Contents

Preface xvii

List of contributors xix

List of figures xxiii

List of tables xxxvii

Abbreviations xli

1 Introduction *(Matthias Jakob and Oldrich Hungr)* 1
    1.1   Debris flows – a global phenomenon. 1
    1.2   Coping with debris-flow hazards. 2
    1.3   The need for this book. 3
    1.4   Structure of the book. 5
    1.5   References 7

2 Classification and terminology *(Oldrich Hungr)* 9
    2.1   Classification of debris flows and avalanches. 9
            2.1.1   Debris-flow materials 11
            2.1.2   Types of flow-like landslides 12
    2.2   Terminology. 16
    2.3   Acknowledgments 21
    2.4   References 21

vi Contents

3 Debris flows in history *(Nigel A. Skermer and Douglas F. VanDine)* .. 25
- 3.1 Introduction ... 25
- 3.2 Historical perspective and its limitations ... 27
- 3.3 Alpine debris flows ... 28
  - 3.3.1 European Alps ... 29
  - 3.3.2 Asia ... 33
- 3.4 Debris flows in semi-arid regions ... 36
  - 3.4.1 South America ... 36
  - 3.4.2 Western North America ... 37
  - 3.4.3 Kazakhstan ... 38
  - 3.4.4 Spain ... 39
- 3.5 Lahars ... 39
  - 3.5.1 Classical times ... 40
  - 3.5.2 The Americas ... 40
  - 3.5.3 Asia ... 42
- 3.6 Jökulhlaups and GLOFs ... 43
- 3.7 Mine waste flows ... 44
- 3.8 Bog bursts ... 45
- 3.9 Summary remarks ... 45
- 3.10 Acknowledgements ... 46
- 3.11 References ... 47

4 Instability of steep slopes *(William Savage and Rex Baum)* ... 53
- 4.1 Introduction ... 53
- 4.2 Stress–strain and strength properties of soil ... 55
- 4.3 Limiting equilibrium methods of slope stability analysis ... 62
- 4.4 Darcy's law and pore pressure in soil media ... 65
- 4.5 Slope stability in partially saturated soils ... 71
- 4.6 Mapping regional slope stability ... 73
- 4.7 Summary ... 75
- 4.8 References ... 77

5 Mechanism of landslide-triggered debris flows: Liquefaction phenomena due to the undrained loading of torrent deposits *(Kyoji Sassa and Gong hui Wang)* ... 81
- 5.1 Introduction ... 81
- 5.2 Undrained loading of the torrent deposits ... 82
  - 5.2.1 Consolidation–shear model ... 82
  - 5.2.2 Stress path ... 83
- 5.3 Liquefaction and sliding surface liquefaction ... 85
- 5.4 Typical cases of landslide-triggered debris flows ... 88
  - 5.4.1 The 1996 Gamahara debris flow ... 88
  - 5.4.2 The 1999 Kameyama debris flow ... 95
- 5.5 Conclusions ... 101
- 5.6 References ... 101

Debris-flow Hazards and Related Phenomena

Matthias Jakob and Oldrich Hungr

Debris-flow Hazards and Related Phenomena

Springer

Published in association with
Praxis Publishing
Chichester, UK

Dr Matthias Jakob
Senior Geoscientist
BGC Engineering Inc.
Vancouver
British Columbia
Canada

Professor Oldrich Hungr
Earth and Ocean Sciences Department
University of British Columbia
Vancouver
British Columbia
Canada

SPRINGER–PRAXIS BOOKS IN GEOPHYSICAL SCIENCES
SUBJECT *ADVISORY EDITOR*: Dr Philippe Blondel, C.Geol., F.G.S., Ph.D., M.Sc., Senior Scientist, Department of Physics, University of Bath, Bath, UK

ISBN 978-3-642-05852-3   e-ISBN 978-3-540-27129-1

Springer is part of Springer-Science + Business Media (springeronline.com)

Bibliographic information published by Die Deutsche Bibliothek

Die Deutsche Bibliothek lists this publication in the Deutsche Nationalbibliografie; detailed bibliographic data are available from the Internet at http://dnb.ddb.de

Apart from any fair dealing for the purposes of research or private study, or criticism or review, as permitted under the Copyright, Designs and Patents Act 1988, this publication may only be reproduced, stored or transmitted, in any form or by any means, with the prior permission in writing of the publishers, or in the case of reprographic reproduction in accordance with the terms of licences issued by the Copyright Licensing Agency. Enquiries concerning reproduction outside those terms should be sent to the publishers.

© Praxis Publishing Ltd, Chichester, UK, 2005
Softcover reprint of the hardcover 1st edition 2005

The use of general descriptive names, registered names, trademarks, etc. in this publication does not imply, even in the absence of a specific statement, that such names are exempt from the relevant protective laws and regulations and therefore free for general use.

Cover design: Jim Wilkie
Project Management: Originator Publishing Services, Gt Yarmouth, Norfolk, UK

Printed on acid-free paper

6	**Debris-flow mechanics** *(Richard M. Iverson)*		105
	6.1	Introduction	105
	6.2	Mechanical definition of debris flow	105
	6.3	Macroscopic dynamics	106
		6.3.1 Continuum conservation laws	107
		6.3.2 Depth integration and mass-change effects	109
		6.3.3 Scaling and shallow flow with non-hydrostatic stress	112
	6.4	Stress estimation	115
		6.4.1 Stress partitioning	116
		6.4.2 Stress due to solid–fluid interaction	117
		6.4.3 Stress due to interactions of solid grains	120
		6.4.4 Stress due to fluid shear	124
		6.4.5 Lumped rheology and calibrated resistance formulas	125
	6.5	Surge dynamics	127
	6.6	Concluding remarks	130
	6.7	Acknowledgements	131
	6.8	References	131
7	**Entrainment of material by debris flows** *(Oldrich Hungr, Scott McDougall, and Michael Bovis)*		135
	7.1	Introduction	135
	7.2	Mechanisms of material entrainment	135
	7.3	Theoretical approach: bed stability	138
	7.4	The yield rate and erosion depth concepts	142
	7.5	Consideration of frequency	149
	7.6	Examples of empirical calibration of the yield rate concept	149
	7.7	The British Columbia debris-flow database	150
	7.8	Conclusions	154
	7.9	References	155
8	**Hyperconcentrated flow – transitional process between water flow and debris flow** *(Thomas C. Pierson)*		159
	8.1	Introduction	159
	8.2	Defining boundaries of hyperconcentrated flow	161
		8.2.1 Rheological criteria	162
		8.2.2 Sand suspension and settling criteria	164
	8.3	Characteristics of hyperconcentrated flow	164
		8.3.1 Initiation	164
		8.3.2 Rheology	165
		8.3.3 Turbulence and velocity distribution	167
		8.3.4 Sediment transport	168
		8.3.5 Bed material, bed forms, and flow resistance	176
	8.4	Examples of observed hyperconcentrated flows	178
		8.4.1 1982 Hyperconcentrated lahar, Toutle River, Mount St. Helens, USA	178

	8.4.2	1993 Hyperconcentrated lahar, Sacobia River, Mount Pinatubo, Philippines.	179
	8.4.3	1995 Hyperconcentrated lahar, Ruapehu volcano, New Zealand.	185
8.5	Sediment deposition.		186
8.6	Hazards		190
8.7	Discussion.		191
8.8	Conclusions.		193
8.9	Acknowledgements		195
8.10	References.		196

9 Subaqueous debris flows *(Jacques Locat and Homa J. Lee)* 203

9.1	Introduction.		203
9.2	Formation of subaqueous mass flow		204
	9.2.1	Triggering mechanism	206
	9.2.2	Transformation.	210
9.3	Parameters and properties of debris flows		212
	9.3.1	Direct measurements	213
	9.3.2	Back analyses.	214
	9.3.3	Empirical relationships.	215
9.4	Physical aspects of subaqueous debris flows		218
	9.4.1	Boundary conditions	218
	9.4.2	Fluid types	219
	9.4.3	Mobility	220
9.5	Modeling tools.		222
	9.5.1	Physical modeling.	222
	9.5.2	Numerical modeling	222
9.6	Examples.		225
	9.6.1	Debris flows on the Mississippi fan, Gulf of Mexico	225
	9.6.2	The Palos Verdes debris avalanche, California	229
9.7	Subaqueous debris flows hazards and consequences		234
	9.7.1	Subaqueous debris-flow hazards.	234
	9.7.2	Consequences of debris flows	237
9.8	Conclusions.		238
9.9	Acknowledgements		239
9.10	References.		239

10 Volcanic debris flows *(James W. Vallance)*. 247

10.1	Introduction.		247
10.2	Definitions.		248
10.3	Origins and downstream behaviour.		253
	10.3.1	Lahars induced by sudden melting of snow and ice, voluminous floods, or torrential rains	254
	10.3.2	Edifice or flank-collapse-induced lahars	256

10.4	Downstream processes and behaviour		257
	10.4.1	Erosion and bulking	257
	10.4.2	Particle-size and particle-density segregation processes	258
	10.4.3	Downstream dilution and transformation	259
	10.4.4	Depositional processes	260
10.5	Selected case studies		261
	10.5.1	Lahars related to crater lakes	261
	10.5.2	Lahars caused by melting of ice and snow	263
	10.5.3	Lahars derived from rain after eruptions	264
	10.5.4	Lahars that derive from debris avalanches	267
10.6	Lahar-hazard assessments		268
	10.6.1	Rationale for lahar-hazard assessments	268
	10.6.2	An automated method of lahar-hazard delineation	269
	10.6.3	Mitigation of lahar hazards	270
10.7	References		271

11 Application of airborne and spaceborne remote sensing methods (Robert T. Pack) — 275

11.1	Introduction		275
11.2	Definitions		275
11.3	Remote sensing methods		277
	11.3.1	Aerial cameras	277
	11.3.2	Digital imagers	280
	11.3.3	Passive microwave sensors	281
	11.3.4	Lidar sensors	282
	11.3.5	Synthetic aperture radar (SAR) sensors	283
11.4	Remote sensing methods in hazard/risk assessment		283
	11.4.1	Event inventories and characterization	284
	11.4.2	Terrain mapping	285
	11.4.3	Hazard/risk assessment	287
11.5	References		287

12 Debris-flow instrumentation (Richard LaHusen) — 291

12.1	Sensors		291
	12.1.1	Sensing debris-flow antecedents	292
	12.1.2	Measuring debris-flow dynamics	296
12.2	Data acquisition systems		300
	12.2.1	Data loggers and other devices	300
	12.2.2	Real-time monitoring from remote sites	301
	12.2.3	Power systems	302
12.3	Summary		303
12.4	References		303

x Contents

13 Runout prediction methods *(Dieter Rickenmann)* 305
- 13.1 Introduction. 305
- 13.2 Total travel distance 306
  - 13.2.1 Travel distance and event magnitude. 306
  - 13.2.2 Volume balance approach. 310
  - 13.2.3 Mass point models (dynamic models) 311
  - 13.2.4 Limiting criteria method. 313
- 13.3 Runout length in the depositional zone 313
  - 13.3.1 Critical slope and deposition on the fan 313
  - 13.3.2 Volume balance approach. 314
  - 13.3.3 Analytical approches 315
- 13.4 Continuum based dynamic simulation models. 318
- 13.5 Summary and conclusions. 320
- 13.6 Acknowledgement 321
- 13.7 References. 321

14 Climatic factors influencing occurrence of debris flow *(Gerald F. Wieczorek and Thomas Glade)* 325
- 14.1 Introduction. 325
- 14.2 Primary climatic factors 325
  - 14.2.1 Intense rainstorms. 326
  - 14.2.2 Rapid snowmelt 330
- 14.3 Secondary climatic factors. 332
  - 14.3.1 Antecedent rainfall 332
- 14.4 Use of climatic data for forecasts and warnings 335
  - 14.4.1 Rainfall thresholds 335
  - 14.4.2 Forecasts and early warning 343
- 14.5 Effects of climate change on debris-flow activity 349
  - 14.5.1 El Niño-Southern Oscillation. 349
  - 14.5.2 Warmer and dryer climates. 349
  - 14.5.3 Forecasting effects of climate change 350
- 14.6 Conclusions 352
- 14.7 Acknowledgements 352
- 14.8 References. 352

15 Wildfire-related debris flow from a hazards perspective *(Susan H. Cannon and Joseph E. Gartner)* 363
- 15.1 Introduction. 363
- 15.2 Fire-related debris-flow initiation processes. 364
  - 15.2.1 Runoff-dominated erosion by surface overland flow. . . 365
  - 15.2.2 Infiltration-dominated landslide failure and mobilization 369
- 15.3 Fire-related debris-flow susceptibility. 371
  - 15.3.1 Bedrock lithology and surficial materials 372
  - 15.3.2 Basin area and average gradient. 373

		15.3.3 Burn extent and severity.	374
		15.3.4 Water repellent soils	375
	15.4	Magnitude of debris-flow response	377
		15.4.1 Relations between peak discharge, area of basin burned, and lithology	377
		15.4.2 Relations between peak discharge, area of basin burned, average basin gradient, and storm rainfall	378
	15.5	Summary and conclusions.	378
		15.5.1 Future research needs	380
	15.6	References	381

16 Influence of forest harvesting activities on debris avalanches and flows (Roy C. Sidle) ... 387

	16.1	Introduction.	387
		16.1.1 Environmental damage.	387
		16.1.2 Triggering mechanisms.	388
	16.2	How do forests enhance slope stability?	389
	16.3	Interaction of forest ecosystems with geomorphic processes	392
	16.4	Effects of management practices in forests	394
		16.4.1 Silvicultural practices and forest stand management	394
		16.4.2 Effects of logging systems.	396
		16.4.3 Harvesting – soil moisture interactions affecting slope stability.	397
		16.4.4 Vegetation conversion	399
		16.4.5 Roads.	400
	16.5	Summary.	402
	16.6	References	403

17 Debris-flow hazard analysis (Matthias Jakob) ... 411

	17.1	Introduction.	411
	17.2	Debris-flow hazard recognition.	412
		17.2.1 Significance of fans	413
		17.2.2 Geomorphic evidence.	413
		17.2.3 Aerial photographs, satellite imagery, and topography	416
		17.2.4 Historical accounts and records	418
	17.3	Debris-flow probability and magnitude	418
		17.3.1 Debris-flow occurrence probability	418
		17.3.2 Debris-flow magnitude.	423
		17.3.3 Debris-flow intensity	434
	17.4	Frequency–magnitude relationships.	434
		17.4.1 Process recognition	435
		17.4.2 The design magnitude approach.	436
	17.5	Debris-flow hazard mapping	436
	17.6	Conclusions	437

	17.7	Acknowledgements	438
	17.8	References	438

18 Debris-flow mitigation measures *(Johannes Huebl and Gernot Fiebiger)* . 445
    18.1   Strategy of protection 445
    18.2   Mitigation measures 446
            18.2.1   Active mitigation measures 446
            18.2.2   Passive mitigation measures 478
            18.2.3   Documentation and control 483
    18.3   Conclusion 483
    18.4   References 485

19 Debris avalanches and debris flows of the Campania Region (southern Italy) *(Francesco M. Guadagno and Paola Revellino)* 489
    19.1   Introduction 489
    19.2   Historic and recent phenomena: an overview 490
    19.3   A fragile geological environment 494
    19.4   Special geotechnical properties and behaviour of pyroclastic soils 501
    19.5   Characteristics of the instabilities 503
    19.6   Instability mechanisms and triggering condition 508
    19.7   Hazard evaluation: some notes on the present situation and future prospective 513
    19.8   Conclusions 515
    19.9   References 515

20 Debris flows of December 1999 in Venezuela *(Reinaldo García-Martínez and J.L. López)* 519
    20.1   Introduction 519
    20.2   Debris-flow events in Latin America 520
    20.3   Geographic characteristics of the disaster area 520
    20.4   The December 1999 storm 522
    20.5   Geological Aspects 524
            20.5.1   Lithological units 524
            20.5.2   Erosion and weathering 524
            20.5.3   Geologic fault system 527
    20.6   Terrain changes 527
            20.6.1   Longitudinal profiles 527
            20.6.2   Changes in bed profiles 527
    20.7   Sediment characteristics 531
            20.7.1   Sediment accumulation 531
            20.7.2   Characteristics of the bed material 531
    20.8   Estimated debris-flow discharges 532
    20.9   Strategies for mitigation 534
            20.9.1   Methodology for hazard maps 534
            20.9.2   Structural measures 536

20.10	Other research efforts.	537
20.11	Conclusions	537
20.12	Acknowledgements	537
20.13	References	538

21 Debris flows caused by Typhoon Herb in Taiwan
(Chyan-Deng Jan and Cheng-lung Chen) 539

21.1	Introduction.	539
21.2	Terrestrial factors triggering debris flows in Taiwan	540
	21.2.1 Geologic and geomorphologic settings.	540
	21.2.2 Recurrence of typhoons	540
	21.2.3 Anthropogenic disturbances.	541
	21.2.4 Typhoons as agents of debris flows	541
21.3	Debris flows triggered by Typhoon Herb	542
	21.3.1 Rainfall.	542
	21.3.2 Debris flows generated in Chenyoulan stream watershed	543
	21.3.3 Debris-flow initiation threshold	551
21.4	Devastation and aftermath of Typhoon Herb	553
	21.4.1 Casualties and property loss (or damage) caused by debris flows	553
	21.4.2 Unjustified engineering structures and land use	555
21.5	Debris-flow mitigation strategies.	557
	21.5.1 Vitalization of debris-flow research	557
	21.5.2 Structural debris-flow countermeasures and warning systems.	557
	21.5.3 Non-structural debris-flow countermeasures	558
21.6	Conclusions	559
21.7	Acknowledgements	560
21.8	References	560

22 Jiangjia Ravine debris flows in south-western China
(Peng Cui, Xiaoqing Chen, Yuyi Wang, Kaiheng Hu, and Yong Li) . . . 565

22.1	Introduction.	565
22.2	Relationships between rainfall and debris flow	569
22.3	Dynamics	571
	22.3.1 Velocity	571
	22.3.2 Impact forces.	574
	22.3.3 Superelevation in channel bends.	576
22.4	Static properties	576
	22.4.1 Density and concentration	576
	22.4.2 Composition	578
22.5	Sediment transportation and influence on the main river.	582
	22.5.1 Sediment transportation	582
	22.5.2 Influence on main river	583
22.6	Deposition.	588

xiv Contents

22.7	Hazard mitigation		589
	22.7.1	Structured mitigation	589
	22.7.2	Forecasting and warning system	591
22.8	Conclusions		592
22.9	Acknowledgements		592
22.10	References		592

23 Debris flows and debris avalanches in Clayoquot Sound
(Terrence P. Rollerson, Thomas H. Millard, and Denis A. Collins) ... 595

23.1	Introduction		595
23.2	Location and setting		597
	23.2.1	Bedrock	598
	23.2.2	Climate and vegetation	598
23.3	Debris avalanche and debris-flow initiation factors in Clayoquot Sound		600
23.4	Landslide management in Clayoquot Sound		607
	23.4.1	Mapping, planning, and avoidance	607
	23.4.2	Remote sensing	609
	23.4.3	Road deactivation and landslide rehabilitation	610
23.5	Summary		610
23.6	Acknowledgements		611
23.7	References		611

24 Analysis and management of debris-flow risks at Sörenberg (Switzerland)
(Markus N. Zimmermann) ... 615

24.1	Introduction		615
24.2	The Swiss strategy for natural disaster reduction		616
	24.2.1	Rationale	616
	24.2.2	Hazard assessment	617
	24.2.3	Definition of protection goals	617
	24.2.4	Priority for mitigation measures	618
24.3	The 1999 natural disasters in Switzerland		618
	24.3.1	Overall conditions	618
	24.3.2	The 1999 landslides and debris flows: Causes and effects	619
	24.3.3	The Sörenberg case	620
24.4	Short and long-term debris-flow risk management		620
	24.4.1	Immediate response	620
	24.4.2	Debris-flow risks assessment	622
	24.4.3	Long-term safety concept for Sörenberg	627
24.5	Unresolved issues		630
24.6	Conclusions		632
24.7	Acknowledgements		632
24.8	References		633

25 Engineering for debris flows in New Zealand (*Mauri J. McSaveney and Tim R.H. Davies*) ... 635
- 25.1 Introduction ... 635
- 25.2 Debris flows at Walter Peak ... 637
  - 25.2.1 Setting ... 638
  - 25.2.2 History ... 639
  - 25.2.3 Meteorology and return period of the event ... 639
  - 25.2.4 Remedial work undertaken ... 640
  - 25.2.5 Conditional probability ... 640
- 25.3 The Rees Valley debris flows ... 640
  - 25.3.1 Background ... 641
  - 25.3.2 Meteorology and return period of the event ... 641
  - 25.3.3 The fatal debris flow ... 644
- 25.4 Waterfall Creek debris flows ... 644
  - 25.4.1 Background ... 644
  - 25.4.2 Bridge #1 ... 645
  - 25.4.3 Bridge #2 ... 646
  - 25.4.4 The Waterfall Creek landslide ... 649
  - 25.4.5 Bridge #3 ... 649
  - 25.4.6 Bridge #4 ... 650
  - 25.4.7 Was it necessary? ... 650
- 25.5 Glencoe Stream and Aoraki/Mount Cook village ... 650
  - 25.5.1 The brief ... 652
  - 25.5.2 Historical background ... 652
  - 25.5.3 Black Birch fan ... 653
  - 25.5.4 Glencoe fan ... 654
  - 25.5.5 Kitchener fan ... 654
- 25.6 Some notable North Island debris-flow events ... 654
  - 25.6.1 Debris flows at Waihi ... 655
  - 25.6.2 Lahars at Mount Ruapehu ... 655
- 25.7 Concluding remarks ... 657
- 25.8 Acknowledgements ... 657
- 25.9 References ... 657

26 Multifaceted hazard assessment of Cheekye fan, a large debris flow fan in south-western British Columbia (*Pierre A. Friele and John J. Clague*) ... 659
- 26.1 Introduction ... 659
- 26.2 Setting and early research ... 660
- 26.3 Basin analysis ... 663
- 26.4 Geochronology and fan evolution ... 665
- 26.5 Fan architecture, sedimentology, and holocene sediment budget ... 673
- 26.6 Debris-flow frequency and magnitude ... 675
- 26.7 Modeling ... 678
- 26.8 Discussion ... 679

26.9	Acknowledgements		681
26.10	References		681

27 Debris flows at Mount St. Helens, Washington, USA
(Jon J. Major, Thomas C. Pierson, and Kevin M. Scott) 685

27.1	Introduction		685
	27.1.1	The significance of volcanic debris flows	685
	27.1.2	Terminology	688
27.2	Eruptions and debris flows at Mount St. Helens – an overview		688
27.3	Initiation processes		690
	27.3.1	Transformation from landslides	690
	27.3.2	Scour, melting, and mixing of snowpack during eruptions	694
	27.3.3	Sediment entrainment (bulking) and transformation of floods	698
27.4	Flow characteristics		700
	27.4.1	Hydrographs, peak discharges, and flow volumes	700
	27.4.2	Sediment entrainment (bulking)	702
	27.4.3	Distal transitions of flow character	703
27.5	Deposit characteristics		704
	27.5.1	Thicknesses	704
	27.5.2	Megaclasts	706
	27.5.3	Textures	707
	27.5.4	Relations among deposit textures and initiation and flow processes	714
27.6	Geomorphic impacts of the 1980 debris flows		716
27.7	Social and economic impacts of the 1980 debris flows		718
27.8	Frequency and magnitude of large debris flows		720
27.9	Hazard assessments and mitigation responses		721
27.10	Conclusions		723
27.11	Acknowledgments		725
27.12	References		725

Index 733

Preface

The dramatic 2004 Boxing Day tsunami in the Indian Ocean, having claimed over a quarter million lives, once again painfully reminded us of our society's vulnerability to natural disasters. If there are any positive aspects to such a terrible and devastating event, it is the World's compassionate response and the knowledge that our awareness of our vulnerability and susceptibility has increased. Undoubtedly, affected governments will now attempt to reduce the consequences of similar events in the future and implement a tsunami-warning system in the region.

Nature is varied and in many ways poorly predictable. Quite possibly, the next major global or regional disaster may not be a tsunami or an earthquake, but a climatic convulsion such as a typhoon or a Class 5 hurricane over a densely populated region. Those who follow the news will realize that a large proportion of life loss suffered in climate-driven natural disasters is due to landslides. More specifically, many people are injured or killed by clusters of debris flows, debris avalanches, and debris floods. Apart from such major events, isolated occurrences of debris flows claim a steady, continuing toll of lives and property in all parts of the world with steep terrain. Every year, approximately 78 million people are added to the world's population. With an ever-increasing population density in high-hazard areas, the consequences of debris flows and related phenomena will grow proportionally.

Risks arising from these landslide processes cannot be eliminated and are difficult to reduce. This can be attributed to globally widespread occurrence of conditions favoring landslides, increased population density in vulnerable areas, and, above all, the unpredictability of the hazard and the difficulty to provide advance warning. Clearly, a better understanding of the character of such phenomena and better predictive tools are urgently needed.

The present book attempts to systematically review significant and up-to-date knowledge of debris flows, debris avalanches, and debris floods, while emphasizing predictive techniques. To accomplish this, we invited leading specialists from five continents, who addressed aspects of debris-flow science ranging from classification to analysis, prediction and mitigation. In addition, we asked renowned scientists to provide technical summaries of notable occurrences from various corners of the world. The invited authors responded enthusiastically and brilliantly. Thanks to their efforts, we believe that this book will be a valuable tool for any scientist or engineer working with debris flows and related hazards. The professionalism of Praxis and Springer publishers and their contractors and consultants was crucial in bringing the book to a successful and timely realization. We are particularly grateful to Clive Horwood, Dr Philippe Blondel (technical editor) and Neil Shuttlewood.

We sincerely hope that the information contained herein will help to save lives and reduce future losses attributed to debris flows and related phenomena.

Matthias Jakob and *Oldrich Hungr*
Vancouver
January 31, 2005

List of contributors

Dr. Rex Baum US Geological Survey, P.O. Box 25046, Denver, CO 80225-0046, USA

Dr. Michael Bovis Geography Department, University of British Columbia, 1984 West Mall, Vancouver, BC, V6T 1Z2, Canada

Dr. John Clague Department of Earth Sciences, Simon Fraser University, Burnaby, British Columbia, V5A 1S6, Canada

Dr. Susan H. Cannon US Geological Survey, Box 25046, DFC, MS 966, Denver, CO 80225, USA

Dr. Xiaoqing Chen Institute of Mountain Hazards and Environment, Chinese Academy of Sciences, P.O. Box 417, Chengdu, Sichuan 610041, P.R. China

Dr. Peng Cui Institute of Mountain Hazards and Environment, Chinese Academy of Sciences, P.O. Box 417, Chengdu, Sichuan 610041, P.R. China

Dr. Cheng-lung Chen 20886 Jollyman Lane, Cupertino, CA 95014-4349, USA

Dr. Denis Collins Ministry of Forests, Coast Forest Region, 2100 Labieux Road, Nanaimo, BC, V9T 6E9, Canada

Dr. Tim Davies Department of Geological Sciences, University of Canterbury, Private Bag 4800, Christchurch, New Zealand

Dr. Gernot Fiebiger Austrian Foresttechnical Service in Torrent Avalanche & Erosion Control, Bergheimerstrasse 57, 5021 Salzburg, Austria

Mr. Pierre Friele Box 612, Squamish, BC, V0N 3G0, Canada

Dr. Reinaldo Garcia Department of Civil, Architectural and Environmental Engineering, University of Miami, 1251 Memorial Drive, Office EB-315, Coral Gables, FL 33124, USA

Dr. Joseph E. Gartner US Geological Survey, Box 25046, DFC, MS 966, Denver, CO 80225, USA

Dr. Thomas Glade Geography Department, Friedrich-Wilhelms-University Bonn, Meckenheimer Allee 166, 53115 Bonn, Germany

Dr. Francesco Maria Guadagno Dipartimento di Studi Geologici e Ambientali, Facoltà di Scienze – Università del Sannio, Via Port'Arsa, 11, 82100 Benevento, Italy

Dr. Kaiheng Hu Institute of Mountain Hazards and Environment, Chinese Academy of Sciences, PO Box 417, Chengdu, Sichuan 610041, P.R. China

Dr. Hannes Huebel Institut für Alpine Naturgefahren, Department Bautechnik + Naturgefahren, Universität für Bodenkultur, Peter Jordan Str. 82, 1190 Wien, Austria

Dr. Oldrich Hungr Earth and Ocean Sciences, University of British Columbia, 6339 Stores Rd, Vancouver, British Columbia, V6T 1Z4, Canada

Dr. Richard M. Iverson US Geological Survey, 1300 SE Cardinal Ct., #100, Vancouver, WA 98683, USA

Dr. Matthias Jakob BGC Engineering Inc., 500-1045 Howe Street, Vancouver, BC, V6Z 2A9, Canada

Dr. Chyan-Deng Jan Department of Hydraulic and Ocean Engineering, National Cheng Kung University, 1 Ta-Hsueh Road, Tainan 70101, Taiwan

Dr. Rick LaHusen US Geological Survey 1300 SE Cardinal Ct., #100, Vancouver, WA 98683, USA

Dr. Homa Lee US Geological Survey, 345 Middlefield Road, Menlo Park, CA 94025, USA

Mr. Yong Li Institute of Mountain Hazards and Environment, Chinese Academy of Sciences, PO Box 417, Chengdu, Sichuan 610041, P.R. China

Dr. Jacques Locat Géologie et de Génie Géologique, Département de Faculté des sciences et de genie, Pavillion Adrien-Pouliot, Bureau 4309, Université Laval, Québec G1K 7P4, Canada

Dr. Jon J. Major US Geological Survey, 1300 SE Cardinal Ct., #100, Vancouver, WA 98683, USA

Mr. Scott McDougall Earth and Ocean Sciences, University of British Columbia, 6339 Stores Rd., Vancouver, BC, V6T 1Z4, Canada

Dr. Mauri McSaveney Institute of Geological & Nuclear Sciences Ltd, PO Box 30368 Lower Hutt, New Zealand

Mr. Tom Millard Ministry of Forests, Coast Forest Region, 2100 Labieux Road, Nanaimo, BC, V9T 6E9, Canada

Dr. Thomas C. Pierson US Geological Survey, Cascades Volcano Observatory, 1300 S.E. Cardinal Ct. #100, Vancouver, WA 98683-9589, USA

Dr. Paola Revellino Dipartimento di Studi Geologici e Ambientali, Facoltà di Scienze – Università del Sannio, Via Port'Arsa, 11, 82100 Benevento, Italy

Dr. Dieter Rickenmann BOKU – University of Natural Resources and Applied Life Sciences, Department of Structural Engineering and Natural Hazards, Institute of Mountain Risk Engineering, Peter Jordan-Strasse 82, A-1190 Vienna, Austria
and
Swiss Federal Research Institute WSL, Department of Natural Hazards, Division of Water, Soil and Rock Movements, CH-8903 Birmensdorf, Switzerland

Dr. Robert T. Pack Research Associate Professor, Utah State University, Logan, UT 84322-4110, USA

Mr. Terry Rollerson Golder Associates Ltd., 1462 Broadview Road, Gabriola, BC, V0R 1X0, Canada

Dr. Kjoji Sassa Kyoto University, Disaster Prevention Research Institute, Research Centre on Landslides, Gokasho, Uji, 611-0011 Kyoto, Japan

Dr. William Savage US Geological Survey, PO Box 25046, Denver, CO 80225-0046, USA

Dr. Kevin M. Scott US Geological Survey, 1300 SE Cardinal Ct., #100, Vancouver, WA 98683, USA

List of contributors

Dr. Roy Sidle Kyoto University, Disaster Prevention Research Institute, Kyoto University, Gokasho, Uji 611-0011, Kyoto, Japan

Nigel Skermer Skaha Consultants, PO Box 287, 418 Lakehill Road, Kaleden, BC, V0H 1K0, Canada

Doug VanDine VanDine Geological Engineering Limited, 267 Wildwood Avenue, Victoria, BC, V8S 3W2, Canada

Dr. Jim Vallance US Geological Survey, 1300 SE Cardinal Ct., #100, Vancouver, WA 98683, USA

Dr. G. Wang Kyoto University, Disaster Prevention Research Institute, Research Centre on Landslides, Gokasho, Uji 611-0011, Kyoto, Japan

Dr. Yuyi Wang Institute of Mountain Hazards and Environment, Chinese Academy of Sciences, PO Box 417, Chengdu, Sichuan 610041, P.R. China

Dr. Gerald F. Wieczorek US Geological Survey, 345 Middlefield Rd., MS 910, Menlo Park, CA 94025, USA

Mr. Yingyan Zhu Institute of Mountain Hazards and Environment, Chinese Academy of Sciences, PO Box 417, Chengdu, Sichuan 610041, P.R. China

Mr. Markus Zimmermann NDR Consulting Zimmermann, Riedstrasse 5, 3600 Thun, Switzerland

Figures

1.1	Damaged buildings on the Caraballeda Fan in the Vargas Province, Venezuela	2
1.2	Freight train derailed by the direct impact of a small debris flow in high mountains.	3
1.3	An aerial view of a debris fan in western Honshu Island, Japan, shortly after the impact of a debris flow, which destroyed several houses and killed 10 persons	4
2.1	A discrete shear surface in earth material (the Big Slide, Quesnel, British Columbia, Canada)	11
2.2	Debris deposits in the fan of Pierce Creek near Chilliwack, British Columbia, Canada	12
2.3	A debris flow in the Khumbu Himalayas, Nepal	14
2.4	A group of debris avalanches triggered by logging and road construction on the west coast of Vancouver Island, Canada	15
2.5	A vehicle surrounded by debris flood deposits, Britannia Beach near Vancouver, Canada	16
2.6	Diagram of a debris-flow surge with a boulder front	18
2.7	A debris-flow channel flanked by levee deposits (Cathedral Mountain, Rocky Mountains, Canada)	19
3.1	Destruction by debris flood on 28 October, 1921, of the mining community of Britannia Beach, north of Vancouver, British Columbia, Canada	26
3.2	Meiringen, Switzerland after a debris flood at the turn of the last century	30
3.3	The village of Inzing in the Tyrol, near Innsbruck, Austria	30
3.4	The side entrance to the library at Inzing is below ground, the level having been built up by earlier debris-flow depositions	31
3.5	Central part of the Annapurna Massive with Machapuchre (Nepal's "Matterhorn") (6,993 m) in the centre and the Sabche Cirque to the right (indicated by an arrow)	34

xxiv **Figures**

4.1	Photograph of a shallow, irregular landslide scar on a steep slope near Clear Creek, Utah	54
4.2	Sketch illustrating the general mechanisms involved in rainfall-induced failure of thin colluvium on a steep hillside	55
4.3	(a) Cross-sectional sketch of a direct shear apparatus. (b) Cross-sectional sketch of a triaxial test apparatus	56
4.4	Stress–strain and void ratio–strain curves for drained triaxial test on loose (contractive) and dense (dilatant) medium-fine sand	58
4.5	(a) Sketch showing Mohr–Coulomb failure criterion. (b) Construction of a Mohr–Coulomb failure envelope for a soil. (c) Mohr–Coulomb strength parameters determined by direct-shear testing of two soils from the Alani–Paty landslide in Hawaii	59
4.6	Role of effective stress in reducing the shear strength of soil	60
4.7	Law of static friction for a rigid block on a rough inclined plane	63
4.8	Variation of the factor of safety for an infinite slope with slope angle for typical values of dimensionless soil cohesion, and the pore pressure ratio	64
4.9	Darcy's law for fluid flow through an inclined porous medium	66
4.10	Groundwater flow boundary conditions	67
4.11	Seepage conditions in an idealized steep slope	69
4.12	Partially saturated cohesion and factor of safety for slopes at Serra do Mar, Brazil	72
4.13	Results of an analysis using TRIGRS (Baum et al., 2002; Savage et al., 2003) of time-dependent slope stability in an area along Puget Sound north of Seattle, Washington	76
5.1	Consolidation–shear model	82
5.2	Drained or undrained loading during the motion of a displaced debris mass and the resulting mobilized apparent friction angle	84
5.3	Torrent deposit liquefied by stamping some hours before the initiation of a debris flow	86
5.4	Conditions for triggering liquefaction on a sample from torrent deposits of Mount Usu	86
5.5	Illustration of the initiation of debris flows	87
5.6	Liquefaction and sliding-surface liquefaction	88
5.7	Aerial photos before and after the debris-flow disaster	89
5.8	Plan of the initial landslide and the Gamahara torrent	90
5.9	A–A' sections made from aerial photos taken on 19 July, 1996, 7 December, 1996, and 17 April, 1997	91
5.10	Schematic illustration of the landslide-triggered Gamahara debris flow, Japan	92
5.11	Control signals for the mechanical simulation test in terms of time series and stress path	94
5.12	Undrained ring-shear test results for the Gamahara debris flow	94
5.13	Aerial oblique views of the landslide	95
5.14	Photos of Kameyama landslide	96
5.15	Central longitudinal section from the source area to the Omoji River and of the source area and cross section of Failure T	color
5.16	Results of a test simulating the undrained loading on the saturated sample	99

5.17	Test results simulating the undrained loading on a sample with natural water content.	100
6.1	Definition of a REV within a debris flow	107
6.2	Definition of the Cartesian coordinate system, bed elevation, debris thickness, and velocity vectors in a depth-averaged debris-flow model.	109
6.3	Typical, debris-flow geometry, illustrating the length scales H and L.	113
6.4	Coulomb yield surface in 3-D stress space.	122
6.5	Geometry used to calculate the basal normal traction due to the apparent weight of a column of debris	123
6.6	Grain trajectories and grain-size segregation in a debris-flow surge	129
6.7	Aerial photographs of an experimental $\sim$10 m^3 debris flow discharging from the mouth of the USGS debris-flow flume.	130
7.1	The Tsing Shan debris flow that started as a small debris slide of 400 m^3 and grew to a total magnitude of 20,000 m^3 through material entrainment	136
7.2	Tsing Shan debris flow.	137
7.3	Schematic diagram of an eroded vertical cross section of a debris-flow channel	138
7.4	An eroded debris-flow channel in the Columbia Mountains, British Columbia	139
7.5	(a) Schematic representation of a saturated bed over-ridden by a debris flow, showing a slope-normal column of unit length and width. (b) Forces acting on the column in (a) include the weight of the column and the shear resistance at its base.	139
7.6	Erosion depths	141
7.7	A debris-flow channel with multiple branches in Banff National Park, Canada	144
7.8	A photo of a firm-base channel, with a substrate of igneous bedrock, near Chilliwack, British Columbia, Canada.	146
7.9	Two debris avalanches widening with distance downslope. (Quindici, Campania Region, southern Italy.).	147
7.10	Queen Charlotte Islands database: histogram of erosion depths for all 1,073 reaches, both confined and unconfined	151
7.11	Queen Charlotte Islands database: histogram of yield rates for the 340 confined reaches.	152
7.12	Queen Charlotte Islands database: histogram of yield rates for the 733 unconfined reaches	153
7.13	Queen Charlotte Islands database: dependence of yield and lag rates on slope angle and the amount of debris-flow material entering the reach for "confined" reaches.	154
7.14	Queen Charlotte Islands database: dependence of yield and lag rates on slope angle and the amount of debris-flow material entering the reach for "unconfined" reaches.	155
8.1	Yield strength of sediment–water mixtures as a function of suspended sediment concentration	163
8.2	Vertical velocity profiles of experimental hyperconcentrated flow.	168
8.3	Examples of smooth and rough flow surfaces of hyperconcentrated flows in Pasig-Potrero River, Mount Pinatubo, Philippines	169

xxvi Figures

8.4	Surface of hyperconcentrated flow revealing reticulate pattern apparently resulting from elutriation of fines at margins of convection cells	172
8.5	Generalized models for movement of bedload beneath overlying turbulent, shearing flow	174
8.6	Vertical distribution of suspended sediment concentration in hyperconcentrated flows	175
8.7	Relation between friction factor (f) and suspended sediment concentration for experimental flume runs	177
8.8	Sediment concentration and flow discharge for the 1982 hyperconcentrated lahar on the Toutle River downstream from Mount St. Helens	182
8.9	Sediment concentration vs. time during passage of the 1982 hyperconcentrated lahar on the Toutle River downstream from Mount St. Helens	183
8.10	Hyperconcentrated lahar on Sacobia River, Mount Pinatubo, Philippines, 26 September, 1993, showing transition from plane bed to antidunes	184
8.11	Approximate stage hydrograph and suspended sediment concentration variation with time for Sacobia River lahar (hyperconcentrated flow), 26 September, 1993, Mount Pinatubo (Philippines)	185
8.12	Examples of hyperconcentrated flow deposits	188
8.13	Differences in grain-size distribution between samples of hyperconcentrated flow (dip samples) and the sediment deposited at the same site during the 19–20 March, 1982 hyperconcentrated lahar on the Toutle River	189
8.14	Burial of house and trees downstream of Mount Pinatubo, Philippines, due to river bed aggradation	191
8.15	Destruction of buildings along a river bank downstream from Mount Pinatubo, Philippines, caused by lateral erosion of channel banks by hyperconcentrated flows	192
8.16	Approximate ranges in total suspended sediment concentrations for natural hyperconcentrated flows and laboratory-mixed hyperconcentrated suspensions of varying grain-size distributions and sediment compositions	194
9.1	Conceptual model of mass movements involving solids and water considering granular and cohesive material	205
9.2	Boundary conditions during a subaqueous mass flow event	205
9.3	The factor of safety and the potential natural and man-made triggers of seabed instability	206
9.4	Mechanics involved in a lateral spread failure	211
9.5	Schematic view of strength evolution with strain rate (displacement or deformation) and the corresponding state of the sediment and its transformation into a fluidized sediment by an increase in water content	212
9.6	Overall approach to the analysis of subaqueous debris flows and their consequences	213
9.7	Soil phases and the use of the liquidity index and the illustration of the normalisation potential of the liquidity index in comparing the remoulded strength profile for three different clayey soils in terms of their sediment concentration and water content	216
9.8	Using the liquidity index–yield strength relationship to estimate rheological properties at the time of debris-flow formation	217
9.9	Bilinear model of Locat (1997) with the boundary conditions used in the 1-D numerical model Bing	220

Figures xxvii

9.10	The mobility of a mass movement	221
9.11	Example of flume modeling of debris flows using kaolin and fine sand mixtures	223
9.12	(a) Location of the distal Missippi fan showing the lobe that was selected for a study of the far-reaching debris flows (b) Relationship between the initial flow thickness and the average flow thickness	226
9.13	(a) Flow thickness and steady-state velocity reached for various slope angles corresponding to those along the slope profile obtained using VIFLOW. (b) Velocity run-out distance simulation using SKRED with an initial thickness varying from 5–50 m	228
9.14	(a) Shaded relief multibeam imagery of the Palos Verdes margin near Los Angeles, California, showing landslide features. (b) An oblique view of the slide	230
9.15	Using the flow deposit thickness in the run-out zone to estimate the yield strength using the relationship proposed by Johnson (1984) for the Palos Verdes debris avalanche	231
9.16	Parametric analysis on the values of the strain rate and the ratio of strain rates using the run-out distance for calibration for the Palos Verdes debris avalanche	231
9.17	Effect of both inital thickness and initial width of the failing mass on the run-out distance for the Palos Verdes debris avalanche	232
9.18	Evolution of the velocity of the frontal element and shape of the flowing mass during the flow event for the Palos Verdes debris avalanche	233
9.19	Velocity and acceleration of the flowing mass for the conditions given in Figure 9.16 for the Palos Verdes debris avalanche	234
9.20	Risk management procedure for submarine mass movements	236
10.1	Plot of VEI vs. volume of primary lahars having three diverse origins	253
10.2	Schematic model of a lahar moving down a river undergoing downstream dilution from debris-flow phase to hyperconcentrated-flow phase and deposit facies derived from it	253
10.3	Schematic diagram showing how segregation of coarse particles to the surface of a debris-flow results in segregation of the coarse particles to the front and margins of the flow	259
10.4	Lahar-hazard map of Hall and von Hillebrand (1988) superimposed on Landscan population data courtesy of John Ewert	color
10.5	Photograph of Armero after the 13 November, 1985 lahars from Nevado del Ruiz, Columbia	color
10.6	Shuttle image of Pinatubo	265
10.7	Plots of sediment yield vs. time for 1991–1997 at Mount Pinatubo and surrounding drainages	266
10.8	Distribution map of the 5,600-year-old, 4-km^3 Osceola Mudflow and coeval Paradise lahar and of the 0.25-km^3, 500-year-old Electron Mudflow	color
10.9	Lahar-hazard zones for Mount Rainier	color
10.10	Idealized lahar path and geometric relationships between A and B describing the extent of the lahar-inundation hazard zone	color
10.11	Lahar-inundation hazard map constructed by applying the computer model for Mount Rainier and vicinity	color
11.1	US Air Force 1951 lens test chart illustrating black and white line pairs	276
11.2	Spectral ranges of the electromagnetic spectrum expressed in nanometers	color

Figures

11.3	Trace of debris flow near Lowman, Idaho, shown on a panchromatic photograph.	278
11.4	Trace of debris flow showing remnant levees in the Chetwynd area of British Columbia.	278
11.5	CIR photograph showing seepage as dark blue patches in the debris slide scar	color
11.6	Lidar range image of a hillslope subject to debris slides and raveling	color
11.7	Three-dimensional photograph of the same hillslope shown in Figure 11.6	color
11.8	Photo showing the use of a ground-based lidar to monitor movement on a debris slide face.	color
11.9	Isoline plot of landslide movements detected by a ground-based lidar system in Austria.	color
11.10	Map of the maximum landslide-induced ground deformation provided by a ground-based D-InSAR and photo of the debris slide headscarp analysed	color
12.1	Cross-section diagram of a snowmelt lysimeter showing collection pan and tipping bucket flowmeter that is buried in a well-drained pit to avoid freezing	293
12.2	Loowit AFM station and precipitation gauge, looking upstream towards Mount St. Helens' crater	297
12.3	Graph of rainfall and AFM ground vibration showing characteristic signal of a debris flow with a rapid rise to more than 10 times background storm flow levels and a more gradual cessation of flow	298
12.4	Illustration of a multiparameter debris-flow monitoring station at Acquabona channel, Italy	299
13.1	Travel angle vs. volume of mass movement	307
13.2	(a) Observed runout distance of debris flows and rock avalanches compared to predictions with (13.1b). (b) Observed runout distance of debris flows and rock avalanches compared to predictions with (13.2)	309
13.3	Travel angle of debris flows vs. watershed area	310
13.4	Comparison between observed and calculated runout lengths for 14 debris flows at the fan of the Kamikamihori valley in Japan	316
13.5	Comparison between observed and calculated runout lengths for 12 debris flows in Switzerland.	316
14.1	Conceptual model for rainfall, infiltration, and aquifers in soils.	326
14.2	Distribution and density of debris flows during the 1982 storm in the San Francisco Bay region.	328
14.3	Timing of debris-flow warnings in San Francisco Bay region during the storms of 12–21 February, 1986	329
14.4a	Debris flows and other landslides following rainstorm events in New Zealand	330
14.4b	Debris flows from the rainstorm in 2002, Gisborne, East Coast, New Zealand	color
14.5	Temperature patterns for snowmelt triggering of a debris flow	331
14.6	Comparison of rainfall thresholds for triggering of debris flows.	336
14.7	Rainfall probability thresholds established by applying the "*Daily Rainfall*" model for the period from 1862 to 1995 for Wellington, New Zealand	340
14.8	Number and spatial density of debris flows for rainstorms of different intensity and duration in the Alps, Prealps, and the Sarno regions of Italy	341

Figures xxix

14.9	Worldwide rainfall thresholds from the literature......................	342
14.10	Rainfall probability thresholds established by applying the "*Antecedent Daily Rainfall*" model for the period 1862–1995 for Wellington, New Zealand	343
14.11	Piezometric response and "*Leaky Barrel*" modeling vs. rainfall intensity for storm...	344
14.12	Rainfall thresholds established by applying the "*Soil Water Status*" model for the Wellington region, New Zealand...............................	345
14.13	Distribution of debris-flow events among calendar months	345
14.14	Landslide occurrence, daily rainfall, and rainfall required to trigger landslides based on calculations applying the "*Antecedent Soil Water Status*" model ...	347
14.15	Slope stability maps from simulation with different hydrologic models......	348
14.16	Grid points of the CRU Global Climate Dataset and duration in days in the Caucasus region	351
14.17	The Soil Water Status index calculated for Wellington, New Zealand.......	color
14.18	Changes of total number of landslide events within each probability class for Wellington, New Zealand...	color
15.1	Photograph of debris-flow path and deposits generated from a burned basin following the 2002 Coal Seam Fire in Colorado............................	color
15.2	Map showing locations of documented fire-related debris-flow events in the western USA	color
15.3	Photograph of debris-flow path generated through process of progressive bulking of runoff with sediment eroded from hillslopes and channels in the 2002 Coal Seam Fire in Colorado...	color
15.4	Frequency of debris-flow initiation processes for 210 debris-flow producing basins...	367
15.5	Average storm rainfall intensity and duration for debris-flow producing storms	368
15.6	Rainfall intensity–duration threshold for the generation of fire-related debris flows from recently burned, steep basins underlain by sedimentary rocks	369
15.7	Photograph of landslide scar on hillslopes burned by the 2002 Missionary Ridge Fire in Colorado..	color
15.8	Frequency of lithologies identified by initiation process underlying 210 debris-flow producing basins	372
15.9	Relations between basin area and average basin gradient, relative to lithology, of basins reported to have produced fire-related debris flows..............	374
15.10	Frequency distributions of percentage of basin area burned at moderate and high severity for 108 debris-flow producing basins	376
15.11	Relation between peak debris-flow discharge estimates and area of basins burned at high and moderate severities, identified by lithology	377
16.1	Alder corridors along headwater channels in southeast Alaska caused by debris flows	388
16.2	Typical changes in forest vegetation rooting strength after timber harvesting (clear-cutting)...	391
16.3	Examples of two types of debris-flow conditions that predominate in steep headwaters: Type 1, landslides immediately or rapidly mobilize into debris flows; Type 2, sediment from hillslope landslides accumulates in channel heads, later to be mobilized into a debris flow..	393

16.4	Simulated effects of clear-cutting vs. partial cutting on landslide volume for different rotation lengths and conditions, Carnation Creek, British Columbia.	396
16.5	Simulations of the effects of timber harvesting on soil moisture, Malaysia ...	398
16.6	Tree root strength deterioration and recovery following timber harvesting with regeneration, as well as for forest conversion to weaker rooted cover.	400
16.7	A landslide initiating on a constructed agricultural terrace in the Loess Plateau, China.	401
17.1	Well-defined debris lobes and levees at Gunbarrel 1 Gully, Lillooet, British Columbia.	414
17.2	The Hinkelstein Boulder rolled at least 200 m by a debris flows at Hummingbird Creek, British Columbia	415
17.3	Scour marks and impact scars in channels well above the flood limit at Canyon Creek, Washington State	416
17.4	Aerial photograph stereopair of a typical debris-flow-prone watershed in south-western British Columbia.	417
17.5	Example of a supply-unlimited watershed in Japan (Kamikamihori Gully)...	419
17.6	Example of a channel with slow recharge rates in the southern Coast Mountains of British Columbia.	420
17.7	Conceptual sketch illustrating the difference between supply-limited and supply-unlimited watersheds.	421
17.8	Summary of stratigraphic evidence of debris flows during the past 7,000 years at Jones Creek, Washington State.	422
17.9	Summary of peak discharge – total volume correlations compiled by Rickenmann (1999).	432
17.10	Frequency–magnitude graph of debris flows at Mount Garibaldi.	435
17.11	Hazard map of Percy Creek at Indian Arm, Vancouver, British Columbia...	color
18.1	Strategy of protection	446
18.2	Schmittenbach in 1887, Salzburg, Austria	449
18.3	Schmittenbach in 1976 after afforestation, Salzburg, Austria	449
18.4	Soil bioengineering at Fendlermure in 1995, Tyrol, Austria.	450
18.5	Slope stabilization at Filprittertobel in 1898, Vorarlberg, Austria.	451
18.6	Transverse toe slope stabilization, Bretterwandbach, Tyrol, Austria.	453
18.7	Longitudinal toe slope stabilization, Eugenbach, Austria	454
18.8	Niedernsiller Muehlbach, Salzburg, Austria.	455
18.9	Sketch of a typical check dam	455
18.10	Upstream wing walls, Duernbach, Salzburg, Austria	456
18.11	Downstream wing walls, Wagrainer Ache, Salzburg, Austria.	457
18.12	Niedernsiller Muehlbach in 1970 after a debris-flow event, Salzburg, Austria.	457
18.13	Gimbach in 2002 after a debris-flow event, Upper Austria, Austria	458
18.14	Anchored longitudinal structure, Kirchbachgraben, Carinthia, Austria.	459
18.15	Bypass tunnel at St. Julien, France.	460
18.16	Bypass at Buergerbach, Salzburg, Austria	461
18.17	Flood-control reservoir in the middle reach of Wartschenbach, Tyrol, Austria	462
18.18	Solid body barrier, Einachgraben, Austria.	462
18.19	Small slot barrier, Koednitzbach, Tyrol, Austria	463
18.20	Sketch of an open barrier.	464
18.21	Large slot barrier, Nieschenbach, Tyrol, Austria	466

Figures xxxi

18.22	Small slot barrier, Truebenbach, Carinthia, Austria	466
18.23	Slit barrier with vertical slits, Zinkenbach, Salzburg, Austria	467
18.24	Slit barrier with horizontal slits, Schnannerbach, Tyrol, Austria	467
18.25	Sectional barrier with fins and beams, Maerzenbach, Tyrol, Austria	468
18.26	Sectional barrier with fins, Sallabach, Styria, Austria	468
18.27	Sectional barrier with piles, Waldbach, Styria, Austria	469
18.28	Beam barrier, Truebenbach, Carinthia, Austria	469
18.29	Sectional rake barrier, Loehnersbach, Salzburg, Austria	470
18.30	Beam barrier, Istalanzbach, Tyrol, Austria	470
18.31	Cross slit barrier, Luggauerbach, Salzburg, Austria	471
18.32	Cross slit barrier in 2000 after a debris flow, Luggauerbach, Salzburg, Austria	471
18.33	Sectional barrier with fins, Fong-Chiu, Nan-Tou County, Taiwan	472
18.34	Frame barrier, Ashiya River, Japan	472
18.35	Net barrier, Gleiersbach, Tyrol, Austria	473
18.36	Sectional barrier with fins, Luggauerbach, Salzburg, Austria	473
18.37	Sectional barrier with fins in 2000 after a debris-flow event, Luggauerbach, Salzburg, Austria	474
18.38	Sectional barrier with fins, Rastelzenbach, Salzburg, Austria	474
18.39	Sectional barrier with fins, Ellmaubach, Salzburg, Austria	475
18.40	Debris-flow grill, Furano River, Japan	475
18.41	Debris-flow grill, Dorfbach Randa, Switzerland	476
18.42	Deflection walls at Luggauerbach, Salzburg, Austria	478
18.43	Deflection wall at Niedernsiller Muehlbach, Salzburg, Austria	479
18.44	Contactless warning by ultrasonic sensors at Lattenbach, Tyrol, Austria	481
18.45	Contact warning by DLT (détecteur lave torrentielle) sensors at Ravoire de Pontamafrey, France	482
18.46	Log pod Mangrtom, Slovenia	483
18.47	Log pod Mangrtom, Slovenia	484
19.1	Debris avalanche paths reaching the town of Sarno and Episcopio (aerial view)	490
19.2	Schematic geo-structural map, showing the location of the destructive debris flows and avalanches in the Campania Region	491
19.3	The 1841 Gragnano landslides, illustrated by Ranieri (1841)	492
19.4	The Salerno area impacted by the landslides of 1954	492
19.5	(a) The extension of the debris fan of Vietri following the 1954 landslide event. (b) Some source areas of 1954 landslides. (c) Spatially diffused slope instabilities (1954 landslide event). (d) The area of Mt. S. Costanzo involved in debris flows triggered by a rock fall. (e) Palma Campania landslide. (f) Pozzano landslides	493
19.6	Patterns of the 1998 flows along the slopes of the Pizzo d'Alvano ridges	495
19.7	Two couples of vertical aerial photographs of the Siano and Quindici landslides respectively	496
19.8	The major landslide event in the Cervinara area	497
19.9	The landslide of the San Martino area in the city of Naples	497
19.10	Cross section of Pizzo d'Alvano	498
19.11	A block diagram showing the peculiar geomorphological setting in the Campanian Apennines	499
19.12	(a) A detail of a typical pyroclastic sequence. (b) A track during the winter. (c) Landslide deposits and a large boulder transported by the flow. (d) Aerial	

xxxii **Figures**

	view of Quindici depositional fans. (e) Initial failure at short distance above the natural scar. (f) Initial failure at short distance above the cut slope of a trackway	500
19.13	Schematic cross sections along a pyroclastics-mantled slope	505
19.14	The morphometrical parameters used in the analysis of debris avalanches	506
19.15	Apex angle vs. the ratio between slope angle tangent and scarp height	506
19.16	Ratio between initial volume and avalanche volume vs. slope length	507
19.17	Example of a DAN back-analysis	508
19.18	A typical source area on the Sarno slopes	509
19.19	Detailed map and section of the source area of the Cervinara landslides	510
19.20	A schematic section along a trackway	512
19.21	The slope involved by debris avalanches in the area of San Francisco gullies	513
20.1	Location of the disaster area in the north coastal range of Venezuela	521
20.2	Aerial view of Avila Mountain and the north coastal range of Venezuela (Ikonos satellite image). Longitudinal distance is approximately 30 km.	521
20.3	Location map of the main basins in the State of Vargas	522
20.4	Rainfall data during December 1999 in the disaster area	523
20.5	Lithological units in the State of Vargas	525
20.6	(*top*) Massive land slides in Cerro Grande and Quebrada Seca basins. (*bottom*) Debris avalanches and debris-flow scars of the 1999 storm in the upper part of San Jose de Galipan river basin (photograph taken in January, 2001)	526
20.7	Longitudinal profiles of main streams in the disaster area	529
20.8	Changes in the bed profiles of Cerro Grande River	529
20.9	Changes in the bed profiles of Uria River	530
20.10	Modification of the river channel and coastline in Uria. (*left*) Picture taken in March 1999. (*right*) Picture taken in December 1999 after the disaster (solid line indicates river channel)	530
20.11	Natural dams formed by debris (tree trunks) in Los Corales (*top*) and Macuto (*bottom*). See man for scale	533
20.12	Discrete hazard levels (PREVENE 2001 project)	color
20.13	Hazard map for San Julian alluvial fan. Squares are 500×500 m	color
21.1	Geographical location and satellite image of Taiwan Island	color
21.2	Number of debris flows per year in Taiwan	542
21.3	Isohyetal map of the 96-hour rainfall during Typhoon Herb	color
21.4	Rainfall magnitude–duration graph for Typhoon Herb at 4 rain-gauge stations in the Chenyoulan stream watershed	544
21.5	Rainfall intensity–duration graph for Typhoon Herb at Alishan Climate Station	545
21.6	Hourly rainfall measured at 4 rain-gauge stations in the Chenyoulan stream watershed during Typhoon Herb	545
21.7	Hourly rainfall hydrograph at Alishan Climate Station during Typhoons Gloria and Herb	546
21.8	Map of the Chenyoulan stream watershed, NCCI Highway, 4 rain-gauge stations, and 47 debris-flow gullies caused by Typhoon Herb	color
21.9	"Bouldery" (rock-rich) debris flow at Fengqiu	color
21.10	"Sand-silt-clayey" (matrix-rich) debris flow at Longhua	color
21.11	"Cobble-gravely" (mixed evenly with rock and matrix) debris flow at Shenmu	color

Figures xxxiii

21.12	Empirical relation between debris-flow volume and debris-flow watershed area	551
21.13	Empirical relation between duration and intensity of debris-flow causing rainstorms for the entire Chenyoulan stream watershed	552
21.14	Cross section of a "cobble-gravely" debris-flow deposit at Junkengqiao. The debris flow destroyed one home except for the wall foundation	color
21.15	Damaged bridge over the Chushui River at Shenmu. The bridge's clearance was designed for water floods and did not have sufficient conveyance for debris flows	color
21.16	Number of debris-flow papers published by Taiwanese researchers in Taiwanese journals and conferences proceedings between 1985 and 2001	558
21.17	Debris-flow structural countermeasures, such as Sabo dams, debris barriers, and debris basins, at Fengqiu in the Chenyoulan stream watershed	color
21.18	A slit dam built at the outlet of a debris-flow-prone stream at Jianqing in Hualien County. The photo inset shows a series of check dams constructed to stabilize the channel bed and banks along the stream channel upstream of the slit dam	color
21.19	A debris-flow warning sign installed at the crossing of a road and a debris-flow-prone channel	color
22.1	Map of debris flows distribution in China	566
22.2	Location of Jiangjia Ravine	567
22.3	Debris-flow source region of Jiangjia Ravine	568
22.4	Plane view of Jiangjia Ravine	569
22.5	10-min rainfall and debris flow of discharge on 8 July, 2001	570
22.6	1-hour rainfall intensity and related debris flows	570
22.7	10-minute rainfall intensity and related debris flows	571
22.8	Four types of rainfall concerning debris flows in the Jiangjia Ravine	572
22.9	Turbulence in a debris-flow front	573
22.10	Concentration of air entrained in debris flow vs. debris-flow velocity ...	574
22.11	Armoured concrete pillar for measuring impact force, as used in 2003 ...	575
22.12	Diagram of superelevation in channel bends	576
22.13	Relationship between suspended sediment and hydraulic strength	578
22.14	Grain-size distribution for eight debris-flow samples	580
22.15	Grain weight vs. grain size for eight debris-flow samples	580
22.16	Rheological curves of debris-flow slurry samples	582
22.17	Annual sediment load caused by debris flows in the Jiangjia Ravine	583
22.18	Simulation of confluence of branch and main river	585
22.19	Cross section nearby the confluence spot of the Jiangjia Ravine and Xiaojiang River	585
22.20	The longitudinal profile of the Xiaojiang River	586
22.21	Sketch map of the four modes	586
22.22	Debris flow from the Jianjgia Ravine entering the Xiaojiang River	587
22.23	Two types of drainage groove	590
22.24	Main structure mitigation in the Jiangjia Ravine	591
23.1	Location map	596
23.2	Landslides at Rae Lake, Clayoquot Sound	596
23.3	Landsat image of Clayoquot Sound study area	599
23.4	Decision-tree diagram for Clayaquot Sound landslides	608

xxxiv **Figures**

23.5	Potential timber salvage associated with landslide activity	color
23.6	Landslides features, Escalante River	color
24.1	Tourist resort of Sörenberg in Central Switzerland	616
24.2	Hazard matrix. Specific quantitative criteria are provided for the magnitude of the processes	color
24.3	Differential safety concept for floods and debris flows	619
24.4	Aerial photograph of the landslide complex (July, 1999)	621
24.5	Displacement measurement on the Sörenberg landslide complex	623
24.6	Sörenberg topographic map, 1918	625
24.7	Cumulative precipitation amount (mm) for 6 months prior to the event or prior to the end of May	626
24.8	Simulation of debris flows at Sörenberg	color
24.9	Hazard map for Sörenberg	color
24.10	Structural measures in the vicinity of Sörenberg Village	629
24.11	Hazard map after the implementation of a set of structural measures	color
25.1	Location on the New Zealand continental land mass of the six debris-flow examples	636
25.2	Aerial view of the Walter Peak debris flows of November 1999	637
25.3	Sagging slopes on the Lake Wakatipu face of Walter Peak	638
25.4	Map and aerial photograph of the upper Rees River Valley where a debris-flow fatality occurred in 2002	642
25.5	The upper fan of the Rees River tributary where the fatality occurred	643
25.6	Large boulders moved hundreds of metres by the fatal debris flow, Rees River Valley	643
25.7	Aerial view of the drainage basin of Waterfall Creek	645
25.8	Aerial view of the Waterfall Creek bridge site	646
25.9	Profiles of the channel of Waterfall Creek, showing changes now implemented	647
25.10	Trajectory of debris flows from the crest of the waterfall to the bridge deck	648
25.11	Aerial view of Aoraki/Mount Cook village on Black Birch and Glencoe fans	651
25.12	Mount Ruapehu in Tongariro National Park, showing the Whangaehu River Valley lahar route	656
26.1	Map of the study area	660
26.2	Digital elevation model of the Cheekye River basin	661
26.3	Map of the Cheekye basin and fan, showing landforms and other features	662
26.4	Younger Dryas ice margin in Howe Sound about 10,600 ^{14}C yr BP	667
26.5	Ground-penetrating radar profile trending south-west across the lower Cheekye fan	669
26.6	Depth and distribution of radiocarbon ages recovered from excavated pits and natural exposures	670
26.7	Evolution of the lower Cheekye fan	671
26.8	Architectural model of the lower Cheekye fan	672
26.9	Two sections of fan sediment on the lower Cheekye fan	672
26.10	Changes in sediment delivery to the lower Cheekye fan and precipitation in south-western British Columbia during the Holocene	674
26.11	Debris-flow probability-of-exceedence plots for the lower Cheekye fan	676

26.12	Core from Stump Lake showing radiocarbon-dated stratigraphy	677
26.13	Locations of sampled trees scarred by debris flows.	678
26.14	Model run for a debris flow with a volume of $3 \times 10^6 \, \text{m}^3$ and a peak discharge of $9{,}600 \, \text{m}^3/\text{s}$.	color
26.15	F/N plot.	680
27.1	Distribution of major volcaniclastic deposits of the cataclysmic 1980 Mount St. Helens eruption and location of gauging stations	687
27.2	Oblique aerial view looking south to the crater of Mount St. Helens	694
27.3	Oblique aerial photograph of the 1980 debris-avalanche deposit	695
27.4	Oblique aerial photograph of the debris-avalanche deposit at a constriction in the North Fork Toutle River valley near Maratta Creek.	696
27.5	Oblique aerial photographs of debris flows triggered by snowmelt on 18 May, 1980.	697
27.6	Oblique aerial photograph of March 1982 snowmelt-triggered debris flow . . .	699
27.7	Relationships between flow volume and runout distances for volcanic debris flows at Mount St. Helens	701
27.8	Reconstructed hydrographs for the 18 May, 1980, South Fork Toutle and North Fork Toutle debris flows	702
27.9	Mudlines on trees along Muddy River valley illustrating the great difference between flow depth and deposit thickness	705
27.10	Megaclasts in ancient debris-flow deposits.	707
27.11	Typical textures of debris-flow deposits at Mount St. Helens.	709
27.12	Morphologic features of debris-flow deposits at Mount St. Helens.	710
27.13	Examples of "sole layers" at bases of debris-flow deposits	711
27.14	Schematic representation of deposit facies of the March 1982 snowmelt-triggered debris flow showing distal transformation from debris flow to hyperconcentrated streamflow.	712
27.15	Transitional and hyperconcentrated-flow facies of the deposit of the March 1982 debris flow along Toutle River	713
27.16	Grain-size distribution of debris-flow deposits at Mount St. Helens	715
27.17	Annual suspended-sediment yields at Mount St. Helens from 1980–2000	717
27.18	Mitigation of post-1980 sediment transport along the Toutle–Cowlitz River system	719

Tables

2.1	Key terms for mass movement in the "flows" category according to Varnes (1978).	10
2.2	Classification of flow type landslides	13
2.3	Landslide velocity scale	21
7.1	Parameters relevant to the yield rate Y_i	143
7.2	A collection of debris-flow entrainment rates and erosion depths reported in the literature.	148
7.3	Queen Charlotte database: yield rates in relation to channel slope and confinement	152
7.4	Queen Charlotte database: erosion depths (in meters) in relation to channel slope and confinement.	153
8.1	Minimum "fines" concentrations required to permit hyperconcentrated flows to suspend large quantities of sand	170
8.2	Directly observed or measured characteristics of natural hyperconcentrated flows	180
9.1	List of parameters required in various slope failure and post-failure analysis (X – primary; x – secondary relevance; RO – run out analysis; Tur – turbidity currents; Sub. – subsidence analysis; D – drained case; U – undrained case).	214
9.2	Example of parameters and properties which can be derived from various field and laboratory observations.	215
9.3	Rheological and geotechnical properties of sediment, at flow conditions, used for numerical models of debris flows on the Mississippi fan	227
9.4	Parameters used for modeling the debris flows on the Mississippi fan	227
10.1	Notable examples of volcanic debris flows (lahars) and their triggering mechanisms. Adapted from Rodolfo (2000), data from Simkin and Siebert (1994).	250

11.1	Morphologic, vegetation, and drainage features characteristic of debris avalanches and debris flows, and their photographic characteristics	285
11.2	Photo interpretable characteristics of debris avalanches and debris flows	286
13.1	Overview of runout prediction methods discussed in the text	306
14.1	A selection of worldwide criteria for debris-flow triggering threshold	338
17.1	Debris-flow hazard analysis	412
17.2	Suggested semi-quantitative probability scale for debris flows	424
17.3	Suggested debris-flow magnitude classification	426
17.4	Equations for indirect determination of debris-flow velocities	429
17.5	Equations for indirect determination of debris-flow peak discharge, Q_p	431
18.1	Active measures	447
18.2	Types of open barriers	465
18.3	Passive mitigation measures	480
18.4	Documentation and control	485
19.1	Key geotechnical Parameter ranges of a typical pyroclastic sequence	502
19.2	Recurrence of the morphological settings recognized in the failure areas	505
19.3	Summary of the analysis of measured cross sections, velocity, runout, and deposit thickness and their comparison with DAN model analysis. (slide numbers correspond to landslides in Figure 19.6. For location of cross sections in flow bends and deposit thickness measurements see Figure 19.6.)	507
20.1	Lithological units in the Avila Mountain	524
20.2	Geometric and physiographic characteristics of main basins in the disaster area	528
20.3	Volume of sediment deposition in alluvial fans of main streams	531
20.4	Comparison of maximum flow discharges estimated by different methods	532
20.5	Definition of hazard level	535
20.6	Event intensities for water flooding	536
20.7	Event intensities for mud and debris flow	536
21.1	Site and rainfall factors affecting 47 debris flows, which were caused by Typhoon Herb in the Chenyoulan stream watershed	547
21.2	Characteristics of catchments with 9 major debris flows caused by Typhoon Herb in the Chenyoulan stream watershed	549
21.3	Characteristics of rainfall and debris flows caused by Typhoon Herb at 9 locations in the Chenyoulan stream watershed	549
21.4	Consequences of Typhoon Herb in Taiwan	554
21.5	Consequences of Typhoon Herb in the 9 debris-flow gullies in the Chenyoulan stream watershed	556
22.1	Morphologic characteristics of the Jiangjia Ravine	568
22.2	Rainfall grade	571
22.3	Ratio of debris-flow events under different rainfall patterns	572
22.4	Sediment and entrained air concentrations of debris-flow samples	574

22.5	Density of a sample and its slurry from the debris flows	577
22.6	Classification for concentration	579
22.7	Particle size of debris flows and associated slurries	581
22.8	Yield stress and Bingham number of the slurries	582
22.9	Sediment transporting capacity of debris flow in the Jiangjia Ravine	583
22.10	Measured annual sediment load and erosion module of debris flows	584
23.1	Comparison of terrain features with post-harvesting landslide density	603
23.2	Terrain features: clear-cut landslide summary statistics	604
24.1	Past events in the Sörenberg landslide complex	624
24.2	Total precipitation prior to the 6 events. The rank indicates the position within the 100-year observation period	626
24.3	Adjustment of the legal framework for future land use in Sörenberg (land-use plan, building code)	628
24.4	Long-term and short-term measurements and observation	629
24.5	Structural measures to control debris-flow hazards in the Sörenberg area	630
24.6	Emergency preparedness for the debris-flow threat at Sörenberg	631
26.1	Radiocarbon ages from lower Cheekye fan sediments	664
26.2	Radiocarbon ages from Stump Lake sediments	668
27.1	Fatalities caused by volcanic debris flows between 1950 and 2001	686
27.2	Characteristics of deposits from the 18 May, 1980, Mount St. Helens eruption	689
27.3	Characteristics of debris flows and hyperconcentrated flows at Mount St. Helens	692

Abbreviations

AFM	Acoustic Flow Monitor
ATM	Airborne Thematic Mapper
BAF	bend-angle function
CCD	charge-coupled devices
CFM	Cumulative/Frequency Magnitude
CIR	colour infrared
CMOS	metal oxide semiconductor
COA	Council of Agriculture
COE	Corps of Engineers (US Army)
CRU	Climatic Research Unit
DAS	data acquisition system
DDFORS	Dongchuan Debris Flow Observation and Research Station
DEM	digital elevation model
DFDP	debris-flow danger period
DOQ	Digital Orthophoto Quadrangle
DPRI	Disaster Prevention Research Institute
DTL	détecteur lave torrentielle
EDFDP	extreme debris-flow danger period
ENSO	El Niño Southern Oscillation
ERLAWS	East Ruapehu Lahar Alarm and Warning System
ESP	effective stress path
FIR	far-infrared
GCM	General Circulation Model
GIO	Government Information Office
GIS	Geographic Information Systems
GPR	ground-penetrating radar
LB	Leaky Barrel
MAP	mean annual precipitation

MDFDP	main debris-flow danger period
MIR	mid-infrared
MMW	millimetre wave
MTF	modulation transfer function
NCCI	New Central Cross-Island
NCKU	National Cheng Kung University
NIR	near-infrared
NSC	National Science Council
NWS	National Weather Service
OTF	optical transfer function
PDO	Pacific Decadal Oscillation
PZ	piezometer
REV	representative elemental volume
RF	radio frequency
RP	resolving power
RTU	remote terminal unit
SAR	synthetic aperture radar
SCADA	supervisory control and data acquisition
SFBA	San Francisco Bay Area
SIP	strongly implicit procedure
SRS	sediment retention structure
SWCB	Soil and Water Conservation Bureau
SWIR	short-wave infrared
TDR	time domain reflectometry
TGRU	Taiwan Geomorphological Research Unit
TSP	total stress path
VEI	volcanic explosivity index
VNIR	visible/near-infrared
VSAT	very small aperture terminals

1

Introduction

Matthias Jakob and Oldrich Hungr

1.1 DEBRIS FLOWS – A GLOBAL PHENOMENON

Debris flows occur in all regions with steep relief and at least occasional rainfall. They are one of the most frequent mass movement processes and play an important role in moving sediment from steep lands and into river systems (e.g., Rapp, 1960). Their high flow velocity, impact forces, and long runout, combined with poor temporal predictability, cause debris flows and debris avalanches to be one of the most hazardous landslide types. Many of the world's most devastating landslide disasters, as measured by loss of life and/or economic value, are attributed to debris flows and their volcanic counterparts called lahars (e.g., Northern Venezuela, 1999; Taiwan, 1996; Columbia, 1985, 1998; see Figure 1.1 and Chapters 20 and 21). A significant portion of the death toll during regional disasters, caused by major storms and earthquakes, is due to debris flows and debris avalanches on steep terrain.

Direct debris-flow damage includes loss of human life, destruction of houses and facilities, damage to roads, rail lines and pipelines, vehicle accidents and train derailments, environmental damage from product spills, damage to agricultural land, livestock, and forest stands, disruption of water supply systems, devaluation of fisheries, and many other losses that are difficult to quantify (see Figure 1.2).

The most important indirect cost of debris flows includes disruption of traffic and communication along linear facilities. It is estimated, for example, that only one day of traffic closure on one of the North American transcontinental rail lines may cost about US $10 million. Another indirect cost is the sterilization of potentially developable land on debris fans. In many mountainous regions these landforms represent the most suitable terrain segment for development in terms of slope gradient and exposure to river flooding; the remainder being steep lands and floodplains (see Figure 1.3).

Figure 1.1. Damaged buildings on the Caraballeda Fan in the Vargas Province, Venezuela, caused by a series of debris flows descending the slopes of the El Avila range (background) in December, 1999.

1.2 COPING WITH DEBRIS-FLOW HAZARDS

Earlier societies dealt with debris-flow hazards through experience, and by avoiding the most exposed areas (Chapter 3). Such an approach is no longer satisfactory, as development proceeds at a fast rate and encroaches quickly into hazard areas, without allowing sufficient time to build experience or understanding. Thus, increasing reliance is being placed on prediction and prevention of debris-flow hazards and related phenomena, which is the task of geologists, geotechnical engineers, and hydrologists skilled in debris-flow studies. The crossing of debris-flow channels and the continued use of debris fans must be based on reliable understanding of the hazardous processes.

Prevention can be achieved by careful siting of facilities away from hazard zones, or by building protective measures (Chapter 18). In many cases, however, it is possible to tolerate existing or new development in hazard zones, based on the concept of acceptable risk. These cases represent the greatest challenge to the debris-flow specialist. Ultimately, it is not the task of the debris-flow expert to recommend an acceptable level of risk but to predict the probability of occurrence, magnitudes,

Sec. 1.3]	The need for this book	3

Figure 1.2. Freight train derailed by the direct impact of a small ddebris flow in high mountains. Although infrequent, such accidents cause property damage and could result in loss of life.

runout distances, velocities, impact forces, and associated potential damage and a range of other parameters necessary to quantify risk.

1.3 THE NEED FOR THIS BOOK

In recent decades, the scientific community responded enthusiastically to the challenges of debris-flow hazard assessment. Most affected countries now have their own specialists in debris-flow science. Hundreds of research articles appear each year in conferences such as the International Symposia on Landslides, meetings of the North American and European Geophysical Unions, conferences on Geotechnical Engineering and various geological meetings and congresses. Relevant articles can be found in journals devoted to natural hazards, geotechnical engineering, geomorphology, geophysics, hydraulics, hydrology, Quaternary sciences, and stratigraphy. Three specialized international conferences on debris flows have been held so far (Chen, 1997; Wieczorek and Naeser, 2000; Rickenmann and Chen, 2003). The next conference is planned in Chengdu, China in 2006. Together with a special volume in the series *Reviews in Engineering Geology* (Costa and Wieczorek, 1987), these publications represent an excellent overview of the debris-flow science.

Figure 1.3. An aerial view of a debris fan in western Honshu Island, Japan, shortly after the impact of a debris flow, which destroyed several houses and killed 10 persons. Debris fans often constitute highly desirable development land, free of river flooding and steep land hazards, but at the price of exposure to debris flows.

What is missing, however, is a volume offering a more systematic review of the state-of-the-art, thematically organized to serve both as an organized repository of some of the main topics and recent advances in the field, and as a key to the overwhelming literature on the subject. To provide such a volume was the main objective of the editors. Individual contributors, who have been selected from a cross section of worldwide leaders in debris-flow science were asked to summarize the present state of their respective specialty focusing on practicality and predictive techniques. Such a task is easier in some parts of the science than in others and we do not anticipate the book will become a "manual" for practice. Nevertheless, we believe that it is a fair summary of the existing knowledge and that it will serve as a good starting point for anyone wishing to understand debris flows and related phenomena and to help reduce their damaging impact on society.

Regarding the format of the book, the editors attempted to strike a balance between exact, theoretical approaches and practical, often approximate, empirical methods. The reader should thus be able to choose whatever approach suits her/his needs. Another balance was attempted between general analytical studies and case histories. The editors believe that no general understanding in the debris-flow science is valid before its application to field cases. For this reason our knowledge base must be continuously measured and re-evaluated against field experience.

1.4 STRUCTURE OF THE BOOK

The first part of the book focuses on debris-flow processes and related phenomena while the second part presents selected case studies. Part I begins with an annotated glossary of terminology (Chapter 2). In the past, a puzzling variety of names have been used creating substantial confusion amongst scientists and practitioners. This chapter is meant to clarify the terms and suggest standards and definitions. In cases where authors chose to use other definitions we have asked them to discuss the differences. Chapter 3 is an account of the history of global debris-flow research and its significance. This chapter cannot be all-inclusive, but provides an anecdotal summary of global debris-flow recognition, analysis, and mitigation.

Chapter 4 offers an overview of the currently accepted principles of slope stability to clarify terms that are used in later chapters and to provide a basic understanding of landslide initiation in shallow soils. Chapter 5 focuses on the initiation of debris flows in a wide variety of settings and on liquefaction as well. Chapter 6 summarizes the current understanding of debris-flow mechanics. This chapter contains many equations describing complex flow behaviour that may ultimately lead to predictive models. Though most of this chapter may not be very helpful to the practitioner a complete treatment of the current understanding of debris-flow mechanics is warranted. Chapter 7 sheds some light on one of the most poorly understood topics in debris-flow science – debris entrainment. In practice the volume estimation of debris flows is necessary to compute runout distance, peak discharge, and area inundated by debris on fans, parameters that will often be used for zoning or the design of debris-flow mitigation structures.

Chapter 8 discusses hyperconcentrated flows (debris floods) as an important and often neglected part of the continuum between water and flowing debris. Although hyperconcentrated flows are typically observed in areas with fine grained surficial materials such as the loess plateaus in southern China, or in the vicinity of volcanoes, a hyperconcentrated flow phase may be present in most debris flows observed worldwide. Chapter 9 details the current scientific understanding of submarine debris flows, which pose an indirect hazard to nearby coastlines and which may damage submarine installations such as pipelines, telecommunication equipment, or foundations of oil and gas drilling platforms.

Some of the most devastating debris flows have originated near volcanoes and have travelled over very large differences transferring their destructive potential into urbanized areas. Research on lahars has advanced greatly in the past 20 years and is summarized in Chapter 10.

The latter chapters of Part I then concentrate on the identification of the hazard using remote-sensing techniques (Chapter 11), debris-flow instrumentation (Chapter 12), debris-flow runout prediction (Chapter 13), the use of hydro-climatic data to forecast debris-flow occurrence (Chapter 14), and the increasing importance of wildfires and logging for the occurrence of debris flows (Chapters 15 and 16). The penultimate chapter addresses hazard assessment on individual debris-flow fans (Chapter 17). Part I concludes with a detailed documentation of active and passive methods to mitigate debris-flow hazards and risk (Chapter 18).

Part II is a selection of relevant case studies from eight countries with a variety of physiographic and climatic settings. Of course, the definition of "relevant" is difficult and there are many more interesting case studies that deserve inclusion. The case studies serve to illustrate the use of some methods described in Part I and to show the limitations of their application. Examples include a series of catastrophic debris avalanches and debris flows in southern Italy (Chapter 19), one of the most devastating debris-flow clusters in northern Venezuela (Chapter 20), and an account of a record-breaking typhoon in Taiwan (Chapter 21). Chapter 22 describes one of the world's most fascinating natural debris-flow laboratories in southern China. Chapter 23 describes a regional study of debris-flow hazards associated with clear-cut logging and road building on the west coast of Vancouver Island, Canada, that serves as an example of the consequences of various forestry techniques in humid maritime climates. Chapter 24 summarizes a detailed debris-flow hazard and risk assessment in Switzerland, while Chapter 25 describes a collection of interesting case studies from New Zealand. Chapter 26 provides a summary of Canada's most detailed debris-flow hazard and risk analysis of a large fan at the foot of a Quaternary volcano. Part II ends with a comprehensive summary of work completed on one of the world's best-studied volcanoes – Mount St. Helens in Washington, USA (Chapter 27).

Our intention then is that this book will find its way onto the shelves of academics in the earth sciences and engineering fields, consultants, specialists in governmental agencies and international organizations, and those local decision makers whose jurisdiction encompasses debris-flow-prone areas. It is our hope that this book will contribute to a propagation of state-of-the-art techniques into everyday engineering and geoscience practice, which ultimately will decrease the number of deaths and socio-economic losses due to debris flows and related phenomena.

1.5 REFERENCES

Chen, C.L. (ed.) (1997) *Proceedings of the 1st International Conference on Debris Flow Hazards Mitigation: Mechanics, Prediction and Assessment*. ASCE, San Francisco.

Costa, J.E. and Wieczorek, E. (eds) (1987) Debris flow: Process, description and mitigation. *GSA Reviews in Engineering Geology*, **VII**.

Rapp, A. (1960) Recent development of mountain slopes in Karkevagge and surroundings, Northern Scandinavia. *Geografiska Annaler*, **42**, 55–200.

Rickenman, D. and Chen L.C. (eds) (2003) *Proceedings of the 3rd International Conference on Debris Flow Hazards Mitigation: Mechanics, Prediction and Assessment*. Millpress, Rotterdam.

Wieczorek, G.F. and Naeser, N.D. (eds) (2000) *Proceedings of 2nd International Conference on Debris Flow Hazards Mitigation: Mechanics, Prediction and Assessment, Taipei*. A.A. Balkema, Rotterdam.

2

Classification and terminology

Oldrich Hungr

2.1 CLASSIFICATION OF DEBRIS FLOWS AND AVALANCHES

The book *Die Muren* by Stiny (1910) was one of the first monographs ever devoted to a specific type of landslide. His definition of *debris flow* begins with the description of a flood in a mountain torrent, carrying suspended load and transporting quantities of bedload. As the amount of sediment carried by the flow increases, "at a certain limit it has changed into a viscous mass consisting of water, soil, sand, gravel, rocks and wood mixed together, which flows like a lava into the valley".

In the USA, Sharpe (1938) made a distinction between debris flows and *debris avalanches*. His definition of the first is similar to that of Stiny and can be paraphrased as rapid flow of saturated, unsorted debris in a steep channel. The second was a rapid shallow landslide from a steep slope, whose morphology resembles that of a snow avalanche (no channel or flood flow is mentioned). The same basic division was retained by Varnes in his influential works on landslide classification carried out for the US Transportation Research Board (Varnes, 1954, 1978), and thus became fairly well established in the North American usage (Table 2.1).

In England, Hutchinson (1968) also recognized debris flow, later dividing it into *channelized* and *hillslope* varieties (Hutchinson, 1988), corresponding respectively to the debris flow and debris avalanche of Varnes. At the same time, some authors preferred the term mudflow (Blackwelder, 1928; Bull, 1964; Crandell, 1957), referring presumably to relatively fine-grained debris flows occurring on rapidly eroding semi-arid slopes in sedimentary rocks and on volcanoes. Broscoe and Thomson (1969) referred to an "alpine mudflow", describing a phenomenon essentially similar to a debris flow. The term mudflow was once also applied to slow-moving, clayey earth flows in England (Skempton and Hutchinson, 1969), although this usage has since been abandoned (pers. commun., J.N. Hutchinson, 2003).

Like Stiny, most practitioners and researchers recognize that in steep granular channels, there is a continuum between sediment-charged flooding and debris flow,

Table 2.1. Key terms for mass movement in the "flows" category according to Varnes (1978).

Rate of movement	Bedrock	Debris (<80% sand and finer)	Earth (>80% sand and finer)
Rapid and higher (>1.5 m/day)	Rock flow (creep, slope sagging)	Debris flow Debris avalanche	Wet sand and silt flow Rapid earth flow Loess flow Dry sand flow
Less than rapid (<1.5 m/day)		Solifluction Soil creep Block stream	Earth flow

as normal bedload transport processes such as rolling and saltation are supplanted by massive bed instability with increasing slope angle (cf. Hutchinson, 1992). Beverage and Culberson (1964) drew attention to the transitional process, coining the term *hyperconcentrated flood* (Chapter 8). Costa and Jarett (1981) proposed to distinguish between debris flow and hyperconcentrated flood on the basis of (presumably mean) sediment concentration. A similar distinction between *"debris flood"* and debris flow was suggested by Aulitzki (1980) in Austria, who pointed out the contrast in the physical character of the two phenomena, one characterized by large clasts and heavy impact, the other by finer material and less vigorous motion. The same distinction is reflected in the French terms "coulée de boue" and "lave torrentielle" (pers. commun., E. Leroi, 2000).

Most debris flows carry some amount of organic matter. Those originating in forested steep lands may contain as much as 60% by volume of large organic debris (timber remnants). Swanston (1974) described such organic-rich landslides from the north-west Pacific Coast of North America and coined a term *"debris torrent"* for them, which is still in use in western USA and Canada. This term is linguistically questionable and its use is declining (cf. Slaymaker, 1988).

In an extensive modification of the Varnes landslide classification, Cruden and Varnes (1996) proposed to restrict the use of the term debris flow to its literal meaning (i.e., a phase of a landslide, during which flowage of coarse material (debris) is occurring). In this usage, many landslide types, including rock avalanches, dry granular flows, and liquefaction flow slides could exhibit a phase termed debris flow. The present author feels that it is better to preserve the term debris flow as an established keyword representing the entire phenomenon, including an initiating slide on a steep slope, rapid flow along a steep confined channel, and deposition on a debris fan. In some cases, of course, the term may be understood in its literal sense (e.g., Chapter 6). In an attempt to reconcile the confused state of terminology for flow-like landslides, while at the same time preserving established concepts and keywords, Hungr et al. (2001) proposed formal definitions of several types of flow-like landslides including debris flows, mud flows, debris avalanches, and debris floods as described below.

2.1.1 Debris-flow materials

Before further discussion of debris-flow classification, it is necessary to review terms describing materials involved in debris flow. The well-established North American landslide classification system of Varnes (1978) and its modification by Cruden and Varnes (1996) distinguish only two types of material: *debris*, a soil containing more than 20% gravel and coarse sizes and *earth* with less than 20% coarse sizes. Hungr et al. (2001) proposed replacing these rather arbitrary definitions with new ones, derived from geomorphology.

The term "*earth*" should refer to unsorted clayey (plastic) colluvium derived from clays or weathered clay-rich rocks, with a consistency closer to the plastic limit than the liquid limit. Such low-sensitivity clay-rich materials of intermediate consistency produce slow to rapid sliding movements along distinct slickensided shear surfaces (earth flows, Figure 2.1).

Geologically, the term "*mud*" refers to liquid or semi-liquid clayey material (Bates and Jackson, 1984). Some mud flows derived from volcanic sources may have clay contents and plasticity indices of more than 10% (Jordan, 1994; Scott et al., 1992). For example, mud flows derived from montmorillonitic shales may have clay contents exceeding 50% (Bull, 1964). Hungr et al. (2001) proposed that the term "mud" be used for soft, remoulded clayey soils whose matrix (sand and finer) is

Figure 2.1. A discrete shear surface in earth material (the Big Slide, Quesnel, British Columbia, Canada).

Figure 2.2. Debris deposits in the fan of Pierce Creek near Chilliwack, British Columbia, Canada.

significantly plastic (plasticity index >5%) and whose liquidity index during motion is greater than 0.5. In order to convert insensitive stiff or dry clayey earth material at a landslide source into mud, rapid mixing with surface water is required to raise the water content to, or above, the liquid limit.

"*Debris*" was defined by Hungr et al. (2001) as loose unsorted material of low plasticity such as that produced by mass wasting processes (*colluvium*), weathering (residual soil), glacier transport (till or ice contact deposits), explosive volcanism (granular pyroclastic deposits), or human activity (e.g., mine spoil). Texturally, debris is a mixture of sand, gravel, cobbles, and boulders, often with varying proportions of silt and a trace of clay (Figure 2.2). Debris may also contain a significant proportion of organic material, including logs, tree stumps, and organic mulch (e.g., Swanston, 1974). Debris is usually non-plastic or weakly plastic and it is characteristically unsorted, sometimes gap-graded (*diamicton*). Many descriptions make reference to *coarse clasts* and *matrix*, although no set separation between these two phases has yet been established. Most often, matrix is considered to be material of sand size or finer, although gravel sizes are sometimes included.

2.1.2 Types of flow-like landslides

Using these material components, Hungr et al. (2001) proposed the following definitions (see also Table 2.2):

Table 2.2. Classification of flow type landslides.
Hungr et al. (2001).

Material	Water content[1]	Special condition	Velocity	Name
Silt, sand, gravel, and debris (talus)	Dry, moist, or saturated	No excess pore-pressure Limited volume	Various	*Non-liquefied sand (silt, gravel, debris) flow*
Silt, sand, debris, and weak rock[2]	Saturated at rupture surface	Liquefiable material[3] Constant water content	Extremely rapid	*Sand (silt, debris, rock) flow slide*
Sensitive clay	At or above liquid limit	Liquefaction *in situ*[3] Constant water content[4]	Extremely rapid	*Clay flow slide*
Peat	Saturated	Excess pore-pressure	Slow to very rapid	*Peat flow*
Clay or earth	Near plastic limit	Slow movements Plug flow (sliding)	Less than rapid	*Earth flow*
Debris	Saturated	Established channel[5] Increased water content[4]	Extremely rapid	*Debris flow*
Mud	At or above liquid limit	Fine-grained debris flow	Greater than, very rapid	*Mud flow*
Debris	Free water present	Flood[6]	Extremely rapid	*Debris flood*
Debris	Partly or fully saturated	No established channel[5] Relatively shallow, steep source	Extremely rapid	*Debris avalanche*
Fragmented rock	Various, mainly dry	Intact rock at source Large volume[7]	Extremely rapid	*Rock avalanche*

[1] Water content of material in the vicinity of the rupture surface at the time of failure.
[2] Highly porous, weak rock (examples: weak chalk, weathered tuff, pumice).
[3] The presence of full or partial *in situ* liquefaction of the source material of the flow slide may be observed or implied.
[4] Relative to *in situ* source material.
[5] Presence or absence of a defined channel over a large part of the path, and an established deposition landform (fan). *Debris flow* is a recurrent phenomenon within its path. *Debris avalanche flow* is not.
[6] Peak discharge of the same order as that of a major flood or an accidental flood. Significant tractive forces of free flowing water. Presence of floating debris.
[7] Volume greater than 10,000 m³ approximately. Mass flow, contrasting with fragmental rock fall.

- *Debris flow* is a very rapid to extremely rapid flow of saturated non-plastic debris in a steep channel. Plasticity index is less then 5% in sand and finer fractions (Figure 2.3).
- *Mud flow* is a very rapid to extremely rapid flow of saturated plastic debris in a channel, involving significantly greater water content relative to the source material (plasticity index >5%).

Figure 2.3. A debris flow in the Khumbu Himalayas, Nepal.

Figure 2.4. A group of debris avalanches triggered by logging and road construction on the west coast of Vancouver Island, Canada.

- *Debris flood* is a very rapid, surging flow of water, heavily charged with debris, in a steep channel.
- *Debris avalanche* is a very rapid to extremely rapid shallow flow of partially or fully saturated debris on a steep slope, without confinement in an established channel (Figure 2.4).

These definitions do not stray far from established North American or British terminology. Being based on readily observable characteristics, they also permit translation of the given concepts into other languages. It is not necessary to use the terms proposed here, but the concepts behind the given definitions are useful in order to develop a typological scheme suitable for research and practical application. The definitions stress aspects of practical significance. For example, the distinction between debris flow and debris avalanche is useful in hazard studies. In case of a debris-flow hazard, the study may concentrate on a given established path (valley or gully) and deposition area (fan). In contrast, a debris avalanche hazard may potentially affect any of a number of steep slopes in a given area. (Of course, debris avalanches will often enter an established confined path and become debris flows.)

It is difficult to base the distinction between a debris flow and debris flood on sediment concentration, a quantity that varies both spatially and in time. It can instead be based on the observed or potential *peak discharge* of an *event*. Discharge limited at most to 2–3 times that of a major flood is the most

Figure 2.5. A vehicle surrounded by debris flood deposits, Britannia Beach near Vancouver, Canada.
Photo K. Fletcher.

important aspect of a debris flood, as it results in relatively limited impacts and relatively low flow depth, in other words, a limited destructive potential (Figure 2.5). The exceptions here are debris floods produced by unusual (catastrophic) water releases from basins, caused by man-made or landslide dam breaks or glacier outbursts (e.g., Jakob and Jordan, 2001). Debris flows, on the other hand, produce extremely large, peak discharges spontaneously, by means of surge growth processes (e.g., Pierson, 1980; Hungr, 2000). These discharges may be as much as 50 times as large as those of a major flood (VanDine, 1985; Jakob and Jordan, 2001). Their destructive potential is much greater than that of a flood.

2.2 TERMINOLOGY

A typical *debris-flow path* is divided into an *initiation zone, transport*, and deposition zone (Figure 2.3). Most often, the initiation zone is a slope failure in the headwall or side slope of a gully or stream channel. The slope failure may have the character of a shallow *debris slide*, transforming into a debris avalanche. It may also be the failure of a man-made (road) fill, or a natural *rock slide*. Sometimes, the bed of the channel itself may become unstable during extreme discharge and the debris flow initiates spontaneously in the steep bed of the channel (Chapter 15). Generally, the area of debris-flow initiation has a steep slope, ranging between 20° and 45°. There may not be sufficient potential energy on flatter slopes to start a failure of granular soil. On the other hand, slopes steeper than 45° usually have a soil cover too thin or discontinuous to be vulnerable to sliding. The *magnitude* (volume) of the *initiating slide* can vary considerably. Often, an initiating slide of only a few tens of cubic metres will trigger a major debris flow. On the other hand, a major debris flow may simply be the final phase or extension of a large rock slide or debris avalanche.

Sometimes, a single initial landslide occurs and leads to a single debris-flow event. However, often, a debris flow-producing storm causes nearly simultaneous triggering at a number of locations, including debris slides/avalanches, slumps, rock falls, and failures of artificial structures such as road fills and cut slopes. While conducting hazard assessment studies, it is prudent to consider multiple triggers, involving several tributaries of a given basin (Chapter 17).

Once initiated, the rapid initial landslide may continue downslope without confinement. In granular materials this always leads to disintegration, producing *flow-like motion* characterized by nearly complete remoulding of the moving mass and a more-or-less evenly distributed velocity profile (debris avalanche). *Rapid loading* of saturated substrate in the flow path may increase the volume and also introduce a higher level of saturation. This middle part of a debris avalanche can be referred to as the transportation zone. Deposition will begin, once the slope angle decreases below a certain value (Chapter 6). Multiple deposits of debris avalanches may form a *debris ("colluvial") apron* at the foot of the slope, which is often recognizable as a distinct terrain unit.

Many debris avalanches enter established gullies or steep stream channels and continue flowing as debris flows in the transportation zone. The entry into the channel may occur spontaneously, without any significant decrease in velocity of the initiating slide or avalanche. In other cases, the initiating slide may arrest in the channel and build a short-lived *landslide dam*. Subsequent breaching and rapid erosion of this dam forms a *debris-flow surge* in the channel. Transportation zones ("gorges") of debris flows are usually steeper than 10°. They may consist of non-erodible bedrock channels or cascades, or channels with erodible soil banks and bedrock base, or fully-erodible gullies cut in overburden, exhibiting perhaps an armour of lag boulder deposits in the base. The debris flow will entrain loose material from the bed and banks. The amount of material entrained per unit length of the channel (measured in cubic metres per metre) was called the *debris yield rate* (Chapter 7). If the yield rate is divided by the top width of the wetted portion of the channel, the resulting average thickness of entrained material is referred to as the *erosion depth*.

Debris flows commonly move in distinct *surges* or slugs of material, separated by watery *intersurge flow*. A *debris-flow event* may consist of one surge, or hundreds of successive waves. The mechanics of surge growth can vary (e.g., Takahashi, 1991). Some surges result from *flow instability* caused by the *longitudinal sorting* of debris-flow material. Such surges are characterized by *boulder fronts* that are relatively free of matrix (e.g., Pierson, 1980, 1986). The main body of the surge is a finer mass of liquefied debris (Figure 2.6). The tail (or "*afterflow*") is a dilute, turbulent flow of sediment-charged water, similar to a debris flood. The growth of the boulder front

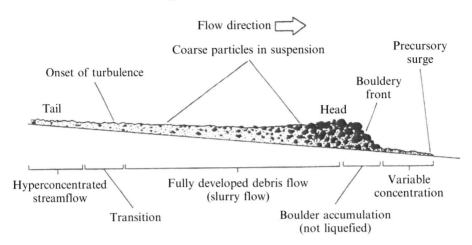

Figure 2.6. Diagram of a debris-flow surge with a boulder front.
From Pierson (1986).

causes a rise in the level of flow behind it, leading to a proportional magnification of the *peak discharge* (Iverson, 1997; Hungr, 2000; Chapter 6).

Another type of surging occurs in fine-grained debris flows or debris floods that lack significant boulder content. Here, the main flow at peak discharge can be of a laminar nature, but it is retained behind a *turbulent front* (Davies, 1986; Takahashi, 1991). Similar flow instability and surging can be observed in thin sheet flows of water over smooth pavement. It is characterized by uniform surge magnitude and spacing. The famous debris flows in the Jiang-jia Ravine, western China, are of this kind (Chapter 22).

Alternative reasons for surging behaviour may include spontaneous surge growth by flow instability as mentioned above, by non-simultaneous triggering of separate initiating landslides, or by periodic stalling of sluggish flow in a channel, followed by a boulder *dam breach* and re-mobilization. Consequently, the individual surges may be of approximately equal volume, or may each be different. Time intervals between surges can be seconds or hours. The total number of surges in a debris-flow event can range from one to many tens. Debris avalanches also surge, usually as a result of *retrogressive failure* or *enlargement* of the source slide.

The main deposition area of a debris flow usually occurs on an established fan, referred to as a *debris fan* or *colluvial fan* or *cone*. The deposition usually occurs as a result of a combination of slope reduction and a loss of confinement. As confinement is lost, the main part of the surge behind the front collapses and the front is deprived of the hydraulic thrust propelling it forward. The front slows down, steepens and may be partly expelled to the margins of the channel, where it builds elongated ridges or *levees* of coarse material (Figure 2.7). Such levees may sustain a moderate degree of confinement beyond what is provided by the morphology of the channel. Eventually, however, the dilute material from the main body of the surge breaks through

Figure 2.7. A debris-flow channel flanked by levee deposits. Note pronounced superelevation of the flow in the two bends, as marked by lines (Cathedral Mountain, Rocky Mountains, Canada).

and overtakes the boulder front, which will then stall and form a lobate *boulder train* deposit, usually on the outside of a bend. Such a process may repeat itself many times on the surface of the fan, thus gradually reducing the content of coarse clasts in the debris surge as well as its peak discharge along its travel down the fan. This behaviour is often the cause of sudden channel abandonement and flow re-direction (*avulsion*) on debris fans and formation of *distributary channels*. Short temporary levees and boulder pockets may sometimes be observed at points of channel widening above the fan as well.

As a result of the deposition process described in the previous paragraph, debris-flow behavior often varies with distance downstream from the debris *fan apex*. In the *proximal part* of the fan, coarse debris forms high discharges and thick deposits. In the *distal part*, finer and thinner deposits form and flow velocities may be reduced. An *afterflow* of heavily sediment-charged water will always reach the *margin* of the debris-flow fan and may continue into the stream channel system below, with the character of a debris flood. Many debris-flow deposits on fans are significantly reworked by water flow immediately following a debris-flow event. Such modifications must be taken into consideration when conducting field studies of debris-flow fans and deposits.

Due to the requirement of steep slopes, debris-flow activity in regions dominated by granular soils is usually confined to basins of zero (gully) to second order. Higher order channels are more likely to produce debris flood events. Basins supplied with large quantities of finer-grained debris from volcanic or residual soil sources may experience debris flows in basins of higher order (Chapter 20). A debris-flow basin may thus contain only one dominant channel, or a number of channels, several of which may be active simultaneously. Morphological parameters such as mean slope, rugosity, and channel density have been correlated with the potential to produce debris flows (e.g., Jakob, 1996; Wilford et al., 2004).

Sources of debris include soil blankets and veneers on steep slopes, colluvial gully fills, channel bedload material, zones of weathered or altered rocks, residual soils, headwalls and side slopes of steep gullies, talus deposits, man-made fills, and similar accumulations of unstable or erodible material. The debris is delivered to the debris-flow channel by shallow or deep-seated sliding from the initiation slide and contributing slides triggered by undercutting of slope toes along the path. The most prolific debris sources are strongly undercut unstable channel sides, which may sometimes represent a continuous line of slumps or slides delivering debris directly into the channel. Direct erosion of bed and banks is another debris generation mechanism. Timber from trees toppled into the channel will add an organic component. Often, bank instabilities operate for a period of time to "prime" the channel with erodible material, which is eventually removed by a debris-flow event.

Velocity of movement is an important parameter of all landslides, as it relates to hazard intensity (e.g., Morgenstern, 1985; Hungr, 1997). According to the velocity classification proposed by Varnes (1978) and Cruden and Varnes (1996), as shown in Table 2.3, most of the phenomena treated in this book are capable of reaching the "*extremely rapid*" class (i.e., speeds exceeding that of a person running). Therefore, with the possible exceptions of some debris floods, these phenomena are of a catastrophic nature, being highly capable of inflicting loss of life. Debris-flow velocities are often estimated during field inspections by observing and analysing the *superelevation* of the flow in channel bends, as evidenced by *mud lines* or *trim lines* (lines of vegetation damage) on the banks (Costa, 1984). Similar observations serve to establish the *maximum flow cross section* and *peak discharge* of the largest surge. A difficulty often arises here, as the channel observed during a field inspection may have been enlarged by *water erosion* in the final stages of the debris-flow event.

The objective of *hazard assessment studies* is to provide estimates of the likely behaviour of debris flows, avalanches, or floods at a given site (cf. Chapter 17). Because the impact of debris flows varies with location on the fan, the concept of *hazard intensity*, as a spatial and temporal function, is useful (Hungr, 1997). In short, a hazard study should estimate the type of events, their magnitudes and probabilities of occurrence, and hazard intensity parameters. The hazard probability is best assessed by means of *Cumulative Frequency/Magnitude* (CFM) curves, derived from historical data, stratigraphic observations of the fan, or dendrochronology. Intensity parameters include estimated *velocity*, *flow depth*, and *thickness of deposits*, together with derived parameters such as *impact forces* on structures and *runup* against barriers and dykes. Of course, the intensity parameters are distributed across a hazard area, whose extent is determined by a *runout analysis*. A given location on a debris fan may not necessarily be impacted each time a debris flow

Table 2.3. Landslide velocity scale.
After Cruden and Varnes (1996).

Velocity class	Description	Velocity (mm/sec)	Typical velocity	Typical human response
7	Extremely rapid			Nil
	↓	5×10^3	5 m/sec	
6	Very rapid			Nil
	↓	5×10^1	3 m/min	
5	Rapid			Evacuation
	↓	5×10^{-1}	1.8 m/hr	
4	Moderate			Evacuation
	↓	5×10^{-3}	13 m/month	
3	Slow			Maintenance
	↓	5×10^{-5}	1.6 m/year	
2	Very slow			Maintenance
	↓	5×10^{-7}	16 mm/year	
1	Extremely slow			Nil

occurs, as the effect of a given event will normally be concentrated along a *damage corridor* of a given width. The ratio of damage corridor width and the total width of the hazard zone (often the circumference of the fan) is sometimes equated with the conditional *probability of intensity* (Hungr, 1997).

2.3 ACKNOWLEDGMENTS

Drs. M. Bovis, M. Jakob, T. Pierson, D. Rickenmann, and J. Vallance provided useful comments.

2.4 REFERENCES

Aulitzky, H. (1980) Preliminary two-fold classification of debris torrents. In: *Proceedings of "Interpraevent" Conference, Bad Ischl, Austria* (Vol. 4, pp. 285–309, translated from German by G. Eisbacher). Internationale Forschungsgesellschaft, Interpraevent, Klagenfurt.

Bates, R.L. and Jackson, J.A. (1984) *Dictionary of Geological Terms* (3rd edn, 571 pp.). Doubleday, New York.

Beverage, J.P and Culbertson, J.K. (1964) Hyperconcentrations of suspended sediment. *ASCE Journal of the Hydraulics Division*, **90**(HY6), 117–126.

Blackwelder, E. (1928) Mudflow as a geologic agent in semi-arid mountains. *Bulletin of the Geological Society of America*, **39**, 465–484.

Broscoe, A.J. and Thomson, S. (1969) Observations on an alpine mudflow, Steele Creek, Yukon. *Canadian Journal of Earth Science*, **6**, 219–229.

Bull, W.B. (1964) *Alluvial Fans and Near-surface Subsidence in Western Fresno County, California* (USGS Professional Paper 437-A). US Geological Survey, Reston, VA.

Costa, J.E. (1984) Physical geomorphology of debris flows. In: J.E. Costa and P.J. Fleisher (eds), *Developments and Applications in Geomorphology* (pp. 268–317). Springer Verlag, New York.

Costa, J.E. and Jarrett, R.D. (1981) Debris flows in small mountain stream channels of Colorado and their hydrologic implications. *Bulletin of the Association of Engineering Geology*, **18**, 309–322.

Crandell, D.R. (1957) Some features of mudflow deposits (Abstract). *Bulletin of the Geological Society of America*, **68**, 18–21.

Cruden, D.M. and Varnes, D.J. (1996) Landslide types and processes. In: A.K. Turner and R.L. Schuster (eds), *Landslides Investigation and Mitigation* (Special Report 247, pp. 36–75). Transportation Research Board, US National Research Council, Washington, DC.

Davies, T.R.H. (1986) Large debris flows: A macroviscous phenomena. *Acta Mechanica*, **63**, 161–178.

Hungr, O. (1997) Some methods of landslide hazard intensity mapping (Invited paper). In: R. Fell and D.M. Cruden (eds), *Proceedings of the Landslide Risk Workshop* (pp. 215–226). A.A. Balkema, Rotterdam.

Hungr, O. (2000) Analysis of debris flow surges using the theory of uniformly progressive flow. *Earth Surface Processes and Landforms*, **25**, 1–13.

Hungr, O., Evans, S.G., Bovis, M., and Hutchinson, J.N. (2001) Review of the classification of landslides of the flow type. *Environmental and Engineering Geoscience*, **VII**, 221–238.

Hutchinson, J.N. (1968) Mass movement. In: R.W. Fairbridge (ed.), *Encyclopedia of Geomorphology* (pp. 688–695). Reinhold, New York.

Hutchinson, J.N. (1988) General report: Morphological and geotechnical parameters of landslides in relation to geology and hydrogeology. In: C. Bonnard (ed.), *Proceedings of the 5th International Symposium on Landslides* (Vol. 1, pp. 3–36). A.A. Balkema, Rotterdam.

Hutchinson, J.N. (1992) Keynote paper: Landslide hazard assessment. In: D.H. Bell (ed.), *Proceedings of the 6th International Symposium on Landslides, Christchurch, New Zealand* (Vol. 3, pp. 1805–1841).

Iverson, R.M. (1997) The physics of debris flows. *Reviews of Geophysics*, **35**(3), 245–296.

Jakob, M. (1996) Morphometric and geotechnical controls of debris flow frequency and magnitude (233 pp). Ph.D. thesis, Department of Geography, University of British Columbia, Vancouver.

Jakob, M. and Jordan, P. (2001) Design floods in mountain streams: The need for a geomorphic approach. *Canadian Journal of Civil Engineering*, **28**(3), 425–439.

Jordan, R.P. (1994) Debris flows in the southern Coast Mountains, British Columbia: Dynamic behaviour and physical properties (258 pp.). Ph.D. thesis, Department of Geography, University of British Columbia, Vancouver.

Morgenstern, N.R. (1985) Geotechnical aspects of environmental control. *Proceedings of the International Conference on Soil Mechanics and Foundation Engineering, San Francisco* (Vol. 1, pp. 155–185).

Pierson, T.C. (1980) Erosion and deposition by debris flows at Mt. Thomas, North Canterbury, New Zealand. *Earth Surface Processes*, **5**, 227–247.

Pierson, T.C. (1986) Flow behaviour of channelized debris flows, Mount St. Helens, Washington. In: A.D. Abrahams (ed.), *Hillslope Processes* (pp. 269–296). Allen & Unwin, Boston.

Scott, K.M., Pringle, P.T., and Vallance, J.W. (1992) *Sedimentology, Behavior and Hazards of Debris Flows at Mt. Rainier, Washington* (Open-File Report 90-385). US Geological Survey, Reston, VA.

Sharpe, C.F.S. (1938) *Landslides and Related Phenomena*. Columbia University Press, New York.

Skempton, A.W. and Hutchinson, J.N. (1969) Stability of natural slopes and embankment foundations. *Proceedings of the 7th International Conference on Soil Mechanics and Foundation Engineering* (State-of-the-art volume, pp. 291–340).

Slaymaker, O. (1988) The distinctive attributes of debris torrents. *Hydrological Sciences Journal*, **33**, 567–573.

Stiny, J. (1910) *Die Muren*. Verlag der Wagner'schen Universitäts-buchhandlung, Innsbruck. [*Debris Flows* (English translation by M. Jakob and N. Skermer, 1997, 106 pp.), EBA Engineering Consultants, Vancouver, Canada.]

Swanston, D.N. (1974) *Slope Stability Problems Associated with Timber Harvesting in Mountainous Regions of the Southwestern United States* (Forest Service General Technical Report PNW-021). US Department of Agriculture, Washington, DC.

Takahashi, T. (1991) *Debris Flow* (IAHR Monograph, 165 pp.). A.A. Balkema, Rotterdam.

VanDine, D.F. (1985) Debris flows and debris torrents in the southern Canadian Cordillera. *Canadian Geotechnical Journal*, **22**, 44–68.

Varnes, D.J. (1954) Landslide types and processes. In: E.B. Eckel (ed.), *Landslides and Engineering Practice* (Special Report 28, pp. 20–47). Highway Research Board, National Academy of Sciences, Washington, DC.

Varnes, D.J. (1978) Slope movement types and processes. In: R.J. Schuster and R.J. Krizek (eds), *Landslides, Analysis and Control* (Special Report 176, pp. 11–33). Transportation Research Board, National Academy of Sciences, Washington, DC.

Wilford, D.J., Sakals, M.E., Innes, J.L., Sidle, R.C., and Bergerud, W.A. (2004) Recognition of debris flow, debris flood and flood hazard through watershed morphometrics. *Landslides*, **1**, 61–66.

3

Debris flows in history

Nigel A. Skermer and Douglas F. VanDine

3.1 INTRODUCTION

"*History is bunk*" is a particularly forward-looking remark attributed to Henry Ford, the inventor of the Ford automobile. Yet, if we choose we can benefit a great deal by looking back into the historical experience in any field. With regards to debris flows, a considerable amount of research has already been carried out. For example, Montandon (1933) and Eisbacher and Clague (1984) have described events in the European Alps as far back as Roman times, and Matsushita (1999) has illustrated events in Japan back to the seventeenth century. In North America, records of debris flows exist for only a few centuries and then for only a few locations (e.g., the Appalachian Mountain region (Clark, 1987; Kochel, 1987)). Without doubt, fascinating histories could also be written of debris flows elsewhere in the world. For example, the *International Symposium on Debris Flow and Flood Disasters Protection*, held in China in 1991, only hinted at the scale of the debris-flow problem facing the Chinese railway system (see the technical note summarizing the meeting and field trips by Jordaan, 1992). A sense of the worldwide economic impacts, at least for recent debris-flow events, can be found in the review by Brabb and Harrod (1989).

A broad survey of the gradual awareness of debris flows in the relatively recently settled province of British Columbia, Canada, has been prepared by VanDine (1990). The earliest record of a debris flow unearthed so far in British Columbia is that for Rubble Creek, near the present town of Whistler, when Major W. Downie, while travelling through the region in 1858, was told that a massive *overflow* occurred three years prior (Downie, 1858). The date of this event was later confirmed by dendochronology (Moore and Mathews, 1978). Sandford Fleming, while surveying the route for the Canadian Pacific Railway in 1872, observed the effects of debris flows on the Snaring River in the Rocky Mountains near the present town of Jasper, Alberta: "*It is a foaming mountain torrent, with a bed full of large round*

Figure 3.1. Destruction by debris flood on 28 October, 1921, of the mining community of Britannia Beach, north of Vancouver, British Columbia, Canada. Thirty-seven people died. A similar event recurred in 1991 but without loss of life. Currently, protection measures are being implemented on the main stem of the creek, and the area is to be re-established as a tourist attraction. (Skermer and Russell (1998); Vancouver Public Library, Special Collections, VPL 15286).

boulders which it piles along its banks, or hurls down its bed to the Athabasca" (Grant, 2000). Devastating debris floods occurred along Howe Sound, British Columbia, during the early part of the twentieth century (Figure 3.1). The earliest debris-flow mitigative works in western Canada were constructed in 1976, following substantial debris-flow damage in 1973 and 1975 to Port Alice on the west coast of Vancouver Island (Nasmith and Mercer, 1979). In this case, model studies were used to design a series of deflection dykes, that were not "tested" until 2003, but then worked quite satisfactorily.

We suspect that a great deal more historical debris-flow gems lie hidden away. In an article, *The History of Rain and Fine Weather*, Le Roy Ladurie (1979) made the case for the professional historian to collaborate with the natural scientist to work out the history of climate. The role of the professional historian was to provide the basic archival material. By his training in such things as ancient languages and the *historian's craft*, he alone has "*the key to certain types of data hidden away centuries ago in bundles of illegible old documents*". Le Roy Ladurie (1971) amply demonstrated the method and value of the professional historian to document climate change in the European Alps over the past millennium by using techniques such as dendrochronology, the glaciological method and what he called the *événementiel* method. The last method is the *recorded events method* – the painstaking accumulation of observations recorded at the time by contemporary witnesses in private correspondence, family diaries, parish registers, and the like, such as was done for debris flows near Savoy in France (Mougin, 1914). It seems to make sense that Le Roy Ladurie's argument, and Mougin's example, would hold true for investigating debris flows in history.

Having sidestepped the task of writing even a brief history of debris flows, this chapter introduces some important historical, and some recent, debris-flow events in alpine countries and in semi-arid regions, discusses the delightful sounding terms *lahars*, *jökulhaups*, and *GLOFs*; and touches on the associated phenomena of mine waste flows and bog bursts. Also quoted are some classic, as well as lesser known, eye-witness accounts of debris flows from various sources in the literature, both technical and from mountain lore. While not necessarily precise scientific descriptions, the images conjured up in these old documents are well worth reading. The aim of this chapter is to provide some historical perspective and also perhaps to inspire a future professional historian to work on a history of the topic.

3.2 HISTORICAL PERSPECTIVE AND ITS LIMITATIONS

The historical method has limitations, one of which is the reactions of people to debris-flow events. Documented records from the past have to be read against the background of events and the socio-economic circumstances prevailing at the time. For example, a debris flow causing loss of life of a dozen people at the time of the Black Death, an epidemic that swept Asia and Europe during the fourteenth century, may have seemed of little consequence. Similarly, in times of war, revolution, famine and the like, debris flows may well have gone unrecorded despite being an event of significant magnitude. Expectations from historical records, therefore, need to be tempered in the light of prevailing circumstances. Nevertheless, parish records may still await discovery and analysis. Searching through old records only turns up what the investigator is looking for, while the rest goes unnoticed. This can also happen in relatively recent history. A record may exist in a museum or in an archive that no one thought to look through.

Sometimes clues to the torrential activity of a creek or river are found in the name of the village. In England, for example, where torrential rivers are unusual, Lynmouth in north Devon derives its name from the Anglo-Saxon word for torrent – Llynna (Delderfield, 1953). The town lived up to its name in August 1952 when an enormous debris-laden flood tore through the holiday town killing 39 people. Previously in 1796 the Lyn River experienced a similar debris flood (Dobbie and Wolf, 1953). Close by in the neighbouring county of Cornwall a very similar devastating torrential debris flood struck the village of Boscastle in August 2004, although miraculously without loss of life (*Western Morning News*, 2004).

Whether we like to believe it or not, history depends on memory. For example, how many readers can honestly recall all four, well-publicized, recent geological disasters of the week of 20–26 December, 2003 (two huge mudslides and two magnitude 6.5 earthquakes – remember?) These events were: the devastating mudslide in the Philippines; the debris flood in southern California; the Paso Robies earthquake in central California; followed by the Bam earthquake in southern Iran. Now think how imperfect memory, and hence knowledge, must have been in medieval times (AD 500 to 1500!) Well, one might argue, without pen, paper, and the computer, memory was unencumbered and far better honed. Even so, memory is surely at all times highly selective.

In terms of perspective, reflect on the newspaper coverage of the December 2003 mudslide in the Philippines that killed an estimated 200 people. Whereas the Philippine newspapers dwelt heavily on the story with multiple articles, maps, and photographs for three or four days, *The New York Times* only displayed one modest photograph accompanied by a few dozen words. Naturally US newspapers gave more coverage to the December 2003 southern California debris floods during which 15 people died. The debris floods were partly the result of the wildfires that burned the San Bernardino Mountains hillsides in late October 2003. The foothills of these mountains have a history of damaging debris floods and debris flows as far back as 1884 with reoccurrences in 1938, 1941, and 1980 (Slosson et al., 1991; Chapter 15).

Finally, anyone doubting the true origin of debris flows ought to visit the small village in the eastern European Alps where a marble monument stands depicting a mythical giant hurling boulders down the creek (Eisbacher, 1982). Such fanciful ideas existed until quite recently in many remote mountain villages throughout the world; the mountains being a source of mystery and indeed fear – literally the home of dragons (de Beer, 1930).

This review of debris flows in history is not inclusive or exhaustive. It relies heavily on the readings and memories of the authors. It is selective and subjective. It is definitely not a history of debris flows. It is a selective historical, and at times a not so historical, review carried out with a little help from our friends around the world, and our files.

3.3 ALPINE DEBRIS FLOWS

The word *alpine* is used here to refer to any lofty mountain range modified by glacial action and subject to a wet climate, such as the Alps in Europe, Japan, and New Zealand and the Cordillera of North America. Every now and again in these regions, a normally tranquil mountain creek decides to clear its throat. The German language has a particularly evocative word for such a creek. It is called a *wildbach*. It translates literally into *wild creek*. The French use the word *torrent*.

3.3.1 European Alps

In the European Alps, debris flows are very common where they threaten hundreds of villages that are built on the fans at the mouths of steep mountain creeks. They have done so for centuries. A classic work on the subject was written in 1910 by Stiny, an Austrian geologist. The preface to his book opens thus:

> Scarcely any other force of nature destroys so much of economic value in our high mountains as the disastrous activity of the torrent creeks, which reach their catastrophic highpoint in the release of debris flows.
>
> Stiny (1910)

Stiny called his book *Die Muren*. *Mure* is a purely Austrian term for *debris flow*, although it is seldom found in the Austrian dictionary. Stiny stated that the word may possibly have come from *muren*, meaning *to grumble*, like the roll of thunder. However, he went on to say that the word *mure* was more likely derived from words meaning *broken*, *rotten*, and *friable*, indicating the condition of the hillside where the *muren* originates. In the Swiss dialect, the word *Ruefe* is used and has a similar origin, *ruin*.

Numerous instances of debris flows damaging villages in the European Alps have been documented by Montandon (1933), and more recently, a history dating back 2,000 years was written by Eisbacher and Clague (1984). Some villages have a long history of debris-flow activity. One such case in Switzerland is the village of Meiringen situated north of the Grimsel Pass on the east side of the River Aare. It was buried in 1764 by a debris flow that originated in the Alpbach watershed, a tributary to the Aare River. An old church was buried by mud almost up to the top of the windows, indicating a flow thickness of 8–9 m. The church had been enlarged 80 years earlier, around a smaller fifteenth century building that itself had been built on the foundations of a Roman temple, dating back to the first century AD. A lengthy and interesting description of a debris flow that took place at the end of the nineteenth century in this village, appears in a travel book by Malby (1912) (Figure 3.2). Recently the church basement has been dug out and the public can tour the excavation.

Similar partial burials of buildings can be seen in other European alpine villages. Inzing near Innsbruck, Austria, for example, has been overrun by debris flows at roughly 40-year intervals since 1807, the last event occurring in 1969 (Figure 3.3). In August 1929, a debris flow swept into Inzing, thrusting open the doors of the church where people were celebrating the festival of *Maria Himmelfahrt* (Assumption of the Virgin Mary into Heaven). One can imagine the effect on the minds of the villagers as the mudflow appeared coinciding with the religious ceremony! Today, the west entrance to the church and the windows are at ground level but, by comparison with the adjacent road levels, the entrance has been dug out by 2 m. The library to the north of the church was also buried during the 1929 event, but has not been excavated, and to gain entrance it is necessary to go down steps to the door (Figure 3.4). This debris flow was said to have been very slow moving. It picked

Figure 3.2. Meiringen, Switzerland after a debris flood at the turn of the last century. *"It was a magnificent sight, and the noise was deafening. Water, pebbles, boulders, mud, all as black as could be, were pouring over the giant fall, and sending up clouds of dingy yellow vapour. The crash of the stones as they found the bottom shook the ground, so that they could be felt, as well as seen and heard"* (Malby, 1912).

Figure 3.3. The village of Inzing in the Tyrol, near Innsbruck, Austria. Note how the village is situated on the left-hand third of the fan, the remainder being unoccupied and, in historical times, given over to debris flows. Nowadays the village is protected by a large debris basin at the head of the fan.

Photo: N.A. Skermer taken in 1978.

Figure 3.4. The side entrance to the library at Inzing is below ground, the level having been built up by earlier debris-flow depositions.
Photo: N.A. Skermer taken in 1978.

up trees and carried them along upright. Today the village is protected by a debris storage basin at the head of the fan (Skermer, 1980).

Another European alpine village with a long history of debris-flow activity is Bourg-Saint-Maurice, in the Isère valley in Savoy, France (Mougin, 1914; Montandon, 1933). The town lies on a large debris fan built by Arbonne Creek, a major tributary of the Isère River, south-west of the Little St. Bernard Pass. This pass was well used in Roman times, connecting the Aosta valley to the Isère valley, and hence the Rhone River where fortifications were concentrated. Bourg-Saint-Maurice was the Roman town of Bergintrum. At the old convent of Clarisses, an inscription tells of how the emperor Caesar Lucius Aurelius Verus, in the year AD 163, *"restored in the country of Centrons (Tarentaise) parts of the road carried away by the force of the torrents"*. He built dykes and *"restored the bridges, temples and baths"*. It seems that just prior, the valley was subjected to a debacle that carried away not only roads and bridges, but also public buildings. Montandon (1933) attributed this event to landslides in the Arbonne drainage basin, leading to mudflows (debris flows) that dammed the Isère River, but which subsequently ruptured. The huge Arbonne fan has pushed the Isère River over to the valley side, and the river has cut deeply into the toe of the fan.

Sometime during the second-half of the fourteenth century, tradition has it that a tributary of Arbonne Creek, Nant-Blanc Creek, suddenly dried up. Seven years later, it re-appeared as a gigantic debris flow that destroyed most houses on both sides of Arbonne Creek. The town was re-built where it stands today, on the flank of the fan.

In the years between 1630 and 1636, a series of new landslides occurred either in the Arbonne or Nant-Blanc drainage basin. Debris flows down Arbonne Creek followed, and 52 houses in the town were swept away. The church was destroyed and a new one was built on the central section of the fan. September 1732 saw more debris-flow activity on Arbonne Creek, and the "new" church and part of the town were carried away. The Isère River was blocked by the debris, and when the landslide dam collapsed, a flood wave swept the length of the valley. All the bridges on the Isère River, as far as Grenoble 140 km downstream, were carried away, with the exception of that at Aigue-Blanche. Mills on the banks of the river were inundated. "*The streets of Moutiers became rivers*" (Montandon, 1933).

Over the centuries, a large amount of work has been undertaken in the European Alps in attempts to control debris flows and protect villages. The history of this work in the Tyrol, Austria, for example, has been reviewed by Stacul (1979), illustrating the use of wooden, masonry, and concrete check dams in an attempt to control debris production in the upper watersheds.

Numerous medieval alpine towns have been threatened by debris flows triggered by deforestation and overgrazing, and this has led to attempts to reforest these areas. For example, primitive wooden check dams in the French Alps were described by Thiéry (1891). Similar early French publications that refer to the protective role of the forests in the Alps include those by Surrell (1870) and Mougin (1914). The latter describes in detail the history of the debris flows in the province of Savoy, with extensive coverage from the fourteenth century onwards. A later work by Mougin (1931) thoroughly addresses the reforestation of the Alps. First Mougin traced the destruction of the forests for pasturage, military operations, public works, and industry. He then discussed the gradual establishment of legislation to protect the forests, and finally reviewed the various methods of restoration including engineering works.

An interesting historical perspective with regard to check dams in the Varunna valley, in south-eastern Switzerland, has been summarized by Rickermann and Zimmerman (1993). Debris-flow catastrophes occurred in 1834, followed by smaller events up until 1880. An event occurred in 1927, then nothing for 60 years. Great efforts were expended from 1860 to 1930 to stabilize the creek gorge using masonry check dams. Gradually, however, sediment built up behind these structures until in 1987 when the sediment was completely eroded following intense, but not exceptionally large, rainstorms. The debris-flow volume was estimated to be about $250,000 \, m^3$ and severe damage to the town of Poschiavo resulted.

Various other contributions to debris flows and protection works in the Alps, as far back as the nineteenth century, are discussed in Chapters 18, 19, and 24.

3.3.2 Asia

Debris flows were witnessed in the nineteenth century by European explorers in the Himalayas, and one of the best descriptions was written on the spot by Conway (1893). He was near the Gilgit River, a tributary of the Indus River, in an area presently being claimed by both Pakistan and India. Conway described a debris flow (*shwā*) down the narrow creek (*nala*) as follows:

> Half an hour further on we were approaching the mouth of a deep narrow side nala that crossed our path, when we heard a noise as of thunder, and beheld a vast black wave advancing down it at a rapid pace. Some glacier-lake had broken high above, and the waters were bringing down the hill. When we reached the edge of the nala the main mass of the stuff had gone by, and only a thick black stream of mud was rushing swiftly past. This became by degrees more liquid, until it was no longer mud, but black water. We waited for some time till the waters subsided and the coolies caught up with us. Harkbir found a way across the torrent by leaping from stone to stone, and we were about to follow him when Karbir, who was looking up the nala, shouted to us to come back. We obeyed with the nimblest feet, and were not more than out of the ditch when another huge mud-avalanche came sweeping down.
>
> It was a horrid sight. The weight of the mud rolled masses of rock down the gully, turning them over and over like so many pebbles, and they dammed back the muddy torrent and kept it moving slowly but with accumulating volume. Each of the big rocks that formed the vanguard of this avalanche weighed many tons; the largest were about ten-foot cubes. The stuff that followed them filled the nala to a width of about forty and depth of about fifteen feet. The thing moved down at a rate of perhaps seven miles an hour. When the front of the avalanche was gone, and the moving mass became shallower, the mixture was about half mud, half rocks, and flowed faster. Now and again a bigger rock than the average would bar the way; the mud would pile up behind it, and presently sweep it on. Looking up the nala we saw the sides of it constantly falling in, and their ruins carried down. Half the river was blackened by the precipitation of so much mud into its brown waters, and went thundering along with added violence. Three times did the nala yield a frightful off-spring of this kind, and each time it found a new exit into the main river below, and entirely changed the shape of its fan. The third avalanche was the largest of all and fortunately left a massive cause-way of stones almost across the nala at our very feet. Some big fall must presently have taken place higher up and dammed back the waters, for the stream ran almost dry, and we were enabled to cross the gully without difficulty, coolies and all. The obstruction delayed us for two hours and three-quarters.
>
> Conway (1893)

Many of the early ideas that were developed in the European Alps for engineered remedial measures against debris flows were later applied in other countries in the early twentieth century. For example, European measures were used to protect India's Asam–Bengal Railway from debris flows. This turned out to be an undertaking which the British Viceroy, Lord Curzon, described as a millstone round the neck of the Indian Finance Department because of the numerous problems with landslides including debris flows (Nolan, 1924).

Debris flows in the history of the Pokhara valley, situated at the foot of the Himalayan Annapurna Massive in Nepal, have recently come to light. The Pokhara valley is a fault-controlled basin about 20 km long and 3 km wide. Early investigation interpreted the well-developed terraces along the entire valley as glaciofluvial (Hormann, 1974; Fort and Freytet, 1983; and Fort, 1984), but subsequently these sediments were ascribed to two huge debris flows ranging between 3 and 7 km^3 (Yamanaka et al., 1982). Both debris flows filled the Pokhara valley completely and blocked all tributary rivers, leading to the formation of lakes, most of which have silted up completely. The earlier event has been dated at approximately 13,000–15,000 years before present (ybp) (Koirala et al., 1998), or in other words, at the end of the last glaciation. Age dating of the younger event has produced ages ranging between 1,000 and 300 ybp (Yamanaka et al., 1982; Fort, 1987; Koirala et al., 1998), and is referred to by a native legend somewhat similar to the legend of Noah's Arc (see Fort, 1987). The source of the tremendous amount of mainly glacial sediment was located by Yamanaka et al. (1982) as between Machapuchre, Annapurna III, and Annapurna IV, where there is a huge glacial cirque, the Sabche Cirque, about 6 km wide (see Figure 3.5). According to Fort (1987) this cirque area still contains about 15 km^3 of glacial material that could descend into the valley. The triggering mechanisms for the two debris flows are not known, although several contributing factors have been suggested: heavy monsoon rainfall, glacial lake outburst flood, glacier surge, and/or an earthquake-induced rock/ice fall (Fort, 1987). Another possibility is blockage of the narrow exit of the cirque by a landslide dam and the subsequent outburst of the landslide-dammed lake.

Today, Japan is well known for its debris flows and debris-flow control structures. But its debris-flow history goes quite far back. Laws prohibiting the cutting of

Figure 3.5. Central part of the Annapurna Massive with Machapuchre (Nepal's "Matterhorn") (6,993 m) in the centre and the Sabche Cirque to the right (indicated by an arrow). The dotted lines mark the top of the valley filling by the more recent debris flow. Note the pronounced terraces in the left foreground.

Photo provided by J. Hanisch.

trees in an attempt to reduce the occurrence of sediment-related disasters date back to AD 700. In spite of these laws, numerous debris flood and debris-flow disasters continued to occur. During the Edo Period (1603–1868), the concepts of forest preservation, flood control tree planting, and riparian works first appeared, and regional feudal administrations began to construct rudimentary sediment control works. In 1666, the *Law for Control of Mountains and Rivers in All Provinces in Japan* was passed. In the late 1800s, foreign engineers were invited from Europe to introduce European debris-flow control technology to Japan. In the 1870s and 1880s, De Rijke and Mulders, Dutch engineers, began working in Japan. They were followed by other Europeans, notably Hofmann, an Austrian professor, who began advising and teaching a course on sediment-control at the Tokyo Imperial University in 1903 (Ikeya, 1979).

In 1897 the *Sabo Law*, one of three water-control laws, was enacted to restrict the production of sediment and to dam or control sediment in order to control flooding and prevent sediment-related disasters. It prohibited potentially sediment-producing activities in designated districts and enabled national and prefectural governments to implement projects throughout Japan. *Sabo* is the Japanese term for erosion control, but the term has been broadened to include "*efforts to prevent sediment-related disasters by means of conserving and recreating the natural environment*" and "*torrent and mountain stream control*" (Matsushita, 1999).

For the past 100 years, Japan has continued its history of sediment and debris-flow control. It is now one of the pillars of debris-flow research and engineering, and over the past forty years has actively passed its collective knowledge on to other countries including: Indonesia, Nepal, the Philippines, Costa Rica, Honduras, Venezuela, Peru, Austria, the USA, and Canada (Matsushita, 1999).

In 2000, Japan enacted the *Sediment-related Disaster Prevention Law*. This law promotes non-structual disaster prevention through the dissemination of information of disaster-prone areas, the restriction of new land development for housing and other purposes, and the development of warning and evacuation systems (Onda, 2001; Kondo, 2002).

Although little is known about debris flows in the history of China, at least to the outside world, it is known that there were early debris flows in that country. In 1701, along the Huashanyu Ravine, near Weinan in Shanxi Province, a debris flow occurred and moved, in addition to its other debris, a boulder 17 m in diameter 1.5 km downstream. In 1743, along the Baini Ravine, near Zhaotong, in Yunnan Province, more than 200 households were destroyed by a debris flow. One particular area with a long debris-flow history is the Jiangjia Ravine, also in Yunnan Province, which has produced giant mudflows (3–5 million m^3) over the last 300–400 years since the forests were removed. Recent works by Liang and Li (2001) of the Chinese Bureau of Hydrology, Ministry of Water Resources, indicates that in Yunnan Province 749 debris flows are known to have occurred between 1743 and 1990. They calculated that up to 1950, the average frequency of these events was approximately one event every five years, but since then the average frequency has increased to one event every 1.6 years. The reason for the increase is not given. Other events in China are described in Chapter 22.

In Taiwan, despite having a long history, debris flows were only first recognized in 1988. The phenomena became publicly well known in 1996 as a result of Typhoon Herb, not because of the serious damage caused by the typhoon, but because a television cameraman captured on video a debris flow destroying a portion of Shen-Mu village, in Nan-Tou Province. When the video was telecast, the Taiwanese people realized for the first time the awesome destructive potential of debris flows. For the months following the hurricane and telecast, debris flows were discussed nation wide, and Shen-Mu village became a tourist attraction. See Chapter 21.

3.4 DEBRIS FLOWS IN SEMI-ARID REGIONS

3.4.1 South America

An example of debris flows going back about 3,000 years is located at the World Heritage site in Chavín de Huántar, Peru, 250 km north of Lima (Burger, 1984). The town is believed to have been one of the very important centres of the pre-Inca civilization, with a population at its peak of around 2,000. The site, located on a fan at the junction of the Huachecsa River and the Mosna River, is a major centre for the study of early Andean societies, and archaeological excavations have taken place over the past 80 years. Debris flows have hampered the archaeological excavations that date back to the Chavín culture (1,200–200 BC). One event in 1945 was triggered by the collapse of a moraine dam, and the resulting debris flow torrented down the Huachecsa River. The debris-flow covered the southern part of the town, and the Old Temple site and its associated excavations, with deposits up to 4 m thick. Our suggestion of using an interdisiplinary approach to study debris flows, alluded to in the introduction, should perhaps be expanded to include archaeologists as well as historians and climatologists!

Debris flows are well known throughout Peruvian history. They are referred to by their native words, *huaico* meaning a *mixture of all particle sizes including boulders*, and *llapana* meaning *sand and mud floods*. The modern word debris flows is *flujos detríticos*, and *flujos de lodos* for mudflows. The first mention of *huaicos* in the English language was made by Sutton (1933) who described the conditions he had seen first hand in the Peruvian Andes. Presently, hazard mapping is being carried out in Peru by the Instituto Geológico Minero y Metalúrgico, and some of its bulletins contain references to historical debris flows.

Another example of where the European Alps ideas for engineered debris flow remedial measures were applied to other countries is Bolivia. In the early twentieth century, European measures were adopted to protect the Bolivia Railway from mudflows, termed *mazamorra*, which translates to *porridge* (Morum, 1936).

3.4.2 Western North America

An early summary of some debris-flow events in the western USA was published by Twenhofel (1932). He referred to those events as *mudflows* and indicated how semi-arid conditions favour their development since "*the fall of rain in occasional downpours precludes the development of a vegetation cover of sufficient magnitude to protect and reinforce unconsolidated surface materials.*" Twenhofel also quoted the classic description by Blackwelder of the mudflow that struck the village of Willard in Utah in 1924:

> The churned-up mass of slimy earth, trees, and boulders gathered momentum as it descended the gorge and burst forth upon the plain at the village of Willard with sufficient impetus to carry it half way down the slope of the fan. It covered the former surface with a layer of bouldery mud 3 to 4 feet deep. The flow deployed rapidly through the village, surrounded houses, carried off small outbuildings or crushed them like eggshells and over-spread the concrete roadway for hundreds of feet ... The front of the flow was about 3 feet high and almost as steep as the edge of a lava coulee. The most striking characteristic of the flow is the abundance of boulders which range in diameter from 1 to 15 feet ... The whole mass is as unstratified as glacial till.
>
> Blackwelder (1928)

(An interesting first-hand observation of a mudflow in Peru in 1924 also appears in a discussion by Singlewald contained in Blackwelder, 1928.) Similar early recognition of the debris-flow process in the arid south-west was written by Pack (1923).

In the early 1980s, Utah was revisited by a spate of debris flows, accompanying an El Niño event. In the spring of 1983, Utah sustained direct damages from landslides, debris flows, mudflows, and flooding costing millions of dollars. Records of precipitation and snow surveys, indicated that snow packs were 150–400% above normal. Although prepared for flooding, both scientists and state emergency agencies were taken by surprise with the extent of the mass movements, which were so widespread and extensive that 22 of the 28 counties in the state were declared national disaster areas. The mass movements were attributed to an abrupt increase in precipitation, unprecedented in Utah's more than 100-year history. The events have been documented in considerable detail by Bowles (1985), and Keaton and Lowe (1998) who reviewed the historical record back to 1847.

California also witnessed many hundreds of debris flows caused by severe winter storms in the same El Niño wet period, 1978–1984. Previously southern California suffered damage from debris flows in 1969 (Kenney, 1969). The "modern-day" damage has resulted largely from hillside development that increased rapidly following the end of World War II (1945). The post-war debris flows are well documented in publications by the US Geological Survey and the US National Research Council. According to Brabb and Harrod (1989) "*no other state comes close to California in total landslide damage.*"

Previous years of above average rainfall leading to debris-flow damage in the Los Angeles area of southern California are 1815, 1825, 1861–1862, 1878–1894,

1914, and 1934–1941. The New Year's Day storm of 1934, in the foothills of Los Angeles, caused tremendous floods and mudflows that resulted in more than 30 deaths and millions of dollars in damage. The catchment areas had been completely denuded by wildfires that swept through the hills the previous autumn, and two weeks prior to the winter storm, the ground was saturated by heavy rain. Matters could have been even worse if it had not been for a number of check dams and the Big Tujunga Reservoir, completed by the Los Angeles County Flood Control District just prior to the event (Eaton, 1936). Eaton documented in detail the engineering aspects of the flood and debris-flow problems and mitigation in the Los Angeles area up to 1934, along with extensive discussion (Lippincott's discussion on the early check dams is particularly interesting). It is interesting to note that the average cost of debris basins in those days amounted to about US$0.16 per m^3 of debris storage capacity.

The use of check dams in Los Angeles County began in about 1914, (*Engineering News*, 1916). Many of the early check dams were loose-rock dams or wire-bound rock mattresses (Ferrell, 1959). Mass failure of these type structures occurred during subsequent storm events, and they were not used after 1934. Baumann (1936) reviewed the design of these early structures and indicated that their function was not to dam or store debris, but to stabilize the creek channel and side slopes. He correctly emphasized that check dams must be able to resist both dynamic impact and erosive forces. He also suggested the use of check dams in the Americas may date back as far as the Maya, Aztec, and Inca civilizations.

An unusual problem that arises in southern California is from the result of watershed burning. Further discussion of this problem is described in Chapter 15. McPhee (1989) described the seemingly hopeless efforts by engineers to design practical long-term solutions for the debris-flow problems in southern California. Events in December 2003 seem to confirm this.

3.4.3 Kazakhstan

Kazakhstan lies on the other side of the world from California. A broad review of debris flows in that region has been written by Khegai and Popov (1989). According to Khegai and Popov, Kazakhstan is the most hazardous of the former USSR republics in terms of debris flows and other landslides, and in that respect is similar to California. Kazakhstan's distinction is in part the result of seismic activity in the northern Tien Shan Mountains, and in part because of the local lithology and hydrology.

In Soviet times, the capital city of Kazakhstan was called *Alma-Ata* (Father of Apples), but is now simply Almaty. The city was originally called *Verny* (Vyernyi) when founded in 1854 during the eastward expansion of Russian Imperialism onto the steppe lands. The Kazakhs were nomadic people, and as a result little historical data on debris flows in this region exists prior to Russian settlement. The city is founded on a very large fan built by debris flows, some of enormous size, that descended the Almaatinka River, where it emerges from the Greater (Bolshaya) and Lesser (Malaya) Almaatinka basins. The debris flow associated with the

Verny earthquake of June 1887 is believed to have involved 70 million m^3 (Mushketov, 1890). Prior to that, a very large mudflow was recorded in 1841. Seven other large mudflows, ranging between 0.5 and 5.0 million m^3, occurred at Almaty during the period 1921–1988. Most of them occurred around July, as a result of rainstorms or glacial outburst floods (see Section 3.6). Travel distances for these events extended up to 30 km (Niyazov and Degovets, 1975; Yesenov and Degovets, 1979). Large amounts of money have been expended to protect Almaty. Measures include openwork, steel lattice barriers across the Lesser Almaatinka River, and a 80 m high rockfill mud collector and dam at Medeo. This dam was constructed, in part, by a series of extremely large chemical blasts (3.5–5 kilotonnes) along the flanks of the river. The dam was later raised to 115 m to provide a debris storage basin capacity of 6 million m^3. Despite the mitigative efforts to protect Almaty, many of the works have been either partially or totally destroyed by the passage of subsequent mudflows. Almaty presents the best documented case study of attempts to control large mudflows by constructing barriers directly across river channels (Gagoshidze, 1969; Gagoshidze and Natishvili, 1974).

3.4.4 Spain

Similar semi-arid conditions occur in south-east Spain on the flanks of the Sierra Nevada Mountains facing the Mediterranean Sea and, similarly, major debris flows have occurred. Following 10 months of drought in 1973, over 400 mm of rain fell in the region in 2 days in October, causing numerous debris-laden floods and damage to many towns and villages over a 150 km long stretch east of Granada. In two of the towns, La Rábita and Puerto Lumbreras, the loss of life totalled about 150 (Thornes, 1974). Debris flows exist in other regions of Spain as well, such as the Pyrenees. More than 50 years experience with the design of debris-flow stabilization and mitigation works in Spain has been summarized in a United Nations publication by López Cadenas de Llano (1993). In August 1996, the *Camping de las Nieves* campground in Biescas near the French border was destroyed by a debris flood with the loss of life of 86 people. The fan was supposed to have been stabilized by check dams, but more than 40 of them were destroyed by the event.

3.5 LAHARS

Lahar is a well-known term, but used rather loosely. It is a word from Java and is translated as *mudstream* (Scrivenor, 1929). In general, lahars refer to flows incorporating volcanic debris. A similar local term in the Philippines, where such events are common, is *baha nin dugi*, meaning *floods of mud*. Lahars are described in detail in Chapter 10. Some historical and prehistoric data on lahars have been compiled for the Americas and Japan by Scott et al. (2001), and includes the 1920 destruction of the town of Barranca in Veracruz, Mexico, with the loss of hundreds of lives. Villagers described the sound of the lahar approaching as "*a prolonged muffled thunder clap, many wagons rolling across pavement, or charging cavalry.*"

Lahars can be initiated by melting of snow and ice by a volcanic eruption, descent of pyroclastic flows into streams, or by the ejection of water from a crater lake. Failure of a crater lake and a resulting giant lahar occurred three days prior to the main eruption of Mount Pelee that destroyed the city of Saint-Pierre on the Caribbean Island of Martinique in May 1902. An eye witness described the event:

> Then I heard a noise that I can't compare with anything else – an immense noise – like the devil on Earth! A black avalanche, beneath white smoke, an enormous mass, full of huge blocks, more than 10 m high and at least 150 m wide, was coming down the mountain with a great din.
>
> Scarth (2002)

Unlike a volcanic eruption, which is a short-lived event, lahars can continue for many weeks, months, and even years, and in fact no eruption needs to occur for a lahar to occur. For this and other reasons, lahars have proved more devastating than pyroclastic flows. In the circum-Pacific region, in historical times, records indicate that about 50,000 people have been killed by lahars. It is perhaps remarkable that two of the really large lahar events, Armero (in Columbia in 1985) and Pinatubo (in the Philippines in 1991) have occurred in very recent history.

3.5.1 Classical times

Records of the destructive effects of lahars go a long way back in history. An ancient example is the Late Bronze Age eruptions of Santorini in Greece with its associated pumice ash falls and mudflows (McCoy and Heiken, 2000). The classical example of a volcanic disaster is the AD 79 eruption of Mount Vesuvius near Naples in Italy that devastated the flourishing cities of Pompeii and Herculaneum, the latter being buried beneath three mudflows totalling a depth of 20 m. The volcanic debris was swept along by torrents of rainwater forming a slurry that penetrated everywhere into the city and that later hardened into a tufaceous material very difficult to excavate.

3.5.2 The Americas

In the Americas, and other regions of the circum-Pacific, destructive lahars have been recorded for centuries. Barely historic, yet a fascinating recent discovery, was the finding of mammoth bones buried in lahar deposits from the Popocatépetl volcano near Mexico City. Dating suggests the event occurred about 11,000 ybp. The lahars appear to have originated from a major eruption dated at about 14,000 ypb. The delay of some 3,000 years is thought to have been the result of the onset of more humid conditions (Siebe et al., 1999). This is of more than academic interest since it corresponds with the idea that Popocatépetl, which is currently quiet, may be subject to violent eruptions at intervals of a few thousand years. Currently more than a million people live within a 35-km radius of this volcano. In prehistoric times, ancient settlements principally nearby Cholula, may have been devastated by lahars associated with a major eruption of Popocatépetl about 1,000 ybp (Siebe et al., 1996). Lahars reached the base of the great pyramid of Cholula. This evidence is disturbing because the relatively harmless eruptions of

Popocatépetl volcano since the Spanish Conquest of Mexico may have lulled people into a false sense of security.

In September 1541, Ciudad Vieja, the former capital of Guatemala, was destroyed by a lahar originating from a crater lake on the extinct volcano Mount Agua, and resulted in the death of over 1,300 persons (Meyer-Abich, 1956).

Similar records exist in Costa Rica where *ash-laden floods* have swept through the city of Cartago at least 6 times in the past 300 years. The *mud floods*, as they are called, descend the Reventado River from the ash covered slopes of the Irazu volcano. Attempts in the 1960s to control the floods using gabion check dams proved unsuccessful, as boulders transported by the floods quickly destroyed the dams (Ulate and Corrales, 1966).

Similar high-frequency lahars are associated with the Cotapaxi volcano in Ecuador, at 5,911 m one of the highest volcanoes on earth. From records dating back to the Spanish Conquest in the sixteenth century, this volcano has produced more than 26 lahars, primarily due to snowmelt. Major eruptions of the Cotapaxi volcano occurred in 1877 from January to September, and lahars travelled more than 240 km to damage buildings.

The largest historic volcanic eruption in South America occurred in February 1600 from the Huaynaputina volcano, approximately 70 km east of Arequipa in Peru. Devastation was widespread and a considerable amount was the result of lahars (de Silva et al., 2000).

The most recent destructive lahar in the world occurred in 1985 near the town of Armero, in Columbia. It was a snowmelt lahar off the side of the Nevado del Ruiz volcano. More than 23,000 people were killed. The town is situated on deposits from previous lahars that were known to have occurred in 1595 and 1845, and killed about 600 and 1,000 people, respectively. The fascinating story of Armero, with the social and political sideshows, is told by Bruce (2001).

In the north-western USA, debris-flow hazards posed by the volcanoes in the Cascade Range were appraised by Crandell and Mullineaux (1975) by studying various types of volcanic-related events over the last 10,000–15,000 years. Crandell and Mullineaux concluded that on average one eruption occurs every century in the Cascades. They predicted the Mount St. Helens volcano with eerie precision at an average frequency of one major eruption every 400 years. How true this proved to be was shown by the eruption of, and massive lahars from, Mount St. Helens in May 1980. The previous eruption was dated in the sixteenth century. As Crandell and Mullineaux wrote six years prior: "... *a potential hazard from lahars extends down valley floors for tens of kilometers* ...." There can be little doubt that some people at that time must have considered Crandell and Mullineaux's prediction to be highly speculative, if not outright geo-fantasy!

Besides destroying cities and villages, lahars are very damaging to transportation routes. Rail cars carrying logs were tipped over by lahars triggered by the 1980 Mount St. Helens volcanic eruption. One of the worst railway disasters in history occurred when Mount Ruapehu in New Zealand erupted in 1951 forming a lahar that weakened a bridge, from which a passenger train plunged killing 151 people (Chapter 25).

3.5.3 Asia

If the destruction of Armero in Columbia was the most recent deadly single lahar, the continuing lahars following the June 1991 volcanic eruption of Mount Pinatubo in the Philippines are the most voluminous and have caused the most widespread devastation. Since 1991, many hundreds of millions of m^3 of volcanic debris have been carried off the flanks of Mount Pinatubo by lahars (by 1996 the volume was estimated at 2 billion m^3). The magnitudes and rates of lahar processes has surprised many experienced researchers, but one reason might be the fact that intense rains from Typhoon Yunya began during the eruption period itself.

Most of the sediment that makes up the lahars is from the 1991 volcanic eruption, while some is from deposits of older eruptions. The rate of sediment yield appears to be decaying exponentially, yet lahars are expected to continue to at least the year 2010. (A similar long period of lahar activity, lasting some 20 years, followed the eruption of Santa Maria volcano in Guatemala in 1902.) The area affected by the recent Mount Pinatubo lahars is said to be approaching 1,000 km^2.

Attempts to control the flows using dykes built of lahar sediment have not been successful. Although a ready source of construction material, the low-density sand and gravel pumice used in the dykes is quite capable of remobilization in subsequent floods and lahars. Direct and indirect damage from the lahars have exceeded that from the volcanic eruption several times over. Although the 1991 eruption of Mount Pinatubo was the world's largest volcanic eruption in more than half a century, it was one of the smallest in its own long history stretching back more than 35,000 years, and the 500-year dormant period that preceded the 1991 eruption was comparatively short for Mount Pinatubo (Newhall and Punongbayan, 1996).

Mayon, also in the Philippines, is one of the most active volcanoes in the world. In 1814 and 1875 it generated lahars that killed some 1,200 and 1,500 people respectively.

In Indonesia, there has been a volcanic eruption on one of its 13,677 islands, on average, every year for the past 1,000 years. Not only have these volcanoes resulted in more than 175,000 deaths, they have caused numerous lahars. Over the centuries, the indigenous population developed methods to reduce the affects of sediment-related disasters on their land. The ancient Harinjing Monument, near Pare, a small city located on the slope of Mount Kelut volcano, was constructed in AD 804 to commemorate the completion of an irrigation and sediment-control dam in the area. In the late 1960s this same area was subject to more debris-flow activity. In this instance, Indonesia, for the first time, relied on foreign expertise. The Japanese Ministry of Public Works introduced several modern debris-flow control techniques to the Indonesians (Hasen, 2002).

The lahars in Indonesia are a major hazard not only due to the presence of the many highly active volcanoes, but also due to the heavy rainfall in the region and the pressure of population that is currently approaching 200 million. Historical records show that many thousands have fallen victim to lahars – Awa (Awoe) in the Sangihe Archipelago in 1711 resulting in 3,000 dead, and Galunggung in 1822 resulting in 3,600 dead, are but two examples. Some volcanoes can be deceiving. Mount Agung on the Island of Bali, for example, had records of volcanic activity back to 1808 without any associated lahar activity, but in 1963 lahars killed more than 200 people. Other volcanoes in Indonesia have a long record of regular lahars, the most frequent probably being Mount Kelut with records of 31 lahars since AD 1000.

According to Neumann van Padan (1951), and other Indonesian records, the majority of the damaging lahars have resulted from volcanic eruptions in, or from the collapse of, crater lakes. An eruption of Mount Kelut occurred without warning in May 1919 resulting in more than 5,000 deaths, and was caused by a volcanic eruption through a crater lake (Scrivenor, 1929). Previous lahars had occurred in 1826, 1848, and 1875, and destroyed many villages and houses.

Mount Kelut is interesting because of the strenuous effort made to mitigate the destructive effects of the lahars. Prior to 1919, attempts were made to protect the city of Blitas, on the south-west flank of the volcano, by building a sediment storage dam along the lahar channel. The dam, however, along with 131 km^3 of arable land and 104 villages, was completely destroyed by the 1919 lahars. The extensive damage was caused by the addition of 38 million m^3 of the water stored in the crater lake and, to prevent that from happening again, a system of tunnels was constructed to drain the lake to less than 1.8 million m^3 of stored water capacity. The tunnel system worked well during the 1951 eruption, when most of the water simply evaporated, and for the first time on record no lahar damage occurred (Zen and Hadikusomo, 1965). The tunnel system, however, was badly damaged and despite some rehabilitation, the crater lake had accumulated some 22 million m^3 of water by 1966 when the next eruption occurred. This time lahars developed again and 260 people were killed. A new tunnel was constructed in 1967 lowering the stored water volume to less than one million m^3. Since then, during three successive eruptions, lahars have not occurred.

3.6 JÖKULHLAUPS AND GLOFs

The sudden failure of a dam, whether of a natural landslide dam or an artificial dam, can result in a large discharge of water downstream. *Jökulhlaup* is an Icelandic term for a glacier outburst flood in which flood waters are suddenly released from either behind a glacier, from within a glacier, or from behind a glacial moraine-impounded lake. In Asia, *GLOF* the acronym for *glacial lake outburst flood*, is the term used to describe the latter type of jökulhlaup. The water floods that result from any of these events, namely dam breaks, jökulhlaups, or GLOFs, can easily turn into debris floods and debris flows by incorporating inorganic and organic debris from the valley floors and sides into the flood waters. Because of the large volumes of water involved, debris floods and debris flows triggered by such events can travel for many tens of kilometres downstream.

Jökulhlaups are not restricted to Iceland. They have been documented in most countries where glaciers exist. Since the decline of the Little Ice Age, in the mid-1800s, along with the relatively recent global climate changes, glaciers are retreating and down-wasting at increasing rates. This has resulted, and will continue to result, in an increased probability of jökulhlaups and GLOFs. Nowhere is this more evident than in the Himalayas.

GLOFs were first recognized as a natural hazard in the Himalayas in the 1950s. In 1954 in Tibet, a GLOF flowed downstream approximately 50 km. At that time it was thought to be a 1:500-year event for the region. However, five GLOFs are known to have occurred in neighbouring Nepal alone between 1977 and 1998. In August 1985, a GLOF from the Dig Tsho (Langmoche) glacial lake in Nepal travelled approximately 40 km down the Bhote Kosi and Dudh Kosi Rivers and destroyed 14 bridges, causing roughly US$1.5 million damage to a hydroelectric power plant, as well as serious loss of life and property damage. It is estimated that a peak discharge of 1,600 m^3/s was attained and approximately three million m^3 of sediment was moved downstream. Approximately 85% of that sediment was deposited in the river valley and 15% travelled further downstream as suspended sediment. In this instance, the travel distance was only 40 km because a great deal of sediment was able to deposit along a wide portion of the river valley upstream of an extremely narrow canyon (Galay et al., 2001).

GLOFs are a growing concern for the Himalayas. Raphstreng Tsho glacial lake in Bhutan, for example, was 1.6 km long, 1 km wide, and 80 m deep in 1986, but by 1995 it had grown to 1.94 km long, 1.13 km wide, and 107 m deep. Tsho Rolpa glacial lake in Nepal covered 0.23 km^2 in the late 1950s, but the mid-1990s the lake covered approximately 1.50 km^2. A recent study by UNEP indicates that there are 24 glacial lakes in Bhutan and 20 glacial lakes in Nepal that have the potential to produce GLOFs, and almost all of these would result in associated far-reaching debris flows (UNEP, 2002).

3.7 MINE WASTE FLOWS

Flow failures from mine tailings impoundments and mine waste rock piles (*waste dumps*) are not usually classified as debris flows. Yet mine tailings and waste rock (collectively known as *mine waste*) are very similar to natural debris, and there is similarity between the flow behaviour of mine waste and natural debris flows. Most major mine waste failures are relatively recent events that have occurred in the last 75 years. The reason for this is simply the enormous increase in the scale of mine operations, particularly following the introduction of very large earth-moving equipment after World War II (1945). Mines these days move a great deal of material each day. They can easily put 200,000 tonnes a day through a mill, and as a result tailings impoundments, and the associated pre-mill waste rock piles, have become some of the largest earth structures ever built. Inevitably some of these fail and the resulting mine waste debris flows can travel large distances, often with disastrous environmental consequences and/or loss of life.

More than 35 mine waste failures have occurred since the 1930s. Presently, on average, one mine waste failure of significance occurs every year. In the mining literature these incidents are often referred to as *flowsides* (Lucia et al., 1981). The value of information on mine waste flowslides to researchers who study natural debris flows is not so much the initiating release mechanisms, but the flow and runout characteristics that have often been recorded in considerable detail (Blight, 1997).

3.8 BOG BURSTS

Bog is the Irish term for *peatland*. Bog bursts are flows of purely organic debris and are of some historical interest. They are rapid downhill movements of masses of saturated peat. They usually occur during heavy and prolonged rainfall, and involve the rupture of the peat surface, followed by the release of liquefied peat from the interior of the bog. There appears to be a gradation of types of movements from a peat flow typical of bog bursts, to peat slides involving coherent blocks of material. Bog bursts have been documented in peatlands in Ireland, Scotland, northern England, the Falkland Islands, Tierra del Fuego, and even in British Columbia, Canada (Hungr and Evans, 1985).

The following is a summary of some historical bog bursts in the British Isles:

> In 1697 a bog of forty acres burst at Charleville, near Limerick. In 1745 a bog burst in Lancashire, and speedily covered a space a mile long and half a mile broad. A bog at Crowhill on the moors near Keighley burst in 1824, and colored the river with a peaty stain as far as to the Humber. In December 1896, a bog of 200 acres burst at Rathmore, near Killarney, and the effects were seen ten miles off. Nine persons perished in one cottage.
>
> Miall (1898)

Probably the worst bog burst occurred in the valley of the Owenmore, Erris, County Mayo, Ireland on 1819:

> This is the most disastrous slide on record, as it carried away a whole village and its inhabitants, also a picket of Highlanders, whose bodies were afterwards picked up in Tullaghan Bay.
>
> Kinahan (1897)

Descriptions by Sollas et al. (1897) recount the events that occurred at Stanley in the Falkland Islands in the years 1878 and 1886. A recent paper by Warburton et al. (2003) contains numerous references to bog bursts.

3.9 SUMMARY REMARKS

Assembling historical accounts of debris flows is a bit like collecting postage stamps – the accumulation of unconsolidated trifles. Hundreds of stories exist about historical debris-flow events, some going back thousands of years. Debris flows are anything but rare events, but often do not appear to be newsworthy except locally. Lacking in rarity they also lack what is described as the dread factor – dread of the unknown (Slovic, 1987), except of course to the immediate recipients of the particular charge of debris. Whereas a news item, such as mad cow disease, can occupy the national papers for days on end, very few people die from the human variant of mad cow disease as compared to deaths from debris flows. Dread of the unknown can generate extraordinary interest. In the writers' experience, dread of debris flows ranks a little higher than dread of automobile accidents. Therefore, people in general will learn little from the history of debris flows. The benefits will be to researchers, professionals, and governments wishing to determine such things as frequency, magnitude, vulnerability, and the effectiveness of mitigative measures. Experience in the European Alps, for example, often taught the mountain folk to give at least one-half of a large fan to the creek, while locating the village on the remainder (see Figure 3.3 of Inzing in Austria). Often though, such a luxury of excess land does not exist and engineering mitigative measures are required, and sometimes constructed.

In a study of the history of Persian earthquakes, Ambraseys and Melville (1982) concluded that people never relocated following such events. Sites were seldom abandoned, and the local residents knowingly rebuilt on the same spot. The ancient town of Bam in Iran, for example, is already being rebuilt following the massive earthquake of December 2003. This is probably because of the widespread effect of earthquakes and the minimal benefit to be gained by moving a few kilometres away. Even so, with regards to debris-flow-prone sites, the same psychology seems to operate. Proposed new developments may be abandoned following the identification of the severity of the hazard (e.g., the village of Garibaldi in British Columbia (GAP, 1978)), however, existing developments tend to stay put. This means that one of the important things history can teach is the effectiveness of mitigative measures, either *passive* (avoidance or appropriate location) or *active* (the construction of engineered mitigative works) (Eisbacher and Clague, 1984).

3.10 ACKNOWLEDGEMENTS

The authors would like to thank a number of individuals and organizations who assisted in the preparation of this contribution. These include the International Sabo Network in Japan, who contributed information on debris flows in Japan, China and Indonesia; Jorg Hanisch, of the German Federal Institute for Geosciences and Natural Resources, who provided information on, and the photograph of the Pokhara valley in Nepal; Li Jian, of the Chinese Bureau of Hydrology, Ministry of Water Resources, who added information on the debris slides in Yunnan Province; and Ko-Fei Liu, of the National Taiwan University, who contributed information on Taiwan. The paper was critically reviewed by Philippe Blondel and Matthias Jakob, and we thank them for their suggestions and corrections. The authors take sole responsibility for any errors or omissions.

3.11 REFERENCES

Ambraseys, N.N. and Melville, C.P. (1982) *A History of Persian Earthquakes* (Cambridge Earth Science Series, 219 pp.). Cambridge University Press, Cambridge, UK.

Baumann, P. (1936) The function of check dams. *Civil Engineering*, **6**, 355–358.

Blackwelder, E. (1928) Mudflow as a geologic agent in semi-arid mountains. *Geological Society of America Bulletin*, **39**, 465–483.

Blight, G.E. (1997) Destructive mudflows as a consequence of tailings dyke failures. *Proceedings of Institute of Civil Engineers, Geotechnical Engineering*, **125**, 9–18.

Bowles, D.S. (1985) *Delineation of Landslide, Flash Flood, and Debris Flow Hazards in Utah: Specialty Conference* (592 pp.). Utah State University, Logan.

Brabb, E.E. and Harrod, B. (eds) (1989) *Landslides: Extent and Economic Significance* (385 pp.). A.A. Balkema, Rotterdam.

Bruce, V. (2001) *No Apparent Danger* (239 pp.). HarperCollins, New York.

Burger, R.L. (1984) *The Prehistoric Occupation of Chavín de Huántar, Peru* (Anthropology Series No. 14, 403 pp.). University of California Press, Berkeley.

Clark, G.M. (1987) Debris slide and debris-flow historical events in the Appalachians south of the glacial border. *Geological Society of America, Reviews in Engineering Geology*, **7**, 125–138.

Conway, W.M. (1893) Exploration in the Mustagh Mountains. *Geographical Journal*, **2**(4), 289–303.

Crandell, D.R. and Mullineaux, D.R. (1975) Technique and rationale of volcanic-hazard appraisals in Cascade Range, Northwestern United States. *Environmental Geology*, **1**, 23–32.

de Beer, G.R. (1930) *Early Travellers in the Alps* (204 pp.). Sidgwick & Jackson, London.

de Silva, S., Alzueta, J., and Salas, G. (2000) *The Socioeconomic Consequences of the AD 1600 Eruption of Huaynaputina, Southern Peru* (GSA Special Paper No. 345). Geological Society of America, Boulder, CO.

Delderfield, E.R. (1953) *The Lynmouth Flood Disaster* (160 pp.). ERD Publications, Exmouth, UK.

Dobbie, C.H. and Wolf, P.O. (1953) The Lynmouth Flood of August 1952. *Proceedings of the Institution of Civil Engineers, Part III*, **2**(3), 522–585.

Downie, W. (Major) (1858) *Report to Governor Douglas on a Proposed Route*. British Columbia Archives, Victoria, BC, Canada.

Engineering News (1916) Flood control in Los Angeles County in California. *Engineering News*, **75**(6), 272–275.

Eaton, E.C. (1936) Flood and erosion control problems and their solution. *Transactions of the American Society of Civil Engineers*, **101**, 1302.

Eisbacher, G.H. (1982) Mountain torrents and debris flows. *International Union of Geological Sciences, Episodes*, **4**.

Eisbacher, G.H. and Clague, J.J. (1984) *Destructive Mass Movements in High Mountains: Hazard and Management* (Geological Survey of Canada Paper No. 84-16, 230 pp.). Geological Survey of Canada, Ottawa, Canada.

Ferrell, W.K. (1959) *Debris Reduction Studies for Mountain Watersheds* (130 pp.). Los Angeles County Flood Control.

Fort, M. (1984) Phases d'accumulations sédimentaires internes et phases orogéniques au sud du massif de l'Annapurna: L'exemple du bassin de Pokhara (Népal). *Montagnes et Piédmonts* (47 pp.). Toulouse [in French].

Fort, M. (1987) Sporadic morphogenesis in a continental subduction setting: An example from the Annapurna Range, Nepal Himalaya. *Zeitschrift Geomorphologie Zusatzband*, **63**, 9–36.

Fort, M. and Freytet, P. (1983) The Quaternary sedimentary evolution of the intra-montane basin of Pokhara, in relation to the Himalayan Midlands and their hinterlands (West Central Nepal). In: A.K. Sinha (ed.), *Contemporary Geoscientific Researches in Himalaya* (Vol. 2, pp. 91–96). Singh Hahendra Pal Singh, Dehru Dun, India.

Gagoshidze, M.S. (1969) Mudflows and floods and their control. *Gidrotekhnika i melioratsiya [Hydraulic Engineering and Reclamation]*, **22**, 29–43. (Translated in *Soviet Hydrology: Selected Papers*, **4**, 410–422.)

Gagoshidze, M.S. and Natishvili, O.G. (1974) Nature of the Alma-Ata mudflow. *Gidrotekhnika i melioratsiya [Hydraulic Engineering and Reclamation]*, **9**, 30–38. (Translated in *Soviet Hydrology: Selected Papers*, **5**, 344–350.)

Galay, V., Schreier, H., and Bestbier, R. (2001) *Himalayan Sediments, Issues and Guidelines* (CD-ROM). Canadian International Development Agency, and Canadian International Water and Energy Consultants, Ottawa, Canada.

GAP (1978) *General Report to the Deputy Minister* (Part 1, pp. 1–76). Garibaldi Advisory Panel, Department of Highways, Victoria, BC, Canada.

Grant, G.M. (2000) *Ocean to Ocean* (Sandford Fleming's expedition through Canada in 1872, 371 pp.). Canadian Collection, Prospero, Ontario.

Hasen, M. (2002) Sediment-related disaster control (Sabo Works) in Indonesia. *Proceedings of International Symposium on Sediment-related Issues in Southeast Asian Pacific Region: Indonesia*. Sediment-related Issues Committee, Third World Water Forum, Yogyakarta, Indonesia.

Hormann, K. (1974) Die Terrassen an der Seti Khola: Ein Betrag zur quartären Morphogenese in Zentralnepal. *Erdkunde*, **28**, 161–176 [in German].

Hungr, O. and Evans, S.G. (1985) An example of a peat flow from near Prince Rupert, British Columbia. *Canadian Geotechnical Journal*, **22**, 246–249.

Iyeka, H. (1979) *An Introduction to Sabo Works* (168 pp.). Japan Sabo Association, Tokyo, Japan.

Jordaan, J.M. (1992) Railway debris flow and flood disaster protection works in mainland China. *Die Siviele Ingenieur in Suid-Afrika*, November, 367–370.

Keaton, J.R. and Lowe, M. (1998) Evaluating debris-flow hazards in Davis County, Utah: Engineering versus geological approaches. *Geological Society of America, Reviews in Engineering Geology*, **12**, 97–121.

Kenney, N.T. (1969) Southern California's trial by mud and water. *National Geographic*, **136**(4), 552–573.

Khegai, A.Y. and Popov, N.V. (1989) The extent and economic significance of the debris flow and landslide problem in Kazakhstan. In: E.E. Brabb and B.L. Harrod (eds), *Landslides: Extent and Economic Significance* (pp. 221–225). A.A. Balkema, Rotterdam.

Kinahan, G.H. (1897) Bog slides and debacles (Letters to the Editor). *Nature*, **55**(1421), 268–269.

Kochel, R.C. (1987) Holocene debris flows in central Virginia. *Geological Society of America, Reviews in Engineering Geology*, **7**, 139–155.

Koirala, A., Hanisch, J., and Geyh, M. (1998) Recurrence history of debris flow events in Pokhara valley: A preview. *Nepal Geological Congress, Kathmandu, 1997* (Abstract 2).

Kondo, K. (2002) Sediment control in Japan. *Proceedings of International Symposium on Sediment-related Issues in Southeast Asian Region: Indonesia*. Sediment-related Issues Committee, Third World Water Forum, Yogyakarta, Indonesia.

Le Roy Ladurie, E. (1971) *Times of Feast, Times of Famine: A History of Climate since the Year 1000* (438 pp.). Doubleday, New York.
Le Roy Ladurie, E. (1979) *The Territory of the Historian* (345 pp.). Harvester Press, Hemel Hempstead, UK.
Liang, J. and Li, J. (2001) Sediment-related disasters in Chinese mountainous areas and measures against them. *Proceedings of International Symposium on Sediment-related Issues in Southwest Asian Region: Nepal*. Sediment-related Issues Committee, Third World Water Forum, Kathmandu, Nepal.
López Cadenas de Llano, F. (1993) *Torrent Control and Streambed Stabilization* (FAO Land and Water Development Series No. 9, 166 pp.). United Nations, Rome.
Lucia, P.C., Duncan, J.M., and Seed, H.B. (1981) *Summary of Research on Case Histories of Flow Failures of Mine Tailings Impoundments* (Information Circular No. 8857, pp. 46–53). US Department of the Interior, Bureau of Mines, Washington, DC.
Malby, R.A. (1912) *With Camera and Rucksack in the Oberland and Valais* (310 pp.). Headley Brothers, London.
Matsushita, T. (1999) *Messages for the 21st Century: Sabo Works and Its Achievements* (71 pp.). Sabo Publicity Centre, Tokyo.
McCoy, F.W. and Heiken, F. (2000) *The Late-Bronze Age Explosive Eruption of Thera (Santorini), Greece: Regional and Local Effects* (GSA Special Paper No. 345). Geological Society of America, Boulder, CO.
McPhee, J.A. (1989) *The Control of Nature* (272 pp.). Farrar Strauss Giroux, New York.
Meyer-Abich, H. (1956) *Los Volcanos Activos de Guatemala y El Salvador* (Anales Servicio Geológico Nacional de El Salvador No. 3). Servicio Geológico Nacional de El Salvador, San Salvador [in Spanish].
Miall, L.C. (1898) On a Yorkshire moor. *Proceedings Royal Institution of Great Britain*, **15**, 621–640.
Montandon, F. (1933) Chronologie des grands éboulements alpins, du début de l'ère Chrétienne à nos jours. *Matériaux pour l'Etude des Calamités, Société de Géographie Genève*, **32**, 271–340 [in French].
Moore, D.P. and Mathews, W.H. (1978) The Rubble Creek landslide, southwestern British Columbia. *Canadian Journal of Earth Sciences*, **15**, 1039–1052.
Morum, S.W.F. (1936) The treatment of mud-runs in Bolivia. *Journal Institution of Civil Engineers*, **3**, 426–487.
Mougin, M.P. (1931) *La Restauration des Alpes* (132 pp.). Ministère de l'Agriculture, Direction Générale des Eaux et Forêts, Eaux et Génie Rural, Paris [in French].
Mougin, M.P. (1914) *Les Torrents de la Savoie* (1251 pp.). Société d'Histoire Naturelle de Savoie, Grenoble, France [in French].
Mushketov, I.V. (1890) Vernenskoe zemletryasenic 28 Maya (9 Juniya) 1887. *Trudy Geologiy Komitata*, **10**(1), 1890–1891 [in Russian].
Nasmith, H.W. and Mercer, A.G. (1979) Design of dykes to protect against debris flows at Port Alice, British Columbia. *Canadian Geotechnical Journal*, **16**, 748–757.
Newhall, C.G. and Punongbayan, R.S. (eds) (1996) *Fire and Mud: Eruptions and Lahars of Mount Pinatubo, Philippines* (1126 pp.). Institute of Volcanology and Seismology, University of Washington Press.
Neumann van Padan, M. (1951) *Catalogue of the Active Volcanoes of Indonesia* (Part 1, 275 pp.). International Volcanological Association, The Hague.
Niyazov, B.S. and Degovets, A.S. (1975) Estimation of the parameters of catastrophic mudflows in the basins of the Lesser and Greater Almatinka Rivers. *Gidrotekhnika i*

melioratsiya [Hydraulic Engineering and Reclamation], **1**, 29–37. (Translated in *Soviet Hydrology: Selected Papers*, **2**, 75–80.)

Nolan, T.R. (1924) Slips and washouts on the hill section of the Assam–Bengal Railway. *Proceedings Institute of Civil Engineers*, **243**(II), 2–25.

Onda, Y. (2001) Sediment-related disasters in Japan. *Proceedings of International Symposium on Sediment-related Issues in Southeast Asian Region: Nepal.* Sediment-related Issues Committee, Third World Water Forum, Kathmandu, Nepal.

Pack, F.J. (1923) The torrential potential of desert water. *Pan-American Geology*, **4**, 349–356.

Rickermann, D. and Zimmermann, M. (1993) The 1987 debris flows in Switzerland: Documentation and analysis. *Geomorphology*, **8**, 175–189.

Scarth, A. (2002) *La Catastrophe: The Eruption of Mount Pelee, the Worst Volcanic Eruption in the Twentieth Century* (246 pp.). Oxford University Press, New York.

Scott, K.M., Macias, J.L., Naranjo, J.A., Rodriguez, S., and McGeehin, J.P. (2001) *Catastrophic Debris Flows Transformed from Landslides in Volcanic Terrains: Mobility, Hazard Assessment, and Mitigation Strategies* (USGS Professional Paper No. 1630, 59 pp.). US Geological Survey, Reston, VA.

Scrivenor, J.B. (1929) The mudstreams ("lahar") of Gunong Keloet in Java. *Geological Magazine*, **LXVI**(X), 433–434.

Siebe, C., Abrams, M., Macias, J.L., and Obenholzner, J. (1996) Repeated volcanic disasters in prehispanic time at Popocatepetl, Central Mexico: Past key to the future? *Geology*, **24**(5), 399–402.

Siebe, C., Schaaf, P., and Urrutia-Fucuganchi, J. (1999) Mammoth bones embedded in a late Pleistocene lahar from Popocatepetl volcano, near Tocuila, central Mexico. *Geological Society of America Bulletin*, **III**(10), 1550–1562.

Skermer, N.A. (1980) Alpine landslides: A lesson for British Columbia? *British Columbia Professional Engineer*, **8**, 19–24.

Skermer, N.A. and Russell, S.O. (1998) The 28th of October. *British Columbia Professional Engineer*, **5**, 21–23.

Slosson, J.E., Havens, G.W., Shuirman, G., and Slosson, T.L. (1991) Harrison Canyon debris flows of 1980. *Environmental Geology*, **18**(1), 27–38.

Slovic, P. (1987) Perception of risk. *Science*, **236**, 280–285.

Sollas, W.J., Praeger, R.L., Dixon, A.F., and Delap, A. (1897) Report of the committee appointed by the Royal Dublin Society to investigate the recent bog flow in Kerry. *Scientific Proceedings of the Royal Dublin Society*, **8**, 475–508.

Stacul, P. (1979) Wildbachverbauung in Südtirol Gestern und Heute. *Sonderbetrieb für Bodenschutz, Wildbach- und Lawinenverbauung* (116 pp.). Autonome Provinz, Bozen-Südtirol, Germany [in German].

Stiny, J. (1910) *Die Muren.* Verlag der Wagnerschen Universitats, Buchhandlung, Innsbruck, Austria. (Translated by M. Jakob and N.A. Skermer, *Debris Flows*, EBA Engineering Consultants, 139 pp.)

Surell, A. (1870) *Etude sur les Torrents des Hautes-Alpes* (Vols 1 and 2). Dunod, Paris [in French].

Sutton, C.W. (1933) Andean mud slide destroys lives and property. *Engineering News-Record*, May 4, 562–563.

Thiéry, E. (1891) *Restauration des Montagnes, Correction des Torrents, Reboisement.* Librairie Polytechnique, Paris [in French].

Thornes, J.B. (1974) The rain in Spain. *Geographical Magazine*, **46**, 337–343.

Twenhofel, W.H. (1932) *Treatise on Sedimentation* (2nd edn, 926 pp.). Williams & Wilkins, Baltimore.

Ulate, C.A. and Corrales, M.F. (1966) Mud floods related to the Irazu volcano eruptions. *Proceedings Hydraulics Division Journal, ASCE*, **92**(HY6), 117–129.
UNEP (2002) *UN Environmental Program*. UN, New York. Available at *http://www.rrcap.unep.org/issues/glof/*
VanDine, D.F. (1990) Development of awareness, knowledge and mitigation of channellized debris flows in British Columbia (Abstract and presentation). *Annual Meeting: Special Session on Landslide Hazards in the Cordillera*. Geological Association of Canada, Vancouver.
Warburton, J., Higgilt, D., and Mills, A. (2003) Anatomy of Pennine peat slide, Northern England. *Earth Surface Process Landforms*, **28**, 457–473.
Western Morning News (2004) It's a miracle: Amazing escapes in flood catastrophe but cost runs in tens of millions (Various news articles). *Western Morning News* (Plymouth, UK), August 18, pp. 2–13.
Yamanaka, H., Yoshida, M., and Arita, K. (1982) Terrace landform and Quaternary deposit around Pokhara Valley, central Nepal. *Journal of Nepal Geological Society*, **2**, 133–147.
Yesenov, U.Ye. and Degovets, A.S. (1979) Catastrophic mudflow on the Bol'shaya Almatinka River in 1977. *Gidrotekhnika i melioratsiya [Hydraulic Engineering and Reclamation]*, **8**, 22–24. (Translated in *Soviet Hydrology: Selected Papers*, **18**(2), 158–160.)
Zen, M.T. and Hadikusumo, D. (1965) The future danger of Mt. Kelut (Eastern Java-Indonesia). *Bulletin Volcanologique*, **28**, 276–282.

4

Instability of steep slopes

William Savage and Rex Baum

4.1 INTRODUCTION

Instability of steep slopes commonly results in the occurrence of debris flows. Figure 4.1 illustrates many of the features of debris-flow source areas or initiation zones (see, Chapter 2). The source area consists of unconsolidated material overlying bedrock. Following geotechnical engineering practice, we refer to this unconsolidated material that, in general, may consist of weathered and relatively weak residuum or slope deposits as soil. In Figure 4.1 the failed soil evacuated the source area exposing the concave-upward basal failure surface of the landslide. The shallow landslide completely disaggregated and mobilized into a debris flow. The landslide was thin, oblong, and somewhat irregular in plan view, several meters wide and long, and at most, only a few meters thick along its longitudinal axis. It was thinner around its edges, indicating that the landslide mass was lens-shaped in transverse cross section. The mass was also slightly thinner at the uphill and downhill ends. Few, if any, exposed roots protruded through the basal failure surface, however, broken roots do protrude from the edges of the scar. This landslide site was underlain by gently dipping sedimentary rocks, but little, if any, bedrock was visible in the source area. Rather, boulder and cobble-sized rock fragments in a fine-grained sandy matrix were exposed in the source area.

Water commonly contributes to the instability of steep slopes. As evidenced by the patches of snow in this example, infiltration of meltwater saturated the soil on the slope. This increased the soil weight and decreased soil strength to the point that the mass began sliding downslope. More commonly, infiltration of rainfall produces the same result. This slide occurred at the head of a small swale, or hollow on the slope, a feature of many shallow slides on steep slopes. Increased soil thickness and convergence of groundwater often contributes to instability at such hollows. Although seismicity was not a factor in this particular landslide, strong ground motion from earthquakes can destabilize steep slopes and produce

54 Instability of steep slopes [Ch. 4

Figure 4.1. Photograph of a shallow, irregular landslide scar on a steep slope near Clear Creek, Utah. This slide, which occurred in May 1984, fully evacuated its source area and mobilized into a debris flow.

shallow landslides that mobilize into debris flows or debris avalanches (Jibson et al., 2000).

Although observed failures of steep slopes that have resulted in debris flows and debris avalanches include topples, falls, and deep-seated slides of wedge, ellipsoidal, or complex geometry, shallow soil slides like that depicted in Figure 4.1 seem to be

Sec. 4.2]   **Stress–strain and strength properties of soil**   55

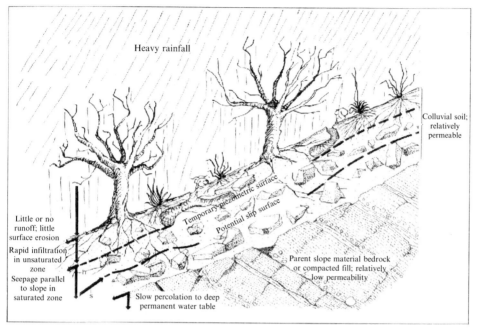

Figure 4.2. Sketch illustrating the general mechanisms involved in rainfall-induced failure of thin colluvium on a steep hillside.
After Campbell (1975).

the most common type. The majority of debris flows and debris avalanches originate on steep slopes that are underlain by relatively strong rocks with a thin mantle of soil. Field experience shows that failure of the unconsolidated mantle (as sketched in Figure 4.2) in response to heavy rainfall is the source of most shallow soil slides. Consequently, our treatment of the instability of steep slopes emphasizes rainfall-induced shallow soil slides.

In what follows, we first review some fundamental aspects of analysis of instability of steep slopes, including stress–strain and strength properties of saturated and unsaturated soils, infinite slope analysis, saturated and unsaturated ground-water flow, and determination of pore pressure in soil media. This is followed by a brief review of methods for assessing instability of steep slopes over wide areas in the context of Geographic Information Systems (GIS) and a final summary.

4.2 STRESS–STRAIN AND STRENGTH PROPERTIES OF SOIL

The stress–strain and strength properties of soils are typically determined in the laboratory by direct shear and triaxial tests. Figures 4.3(a) and (b) respectively show the general configurations of a laboratory direct-shear test apparatus and a triaxial test apparatus.

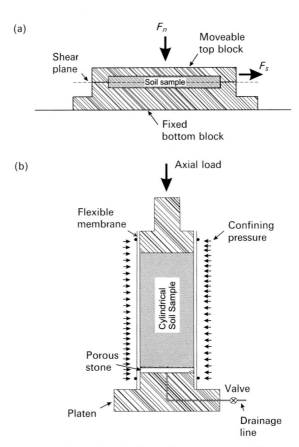

Figure 4.3. (a) Cross-sectional sketch of a direct shear apparatus. The shear box encloses the soil sample and may be rectangular or circular in plan view. (b) Cross-sectional sketch of a triaxial test apparatus. This apparatus is usually circular in plan view.

In a direct-shear test the soil is placed in a steel box that is split along the line of the shear plane as depicted in Figure 4.3(a). A normal force F_n is applied along the top of the box and a shear force F_s is applied to the side of the upper, moveable block to cause shearing in the enclosed soil. Shear and normal stress magnitudes are obtained by dividing the shear and normal force by the original horizontal cross-sectional area of the soil sample. These stresses and the displacements of the upper block are recorded and the results are given in terms of shear τ and normal σ stress on the shear plane and the horizontal and vertical displacements of the upper, moveable block. In addition, changes in pore-water pressure in wet or saturated soils can be recorded.

In the triaxial test apparatus shown in Figure 4.3(b), the soil sample, enclosed in a flexible membrane, is placed between a moveable upper platen and a fixed lower platen and a confining pressure is applied. This is followed by application of an axial load. The axial load is converted to axial stress by dividing the axial load by the original horizontal cross-sectional area of the soil sample. The axial stress is the major (most compressive) principal stress σ_1 and the confining pressure is the minor (least compressive) principal stress σ_3 in a compression test. A porous stone disc connected to an outlet tube allows drainage of fluid from the bottom of the sample. The valve on the outlet tube is left open in a drained test. An undrained test is achieved by shutting the valve on the outlet tube. The stresses σ_1 and σ_3 and the displacements of the upper platen are recorded and, in addition, changes in water pressure in an undrained test for wet or saturated soils are usually recorded. Further details on direct-shear and triaxial tests on soils are found in Lambe and Whitman (1969) and Bishop and Henkel (1957). Modifications to these tests for partially saturated soils involve measurement of pore-air and pore-water pressures in a sample that is in a sealed test chamber; water is allowed to drain freely under the applied air pressure (Bishop et al., 1961; Fredlund and Rahardjo, 1993).

Depending on their initial porosity relative to the applied normal stress, soils initially exhibit either contractive or dilatant behaviour during shearing deformation. Loose soils tend to compact or contract as they deform under load. Dense soils tend to dilate or expand as they deform under a normal load. Both tend toward a "steady" or "critical" state as deformation increases under constant normal load (Lambe and Whitman, 1969). Soil particles deform very little at the low normal stresses present in shallow soils on steep slopes. Rather, soil deformation at low normal stresses mainly involves rearrangement of soil particles and changes in pore space.

Figure 4.4 shows the results of triaxial tests on granular soil in loose and dense states. The dense soil failed in a relatively brittle manner, as indicated by the steep slope of the initial (rising) part of the force-displacement curve. It reached its peak strength after an axial strain of a few %. After reaching the peak, the strength of the dense soil gradually declined toward the ultimate or residual shear strength. As shown by the change in void ratio, the dense soil began dilating after a very small axial strain and continued dilating throughout most of the test (Figure 4.4).

The initial slope of the force displacement curve of the loose, contractive soil is much flatter than that of the dense soil, indicating ductile failure (Figure 4.4). The shearing resistance of the loose soil gradually built to its ultimate strength, which was attained after significant axial strain. Change in void ratio indicates that the soil compacted slightly at the beginning of the test and then dilated to approximately its original void ratio. After large axial strain (about 30%), the "critical state" was reached where the shear strength and void ratio of the two specimens were approximately equal.

The results of direct shear and triaxial tests of soil show that at the point of incipient shear failure there can occur planes along which the shear stress is given by:

$$\tau = c + \sigma \tan \phi \tag{4.1}$$

where τ is the shear stress on a potential failure plane, σ is the normal stress at failure on this plane, c is cohesion, and ϕ is the angle of internal friction of the soil. Equation 4.1, first proposed by Coulomb in 1773, is known as the Coulomb failure criterion. It plots as a straight line in 2-D Mohr stress space (Terzaghi,

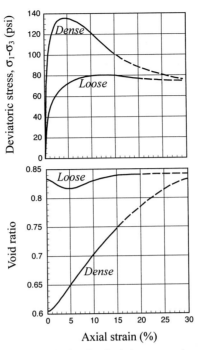

Figure 4.4. Stress–strain curves for drained triaxial test on loose (contractive) and dense (dilatant) medium-fine sand. Solid lines represent actual test data; dashed lines represent extrapolations based on results of other tests. The dense soil reached its peak strength after a small axial strain, but continued to dilate, as axial strain increased. The loose soil began contracting early in the test and reached its ultimate strength after a relatively large axial strain.
From Lambe and Whitman (1969; p. 131).

1943) as shown in Figure 4.5(a). This line separates the Mohr stress space into stable and unstable parts. If a Mohr circle constructed from the major and minor principal stresses, σ_1 and σ_3, lies below the Coulomb failure line, the soil behaves elastically. If, however, a Mohr circle becomes tangent to these lines, failure will ensue. No Mohr circle can lie beyond this limiting line because the shear stress cannot exceed the yield strength of the soil.

The strength parameters cohesion c and the angle of internal friction ϕ of a soil are determined as illustrated in Figures 4.5(a) and (b). Mohr's circles representing the state of stress at failure for three different tests are plotted and a line tangent to the circles is constructed (Figure 4.5(b)). This line is the failure envelope. Commonly, the failure envelope can be treated as a linear, two-parameter model, the Mohr–Coulomb failure criterion (Equation 4.1), over the range of stresses that apply in shallow landslides. Figure 4.5(c) shows Mohr–Coulomb strength parameters determined by drained direct-shear testing of two soils from the Alani–Paty landslide in Hawaii (Baum and Reid, 1995). The expansive clay is similar to material from the basal failure surface of the landslide and the sandy clay represents some of

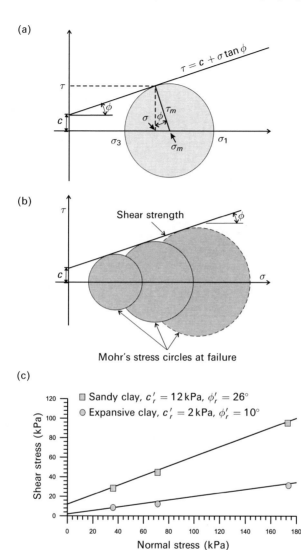

Figure 4.5. (a) Sketch showing Mohr–Coulomb failure criterion. Here, $\tau_m = 1/2(\sigma_1 - \sigma_3)$ is the maximum shear stress and $\sigma_m = 1/2(\sigma_1 + \sigma_3)$ is the mean stress where σ_1 is the major (most compressive principal stress) and σ_3 is the minor (least compressive) principal stress in a triaxial test. The stresses τ and σ, are, respectively, the shear and normal stresses on planes undergoing shear failure, c is cohesion, and ϕ is the angle of internal friction of the soil. (b) Construction of a Mohr–Coulomb failure envelope for a soil. (c) Mohr–Coulomb strength parameters determined by direct-shear testing of two soils from the Alani–Paty landslide in Hawaii. Primes indicate that the strength parameters are determined under effective stress conditions and the subscript r indicates that the strength parameters are residual values.

the material in the body of the landslide above the failure surface. Primes indicate that the strength parameters were determined under effective stress conditions – conditions that we now define.

For water-saturated soils, effective stress is defined by subtracting pore-water pressure (interstitial fluid pressure) p from the total normal stress components (Terzaghi, 1943). In Cartesian xyz coordinates the total normal stress components are σ_x, σ_y, and σ_z and the effective stresses are given by $\sigma'_x = \sigma_x - p$, $\sigma'_y = \sigma_y - p$, and $\sigma'_z = \sigma_z - p$. Static pore pressure does not affect shear stress; however, groundwater flow fields can affect shear stresses as well as normal stresses (Iverson and Reid, 1992). When pore pressure is present the Mohr–Coulomb failure criterion becomes:

$$\tau = c' + (\sigma - p)\tan\phi' \qquad (4.2)$$

where c' and ϕ' are, respectively, the cohesion and the angle of internal friction measured under effective stress conditions. Thus, pore pressure reduces the normal stress on potential failure planes and, in effect, reduces the internal friction resisting failure on these planes. Thus increasing pore pressure reduces the shear strength of a soil mass. This effect is illustrated in Figure 4.6, where we see that the addition of positive pore pressure shifts the Mohr circle to the left towards tangency with the failure envelope.

In partially saturated soils, the concept of effective stress is more complicated than indicated by Equation 4.2, because water is held in the soil pores by surface-tension forces that arise from adsorption and capillary phenomena. Pore pressure in partially saturated soils is negative and is known as matric potential or matric suction (Freeze and Cherry, 1979; Hillel, 1982). The matric suction contributes to the normal stress acting on the soil, which results in greater interparticle friction that increases soil strength. Experimental observations show the relationships between applied normal stress, matric suction, and shear strength of partially saturated soils (Bishop et al., 1961; Escario et al., 1989). Suction-induced shear strength occurs through the mechanism of interparticle friction, but commonly manifests itself as an apparent cohesion that, for example, allows damp sand to stand at steeper angles than dry sand. However, the response of partially saturated soil to changes in matric

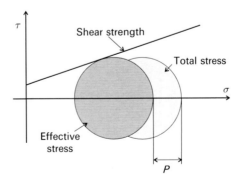

Figure 4.6. Role of effective stress in reducing the shear strength of soil.

suction is more complex than the response of saturated soil to changes in pore pressure. Laboratory observations indicate that the strength contribution by soil suction is non-linear and asymptotically reaches a finite maximum value at high suctions (Escario et al., 1989).

Attempts to interpret shear-strength measurements on partially saturated soils have led to several models including an extended Mohr–Coulomb failure criterion, (Fredlund et al., 1978; Fredlund and Rahardjo, 1993) an extended definition of effective stress (Bishop, 1959), and a Mohr–Coulomb criterion with variable cohesion (Abramento and Carvalho, 1989). The extended Mohr–Coulomb criterion uses an additional parameter ϕ^b which measures the effect of matric suction on soil strength:

$$\tau = c' + (\sigma - p_a)\tan\phi' + (p_a - p_w)\tan\phi^b$$

where p_a is air pressure, p_w is water pressure, $\sigma - p_a$ represents the standard component of total stress used in soil mechanics, and $p_a - p_w$ is the matric suction. Experimental studies have shown that ϕ^b is not constant, but rather varies with the volumetric water content and the matric suction (Abramento and Carvalho, 1989).

Theoretically, the Mohr–Coulomb failure criterion can be applied to partially saturated soils without modification by using an extended definition of effective stress. Lambe and Whitman (1969) define effective stress for partially saturated soils as $\sigma' = \sigma - p^*$, where p^* is an equivalent pore pressure that varies non-linearly with the degree of saturation or matric suction. Bishop (1959) proposed that the effective stress in partially saturated soils could be represented approximately as:

$$\sigma' = (\sigma - p_a) + \chi(p_a - p_w)$$

The term $\chi(p_a - p_w) \approx -p^*$. The parameter χ usually ranges from 1 in saturated soils (including tension-saturated soils) to 0 in dry soils. Tension-saturated soils are fully saturated but subject to matric suction; they occur above the water table, usually in the capillary fringe. Experimental results sometimes yield values of χ that are greater than 1, which seems inconsistent with its definition (Abramento and Carvalho, 1989). Experimental studies have also shown that χ is non-linear with respect to the degree of saturation and matric suction (Bishop et al., 1961). Although experimental results indicate that ϕ' is relatively constant according to the extended effective stress law, results for some soils indicate that ϕ' gradually increases with matric suction (Escario et al., 1989). It is unclear whether this increase results from increased particle interlocking or some other cause.

Abramento and Carvalho (1989) and Wolle and Hachich (1989) used variable apparent cohesion c_a in the Mohr–Coulomb criterion as an alternative to the extended Mohr–Coulomb criterion and the extended effective stress law:

$$\tau = c_a + \sigma\tan\phi'$$

The apparent cohesion c_a is the sum of the cohesion for effective stress c' and a non-linear function of the matric suction $f(p_a - p_w)$ which depends on the character

62 Instability of steep slopes [Ch. 4

of the soil as well as the range of applied stress and stress path(s). For the soil tested by Abramento and Carvalho (1989), the function varied with the square root of matric suction.

The stress–strain and strength properties of soil combined with the effects of pore water have significant implications for the formation, speed, and travel distance of landslides. In particular, dilatant and contractive deformations of soils on steep slopes affect landslide behaviour. Soils that deform contractively do so by reducing their pore space and soils that deform dilatantly increase their pore space. Reduction in pore space of a saturated or nearly saturated soil results in an immediate increase in pore pressure. According to the principle of effective stress (Equation 4.2), this increase in pore pressure causes a loss of soil strength. Consequently, failure of a loose, contractive soil may result in a rapid evolution from soil slide to debris flow.

In contrast, failure of a relatively dense, dilatant soil tends to proceed incrementally; expansion of pore space in a soil reduces pore pressure and may result in matric suction. If sufficient water is available the suction will draw water into the zone of failure in the dilatant soil. The additional water thus drawn into the soil reduces the suction (or increases the pore pressure) and may allow an additional increment of dilatant failure (Moore and Iverson, 2002). The rate at which dilatant failure proceeds depends upon the hydraulic properties of soil; the rate tends to decrease as the soil becomes finer and less permeable. Fleming et al. (1989) showed from field studies that contractive soils tend to fully mobilize into fluid debris flows that travel relatively long distances and that soil slides in dilatant soils will either mobilize only partially into flows of relatively short runout or will remain perched on the hillside as coherent slide masses. Iverson et al. (2000) confirmed these observations experimentally and showed that an initial critical porosity of at least 20.5% is required for rapid failure and debris-flow fluidization in a sandy-loam soil.

4.3 LIMITING EQUILIBRIUM METHODS OF SLOPE STABILITY ANALYSIS

In cases where a thin layer of soil overlays more competent soil or bedrock on a long steep slope, infinite slope stability analysis may be appropriate. Ideally, the length and width of the potential landslide should be much greater than the thickness to minimize edge effects. Infinite slope stability analysis is a simple approximate limiting equilibrium procedure based on a generalization of the law of static friction for a rigid block on a rough inclined plane as illustrated in Figure 4.7. Failure of an infinite slope is characterized by the ratio of resisting Coulomb friction and cohesion on a basal slip surface to gravitationally induced downslope basal driving stress. This ratio defines the factor of safety against slope failure. The effective normal stress in a direction normal to the slope at a depth z in an infinite slope inclined at an angle θ to the horizontal is given by:

$$\sigma' = \gamma_s z \cos\theta - \gamma_w \Psi \tag{4.3}$$

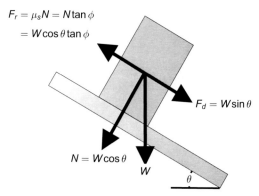

Figure 4.7. Law of static friction for a rigid block on a rough inclined plane. The coefficient of static friction is μ_s and the angle of static friction is ϕ. Movement of the block is imminent when the resisting force of static friction $F_r = \mu_s N = N \tan \phi$ equals the driving force $F_d = W \sin \theta$ (i.e., $F_r/F_d = \tan \phi / \tan \theta = 1$). Thus, the block on the rough plane is in a state of limiting equilibrium when the angle of inclination of the plane θ equals the angle of static friction ϕ.

and the shear stress parallel to the slope is given by:

$$\tau = \gamma_s z \sin \theta \tag{4.4}$$

where γ_s is the saturated soil unit weight, γ_w is the unit weight of groundwater, and Ψ is the pressure head defined by dividing the pore pressure p by γ_w (i.e., $\Psi = p/\gamma_w$). Thus, the factor of safety for an infinite slope F_s is calculated for a slide of thickness T by:

$$F_s = \frac{c' + (\gamma_s T \cos \theta - \Psi \gamma_w) \tan \phi'}{\gamma_s T \sin \theta} \tag{4.5}$$

where, again, c' is effective soil cohesion and ϕ' is the effective soil friction angle. The infinite slope is stable when $F_s > 1$, in a quasi-stable state of limiting equilibrium when $F_s = 1$, and unstable when $F_s < 1$. In engineering practice, an infinite slope with a factor of safety of unity or less is either quasi-stable or unstable and is assumed to fail.

The factor of safety for an infinite slope F_s as a function of slope angle θ computed by Equation 4.5 is plotted in Figure 4.8. For the plots in Figure 4.8, we have followed Bishop and Morgenstern (1960) by choosing ranges of values of slope angle, dimensionless effective soil cohesion $c^* = c'/\gamma_s T$, and pore pressure ratio $r_u = p/\gamma_s T$, that are commonly encountered in practice. The shaded area in the figure indicates where $F_s \leq 1$ and the slope will fail. For $c^* = 0.0$ and $r_u = 0.0$ we see that the infinite slope model is analogous to the block on the rough inclined plane in that it is in a state of limiting equilibrium when the angle of the slope θ equals the angle of internal friction ϕ'. Maintaining $\phi' = 30°$ and $c^* = 0.0$ and setting $r_u = 0.70$ decreases stability in a manner similar to lowering the effective angle of internal friction. Increasing the dimensionless cohesion c^* from 0.0 to 0.05 while maintaining

64 Instability of steep slopes   [Ch. 4

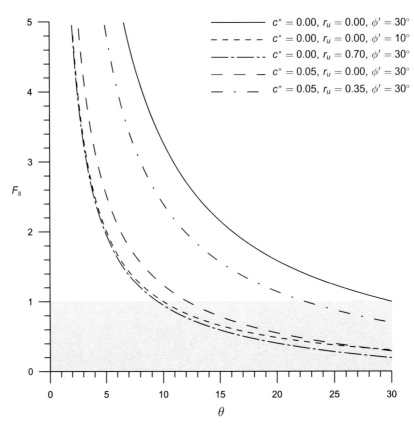

Figure 4.8. Variation of the factor of safety for an infinite slope F_s with slope angle θ for typical values of the dimensionless soil cohesion $c^* = c'/\gamma_s T$, and the pore pressure ratio $r_u = p/\gamma_s T$. The shaded area indicates where $F_s \leq 1$. See text for definition of symbols.

$\phi' = 30°$ and $r_u = 0.70$ leads to increased stability, while maintaining $c^* = 0.05$ and $\phi' = 30°$ and decreasing the dimensionless pore pressure r_u to 0.35 further increases stability. Note that the pore pressure ratio, r_u is constant in each of these examples. The case where r_u is not constant, the steady seepage case, will be discussed below.

A computer-based probabilistic method of infinite-slope stability analysis has been developed to account for measurement uncertainties and the natural variability of the parameters in Equation 4.5 (Hammond et al., 1992a, b). In this method, incorporated in the computer program LISA, the user enters probability distributions for each of the parameters in the infinite-slope equation and a Monte-Carlo simulation is used to estimate the probability of slope failure.

Finally, several workers have shown that the strength and pullout resistance of roots may also contribute to slope stability. Mechanically, roots reinforce soils in a manner analogous to soil anchors (Riestenberg, 1994). However, for convenience most workers generalize the reinforcement provided by roots as an apparent cohesion (Endo and Tsuruta, 1969; O'Loughlin, 1974; Waldron, 1977; Schmidt et al., 2001). Root reinforcement is greatest in the upper few tens of centimeters where the size and quantity of roots is also greatest. Root reinforcement rapidly declines below depths greater than 50 cm in temperate forests of North America; consequently, the greatest contribution of root reinforcement to slope stability is through lateral reinforcement (Schmidt et al., 2001). Incorporating the effects of root cohesion in Equation 4.5 can be accomplished by defining c' in terms of lateral and basal components of soil and root cohesion:

$$c' = (c'_{sb} + c_{rb}) + (c'_{sl} + c_{rl})(A_l/A_b)$$

in which c'_{sb} and c'_{sl} are, respectively, the basal and lateral soil cohesion for effective stress, c_{rb} and c_{rl} are, respectively, the basal and lateral root cohesion, A_l is the lateral area, and A_b is the basal area of the potential landslide. This expanded definition of cohesion introduces a geometric dependence that violates the assumptions of the infinite slope analysis; however, it illustrates that root reinforcement may partially control the dimensions of shallow slope failures. The driving force of a slide must be great enough to overcome the lateral reinforcement of the roots as well as the basal shear resistance. Consequently, the basal area tends to be significantly greater than the lateral area of shallow landslides.

4.4 DARCY'S LAW AND PORE PRESSURE IN SOIL MEDIA

Darcy in 1856 found experimentally that the rate of flow of fluid Q through a porous medium is proportional to the cross-sectional area of the flow A and the gradient in total hydraulic head, that is:

$$Q = KA \left[\frac{h_1 - h_2}{L} \right] \quad (4.6)$$

Here K, the hydraulic conductivity with dimensions of length over time $[L/T]$, is a measure of how easily water flows through the porous matrix. Typical values of hydraulic conductivity for various geologic materials range over many orders of magnitude (Freeze and Cherry, 1979). Figure 4.9, which illustrates Darcy's law for fluid flow through an inclined porous medium of length L, defines the total hydraulic heads h_1 and h_2. The hydraulic gradient in Equation 4.6 is given by $(h_1 - h_2)/L$. Pore pressure p within a saturated soil medium can be obtained from h, the total head, given by $h = \Psi + Z$. Here Ψ is the pressure head and Z is the elevation head defined by elevation above a horizontal datum. Thus $p = \gamma_w(h - Z)$. Relationships between total head, pressure head, and elevation head for an inclined porous medium are shown in Figure 4.9.

Dividing both sides of Equation 4.6 by A, taking appropriate limits, and generalizing to three dimensions yields the formal statement of Darcy's law in Cartesian xyz coordinates:

66 Instability of steep slopes [Ch. 4

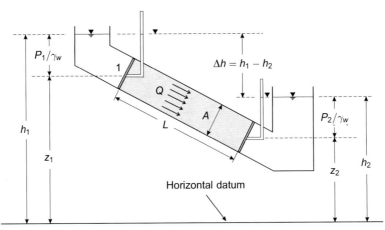

Figure 4.9. Darcy's law for fluid flow through an inclined porous medium. The rate of flow through the inclined porous sample is Q, A is the cross-sectional area of the sample, L is the flow path length, and h_1 and h_2 are total hydraulic heads measured at the entrance and exit ends of the sample.

$$q_x = -K_{xx}\frac{\partial h}{\partial x} - K_{xy}\frac{\partial h}{\partial y} - K_{xz}\frac{\partial h}{\partial z} \quad (4.7a)$$

$$q_y = -K_{yx}\frac{\partial h}{\partial x} - K_{yy}\frac{\partial h}{\partial y} - K_{yz}\frac{\partial h}{\partial z} \quad (4.7b)$$

$$q_z = -K_{zx}\frac{\partial h}{\partial x} - K_{zy}\frac{\partial h}{\partial y} - K_{zz}\frac{\partial h}{\partial z} \quad (4.7c)$$

where q_x, q_y, and q_z are discharges per unit area in the x, y, and z coordinate directions at a point in the porous medium. In the general case given by Equations 4.7, the hydraulic conductivity is a tensor function with nine components and the porous medium is completely anisotropic with respect to flow.

Various degrees of flow anisotropy are possible. For example, layered soils can be taken to be transversely isotropic. However, in many practical applications of Equations 4.7 it is assumed that the soil is isotropic and thus the nine components of the hydraulic conductivity tensor reduce to one K and Equations 4.7 reduce to:

$$q_x = -K\frac{\partial h}{\partial x} \quad (4.8a)$$

$$q_y = -K\frac{\partial h}{\partial y} \quad (4.8b)$$

$$q_z = -K\frac{\partial h}{\partial z} \quad (4.8c)$$

If we assume the pore fluid to be incompressible, then the rate of flow of fluid into a small volume of isotropic porous material will be equal to the rate of increase of fluid in this volume. This leads (Freeze and Cherry, 1979) to a statement of the equation of continuity for an isotropic, homogeneous porous medium as:

$$\frac{\partial \Theta}{\partial t} = K\left[\frac{\partial^2 h}{\partial x^2} + \frac{\partial^2 h}{\partial y^2} + \frac{\partial^2 h}{\partial z^2}\right] \tag{4.9}$$

where Θ is volumetric fluid content.

Equation 4.9, subject to appropriate boundary and initial conditions, governs fluid movement in saturated isotropic, porous media. It is a partial differential equation of diffusion type that defines the value of hydraulic head at any point in a flow field at a given time. Equation 4.9 is analogous to the governing equation of heat conduction and, because of this, solutions for porous fluid movement may be readily obtained from known heat conduction solutions (e.g., Carslaw and Jaeger, 1959).

Equation 4.9 becomes Laplace's equation:

$$\frac{\partial^2 h}{\partial x^2} + \frac{\partial^2 h}{\partial y^2} + \frac{\partial^2 h}{\partial z^2} = 0 \tag{4.10}$$

for steady-state flow. Solution of Equation 4.10, subject to appropriate boundary conditions, provides the values of total head as a function of x, y, and z at any point in a steady flow field. These values of total head define equipotential surfaces in the flow field that are everywhere orthogonal to flow lines in isotropic and homogenous saturated porous media. Alternative boundary conditions for the solutions to Equation 4.10 are shown in Figure 4.10. These conditions require that equipotential surfaces intersect impervious boundaries at right angles (Figure 4.10(a)) and be parallel to boundaries along which the head is constant (Figure 4.10(b)). Equipotential surfaces need not be normal or parallel to a water-table boundary (Figure 4.10(c)) where the total head h equals the elevation head Z. This last condition holds because pore pressure, given in general by $p = \gamma_w(h - Z)$, is zero at the water-table boundary by definition. In addition, a tangent law applies for

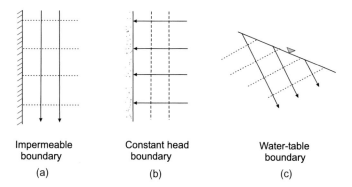

Figure 4.10. Groundwater flow boundary conditions. Dashed lines are equipotentials and arrows are flow lines.
After Freeze and Cherry (1979).

refraction of flow lines at boundaries between media of different hydraulic conductivities. The interested reader is referred to Freeze and Cherry (1979) for further details on methods for solving Equations 4.9 and 4.10.

Unsteady, partially saturated Darcian groundwater flow in response to infiltration of surface water is governed by Richards' equation. In the case of a slope inclined at an angle θ to the horizontal this equation, following Iverson (2000), is written as:

$$\frac{\partial \Psi}{\partial t}\frac{d\Theta}{d\Psi} = \frac{\partial}{\partial x}\left[K_L(\Psi)\left(\frac{\partial \Psi}{\partial x} - \sin\theta\right)\right]$$
$$+ \frac{\partial}{\partial y}\left[K_L(\Psi)\left(\frac{\partial \Psi}{\partial y}\right)\right] + \frac{\partial}{\partial z}\left[K_z(\Psi)\left(\frac{\partial \Psi}{\partial z} - \cos\theta\right)\right] \quad (4.11)$$

where Ψ is groundwater pressure head, Θ is fluid content, and K_L and K_z are the lateral and slope normal pressure-head-dependent hydraulic conductivities referred to an xyz-coordinate system with x tangential to the local surface slope, y tangential to the local topographic contour, and z normal to the xy plane.

Because of the dependence of fluid content and hydraulic conductivities on pressure head, Richards' equation is non-linear and specialized techniques are required to obtain solutions for pressure head, which can then be used to assess the influence of infiltration on slope stability. Anderson and Howe (1985) coupled a 1-D numerical solution of Richards' equation with a slope stability model to compute the factor of safety in response to rain infiltration. Wilkinson et al. (2002) coupled a numerical quasi 3-D infiltration model with a 2-D slope stability model. Some end-member approximate analytical solutions for pressure head can be obtained by using perturbation methods. For example, long and short-term approximate solutions for pressure head variations in response to infiltration have been obtained. Iverson (2000), Morrissey et al. (2001), and Baum et al. (2002) couple these approximate solutions to Richards' equation with an infinite-slope stability model to compute pressure heads and factors of safety as functions of depth and time. In addition, for dry soils, solution of Richards' equation can be shown to predict that the saturated wetting fronts and hence the pressure-head variations move into the soil as kinematic waves (Iverson, 2000). However, in general, for partially saturated flow, numerical solution of Richards' equation for pressure head is generally required (e.g., Rubin and Steinhart, 1963; Freeze, 1969; Havercamp et al., 1971).

In a porous medium, the drag of fluid moving past the soil grains produces a seepage force. Seepage forces have two notable effects on soil stability. A cohesionless saturated soil will liquefy when a steady upward-acting seepage force becomes equal to the submerged unit weight of the soil (Lambe and Whitman, 1969). A sloping (dry) cohesionless soil mass will fail when the slope angle is equal to the soil's angle of internal friction. However, seepage reduces the angle at which the soil mass will fail. For example, if steady slope-parallel seepage is present and the water table coincides with the ground surface, the soil will fail when the slope angle is approximately half the soil's angle of internal friction for effective stress. This well-known result of infinite-slope stability analysis occurs because the buoyant unit weight of soil is about half the saturated unit weight (Lambe and Whitman, 1969).

Mathematically, the seepage force is expressed as:

$$S_j = -\gamma_w \frac{\partial h}{\partial x_j} \qquad (4.12)$$

where the subscript j ranges from 1 to 3 and represents the 3 Cartesian coordinate directions, x, y, and z; S_j represents the x, y, and z components of the seepage force vector; γ_w is the unit weight of water; and the last term $\partial h/\partial x_j$ is the hydraulic gradient in each of the coordinate directions. Note that seepage force is a force per unit volume and has the direction of the hydraulic gradient at each point. In 2-D, the hydraulic gradient is defined by:

$$\frac{\partial h}{\partial x} = -i \sin \lambda \qquad (4.13)$$

and

$$\frac{\partial h}{\partial z} = i \cos \lambda \qquad (4.14)$$

where i is the magnitude of the hydraulic gradient vector and λ is the angular direction of the hydraulic gradient with respect to the normal to the ground surface (Figure 4.11). Analysis by Iverson (1990) shows that $\partial h/\partial x = -\sin\theta$ in infinite slopes that have a water-table boundary condition (i.e., the water table is at the ground surface), regardless of anisotropy. Substituting this formula into Equation 4.13 and solving for i shows that in a homogeneous infinite slope with a water-table boundary condition, the combination of θ and λ uniquely determines the magnitude of the hydraulic gradient i and, therefore, the seepage force (Iverson and Reid, 1992). The hydraulic gradient is indeterminate where it acts normal to the ground surface. Introducing heterogeneity or changing the boundary condition

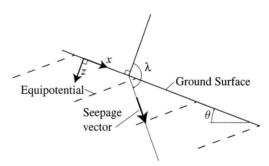

Figure 4.11. Seepage conditions in an idealized steep slope. The seepage-force vector acts parallel to the hydraulic gradient, which may be different than the flow direction in hydraulically anisotropic soils. The slope angle, measured clockwise from horizontal, is θ. The clockwise angle from the normal to the ground surface to the seepage vector is λ. The x and z coordinates are parallel and perpendicular to the ground surface as shown.

allows a wider range of hydraulic gradients and attendant seepage effects on slope stability (Iverson, 1990).

For simplicity, the following analysis of limiting equilibrium of an infinite slope under steady seepage, which follows Iverson and Major (1986), assumes that the slope materials are homogeneous, cohesionless, and water-saturated to the ground surface. For a saturated infinite slope inclined at an angle θ to the horizontal, the gradient of effective normal stress in a direction normal to the slope is given in 2-D by:

$$\frac{d\sigma'}{dz} = (\gamma_s - \gamma_w)\cos\theta - i\gamma_w \cos\lambda \qquad (4.15)$$

and the gradient of shear stress parallel to the slope is given by:

$$\frac{d\tau}{dz} = (\gamma_s - \gamma_w)\sin\theta + i\gamma_w \sin\lambda \qquad (4.16)$$

Substituting Equations 4.15 and 4.16 into the z-derivative of the Mohr–Coulomb failure criterion and solving for the ratio of shearing resistance to shear stress results in the factor of safety F_s:

$$F_s = \left[\frac{(\gamma_s - \gamma_w)\cos\theta - i\gamma_w \cos\lambda}{(\gamma_s - \gamma_w)\sin\theta + i\gamma_w \sin\lambda}\right] \tan\phi' \qquad (4.17)$$

(A slightly more complicated formula applies if soil cohesion is considered.) At limiting equilibrium, the factor of safety is unity, and Equation 4.17 can be reduced to the following general form by algebraic manipulation and application of trigonometric identities (Iverson and Major, 1986):

$$\zeta = \frac{-\sin(\theta - \phi')}{\sin(\lambda + \phi')} \qquad (4.18)$$

in which

$$\zeta = \frac{i\gamma_w}{(\gamma_s - \gamma_w)} \qquad (4.19)$$

Equations 4.18 and 4.19 indicate that the limiting stable slope angle depends on the magnitude and direction of the seepage force, the buoyant unit weight $(\gamma_s - \gamma_w)$, and the soil's angle of internal friction for effective stress. Analysis using Equations 4.18 and 4.19 shows that, in general, minimum slope stability holds when the seepage direction is given by $90° - \phi'$. For typical values of θ and ϕ', this corresponds to approximately horizontal seepage directed out of the slope; as often occurs above an approximately horizontal, low-permeability layer (Iverson and Major, 1986). Also, analysis using Equations 4.18 and 4.19 shows that static liquefaction occurs for some cases of upward seepage. Such liquefaction can lead to debris-flow mobilization (Chapter 5).

Field observations of pore pressures and water levels help define the natural seepage conditions that are common in steep slopes. Observations based on field instrumentation in landslides and unstable slopes indicate that seepage is commonly directed downward and toward the toe of the slope (Baum and Reid, 1995).

Hydraulic heterogeneity and topographic factors strongly influence seepage locally. Iverson and Reid (1992) noted that 1-D flow fields as modeled in infinite slopes cause a much greater reduction in the calculated slope stability than corresponding 2-D flow fields. The lower reduction results from variability in 2-D groundwater-flow and stress fields and the effects of boundary constraints. Thus, while infinite slope analyses with general, 1-D seepage (such as Equation 4.17) are instructive and simple, stability analyses that use observed values of pore pressure are more apt to yield reliable results for practical applications. If the depth of failure is not known, then pore-pressure observations must be recorded at a range of depths and preferably at several locations to account for heterogeneity and subtle topographic effects. In practice, use of Equation 4.5 with observed pore-pressure values should yield reasonably accurate results for long, relatively straight slopes and saturated conditions. Two or three-dimensional methods are needed to analyse the stability of slopes that are curving, irregular, or heterogeneous and where the thickness of the potentially unstable mass is greater than about one-fourth its length or width.

4.5 SLOPE STABILITY IN PARTIALLY SATURATED SOILS

As noted previously, the shear strength of partially saturated soils, and thus their stability on steep slopes, increases with matric suction. Studies at Serra do Mar, Brazil have documented how loss of soil suction during rainstorms, without development of positive pore pressure, can explain failures of shallow soils on steep slopes (Abramento and Carvalho, 1989; Wolle and Hachich. 1989). At Serra do Mar, permeable fractured granitic and gneissic rocks were overlain successively by less permeable layers of saprolite, saprolitic soil and colluvial soil. Field observations over several years indicated that water levels in the fractured bedrock remained 20–30 m below ground surface and the bedrock served to drain the overlying materials, thus perched water was not likely a significant factor in the shallow landslides (Wolle and Hachich, 1989). The landslides had uniform depths of the order of 1 m and failure occurred below the root mat, which was about 0.5–0.8 m thick. Shear strength of the soil increases nonlinearly with matric suction (Figure 4.12(a)). Factor of safety calculations for typical slopes from Serra do Mar illustrate the effects of soil suction on slope stability (Figure 4.12(b)). Tensiometric observations indicated that matric suction at a depth of 0.96 m decreased from about 6 kPa during dry periods to less than 1 kPa following intense rainfall. Suction generally increased with depth below the ground surface and observed matric suction at a depth of 2 m ranged from about 6–21 kPa (Abramento and Carvalho, 1989). Factor of safety calculations based on field and laboratory data (Figure 4.12(b)) indicate that the soil was stable during periods accompanied by moderate to high suction but only marginally stable or unstable after relatively intense rainfall due to greatly decreased

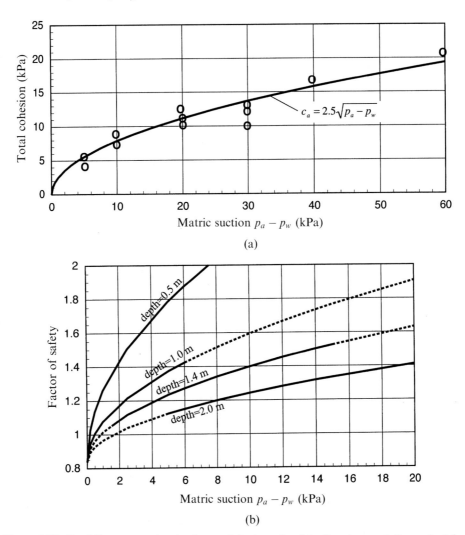

Figure 4.12. Partially saturated cohesion and factor of safety for slopes at Serra do Mar, Brazil. (a) Laboratory measurements of apparent cohesion with variable matric suction. (b) Factor of safety as a function of matric suction $p_a - p_w$; $\theta = 43°$, $\gamma_s = 18.2\,\text{kN/m}^3$, $\phi' = 38°$, $q = 2\,\text{kPa}$, and $c_a = 2.5(P_a - P_w)^{1/2}$, as defined by the best-fit line in (a). Solid lines indicate approximate range of matric suction observed in the field using tensiometers at depth indicated for each curve; low end of each range was associated with rainfall and high end with dry periods.

From Abramento and Carvalho (1989) and Wolle and Hachich (1999).

suction. The factor of safety in Figure 4.12(b) was calculated using the following formula:

$$F_s = \frac{c_a + (\gamma_s T + q)\tan\phi'\cos\theta}{(\gamma_s T + q)\sin\theta} \quad (4.20)$$

where q represents the component of the vegetation surcharge (load) acting normal to the ground surface, c_a is the apparent cohesion as a function of matric suction, and $T/\cos\theta$ is the analysed potential failure depth such as given in Figure 4.12(b). By considering the strength contributed by roots at the edges of shallow slides in a simplified 3-D analysis, Wolle and Hachich (1989) also estimated the critical width of failures at Serra do Mar.

Estimating the factor of safety of partially saturated soils on steep slopes requires information about the partially saturated shear strength as described earlier and either field data on soil suctions (tensiometric measurements, Chapter 12) or soil-moisture characteristics (Freeze and Cherry, 1979; Hillel, 1982; Stephens, 1996) and numerical model results. Therefore, it is more complicated than estimating the F_s for saturated soils. In the absence of field measurements of soil suction, computer programs such as VS2DI (Hsieh et al., 2000) or SUTRA (Voss, 1984) can be used to estimate the soil-suction values if the soil-moisture characteristics and hydraulic conductivity are known in detail. Both programs are available for free download from the US Geological Survey. The public-domain program HYDRUS 6.0 (Simunek et al., 1998) can solve the flow equations in very-nearly saturated soils where other programs encounter serious numerical difficulties. Successful application of such programs requires understanding of partially saturated groundwater flow, the soil-moisture characteristics of soils at the site, and appropriate initial and boundary conditions. Once estimates or observations of the soil suction are obtained, they can be compared to laboratory measurements of soil shear strength at different suction values to estimate the shear strength at field conditions. One can compute the factor of safety by substituting the angle of internal friction and the apparent cohesion appropriate for the soil suction, into Equation 4.20.

4.6 MAPPING REGIONAL SLOPE STABILITY

Modern methods of regional slope stability mapping use regional rainfall data, geologic mapping, digital terrain data, and GIS to estimate spatial and temporal distributions of potential slope instability. These slope stability mapping methods range from purely empirical, to empirically-based probabilistic, to purely analytical. The empirical approach (Guzzetti et al., 1999; Carrara and Guzetti, 1999) relies on multivariate statistical correlation between locations of failures on steep slopes and such factors as slope angle, slope curvature, bedrock lithology, soil type, basin morphometry, or other spatial data. Empirically-based probabilistic methods (Coe et al., 2000; Coe et al., 2004) use historic records of landslide occurrence to predict the temporal and spatial probability of future landslides. The results of the empirically-based probability analyses are portrayed on GIS-based maps. Analytical methods rely on application of simple formulations for seepage or infiltration with infinite slope analysis in a GIS setting to estimate relative stability of slopes. These methods are packaged in computer programs that include SINMAP (Pack et al., 1998), SHALSTAB (Montgomery and Dietrich, 1994; Dietrich et al., 2001), and

TRIGRS (Baum et al., 2002; Savage et al., 2003). In addition, there are GIS-based analytical methods for predicting regionally distributed earthquake-induced landslides (Jibson et al., 2000).

Mechanical, geometric, and hydrologic heterogeneity of natural slopes limits the ability of GIS-based methods to provide detailed assessments of slope-failure hazards. Mapping subsurface variations in soil thickness, degree of saturation, hydraulic properties, shear strength, and other parameters relevant to slope stability is a difficult problem. For example it should be clear from the strong influence of seepage gradients and degrees of saturation, discussed previously in this chapter, that local hydrologic heterogeneities, such as concealed bedrock ledges, can divert seepage and localize failure directly above the ledge. Consequently, modern GIS-based methods of mapping slope stability are suitable for preliminary assessments of large areas; more detailed studies are needed for site-specific assessment of slope stability.

SINMAP and SHALSTAB are designed primarily for assessing stability of steep slopes in areas where significant numbers of shallow landslides initiate in colluvium-filled hollows. Such hollows are common in areas that have well-developed dendritic drainage patterns. SINMAP and SHALSTAB combine a simple hydrologic model for generation of steady-state pore water pressure with infinite slope analysis.

The hydrologic model is based on Darcy's law (Equation 4.6) and the assumption that shallow subsurface flow follows topographic gradients, that is:

$$Q = bdK \sin \theta$$

where Q is discharge in the downslope direction, b is the cross-sectional width of the flow, d is the height of the water table in the soil layer, K is saturated hydraulic conductivity, and θ is the slope angle. In addition, it is assumed that discharge at each point in the flow is in equilibrium with surface infiltration I over an upslope contributing area A, that is:

$$Q = IA$$

Elimination of discharge Q between these last two equations leads to a solution for steady-state water-table depth:

$$d = \left(\frac{I}{K}\right)\left(\frac{A}{b \sin \theta}\right)$$

from which pore pressures used in the infinite-slope factor of safety calculations are obtained from the relationships between total pressure and elevation heads given above.

Both SINMAP and SHALSTAB predict that shallow landslides will concentrate in areas of topographic convergence. This is because the ratio of contributing area A to cross-sectional width b in the equation for water-table height above will increase in hollows. Details, such as the flow-routing methods, treatment of material properties, and use of results of the factor of safety calculations in identifying unstable slopes differ between the two programs, however, their overall approach to assessing slope stability is similar. SINMAP and SHALSTAB are useful for preliminary assessment of stability over broad areas where the assumptions of the underlying models are adequately satisfied.

TRIGRS, which is based on solutions to Iverson's (2000) linearization of Richard's equation for transient partially saturated flow, includes Iverson's original solution for the response of a porous half space to infiltration of rainfall of fixed duration and intensity, as well as solutions for the pressure-head responses to rainfall sequences of varying durations and intensities in a surface layer of finite depth. These calculated groundwater pressures are used in Iverson's (2000) infinite-slope stability formulation to give time and depth-dependent factors of safety that are then incorporated on a cell-by-cell basis in a grid-based GIS framework to provide a regionally distributed and time-dependent model for rainfall-induced initiation of shallow landslides (Figure 4.13). Application of TRIGRS or similar programs based on the work of Iverson (2000) requires the user to specify the initial steady groundwater flow field and is valid for saturated or nearly saturated soils in which the hydraulic conductivity approximately equals the saturated hydraulic conductivity. In addition, because of the assumptions made by Iverson (2000) for linearization of Richard's equation, the infiltration model applies mainly to soil depths that are shallow compared to the upslope groundwater contributing area. Work is presently in progress to assess the applicability of TRIGRS to various field situations and the appropriate level of effort and data needed to specify the initial conditions.

4.7 SUMMARY

Failure of shallow surficial deposits on steep slopes is the source of many debris flows and debris avalanches. Many of these failures have sufficiently simple geometry that they can be analysed adequately by infinite-slope analysis. Water has a significant role in destabilizing surficial deposits on steep slopes, most commonly by direct infiltration of rainfall or snow meltwater or by lateral flow above low-permeability layers in the underlying bedrock. Recent work has helped clarify the mechanisms by which water destabilizes both saturated and partially saturated materials on steep slopes and this work emphasizes the need for accurate field data on subsurface water flow in assessing the instability of these slopes. Finally, combination of simple groundwater flow and slope-stability models with GIS has led to the development of several computer models for analysing slope instability in steep terrain and characterizing its susceptibility to generation of shallow landslides and debris flows.

76 Instability of steep slopes [Ch. 4

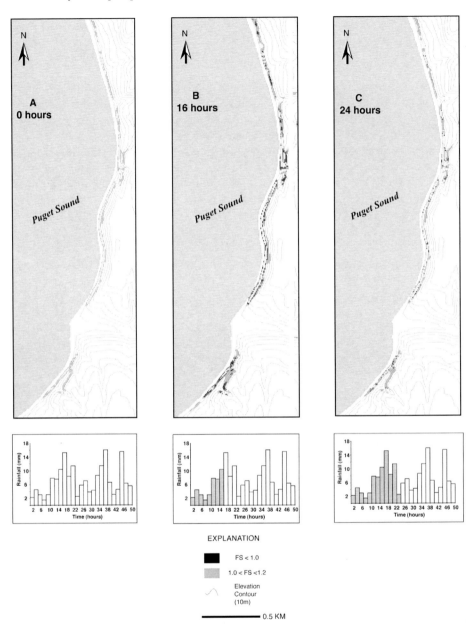

Figure 4.13. Results of an analysis using TRIGRS (Baum et al., 2002; Savage et al., 2003) of time-dependent slope stability in an area along Puget Sound north of Seattle, Washington. Gray cells have a computed factor of safety of greater than 1.0 and less than 1.2. Black cells have a computed factor of safety of less than 1.0. The rainfall histograms below the maps are from the storm of 3–5 April, 1991. The times shown in the upper left-hand corners and by grey bars in the histograms are the elapsed times from the beginning of the storm. Contour interval is 10 m.

4.8 REFERENCES

Abramento, M. and Carvalho, C.S. (1989) Geotechnical parameters for the study of natural slope instabilization at Serra do Mar, Brazil. In: *Proceedings of the 12th International Conference on Soil Mechanics and Foundation Engineering, Rio De Janeiro, 1989* (pp. 1599–1602). A.A. Balkema, Rotterdam.

Anderson, M.G. and Howes, S. (1985) Development and application of a combined soil water-slope stability model. *Quarterly Journal of Engineering Geology*, **18**, 225–236.

Baum, R.L. and Reid, M.E. (1995) Geology, hydrology, and mechanics of a slow-moving, clay-rich landslide, Honolulu, Hawaii. *GSA Reviews in Engineering Geology*, **X**, 79–105.

Baum, R.L., Savage, W.Z., and Godt, J.W. (2002) *TRIGRS A Fortran Program for Transient Rainfall Infiltration and Grid-Based Regional Slope-Stability Analysis* (USGS Open-File Report 02-0424, 27 pp., 2 Appendices). US Geological Survey, Reston, VA.

Bishop, A.W. (1959) The principle of effective stress. *Teknisk Ukeblad*, **39**, 859–863.

Bishop, A.W. and Henkel, D.J. (1957) *The Triaxial Test* (228 pp.). Edward Arnold, London.

Bishop, A.W. and Morgenstern, N.R. (1960) Stability coefficients for earth slopes. *Geotechnique*, **10**, 129–150.

Bishop, A.W., Alpan, I., Blight, G.E., and Donald, I.B. (1961) Factors controlling the strength of partly saturated cohesive soils. In: *Research Conference on Shear Strength of Cohesive Soils, Boulder, Colorado, June 1960* (pp. 503–532). American Society of Civil Engineers, New York.

Campbell, R.H. (1975) *Soil Slips, Debris Flows, and Rainstorms in the Santa Monica Mountains and Vicinity, Southern California* (USGS Professional Paper 851, 51 pp.). US Geological Survey, Reston, VA.

Carrara, A. and Guzzetti, F. (1999) Use of GIS technology in the prediction and monitoring of landslide hazard. *Natural Hazards*, **20**, 117–135.

Carslaw, H.S. and Jaeger, J.C. (1959) *Conduction of Heat in Solids* (510 pp.). Oxford University Press, New York.

Coe, J.A., Michael, J.A., Crovelli, R.A., and Savage, W.Z. (2000) *Preliminary Map Showing Landslide Densities, Mean Recurrence Intervals, and Exceedance Probabilities as Determined from Historic Records, Seattle, Washington* (USGS Open-File Report 00-303, 32 pp.). US Geological Survey, Reston, VA.

Coe, J.A., Michael, J.A., Crovelli, R.A., and Savage, W.Z. (2004) Probabilistic assessment of precipitation-triggered landslides using historic records of landslide occurrence, Seattle, Washington. *Environmental and Engineering Geoscience*, **X**, 103–122.

Dietrich, W.E., Bellugi, D., and Real de Asua, R. (2001) Validation of the shallow landslide model, SHALSTAB, for forest management. *Water Science and Application*, **2**, 195–227.

Endo, T. and Tsuruta, T. (1969) Effects of trees roots upon the shearing strength of soils. In: *18th Annual Report of the Hokkaido Branch, Government Forest Experimental Station. Tokyo* (pp. 167–179). Forest Experiment Station, Saporo, Japan [in Japaneese].

Escario, V, Juca, J.F.T., and Coppe, M.S. (1989) Strength and deformation of partly saturated soils. In: *Proceedings of the 12th International Conference on Soil Mechanics and Foundation Engineering, Rio De Janeiro, 1989* (pp. 43–46). A.A. Balkema, Rotterdam.

Fleming, R.W., Ellen, S.D., and Algus, M.A. (1989) Transformation of dilative and contractive landslide debris into debris flows: An example from Marin County, California. *Engineering Geology*, **27**, 201–223.

Fredlund, D.G. and Rahardjo, H. (1993) *Soil Mechanics for Unsaturated Soils* (517 pp.). John Wiley & Sons, New York.

Fredlund, D.G., Morgenstern, N.R., and Widger, R.A. (1978) Shear strength of unsaturated soils. *Canadian Geotechnical Journal*, **15**, 313–321.

Freeze, R.A. (1969) The mechanism of natural ground-water recharge and discharge: 1. One-dimensional, vertical, unsteady, unsaturated flow above a recharging or discharging ground-water flow system. *Water Resources Research*, **5**, 153–171.

Freeze, R.A. and Cherry, J.A. (1979) *Groundwater* (604 pp.). Prentice Hall, Englewood Cliffs, NJ.

Guzzetti, F., Carrara, A., Cardinali, M., and Reichenbach, P. (1999) Landslide hazard evaluation: A review of current techniques and their application in a multi-scale study, central Italy. *Geomorphology*, **31**, 181–216.

Hammond, C.J., Hall, D.E., Miller, S.M., and Swetik, P.G. (1992a) *Level I Stability Analysis (LISA) Documentation for Version 2.0* (USDA Forest Service General Tech. Report INT-25). US Department of Agriculture, Washington, DC.

Hammond, C.J., Prellwitz, R.W., and Miller, S.M. (1992b) Landslide hazard assessment using Monte-Carlo simulation. In: D.H. Bell (ed.), *Proceedings of the International Symposium on Landslides* (Vol. 6, pp. 959–964). A.A. Balkema, Rotterdam.

Havercamp, R., Vauclin, M., Touma, J., Wierenga, P.J., and Vachaud, G. (1971) A comparison of numerical simulation models for one-dimensional infiltration. *Soil Science Society of America Journal*, **41**, 285–294.

Hillel, D. (1982) *Introduction to Soil Physics* (364 pp.). Academic Press, San Diego, CA.

Hsieh, P.A., Wingle, W., and Healy, R.W. (2000) *VS2DI: A Graphical Software Package for Simulating Fluid Flow and Solute or Energy Transport in Variably Saturated Porous Media* (USGS Water-Resources Investigations Report 99-4130, 16 pp.). US Geological Survey, Reston, VA.

Iverson, R.M. (1990) Groundwater flow fields in infinite slopes. *Géotechnique*, **40**, 139–143.

Iverson, R.M. (2000) Landslide triggering by rain infiltration. *Water Resources Research*, **36**, 1897–1910.

Iverson R.M., and Major, J.J. (1986) Groundwater seepage vectors and the potential for hillslope failure and debris flow mobilization. *Water Resources Research*, **22**, 1543–1548.

Iverson R.M. and Reid, M.E. (1992) Gravity-driven groundwater flow and slope failure potential: 1. Elastic effective-stress model. *Water Resources Research*, **28**, 925–938.

Iverson, R.M., Reid, M.E., Iverson, N.R., LaHusen, R.G., Logan, M., Mann, J.E., and Brien, D.L. (2000) Acute sensitivity of landslide rates to initial porosity. *Science*, **290**, 513–516.

Jibson, R.W., Harp, E.L., and Michael, J.A. (2000) A method for producing digital probabilistic landslide hazard maps. *Engineering Geology*, **58**, 271–289.

Lambe, T.W. and Whitman, R.V. (1969) *Soil Mechanics* (553 pp.). John Wiley & Sons, New York.

Montgomery, D.R., and Dietrich, W.E. (1994) A physically-based model for the topographic control on shallow landsliding. *Water Resources Research*, **30**, 1153–1171.

Moore, P.L. and Iverson, N.R. (2002) Slow episodic shear of granular materials regulated by dilatant strengthening. *Geology*, **30**(9), 843–846.

Morrissey, M.M, Wieczorek, G.F., and Morgan, B.A. (2001) *Regional Application of a Transient Hazard Model for Predicting Initiation of Debris Flows in Madison County, Virginia* (USGS Open-File Report 01-481, 7 pp.). US Geological Survey, Reston, VA.

O'Loughlin, C.L. (1974) The effect of timber removal on the stability of forest soils. *Journal of Hydrology (N.Z.)*, **13**, 121–134.

Pack, R.T., Tarboton, D.G., and Goodwin, C.N. (1998) The SINMAP approach to terrain stability mapping. *Proceedings of 8th Congress of the Association of Engineering Geology* (Vol. 2, pp. 1157–1165). American Society of Civil Engineers, New York.

Riestenberg, M.M. (1994) *Anchoring of Thin Colluvium by Roots of Sugar Maple and White Ash on Hillslopes in Cincinnati* (USGS Bulletin 2059-E). US Geological Survey, Reston, VA.

Rubin, J. and Steinhardt, R. (1963) Soil water relations during rainfall infiltration: 1. Theory. *Soil Science Society Proceedings*, **27**, 246–251.

Savage, W.Z., Godt, J.W., and Baum, R.L. (2003) A model for spatially and temporally distributed shallow landslide initiation by rainfall infiltration. *Proceedings of the 3rd International Conference on Debris Flow Hazards Mitigation. Mechanics, Prediction, and Assessment, September 10–12, 2003, Davos, Switzerland* (pp. 179–187). Millpress, Rotterdam.

Schmidt, K.M., Roering, J.J., Stock, J.D., Dietrich, W.E., Montgomery, D.R., and Schaub, T. (2001) The variability of root cohesion as an influence on shallow landslide susceptibility in the Oregon Coast Range. *Canadian Geotechnical Journal*, **38**, 995–1024.

Simunek, J., Huang, K., and van Genuchten, M.Th. (1998) *The HYDRUS Code for Simulating 1-dimensional Movement of Water, Heat and Multiple Solutes in Variably Saturated Media, Version 6.0* (US Salinity Laboratory Research Report No. 144). US Salinity Laboratory, Riverside, California.

Stephens, D.B. (1996) *Vadose Zone: Hydrology* (347 pp.). Lewis, Boca Raton, FL.

Terzaghi, K. (1943) *Theoretical Soil Mechanics* (510 pp.). John Wiley & Sons, New York.

Voss, C.I. (1984) *A Finite-element Simulation Model for Saturated-unsaturated, Fluid-density-dependent Ground-water Flow with Energy Transport or Chemically-reactive Single-species Solute Transport* (USGS Water-Resources Investigations Report 84-4369, 409 pp.). US Geological Survey, Reston, VA.

Waldron, L.J. (1977) The shear resistance of root-permeated homogeneous and stratified soil. *Soil Science Society of America Journal*, **41**, 843–849.

Wilkinson, P.L., Anderson, M.G., and Lloyd, D.M. (2002) An integrated hydrological model for rain-induced landslide prediction. *Earth Surface Processes and Landforms*, **27**, 1285–1297.

Wolle, C.M., and Hachich, W. (1989) Rain-induced landslides in southeastern Brazil. In: *Proceedings of the 12th International Conference on Soil Mechanics and Foundation Engineering, Rio de Janeiro, 1989* (pp. 1639–1642). A.A. Balkema, Rotterdam.

5

Mechanism of landslide-triggered debris flows: Liquefaction phenomena due to the undrained loading of torrent deposits

Kyoji Sassa and Gong hui Wang

5.1 INTRODUCTION

The long-travelling motion of debris agitated and saturated with water is called a "debris flow". Debris flows can develop during heavy rainfall, when a debris slide changes into a debris flow. They can travel several kilometres from their source and inundate a large area, causing great loss of lives and properties.

Debris flows can be triggered by pyroclastic flows (e.g., Pierson et al., 1990), and accumulated water (e.g., Takahashi, 1978, 1991), but in most cases, they were mobilized from landslides (e.g., Johnson, 1984; Sassa, 1985; Ellen and Fleming, 1987; Iverson et al., 1997). The displaced landslide masses in the source area transform into debris flows while sliding downwards along the slope, a channel, or a gully. In this case, the motion involves not only the displaced soil mass from the initial failure, but also the deposits added along the path of travel (Costa and Williams, 1984; Sassa, 1985, 2000a; Santi and Mathewson, 1988; Jibson, 1989; Wieczorek et al., 1989, 2000; see also Chapter 7).

When the initial landslide mass rides on the torrent bed deposits, an undrained loading process may generate a high pore-water pressure within the torrent deposits and this helps incorporate those deposits into moving mass (Sassa, 2000a).

The process of undrained loading process has been reported by Hutchinson and Bhandari (1971) for earth flows. Through a series of studies on debris flows that occurred on Mount Usu and the Sakurajima volcano, it was pointed out that similar undrained loading on the torrent deposits due to the displaced landslide mass performed a key role on the occurrence and motion of debris flows (Sassa, 1985).

This paper introduces the mechanism of undrained loading on the torrent deposits and describes two landslide-triggered debris flows in which undrained shear loading played a very important role in the downslope volume enlargement of debris flow.

5.2 UNDRAINED LOADING OF THE TORRENT DEPOSITS

When a displaced landslide mass covers the torrent deposits, the loading is applied quickly. Therefore, the saturated torrent deposits suffer undrained loading, with effects similar to the seismic force acting on slope. In the following, let us analyse the response of torrent deposits to the dynamic loading of displaced landslide mass, by utilizing a consolidation–shear model and the corresponding stress path.

5.2.1 Consolidation–shear model

The concept of undrained loading due to the application of normal stress was proposed by Terzaghi and Peck (1948). In this model, the soil element is assimilated to an elastic spring, and this spring lies in a container full of water with a small drainage hole. The model of Terzaghi and Peck was proposed for the consolidation of clay. A different process takes place in loose sands, where contraction due to the application of shear loadings, especially if accompanied by grain crushing, will result in the generation of excess pore-water pressure. To model the undrained shearing on the torrent deposits, both models need to be combined as shown in Figure 5.1

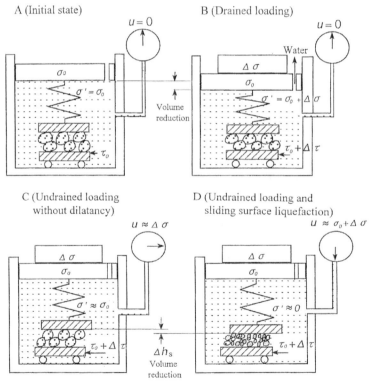

Figure 5.1. Consolidation–shear model (Sassa, 1997, 2000a, b). Here σ' is the normal stress on the spring; $\Delta\sigma$ is added normal stress; u is water pressure within the container; Δh_c is the change in the consolidation model (spring) height; Δh_s is the change in the shear model height; and $\Delta\tau$ is the added shear stress.

(Sassa, 1997). To simplify the calculation of normal stress due to loadings, here a unit cross section of the upper part of the container is assumed.

At first, let us consider the model of Terzaghi and Peck, with normal stress but no shear stress. If the upper drainage hole was closed (i.e., the box would be in an undrained condition), the increment in pore-water pressure (Δu) due to the increment of normal stress ($\Delta \sigma$) could be expressed as follows (Sassa, 1985):

$$\Delta u = B_D \times \Delta \sigma \tag{5.1}$$

$$B_D = 1/(1 + \eta C_W/C_{C1}) \tag{5.2}$$

where η is the porosity of soil; C_w is the compressibility of pore fluid; and C_{C1} is the 1-D compressibility of soil.

In fully saturated soil, the compressibility of water is several orders of magnitude less than that of soil, then B_D could be regarded approximately as 1.0. In a dry soil, it is 0. Next, let us consider the case where both normal stress and shear stress are applied simultaneously to the saturated sand-grain layer shown in Figure 5.1.

(a) *Initial state.* Initial normal stress and shear stress are applied on the sand-grain layer, and the pore-water pressure is zero.
(b) *Case of drained loading.* Excess pore pressure is zero, irrespective of the occurrence of shear failure or grain crushing. Therefore, the increment in the effective normal stress will be equal to $\Delta \sigma$. The spring is compressed due to the consolidation and $\Delta h_c < 0$.
(c) *Case of undrained loading without dilatancy.* If there is no volume change in the soil during the shearing, $\Delta h_c + \Delta h_s = 0$, therefore, $\Delta h_c = 0$. The increment in normal stress is transferred fully into the pore-water pressure, and the effective normal stress remains the same as the initial value.
(d) *Case of undrained loading and sliding-surface liquefaction.* In this case, there is some volume change in the sand grains due to contraction during shearing ($\Delta h_s < 0$). Because the box is under undrained condition (i.e., $\Delta h_c + \Delta h_s = 0$), then $\Delta h_c > 0$, and the initial effective normal stress will decrease. If collapse of soil structure and grain crushing takes place, the initial normal stress could even be fully balanced by the excess pore-water pressure (i.e., $u \cong \sigma_0 + \Delta \sigma$), and the effective normal stress would reduce to a very small value.

In summarizing, the entire increment of normal stress ($\Delta \sigma$) will be carried by the spring in the case of drained conditions, or it will lead to an increment of water pressure (Δu) with $\Delta u = \Delta \sigma$ in the case of undrained conditions (Figure 5.1(c)). However, if there is volume change (contraction) in the sand-grain model as a result of shearing, the spring will compress further, and the effective normal stress will be further reduced (Figure 5.1(d)). Full liquefaction could then result.

5.2.2 Stress path

Figure 5.2(a) shows the common case where a debris mass is descending along a channel or slope covered with debris deposits. The stresses acting on the base of the

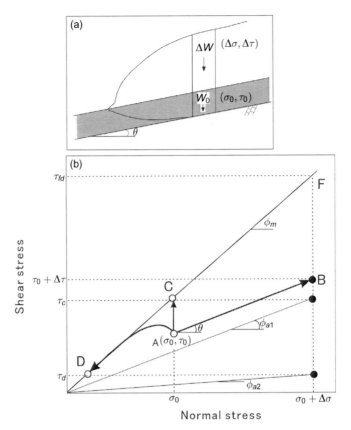

Figure 5.2. Drained or undrained loading during the motion of a displaced debris mass and the resulting mobilized apparent friction angle. (a) Downslope motion of debris mass along channel or slope with debris deposits. (b) Stress path acting on the sliding surface (bottom in (a)) within the debris deposits.
Sassa (1997, 2000a, b).

column with weight of W_0 are presented in Figure 5.2(b). The initial stress state due to W_0 is expressed by point "A". When the displaced landslide mass rides onto the torrent deposits, if no excess pore pressure was generated during this process, the stress point would move to point "B" by adding static normal and shear stresses ($\Delta\sigma = \Delta W \times \cos\theta$; $\Delta\tau = \Delta W \times \sin\theta$) due to the weight of the mass (ΔW). The stress path would thus move from point "A" to point "B". Normally, the inclination of the torrent slope is smaller than the internal friction angle of the soil grains, so shear failure will not occur. A short-lived impact load may temporarily bring the stress path to the failure point "F", but will soon disappear. For this drained condition, the apparent friction angle (ϕ_a) during shear movement is:

$$\tan\phi_a = \tau_{fd}/(\sigma_0 + \Delta\sigma) \tag{5.3}$$

Therefore, if cohesion is 0, ϕ_a will be the same as the internal friction angle of soil grains in motion (ϕ_m).

However, when no dilatancy occurs under undrained conditions (Figure 5.1(c)), there would be no change in the effective normal stress, and the stress path will go vertically upwards. If the stress path reaches the failure line, a shear failure would result. If there were no dilatancy even after the shear failure, the mobilized shear strength during movement would remain the same as that at the failure point (point C). The apparent friction angle (ϕ_{a1}) defined by $\{\tau_c/(\sigma_0 + \Delta\sigma) = \tan\phi_{a1}\}$ could be calculated through the following equation:

$$\tau_c = (\sigma_0 + \Delta\sigma) \times \tan\phi_{a1} = c + (\sigma_0 + \Delta\sigma - \Delta u) \times \tan\phi_m \quad (5.4)$$

where c is true cohesion, usually zero for sandy soils; and Δu is the increment of pore-water pressure due to introduction of $\Delta\sigma$. Now assuming that $\Delta\sigma = \Delta u$, then we have:

$$\tan\phi_{a1} = \frac{\sigma_0}{\sigma_0 + \Delta\sigma} \times \tan\phi_m = \frac{1}{1 + \Delta\sigma/\sigma_0} \times \tan\phi_m \quad (5.5)$$

This hypothesis has been used for the interpretation of the Ontake catastrophic debris flow and another large debris flow that occurred on the Myoko Plateau, Japan. Both of these debris flows occurred on a large scale with high mobility (Kawakami, 1997).

If sliding-surface liquefaction were triggered (Figure 5.1(d)), the shear resistance would decrease to a very small value (τ_D) compared to the shear strength (τ_C) at the initial normal stress. The apparent friction angle at this time (ϕ_{a2}, as shown in Figure 5.2(b)) would be very small, compared to both ϕ_m and ϕ_{a1}.

5.3 LIQUEFACTION AND SLIDING SURFACE LIQUEFACTION

The liquefaction failure phenomenon of torrent deposits is well documented in a video, entitled *Debris Flow Dynamics*, made by Costa and Williams (1984). In this video, the displaced soil mass overrode torrent deposits along its path, triggered a liquefaction failure resulting in the debris flow becoming enlarged in volume. Sassa (1985) investigated the upstream reaches of the Kousu torrent on Mount Usu during 2–12 August, 1981, when a series of heavy rainfalls saturated the area, and many small debris flows ensued. The torrent deposits (about 1 m in depth) were almost submerged, and groundwater flowed at 10 cm below the surface. As reported, "I walked on the deposit and felt soft reaction, then I heavily stamped it. The deposit around me behaved like a water cushion, then water came out as consolidation began" (Figure 5.3) (Sassa, 1985).

To examine the liquefaction initiation under undrained conditions undrained load-controlled triaxial tests were performed on samples from the torrent deposits of Mount Usu (Figure 5.4). While in some tests excess pore pressure did not occur, in others sudden liquefaction failure occurred and the effective normal stress decreased sharply. If the effective normal stress decreased to a value less than 1/10 of the initial

86 Mechanism of landslide-triggered debris flows [Ch. 5

Figure 5.3. Torrent deposit liquefied by stamping some hours before the initiation of a debris flow.

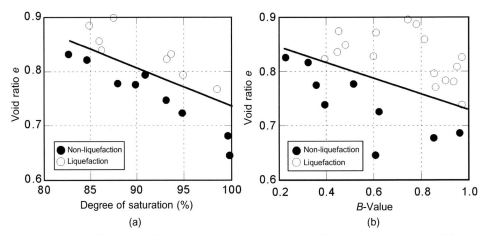

Figure 5.4. Conditions for triggering liquefaction on a sample from torrent deposits of Mount Usu.

stress, liquefaction was said to have occurred. Therefore, a boundary between liquefaction and non-liquefaction phenomena was drawn (Figure 5.4), and defined by the void ratio and degree of saturation that is expressed by the pore pressure coefficient B (Skempton, 1954). Although to some extent the measured void ratios of the torrent deposits were scattered, they were very near the boundary between liquefaction and

Sec. 5.3] Liquefaction and sliding surface liquefaction 87

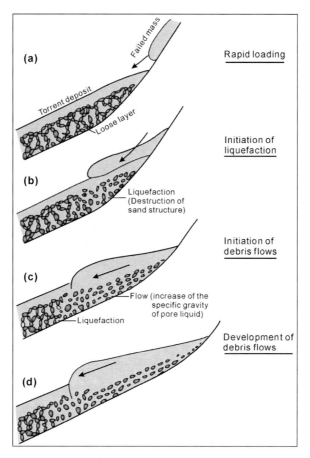

Figure 5.5. Illustration of the initiation of debris flows (Sassa, 1985).

non-liquefaction. If the torrent deposits liquefied, they could quickly change to debris flows.

Debris flows and their development are depicted in Figure 5.5, based on field observations in Japan and laboratory liquefaction tests. Torrent deposits with loose and unstable structure can collapse structurally under rapid loading, and the displaced debris mass becomes seated on a liquefied layer (Figures 5.5(a, b)). The torrent deposit starts to flow, causing liquefaction at its front and increasing its volume (Figures 5.5(c, d)). Similar mechanisms were proposed by Hutchinson and Bhandari (1971), Tabata and Ichinose (1973), Costa and Williams (1984), and Sassa et al. (1985, 1997).

Liquefaction can take place in meta-stable granular soils. The collapse of the soil structure causes volume reduction and results in excess pore-water pressure generation. The existence of sliding-surface liquefaction was found during the study of the Nikawa landslide caused by the 1995 Kobe earthquake (Sassa 1996; Sassa et al.,

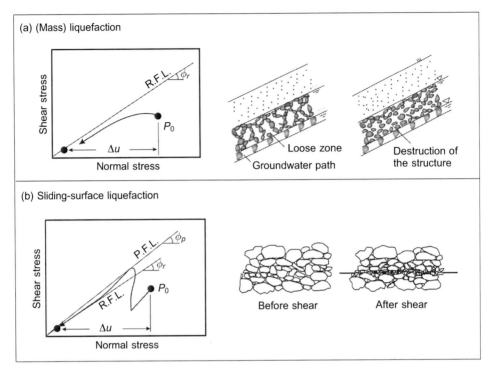

Figure 5.6. Liquefaction and sliding-surface liquefaction. P.F.L. is the peak failure line; R.F.L. the residual failure line; P_0 the initial stress state; Δu the increment in pore-water pressure; $\Delta \tau$ the increment in shear stress; and ϕ_p, ϕ_r the peak and residual friction angles, respectively.

From Sassa (1996, 2000a)

1996). Normally, liquefaction causes landslides, but in this case it is the sliding that triggers liquefaction and results in a mobile landslide (Figure 5.6).

Liquefaction takes place only in soils with very loose meta-stable structure. On the other hand, sliding-surface liquefaction can take place in loose, medium, or even dense soils so long as grain crushing and the resulting volume reduction takes place under the given overburden pressure. When a landslide mass moves onto slope or torrent deposits, drained or undrained loading can be involved, and either liquefaction or sliding-surface liquefaction can occur.

5.4 TYPICAL CASES OF LANDSLIDE-TRIGGERED DEBRIS FLOWS

5.4.1 The 1996 Gamahara debris flow

Outline of the Gamahara landslide-triggered debris flow

A landslide-triggered debris flow occurred in the Gamahara torrent, Otari village, Nagano Prefecture, Japan, on 6 December, 1996 (Kaiyo Syuppan, 1997;

Sec. 5.4] Typical cases of landslide-triggered debris flows 89

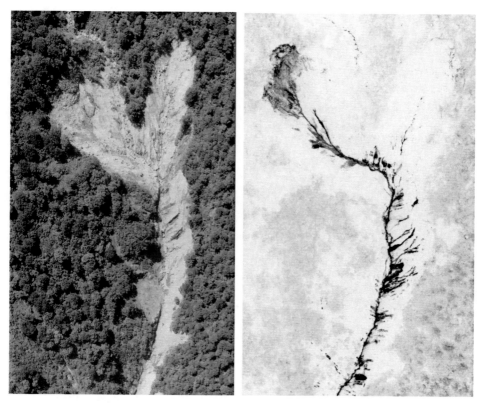

Figure 5.7. Aerial photos before and after the debris-flow disaster. The left-hand photo was taken on 19 July, 1996. The right-hand photo was taken on 7 December, 1996, while snow was covering this area.

Left-hand photo courtesy of Nakanihon Kouku Company. Right-hand photo courtesy of Pasco Company.

Marui et al. 1997; Sassa et al., 1997; Sassa, 1999). It killed 14 people and injured 8 who were constructing check dams. The event occurred during a season not normally known for debris flows. The day of the debris flow saw almost no rainfall (1 mm), and rainfall the day before the debris flow was 49 mm (Marui et al., 1997; Sassa et al., 1997). Occurrence of debris flows is regulated by rainfall intensity and cumulative rainfall. Warnings of debris flows are issued based on the monitoring of rainfall in Japan (JSECE, 1992) and in the USA (Wieczorek et al., 1990). In this case, no warning was issued because of the lack of rainfall.

Figure 5.7 shows two aerial-photos, one before and one after the debris flow. The right-hand photo taken one day after the event shows the trace of the debris flow. A landslide was located at the top of the path; it was apparently the cause of this debris flow. The left-hand photo shows the same area in July 1996 (i.e., before the event). Two previous landslides were observed. Comparing these two photos, the landslide that triggered the 1996 debris flow was a retrogressive failure at the head scarp of a previous landslide (45 m wide and 65 m long).

Field investigation

The Disaster Prevention Research Institute (DPRI), Kyoto University team investigated the initial landslide shown in Figure 5.7, surveyed it using a total station theodolite, and took samples on 4 May, 1997. Figure 5.8 shows the plan of the initial

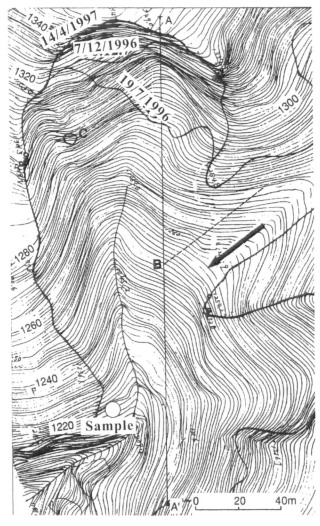

Figure 5.8. Plan of the initial landslide and the Gamahara torrent (made from the aerial photo of 17 April, 1997). The vertical line A–A′ corresponds to the survey line. The arrow represents the main stream of the Gamahara torrent. (○ represents the sampling point.)

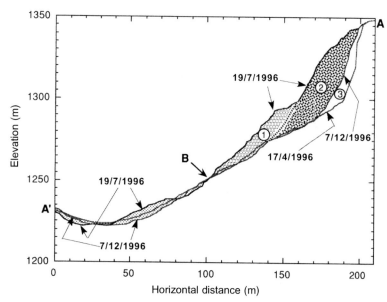

Figure 5.9. A–A′ sections made from aerial photos taken on 19 July, 1996, 7 December, 1996, and 17 April, 1997. ① represents the landslide that occurred during the 11 July, 1995 heavy rainstorm; ② the landslide mass of the 6 December, 1996 debris flow; and ③ the retrogressive slide in early April, 1997.

landslide. Line A–A′ was the central survey line. The section of A–A′ is presented in Figure 5.9, which also shows the results of aerial photo interpretations of 19 July, 1996 (before the 6 December, 1996 debris flow), 7 December, 1996 (one day after the debris flow), and 17 April, 1997 (after another retrogressive slide). The profile of 19 July, 1996 shows the sliding surface of a landslide which took place during the 11 July, 1995 heavy rainstorm and the debris left by it (mass designated by the numeral 1). The mass designated by the numeral 2 is estimated to be the landslide mass of the 6 December, 1996 debris flow. The mass shown as 3 probably formed the retrogressive slide in early April, 1997. Material shown downslope from B in Figures 5.8–5.9 was the torrent bed and the deposits of the Gamahara gully. As revealed by Figure 5.9, about 6 m of torrent deposits between the two ground surface lines of 19 July, 1996 and 7 December, 1996 were transported together with the landslide mass.

Landslide 2 (Figure 5.9) occurred on a steep slope and it might not have been saturated and liquefied at the time of the slope failure. However, the torrent deposits downstream from Point B could have been saturated by streamflow. It is inferred that a landslide occurred in the source area first; then the displaced landslide mass moved onto the saturated torrent deposit, and triggered the liquefaction failure. To examine whether shear failure could be triggered in the torrent deposits by the displaced soil mass from failure 2 and what was the critical condition for this kind of failure, experimental analyses were performed by conducting ring-shear tests on soil samples taken from the torrent deposits downstream of B.

Model concept of the landslide-triggered debris flow at Gamahara

Figure 5.10(a) presents a schematic illustration of the 6 December, 1996 Gamahara debris-flow case. In this event, a retrogressive landslide initially occurred at the headscarp. The displaced soil mass of the *trigger failure* slipped down the slope (step I), and then rode onto the debris deposits at the toe of the slope (step II). The debris deposits were sheared and moved together with the displaced soil mass

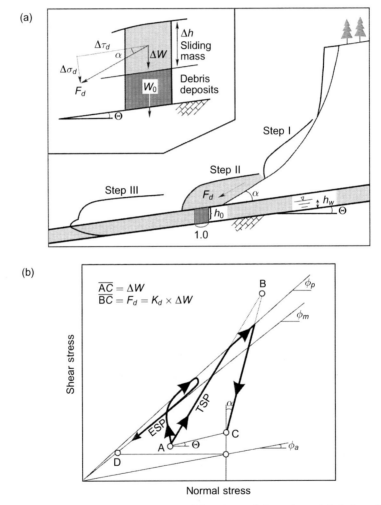

Figure 5.10. Schematic illustration of the landslide-triggered Gamahara debris flow, Japan. (a) The assumed process of debris flow. (b) Stress path of the debris deposit during loading. W_0 represents the self-weight of the column of the debris deposit; ΔW the increment of the static stress due to the self-weight of the sliding mass; α the angle of thrust between the slope and the debris bed; K_d the dynamic coefficient ($F_d/\Delta W$); F_d the dynamic stress; and ϕ_p, ϕ_m, ϕ_a the peak, residual, and apparent friction angles, respectively.

from the *trigger failure* (step III). Consider a column of unit width inside the debris deposit. In step I, the self-weight of the column (W_0) was in effect. When the sliding mass rode onto the debris deposit (step II) with a certain velocity, it caused a dynamic load on the column. Here, assume the applied stress as the sum of the static stress (ΔW) (load due to the self-weight of the displaced soil mass) and the dynamic stress along the movement direction of the displaced mass (F_d).

The dynamic force (F_d) is calculated by using the static stress (ΔW) and the dynamic coefficient (K_d), which was denoted by Sassa et al. (1997) and is similar to a seismic coefficient:

$$F_d = K_d \times \Delta W \tag{5.6}$$

Figure 5.10(b) shows two stress paths caused by the condition of static and dynamic loading as shown in Figure 5.10(a). The total stress path is outlined by ABC, with excess pressure caused by rapid undrained loading or sliding surface liquefaction. Path A–D would be followed in terms of effective stress, leading to substantial strength loss.

Ring-shear test for mechanical simulation

To examine the model concept of Figure 5.10 for the Gamahara debris flow, we took a sample of the torrent deposit downstream of B in Figure 5.9. The sample was taken in the debris derived from Jurassic sedimentary rocks at the valley side slope (marked as ○ in Figure 5.8). Using this sample, the situation was simulated by using a torque-controlled undrained ring-shear apparatus (DRPI-5) (Sassa et al., 2003).

A test condition was adopted corresponding to the field situation as described in Figure 5.10, with the flow parameters: $\Delta h = 18\,\text{m}$, $h_0 = 6\,\text{m}$, $h_w = 1\,\text{m}$, $\theta = 20°$, and $\alpha = 0°$ because the slope near the torrent deposits area was smooth. The total unit weight of the soils was $\gamma = 20\,\text{kN/m}^3$.

A dynamic coefficient was assumed to be $k_d = 0.42$ so that the stress path could reach the failure line of this material. The control signals for normal stress and shear stress are shown in Figure 5.11. The line segment of A–C is the static stress increment due to the self-weight of debris of about 18 m. The line segment of B–C is the dynamic (impact) stress increment component. The real stress increment stress path is from A to B. The loading and unloading speed in the test were set to be 5 seconds for loading time, 5 seconds for unloading time, 5 seconds for constant, and 5 seconds for decreasing shear stress to zero (as shown in Figure 5.11(a)).

Figure 5.12(a) presents variation of normal stress, pore pressure, shear resistance, and shear speed, which were monitored during the test. After failure, pore pressure was much increased and shear resistance became very low. The shear velocity reached the maximum speed of this apparatus of 224 cm/s. Figure 5.12(b) shows the corresponding effective stress path (ESP) and the total stress path (TSP). The ESP reached the failure line, and then decreased to a low stress point along the failure line, showing a typical stress path of sliding-surface liquefaction. The apparent friction angle (ϕ_a) at the steady state was 3.6°. Therefore, rapid motion could continue in a torrent bed steeper than 3.6°.

94 Mechanism of landslide-triggered debris flows [Ch. 5

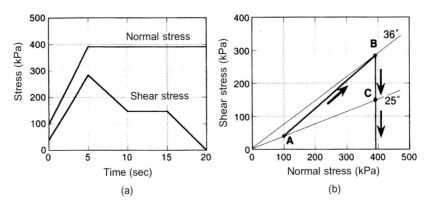

Figure 5.11. Control signals for the mechanical simulation test in terms of time series (a) and stress path (b).

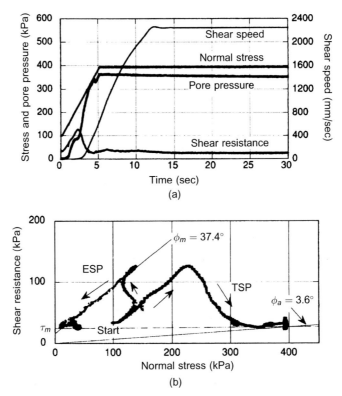

Figure 5.12. Undrained ring-shear test results for the Gamahara debris flow. (a) Time series data. (b) Stress paths. $B_D = \Delta u / \Delta \sigma = 0.79$, $e = 0.67$.

With such low strength, the debris was able to continue moving as far as the alluvial fan of the Himekawa River. The moving debris changed to a debris flow. Calculation based on the concept of Figure 5.10 showed that no dynamic stress (impact force) was necessary to cause shear failure of the torrent deposit.

This test result supports the model concept illustrated in Figure 5.10. The sliding-surface liquefaction was similar to that in the Nikawa landslide. Although earthquake loading in the Nikawa landslide was cyclic, and the Gamahara debris flow was monotonic loading, the basic concept was common in both.

5.4.2 The 1999 Kameyama debris flow

On 29 June, 1999 heavy rainfall in the granitic mountains of Hiroshima Prefecture, Japan, triggered hundreds of landslides and debris flows resulting in 24 fatalities. The most catastrophic was a soil slide that mobilized into a debris flow in the Kameyama area of Hiroshima City (referred to here as the Kameyama landslide), killing four people and destroying several houses. Field and laboratory investigation of the possible mechanism of the Kameyama landslide strongly suggests that it was also a landslide-triggered debris flow (Wang et al., 2003).

Figure 5.13(a) shows an oblique aerial view of the Kameyama landslide and Figure 5.13(b) is a plan view of the landslide. Figure 5.13(b) shows two streams A and B, which originated on the upper slopes and are joined at C. Stream A acted as the main contributor to the disaster. It was an intermittent stream in a gentle

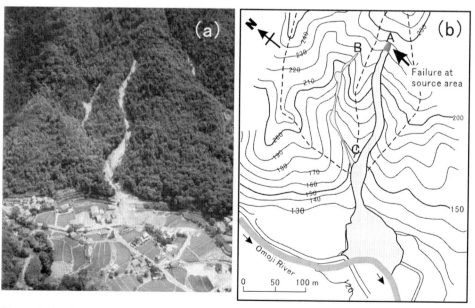

Figure 5.13. Aerial oblique views of the landslide. (a) A total scene of the landslide. (b) Plan of the landslide. The contour lines are labeled in meters.
After Yokota et al. (1999).

96 Mechanism of landslide-triggered debris flows [Ch. 5

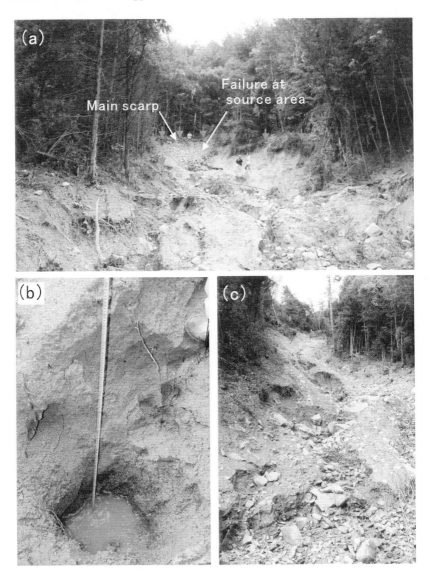

Figure 5.14. Photos of Kameyama landslide. (a) Failure in the source area. (b) Cross section of the excavated pit. (c) Middle part of the travel path.

valley-shaped concave slope, a zero-order basin that appeared to be an upslope extension of a torrent (Tsukamoto, 1973; Dietrich et al., 1987; Sidle, 1987). The main efforts in this study were expended on landslide activity in stream A.

Figure 5.14(a) is a photograph showing the source area of stream A. The sliding mass consisted of decomposed granitic soils. Forest covered this source area. Figure 5.14(b) presents a study pit which was dug 2 days after the heavy rainfall (1 July). As indicated by this photograph, groundwater appeared below the present ground surface (sliding surface) after failure at a depth of about 1.1 m, and the decomposed granitic soil (at a depth of about 80–90 cm) changed in color from brown (due to oxidization) to blue–gray (due to deoxidization). This means that the iron contained in the soil was not oxidized due to the long-term existence of groundwater. Therefore, groundwater could rise rapidly during the rainfall.

Figure 5.14(c) shows the middle parts of the travel path of the displaced soil mass. Most of the newly exposed surface consisted of bedrock and inorganic debris, mainly colluvium that had previously been deposited by sliding, falling, or creeping of the weathered granite.

Using a total station surveying instrument, the following sections were established (through the landslide): (a) a central longitudinal section from the source area to the Omoji River (Figure 5.15(a), see color section); (b) a central longitudinal section through the source area, including two small failures (denoted as Failure T and Failure F) (Figure 5.15(b), see color section); and (c) a cross section through Failure T (Figure 5.15(c), see color section). The distance from the source area to the destroyed houses near the Omoji River was approximately 300 m (Figure 5.15(a), see color section). The original ground surfaces shown in Figure 5.15(b, c) (see color section) were assumed based on the shape of the neighbouring ground surface. From the longitudinal section of the source area (shown in Figure 5.15(b), see color section), it can be seen that Failure T and Failure F have spoon-shaped sliding surfaces. The length of Failure T was approximately 16 m horizontally. Failure T was estimated to be approximately 13 m wide and up to 2 m deep (Figure 5.15(c), see color section).

The total volume of the displaced soil mass from Failure T in the source area was approximately 250 m^3. However, the debris masses that flowed into and were deposited in the residential area near the toe of the slope were approximately 5,000 m^3, more than 10 times as great as that in the source area, even after the possible contribution from stream B and from secondary erosion on the slope by post-landslide rainwater were subtracted (Sassa, 1999).

For the case of the Kameyama landslide, the displaced material increased its volume simultaneously with sliding, indicating the nature of landslide-triggered debris flow. Similar to the Gamahara debris-flow case, undrained ring-shear tests were conducted on the samples taken from the source area of the Kameyama landslide to examine the triggering process.

Test conditions

Two days after the failure, disturbed soil samples were taken from the source area of the Kameyama landslide. The sample in the field had a unit weight (γ_t) of 15.0 kN/m^3 with a water content (at sampling time) of 13% by weight and a void ratio (between the volume of voids and that of solids) of 1.01, and the saturated unit weight (γ_{sat}) for the soil mass in the field was calculated as 17.9 kN/m^3.

The ring-shear apparatus employed in this research was designed to simulate landslides on large scales. Although the estimated thickness for the failed soil mass in the source area and in the debris deposit were approximately 2.0 m, in our laboratory simulation we used a somewhat larger stress corresponding to a thickness of 2.5 m. The stresses acting on the sliding surface, a slope angle (θ in Figure 5.6) was taken as 25° from Figure 5.15(b) (see color section). A dynamic coefficient of 0.9 was assumed and used in the test condition. The initial normal stress and shear stress acting on the potential sliding surface of Failure F for the debris deposit due to W_0 were $\sigma_0 = 40\,\text{kPa}$, $\tau_0 = 20\,\text{kPa}$; the increment in static load due to ΔW of the displaced mass from Failure T were $\Delta\sigma_0 = 40\,\text{kPa}$, $\Delta\tau_0 = 20\,\text{kPa}$; and the dynamic stresses caused by the displaced mass from Failure T (F_d) were assumed to be $\sigma_d = 0\,\text{kPa}$, $\tau_d = 40\,\text{kPa}$. Loads caused by the displaced mass from Failure T were assumed to be applied in 5 seconds, and the dynamic force to be dissipated in 2.5 seconds. Therefore, the loading path shown in Figure 5.16(a) was assumed for simulating the possible impact of loading on the soil unit at the sliding surface of Failure F.

Impact loading test on the saturated sample

After the sample was saturated (the B_D value was checked as 0.96) and normally consolidated in the drained state, it was loaded by the impacting stresses shown in Figure 5.16(a) under undrained conditions. For this test, a rotating speed of 32.3 cm/s was selected.

The test results are presented in Figure 5.16. As can be seen in Figure 5.16(b–d), the pore pressure began to increase almost immediately, and the sample failed at a very small increment of shear stress (approximately 10 kPa). Soon after, the pore pressure increased greatly and the shear resistance reduced to a very small value (about 2 kPa). The ESP shifted from the failure line in the initial loading period, and later dropped along the failure line, showing the characteristic of sliding-surface liquefaction. From Figure 5.16(c) it is seen that the shear velocity reached the maximum value of the selected gear of 32.3 cm/sec, and the shear displacement reached 5 m during the test duration of a few seconds. It is inferred that upon sliding of the displaced soil mass from Failure T, liquefaction failure was triggered in the saturated colluvial deposits along the path of travel, thus including additional deposits in motion. This could explain the tenfold increase in soil volumes as described above.

Impact loading test on a sample with natural water content

As pointed out, full saturation is not always a prerequisite for liquefaction; in some cases, partially saturated soils can also undergo liquefaction failure (Sassa et al., 2001). The test mentioned above was repeated on a sample with a natural water content of 15.3% (without saturation).

Figure 5.17(b) shows the measured normal stress and mobilized shear resistance on the sliding surface; and Figure 5.17(c) indicates the resultant shear velocity and shear displacement. Figure 5.17(d) presents the corresponding stress path. When the load was applied at its initial state (start point (○)), the stress path reached the failure line with increase of load, and then failed. Figure 5.17(c) shows a very small shear displacement (approximately 16 mm). The maximum shear velocity reached

Sec. 5.4] Typical cases of landslide-triggered debris flows 99

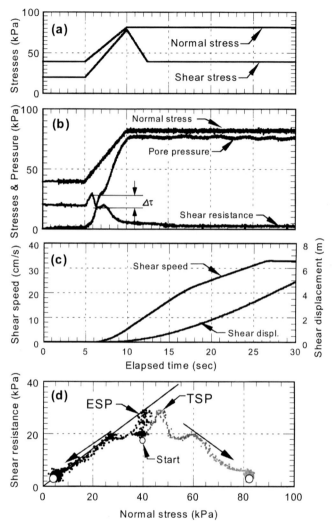

Figure 5.16. Results of a test simulating the undrained loading on the saturated sample ($\gamma_d = 1.50\,\text{gf/cm}^3$, $e = 0.74$). (a) Control signal for normal stress and shear stress. (b) Time series data of normal stress and shear resistance. (c) Time series data of shear speed and shear displacement. (d) ESP and TSP.

was about 4.5 mm/s (the maximum velocity produced by the gear used in this test was 32.3 cm/s). When the soil was in the state of natural water content (15.3%) in this slope, only a very small shear displacement within the soil layer could have been initiated. The soil mass that originated from the source area could not have enlarged its volume during sliding downslope, but in fact decreased in volume by depositing material below the shear zone. The existence of pre-deposited colluvial soil masses along Stream A could be interpreted so that in normal times of little or no rainfall,

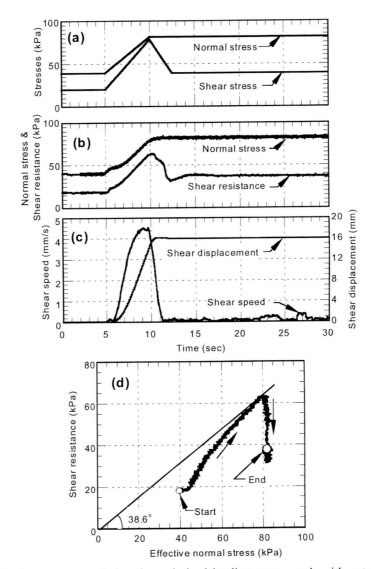

Figure 5.17. Test results simulating the undrained loading on a sample with natural water content ($\gamma_d = 1.54\,\text{gf/cm}^3$, $e = 0.69$, $S_r = 63.9\%$). (a) Control signal for normal stress and shear stress. (b) Time series data of normal stress and shear resistance. (c) Time series data of shear speed and shear displacement. (d) ESP.

the soil mass near the sliding surface was not saturated enough for liquefaction failure to occur. Even taking into account impact loading by a displaced soil mass from upstream failure, the displaced soil mass could have remained in place.

5.5 CONCLUSIONS

This chapter gives an interpretation of landslide-triggered debris flows, and introduces different types of liquefaction failure processes. Two recent debris flows were introduced and examined in detail. The conclusions are:

1. Undrained loading plays a key role in landslide-triggered debris flows. When debris deposits along the travel path were subjected to undrained loading by displaced landslide mass, then: (a) liquefaction could be triggered due to the collapse failure of soil structure, if the void ratio of the debris deposits were greater than the critical void ratio; (b) sliding surface liquefaction could be triggered within the shear zone due to the grain crushing if the grains are crushable at the applied stress level, even if the void ratio of the debris deposits were not so great.
2. For the Gamahara debris flow, it is inferred that the scarp of a previous landslide could have been marginally stable. A small amount of rainfall or snow melt could cause a landslide failure on the scarp. Meanwhile, there probably were some saturated parts within the torrent deposits due to the rainfall and melted snow. The displaced landslide mass from the scarp could cause liquefaction in the torrent deposits, leading to enlargement in the debris-flow volume. This was examined and supported by the undrained ring-shear test results, which showed that the static shear stress due to the weight of displaced landslide mass was great enough to cause shear failure of the torrent deposits.
3. For the Kameyama landslide, it is concluded that a slope failure in the source area was triggered due to a rise in groundwater level. The standing groundwater made it possible for the groundwater level to rise during the rainstorm and to saturate the potential sliding surface. The displaced soil mass of the small failure slid downslope onto colluvial debris deposits. The saturated portions of the debris deposits were sheared, and joined in movement with the original displaced soil mass, enlarging the volume of moving debris flow. Undrained ring-shear tests simulating the triggering process revealed that only a small increment of shear stress due to the impact of displaced soil mass could result in shear failure with the generation of high pore-water pressure.
4. A ring-shear test on a sample from Kameyama at natural water content revealed that saturated soil mass along the travel path is a prerequisite for the triggering of debris flow by landslides with downslope volume enlargement processes mentioned above. Under normal climatic conditions of little or no rainfall, failure by undrained loading and displacement of debris flow would not be possible, even under impact by a displaced soil mass from an upstream failure.

5.6 REFERENCES

Costa, J.E. and Williams, G.D. (1984) *Debris Flow Dynamics* (USGS Water Resources Division, Open-file Report 84/606, Video). US Geological Survey, Reston, VA.

Dietrich, W.E., Reneau, S.L., and Wilson, C.J. (1987) Overview: Zero-order basins and problems of drainage density, sediment transport and hillslope morphology. *Erosion and Sedimentation in the Pacific Rim: Proceedings of the Corvallis Symposium, Oregon* (IAHS Publication No. 165, pp. 27–37). International Association of Hydrological Sciences, Christchurch, New Zealand.

Ellen, S.D. and Fleming, R.W. (1987) Mobilization of debris flows from soil slips, San Francisco Bay region. In: J.E. Costa and G.F. Wieczorek (eds), *Debris Flows/Avalanches: Process, Recognition, and Mitigation* (GSA Reviews in Engineering Geology No. 7, pp. 31–40). Geological Society of America, Boulder, CO.

Hutchinson, J. and Bhandari, R. (1971) Undrained loading, a fundamental mechanism of mudflows and other mass movements. *Géotechnique*, **21**(4), 353–358.

Iverson, R.M., Reid, M.E., and LaHusen, R.G. (1997) Debris-flow mobilization from landslides. *Annual Review of Earth Planetary Science*, **25**, 85–138.

Jibson, R.W. (1989) Debris flows in southern Puerto Rico. In: A.P. Schultz and R.W. Jibson (eds), *Landslide Processes of the Eastern United States and Puerto Rico* (GSA Special Paper No. 236, pp. 29–55). Geological Society of America, Boulder, CO.

Johnson, A.M. (1984) Debris flow. In: D. Brunsden and D.B. Prior (eds), *Slope Instability* (pp. 257–361). John Wiley & Sons, New York.

JSECE (1992) *Prevention of Sediment Disasters* (Japan Society of Erosion Control Engineering Series of Sabo Text No. 6-1, pp. 143–166). Sankaido, Tokyo [in Japanese].

Kaiyo Syuppan Ltd (1997) Gamahara debris-flow disasters (Special issue on landslide-triggered Gamahara debris flows, 10 reports). *Monthly Journal Chikyu [Earth]*, **19**(10), 603–670 [in Japanese].

Kawakami, H. (1997) *Surveillance Study on the 1996 Otanimura, Nagano Prefecture, Debris Flow Disaster* (Report for the Grant-in-Aid for Scientific Research in 1996, Project No. 08300017, pp. 4-1–4-13). Kyoto University, Tokyo.

Marui, H., Sato, O., and Watanabe, N. (1997) Gamahara torrent debris flow on 6 December 1996, Japan. *Landslide News*, **10**, 4–6.

Pierson, T.C., Janda, R.J., Thouret, J.C., and Borrero, C.A. (1990) Perturbation and melting of snow and ice by the 13 November 1985 eruption of Nevado del Ruiz, Colombia, and consequent mobilization, flow, and deposition of lahars. *Journal of Volcanology and Geothermal Research*, **41**, 17–66.

Santi, P.M. and Mathewson, C.C. (1988) What happens between the scar and the fan? The behavior of a debris flow in motion. In: R.J. Fragaszy (ed.), *24th Annual Symposium on Engineering Geology and Soils Engineering* (pp. 73–78). Washington State University, Washington, DC.

Sassa, K. (1985) The mechanism of debris flow. *Proceedings of XI International Conference on Soil Mechanics and Foundation Engineering, San Francisco* (pp. 1173–1176). A.A. Balkema, Rotterdam, the Netherlands.

Sassa, K. (1996) Prediction of earthquake induced landslides (Special lecture). *7th International Symposium on Landslides* (Vol. 1, pp. 115–132). A.A. Balkema, Rotterdam.

Sassa, K. (1997) Landslide triggered debris flows: Mechanism of undrained loading of torrent deposits (Special issue on landslide-triggered debris flows). *Monthly Journal Chikyu [Earth]*, **19**(10), 652–660 [in Japanese].

Sassa, K. (1999) Fluidized landslides on urbanized areas and their mechanisms: Case studies on some recently occurred disasters in the Kameyama area, Hiroshima City, and other areas. Unpublished paper presented at *Symposium of Landslide Research Council of Japan* (pp. 1–12) [in Japanese].

Sassa, K. (2000a) Mechanism of flows in granular soils (Invited paper). *Proceedings of GeoEng2000, Melbourne, 19–24 November* (Vol. 1, pp. 1671–1702). Technomic Publishing Company, Inc., Pennsylvania.

Sassa, K. (2000b) Mechanism of fluidized landslides. *Proceedings of Symposium on Initiation, Motion and Prediction of Fluidized Landslides* (pp. 1–26). Japan Landslide Society, Tokyo [in Japanese].

Sassa, K., Kaibori, M., and Kitera, N. (1985) Liquefaction and undrained shear of torrent deposits as the cause of debris flows. *Proceedings International Symposium on Erosion, Debris Flows and Disaster Prevention* (pp. 231–236). Toshindo, Tokyo.

Sassa, K., Fukuoka, H., Scarascia-Mugnozza, G., and Evans, S. (1996) *Earthquake-induced Landslides: Distribution, Motion and Mechanism: Special issue. Soils and Foundations* (pp. 53–64). Japanese Geotechnical Society, Tokyo.

Sassa, K., Fukuoka, H., and Wang, F.W. (1997) Mechanism and risk assessment of landslide-triggered-debris flows: Lesson from the 1996.12.6 Otari debris flow disaster, Nagano, Japan. In: D. Cruden and R. Fell (eds), *Landslide Risk Assessment* (pp. 347–356). A.A. Balkema, Rotterdam.

Sassa, K., Wang, G., Fukuoka, H., Vankov, D., and Okada, Y. (2001) Evaluation of dynamic shear characteristics in landslides. *Proceedings International Conference on Landslides: Causes, Impacts and Countermeasures, Davos, Switzerland, 17–21 June* (pp. 305–318). Verlag Glückauf, Essen, Germany.

Sassa, K., Wang, G., and Fukuoka, H. (2003) Performing undrained shear tests on saturated sands in a new intelligent type of ring shear apparatus. *Geotechnical Testing Journal*, **26**(3), 257–265.

Sidle, R.C. (1987) A dynamic model of slope stability in zero-order basins. *Erosion and Sedimentation in the Pacific Rim: Proceedings of the Corvallis Symposium, Oregon* (IAHS Publication No. 165, pp. 101–110). International Association of Hydrological Sciences, Christchurch, New Zealand.

Skempton, A.W. (1954) The pore-pressure coefficient A and B. *Géotechnique*, **4**, 143–147.

Tabata, S. and Ichinose, E. (1973) Study on the primary factors of the mud-flow disasters in Owase area. *Shin-sabo [Journal of the Erosion Control Engineering Society, Japan]*, **86**, 20–24.

Takahashi, T. (1978) Mechanical characteristics of debris flow. *Journal of Hydraulics Division, ASCE*, **104**(HY8), 1153–1169.

Takahashi. T. (1991) *Debris Flow* (165 pp.). A.A. Balkema, Rotterdam.

Terzaghi, K. and Peck, R.B. (1948) Consolidation of clay layers. *Soil Mechanics in Engineering Practice* (pp. 74–78). John Wiley & Sons, New York.

Tsukamoto, Y. (1973) Study on the growth of stream channels. I: Relationship between steam channel growth and landslides occurring during heavy storms. *Journal of Japan Erosion Control Society*, **25**(4), 4–13.

Turner, A.K. and Schuster, R.L. (1996) *Landslides: Investigation and Mitigation*. National Academy Press, Washington, DC.

Wang, G., Sassa, K., and Fukuoka, H. (2003) Downslope volume enlargement of a debris slide–debris flow in the 1999 Hiroshima, Japan, rainstorm. *Engineering Geology*, **69**(3/4), 309–330.

Wieczorek, G.F., Lips, E.W., and Ellen, S.D. (1989) Debris flows and hyperconcentrated floods along the Wasatch Front, Utah, 1983 and 1984. *Association of Engineering Geologists' Bulletin*, **26**(2), 191–208.

Wieczorek, G.F., Wilson, R.C., Mark, R.K., Keefer, D.K., Harp, E.L., Ellen, S.D., Brown III, W.M., and Rice, P. (1990) Landslides warning system in the San Francisco Bay Region, California. *Landslide News*, **4**, 5–8.

Wieczorek, G.F., Morgan, B.A., and Campbell, R.H. (2000) Debris-flow hazards in the Blue Ridge of central Virginia. *Environmental and Engineering Geoscience*, **6**(1), 3–23.

Yokota, S., Moriyama, T., Ando, S., Hamazaki, A., and Osaka, S. (1999) *The Geo-disasters around the Omoji River Basin on Asakita Area of Hiroshima City* (Disaster Emergency Research Report, 6 pp.). (unpublished.)

6

Debris-flow mechanics

Richard M. Iverson

6.1 INTRODUCTION

Debris flows involve gravity-driven motion of solid–fluid mixtures with abrupt surge fronts, free upper surfaces, variably erodible basal surfaces, and compositions that may change with position and time. These complications pose great challenges in efforts to understand debris-flow mechanics and predict debris-flow behavior. Recently, however, a combination of observational, experimental, and theoretical research has begun to yield a coherent picture of debris-flow mechanics. To help build a foundation for future research, this chapter emphasizes principles of debris-flow mechanics that are relatively well established and also highlights areas where critical knowledge is lacking. The chapter does not provide a comprehensive review of debris-flow mechanics literature, which has become voluminous during the past decade. An entree to this literature is provided by the proceedings of three International Conferences on Debris-Flow Hazards Mitigation: Mechanics, Prediction, and Assessment (Chen, 1997; Wieczorek and Naeser, 2000; Rickenmann and Chen, 2003).

6.2 MECHANICAL DEFINITION OF DEBRIS FLOW

Debris flows encompass a broad and imprecisely defined range of phenomena intermediate between dry rock avalanches and sediment-laden water floods, but to limit the scope of mechanical analysis, it is necessary to identify some distinguishing traits. Although debris flows are largely saturated with water, they differ from surging water floods in which sediment is held in suspension almost exclusively by fluid mechanical phenomena (e.g., viscous drag, buoyancy, turbulence). In such floods the presence of suspended sediment is mostly incidental to the dynamics of the flood wave as a whole. At the opposite extreme, although debris flows have sediment concentrations comparable to those of rock avalanches, they differ from avalanches in which grains interact almost exclusively through solid-contact phenomena (e.g., collisions, adhesion, friction), perhaps mediated by intergranular air. In such avalanches the presence of water is mostly incidental to the dynamics of the avalanche as a whole. In contrast, strong interactions of the solid and liquid constituents are an essential element of the mechanics of debris flows. The magnitude and character of solid–liquid interactions may vary from flow to flow and within an individual flow, but the interactions always play a definitive mechanical role.

Typically, solid grains and intergranular liquid constitute roughly equal percentages (30–70%) of the volume of a debris flow. Rock avalanches can transform into debris flows through entrainment of water or water-rich sediment, and debris flows that entrain additional water can become so dilute that they transform to surging floods. Subaqueous debris flows can undergo a similar transformation as a result of entrainment of ambient water, thereby forming buoyancy-dominated gravity currents.

This chapter focuses on the mechanics of relatively simple subaerial debris flows in which average compositions remain more-or-less constant. Although the chapter emphasizes debris-flow motion, it presents a mechanical framework that also applies to quasistatic processes such as liquefaction during debris-flow initiation and consolidation of debris-flow deposits (cf. Iverson et al., 1997; Major and Iverson, 1999; Iverson et al., 2000; Iverson and Denlinger, 2001; Denlinger and Iverson, 2001). The conceptual continuity provided by this framework is important because debris-flow motion begins and ends in static states. In this respect debris flows have more in common with rock avalanches than with water floods.

6.3 MACROSCOPIC DYNAMICS

Any mechanical assessment of debris flows must begin with identification of the scale of behaviour of interest. This chapter adopts a continuum perspective, which considers behaviour on scales no smaller than that of representative elemental volumes (REVs) containing large numbers of individual solid grains (see Figure 6.1). The number of grains in an REV must be great enough that spatially and temporally averaged continuum quantities such as stress are meaningful and measurable, and are not subject to significant fluctuations due to the motion of individual grains. Drew and Lahey (1993) discuss mathematical issues regarding continuum averaging of fluctuating phenomena in grain–fluid mixtures. Iverson (1997) presents data that show how continuum stress fluctuations at the base of debris flows diminish as the size of the measurement device (or REV) increases to include the simultaneous effects of many thousands of grains.

An alternative approach to debris-flow mechanics considers behaviour at the scale of individual grains. Advances in computational power have facilitated progress in this area (e.g., Campbell et al., 1995; Asmar et al., 2003), but such a "discrete-body" approach appears unlikely to supplant continuum mechanics in the foreseeable future, as even laboratory-sized debris flows ($\sim 10\,\mathrm{m}^3$) commonly contain

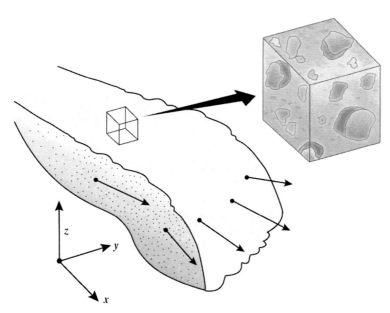

Figure 6.1. Definition of a REV containing a large number of solid grains as well as intergranular muddy fluid within a debris flow. Here the REV is a cube aligned with Cartesian coordinate axes, a geometry that is convenient but not essential.

more than 10^{10} interacting grains (Iverson, 1997). Nevertheless, an intriguing possibility for the future involves melding continuum and discrete-body mechanics to investigate the interaction of isolated large clasts with adjacent, finer-grained debris (e.g., Yamagishi et al., 2003).

6.3.1 Continuum conservation laws

The conservation laws of classical physics provide the fundamental tools for analysis of debris-flow continuum mechanics. The most useful of these laws describe conservation of mass and linear momentum. Conservation of angular momentum also applies to debris flows, but in conventional continuum mechanics angular momentum is conserved implicitly through the use of a symmetric stress tensor, as is used here (e.g., Malvern, 1969). Conservation of energy applies to debris flows, but does not provide additional information if debris flows are treated as isothermal phenomena, as they are here (cf. Iverson, 1997). (In 1-D analyses of debris-flow motion, conservation of energy and linear momentum yield equivalent equations of motion because the vectorial character of momentum reduces to a scalar form like that of energy. However, this equivalence does not extend to multidimensional debris flows, wherein the vectorial character of momentum conservation makes it the most useful principle.)

Differential equations describing mass and linear momentum conservation, valid for each phase of a debris-flow mixture treated as a continuum, are (e.g., Gidaspow, 1994):

$$\frac{\partial \rho_i n_i}{\partial t} + \nabla \cdot \rho_i n_i \vec{v}_i = 0 \tag{6.1}$$

$$\frac{\partial \rho_i n_i \vec{v}_i}{\partial t} + \nabla \cdot \rho_i n_i \vec{v}_i \vec{v}_i = -\nabla \cdot \mathbf{T}_i + \rho_i n_i \vec{g} + \vec{f}_i \tag{6.2}$$

where $\vec{g}$ is gravitational acceleration, t is time, and quantities with a subscript i apply to each phase individually. For each phase, n denotes the volume fraction (such that $\Sigma n_i = 1$), ρ denotes mass density, $\vec{v}$ denotes velocity ($\vec{v}\vec{v}$ is a dyadic product: a 3×3 tensor with Cartesian components of the form $v_x v_y$), and T denotes stress (a 3×3 tensor with Cartesian components of the form τ_{yx}). A minus sign precedes the stress term in (6.2) because stress is defined as positive in compression, as is conventional in soil and rock mechanics. The vector $\vec{f}_i$ in (6.2) denotes the interaction force per unit volume exerted on phase i due to relative motion of the other phase(s).

Terms on the left-hand side of the momentum-conservation equation (6.2) differ from those on the left-hand side of the mass-conservation equation (6.1) only through inclusion of an additional $\vec{v}_i$, reflecting the definition of momentum: mass times velocity. This connection between mass and momentum conservation is clear because (6.2) depicts the "conservative" form of the momentum equation. Some readers may be more familiar with the "primitive" momentum equation wherein, for example, $\vec{v}_i \cdot \nabla \vec{v}_i$ replaces $\nabla \cdot \vec{v}_i \vec{v}_i$ in (6.2). The primitive form is obtained by algebraic rearrangement of (6.2) and elimination of some terms through use of (6.1).

Conservation equations for the debris-flow mixture as a whole can be derived by summing the equations for the individual phases while using appropriately weighted averages to define the mixture density $\rho = \rho_s n_s + \rho_f n_f$ and mixture velocity $\vec{v} = (\rho_s n_s \vec{v}_s + \rho_f n_f \vec{v}_f)/\rho$. Here subscripts s and f denote the solid and fluid phases indicated generically by "i" in (6.1) and (6.2). Summation of (6.1) for the solid and fluid phases yields the mixture mass-conservation equation, which can be written in several alternative forms, including:

$$\frac{\partial \rho}{\partial t} + \nabla \cdot \rho \vec{v} = 0 \tag{6.3a}$$

$$\frac{1}{\rho}\frac{d\rho}{dt} + \nabla \cdot \vec{v} = 0 \tag{6.3b}$$

Similarly, summation of (6.2) for the solid and fluid phases yields the mixture momentum-conservation equation, which can be written in several forms including:

$$\frac{\partial \rho \vec{v}}{\partial t} + \nabla \cdot \rho \vec{v} \vec{v} = -\nabla \cdot \mathbf{T} + \rho \vec{g} \tag{6.4a}$$

$$\frac{d\vec{v}}{dt} = -\frac{1}{\rho} \nabla \cdot \mathbf{T} + \vec{g} \tag{6.4b}$$

In (6.3b) and (6.4b) the total time derivative d/dt represents the differential operator

$\partial/\partial t + \vec{v} \cdot \nabla$, and it denotes differentiation in a frame of reference that moves with the mixture velocity $\vec{v}$. Derivatives of ρ do not appear in (6.4b) because they cancel from (6.4a) through subtraction of (6.3a). The interaction force $\vec{f_i}$ in (6.2) does not appear explicitly in (6.4a) or (6.4b) because the force exerted on the solid grains by the fluid balances the force exerted on the fluid by the grains (thereby satisfying Newton's third law of motion).

Equations (6.3) and (6.4) are identical to the mass and momentum conservation equations for a single-phase continuous medium (cf. Malvern, 1969). However, the summed stress T in (6.4) implicitly includes distinct contributions from solid and fluid phases and from relative motion of the phases (cf. Iverson, 1997; Iverson and Denlinger, 2001).

6.3.2 Depth integration and mass-change effects

A more useful form of the 3-D conservation equations (6.3) and (6.4) can be obtained by integrating the equations through the debris-flow thickness h, measured vertically from the bed at elevation $z = b(x, y, t)$ to the flow surface at elevation $z = \eta(x, y, t)$, where x and y are planimetric coordinates (see Figure 6.2). Alternatively, depth integration can be performed in a direction normal to the bed, as presented, for example, by Savage and Hutter (1989, 1991), Iverson (1997), Gray et al. (1999), and Iverson and Denlinger (2001). Results are similar in either case, although vertical integration serves to highlight the influence of boundary conditions and non-hydrostatic stress states, which are elaborated below.

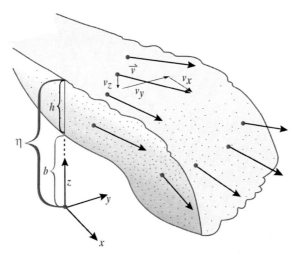

Figure 6.2. Definition of the Cartesian coordinate system, variable bed elevation b, variable debris thickness $h = \eta - b$, and variable velocity vectors in a depth-averaged debris-flow model obtained by vertical integration of the equations of motion. Cartesian velocity components (v_x, v_y, v_z) are depicted for one of the velocity vectors, and v_z is negative when pointing downward as shown.

110   Debris-flow mechanics   [Ch. 6

Prior to depth integration, the vector momentum equation (6.4) is disaggregated into component equations containing the x, y, and z velocity components, v_x, v_y, and v_z (see Figure 6.2). Integration of each component equation makes use of Leibniz' theorem for interchanging the order of integration and differentiation (Abramowitz and Stegun, 1964, p. 11). Most integrals are thereby absorbed into definitions of depth-averaged quantities such as the depth-averaged velocity components (denoted by overbars):

$$\bar{v}_x = \frac{1}{h}\int_{z=b}^{z=\eta} v_x\,dz \qquad \bar{v}_y = \frac{1}{h}\int_{z=b}^{z=\eta} v_y\,dz \qquad \bar{v}_z = \frac{1}{h}\int_{z=b}^{z=\eta} v_z\,dz \qquad (6.5\text{a, b, c})$$

Also absorbed into (6.3) and (6.4) during depth integration are kinematic boundary conditions, which specify volumetric fluxes of debris through the basal and upper surfaces of a debris flow due to erosion and sedimentation. The kinematic conditions may be written as:

$$v_z(\eta) = \frac{\partial \eta}{\partial t} + v_x(\eta)\frac{\partial \eta}{\partial x} + v_y(\eta)\frac{\partial \eta}{\partial y} + A(x,y,t) \qquad (6.6)$$

$$v_z(b) = \frac{\partial b}{\partial t} + v_x(b)\frac{\partial b}{\partial x} + v_y(b)\frac{\partial b}{\partial y} + B(x,y,t) \qquad (6.7)$$

In (6.6), $A(x, y, t)$ specifies the rate of vertical accretion to a debris flow's surface as a result of collapse of adjacent bank material, for example. In (6.7), $B(x, y, t)$ specifies the rate of bed elevation change as a result of sedimentation ($B > 0$) or erosion ($B < 0$). If no mass enters or leaves a debris flow, A and B as well as $\partial b/\partial t$ equal zero. Depth integrations that embed the kinematic conditions (6.6 and 6.7) in the conservation equations (6.3 and 6.4) assume that debris entering or leaving the debris flow locally has the same bulk density as the debris within the flow.

The mathematical details of depth integration of (6.3b) illustrate use of Leibniz' theorem and incorporation of the kinematic boundary conditions (6.6) and (6.7):

$$\int_b^\eta \left(\frac{1}{\rho}\frac{d\rho}{dt} + \nabla \cdot \vec{v}\right)dz = \int_b^\eta \frac{1}{\rho}\frac{d\rho}{dt}dz + \int_b^\eta \left(\frac{\partial v_x}{\partial x} + \frac{\partial v_y}{\partial y} + \frac{\partial v_z}{\partial z}\right)dz$$

$$= \int_b^\eta \frac{1}{\rho}\frac{d\rho}{dt}dz + \frac{\partial}{\partial x}\int_b^\eta v_x\,dz - v_x(\eta)\frac{\partial \eta}{\partial x} + v_x(b)\frac{\partial b}{\partial x}$$

$$+ \frac{\partial}{\partial y}\int_b^\eta v_y\,dz - v_y(\eta)\frac{\partial \eta}{\partial y} + v_y(b)\frac{\partial b}{\partial y} + v_z(\eta) - v_z(b)$$

$$= \int_b^\eta \frac{1}{\rho}\frac{d\rho}{dt}dz + \frac{\partial(\bar{v}_x h)}{\partial x} + \frac{\partial(\bar{v}_y h)}{\partial y}$$

$$- \left[v_x(\eta)\frac{\partial \eta}{\partial x} + v_y(\eta)\frac{\partial \eta}{\partial y} - v_z(\eta)\right] + \left[v_x(b)\frac{\partial b}{\partial x} + v_y(b)\frac{\partial b}{\partial y} - v_z(b)\right]$$

$$= \int_b^\eta \frac{1}{\rho}\frac{d\rho}{dt}dz + \frac{\partial(\bar{v}_x h)}{\partial x} + \frac{\partial(\bar{v}_y h)}{\partial y} + \frac{\partial h}{\partial t} + A - B = 0 \qquad (6.8)$$

The last line of (6.8) is obtained by substituting (6.6) and (6.7) into the fifth line of (6.8) and then using the substitution $\partial \eta/\partial t - \partial b/\partial t = \partial h/\partial t$ (see Figure 6.2). If the debris bulk density ρ is constant, the integral in the last line of (6.8) vanishes, and the last line of (6.8) thereby reduces to the depth-integrated mass-conservation equation conventionally used in shallow-water theory (e.g., Vreugdenhil, 1994).

If ρ is not constant but the solid and fluid constituents of a debris flow are individually incompressible, the bulk density change $d\rho/dt$ that appears in the integral in the last line of (6.8) can be expressed in terms of porosity change (i.e., fluid volume–fraction change), because $d\rho/dt = (\rho_f - \rho_s) dn_f/dt$. Changes in porosity imply relative motion of the solid and fluid constituents and thereby produce solid–fluid interaction stresses, as detailed below.

Mathematical operations similar to those in (6.8) are used to derive from (6.4) the depth-integrated momentum-conservation equations for the x, y, and z directions:

$$\frac{\partial h\bar{v}_x}{\partial t} + \frac{\partial h\bar{v}_x^2}{\partial x} + \frac{\partial h\bar{v}_y\bar{v}_x}{\partial y} = -\frac{1}{\rho}\int_b^\eta \left[\frac{\partial \tau_{xx}}{\partial x} + \frac{\partial \tau_{yx}}{\partial y} + \frac{\partial \tau_{zx}}{\partial z}\right]dz - Av_x(\eta) + Bv_x(b) \quad (6.9)$$

$$\frac{\partial h\bar{v}_y}{\partial t} + \frac{\partial h\bar{v}_y^2}{\partial y} + \frac{\partial h\bar{v}_x\bar{v}_y}{\partial x} = -\frac{1}{\rho}\int_b^\eta \left[\frac{\partial \tau_{yy}}{\partial y} + \frac{\partial \tau_{xy}}{\partial x} + \frac{\partial \tau_{zy}}{\partial z}\right]dz - Av_y(\eta) + Bv_y(b) \quad (6.10)$$

$$\frac{\partial h\bar{v}_z}{\partial t} + \frac{\partial h\bar{v}_x\bar{v}_z}{\partial x} + \frac{\partial h\bar{v}_y\bar{v}_z}{\partial y} = -gh - \frac{1}{\rho}\int_b^\eta \left[\frac{\partial \tau_{zz}}{\partial z} + \frac{\partial \tau_{xz}}{\partial x} + \frac{\partial \tau_{yz}}{\partial y}\right]dz - Av_z(\eta) + Bv_z(b)$$

$$(6.11)$$

Several features of these equations deserve emphasis. First, the equations apply even if the debris bulk density varies, provided that A and B represent boundary fluxes of debris with a bulk density equal to the local flow bulk density. Second, the gravitational forcing term $-gh$ appears only in the z momentum equation (6.11), because gravity is assumed to act vertically downward. Motion in the x and y directions is driven by stress gradients that arise in reaction to this gravitational forcing. Third, the non-linear advective acceleration terms (which contain velocity products such as $\bar{v}_x^2$ and $\bar{v}_y\bar{v}_x$) neglect the effects of non-uniform vertical velocity profiles, which produce differential advection of momentum (cf. Vreugdenhil, 1994). Compensation for this neglect involves adding momentum "correction coefficients" to the advective acceleration terms. However, such coefficients are omitted here because vertical velocity profiles in debris flows are poorly constrained and are likely variable (Iverson and Vallance, 2001). Therefore, at this juncture, addition of momentum correction coefficients to the momentum-conservation equations would add no mechanical insight.

The presence of the erosion and sedimentation terms involving A and B on the right-hand side of (6.9), (6.10), and (6.11) distinguishes these equations from typical depth-integrated momentum equations, such as those of Denlinger and Iverson (2004). Mathematically, these terms arise from use of (6.6) and (6.7) during depth integration. Physically, these terms represent the momentum change associated with accelerating newly added mass (assumed to have no initial velocity) to the speed of the debris flow or with expelling debris-flow mass to create a static deposit. The terms are exact insofar as mathematical book-keeping is concerned, but they do not account fully for the mechanics of the erosion or deposition processes. Erlichson (1991) likened mass-change terms such as those in (6.9), (6.10), and (6.11) to mass-change terms in rocket equations: only in systems in which mass change occurs independently of external forces do such terms account completely for momentum change produced by mass change. Otherwise, it is necessary to account explicitly for the external forces that *cause* the mass change, as well as for the *effects* of mass change as represented in (6.9), (6.10), and (6.11).

Conservation equations such as (6.9), (6.10), and (6.11) provide a starting point for investigation of erosion and sedimentation by debris flows, but characterization of forces that cause such mass change remains largely speculative and requires further research. For this reason, and to streamline the mathematics in the remainder of this chapter, the assumption $A = B = 0$ will be used hereafter.

6.3.3 Scaling and shallow flow with non-hydrostatic stress

Virtually all computational models of debris-flow motion use some form of shallow-flow approximation. Shallow-flow approximations of 4-D conservation laws (involving three space coordinates plus time, as in (6.8)–(6.11)) reduce the number of governing equations and dependent variables from 4 to 3, thereby facilitating computation of solutions (e.g., Denlinger and Iverson, 2001). Shallow-flow approximations also simplify evaluation of stresses – a particularly significant advantage when stress states are poorly constrained (e.g., Iverson and Denlinger, 2001). Below, scaling of (6.8)–(6.11) is used to obtain a shallow debris-flow approximation that is valid on both steep and gentle slopes. This approximation subsumes as a special case the approximation commonly used in shallow-water theory (cf. Vreugdenhil, 1994).

For either debris flows or shallow-water flows of finite extent, the pertinent length scale in the z direction is the typical flow thickness H, whereas the length scale in the x and y directions is a typical planimetric dimension (length, width) of the flow L (Figure 6.3). Velocity components in the x, y, and z directions scale with the product of gravitational acceleration and the pertinent length scale (L or H) raised to the $\frac{1}{2}$ power, and time scales with $(L/g)^{1/2}$ because gravity drives time-dependent motion dominantly in the x and y directions. All stress components in (6.9), (6.10), and (6.11) scale with the static stress due to gravity, $\rho g H$, because gravity and its effect on debris weight are the fundamental phenomena driving motion. These scalings are summarized by (cf. Savage and Hutter, 1989):

$$x, y \sim L \qquad z \sim H \qquad \text{(6.12a, b)}$$

$$T_{xx}, T_{yy}, T_{zz}, T_{yx}, T_{zx}, T_{yz} \sim \rho g H \qquad \text{(6.12c)}$$

$$\bar{v}_x, \bar{v}_y \sim (gL)^{1/2} \qquad \bar{v}_z \sim (gH)^{1/2} \qquad t \sim (L/g)^{1/2} \qquad \text{(6.12d, e, f)}$$

Typically $H/L \ll 1$ because debris flows generally have more-or-less tabular geometries like that shown in Figure 6.3. As a consequence of this geometry and

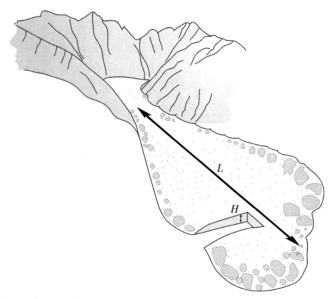

Figure 6.3. Schematic of typical, tabular debris-flow geometry, illustrating the length scales H and L.

the scalings in (6.12a, b, c, d, e, f), the stress derivatives with respect to x and y in (6.9), (6.10), and (6.11) are inferred to be significantly smaller than the stress derivatives with respect to z (by a factor $\sim H/L$). Similarly, the velocity component in the z direction is inferred to be smaller than the velocity components in the x and y directions (by a factor $\sim (H/L)^{1/2}$). Thus, a rational approximation of the vertical momentum equation (6.11) can be obtained by omitting the relatively small terms involving z velocity components and x and y derivatives of stress components. Omission of these terms reduces (6.11) to a simple hydrostatic stress balance $\int_b^\eta (\partial \tau_{zz}/\partial z) dz = -\rho g h$, and integration of this equation using the free-surface boundary condition $\tau_{zz}(\eta) = 0$ yields a hydrostatic basal normal stress $\tau_{zz}(b) = \rho g h$. This is the stress state (and supporting rationale) assumed in conventional shallow-water theory (e.g., Vreugdenhil, 1994).

In debris flows the stress state may differ significantly from the hydrostatic state assumed in conventional shallow-water theory, in part because vertical accelerations (involving $\bar{v}_z$) may not be negligible. Vertical accelerations effectively change the weight of a moving debris mass, and their consequent effect on stresses is apt to be particularly important where debris flows encounter steep slopes or irregular terrain that deflects the flow. If vertical accelerations are significant, a suitable approximation of the vertical momentum equation (6.11) can be obtained by neglecting the terms involving x and y derivatives of stress (of order H/L) but retaining the vertical velocity terms (of order $(H/L)^{1/2}$), yielding:

$$\frac{\partial (h\bar{v}_z)}{\partial t} + \frac{\partial (h\bar{v}_x\bar{v}_z)}{\partial x} + \frac{\partial (h\bar{v}_y\bar{v}_z)}{\partial y} = -\frac{1}{\rho}\int_b^\eta \frac{\partial \tau_{zz}}{\partial z} dz - gh \qquad (6.13)$$

Evaluating the integral in (6.13) and inferring that the flow surface is stress-free (i.e., $T_{zz}(\eta) = 0$) yields an expression for the basal normal stress, which includes dynamic terms due to z momentum in addition to the hydrostatic term $\rho g h$:

$$T_{zz}(b) = \rho g h + \rho \bar{v}_z \left[\frac{\partial h}{\partial t} + \frac{\partial (h \bar{v}_x)}{\partial x} + \frac{\partial (h \bar{v}_y)}{\partial y} \right] + \rho h \frac{d \bar{v}_z}{dt} \quad (6.14)$$

For debris flows with no mass exchange with boundaries ($A = B = 0$), (6.14) simplifies because the terms in brackets reduce to a single term through application of the depth-averaged mass-conservation equation (6.8). (Note that when d/dt operates on a depth-averaged quantity, such as $\bar{v}_z$ in (6.14), it represents a total time derivative advected with the depth-averaged horizontal motion, $d/dt = \partial/\partial t + \bar{v}_x \partial/\partial x + \bar{v}_y \partial/\partial y$.)

After applying (6.8) with $A = B = 0$ to the terms in brackets (6.14) can be rewritten as:

$$T_{zz}(b) = \rho g' h - \rho \bar{v}_z \int_b^\eta \frac{1}{\rho} \frac{d\rho}{dt} dz \quad (6.15)$$

where g' denotes a total vertical acceleration defined by:

$$g' = g + \frac{d\bar{v}_z}{dt} \quad (6.16)$$

Equation (6.15) constitutes the central approximation in a shallow-flow theory for motion of debris flows on steep, irregular slopes, and it reduces to the hydrostatic approximation if $d\bar{v}_z/dt = 0$ and $d\rho/dt = 0$. For debris flows in which ρ is nearly constant, the magnitude of the term containing g' in (6.15) greatly exceeds the magnitude of the term containing $d\rho/dt$, and the integral in (6.15) can be neglected. Ramifications of (6.15) and (6.16) for evaluation of stress components in addition to T_{zz} are discussed below.

For (6.15) and (6.16) to be useful, estimation of $\bar{v}_z$ is necessary. A suitable estimate results from approximating the value of $\bar{v}_z$ with the mean of the surface and basal values of v_z, which is obtained by inserting the depth-averaged velocity components $\bar{v}_x$ and $\bar{v}_y$ in the kinematic boundary conditions (6.6) and (6.7):

$$\bar{v}_z \approx \frac{1}{2} \left[\frac{\partial \eta}{\partial t} + \bar{v}_x \frac{\partial \eta}{\partial x} + \bar{v}_y \frac{\partial \eta}{\partial y} \right] + \frac{1}{2} \left[\frac{\partial b}{\partial t} + \bar{v}_x \frac{\partial b}{\partial x} + \bar{v}_y \frac{\partial b}{\partial y} \right] \quad (6.17)$$

Through use of (6.15), (6.16), and (6.17), effects of vertical accelerations can be included in stress evaluations without including $\bar{v}_z$ as an explicit dependent variable (Denlinger and Iverson, 2004).

An alternative way of representing the effects of vertical accelerations on stresses involves use of curvilinear coordinate systems fitted to topography. In this case, vertical accelerations are represented by the combined effects of downslope accelerations and centripetal accelerations that are induced as flows traverse the topography (cf. Savage and Hutter, 1991; Hungr, 1995; Gray et al., 1999; Iverson and Denlinger, 2001). However, such an approach is difficult to implement if topographic

6.4 STRESS ESTIMATION

Whereas development of applicable conservation equations for shallow debris flows (i.e., equations 6.8, 6.9, 6.10, and 6.15) is tightly constrained by universal physical laws, mathematical rules, and scaling principles, stress estimation is more ambiguous. Stresses are important because they do the irreversible small-scale work that is responsible for continuum-scale energy dissipation and resistance to debris-flow motion. Of course, stress is simply a surrogate for the effects of momentum transport at scales too small to be resolved at a continuum scale. Therefore, momentum transport at scales much smaller than that of a continuum REV (Figure 6.1) can in principle be analysed to gain insight about stress and rheology, which relates stress to deformation. However, this chapter focuses on the continuum viewpoint and depth-averaged modeling described above, and does not present such analyses.

An advantage of depth-averaged debris-flow modeling is that the magnitude of one crucial stress component can be estimated without ambiguity: $\tau_{zz}(b)$ is given by (6.15) regardless of flow rheology. Estimation of the magnitudes of other individual stress components depends on rheology but can be simplified by first applying Leibniz' theorem to the integrals on the right-hand sides of (6.9) and (6.10). From (6.9) this operation yields:

$$\int_b^\eta \left[\frac{\partial \tau_{xx}}{\partial x} + \frac{\partial \tau_{yx}}{\partial y} + \frac{\partial \tau_{zx}}{\partial z} \right] dz = \frac{\partial \bar{\tau}_{xx} h}{\partial x} + \tau_{xx}(b)\frac{\partial b}{\partial x} + \frac{\partial \bar{\tau}_{yx} h}{\partial y} + \tau_{yx}(b)\frac{\partial b}{\partial y} - \tau_{zx}(b) \quad (6.18)$$

and an exactly analogous expression arises from (6.10). When Leibniz' theorem is used to obtain (6.18), some terms vanish because the upper surface of the flow (at $z = \eta$) is assumed free of all stresses. Stress components with overbars in (6.18) denote depth-averaged quantities defined by integrals analogous to those in (6.5).

Terms on the right-hand side of (6.18) can be grouped into two categories. The collection of terms $-\tau_{zx}(b) + \tau_{xx}(b)[\partial b/\partial x] + \tau_{yx}(b)[\partial b/\partial y]$ describes basal resistance to motion, and includes both a shear stress term $\tau_{zx}(b)$ and two "form drag" terms that are non-zero if the local components of bed slope $(\partial b/\partial x, \partial b/\partial y)$ result in a component of horizontal force directed into or out of the bed. These form drag terms vanish if depth integration is performed normal to the bed, rather than vertically, as is performed above (cf. Iverson and Denlinger, 2001). However, if the bed topography is irregular, depth integration normal to the bed results in spatial variation of the integration direction, which leads to other mathematical complications (Keller, 2003).

Additional terms on the right-hand side of (6.18), $\partial(\bar{\tau}_{xx} h)/\partial x + \partial(\bar{\tau}_{yx} h)/\partial y$, express the influence of depth-averaged horizontal stress gradients. These terms are non-zero even in steady, uniform flows on slopes, because the flow depth at a fixed z varies as a function of x and y in such flows. If depth integration is performed normal to the bed, the terms $\partial(\tilde{\tau}_{xx}h)/\partial x + \partial(\tilde{\tau}_{yx}h)/\partial y$ have the same form as in (6.18), but the terms vanish if flow is steady and uniform.

A lowest order approximation of the stress terms on the right-hand side of (6.18) can be identified by using the scalings summarized in (6.12). The scalings indicate that the basal resistance is of the order of $\rho g H$, whereas the terms involving the depth-averaged horizontal stresses $\tilde{\tau}_{xx}$ and $\tilde{\tau}_{yx}$ are smaller, of the order of $\rho g H^2/L$ (where $H/L \ll 1$). Thus, basal shear resistance has primary importance in depth-averaged debris-flow models, although the effects of lateral stress gradients cannot be ignored when modeling surge-like motion (Savage and Hutter, 1989).

6.4.1 Stress partitioning

As noted above, stresses in debris flows include distinct contributions from solid grains, intergranular fluid, and solid–fluid interactions. This partitioning of stress motivates two fundamental questions: In a debris mixture comprising a great diversity of grains and fluids, how are the solid and fluid phases distinguished? Once a distinction between solids and fluids is made, how is partitioning of stress determined?

Definition of the fluid phase in debris flows is not as simple as it might seem. The most straightforward definition (that the fluid consists of pure liquid water and pure gaseous air) is not the most useful for analysing debris-flow mechanics. To a large degree, air can be excluded from consideration because its low density, its low viscosity, and its large compressibility make its mechanical effects very small compared to those of liquid water. Furthermore, liquid water in debris flows generally carries small solid grains that can remain suspended solely as a consequence of buoyancy, viscosity, and turbulence. Because suspension of these small grains can occur in the absence of direct grain-to-grain contacts, it is not appropriate to treat the suspended grains as solids that transfer momentum only through direct contacts. Therefore, the discussion below defines the fluid phase of debris flows as water plus suspended small grains, which can in turn influence the effective fluid properties.

A scaling criterion can be used to distinguish the sizes of grains that are treated as part of the debris-flow fluid (Iverson, 1997). If the duration t_D of a debris flow is long in comparison with the time required for settling of a grain in static, pure water, the grain must be considered part of the solid fraction. On the other hand, if a grain can remain suspended for times that exceed t_D as a result of only water viscosity and buoyancy, the grain acts as part of the fluid. Durations of debris flows range from about $t_D = 10$ s for small but significant events to 10^4 s for the largest. The time scale for grain settling (in the absence of other grains) can be estimated by dividing the characteristic settling distance or half thickness of a debris flow $H/2$ by the grain settling velocity v_{set} estimated from Stokes law or a more general equation that accounts for grain inertia (Vanoni, 1975). Thus if $H/(2t_D v_{set}) < 1$ the debris-flow duration is large compared with the time scale for grain settling. The half thickness of debris flows ranges from about 0.01 m for small flows to 10 m for large ones. Thus $H/(2t_D) \sim 0.001$ m/s is typical for both small and large debris flows, which implies that $v_{set} < 0.001$ m/s is required for grains to act as part of the fluid. In water, grain settling velocities of ~ 0.001 m/s or less occur if grain diameters are less than about 0.05 mm (Vanoni, 1975). This critical grain size corresponds reasonably well with the conventional silt–sand boundary (0.062 5 mm), and it also falls in the range where settling is characterized by grain Reynolds numbers much smaller than 1, such that viscous and buoyancy forces dominate grain motion. By this rationale, a useful but inexact guideline states that grains larger than silt-size generally constitute the solid phase in debris flows, whereas grains in the silt–clay (i.e., "mud") size fraction act as part of the fluid. Size distributions of grains in muddy fluids drained from freshly emplaced debris-flow deposits support this interpretation (Iverson, 1997).

Stress partitioning between the solid and fluid phases in mixtures can be accomplished in a variety of ways, but for debris flows it is convenient to employ a partitioning that is consistent with well-established conventions of soil mechanics (cf. Passman and McTigue, 1986). Thus, the total mixture stress T can be partitioned as (cf. Iverson, 1997; Iverson and Denlinger, 2001):

$$\mathbf{T} = \mathbf{T}_e + \mathbf{I}p + n\mathbf{T}_{vis} \tag{6.19}$$

where $\mathbf{T}_e$ is the effective stress, p is the pore-fluid pressure, $\mathbf{I}$ is the identity tensor (which in equation 6.19 indicates that fluid pressure acts isotropically), $\mathbf{T}_{vis}$ is the viscous or deviatoric fluid stress (total fluid stress minus pressure), and n is the mixture porosity (or fluid volume fraction). This stress partitioning treats $\mathbf{T}_e$ and p as stresses that effectively act throughout the mixture (just as in conventional soil mechanics), whereas $\mathbf{T}_{vis}$ acts only within the fluid phase. For the special case in which the fluid is essentially static and the state of stress is 1-D (6.19) reduces to the familiar effective-stress definition of Terzaghi (1936):

$$\sigma = \sigma_e + p \tag{6.20}$$

where σ denotes a normal-stress component and σ_e is effective normal stress. However, definitions such as (6.19) and (6.20) imply nothing about the mechanical roles of pore pressure or effective stress; they merely provide a convenient means of partitioning the total stress.

6.4.2 Stress due to solid–fluid interaction

Practical application of (6.19) and (6.20) requires specification of the mechanical roles of pore-fluid pressure and effective stress in debris flows. Clearly, these roles may be very complicated, because solid and fluid constituents in grain–fluid mixtures may interact in diverse ways (Iverson, 1997; Koch and Hill, 2001). However, to provide the simplest viable theory and establish a link with classical soil mechanics, the analysis below employs three postulates about the bulk interactions of solids and fluids in continuum REVs like that shown in Figure 6.1: the fluid pressure p mediates solid–fluid interactions; a linear drag equation specifies how fluid pressure gradients are coupled to relative motion of solid grains and intergranular fluid; and effective stress governs solid-contact friction. As shown below, use of these postulates in conjunction with mass conservation laws yields a theoretical framework that can be applied to rapid flows with large deformations as well as to the special cases of quasistatic liquefaction and consolidation of soils.

For a debris-flow mixture with porosity n fully saturated with liquid, the mass-conservation equations for the fluid and solid phases can be inferred directly from (6.1). The equations are:

$$\frac{\partial [n\rho_f]}{\partial t} + \nabla \cdot [\vec{v}_f n \rho_f] = 0 \qquad (6.21)$$

$$\frac{\partial [(1-n)\rho_s]}{\partial t} + \nabla \cdot [\vec{v}_s (1-n)\rho_s] = 0 \qquad (6.22)$$

where subscripts f and s denote the fluid and solid phases, respectively. If the densities of the solid and fluid phases are constant (a reasonable assumption for the stress magnitudes $<100\,\text{kPa}$ typical of debris flows) then (6.21) and (6.22) reduce to:

$$\partial n/\partial t + \nabla \cdot (\vec{v}_f n) = 0 \qquad (6.23)$$

$$-\partial n/\partial t - \nabla \cdot (n\vec{v}_s) + \nabla \cdot \vec{v}_s = 0 \qquad (6.24)$$

Addition of (6.23) and (6.24) yields a special form of the mixture mass-conservation equation, which shows how the divergence of solid grain velocities $\nabla \cdot \vec{v}_s$ must be balanced by flow of fluid relative to the grains:

$$\nabla \cdot \vec{v}_s = -\nabla \cdot n(\vec{v}_f - \vec{v}_s) \qquad (6.25)$$

An additional equation shows how the same divergence must be balanced by changes in porosity. The equation is obtained by rearranging (6.24) as:

$$\nabla \cdot \vec{v}_s = \frac{\partial n}{\partial t} + \nabla \cdot (n\vec{v}_s) = \frac{d_s n}{dt} + n(\nabla \cdot \vec{v}_s) = \frac{1}{1-n} \frac{d_s n}{dt} \qquad (6.26)$$

where $d_s/dt = \partial/\partial t + \vec{v}_s \cdot \nabla$ denotes a total time derivative in a frame of reference advected with the velocity of the solid grains $\vec{v}_s$. Equating (6.25) and (6.26) yields a particularly useful form of the mixture mass-conservation equation (6.3b), which is exact if the solid and fluid phases are individually incompressible:

$$\frac{d_s n}{dt} = -(1-n)\nabla \cdot n(\vec{v}_f - \vec{v}_s) \qquad (6.27)$$

This equation shows that local porosity changes necessarily are accompanied by differences in the local solid and fluid velocities. Even slight velocity differences have large mechanical ramifications if they result in significant solid–fluid drag.

Equation (6.27) can be converted to a form that uses a total time derivative advected with the mean mixture velocity, $d/dt = \partial/\partial t + \vec{v} \cdot \nabla$. This conversion is important because d/dt is the total time derivative used in (6.3b) and (6.4b) to describe mixture mass and momentum change, and it is helpful to express porosity change in the same frame of reference. Some simple algebraic substitutions and cancellations show that:

$$\frac{dn}{dt} - \frac{d_s n}{dt} = \left(\frac{\rho_s[1-n]}{\rho} - 1\right)\vec{v}_s \cdot \nabla n + \left(\frac{\rho_f n}{\rho}\right)\vec{v}_f \cdot \nabla n = \frac{\rho_f n}{\rho}(\vec{v}_f - \vec{v}_s) \cdot \nabla n \qquad (6.28)$$

Therefore, by utilizing (6.28), (6.27) can be rewritten as:

$$\frac{dn}{dt} = -(1-n)\nabla \cdot n(\vec{v}_f - \vec{v}_s) + \frac{\rho_f n}{\rho}(\vec{v}_f - \vec{v}_s) \cdot \nabla n \qquad (6.29)$$

The solid–fluid velocity difference $\vec{v}_f - \vec{v}_s$ that appears in (6.27), (6.28), and (6.29) implies the existence of drag due to relative motion of solid grains and adjacent fluid. A simple estimate of this drag for continuum REVs (Figure 6.1) assumes that it is proportional to the gradient in excess fluid pressure p_e that arises in reaction to relative motion of grains and fluid (where p_e is defined as total fluid pressure minus hydrostatic fluid pressure). This reasoning yields a linear drag equation with the same form as Darcy's law, which may be written as:

$$\vec{v}_f - \vec{v}_s = \frac{\vec{q}}{n} = -\frac{k}{n\mu}\nabla p_e \qquad (6.30)$$

Here $\vec{q}$ is the specific discharge of fluid (the flux relative to the adjacent granular aggregate), k is the intrinsic hydraulic permeability of the granular aggregate, μ is the fluid viscosity, and the coefficient group $k/n\mu$ may be viewed as a drag parameter. Bear (1972) provides an intensive discussion of Darcy's law and its interpretation for quasistatic rocks and soils. Experimental data indicate that Darcian drag is probably prevalent even in liquefied debris-flow mixtures (Iverson, 1997; Major et al., 1997).

Substitution of (6.30) into (6.27) and (6.29) yields alternative forms of an equation describing diffusive redistribution of excess pore pressure that occurs in response to porosity change:

$$\frac{1}{1-n}\frac{d_s n}{dt} = \nabla \cdot \frac{k}{\mu}\nabla p_e \qquad (6.31a)$$

$$\frac{dn}{dt} = \left[(1-n)\nabla - \frac{\rho_f}{\rho}\nabla n\right] \cdot \frac{k}{\mu}\nabla p_e \qquad (6.31b)$$

Slight changes in porosity n can produce very significant changes in excess pore pressure p_e because plausible values of the coefficient k/μ in (6.31a, b) range from about 10^{-16} to 10^{-6} m^3 kg^{-1} s for debris-flow mixtures (Iverson, 1997; Major et al., 1997).

A more familiar form of the excess pore-diffusion equation arises from defining a debris bulk compressibility α such that:

$$\frac{1}{1-n}\frac{d_s n}{dt} = -\alpha \frac{d_s \bar{T}_e}{dt} \qquad (6.32)$$

where $\bar{T}_e$ is the mean effective normal stress (cf. Savage and Iverson, 2003). Substitution of (6.32) in (6.31a) yields an equation that shows how excess pore pressure changes in response to changes in effective stress:

$$\frac{d_s \bar{T}_e}{dt} = -\frac{1}{\alpha}\nabla \cdot \left(\frac{k}{\mu}\nabla p_e\right) \qquad (6.33)$$

Alternatively, the definition of effective stress (equation 6.19) can be used to rewrite (6.33) as a forced diffusion equation for excess pore pressure p_e:

$$\frac{d_s p_e}{dt} = \frac{1}{\alpha} \nabla \cdot \left(\frac{k}{\mu} \nabla p_e \right) + \frac{d_s}{dt} [\bar{T} - \rho_f g(\eta - z)] \qquad (6.34)$$

where $\rho_f g(\eta - z)$ is the hydrostatic component of pore pressure.

Several attributes of (6.34) are noteworthy. Equation (6.34) is similar to pore-pressure diffusion equations used in standard theories of soil consolidation and groundwater motion, except that (6.34) includes an advected time derivative which accounts for the fact that pressure diffusion occurs in debris that may move at significant rates. Equation (6.34) also includes a forcing term that accounts for evolution of the mean total stress $\bar{T}$ and the hydrostatic component of pore pressure, quantities that change as debris-flow geometry changes. Finally, the equation contains a group of parameters $(k/\alpha\mu)$ that plays the role of a pore-pressure diffusivity or consolidation coefficient. Values of these parameters may, of course, evolve as debris-flow composition and bulk density evolve.

Savage and Iverson (2003) showed how (6.34) may be solved in conjunction with debris-flow dynamics equations for cases in which debris-flow motion is 1-D, excess pore pressure diffuses only normal to the bed, and pore-pressure diffusivity is a simple function of position within the flow. Denlinger and Iverson (2001) computationally solved an equation similar to (6.34), but lacking the forcing term, in conjunction with multidimensional debris-flow equations. However, completely general models that couple 3-D pore-pressure diffusion to porosity changes caused by debris-flow motion remain to be developed. Such models require consideration of the coupling between debris agitation and porosity change, perhaps through use of the "granular temperature" concept commonly used in grain–flow dynamics (e.g., Goldhirsh, 2003).

Despite the current lack of a complete model, the most important implications of pore-fluid pressures in debris flows are well established on the basis of both theory and experiments: pore pressure co-evolves with debris-flow deformation; significant pore-pressure changes can result from small changes in debris porosity; and pore-pressure changes imply commensurate changes in intergranular effective stress, which plays an important role in debris-flow mechanics owing to its influence on intergranular friction (Iverson, 1997, 2003a; Major and Iverson, 1999; Savage and Iverson, 2003).

6.4.3 Stress due to interactions of solid grains

Field observations and laboratory experiments indicate that contacts of solid grains against the bed and one another transfer much momentum, dissipate much energy, and therefore produce much of the stress during debris-flow motion (Iverson, 2003a). Grains can interact with the bed and one another through both enduring contacts (i.e., frictional sliding, rolling, and locking) and brief inelastic collisions. A wealth of experimental and theoretical evidence indicates that both types of interaction tend to produce intergranular shear stresses directly proportional to intergranular normal stresses, as summarized by the Coulomb equation (e.g., Bagnold, 1954; Hungr and Morgenstern, 1984; Savage and Sayed, 1984). Therefore, estimation of intergranular shear stresses in depth-averaged debris-flow models can be based on (6.15), which provides a basis for estimating intergranular normal stress.

Coulomb (1776) proposed his well-known equation describing bulk stresses in failing masses of grains through analogy with frictional behaviour of discrete solid bodies in contact. The Coulomb equation:

$$\tau_{shear} = \sigma_{norm} \tan \varphi \tag{6.35}$$

has an apparent simplicity that belies its subtle (and sometimes complicated) implications. The equation states that the bulk intergranular shear stress τ_{shear} on a plane of shearing is directly proportional to the bulk intergranular normal stress σ_{norm} acting on the same plane, irrespective of the area of grain contacts, rate of shearing, and magnitudes of stress components not acting on the plane of shearing. The proportionality constant is specified by the tangent of the friction angle φ.

By measuring and calculating stresses produced by collisions in a shearing mixture of neutrally buoyant spherical grains, Bagnold (1954) provided the first evidence that a Coulomb proportionality applies even in rapid, collisional grain flows. Subsequent analyses and experiments with diverse materials have generally supported Bagnold's findings and lent credibility to the Coulomb proportionality (e.g., Brown and Richards, 1970; Savage and Sayed, 1984; Hunt et al., 2002).

The Coulomb equation (6.35) may be generalized by replacing σ_{norm} with the intergranular effective normal stress σ_e, defined as $\sigma_e = \sigma_{norm} - p$, and by adding a cohesive strength component c. These generalizations produce the Coulomb–Terzaghi equation typically used to describe stresses during shear failure of rocks and soils (e.g., Terzaghi, 1936; Lambe and Whitman, 1979):

$$\tau_{shear} = (\sigma_{norm} - p) \tan \varphi + c \tag{6.36}$$

For granular materials subject to large deformations (as in debris flows) cohesive forces are generally negligible, and (6.36) reduces to $\tau_{shear} = (\sigma_{norm} - p) \tan \varphi$. Importantly, intergranular stresses are coupled to the solid-fluid interaction stresses described in the previous section though inclusion of p in (6.36).

Equation (6.36) with $c = 0$ provides information about the state of intergranular stress in shearing debris, but does not constitute a rheological model in the usual sense, because it implies no one-to-one correspondence between stress and deformation or deformation rate. Moreover, (6.36) is a 1-D equation, and generalizing the equation to 3-D results in a complicated mathematical formulation (e.g., Desai and Siriwardane, 1984). Coulomb stress states in a 3-D medium can be represented relatively simply, however, by depicting them geometrically in a 3-D stress space in which Cartesian axes denote principal stresses (Figure 6.4). In this depiction, Coulomb stress states lie on the surface of an irregular hexagonal cone, but these states cannot be determined without some independent knowledge or concurrent calculation of deformation (e.g., Denlinger and Iverson, 2004). In this sense, the Coulomb stress model shares a property with traditional rheological models (e.g.,

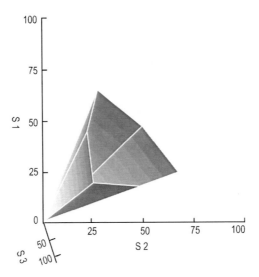

Figure 6.4. A Coulomb yield surface in 3-D stress space. In this space the coordinate axes represent principal stresses, but the relative magnitudes of the three principal stresses S1, S2, and S3 are unspecified. See Denlinger and Iverson (2004) for further details.

viscosity, elasticity): calculation of multidimensional states of stress cannot be accomplished without calculation of deformation.

In cases where debris-flow thickness varies only gradually, however, the 1-D form of the Coulomb–Terzaghi equation (6.36) suffices to estimate basal intergranular shear stresses that resist debris-flow motion. In such cases the total stress normal to the bed can be estimated from the apparent weight of the superincumbent debris, and lateral stresses are assumed to have negligible influence on the basal shear stress. With reference to Figure 6.5, the apparent weight of a moving column of debris of constant density ρ, mean vertical height h, and horizontal planimetric area $\Delta x \Delta y$ is $\rho g' h \Delta x \Delta y$, where g' is defined as in (6.16). Normal to the bed the component of apparent weight is $\rho g' h \Delta x \Delta y \cos\theta$, where θ is the angle between the bed and a horizontal reference surface. The basal traction (defined as force per unit of bed area) due to this component of apparent weight is $\rho g' h \cos^2\theta$. Therefore, according to (6.36), the basal Coulomb shear resistance acting parallel to the base of the debris column is approximately:

$$\tau_{shear}(b) = (\rho g' h \cos^2\theta - p_{bed})\tan\varphi_{bed} \qquad (6.37)$$

where φ_{bed} denotes a friction angle appropriate for the grain–bed interface and p_{bed} is the pore-fluid pressure at this interface. For simple Coulomb sliding parallel to the bed, the basal resistance equation (6.37) takes the place of the collection of basal stress terms that appear in (6.18), $-\tau_{zx}(b) + \tau_{xx}(b)[\partial b/\partial x] + \tau_{yx}(b)[\partial b/\partial y]$. Moreover, (6.37) provides a first (or "lowest order" in H/L) approximation of resistance to debris-flow motion even in more complicated cases (cf. Iverson and Denlinger, 2001).

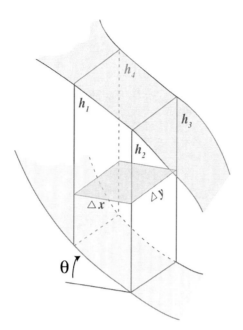

Figure 6.5. Geometry used to calculate the basal normal traction due to the apparent weight of a vertical column of debris with mean height $h = (h_1 + h_2 + h_3 + h_4)/4$. The apparent weight depends on the total vertical acceleration g' defined in (6.16).

Typical values of basal Coulomb shear stresses in debris flows may be estimated by using (6.37) in conjunction with typical parameter values such as $h = 1$ m, $\rho = 2{,}000$ kg/m^3, $\varphi_{bed} = 30°$, and $\theta = 5°$. With these values and $p_{bed} = 0$, (6.37) indicates a basal shear stress of roughly 10,000 Pa. Of course, this Coulomb shear stress will be reduced as p_{bed} increases, and will be zero in the extreme case where p_{bed} balances the total basal normal stress.

Additional Coulomb stress components that appear in (6.18) can be estimated with varying degrees of sophistication, although all such estimations require analyses that are too involved to be presented in detail here. The simplest approach entails use of lateral earth-pressure coefficients similar to those used in quasistatic soil mechanics (e.g., Lambe and Whitman, 1979). In this case depth-averaged lateral stresses such as $\bar{\tau}_{xx}$ are related to the vertical stress $\bar{\tau}_{zz}$ through a simple proportionality:

$$\bar{\tau}_{xx} = k_{act/pass}\bar{\tau}_{zz} \qquad (6.38)$$

Values of the lateral pressure coefficient $k_{act/pass}$ are computed by assuming Coulomb limiting equilibrium in a granular slab deforming uniformly in compression or extension (Savage and Hutter, 1989; Iverson and Denlinger, 2001).

A more accurate approach estimates all components of Coulomb stresses acting on vertical planes by using deformation kinematics to infer the location of principal stresses on the Coulomb cone of Figure 6.4 (Denlinger and Iverson, 2004). Then standard mathematical rules for tensor transformations can be employed together with (6.15) and (6.18) to obtain all stress components. This approach may be necessary to accurately resolve Coulomb stresses where debris-flow thicknesses change abruptly as a result of flow interaction with irregular terrain (Iverson et al., 2004).

More elaborate models may someday supplant the simple Coulomb model of intergranular stresses in debris flows. At present, however, no data convincingly demonstrate the inadequacy of the Coulomb model, and as noted above, the model is particularly useful in the context of depth-averaged flow computation.

6.4.4 Stress due to fluid shear

Stress also results from shear deformation of the fluid phase in debris flows. Magnitudes of shear stresses in debris-flow fluids that consist of water plus silt and clay carried in suspension have been evaluated in rheometric tests. Such tests approximate steady, uniform, 1-D flow in devices (typically rotational rheometers) in which the stress and deformation fields can be calculated exactly or measured directly. Rheometric tests indicate that viscosities of fine-grained slurries range from about 0.1 to 50 Pa-s (about 100 to 50,000 times greater than the viscosity of pure water), depending on the sediment concentration (cf. Hunt et al., 2002). Given that shear rates in debris flows are typically of the order of $10\,\text{s}^{-1}$, this range of viscosities implies the existence of viscous shear stresses no larger than about 500 Pa.

Rheometric measurements of muddy debris-flow slurries also reveal the existence of finite shear strengths, which contribute to stress by resisting deformation. Strengths of mud slurries typically range from about 10–400 Pa (e.g., Kang and Zhang, 1980; O'Brien and Julien, 1988; Phillips and Davies, 1991; Major and Pierson, 1992; Coussot and Piau, 1995; Locat, 1997; Parsons et al., 2001). To gain some intuitive grasp of the size of these strengths, it is instructive to slide a book across a tabletop. For diverse books and tabletops, basal shear stresses that resist sliding are comparable to the strengths in fine-grained debris-flow slurries.

Although debate continues about the best mathematical model for representing fluid shear stresses in debris-flow mixtures, the mechanical effects described above can be represented by a simple Bingham model. The 1-D form of this model can be expressed as:

$$\tau_{shear} = s_f + \mu \frac{dv_x}{dz} \qquad (6.39)$$

where s_f is the fluid (slurry) shear strength and dv_x/dz is the 1-D shear deformation rate. Iverson (1985) showed how (6.39) and related rheological equations may be generalized to three space dimensions.

The basal shear stress implied by (6.39) can be expressed as a function of the depth-averaged debris-flow velocity $\bar{v}_x$ and flow thickness h by assuming the shear stress results only from the weight of the superincumbent debris ($\tau_{shear} = \rho g'(\eta - z)\sin\theta\cos\theta$, analogous to the assumption used to obtain the basal Coulomb shear stress equation (6.37)). Then integration of (6.39) from $z = 0$ to

$z = h$ shows that the basal resistance due to fluid shear can be expressed as (cf. Bird et al., 1960, pp. 37–40):
$$\tau_{shear}(b) = s_f + 3\mu \bar{v}_x/h \tag{6.40}$$
It is useful to compare the magnitude of this fluid basal shear stress with that of the Coulomb basal shear stress described by (6.37) for a typical debris flow with bulk density $\rho = 2{,}000\,\text{kg/m}^3$, thickness $h = 1$ m, and depth-averaged velocity $\bar{v}_x = 10\,\text{m/s}$ descending a planar surface with slope $\theta = 5°$. If the fluid slurry has typical properties $s_f = 100$ Pa and $\mu = 10$ Pa-s, the basal shear stress described by (6.40) then has a value of 500 Pa. In contrast, if the basal friction angle of the granular debris has a typical value $\varphi_{bed} = 30°$ and the basal pore pressure has a value $p_{bed} = 0$, the Coulomb basal shear stress described by (6.37) has a value of roughly 10 kPa, 20 times larger than the fluid stress. This stress ratio is reduced to 1:1 if the basal pore pressure nearly liquefies the mixture by balancing 95% of the total basal normal stress. In such instances fluid resistance to shear is comparable to the Coulomb shear resistance.

The foregoing discussion intentionally omits any mention of fluid turbulence. Although turbulent fluid flow might occur in debris flows, turbulence is suppressed by the presence of high concentrations of solid grains (Koch and Hill, 2001). Indeed, a crucial difference exists between fluid turbulence and the generally agitated state commonly evident in debris flows. Agitation associated with bobbing and jostling of grains in debris flows indicates the presence of disorganized kinetic energy, commonly called "granular temperature" by analogy with thermodynamic temperature in the kinetic theory of molecular gases (e.g., Haff, 1983; Iverson, 1997; Goldhirsch, 2003). Granular temperature can exist in the absence of any fluid or fluid turbulence, and it involves energy dissipation so intense that it hinders development of coherent vorticity structures like those associated with eddies in turbulent fluid flow. Further research is needed to understand the relationships between granular temperature and small-scale turbulent fluctuations of intergranular fluid in debris-flow mixtures. At present, however, inclusion of such relationships in debris-flow models would entail almost pure conjecture (cf. Koch and Hill, 2001).

6.4.5 Lumped rheology and calibrated resistance formulas

As an alternative to separating the stress contributions of solid and fluid constituents and their interactions, many investigators use lumped-rheology models or calibrated resistance formulas to represent the effects of stresses throughout debris flows. In the lumped-rheology approach, debris is treated as a single-phase continuum, and the stress in (6.4a, b) is specified explicitly as a function of debris deformation or deformation rate (e.g., Iverson, 1985; Chen, 1988). Although the mathematical simplicity of this approach is appealing, the approach cannot represent evolution of stress-generation processes. Lumped-rheology equations assume that stress-generation processes remain essentially constant in space and time, whereas field observations and experimental data indicate that dominant stress-generation processes differ in the coarse granular surge fronts and nearly liquefied interiors that develop in debris flows (Iverson, 2003a).

A related factor limiting the utility of the lumped-rheology approach is the complexity and poor reproducibility of rheological properties of debris-flow materials treated as single-phase continua (Phillips and Davies, 1991; Major and Pierson, 1992). No standard devices or established protocols exist for measuring rheologies of mixtures consisting of both muddy slurry and coarse granular debris that may include gravel, cobbles, and boulders. (On the other hand, the distinct solid and fluid constituents of debris flows have mechanical properties that are clearly defined and readily measured in standard tests. The most important solid–fluid coupling parameter, the mixture permeability k, is also readily measured (Major et al., 1997).) Therefore, the relative simplicity of the lumped-rheology approach is largely illusory. Lumping rheological effects into a single equation can simplify mathematical and computational tasks, but it complicates the task of measuring relevant parameters.

Measurement difficulties can be circumvented by using calibrated resistance formulas rather than explicit rheological equations to represent the net effect of stresses in debris flows. Calibrated resistance formulas often can provide good agreement between model results and field observations, because values or even the functional forms of resistance terms can be adjusted with the explicit aim of achieving good fits. However, the appeal of good fits must be weighed against loss of the ability to perform conclusive hypothesis tests. From a scientific perspective, a mechanical model represents a hypothesis cast in precise mathematical form, but the hypothesis can be tested only if it makes unequivocal predictions. If a mechanical model is calibrated to fit data rather than tested against data, no unequivocal predictions are made, and it becomes difficult to distinguish whether good model performance reflects inherent model accuracy or merely model adaptability that is accommodated by calibration (Iverson, 2003b).

The resistance-formula approach generally focuses on adjustment of the basal shear stress (i.e., $\tau_{zx}(b)$ in the depth-averaged stress equation (6.18)), with the aim of fitting observations of debris-flow travel times and distances. Commonly modellers assume that $\tau_{zx}(b)$ is some function of the depth-averaged velocity $\bar{v}_x$ (e.g., O'Brien et al., 1993; Hungr, 1995). Any such function can be represented by a power-series expansion of the form:

$$\tau_{zx}(b) = a_0 + a_1 \bar{v}_x + a_2 \bar{v}_x^2 + \cdots + a_N \bar{v}_x^N \tag{6.41}$$

and if the coefficients a_0 through a_N are freely adjustable, (6.41) can be calibrated to any desired precision (i.e., if the power series contains N terms, it can be fitted to $N+1$ data points exactly).

Typically debris-flow modellers restrict attention to the first three terms on the right-hand side of (6.41), and also ascribe some rheological significance to the coefficients a_0, a_1, and a_2 (e.g., O'Brien et al., 1993; Hungr, 1995). The basis for this reasoning becomes clearer if some additional factors and constants are inserted in (6.41). For example, if some appropriate factors and constants are inserted and the series expansion is truncated to two terms, (6.41) takes the form:

$$\tau_{zx}(b) = \rho g h a_0 + (3/h) a_1 \bar{v}_x \tag{6.42}$$

Here it is clear that a_0 can be regarded as analogous to the basal Coulomb friction coefficient $\tan\varphi_{bed}$ in (6.38) and a_1 can be regarded as analogous to the viscosity coefficient μ in (6.40). Thus, in principle it should be possible to determine applicable values of these coefficients using appropriate rheometric tests. Instead, in the calibrated-resistance approach, the values of coefficients such as a_0 and a_1 are adjusted to fit model results to field data. This adjustment constitutes the most fundamental distinction between lumped-rheology and calibrated-resistance approaches.

Another distinction between the mixture-theory, lumped-rheology, and calibrated-resistance approaches involves evaluation of stress components other than $\tau_{zx}(b)$ in (6.18). Mixture theory assumes that solid and fluid constituents and their interactions can influence all stress components (Iverson, 1997; Iverson and Denlinger, 2001). The lumped-rheology approach similarly assumes that all stress components depend on rheology of a mixture idealized as a one-phase material. In contrast, the calibrated-resistance approach typically assumes that stresses such as $\bar{\tau}_{xx}$ in (6.18) have a fixed form that is independent of the calibrated form of $\tau_{zx}(b)$ (e.g., O'Brien et al., 1993). This dissociation of stress components impedes efforts to interpret calibrated resistance formulas rheologically.

6.5 SURGE DYNAMICS

A conspicuous and important trait of debris flows involves their tendency to move as a discrete surge or series of surges, with each surge typically exhibiting a coarse-grained head and finer grained, more-liquefied tail (Sharp and Nobles, 1953; Davies, 1990; Iverson, 1997; Hungr, 2000). The head-and-tail morphology results from mass and momentum conservation operating in conjunction with solid–fluid stress partitioning and grain-size segregation. Other phenomena, such as buoyancy and inertia due to ambient fluid surrounding a flow, can contribute significantly to head-and-tail surge morphology in dilute density currents and subaqueous debris flows, but are relatively unimportant in subaerial debris flows (Iverson, 2003c).

To illustrate some key aspects of surge dynamics, it is useful to consider a simplified depth-integrated momentum-conservation equation, which applies to an infinitely wide debris flow travelling in the x direction across a horizontal surface without erosion or deposition. In this case the y and z velocity components are zero (Figure 6.2), and the momentum conservation equation for the x direction reduces to:

$$\frac{\partial(h\bar{v}_x)}{\partial t} + \frac{\partial(h\bar{v}_x^2)}{\partial x} = -\frac{1}{\rho}\left[\frac{\partial(\bar{\tau}_{xx}h)}{\partial x} - \tau_{zx}(b)\right] \quad (6.43)$$

which is obtained by combining (6.9) and (6.18) and eliminating terms that equal zero in this special case.

To facilitate interpretation of (6.43), the depth-averaged longitudinal normal stress $\bar{\tau}_{xx}$ can be approximated as a gravity-induced stress that is proportional to the local debris thickness h, such that $\bar{\tau}_{xx} = (1/2)\rho g h k_{act/pass}$, where $k_{act/pass}$ is the same proportionality coefficient as in (6.38) (cf. Savage and Hutter, 1989; Hungr,

1995; Iverson, 1997). Substituting this expression and the applicable form of (6.37) (i.e., $\tau_{zx}(b) = -(\rho g h - p_{bed})\tan\varphi_{bed}$) into (6.43) yields a 1-D momentum equation in which physical aspects of surge dynamics are especially transparent:

$$\rho\left(\frac{\partial(h\bar{v}_x)}{\partial t} + \frac{\partial(h\bar{v}_x^2)}{\partial x}\right) = -\rho g h\left[k_{act/pass}\frac{\partial h}{\partial x} + \left(1 - \frac{p_{bed}}{\rho g h}\right)\tan\varphi_{bed}\right] \qquad (6.44)$$

In this simplified momentum equation, longitudinal normal stresses due to the solid and fluid constituents have been lumped into a single term, but separation of these stresses is straightforward (Iverson, 1997).

All the terms in (6.44) have clear physical implications that can be couched in terms of Newton's second law of motion. The terms on the left-hand side of (6.44) express the change in momentum of the debris per unit of bed area $\Delta x \Delta y$, and the terms on the right-hand side express the net force on the debris per unit of bed area. These forces result from gravity and are therefore proportional to the lithostatic stress $\rho g h$. The term $-\rho g h(1 - p_{bed}/\rho g h)\tan\varphi_{bed}$ expresses the basal resisting stress due to Coulomb friction, and this resistance is modulated by the basal pore pressure p_{bed}, which is generally a function of x and t. The term $-\rho g h k_{act/pass}(\partial h/\partial x)$ expresses the variation in longitudinal normal stress due to variation in debris thickness. This term acts to drive debris forward if $\partial h/\partial x < 0$, as in the head of an advancing surge, whereas it acts to drive debris backward if $\partial h/\partial x > 0$, as in the tail of a surge. Thus, the longitudinal stress term indicates that gravitational spreading should cause surges to elongate and thereby attenuate. In contrast, observations and data show that, although debris-flow surges may elongate, attenuation of surge fronts is by no means pervasive. Instead, debris-flow surge fronts tend to grow large and steep, and secondary surge fronts tend to appear. Therefore, understanding the development and persistence of debris-flow surges and surge fronts requires delving deeper into the implications of (6.44).

Stability analysis of an equation similar to (6.44) shows that development of small-amplitude surges can result from an interaction of inertial and gravitational effects that are present in shallow flows of any fluid-like substance, irrespective of its rheology (Forterre and Pouliquen, 2003). However, surge fronts in debris flows appear to grow to large amplitudes as a consequence of non-uniform frictional resistance that results from grain-size segregation and pore-pressure diffusion (cf. Iverson, 1997; Savage and Iverson, 2003). In (6.44), non-uniform frictional resistance is represented by a non-uniform distribution of $-\rho g h(1 - p_{bed}/\rho g h)\tan\varphi_{bed}$, which results mostly from variation in basal pore-fluid pressure p_{bed}. Recall from (6.34) that variation of pore-fluid pressure about an equilibrium (hydrostatic) distribution obeys a diffusion equation, and that the diffusivity $k/\alpha\mu$ includes the permeability k, which can vary by many orders of magnitude as a consequence of variations in grain-size distributions (Iverson, 1997; Major et al., 1997). Thus, grain-size segregation in debris flows can result in great variations in dissipation of excess (non-equilibrium) pore-fluid pressure, which causes great variation in frictional resistance.

Figure 6.6 depicts schematically the means by which grain-size segregation in debris flows appears to develop and persist. As a consequence of grain-size

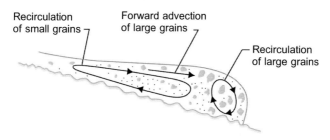

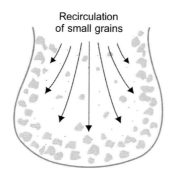

Figure 6.6. Schematic of grain trajectories and resulting grain-size segregation in a debris-flow surge.

segregation, debris-flow surges typically have steep, coarse-grained, high-resistance heads with little excess pore-fluid pressure and relatively fine-grained, low-resistance, tapering surge tails nearly liquefied by high pore-fluid pressure (Figure 6.7). The liquefied tail tends to push against the high-friction head, which can act somewhat like a moving dam. Thus, a disparity in frictional resistance between the surge head and tail can amplify the surge waveform despite the tendency for the term $-\rho g h k_{act/pass}(\partial h/\partial x)$ in (6.44) to attenuate the surge (Savage and Iverson, 2003).

The important role of grain-size segregation in growth of debris-flow surges is indicative of an emergent phenomenon. ("Emergent" is a term commonly used in non-linear dynamics to describe phenomena or structures that arise from dynamical feedbacks rather than from physical properties that can be specified a priori.) Indeed, any a priori specification of debris-flow rheology or flow resistance disregards a fundamental fact of debris-flow mechanics: the form of the chief macroscopic flow structure (i.e., a blunt, large-amplitude surge head followed by a tapering, more dilute tail) is contingent on a non-uniform distribution of flow resistance that evolves as a *consequence* of flow dynamics. Future progress in debris-flow mechanics may therefore depend on the degree to which grain-size segregation processes and the consequent emergence and persistence of surges can be successfully represented in continuum models.

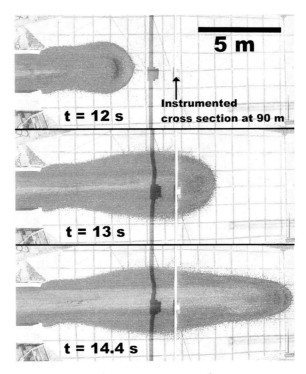

Figure 6.7. Aerial photographs of an experimental ~10 m³ debris flow discharging from the mouth of the USGS debris-flow flume. Grain-size segregation produced a surge front composed almost entirely of gravel (dark toned) and a surge tail composed of liquefied mud (light toned). Time stamps of photo frames are referenced to $t = 0$ when the debris flow was released from a headgate 82.5 m upslope from the flume mouth. Iverson (2003a) provides further details and data on this experiment.

6.6 CONCLUDING REMARKS

Continuum mechanical models of debris flows are founded partly on well-established physical laws and mathematical rules, and partly on empirical and theoretical inferences about stresses that are responsible for energy dissipation. Therefore, a primary objective in mechanistic debris-flow modelling is to honor physical laws and mathematical rules as faithfully as possible, and a second objective is to link representation of stresses to data and theory in a manner that is direct and transparent.

Mechanical models of debris flows can have value as both hazard-assessment tools and precise conceptual frameworks. The conceptual value depends largely on the degree to which models link debris-flow behavior to universal scientific principles, such as conservation of momentum and mass, and to replicable experimental data. Models with weak linkages to replicable data and universal principles have limited conceptual value but may offer useful methodology for practical hazard assessment in some circumstances.

Clear tests of the predictive power of mechanistic models can be achieved only in controlled experiments in which all parameter values and boundary conditions are independently constrained. On the other hand, natural debris flows typically have indeterminate parameter values and unconstrained initial and boundary conditions, which preclude decisive model tests. As a consequence, models commonly are calibrated by adjusting stress or resistance terms to fit the observed behaviour of natural debris flows. The differences between model testing using experimental data and model calibration using field observations can have a synergistic effect, however. The two procedures can be used together to reveal model weaknesses and thereby lead to model improvements.

Predictions of debris-flow models should be regarded with ample skepticism by both model developers and model users. Scientific interpretations derived from model output are only as valid as the assumptions used in model formulation. Model developers should, therefore, make painstaking efforts to thoroughly document all physical, mathematical, and computational aspects of their models, as well as the sources and reliability of data that serve as model inputs. Model users should demand this thorough documentation and should make their own painstaking efforts to understand model limitations. Through such combined efforts of model developers and model users, mechanistic models of debris flows can be expected to improve.

6.7 ACKNOWLEDGEMENTS

I thank James Vallance, Joseph Walder, and the editors for their constructive criticism of a preliminary draft of this chapter. Dr. Vallance provided a draft version of Figure 6.6.

6.8 REFERENCES

Abramowitz, M. and Stegun, I.A. (eds) (1964) *Handbook of Mathematical Functions with Formulas, Graphs, and Mathematical Tables* (1046 pp.). US Department of Commerce, National Bureau of Standards, Washington, DC.

Asmar, B.N., Langston, P.A., and Ergenzinger, P. (2003) The potential of the discrete element method to simulate debris flow. In: D. Rickenmann and C-L. Chen (eds), *Debris-flow Hazards Mitigation: Mechanics Prediction and Assessment* (Vol. 1, pp. 435–445). Millpress, Rotterdam.

Bagnold, R.A. (1954) Experiments on a gravity-free dispersion of large solid spheres in a Newtonian fluid under shear, *Proceedings of the Royal Society of London, Ser. A*, **225**, 49–63.

Bear, J. (1972) *Dynamics of Fluids in Porous Media* (764 pp.). Dover, New York.

Bird, R.B., Stewart, W.E., and Lightfoot, E.N. (1960) *Transport Phenomena* (780 pp.). John Wiley & Sons, New York.

Brown, R.L. and Richards, J.C. (1970) *Principles of Powder Mechanics* (221 pp.). Pergamon Press, Oxford, UK.

Campbell, C.S., Cleary, P.W., and Hopkins, M.A. (1995) Large-scale landslide simulations: Global deformation, velocities and basal friction. *Journal of Geophysical Research B*, **100**, 8267–8283.

Chen, C-L. (1988) Generalized viscoplastic modeling of debris flow. *Journal of Hydraulic Engineering*, **114**, 237–258.

Chen, C-L. (ed.) (1997) *Debris-flow Hazards Mitigation: Mechanics, Prediction, and Assessment* (817 pp.). American Society of Civil Engineers, New York.

Coulomb, C.A. (1776) Sur une application des règles de maxima & minima à quelques problèmes de statique, rélatifs à l'architecture. In: *Mémoires de Mathématique et de Physique, Presented at the Royal Academy of Sciences, 1773* (pp. 343–384). Imp. R. Acad. Sci., Paris.

Coussot, P. and Piau, J-M. (1995) A large-scale field coaxial cylinder rheometer for the study of the rheology of natural coarse suspensions. *Journal of Rheology*, **39**, 105–124.

Davies, T.R.H. (1990) Debris-flow surges: Experimental simulation. *Journal of Hydrology (N.Z.)*, **29**, 18–46.

Denlinger, R.P. and Iverson, R.M. (2001) Flow of variably fluidized granular masses across three-dimensional terrain: 2. Numerical predictions and experimental tests. *Journal of Geophysical Research B*, **106**, 553–566.

Denlinger, R.P. and Iverson, R.M. (2004) Granular avalanches across irregular three-dimensional terrain: 1. Theory and computation. *Journal of Geophysical Research F*, **109**, doi:10.1029/2003JF000085, 14 pp.

Desai, C.S. and Siriwardane, H.J. (1984) *Constitutive Laws for Engineering Materials, with Emphasis on Geologic Materials* (468 pp.). Prentice Hall, Englewood Cliffs, NJ.

Drew, D.A. and Lahey, R.T. (1993) Analytical modeling of multiphase flow. In: M.C. Roco (ed.), *Particulate Two-Phase Flow* (pp. 509–566). Butterworth-Heinemann, Boston.

Erlichson, H. (1991) A mass-change model for the estimation of debris-flow runout, a second discussion: Conditions for the application of the rocket equation. *Journal of Geology*, **99**, 633–634.

Forterre, Y. and Pouliquen, O. (2003) Long-surface-wave instability in dense granular flows. *Journal of Fluid Mechanics*, **486**, 21–50.

Gidaspow, D. (1994) *Multiphase Flow and Fluidization* (467 pp.). Academic Press, Boston.

Goldhirsh, I. (2003) Rapid granular flows. *Annual Review of Fluid Mechanics*, **35**, 267–293.

Gray, J.M.N.T., Wieland, M., and Hutter, K. (1999) Gravity driven free surface flow of granular avalanches over complex basal topography *Proceedings of the Royal Society of London, Ser. A*, **455**, 1841–1874.

Haff, P.K. (1983) Grain flow as a fluid-mechanical phenomenon. *Journal of Fluid Mechanics*, **134**, 401–430.

Hungr, O. (1995) A model for the runout analysis of rapid flow slides, debris flows, and avalanches. *Canadian Geotechnical Journal*, **32**, 610–623.

Hungr, O. (2000) Analysis of debris flow surges using the theory of uniformly progressive flow, *Earth Surface Processes and Landforms* **25**, 483–495.

Hungr, O. and Morgenstern, N.R. (1984) Experiments on the flow behaviour of granular materials at high velocity in an open channel. *Géotechnique*, **34**, 405–413.

Hunt, M.L., Zenit, R., Campbell, C.S., and Brennen, C.E. (2002) Revisiting the 1954 suspension experiments of R.A. Bagnold. *Journal of Fluid Mechanics*, **452**, 1–24.

Iverson, R.M. (1985) A constitutive equation for mass-movement behavior. *Journal of Geology*, **93**, 143–160.

Iverson, R.M. (1997) The physics of debris flows. *Reviews of Geophysics*, **35**, 245–296.

Iverson, R.M. (2003a) The debris-flow rheology myth. In: D. Rickenmann and C.L. Chen (eds), *Debris-flow Hazards Mitigation: Mechanics, Prediction, and Assessment* (Vol. 1, pp. 303–314). Millpress, Rotterdam.

Iverson, R.M. (2003b) How should mathematical models of geomorphic processes be judged? In: P.R. Wilcock and R.M. Iverson (eds), *Prediction in Geomorphology* (Geophysical Monograph 135, pp. 83–94). American Geophysical Union, Washington, DC.

Iverson, R.M. (2003c) Gravity-driven mass flows. In: G.V. Middleton (ed.), *Encyclopedia of Sediments and Sedimentary Rocks* (pp. 347–353). Kluwer, Dordrecht, The Netherlands.

Iverson, R.M. and Denlinger, R.P. (2001) Flow of variably fluidized granular masses across three-dimensional terrain: 1. Coulomb mixture theory. *Journal of Geophysical Research B*, **106**, 537–552.

Iverson, R.M. and Vallance, J.W. (2001) New views of granular mass flows. *Geology*, **29**, 115–118.

Iverson, R.M., Reid, M.E., and LaHusen, R.G. (1997) Debris-flow mobilization from landslides. *Annual Review of Earth and Planetary Sciences*, **25**, 85–138.

Iverson, R.M., Reid, M.E., Iverson, N.R., LaHusen, R.G., Logan, M., Mann, J.E., and Brien, D.L. (2000) Acute sensitivity of landslide rates to initial soil porosity. *Science*, **290**, 513–516.

Iverson, R.M., Logan, M., and Denlinger, R.P. (2004) Granular avalanches across irregular three-dimensional terrain: 2. Experimental tests. *Journal of Geophysical Research F*, **109**, doi:10.1029/2003JF000084, 16 pp.

Kang, Z. and Zhang, S. (1980) A preliminary analysis of the characteristics of debris flow. *Proceedings of the International Symposium on River Sedimentation* (pp. 225–226). Chinese Society for Hydraulic Engineering, Beijing.

Keller, J.B. (2003) Shallow-water theory for arbitrary slopes of the bottom. *Journal of Fluid Mechanics*, **489**, 345–348.

Koch, D.L. and Hill, R.J. (2001) Inertial effects in suspension and porous-media flows. *Annual Review of Fluid Mechanics*, **33**, 619–647.

Lambe, T.W. and Whitman, R.V. (1979) *Soil Mechanics, SI Version* (553 pp.). John Wiley & Sons, New York.

Locat, J. (1997) Normalized rheological behavior of fine muds and their properties in a pseudoplastic regime. In: C-L. Chen (ed.), *Debris-flow Hazards Mitigation: Mechanics, Prediction, and Assessment* (pp. 260–269). American Society of Civil Engineers, New York.

Major, J.J. and Iverson, R.M. (1999) Debris-flow deposition: Effects of pore-fluid pressure and friction concentrated at flow margins. *Geological Society of America Bulletin*, **111**, 1424–1434.

Major, J.J. and Pierson, T.C. (1992) Debris-flow rheology: Experimental analysis of fine-grained slurries. *Water Resources Research*, **28**, 841–857.

Major, J.J., Iverson, R.M., McTigue, D.F., Macias, S., and Fiedorowicz, B.K. (1997) Geotechnical properties of debris-flow sediments and slurries. In: C-L. Chen (ed.), *Debris-flow Hazards Mitigation: Mechanics, Prediction, and Assessment* (pp. 249–259). American Society of Civil Engineers, New York.

Malvern, L.E. (1969) *Introduction to the Mechanics of a Continuous Medium* (713 pp.). Prentice Hall, Englewood Cliffs, NJ.

O'Brien, J.S. and Julien, P.Y. (1988) Laboratory analysis of mudflow properties. *Journal of Hydraulic Engineering*, **114**, 877–887.

O'Brien, J.S., Julien, P.Y, and Fullerton, W.T. (1993) Two-dimensional water flood and mudflow simulation. *Journal of Hydraulic Engineering*, **119**, 244–261.

Parsons, J.D., Whipple, K.X., and Simioni, A. (2001) Experimental study of the grain-flow, fluid-mud transition in debris flows. *Journal of Geology*, **109**, 427–447.

Passman, S.L. and McTigue, D.F. (1986) A new approach to the effective stress principle. In: S.K. Saxena (ed.), *Compressibility Phenomena in Subsidence* (pp. 79–91). Engineering Foundation, New York.

Phillips, C.J. and Davies, T.R.H. (1991) Determining rheological parameters of debris flow material. *Geomorphology*, **4**, 101–110.

Rickenmann, D. and Chen, C-L. (eds) (2003) *Debris-flow Hazards Mitigation: Mechanics, Prediction, and Assessment* (1335 pp.). Millpress, Rotterdam.

Savage, S.B. and Hutter, K. (1989) The motion of a finite mass of granular material down a rough incline. *Journal of Fluid Mechanics*, **199**, 177–215.

Savage, S.B. and Hutter, K. (1991) The dynamics of avalanches of granular materials from initiation to runout Part I: Analysis. *Acta Mechanica*, **86**, 201–223.

Savage, S.B. and Iverson, R.M. (2003) Surge dynamics coupled to pore-pressure evolution in debris flows. In: D. Rickenmann and C-L. Chen (eds), *Debris-flow Hazards Mitigation: Mechanics, Prediction, and Assessment* (Vol. 1, pp. 503–514). Millpress, Rotterdam.

Savage, S.B. and Sayed, M. (1984) Stresses developed in dry cohesionless granular materials sheared in an annular shear cell. *Journal of Fluid Mechanics*, **142**, 391–430.

Sharp, R.P. and Nobles, L.H. (1953) Mudflow of 1941 at Wrightwood, southern California. *Geological Society of America Bulletin*, **64**, 547–560.

Terzaghi, K. (1936) The shearing resistance of saturated soils. *Proceedings of the 1st International Conference on Soil Mechanics* (Vol. 1, pp. 54–56).

Vanoni, V.A. (ed.) (1975) *Sedimentation Engineering* (745 pp.). American Society of Civil Engineers, New York.

Vreugdenhil, C.B. (1994) *Numerical Methods for Shallow-water Flow* (261 pp.). Kluwer Academic, Dordrecht, The Netherlands.

Walton, O.R. (1993) Numerical simulation of inelastic, frictional particle–particle interactions. In: M.C. Roco (ed.), *Particulate Two-Phase Flow* (pp. 884–911). Butterworth-Heinemann, Boston.

Wieczorek, G.F. and Naeser, N.D. (eds) (2000) *Debris-flow Hazards Mitigation: Mechanics, Prediction, and Assessment* (608 pp.). A.A. Balkema, Rotterdam.

Yamagishi, M., Mizuyama, T., Satofuka, Y., and Mizuno, H. (2003) Behavior of big boulders in debris flow containing sand and gravel. In: D. Rickenmann, and C-L. Chen (eds), *Debris-flow Hazards Mitigation: Mechanics, Prediction, and Assessment* (Vol. 1, pp. 411–420). Millpress, Rotterdam.

7

Entrainment of material by debris flows

Oldrich Hungr, Scott McDougall, and Michael Bovis

7.1 INTRODUCTION

Debris-flow magnitude can be defined as the total volume of material moved to the deposition area during an event. It is an important quantity as it serves to scale the event and correlates with other parameters such as maximum discharge and runout distance (Chapters 13 and 17).

From descriptions in many other chapters of this book, it is clear that debris-flow magnitude is rarely determined by the volume of the initiating landslide. Often, the initiating slide is small and the bulk of the volume transported to the deposition area results from entrainment of material along the path. An excellent example is the 1990 Tsing Shan debris flow shown in Figure 7.1, the largest natural debris flow observed in Hong Kong (King, 1996). Here, a small slip of $400 \, m^3$ enlarged to a final volume of $20,000 \, m^3$ by entraining colluvium from the flow path. Figure 7.2 shows a reconstruction of the mass balance curve of this debris flow, based on data collected by King (1996).

Thus, it is the efficiency of the entrainment mechanism that primarily determines the total volume of a debris flow.

7.2 MECHANISMS OF MATERIAL ENTRAINMENT

Bedload in a stream channel lined with granular bed material can be transported by suspension, rolling, sliding, or saltation (e.g., Easterbrook, 1999). As shown in flume experiments, once the slope of the channel increases beyond approximately 10°, the bed itself may become unstable under the combination of gravity and drag forces imposed by the over-riding water flow (Bagnold, 1966). If the surface fluid is saturated debris instead of water, even greater drag forces result and the bed material can be massively mobilized and entrained into the flow.

M. Jakob and O. Hungr (eds), *Debris-flow Hazards and Related Phenomena.*
© Praxis. Springer Berlin Heidelberg 2005.

Figure 7.1. The Tsing Shan debris flow that started as a small debris slide of $400\,m^3$ and grew to a total magnitude of $20{,}000\,m^3$ through material entrainment.

Photo courtesy, J. King, Geotechnical Engineering Office, Hong Kong.

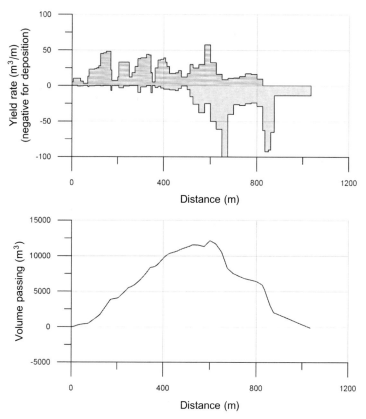

Figure 7.2. Tsing Shan debris flow. (a) Distribution of observed yield rate. (b) Approximate mass balance curve.

Based on data reported by King (1996).

The process of surge formation, resulting from longitudinal sorting and the emergence of boulder fronts (e.g. Iverson, 1997) or from the formation of turbulent fronts (e.g. Davies, 1986), magnifies the peak discharge of debris-flow surges (Hungr, 2000) and is thus likely responsible for an increase in drag forces and further enhancement of the entrainment intensity.

One of the mechanisms causing material entrainment in debris flows is bed destabilization and erosion. Destabilization of bed material is the result of drag forces acting at the base of the flow, but may be aided by strength loss due to rapid undrained loading (Hutchinson and Bhandari, 1971), impact loading, and liquefaction of the saturated channel fill (Sassa, 1985, see also Chapter 5). As shown schematically in Figure 7.3, bed destabilization during a debris flow may affect not only bedload, but any erodible bed substrate. This process of bed destabilization can be quantified to some degree as shown in the following section, although the necessary data regarding bed stratigraphy, bed material, substrate strength, and drainage characteristics is difficult to obtain. Knowledge of erosion

138 Entrainment of material by debris flows   [Ch. 7

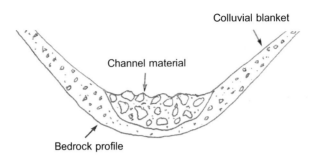

Figure 7.3. Schematic diagram of an eroded vertical cross section of a debris-flow channel.

depth is also useful for practical purposes, such as the protection of pipelines crossing a debris-flow channel (Jakob et al., 2004b).

The second important mechanism of material entrainment results from instability of stream banks undercut by bed erosion, as also shown in Figure 7.3. It is important to consider that steep stream and gully channels are often being actively incised. Thus, their banks may exist in a state of marginal equilibrium that is easily disturbed by lowering of the bed, such as often occurs during passage of a debris-flow surge. The bank may respond immediately and release a shallow landslide directly into the body of the surge, or may release with a delay, to provide material available for incorporation into the next surge. Thus, some debris may form transient deposits in the channel, only to be re-mobilized later during the same event, or in a later event. Eyewitness reports mention debris from bank slides that briefly dam the channel and are then rapidly eroded by overtopping water or debris flow, and are readily liquefied by mixing with stream water (Johnson, 1970). Such processes are complex and defy mechanistic quantification, since the required data on side slope stability, including strength provided by vegetation, and the temporal relationships between surging flow discharge, bed erosion, bank slope failure, and mixing of water and debris, cannot normally be obtained. A photo of a channel combining highly erodible bed and banks is shown in Figure 7.4.

7.3 THEORETICAL APPROACH: BED STABILITY

The process of bed destabilization during a debris flow can be represented by a simple extension of the infinite slope stability theory (e.g. Morgenstern and Sangrey, 1978). However, much depends on the assumptions made concerning pore-water pressure in the bed materials. In the first attempt at deriving a formula for the depth of bed instability, Takahashi (1978, 1991) assumed slope-parallel seepage in a saturated bed. In the following derivation we use different symbols than Takahashi (1978), but the same physical concept.

The problem configuration is shown schematically in Figure 7.5. A sheet of debris, of thickness z_d, flows over a bed of cohesionless material inclined at an angle β. As a result of the added tractive force of the debris, the bed becomes

Sec. 7.3] Theoretical approach: bed stability 139

Figure 7.4. An eroded debris flow channel in the Columbia Mountains, British Columbia. Material was derived both from vertical erosion of the bed and from instability of the banks.

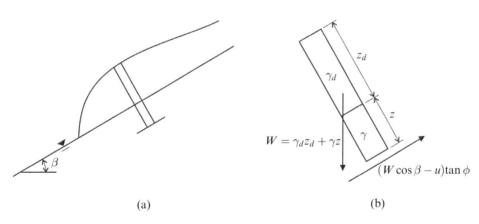

Figure 7.5. (a) Schematic representation of a saturated bed over-ridden by a debris flow, showing a slope-normal column of unit length and width. (b) Forces acting on the column in (a) include the weight of the column and the shear resistance at its base. Loading due to the over-riding flow destabilizes the bed to depth z. See text for details.

unstable to an unknown depth z below the original bed surface. As is standard in the infinite slope approach, only the stability of a typical column of a unit length in the downslope direction is considered (the width perpendicular to the flow is unity).

Based on the diagram of the column shown in Figure 7.5, its weight equals:

$$W = \gamma_d z_d + \gamma z \qquad (7.1)$$

where γ is the saturated unit weight of the bed material (typically 20–23 kN/m^3) and γ_d is the bulk unit weight of the debris (18–20 kN/m^3).

From Figure 7.5, the normal total stress at the column base equals:

$$\sigma = W \cos \beta \qquad (7.2)$$

The shear stress equals:

$$\tau = W \sin \beta \qquad (7.3)$$

Takahashi (1978) assumed slope-parallel seepage and uniform flow, combined with instant drainage, so that the pore fluid is hydrostatically pressurized and flowing in a steady-state regime, with no excess pore pressure. Such assumptions may not be justified, as discussed below. However, to continue the analysis, with these assumptions the pore pressure at the base of the column is:

$$u = \gamma_w (z_d + z) \cos \beta \qquad (7.4)$$

where γ_w is the unit weight of water and $(z_d + z) \cos \beta$ is the elevation difference measured along an equipotential line.

The shear strength of the bed material is given by the cohesionless Mohr–Coulomb shear strength equation, in which ϕ is the friction angle:

$$S = (\sigma - u) \tan \phi \qquad (7.5)$$

At the point of shear failure $S = \tau$. Hence, from (7.2), (7.3), and (7.5):

$$(W \cos \beta - u) \tan \phi = W \sin \beta \qquad (7.6)$$

Substituting further using (1) and (4) and solving for z:

$$z = z_d \frac{\left[\dfrac{\gamma_d}{\gamma} \left(1 - \dfrac{\tan \beta}{\tan \varphi} \right) - \dfrac{\gamma_w}{\gamma} \right]}{\left[\dfrac{\gamma_w}{\gamma} - \left(1 - \dfrac{\tan \beta}{\tan \varphi} \right) \right]} \qquad (7.7)$$

This equation is equivalent to Takahashi's (1978, Equation 22). Its results are represented by the dashed lines in Figure 7.6. The saturated unit weight of the bed material was chosen as 20 kN/m^3 (i.e., about twice the unit weight of water). Therefore, by application of the simple infinite slope stability equation, the ratio $\tan \beta / \tan \phi$ must be less than about 0.5, or the bed itself would be inherently unstable. The diagram shows that a certain amount of entrainment is possible for any value of γ_d less than $\gamma_w / (1 - \tan \beta / \tan \phi)$, with more dilute flows causing instability to greater depths. For fully developed debris surges, whose bulk density approximates the density of the bed material (i.e., $\gamma_d / \gamma = 1$), no entrainment will be predicted with these assumptions, except if the bed itself is inherently unstable.

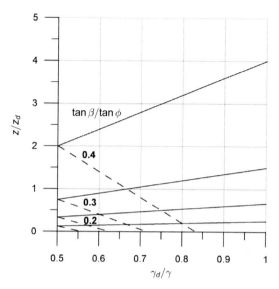

Figure 7.6. Erosion depths predicted by solution of (7.7) (dashed lines) and (7.9) (solid lines).

Debris-flow surges travel many metres per second and, even with a relatively coarse and pervious bed material, it is unlikely that a steady seepage condition can be achieved in the short time while a surge peak is passing. A more realistic assumption is that the full bulk weight of the debris flow $\gamma_d z_d$ will be transferred to pore-water by undrained loading, generating geostatic pressure within the bed materials. Thus, instead of (7.4) we have:

$$u = (z_d \gamma_d + z \gamma_w) \cos \beta \tag{7.8}$$

Following the same steps that have led to (7.7) we obtain:

$$z = z_d \left[\frac{-\dfrac{\gamma_d \tan \beta}{\gamma \tan \varphi}}{\dfrac{\gamma_w}{\gamma} - \left(1 - \dfrac{\tan \beta}{\tan \varphi}\right)} \right] \tag{7.9}$$

This equation is plotted using the full lines in Figure 7.6. Here, a very different trend is observed. The unstable depth *increases* with the bulk density of the debris flow and entrainment is predicted for all values of γ_d and β, as long as the bed is not horizontal. The actual value of the unstable depth may lie somewhere between the two extremes depicted in Figure 7.6 although it is likely closer to the undrained condition (full lines) than those for the drained condition.

While these results are conceptually interesting, they are of little value for practical application. One reason is that little is known about the shear strength of materials comprising the bed of a debris-flow stream and its variation with depth. Often, there will be a layer of cohesionless, coarse material, underlain by a substrate possessing either true or apparent cohesion, such as glacial till or residual soil.

Pore-pressures are also unlikely to be easily predictable, due to possible discharge gradients at the base of a steep-sided path segment or excess gradients generated by rapid loading and vibration due to the debris flow. Three-dimensional effects (i.e., the strength of lateral surfaces at the channel edge), are also likely to be important. Although the equations may help predict when erosion will begin to occur, the rate at which material is entrained into the flow requires further analysis and further difficult assumptions. Thus, the above equations can be regarded merely as conceptual guidelines. As argued in the preceding section, the process of entrainment involving bank instability is even less amenable to mechanistic analysis. Consequently, the remainder of this chapter concentrates on empirical approaches to the problem.

7.4 THE YIELD RATE AND EROSION DEPTH CONCEPTS

Several early empirical algorithms for prediction of debris-flow magnitude were reviewed by VanDine (1985). Some of these methods attempt to correlate magnitude with the drainage area, but the results tend to be widely scattered (see Chapter 17). Other methods concentrate on erosion of material along the length of channels. The first such published attempt was Ikeya (1981), who suggested that potential magnitude can be calculated as a product of the channel length L, mean width B, and mean erosion depth D. He used empirical relationships involving drainage area for L and B and estimated D as ranging between 0.5 and 3.2 m.

A more direct method was developed by Thurber Consultants (1983) and Hungr et al. (1984), based on the concept of yield rate Y_i. The yield rate is defined as the volume eroded per metre of channel length (Hungr et al., 1984). With reference to Figure 7.3, it is the area of the vertical cross section of the eroded space, multiplied by the cosine of the channel slope angle β to convert vertical depth to thickness. To apply the concept, the channel system of a debris-flow watershed is divided into channel reaches considered to be approximately constant in terms of the parameters critical for material entrainment, as listed in Table 7.1.

Once the applicable yield rates are estimated, the debris-flow magnitude (volume) V (in m^3) can be estimated by the formula:

$$V = V_{initial} + \sum V_{point} + \sum_{i=1}^{n} Y_i L_i \qquad (7.10)$$

Here, $V_{initial}$ represents the volume of the initiating landslide, V_{point} is the volume of any "point sources" (i.e., tributary landslides that may be destabilized by a passing debris flow and add volume to the flow), and L_i and Y_i are the length and yield rate of n channel reaches as defined above.

While the concept represented by (7.10) is simple, several problems remain. First is the optimal number of tributaries to the main debris-flow channel to include in the summation. Some debris flows affect only one branch at a time. Other events, especially those triggered by a major regional storm, may mobilize nearly every

Table 7.1. Parameters relevant to the yield rate Y_i

Slope angle
Existing channel width and depth
Bed material
Bank slope angle
Bank slope height
Bank slope material
Bank slope stability rating
Tributary drainage area or discharge

tributary of the drainage network, right down to zero-order colluvial hollows (Figure 7.7). To the authors' knowledge, no research clarifies this issue, and a judgmental decision needs to be made based on local experience. Knowledge of the time elapsed since the last debris flow may provide some guidance, since this controls the amount of debris replenishment in formerly scoured channel segments. This issue is further discussed in Chapter 17.

In exceptional cases of catastrophic rainstorms, the extent of instability of steep slopes in the headwaters may be so large as to make the yield rate approach impractical. For example, the December 1999 storm in the Vargas State, Venezuela, caused the failure of up to 30% of the steep slopes in certain basins (Lopez et al., 2003; see Chapter 20). The magnitude of debris produced by such events may best be estimated by multiplying the predicted area of landslide scars by the average erosion depth.

The second problematic issue is determination of the downstream limits of debris-flow erosion. It is well known that debris-flow surges erode material on steep slopes, but deposit material on slopes flatter than a certain limit. Several suggestions for setting the deposition slope can be found in the literature:

1. Ikeya (1981), followed by Okubo and Mizuyama (1981), suggested a slope of 10°.
2. Hungr et al. (1984), referring to relatively coarse-grained non-volcanic debris flows from Southern British Columbia, Canada, suggested a deposition slope of 8–12° for "confined" channels and 10–14° for "open" channels, the former being characterized by a maximum flow depth/width ratio of over 1 : 5.
3. Small debris flows and debris avalanches in Hong Kong often deposit on slopes exceeding 30–40° (Wong et al., 1997).
4. Referring to debris flows from the Pacific North-west, USA, Benda and Cundy (1990) placed the downstream end of erosion at a 10° slope. Deposition is said to begin at a slope of 3.5°, or earlier if a sharp change of flow direction occurs at a stream junction.
5. Fannin and Wise (2001), using data from coastal and insular British Columbia, found both deposition and entrainment occurring on slopes of 10–22° in confined reaches and 19–24° in unconfined ones (e.g., open slopes, gully sidewalls or headwalls).

Figure 7.7. A debris flow channel with multiple branches in Banff National Park, Canada.
Photo D. Ayotte.

6. Jordan (1994) made a comparison between coarse-grained debris flows derived from igneous rocks of the Coast Plutonic Complex, southern British Columbia, Canada and fine-textured volcanogenic debris flows from the same region. The average slope angles of the deposits ranged between 7 and 15° for the former and 0.5 to 5° for the latter, showing a strong inverse relationship to volume (ranging up to $10^7 \, m^3$ for the largest lahars).
7. Large debris flows from metamorphic rock sources, triggered during the 1999 Vargas State disaster in Venezuela, eroded fan and floodplain deposits and carried large boulders to slopes as low as 2° without substantial confinement (Lopez et al., 2003, see also Chapter 20).
8. Large, eruption-triggered debris flows on volcanoes may erode substantially on slopes as flat as 1° (Pierson, 1995, see also Chapters 10 and 27).

This collection of deposition criteria confirms the disquieting fact that no general guidelines for the determination of the deposition angle exist. Furthermore, experience shows that smaller debris flow and debris avalanche events can deposit on considerably steeper angles than larger events on the same path. Mean water content of individual surges also seems to play an important role as does the composition and particle size of the surge front. Thus, the downstream limit of erosion may vary from one event to another, or even between individual surges.

Research is needed to establish two criteria: the slope where substantial erosion ends ("limit of erosion slope") β_e, and the slope where deposition begins ("slope of deposition") β_d (neither may be a unique value for a given path). The factors likely to influence both limits are probably those listed in Table 7.1. Both angles will probably be strongly affected by the average solids concentration of a debris-flow surge. Flows with lower solids concentrations by volume should be more erosive and also should have the lowest deposition angles. The implication of this is that relatively steep fan accumulations of material laid down by one flow may subsequently be eroded and remobilized by later flows having lower solids concentrations. This type of behavior is observable on many debris fans, where complex patterns of filling and cutting tend to be the rule rather than the exception.

The third problem is the estimation of the yield rate itself. Some channels and gullies are formed in substrate of low erodibility (e.g., bedrock or dense or very coarse granular soil or stiff cohesive soil). Bedload material and colluvial wedges at the base of stable banks are ephemeral in such channels and likely to be eroded by a climax debris-flow surge. An example of such a "firm-base channel", formed in igneous bedrock, is shown in Figure 7.8. The corresponding term in sediment hydraulics is a "supply-limited" channel (cf. Bovis and Jakob, 1999; Jakob et al., 2004). It is possible to estimate the amount of debris available in firm-base channels by a direct visual inspection (Thurber, 1983; Hungr et al., 1984; VanDine, 1985).

Estimation of the yield rate is more difficult and subjective for "erodible-base channels", where no shallow, firm substrate exists and the entrainment process is transport-limited. As stated earlier, a theoretical means of estimating erosion depths in such cases is not practically useful, and recourse must be taken to subjective judgment, or empirical relations as discussed later in this chapter.

Figure 7.8. A photo of a firm-base channel, with a substrate of igneous bedrock, near Chilliwack, British Columbia, Canada.

Sec. 7.4] The yield rate and erosion depth concepts 147

The yield rate approach is difficult to use for unchannellized debris avalanches, where the yield rate depends strongly on the width of the path. In this case, the "erosion depth" parameter is a more suitable index for estimating volumes, provided that the width of the path is known. The relation between yield rate (Y_i, in m^3/m), erosion depth (D_i in m), and path width (B_i in m) is:

$$Y_i = B_i \times D_i \quad (7.11)$$

Here the index i represents a particular reach of a path. The erosion depth in open-slope debris slides depends primarily on the depth of any loose layer such as a colluvial veneer, an organic rich soil horizon, or a loosened surficial layer. In some cases, it can be estimated directly from field observation or by subsurface investigations such as test pits or geophysical surveys.

Path width is another parameter that is difficult to estimate, except in the case of firm-based channels. Width estimation is particularly difficult on open slopes, where debris avalanche scars are often observed to widen with distance downslope (Figure 7.9). Guadagno et al. (2003) suggested empirical means to estimate the angle of spreading of debris avalanche scars in the Campania Region, Italy (Chapter 19).

A collection of reported values of erosion depth and yield rate from several parts of the world is shown in Table 7.2, which shows that depth ranges up to about 6 m and yield rate up to about 30 m^3/m, although much larger values probably occur in

Figure 7.9. Two debris avalanches widening with distance downslope. (Quindici, Campania Region, southern Italy.)

Table 7.2. A collection of debris flow entrainment rates and erosion depths reported in the literature.

Reference	Location	No. events	Confinement	Erosion depth (m)	Yield rate (m³/m)
Hungr et al. (1984)	B.C. Coast, Canada	5	C	–	6–18
Jakob et al. (1997)	B.C. Coast, Canada	2	C	–	(23)
Campbell and Church (2003)	B.C. Coast, Canada	37	C	0.5–1	–
Fannin and Rollerson (1993)	Queen Charlotte Island, Canada	253	C	–	(12.6)
		196	U		(24)
Jakob et al. (2000)	B.C. Interior, Canada	1	C	0.5–1.5	(28)
Cenderelli and Kite (1998)	The Appalachians, USA	4	U	0–2.5	0–42 (4.2)
Springer et al. (2001)	The Appalachians, USA	2	C	–	2–18
Agostino and Marchi (2003)	Southern Alps, Italy	1	C	0.1–6 (1.0)	–
Revellino et al. (2003)	Campania, Italy	17	C, U	(1.5)	–
Li and Yuan (1983)	South-west China	1	C	5–8	–
Franks (1999)	Hong Kong	40	C	–	0.2–5 (3.6)
King (1996)	Hong Kong	1	U	0–3	0–50
Okuda et al. (1980)	The Alps, Japan	1	C	0–5	–
Rickenmann et al. (2003)	Kazakhstan	1	C	–	8–300

C = mainly confined events (debris flows), U = mainly unconfined events (debris avalanches).
Values in brackets are averages.

catastrophic events (see Chapter 27). The largest values in the table relate to reaches with large bank failures (Rickenmann et al., 2003).

7.5 CONSIDERATION OF FREQUENCY

The above discussion has attempted to present a deterministic picture. However, in reality, all of the parameters contributing to (7.10), including length of the eroded channels, location of the end of erosion, and the yield rate or erosion depth, are stochastic in nature. Thus, we are faced not only with the need to estimate the mean values of these parameters, but also their variance. Some attempts have been made to take a stochastic approach to this problem as discussed below. However, in many practical applications, the analysis focuses on determination of a flow "design magnitude" for a particular channel, which could be defined as the magnitude of an event whose probability of occurrence approximates the inverse of the expected lifetime of a given structure – such as a debris-flow basin – reduced by a suitable factor of safety (Thurber, 1983). The empirical calibration process could then be based on previous events with similar return periods.

Event return periods may be estimated more reliably for firm-base channels. If the path is underlain by a firm base, a debris flow may remove most of the loose material accumulated on it. Following the debris flow, there is no unconsolidated material available and a second event cannot happen. Gradually, more debris accumulates by erosion and mass movement from the hillsides adjacent to the path and by bedload deposition, thus "priming" the channel for another event (Bovis and Dagg, 1988). Bovis and Jakob (1999), found that debris flows can reoccur in some gullies of coastal British Columbia at intervals as short as a few decades. Benda and Dunne (1997), on the other hand, estimated that recharge times for gullies in the US Pacific North-west may be as high as several thousand years. Thus, no simple rule can be given at present.

7.6 EXAMPLES OF EMPIRICAL CALIBRATION OF THE YIELD RATE CONCEPT

The use of the yield rate concept was extended to the simulation of the deposition behavior of debris flows and avalanches on Oahu Island, Hawaii by Cannon (1993). She assumed that each event begins as a discrete debris slide, the volume of which can be estimated beforehand by independent means. A constant "lag rate" is then assumed, being the equivalent of the yield rate, but negative in this case, since material is gradually discarded along the path in levees and sheets. The runout distance is determined by dividing the slide volume by the lag rate. Using multiple regression analysis, Cannon (1993) found an empirical relation connecting lag rate with slope and width of the path (lateral confinement). Cannon's approach is difficult to apply in many cases of debris flows and avalanches, in which the volume of the initiating slide tends to be only a small fraction of the total volume mobilized along the path, and where runout distance is usually significantly controlled by lateral spreading.

Fannin and Wise (2001), using data from the Queen Charlotte Islands of British Columbia, combined both the yield rate and lag rate approaches. Erosion tends to be dominant in the steeper reaches of the path, causing down-channel increases in flow volume. As slope angle decreases, deposition begins and volume is discarded according to a negative lag rate. Runout is predicted when the deposited volume equals the total entrained (i.e., a volume balance is established).

Fannin and Wise's (2001) approach was different from the concepts discussed here, in that neither yield rate nor erosion depth were used explicitly. Instead, regression equations were developed for incremental volumes of erosion and deposition per reach. Separate correlation equations were developed for reaches dominated by erosion and deposition and for transitional reaches. The primary predictor variables in the regressions included reach length and path width. Their correlation coefficients were close to 1, suggesting that these two quantities serve primarily as scaling factors. Thus, the use of incremental volumes instead of yield rates does not seem to be advantageous. Weaker, highly scattered correlations were also found for some path types with slope angle, lateral confinement, cumulative passing volume, and angular changes in flow direction.

The yield rate method was also examined by Lau and Woods (1997) for debris flows and avalanches in Hong Kong. They developed stepwise regression equations for the average yield rate of a channel, based on material type (colluvium, residual soil), slope morphology (planar, concave, convex), vegetation type, slope angle, and radius of channel cross section. However, they concluded that the correlations were very weak, and that the resulting model was no more consistent than the routine empirical runout estimation method based on the travel angle.

7.7 THE BRITISH COLUMBIA DEBRIS-FLOW DATABASE

T. Rollerson and T. Millard, scientists employed by the MacMillan Bloedel Company in British Columbia, investigated 449 debris flow and debris avalanche events from the Queen Charlotte Islands, British Columbia. The region is a heavily dissected plateau with a relief of sea level to approximately 700 m, composed of metamorphosed volcanics and sediments mantled by Pleistocene glacial soils and colluvium (Fannin and Rollerson, 1993). It has a cool, perhumid maritime climate, with an annual precipitation of 1,000–4,000 mm.

The full length of each landslide was traversed on the ground and estimates were made of material eroded and deposited in reaches judged to be homogeneous. An additional 39 events were studied subsequently on the south-western British Columbia coast by Wise (1997) and Yonin and Hungr (unpublished). After eliminating internally inconsistent and incomplete records, a database of 174 debris flow and debris avalanche events, comprising 1,073 channel reaches, was compiled. Unfortunately, the database does not contain any descriptive information on the reaches,

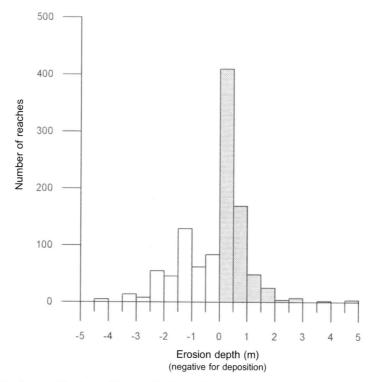

Figure 7.10. Queen Charlotte Islands database: histogram of erosion depths for all 1,073 reaches, both confined and unconfined.

Data courtesy Messrs. T. Rollerson, Golder Associates Ltd and M. Wise, Vancouver.

except for slope, channel width, presence or absence of lateral confinement, path azimuth and passing volume.

The average magnitude of events in the database is only 1,100 m^3, and ranges from 50–30,000 m^3. The total estimated volume of debris eroded in all of the events is 200,000 m^3, while that deposited is 17% greater, indicating a measure of inaccuracy in the field estimates.

Figure 7.10 shows the distribution of erosion depths compiled from the database. Table 7.3 gives average erosion depths for various slope categories. The following observations can be made:

- The erosion depths vary from −3 m (deposition) to 3 m (erosion), with a few exceptional cases reaching the range of −5 to 5 m.
- The erosion depth exhibits a Poisson distribution within the erosional domain, while being rather uniform in the depositional domain.
- The average value is approximately 0.5 m for all eroding reaches and −1.0 for all depositing reaches. However, the standard deviation exceeds the mean at all slopes, indicating significant skewness in the distributions. The range of observed values is always at least ±2 m (except at angles exceeding 40°).

Table 7.3. Queen Charlotte Islands database: yield rates in relation to channel slope and confinement.

Data courtesy Messrs. T. Rollerson, Golder Associates Ltd. and M. Wise, Vancouver.

Slope angle	Unconfined		Confined	
	Average	Shaded deviation	Average	Standard deviation
All	−0.94	39.50	0.26	11.62
0–15	−22.33	58.66	−10.89	18.99
16–20	1.55	33.13	2.07	5.57
21–30	5.53	30.75	3.41	6.55
31–50	11.10	12.96	3.88	3.23

Scatter plots correlating erosion depth with path width, cumulative volume of debris passing into the reach, slope angle, and other variables, showed very weak trends and extreme scatter. It was concluded that those variables available in the database (i.e., confinement, path width, slope angle, path azimuth, and passing volume) do not have a significant systematic influence on the erosion depth. Any model using these data must therefore treat D_i as a random variable, with a range of at least −3 to 3 m.

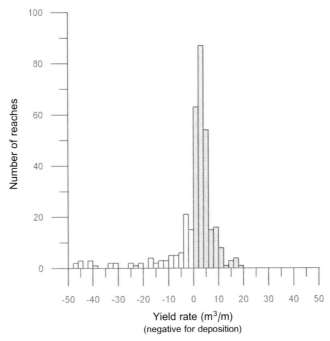

Figure 7.11. Queen Charlotte Islands database: histogram of yield rates for the 340 confined reaches.

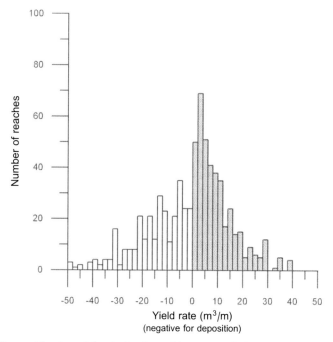

Figure 7.12. Queen Charlotte Islands database: histogram of yield rates for the 733 unconfined reaches.

The distributions of yield rates in all confined and unconfined reaches of the database are shown in Figures 7.11 and 7.12 respectively. In both instances the distributions are approximately normal, centered at a slightly positive value. A summary of means and standard deviations for these data is given in Table 7.4. It is interesting to note that the distributions show a similar trend to the worldwide values compiled in Table 7.2. Thus, the distributions appear to represent typical conditions for relatively small debris flows and debris avalanches.

Figures 7.13 and 7.14 represent an attempt to delineate zones of erosion (full symbols) and deposition (open symbols) in terms of slope angle and the volume of

Table 7.4. Queen Charlotte Islands database: erosion depths (in meters) in relation to channel slope and confinement.

Slope angle	Unconfined		Confined	
	Average	Shaded deviation	Average	Standard deviation
All	0.0	1.2	0.2	1.1
0–15	−1.2	0.9	−0.9	1.4
16–20	−0.1	1.1	0.2	0.8
21–30	0.3	1.0	0.5	0.8
31–50	0.7	0.7	0.6	0.7

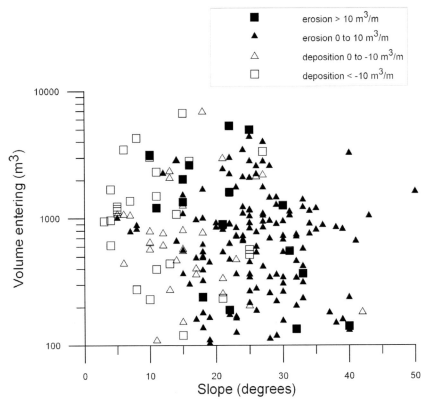

Figure 7.13. Queen Charlotte Islands database: dependence of yield and lag rates on slope angle and the amount of debris flow material entering the reach for "confined" reaches.

debris-flow material entering a given reach. In the case of confined reaches, there appears to be a crude trend, indicating that erosion can occur on slopes as low as 10°, provided that the flow is already fully developed. On the other hand, the volume entering a reach seems to have no effect in unconfined reaches (Figure 7.14). In both cases, there is a transitional zone, at least 10° wide, as well as numerous outliers. Indeed, the occurrence of erosion and deposition in these small magnitude events appears to be a largely random process.

7.8 CONCLUSIONS

The ability to determine entrainment is a crucial step in prediction of debris flow and debris avalanche magnitude and behavior. As shown in this chapter, analytical techniques are unlikely to be useful in the foreseeable future. Empirical relations must be developed, but this task is made complex by the wide scatter in the available data sets, combined with the difficulty of acquiring such data and their generally low level

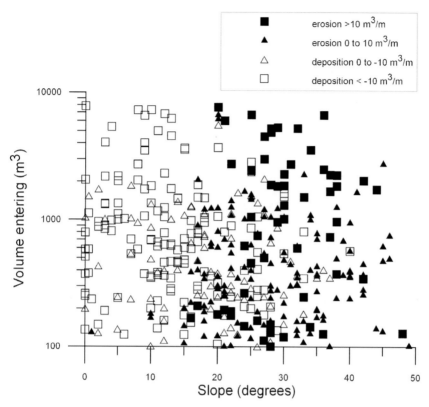

Figure 7.14. Queen Charlotte Islands database: dependence of yield and lag rates on slope angle and the amount of debris flow material entering the reach for "unconfined" reaches.

of reliability. Although difficult, the approach of collecting data on entrainment depth and yield rate, then correlating these data with well-chosen descriptive parameters in a statistical treatment seems to be the only course available. Any such methodology will probably always need to be complemented by judgment.

7.9 REFERENCES

Agostino, V.D. and Marchi, L. (2003) Geomorphological estimation of debris-flow volumes in alpine basins. In: D. Rickenman and L-C. Chen (eds), *Proceedings of the 3rd International Conference on Debris-flow Hazards Mitigation: Mechanics, Prediction and Assessment* (pp. 1097–1106). Millpress, Rotterdam.

Bagnold, R.A. (1966) *An Approach to the Sediment Transport Problem from General Physics: Physiographic and Hydraulic Studies* (USGS Professional Paper 422-I). US Geological Survey, Washington, DC.

Benda, L.E. and Cundy, T.W. (1990) Predicting deposition of debris flows in mountain channels. *Canadian Geotechnical Journal*, **27**, 409–417.

Benda, L. and Dunne, T. (1997) Stochastic forcing of sediment supply to channel networks from landsliding and debris flow. *Water Resources Research*, **33**(12), 2849–2863.

Bovis, M.J. and Dagg, B.R. (1988) A model for debris accumulation and mobilization in steep mountain streams. *Hydrological Sciences Journal*, **33**(6), 589–604.

Bovis, M. and Jakob, M. (1999) The ratio of debris supply in predicting debris flow activity. *Earth Surface Processes and Landforms*, **24**, 1039–1054.

Campbell, D. and Church, M. (2003) Reconnaissance sediment budgets for Lynn Valley, British Columbia: Holocene and contemporary time scales. *Canadian Geotechnical Journal*, **40**, 701–713.

Cannon, S.H. (1993) An empirical model for the volume-change behavior of debris flows. In: H.W. Shen, S.T. Su and F. Wen (eds), *Proceedings of Hydraulic Engineering '93, San Francisco* (Vol. 2, pp. 1768–1773). American Society of Civil Engineers, New York.

Cenderelli, D.A. and Kite, J.S. (1998) Geomorphic effects of large debris flows on channel morphology at North Fork Mountain, eastern West Virginia, USA. *Earth Surface Processes and Landforms*, **23**, 1–19.

Davies, T.R.H. (1986) Large debris flows: A macroviscous phenomena. *Acta Mechanica*, **63**, 161–178.

Easterbrook, D.J. (1999) *Surface Processes and Landforms* (2nd edn, 546 pp.). Prentice Hall, Englewood Cliffs, NJ.

Ellen, S.D., Mark, R.K., Cannon, S.H., and Knifong, D.C. (1993) *Map of Debris Flow Hazards in the Honolulu District of Oahu, Hawaii* (USGS Open File 93-213). US Geological Survey, Reston, VA.

Fannin, R.J. and Rollerson, T.P. (1993) Debris flows: Some physical characteristics and behaviour. *Canadian Geotechnical Journal*, **30**, 71–81.

Fannin, R.J. and Wise, M.P. (2001) An empirical-statistical model for debris flow travel distance. *Canadian Geotechnical Journal*, **38**, 982–994.

Franks, C.A.M. (1999) Characteristics of some rainfall-induced landslides on natural slopes, Lantau Island, Hong Kong. *Quarterly Journal of Engineering Geology*, **32**, 247–259.

Guadagno, F.M., Martino, S., and Scarascia-Magnozza, G. (2003) Influence of man-made cuts on the stability of pyroclastic covers (Campania, southern Italy): A numerical approach. *Environmental Geology*, **43**, 371–384.

Hungr, O. (2000) Analysis of debris flow surges using the theory of uniformly progressive flow. *Earth Surface Processes and Landforms*, **25**, 1–13.

Hungr, O., Morgan, G.C., and Kellerhals, R. (1984) Quantitative analysis of debris torrent hazards for design of remedial measures. *Canadian Geotechnical Journal*, **21**, 663–677.

Hutchinson, J.N. and Bhandari, R.K. (1971) Undrained loading, a fundamental mechanism of mudflows and other mass movements. *Géotechnique*, **21**, 353–358.

Ikeya, H. (1981). A method of designation for area in danger of debris flow. *Erosion and Sediment Transport in Pacific Rim Steeplands* (IAHS Publication No. 132, pp. 576–588). International Association of Hydrological Sciences, Wallingford, UK.

Iverson, R.M. (1997) The physics of debris flows. *Reviews of Geophysics*, **35**(3), 245–296.

Jakob, M., Hungr, O., and Thomson, B. (1997) Two debris flows with anomalously high magnitude. In: C-L. Chen (ed.), *Proceedings of the 1st International Conference on Debris-flow Hazards Mitigation: Mechanics, Prediction and Assessment* (pp. 382–394). American Society of Civil Engineers, New York.

Jakob, M., Anderson, D., Fuller, T., Hungr, O., and Ayotte, D. (2000). An unusually large debris flow at Hummingbird Creek, Mara Lake, British Columbia. *Canadian Geotechnical Journal*, **37**, 1109–1125.

Jakob M., Bovis, M., and Oden, M. (2004a) Estimating debris flow magnitude and frequency from channel recharge rates. *Earth Surface Processes and Landforms*, in print.

Jakob, M., Porter, M., Savigny, K.W., and Yaremko, E. (2004b) A geomorphic approach to the design of pipeline crossings of mountain streams. *Proceedings of IPC 2004 International Pipeline Conference, October 4–8, 2004, Calgary* (in print).

Johnson, A.M. (1970) *Physical Processes in Geology* (577 pp.). W.H. Freeman, New York.

Jordan, M. (1994) Debris flows in the Southern Coast Mountains, British Columbia: Dynamic behaviour and physical properties (250 pp.). Ph.D. Thesis (Geography), University of British Columbia, Vancouver.

King, J. (1996) *Tsing Shan Debris Flow* (Special Project Report SPR 6/96, 133 pp.). Geotechnical Engineering Office, Hong Kong Government.

Lau, K.C. and Woods, N.W. (1997) Review of methods for predicting travel distance of debris from landslides on natural terrain (Geotechnical Engineering Office, Hong Kong, Technical Note, TN 7/97, 48 pp.).

Li, J., Jianmo, Y., Cheng, B., and Defu, L. (1983) The main features of the mudflow in Jiang-Jia Ravine. *Zeitschrift für Geomorphologie*, **27**(3), 325–341.

Li, J. and Yuan, J. (1983) The main features of the mudflow in Jiang-Jia Ravine. *Zeitschrift Geomophologie*, **27**, 325–341.

Lopez, J.L., Perez, D., and Garcia, R. (2003) Hydrologic and geomorphological evaluation of the 1999 debris flow event in Venezuela. *Third International Conference on Debris Flow Hazards Mitigation: Mechanics, Prediction and Assessment* (Davos, Switzerland), **2**, 989–1000

Morgenstern. N.R. and Sangrey (1978) Methods of slope stability analysis. In: R.J. Schuster and R.J. Krizek (eds), *Landslides, Analysis and Control* (Special Report 176, pp. 155–171). Transportation Research Board, National Academy of Sciences, Washington, DC.

Okubo, S. and Mizuyama, T. (1981) Planning of countermeasures against debris flow. *Civil Engineering Journal*, **23**(9) [in Japanese].

Okuda, S., Suwa, H., Okunishi, K., Yokohama, K., and Ogawa, K. (1980) Synthetic observation on debris flow. *Annals of the Disaster Prevention Research Institute, Kyoto University*, **24**, 411–448.

Pierson, T.C. (1995) Flow characteristics of large eruption-triggered debris flows at snow-clad volcanoes: Constraints for debris-flow models. *Journal of Volcanology and Geothermal Research*, **66**, 283–294.

Revellino, P., Hungr, O., Guadagno, F.M., and Evans, S.G. (2003) Velocity and runout prediction of destructive debris flows and debris avalanches in pyroclastic deposits, Campania Region, Italy. *Environmental Geology*, **45**, 295–311.

Rickenmann, D., Weber, D., and Stepanov, B. (2003) Erosion by debris flows in field and laboratory experiments. In: D. Rickenman and L-C. Chen (eds), *Proceedings of the 3rd International Conference on Debris-flow Hazards Mitigation: Mechanics, Prediction and Assessment* (pp. 883–893). Millpress, Rotterdam.

Sassa, K. (1985). The mechanism of debris flows. In: *Proceedings of the XI International Conference on Soil Mechanics and Foundation Engineering, San Francisco* (Vol. 1, pp. 1173–1176).

Springer, G.S., Dowdy, H.S., and Eaton, L.S. (2001) Sediment budgets for two mountainous basins affected by a catastrophic storm: Blue Ridge Mountains, Virginia. *Geomorphology*, **37**, 135–148.

Takahashi, T. (1978) Mechanical characteristics of debris flow. *Journal of the Hydraulics Division, ASCE*, **104**(HY8), 1153–1169.

Takahashi, T. (1991). *Debris Flow* (IAHR Monograph, 165 pp.). A.A. Balkema, Rotterdam.

Thurber Consultants Ltd (1983) *Debris Torrent and Flooding Hazards, Highway 99, Howe Sound* (Report, 42 pp.). British Columbia Ministry of Transportation and Highways, Victoria, Canada.VanDine, D.F. (1985) Debris flows and debris torrents in the Southern Canadian Cordillera. *Canadian Geotechnical Journal*, **22**, 44–68.

Wise, M.P. (1997) Probabilistic modelling of debris flow travel distance using empirical volumetric relationships. Master of Applied Science thesis, University of British Columbia, Vancouver.

Wong, H.N., Ho, K.K.S., and Chan, Y.C. (1997) Assessment of consequence of landslides. In: R. Fell and D.M. Cruden (eds), *Proceedings of the Landslide Risk Workshop* (pp. 111–126). A.A. Balkema, Rotterdam.

8

Hyperconcentrated flow – transitional process between water flow and debris flow
Thomas C. Pierson

8.1 INTRODUCTION

Naturally occurring high-discharge flows of water and sediment in open channels (i.e., "floods"), vary over a wide and continuous spectrum of sediment concentration and particle-size distribution. Water floods normally transport mostly fine sediment and in relatively small quantities (as a proportion of total flow volume), with suspended sediment having little effect on flow behaviour and concentrations generally less than 4% by volume (vol%) or 10% by weight (wt%) (e.g., Waananen et al., 1970). At the other end of the spectrum, especially in favourable geologic or geomorphic settings, high-discharge *debris flows* and *mudflows*[1] may transport more sediment than water. In these flows sediment concentrations are often in excess of 60 vol% (80 wt%)[2] (Costa, 1984, 1988; Pierson and Costa, 1987), and the sediment plays an integral role in the flow behaviour and mechanics (Wan and Wang, 1994; Coussot and Piau, 1994; Coussot, 1995; Iverson, 1997; see also Chapter 6). The term *hyperconcentrated flow* is most often applied to flows intermediate between these two end-members, although *debris flood* and *mud flood* have also been used. There is a subtle but important distinction to be considered in choosing the term, however. Debris floods and mud floods, discussed below, are extreme-magnitude, sediment-rich flow events, which may or may not

---

[1] Debris flows and mudflows are similar processes – complex, highly concentrated, pseudo-one-phase gravitational flows of sediment and water (Pierson and Costa, 1987; Wan and Wang, 1994). Debris flows are generally considered to contain more than 50% particles larger than sand size (Varnes, 1978), and varying physical interactions between coarse clasts and between clasts and fluid play significant roles in the flow mechanics (Iverson, 1997; see also Chapter 6). Mudflows (of the type studied by Chinese researchers and discussed in this chapter) are composed predominantly of silt, with some clay and fine sand. A combination of concentration-dependent and shear-rate-dependent, electrochemical and frictional interactions between particles largely determine flow behaviour (Coussot and Piau, 1994).
[2] A particle density of 2.65 g/cm³ is used to compute volumetric concentrations from weight concentrations for kaolinite clay and silicate mixtures of silt, sand, and gravel; 2.3 g/cm³ is used for smectite clay.

involve the physical mechanisms characteristic of hyperconcentrated flows. It will be shown in this chapter that hyperconcentrated flow is a distinct flow process that can occur at low, as well as high, discharges.

Debris flood has been defined in a general sense as "a flood intermediate between the turbid flood of a mountain stream and a true mudflow" (Bates and Jackson, 1987). The approximately synonymous term *mud flood* (Gagoshidze, 1969; Committee on Methodologies for Predicting Mudflow Areas, 1982) has also appeared a few times in the literature. More specifically, a debris flood has been defined as "a very rapid, surging flow of water, heavily charged with debris, in a steep channel" (Hungr et al., 2001; see also Chapter 2), differentiated from debris flow in that it presumably maintains the characteristics of a Newtonian fluid and does not exhibit a surging or pulsating behaviour (Aulitsky, 1980). Although the term is discussed in classification schemes, it is generally not used in the English-language literature to describe actual events.

Floods that would qualify as debris floods or mud floods by the above definitions (but not necessarily labeled as such by the researchers who studied them) might include landslide-dam or log-jam breakout floods or thunderstorm-generated flash floods in relatively steep, narrow canyons. Such floods can move prodigious quantities of mud, sand, and gravel, including very large boulders, owing to high discharge, steep water-surface slopes, and valley constrictions, which keep flows deep and fast enough to produce high bed shear stress and strong turbulence. These floods are typically described by eye-witnesses (e.g., Glancy and Harmsen, 1975; Glancy and Bell, 2000) as: (a) arriving initially as a steep, fast-moving "wall" of boulders, logs, and other debris; (b) being very muddy in appearance and extremely turbulent; and (c) appearing overall as very muddy water, but occasionally having a consistency similar to fresh, wet concrete near the flow front (i.e., possibly including a debris-flow phase). Published examples of such floods include the 1964 dam-failure flood on the Rubicon River, California (Scott and Gravlee, 1968), the 1974 Eldorado Canyon flash flood, Nevada (Glancy and Harmsen, 1975), the 1976 Big Thompson River flash flood, Colorado (Balog, 1978; Costa, 1978), the 1982 Lawn Lake dam-failure flood, Colorado (Jarrett and Costa, 1986), the 1983 Ophir creek flood, Nevada (Glancy and Bell, 2000), and the 1996 Barranco de Aras flood, Spain (Alcoverro et al., 1999; Batalla et al., 1999). While debris floods or mud floods may achieve debris-flow characteristics toward the front of the flood wave or may include a submerged debris-flow phase along the channel bed (e.g., Scott and Gravlee, 1968), debris floods are primarily normal water floods or hyperconcentrated floods that are able to move large quantities of coarse sediment because of high discharges and/or steep channel slopes. Their deposits typically are composed of poorly stratified, loose mixtures of coarse sand and gravel, with the gravels commonly showing both an openwork texture throughout and an imbricated orientation of cobbles and boulders – characteristics not typical of debris-flow deposits. With the exception of the minor debris-flow phases, there are as yet no data nor compelling arguments to suggest that (a) basic bedload transport mechanics are not sufficient to account for the large-scale bedload movement in debris floods or mud floods, or (b) debris floods or mud floods should be classified as a separate process. Bedload transport is proportional to flow velocity raised to the exponent of 3–5, so dramatic increases in coarse-sediment transport can be expected in normal floods where high velocities are maintained and abundant coarse sediment is available (Mizuyama, 1981; Komar, 1988). Thus, the terms *debris flood* and *mud flood* may be descriptively useful to characterize a specific flood event, but they will not be discussed further in this chapter.

Evidence presented in this chapter from field observations and laboratory experiments strongly supports the argument that hyperconcentrated flow should be considered a separate flow process, not just a flow event. Flows with unusually high concentrations of suspended sediment were noted decades ago in streams and rivers in the semiarid and arid areas of the western USA (Lane, 1940; Bondurand, 1951; Nordin, 1963; Richardson and Hanly, 1965), in rivers draining the loess plateau of central China (Todd and Eliassen, 1940), and in freshly disturbed volcanic landscapes (Segerstrom, 1950). However, the term *hyperconcentrated flow* was first coined in 1964 (Beverage and Culbertson, 1964) to distinguish these muddy flows from normal streamflow, because of their tendency to clog irrigation canals and aggrade natural stream beds. Subsequent field and experimental studies have shown that natural hyperconcentrated flows are turbulent, two-phase, gravity-driven flows of water and sediment, intermediate in suspended-sediment concentration between normal sediment-laden streamflow and debris flow or mudflow. They appear to be fundamentally different from these other flow types on the basis of how sediment is transported, although more research is needed. Likewise, they present hazards that are somewhat different from those presented by water flows or debris flows and mudflows. This chapter reviews the various ways in which hyperconcentrated flows have been defined, explores ways in which they are different from other flow types, and synthesizes results from disparate studies to obtain a workable conceptual definition of hyperconcentrated flow.

8.2 DEFINING BOUNDARIES OF HYPERCONCENTRATED FLOW

Problems arise in trying to define hyperconcentrated flow. In water flow, the fluid mechanics are dominated by viscous fluid forces acting on channel boundaries and on individual entrained sediment grains that have little meaningful interaction with each other. In debris flow and mudflow, the flow mechanics involve complex combinations of physical particle interactions (friction and momentum transfer between coarse particles) and electrochemical particle interactions (double-layer and van der Waals attractions between fine particles), physical interactions between the sediment load and the bed, and strong but varying interactions between sediment grains and the fluid (Coussot and Piau, 1994; Coussot, 1995; Iverson, 1997; see also Chapter 6). The spectrum of physical processes at work in water flow and in debris flow or mudflow represents a continuum, and sharp, discrete demarcations between flow types probably do not exist.

Beverage and Culbertson (1964) initially defined hyperconcentrated flow as having a suspended sediment concentration of at least 20 vol% (40 wt%) and not more than 60 vol% (80 wt%). While these authors described some of the unique properties of hyperconcentrated flow, the boundary values were assigned without objective criteria that could be transferred to other settings. Since then, many other authors have simply taken these concentration limits and applied (or misapplied) them. In addition to sediment concentration, two other types of criteria have been used to distinguish hyperconcentrated flow: (1) criteria based on the bulk rheological properties of the suspensions, and (2) criteria based on how (and how much) sand is suspended and deposited in the flow. These approaches are discussed below. So far, a single, precise, comprehensive, and commonly accepted definition of hyperconcentrated flow has remained elusive.

8.2.1 Rheological criteria

Rheology, the study of the deformation and flow of materials, examines the bulk behaviour of two-phase sediment water mixtures, within which complicated interactions between solid and fluid forces take place during flow. So while rheology may ignore the mechanistic details, rheological definitions of flow type have the advantage that tests sometimes can be performed in the laboratory (or observations made in the field) to define flow type.

Water behaves as a Newtonian fluid,[3] even when mixed with up to 35 vol% sand or gravel-size particles (Fei, 1983; Wan and Wang, 1994). However, smaller amounts of *fines*,[4] especially clay, added to a suspension will cause the onset of measurable yield strength[5] (Figure 8.1), marking the onset of non-Newtonian fluid behaviour. This transition from a Newtonian to a non-Newtonian fluid has been used by some authors to define the lower threshold of hyperconcentrated flow (Qian et al., 1981; Pierson and Costa, 1987; Rickenmann, 1991; Xu, 2002b, 2003). The upper threshold, namely the transformation to debris flow or mudflow, can also be defined in terms of yield strength – the point at which mixture yield strength, combined with buoyancy, is sufficient to fully suspend particles recognizable in the field as gravel,[6] whether or not the flow is moving (Pierson and Costa, 1987). A minimum yield strength of about 60 Pa is required, in addition to the buoyancy provided by the fluid, to suspend a 4 mm mineral grain in a static mixture of clay and water (Hampton, 1975), although turbulence in a flowing mixture can decrease the yield strength by breaking flocculent structures (Coussot and Piau, 1994; Wan and Wang, 1994). Support of coarse gravel in static debris-flow mixtures requires the additional support mechanism of grain-to-grain frictional contact (Pierson, 1981), and

---

[3] Newtonian fluids have a constant viscosity with respect to shear rate and will flow in response to any applied shear stress (i.e., they do not have internal strength to resist flow).
[4] "Fines" are considered by North American researchers to be particles <0.62 mm (i.e., the full range of silt and clay-size grains). In China, however, where much of the work on hyperconcentrated flow has been done, "fines" are considered to be only particles <0.01 mm (i.e., fine silt and clay). Unless otherwise stated, the North American convention will be used in this chapter.
[5] Yield strength (also termed shear strength) is the internal resistance of the sediment mixture to shear deformation; it is the result of friction between grains and cohesion.
[6] Gravel here refers to particles >4 mm diameter, because particles 2–4 mm are sometimes labeled "grus" and are hard to differentiate from sand.

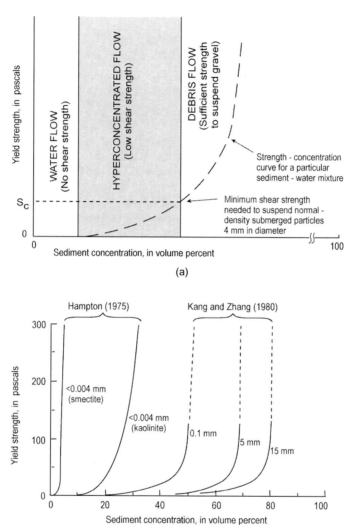

Figure 8.1. Yield strength of sediment–water mixtures as a function of suspended sediment concentration. (a) Definitions of flow type based on an idealized yield-strength curve of a poorly sorted sediment-water mixture. (b) Range in measured rheological properties of various sediment–water mixtures resulting from differences in grain-size distribution (average grain size indicated, in mm).

(b) From Pierson and Scott (1985).

increasing granular friction between more closely packed grains may explain the exponential steepening of yield strength curves (Figure 8.1) in response to increasing sediment concentration. Moving debris flows, however, involve a complex and variable interplay between solid grain forces and viscous fluid (Iverson, 1997, 2003; Iverson and Vallance, 2001, see also Chapter 6).

8.2.2 Sand suspension and settling criteria

Hyperconcentrated flows also have been characterized as flows in which large quantities of sand are transported in full dynamic suspension once minimum concentrations of fines (clay and fine silt) are achieved (Cao and Qian, 1990; Rickenmann, 1991; Dinehart, 1999). Field observations (Cronin et al., 1999) indicate that some fine gravel may be included with the sand. The point at which the proportion of sand in suspension abruptly increases relative to the suspended fines can also be used to define the lower boundary of hyperconcentrated flow. Hyperconcentrated flows characteristically have sand concentrations that greatly exceed the fines concentrations.[7] This sudden increase in the effectiveness of sand suspension is in agreement with a model for transition from low-concentration suspensions to intermediate or moderate-concentration suspensions (Druitt, 1995; Major, 2003), whereby upwelling of fluid displaced by the downward settling of some of the grains adds an additional buoyancy mechanism to the mixture, allowing the suspension to hold more sand. A critical part of this definition is that the sand and gravel still can settle out of the water column whenever energy of the flow decreases. Hyperconcentrated-flow deposits, therefore, commonly acquire some stratification and better sorting than the suspensions from which they settle out.

If sediment continues to be added to a hyperconcentrated mixture, however, a point is eventually reached whereby sand and gravel grains in the suspension begin to significantly interact with each other and frictional forces between grains hinder selective settling from the fluid suspension when the flow slows or stops (Druitt, 1995; Major, 2003). Frictional contact between grains prevents the larger and denser grains from settling faster than the surrounding finer and lighter grains. Consequently all the grains settle at the same rate and the result is an unsorted, unstratified deposit. Chinese researchers refer to this as the transition from "heterogeneous hyperconcentrated flow" to "homogeneous hyperconcentrated flow" or mudflow (Qian et al., 1981; Cheng et al., 1999; Cao and Qian, 1990). For suspensions of sediment from the Yellow River, this upper boundary is at about 19 vol% (Qian et al., 1981), but for coarser, more poorly sorted mixtures the boundary is between 50 and 55 vol% (Pierson, 1986; Cronin et al., 1999; Major, 2003).

8.3 CHARACTERISTICS OF HYPERCONCENTRATED FLOW

8.3.1 Initiation

Hyperconcentrated flows can initiate when water floods acquire added suspended sediment through erosion and entrainment or when debris flows lose coarse sediment through dilution and selective deposition. Documented initiation mechanisms include hillslope and channel erosion during intense rainstorms (Beverage and Culbertson, 1964; Major et al., 1996; Pierson et al., 1996), lake-breakout floods

---

[7] Some water floods may carry more sand than fines in suspension, but generally the fines fraction is significantly greater than the sand fraction (cf. Waananen et al., 1970).

(Rodolfo et al., 1996; O'Connor et al., 2001), glacier-outburst floods (Maizels, 1989), dilution and/or selective deposition at the heads and tails of debris flows (Pierson, 1986; Pierson and Scott, 1985; Cronin et al., 1999, 2000), and inputs of large sediment volumes to water floods by landslides (Kostaschuk et al., 2003). No matter what the water source is, an ample supply of easily erodible, relatively fine-grained sediment is critical. Most naturally occurring hyperconcentrated flows occur as floods (i.e., have higher than normal discharges), but in basins where sediment is extremely erodible they readily occur under low-flow conditions as well (Beverage and Culbertson, 1964; Montgomery et al., 1999). Studies have shown that there is often no direct relation between discharge and suspended-sediment concentration in natural occurrences of hyperconcentrated flow (Beverage and Culbertson, 1964; Komar, 1988; Alexandrov et al., 2003).

Hyperconcentrated flows occur commonly in semiarid and arid regions, particularly where basins are steep, hillslopes are erodible, channel banks are fragile, and channel beds are unarmored and erodible (Gerson, 1977; Laronne and Reid, 1993; Laronne et al., 1994). Although suspended sediment concentrations during floods in this terrain normally are in the concentration range of 0.5–5 vol%, much higher concentrations (in excess of 40 vol%) occur during extreme discharge events, during first runoff events of the season, or when landslides enter streams during floods (Lane, 1940; Beverage and Culbertson, 1964; Wannanen et al., 1970; Gerson, 1977). The loess plateau of central China has an especially high incidence of hyperconcentrated flows due to the combination of abundantly available fine sediment (thick deposits of eolian silt and fine sand), a subhumid to semiarid climate, and relatively steep, deeply incised channels resulting from regional tectonic uplift (Todd and Eliassen, 1940; Cao and Qian, 1990; Wang, 1990; Cheng et al., 1999; Xu, 1999). This region produces rainfall-runoff floods with maximum concentrations occasionally exceeding 50 vol%. Disturbed watersheds on volcanoes that have had recent explosive eruptions also tend to produce hyperconcentrated flows, owing to the widespread distribution of uncompacted fine tephra deposits (Segerstrom, 1950; Waldron, 1967; Scott, 1988; Major et al., 1996; Pierson et al., 1996; Rodolfo et al., 1996).

Two factors (found primarily at volcanoes) appear to be prerequisite for hyperconcentrated flows to evolve from debris flows: (1) the debris flows need to be sufficiently large to flow for long enough distances for flow transformations to occur, and (2) the debris mixtures must be relatively poor in fines (clay-rich debris flows tend not to transform). Where these conditions have been met, transformation from debris flow to hyperconcentrated flow has been inferred or documented (Janda et al., 1981; Pierson and Scott, 1985; Scott, 1988; Major and Newhall, 1989; Smith and Lowe, 1991; Scott et al., 1995; Major et al., 1996; Cronin et al., 1999, 2000; O'Connor et al., 2001).

8.3.2 Rheology

Viscosity is the property of a fluid that slows the settling of suspended particles and allows it to resist shear deformation, thus controlling the rate of shear or flow.

Viscosity increases with increasing sediment concentration in aqueous suspensions of silt and clay (Cao and Qian, 1990; Julian and Lan, 1991), and the effect can be dramatic. Suspensions of silt and clay containing between 15 and 45 vol% solids can have dynamic viscosities 1.5–4 orders of magnitude greater than the viscosity of clear water (Julian and Lan, 1991).

While Newtonian fluids such as clear water are deformable under any applied stress and have a constant viscosity that is independent of shear rate, suspensions of sediment and water commonly do not show either of these traits – viscosity can change with shear rate, and some minimum stress may be required to deform some mixtures. There is an extensive literature on the rheometry of fine-particle suspensions and a number of rheologic models have been proposed to explain their behaviour under applied stress (see Major, 1993 for a summary). Despite reported complexities in rheologic behaviour, flow of natural silt/clay suspensions is commonly considered to be reasonably well approximated by the Bingham model over naturally occurring ranges of grain-size distribution, sediment concentration, and shear rate (including rates representative of natural open-channel flows) (Fei, 1983; Wang et al., 1983; Engelund and Wan, 1984; Cao and Qian, 1990; Phillips and Davies, 1991; Major and Pierson, 1992; Wan and Wang, 1994). The Bingham model predicts a constant Bingham viscosity (μ_B), after a finite yield stress has been exceeded and can be represented as:

$$\tau = s + \mu_B \frac{du}{dy} \qquad (8.1)$$

where τ is shear stress, s is yield stress, and du/dy is the velocity gradient. Barnes and Walters (1985) argue, however, that a true yield stress does not exist in fine-particle suspensions and that apparent yield stress disappears at very low shear rates ($<10^{-4}\,\mathrm{s}^{-1}$), although such low shear rates may be irrelevant for natural flows.

In most natural fine-grained mixtures, laboratory measurement of shear stress over a wide range in shear rate (>3 orders of magnitude) results in demonstrable shear thinning (i.e., decrease in viscosity with increase in shear rate), which is probably due to the shear-induced breakdown of interconnected floc networks (Coussot, 1995). At relatively high shear rates (e.g., $100\,\mathrm{s}^{-1}$), even in mixtures containing some sand, particle interactions are minimized, and the viscosity of the interstitial fluid controls energy dissipation. At low shear rates ($<5\,\mathrm{s}^{-1}$), disrupted floc networks can reconnect (Coussot, 1995) and sand grains begin to physically interact (Major and Pierson, 1992). Both of these phenomena also cause the mixture rheology to deviate from the Bingham model by increasing apparent viscosity. This control of viscosity by shear rate can be better modeled by the Herschel–Bulkley model (Coussot and Piau, 1994; Coussot, 1995):

$$\tau = s + \mu \left(\frac{du}{dy}\right)^n \qquad (8.2)$$

where μ is variable viscosity and the exponent n ($0 \leq n \leq 1$) defines the rate of shear thinning. When $n = 1$, this equation becomes the Bingham model.

8.3.3 Turbulence and velocity distribution

Much of the research into the effect of sediment loads on turbulence structure and velocity distribution in open-channel flow has been carried out on clear-water flows transporting large quantities of relatively coarse sand (no fines in suspension), and the conclusions are not always in agreement (discrepancies summarized by Wang and Qian, 1989; and Cao et al., 2003).

Limited experimental work, specifically with turbulent hyperconcentrated flows (large quantity of fines in suspension), suggests that vertical velocity distributions are essentially logarithmic in shape as they are in clear-water streamflow (Zhang and Ren, 1982; Yang and Zhao, 1983; Zhou et al., 1983), although experimental work by Bradley (1986) indicates that a power-law function may also adequately describe velocity profiles in hyperconcentrated flow. Overall, turbulence is somewhat dampened in hyperconcentrated flows by the higher fluid viscosity, resulting in less frictional loss of energy (van Rijn, 1983; Yang and Zhao, 1983), although flows transporting coarse bed load experience greater turbulence intensity near the bed and thus greater energy loss (Wang and Larsen, 1994). This effect of bedload intensifying near-bed turbulence has also been noted in clear-water flows (at normalized depths < 0.2). Causes of near-bed turbulence have been ascribed to greater bed roughness height, eddy shedding from large grains, grain inertial effects, and grain/bed form interactions (Best et al., 1997). Relative size of the bedload particles is an important variable (Cao et al., 2003). Turbulence has also been shown to increase when shallow hyperconcentrated flow moves over bed forms that have been immobilized by clay impregnation (Simons et al., 1963, p. G31). Decrease in flow velocity or increase in sediment concentration or suspended-particle size can alter the turbulence structure of flow and cause flow stratification to develop, whereby the upper part of the flow may become laminar and re-acquire a higher yield strength (floc networks re-establish) while flow near the bed remains turbulent (Wan and Wang, 1994; Wang and Larsen, 1994; Cao et al., 2003). Turbulence induced by coarse bed load can disrupt this stratified structure, however, and re-establish turbulence throughout the fluid column (Wang and Larsen, 1994). Rigid plugs (zones with zero velocity gradient) may develop in the upper, laminar part of the flow if concentrations approach the debris-flow/mudflow concentration threshold or the flow slows (Zhang and Ren, 1982; Yang and Zhao, 1983; Wilson, 1985) (Figure 8.2). Velocity distributions beneath such plugs remain logarithmic but with an increased velocity gradient (Yang and Zhao, 1983; Wan and Wang, 1994).

Field observations support the conclusion that many natural hyperconcentrated flows move principally as fully turbulent flows but with velocities near the river bed deviating from the logarithmic distribution (Zhou et al., 1983). Some damping of turbulence is noted, however (Pierson and Scott, 1985; Dinehart, 1999; Pringle and Cameron, 1999; Cronin et al., 1997, 1999, 2000). Large-scale boils, eddies, hydraulic jumps, and breaking antidune waves can be seen, but the small-scale choppy waves and splashing common waves in clear-water flows of equivalent discharge are commonly diminished or absent in relatively deep flows, and flow surfaces are often described as having an oily sheen (Figure 8.3(a)). One early observer (Pierce,

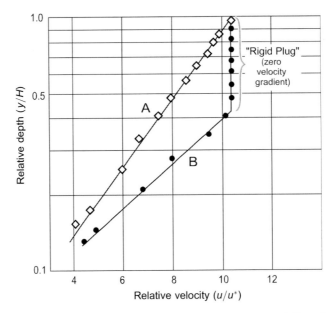

Figure 8.2. Vertical velocity profiles of experimental hyperconcentrated flow showing both a fully logarithmic profile (a) and a profile with a rigid plug above and a logarithmic profile below (b).
From Yang and Zhao (1983).

1917, p. 41) described one such flow at $Q = 370\,\text{m}^3/\text{s}$ and having close to 50 vol% suspended sediment: "... the river ran with a smooth, oily movement and presented the peculiar appearance of a stream of molten red metal instead of its usual rough, choppy surface." Rapid shallow flows do not exhibit this oily smooth surface, probably due to turbulence induced by coarse bedload or clay-encrusted bed forms (Figure 8.3(b)). Stratified flows probably also occur in the field, particularly at high concentrations, where turbulence near the surface vanishes while a turbulent layer is maintained near the bed. In Chinese rivers rigid plugs have been observed to form and thicken sufficiently to interact with the bed and bring flow in the channel to a standstill (Qian et al., 1981).

8.3.4 Sediment transport

Suspended load and the role of fines

There are two types of suspended load in hyperconcentrated flow, as in water flow: (1) fines (*wash load*) which form a stable suspension relative to the duration of flow and remain in suspension independent of concentration, flow velocity, or flow discharge (Cao and Qian, 1990; Xu, 1999), and (2) coarser particles (*intermittently suspended bed-material load* or simply *intermittently suspended load*) which remain temporarily in dynamic suspension for long periods of time relative to

Sec. 8.3] **Characteristics of hyperconcentrated flow** 169

Figure 8.3. Examples of smooth and rough flow surfaces of hyperconcentrated flows in Pasig-Potrero River, Mount Pinatubo, Philippines. Both were identified as hyperconcentrated by the high concentrations of suspended sand, but concentrations were not measured. (a) Hyperconcentrated flow (lahar) spilling over highway and creating a hydraulic jump on downstream side; (b) shallow hyperconcentrated flow on a sand bed with randomly choppy surface, possibly due to fines-crusted bed forms resisting transformation to a higher order form as noted in experiments by Simons et al. (1963, p. G31).

their size during flow (Komar, 1988; Bridge, 2003). Sand grains may only occasionally contact the bed while saltating cobbles may stay suspended for only seconds. However, all of the coarser particles that make up the intermittently suspended load will settle out of suspension as flow slows or stops. Although these two types of suspended load are present in both hyperconcentrated flow and water flow, the mean grain size of the intermittently suspended load is coarser in hyperconcentrated flow, much higher concentrations of intermittently suspended load are achieved in hyperconcentrated flow, and the relative concentration of intermittently suspended sand with distance from the bed is typically uniform in hyperconcentrated flow but strongly stratified in water flow. Intermittently suspended particles may be as large as small boulders in some hyperconcentrated flows (Cronin et al., 1999).

Concentration of fines appears to play a dominant role in the mechanics of transport of the intermittently suspended load. Field and experimental evidence suggest that high concentrations of sand cannot be transported in suspension in hyperconcentrated flows *unless* a minimum fines concentration is first achieved (Beverage and Culbertson, 1964; Gerson, 1977; Wan and Song, 1987). In this sense, the water plus wash load mixture can be considered to be the "carrier fluid" for the coarser suspended load (Wilson, 1985; Cao and Qian, 1990). Minimum wash-load concentrations appear to vary with the grain-size distribution of both types of suspended load (Table 8.1). At Mount St. Helens for example, a

Table 8.1. Minimum "fines" concentrations required to permit hyperconcentrated flows to suspend large quantities of sand. It should be noted that Cronin et al. (1999) measured 5 vol% suspended fines in measured flow LH7, which did not make the transition to hyperconcentrated flow (i.e., the relative amount of suspended sand remained low).

Location	Minimum fines required[a] (in wt% solids)	Minimum fines required[b] (in wt% solids)	Minimum fines required[c] (in vol% solids)	Reference
New Mexico and Arizona (Rio Puerco, Rio Salado, Paria River, Little Colorado River)		~15	~6	Beverage and Culberson (1964)
Toutle River, Mount St. Helens, Washington		~18*	~8	Dinehart (1999)
Yellow River basin, China	3–5*	Estimated 7–23 from range of grain-size curves (Qian et al., 1981)	3–10	Xu (2002b)

[a] "Fines" = particles ≤ 0.01 mm.
[b] "Fines" = particles ≤ 0.062 mm.
[c] "Fines" = particles ≤ 0.062 mm.
* Concentration values read from graphs of fines concentration with corresponding sand concentration, at the point at which sand concentration begins to increase significantly.

fines concentration of about 8 vol% is required before significant increases in suspended sand transport will occur, while at fines concentrations below 6 vol%, relatively little sand is suspended (Dinehart, 1999). During the passage of hyperconcentrated flow flood waves, fines concentrations tend to reach a maximum and fairly constant "plateau level" (even though sand concentrations may vary greatly), which may be due to supply limitations for the fines (Dinehart, 1999; Xu, 2002b). These maximum levels vary for different flows in different regions.

The roles played by the fines in helping keep sand in suspension during hyperconcentrated flow are complex. For decades it has been observed that high concentrations of suspended fines decrease fall velocity[8] of sand grains by increasing both fluid density (i.e., increasing buoyancy) and fluid viscosity (Simons et al., 1963; Nordin, 1963; Beverage and Culbertson, 1964; Xu, 1999). Dynamic viscosity can increase by as much as four orders of magnitude in going from clear water to a fine-sediment suspension of 45 vol% solids (Julian and Lan, 1991). Fall velocities for fine sand (0.125 to 0.25 mm) are reduced by an order of magnitude, whereas those for coarse sand (0.5 to 1.0 mm) are cut in half by fines concentrations typically found in hyperconcentrated flows in Arizona and New Mexico (Beverage and Culbertson, 1964). Bentonite suspensions of 4 vol% solids reduced fall velocity of various natural sands (0.1 to 1 mm diameter) 2.5 to 4 times (Simons et al., 1963). Thus, both yield strength and viscosity play important roles. Yield strength maintains the integrity of the carrier fluid by keeping fines permanently in suspension (Wan and Wang, 1994), and increased viscosity slows the settling of sand and coarser particles (Wang et al., 1983; Wan, 1985). Not coincidentally, yield strength is acquired at about the same point that sand concentrations begin to increase dramatically (Wang et al., 1983).

Suspension of coarse sand and gravel must rely on dynamic processes in addition to the fluid properties that hinder grain settling. These have not been studied in detail, but probably include increased drag provided by turbulent eddies in the carrier fluid, and upward dispersive stress provided by collisions between saltating grains and the bed and between grains themselves (Bridge, 2003). Also, experiments have shown that the quantity of suspended coarse load increases with increasing discharge (i.e., with increasing velocity and turbulence) (Wang, 1990), suggesting that the dynamic mechanisms of turbulence and grain collisions must play very important roles.

Coarse particles in hyperconcentrated flows will stay dynamically suspended so long as the sum of fluctuating upward-directed fluid flow and momentum transfer from grain collisions is greater than the particle fall velocity. Some of the upward flow results from turbulence and some from convection. Upward-directed fluid turbulence in hyperconcentrated flows, especially the action of macroturbulent vortices, observed in natural flows (Pierson and Scott, 1985; Dinehart, 1999), probably is fundamental in getting sand into suspension and helping to keep it there (Wilson, 1985; Komar, 1988). Convection occurs as a direct result of particle

---

[8] Constant fall velocity is reached when the submerged weight of a particle is balanced by the drag force imposed by the fluid (Wan and Wang, 1994).

Figure 8.4. Surface of hyperconcentrated flow revealing reticulate pattern apparently resulting from elutriation of fines at margins of convection cells caused by downward sinking coarse suspended load. Flow was the LH5 lahar in Whangaehu River (Ruapehu volcano, New Zealand), 27 September, 1995, 0945 hr, 42 km from source. Sediment concentration was 46 vol% (see Cronin et al., 1999).

Photo by Shane J. Cronin (Massey University).

settling – irregular cylindrical or planar zones of fluid are displaced upward by downward settling of larger, denser particles (Major, 2003). Such upward counterflow can elutriate fines from within the suspension and carry them to the flow surface (Cronin et al., 1999; Major, 2003), sometimes resulting in the formation of swirling reticulate patterns on the flow surface (Figure 8.4).

Bedload

Bedload (or contact load) is the sum of all sliding or rolling particles that stay in more or less continuous contact with the bed, as well as the saltating particles that move close to the bed and are frequently in contact with it. In hyperconcentrated flow, the boundary between bedload and intermittently suspended load cannot be sharply defined. Little is known about bedload transport during natural hyperconcentrated flows, because standard bedload samplers cannot be deployed in flows with such high drag forces. It has been noted that the sound of boulders moving on the bed increases as the total suspended load increases (Pierson and Scott, 1985;

Dinehart, 1999). Cronin et al. (1997) observed boulders up to 2 m in diameter being rolled by hyperconcentrated flow having a sediment concentration of 37 vol%.

Recent theoretical and experimental work on high-concentration particle transport suggests that bedload sediment in hyperconcentrated flow may be transported in a concentrated zone of intense bed shear that has been referred as a *traction carpet* (Hanes and Bowen, 1985; Todd, 1989; Sohn, 1997). Observers who have reached hands into shallow (≤ 1 m) hyperconcentrated flows have verified that an increasingly dense zone of moving bed material can be felt near the bed (Dinehart, 1999). A dense basal underflow layer (akin to a submerged debris flow beneath a more dilute flow) has also been inferred from field relationships (Scott and Gravlee, 1968; Cronin et al., 2000; Manville et al., 2000). In water flows, this dense zone of moving bed material has been subdivided into two zones: an upper *saltation* zone and a lower *collisional grain flow* zone (Hanes and Bowen, 1985). A somewhat similar model is used to describe motion within dry grain flows (e.g., Drake, 1990); it includes (1) an upper *collisional zone*, which is characterized by large gradients in particle concentration and velocity, active grain collision, high granular temperature, and generation of dispersive pressure; and (2) a lower *frictional* zone, which is a compact layer of slowly moving grains that are entirely in frictional contact with each other. Both models have been combined by Sohn (1997) in a conceptual model for bedload movement beneath turbulent overlying flows (Figure 8.5). It has been noted that variation in thickness of the total bedload shear layer, regardless of its internal mechanics, affects flow resistance and may lead to flow instability for hydraulic reasons (Wilson, 1985).

Vertical distribution of sediment

The vertical distribution of suspended sediment concentration is relatively uniform in turbulent hyperconcentrated flow – much more so than in normal streamflow (Qian et al., 1981; Yang and Zhao, 1983; McCutcheon and Bradley, 1984; Bradley, 1986; Rickenmann, 1991), and this uniformity of distribution depends on turbulence (Durand and Condolios, 1952). The concentration distribution is size-dependent (Figure 8.6) and can be predicted by the Rouse equation for distribution of relative concentration (Nordin, 1963, p. C10). Nevertheless, concentration will increase near the bed, and there may be no real demarcation between bedload and suspended load in the near-bed region, where there should be a continuous exchange of particles between suspended load and bedload (Wilson, 1985). As flow velocity decreases: (a) particles settle out of suspension according to size, (b) the average size of the intermittently suspended load becomes finer, (c) suspended sediment becomes progressively more concentrated near the bed, and (d) flow becomes more stratified (Wilson, 1985).

Transport rates

It has been well demonstrated that suspended-load transport rates for turbulent water flow and hyperconcentrated flow increase with increasing fine-material concentration (Simons et al., 1963; Kikkawa and Fukuoka, 1969; Wan, 1982; Bradley,

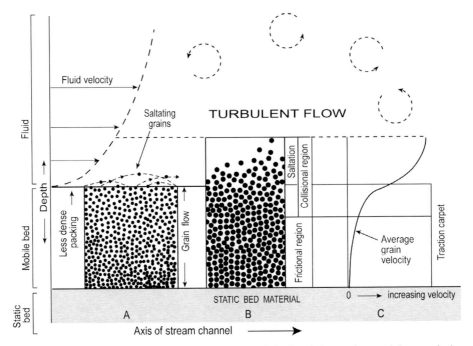

Figure 8.5. Generalized models for movement of bedload beneath overlying turbulent, shearing flow, which are presumed to describe mechanisms operative in a traction carpet in hyperconcentrated flow. (a) Model of Hanes and Bowen (1985) that postulates an internally shearing, collision-dominated granular fluid region beneath a turbulent fluid shear region. (b) Model of Drake (1990), based on high-speed photography of dry grain-flow experiments, that identifies (1) a basal frictional region with enduring frictional contacts between grains, which has a lower quasistatic zone and an upper block-gliding zone with groups of grains moving as coherent blocks, and (2) an overlying collisional region that is subdivided into a lower grain-layer gliding zone, a middle chaotic zone, and an upper saltational zone. (c) The conceptual synthesis of models A and B (Sohn, 1997) that predicts non-linearly increasing average velocities of grains with height above the static bed, up to the top of the traction carpet.

1986; Wan and Song, 1987; Dinehart, 1999). The average particle size of transported sand also increases with increasing fines concentration (Cao and Qian, 1990; Xu, 1999). Beverage and Culbertson (1964) predicted overall transport rate increases of about 500% for hyperconcentrated flow above water-flow transport rates. For over an hour during the 1982 hyperconcentrated flow at Mount St. Helens, the quantity of sediment in suspension was between 15 and 40 times more than during normal floods of similar magnitude (Pierson and Scott, 1985; Dinehart, 1999). Wan and Song (1987) predict overall increases in sediment transport rate of up to nearly 100 times more than normal streamflow.

Bedload transport rates are higher for hyperconcentrated flow than normal streamflow, as well. Bradley and McCutcheon (1985) calculated that bedload transport rate should increase between 50–70% for suspended sediment concentrations of 20 vol%, using well-known sediment transport formulas. Rickenmann (1991)

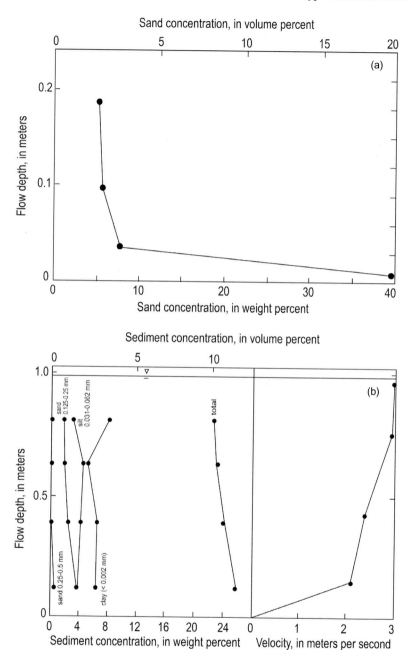

Figure 8.6. Vertical distribution of suspended sediment concentration in hyperconcentrated flows. (a) Experimental flow using sand ($D_{50} = 0.18$ mm) added to bentonite clay suspension of 3.8 vol% concentration. (b) Apparent natural hyperconcentrated flow on Rio Puerco near Bernardo, New Mexico, station 185, 20 September, 1961.

(a) From Bradley (1986). (b) From Nordin (1963).

experimentally achieved bedload transport rates up to about 300% above clear-water rates, as fines concentration was increased up to a limiting value of 17 vol%. At concentrations above this limiting value, the fluid was less effective in transporting bedload. In turbulent flow, Rickenmann attributed this increase in bedload transport capacity to the efficacy of increased fluid density. Bradley's (1986) experiments showed an increase of two orders of magnitude in bed-material transport rates.

8.3.5 Bed material, bed forms, and flow resistance

Resistance to flow in an open channel is affected both by the internal properties of the fluid itself (fluid density and viscosity), by the shape and sinuosity of the channel, and by the drag imposed by the roughness of the bed. In alluvial channels, bed roughness can be further subdivided between particle roughness (size and shape of individual particles on the bed, mobile or stationary) and form roughness (shape and size of bed forms such as ripples, dunes, and antidunes).

Flume experiments have indicated that the roughness of the stationary bed and the presence or absence of relatively coarse bedload are important to flow resistance in hyperconcentrated flows. Hyperconcentrated flows flowing over rough beds experience less drag than clear-water flows at the same flow rate on the same beds, because the fines reduce turbulence intensity and suppress small eddies, therefore less energy is consumed and flow velocities are higher (Wan and Wang, 1994). Increasing the sediment concentration (in the range 3–9 vol%) can reduce flow resistance further. Even on smooth (sand) beds, flow resistance of the hyperconcentrated flow without coarse bedload can be less than that of the clear-water flow by about a factor of two (Shu et al., 1999). Also, it was found that flow resistance decreased with increasing suspended sediment concentration, but only until a minimum flow-resistance value was reached at a concentration of about 11–15 vol%; at concentrations above this threshold, flow resistance increased with increasing concentration (Figure 8.7). Yang and Zhao (1983) also concluded that flow resistance was commonly less in hyperconcentrated flow than in clear-water flow. With coarse bedload present that requires energy to transport, however, flow resistance is greater than in a similar flow without the bedload. The presence of coarse bed load increases turbulence intensity throughout the flow depth, transfers energy from mean motion to turbulence, and slows mean velocity (Wang and Larsen, 1994).

The effect of bed forms (form roughness) on flow resistance has been investigated for flows of clay–water suspensions at various concentration ranges: 0.4–1.0 vol% (Wan, 1985), 3–4 vol% (Simons et al., 1963), and 1–16 vol% (Wan and Song, 1987). In the lower flow regime (subcritical flow), a higher discharge is required to initiate particle motion on the bed for clay–water suspensions than for clear water, but the transformation from dunes to plane bed occurs at progressively smaller discharges as fines concentration is increased (Simons et al., 1963; Wan, 1985). In addition, the dunes that formed in flows with high concentrations of suspended clay were lower, smoother, more spaced out, and more symmetric, thus providing less

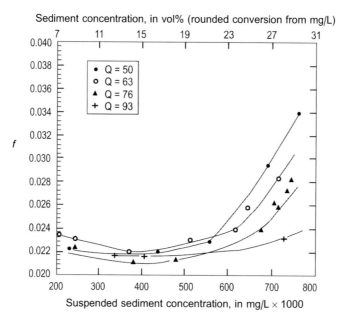

Figure 8.7. Relation between friction factor (f) and suspended sediment concentration for experimental flume runs at flow discharges between 50 and 93 L/s.
From Shu et al. (1999).

form resistance than those formed under similar flow conditions by clear water (Simons et al., 1963; Wan, 1985; Wan and Song, 1987). Thus in the lower flow regime, form roughness decreased with increasing suspended sediment concentration. However, the opposite effect was observed in supercritical flow – form roughness and flow resistance increased with increasing suspended sediment concentration. At constant discharge, the addition of fines caused standing waves to transform to breaking antidunes (with accompanying increase in turbulence and sediment transport) (Simons et al., 1963). This transition can increase flow resistance by a factor of 2–3 (Simons and Richardson, 1966). It was noted that the high fines concentrations could cause stabilization (crust formation) of ripple and dune bed surfaces due to bed impregnation by the fines. These clay-encrusted bed forms had to be broken up by higher flow discharges before higher order bed forms could be constructed (Simons et al., 1963).

Transverse and downstream-pointing V-shaped standing waves (non-breaking and breaking) have been observed in natural hyperconcentrated flows (Pierson and Scott, 1985; Dinehart 1999; Cronin et al., 1999, 2000). A progression in bed-form development was noted during flow (Dinehart, 1999), where apparent plane-bed conditions changed to standing waves and then to breaking antidunes while suspended sediment concentration increased and flow discharge decreased, supporting the experimental findings noted in the previous paragraph. Development of chutes and pools (bed forms providing still greater flow resistance) has also been observed in natural hyperconcentrated flows (cf. Pringle and Cameron, 1999).

The experimental work cited above predicts that hyperconcentrated flow should flow faster than clear-water flow for the same depth and slope, so long as flow remains predominantly in the lower flow regime and transported bedload is not excessively coarse. Furthermore, other experiments have shown that mean flow velocity in open-channel flow increases with increasing suspended-sediment concentration even before hyperconcentration levels are reached (Vanoni, 1946; Vanoni and Nomicos, 1960). These experimental conclusions are supported by field measurements at Mount St. Helens, where a high-discharge hyperconcentrated flow in 1982 had a mean velocity approximately 10% faster than several previous normal-concentration floods of approximately equivalent depth and discharge at the same gauging station (Pierson and Scott, 1985).

8.4 EXAMPLES OF OBSERVED HYPERCONCENTRATED FLOWS

Hyperconcentrated flows are relatively common in volcanic terrains recently impacted by explosive eruptions, and the best documented examples have been observed in volcanic terrains. In such settings rapid releases of water from a variety of sources will typically erode and incorporate exceptionally large volumes of unconsolidated deposits on hillslopes and in channels, and then flow downstream as *lahars*.[9] The following are three examples of lahars that were, for much of their flow paths, hyperconcentrated flows (Table 8.2), identified as such by their dynamic suspension of sand and fine gravel in large quantities. Equivalent data for non-volcanic hyperconcentrated flows have not been published.

8.4.1 1982 Hyperconcentrated lahar, Toutle River, Mount St. Helens, USA

The best documented example of a hyperconcentrated flow in North America is an eruption-triggered lahar that occurred on 19–20 March, 1982, in the Toutle River downstream from Mount St. Helens (Pierson and Scott, 1985; Scott, 1988; Dinehart, 1999; Pierson, 1999). It was caused by an explosive eruption that rapidly produced about $4 \times 10^6 \, \text{m}^3$ of meltwater, forming a temporary lake in the crater. Breakout of the lake produced a pumice-charged water flood that eroded deeply into crater-floor volcaniclastic sediment and bulked to debris-flow concentration as it flowed out of the crater. Dip samples collected from the surface of the flow at three US Geologic Survey gauging stations 49, 73, and 81 km downstream from the source verified that it had transformed from debris flow to hyperconcentrated flow by the time it reached these stations. Estimated and measured peak discharges were about 960, 650, and 450 m³/s, respectively – a flood peak attenuation of more than 50% over this 32-km reach. Average peak-flow depths ranged from 2.0 to 3.5 m. This example differs significantly from examples in central China in that the input of fines occurred primarily at or near source, and that the initial sediment had an extremely wide

---

[9] Lahars are "rapidly flowing mixtures of rock debris and water (other than normal streamflow) from volcano" (Smith and Fritz, 1989; Smith and Lowe, 1991). They can be hyperconcentrated flows or debris flows, and they commonly transform from one to the other. This definition is now gaining wide acceptance over previous usage that equated "lahar" with "volcanic debris flow".

range of particle sizes. High discharge hyperconcentrated flows in central China involve primarily silt and fine sand, and sediment inputs occur throughout the drainage basin wherever rainfall is contributing runoff.

Flood hydrographs and sediment sampling from the three stations (Figure 8.8) and field relationships show that: (1) the flood wave peaked quickly at each gauging station but peak magnitude attenuated as the flood progressed downstream and deposited large volumes of sand and gravel; (2) sediment concentration peaks lagged the flow discharge peaks with lag times increasing in the downstream direction; (3) fines concentrations plateaued at 250,000–350,000 mg/L (9–13 vol%); (4) suspension of large volumes of sand did not begin occurring until fines concentrations reached about 200,000 mg/L (8 vol%); (5) at peak concentration, the amount of sand in suspension far exceeded the quantity of suspended fines; and (6) peak total sediment concentration steadily decreased with distance downstream from at least 46 vol% (actual peak value not recorded at first station) to 41 vol% at the second station to 36 vol% at the third. Over this 32-km-long reach (average slope of 0.006 5 m/m), the flood deposited 37% of its sediment load. Of the total sediment in suspension at the time of peak sediment concentration, 70–80% was sand and fine gravel, although the suspended gravel was primarily lower density vesicular dacite and pumice (Dinehart, 1999). The coarsest suspended particles sampled at the flow surface were 16–32 mm in diameter. Plots comparing suspended sediment concentrations by size class with time (Figure 8.9) show that the coarser the grain size, the more quickly it settled out of suspension.

8.4.2 1993 Hyperconcentrated lahar, Sacobia River, Mount Pinatubo, Philippines

Monsoonal rain triggered a small (of the order of 10 m^3/s peak discharge) hyperconcentrated lahar lasting about 2 hours in a broad sand and pumice-gravel-bedded braided alluvial channel on the lower east flank of Mount Pinatubo on 26 September, 1993 (Figures 8.10 and 8.11). It arrived at the observation point (20–25 km downstream from estimated initiation point) as a single low bore, 5–10 cm high, moving slightly faster than the ambient, already highly concentrated streamflow. Peak flow depth (in the distributary channel closest to bank), about 0.7 m, was reached in about 10 min, and flow discharge appeared to gradually recede as the thalweg moved to a mid-valley channel. The photographs in Figure 8.10 show both apparent plane-bed conditions accompanied by bank-parallel small waves and development of breaking antidune waves only 2 min later (slight increase in depth and possibly velocity). Total active floodway width was estimated at 30 m, although flow was concentrated in 3 or 4 distributary channels.

Crude depth-integrated sampling was carried out at intervals along the right bank using heavy-duty Ziplock®-type[10] plastic bags. The highest sampled suspended sediment concentration was 30 vol% (average particle density = 1.84 g/cm^3 from pumice content), which lagged about 10 min behind the apparent flow peak. When flow was at or near peak depth, sand and pumice gravel could be felt in

---

[10] For foreign readers, these plastic bags have a zipper-type closure that permits them to be sealed watertight; samples can be carefully transported to a laboratory for analysis without loss of fluid.

Table 8.2. Directly observed or measured characteristics of natural hyperconcentrated flows at approximate times of peak sediment concentration.

	Eruption-triggered lahar (3/19–20/82), Toutle River near Mount St. Helens (USA)	Rainfall-triggered lahar (9/26/93), Sacobia River on flank of Mount Pinatubo (Philippines)	Eruption-triggered lahar, LH6 (9/29/95), Whangaehu River on flank of Ruapehu volcano (New Zealand)
Observation location	73 km downstream from source	20–25 km downstream from source	42 km downstream from source
Flow discharge (m^3/s)	450	10 (estimate)	42 (peak, measured 14 km farther downstream)
Flow depth (m)	2.1 (mean estimated)	0.5–0.7 m (where sampled)	2–3 (estimated)
Flow velocity (m/s)	3.5–4 (estimated)	2.6–3.1 (surface)	2.7–3.0 (estimated)
Concentration peak lag behind flow peak (min)	38	Not measured	50
Suspended sediment concentration (vol%)	43	30	52
Silt + clay (fraction of total suspended sediment)	0.22	0.34	0.19
Maximum clast size in suspension	30–40 mm observed; 16–32 mm in dip samples (mostly low-density rock)	16–32 mm (low-density rock)	Up to 40 mm (normal-density rock)
Bedload movement	Boulder movement heard; observers described bed being "in motion"	Pebbles (coarser clasts not available)	Boulders; 0.5 m boulders observed saltating (surfacing momentarily)

Flow surface appearance	Oily, glassy-smooth; floating debris and pumice clasts	Slight smoothing of surface	Sandy-textured; had appearance of flowing quicksand; swirling reticulate pattern of fines; damped turbulence
Vertical sediment distribution	Sand and fine gravel in suspension near surface	Sand and fine gravel in suspension near surface; no sharp contact between bedload and suspended load	Fine gravel in full dynamic suspension; small boulders rising occasionally to surface
Bed forms	Oblique standing waves (antidunes) up to 2.5 m (estimate) high, generally breaking; alternating with plane-bed conditions	Transverse antidunes, non-breaking (0.3–0.5 m high); alternating with plane-bed conditions	Oblique standing waves up to 0.5 m
Sediment deposition	Channel bed aggraded 0.5–0.7 m; benches ~1 m thick formed at channel margins as stage dropped; deposits firm within minutes of deposition	Estimated ~10 cm bed aggradation	Benches with silty tops formed at channel margins
References	Pierson and Scott (1985); Dinehart (1999)	Pierson et al. (1996); T.C. Pierson, unpublished data	Cronin et al. (1999); written commun. S.J. Cronin (2004)

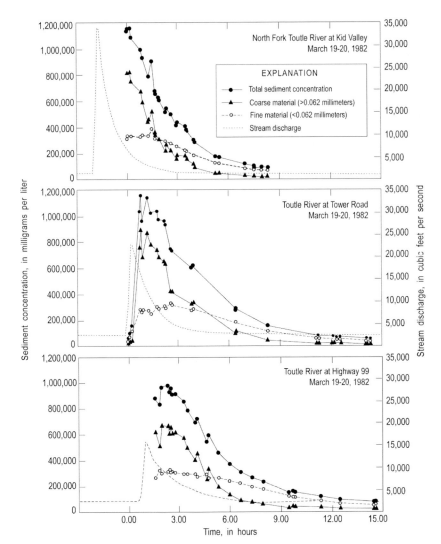

Figure 8.8. Sediment concentration and flow discharge for the 1982 hyperconcentrated lahar on the Toutle River downstream from Mount St. Helens, resulting from direct measurements of the flow at three USGS gauging stations.
From Dinehart (1999).

suspension throughout the vertical water column, and a denser layer of moving sand and gravel could be felt near the bed. In fact, it was difficult to identify by feel the boundary between the moving bed material and the static bed. Interestingly, fines concentrations in this hyperconcentrated flow were very low – all under 1 vol%. However, it is likely the sand-size and some gravel-size pumice grains would have fall

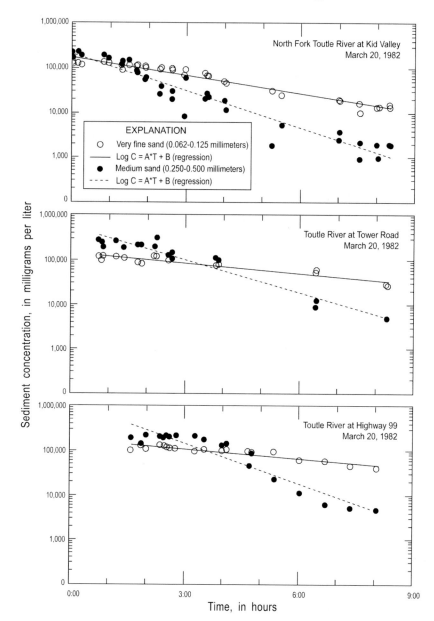

Figure 8.9. Sediment concentration vs. time during passage of the 1982 hyperconcentrated lahar on the Toutle River downstream from Mount St. Helens at different gauging stations, showing the more rapid settling out of the coarser size fraction as total concentration is decreasing in the tail of the flow.
From Dinehart (1999).

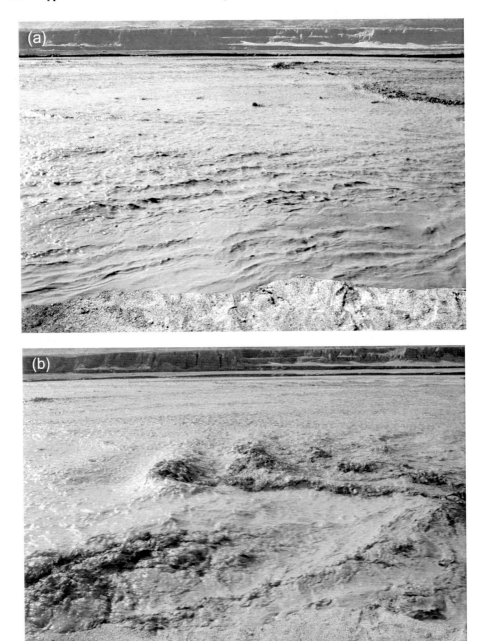

Figure 8.10. Hyperconcentrated lahar on Sacobia River, Mount Pinatubo, Philippines, 26 September, 1993, showing transition from plane bed to antidunes. Flow is left to right, and both photographs were taken from same position; time shown in upper right of each. For both, sediment concentration was about 23 vol%, flow velocity was about 2.5 m/s, flow depth was 30–50 cm; the transition to antidunes accompanied a slight increase in stage.

Sec. 8.4] Examples of observed hyperconcentrated flows 185

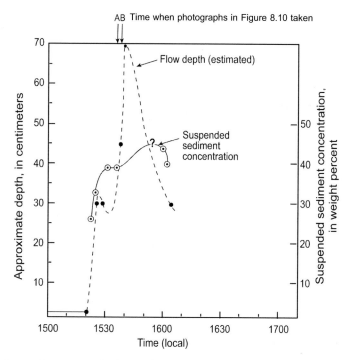

Figure 8.11. Approximate stage hydrograph and suspended sediment concentration variation with time for Sacobia River lahar (hyperconcentrated flow), 26 September, 1993, Mount Pinatubo (Philippines).

diameters in the silt range (particle densities as low as $0.7\,\text{g/cm}^3$) and thus would be acting like fines and supplementing the silt-size particles in forming the wash load.

8.4.3 1995 Hyperconcentrated lahar, Ruapehu volcano, New Zealand

Numerous lahars were generated on the eastern flank of Ruapehu volcano in late 1995 by pyroclastic debris erupting explosively up through the crater lake (Cronin et al., 1997, 1999, 2000). Several single-peaked and relatively sustained flows were generated on 29 September that were observed, measured, photographed, and sampled in detail (Cronin et al., 1999). The LH6 flow (Table 8.2) was identified as fully hyperconcentrated (by the criterion of dynamic suspension of sand and fine gravel) at observation stations 42 and 56.5 km from source, having transformed from a debris flow farther upstream.

This single-peaked lahar moved downstream in four phases, as did the smaller and more dilute (probably not hyperconcentrated) LH7 flow. The phases could be distinguished by sampling the relative mix of chemically distinctive (highly acidic) lake water and the stream water as the flows passed by the observation points (Cronin, 1999):

1. *Rising limb of the flood wave.* At 42 km from source, LH6 stage rose relatively rapidly – 0.5 m in 50 min, and the rise to peak stage/discharge was accompanied by only a slight increase in suspended sediment comprising mostly silt and clay (up to 4 vol%), by an increase in floating woody debris, but by essentially no change in water chemistry. This last observation suggests that this part of the flood wave was only stream water that was "piling up" and being pushed ahead by the actual lahar that was moving faster than the stream.
2. *Arrival of lahar front as stage recedes.* Within the next hour stage dropped to about 60% of peak, but a large and fairly abrupt jump in dissolved sulfate, chloride, and magnesium, and a drop in pH, indicated that the packet of ejected water from the crater lake had arrived, accompanied by an increase in suspended sediment concentration that reached a peak of 52 vol%. Increases in both sediment concentration and ion concentration were more gradual in another lahar (LH7), indicating that lahar fronts can mix with stream water as well as push it ahead. Flow turbulence was dampened but nonetheless active. Slowly saltating boulders surfaced for a few seconds and then sank back down, and the flow surface had the appearance of flowing quicksand.
3. *Lahar recessional limb.* Suspended sediment concentration decreased to 33 vol% during this phase of the flow, but dissolved ions and acidity continued to increase, indicating that the flood wave of ejected lake water was being diluted by stream water at the front of the lahar. Gravel and sand (normal-density and scoriaceous) remained in dynamic suspension, and small boulders continued to saltate up to the flow surface.
4. *Highly erosive lahar tail.* Both water acidity and sediment concentration decreased during this phase, and flow transitioned back to normal streamflow. However, vigorous turbulence while sediment concentration remained above 10 vol% apparently triggered an acceleration of bank erosion in the channel.

The observations of Cronin et al. (1999) are important in clarifying the difference between debris flows and hyperconcentrated flows in coarse sediment mixtures. At their upper-concentration range, hyperconcentrated flows superficially can look like debris flows, with gravel in suspension and even having boulders bobbing at the surface. The key, however, is that the suspension of sand and gravel depends on dynamic processes that occur during flow. Once the fluid motion stops, sand and gravel readily settle out of suspension.

8.5 SEDIMENT DEPOSITION

Sediment is deposited in two ways in hyperconcentrated flows – by settling out of suspension (suspension fallout) and by traction-carpet accretion. Although a variety of depositional features have been attributed to hyperconcentrated flows and described in the sedimentological literature, the brief discussion here is limited to observations of deposits immediately after hyperconcentrated flows have passed.

Deposition by suspension fallout should occur (1) where velocity, and hence turbulence, are decreasing, and (2) where dilution by addition of stream water at flow-front mixing interfaces or at tributary inflows result in loss of suspension competence (Pierson and Scott, 1985; Cronin et al., 2000). At energy-drop locations (such as along channel margins, where flow depth decreases, in eddies, or below hydraulic jumps), the fallout from suspension should be a function of flow energy at that point. A relatively narrow range of grain sizes should rain down if discharge remains fairly constant and stratification should be only weakly developed or absent. Relatively massive and well-sorted deposits have been observed along some channel margins, for example (Pierson and Scott, 1985; Cronin et al., 2000). In suspension fallout zones where flow discharge fluctuates or where turbulence intensity varies (e.g., at mixing fronts), deposits can show weak to strong horizontal bedding (Cronin et al., 2000) (Figure 8.12(a)). Mud layers commonly form on the tops of these depositional units after subaerial exposure, due to vertical dewatering of the deposits with transport of fines to the surface, and upper portions of the deposits are commonly normally graded (Cronin et al., 2000).

Deposition by traction-carpet accretion should be greatest where flow velocities, and hence bed shear stresses, are high. Therefore, deposition along channel thalwegs should comprise a high proportion of traction-carpet accretion deposits. Here, bed material is deposited largely as layers or sheets of grains accreted from the base of the mobile traction carpet, and deposits are coarser than at channel margins (Cronin et al., 2000). These deposits commonly show more pronounced horizontal stratification than channel margin-deposits, particularly when contrasting grain types are available, but laminations typically occur without high-angle cross-bedding (Figure 8.12(b), lower part). Coarse bedload (lenses or single outsized clasts of cobbles or boulders) is commonly enveloped by finer accretionary strata or is left stranded on surfaces of berms or terraces. In distal depositional areas where flow sediment concentrations are decreasing significantly and progressively more sediment is moving as bedload, deposits become more distinctly stratified and high-angle cross-bedding can be formed and preserved (Cronin et al., 2000).

A third general type of deposit is also sometimes observed in channels after the passage of highly concentrated hyperconcentrated flows: massive, very poorly sorted, but only moderately compacted diamicts of sand and gravel, which are more friable than typical debris-flow deposits (Pierson and Scott, 1985; Cronin et al., 2000; see also Figure 8.12(b), upper part). It is uncertain whether these deposits represent a submerged debris-flow phase that lags behind peak discharge and flows beneath more dilute surface flow (i.e., a stratified lahar (the explanation favoured by Cronin et al., 2000)), or whether these deposits result simply from very rapid and chaotic suspension fallout in the rapidly fluctuating, high-energy environment found in the deepest, fastest, most concentrated part of the flow.

Overall, median grain size in all types of deposits typically decreases progressively with distance downstream (Pierson and Scott, 1985), and deposits are better sorted than the fluid mixtures that deposit them (Pierson and Scott, 1985; Cronin et al., 1997, 2000; see also Figure 8.13) – most of the silt and clay remains in suspension while the sand and gravel are deposited. In addition, hyperconcentrated flow deposits are relatively similar over hundreds of metres along a channel (Cronin

Figure 8.12. Examples of hyperconcentrated flow deposits. (a) Laminated deposit from 26 September, 1993 Sacobia River lahar, Mount Pinatubo (see Figure 8.11). Laminations emphasized by contrasting sediment grains – dark lithic medium sand and light pumiceous very coarse sand/fine gravel. (b) Section near main axis of channel showing two contrasting deposit types from the 1982 Toutle River lahar at Mount St. Helens. Lower part of section – faintly, horizontally stratified medium to coarse lithic sand; upper part of section – massive, poorly sorted diamict (resembling a debris-flow deposit but not as indurated because few fines are deposited with the sand and gravel).

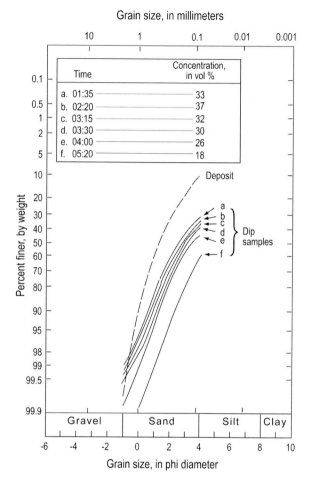

Figure 8.13. Differences in grain-size distribution between samples of hyperconcentrated flow (dip samples) collected 81 km downstream from source and the sediment deposited at the same site during the 19–20 March, 1982 hyperconcentrated lahar on the Toutle River. Dip samples were collected at different times during passage of the flow; corresponding sediment concentrations are shown.
From Pierson and Scott (1985).

et al., 2000), whereas fluvial deposits laid down by low-concentration water floods typically exhibit abrupt changes in mean grain size and stratification over relatively short distances.

Thalwegs tend to straighten during hyperconcentrated flow, moving away from the outsides of bends in sinuous reaches as a result of the formation of massive lateral berms (Zhou et al., 1983; Wan and Wang, 1994). Such berm formation tends to narrow the flow cross section, which may lead to bed degradation in the thalweg (Wan and Wang, 1994). Dinehart (1999) described such berm formation on the outside of a sweeping channel bend, where the berm formed a shallow beach (deposition occurring just below the water surface) that extended as far as 30 m out from the original bank. During more dilute flow recession, large parts of this berm were re-eroded.

8.6 HAZARDS

Hyperconcentrated flows at high discharge present significant hazards in addition to those of normal water floods and different from those of debris flows or mudflows. These hazards are commonly exhibited in watersheds where prodigious quantities of loose, erodible sediment become available for transport – in active volcanic areas, where eruptions can deposit large volumes of erodible material over broad areas and destroy vegetation cover (Major et al., 1996; Pierson et al., 1996; Rodolfo et al., 1996; Scott et al., 1996), and in mountainous areas subjected to wildfire, where burning of vegetation cover leaves soils loose and extremely vulnerable to sheetwash and rill erosion (Meyer and Wells, 1997; Cannon, 2001; see also Chapter 15).

Hyperconcentrated flows can be highly erosive, especially where channels are relatively steep (Waldron, 1967; Wang, 1990; Rickenmann, 1991; Xu, 1999, 2002a), but degree of scour is also a function of sediment concentration (Xu, 2002a). Streamflow tends to become more erosive when it transitions into the hyperconcentrated range (Xu, 2002a), but after the concentration limit (a function of grain-size distribution) has been reached for a flow, erosivity begins to decline (Wang, 1990). Debris flows and mudflows are comparatively less erosive, although the low-discharge "tails" of debris flows commonly transform to hyperconcentrated flow and can accomplish more erosion than the main bodies of the debris flows themselves (Pierson, 1986). Tens of metres of vertical scour have been observed in hyperconcentrated flow in China within a period of 10 hours (Kuang et al., 1999).

Hyperconcentrated flows are not always erosive, however. They can cause rapid deposition and river bed aggradation at places where channel gradients decrease or channels widen (Pierson et al., 1996; Rodolfo, 1996; Scott et al., 1996). With time and repeated flood events, deposition by hyperconcentrated flows can lead to incremental infilling of channels and channel shifting, reduction of flood-conveyance capacity, and burial of low-lying areas and structures in sediment (Figure 8.14). Hazardous river bed aggradation was well demonstrated along the rivers draining Mount Pinatubo (Philippines) following the large 1991 eruption of that volcano. Up to 25 m of channel aggradation occurred in rivers within the first three months following the June 1991 eruptions, and more has occurred since, due to deposition by thousands of rainfall-generated lahars (some debris flows but mostly hyperconcentrated flows) (Major et al., 1996; Pierson et al., 1996; Scott et al., 1996; Rodolfo et al., 1996). This activity has led to widespread burial of towns, roads, farms, and prime agricultural land by sediment.

Rapid lateral migration of channels is also common in low-gradient channels during aggradation by hyperconcentrated flows. This can lead to extreme rates and amounts of lateral erosion of river banks and destruction of buildings, roads, and bridges located on floodplains or alluvial fans (Major et al., 1996; Rodolfo et al.,

Sec. 8.7]   Discussion   191

Figure 8.14. Burial of house and trees downstream of Mount Pinatubo, Philippines, due to river bed aggradation resulting primarily from repeated, relatively small hyperconcentrated flows that spread out on the alluvial fan after the stream channel was filled with sediment.

1996; see also Figure 8.15). Laterally eroded channels are typically rectangular in cross section, due to undercutting and the formation of near-vertical banks in unconsolidated alluvial fill. Width/depth ratios of channels cut in alluvium can be correlated to suspended sediment concentration; the higher the concentration, the deeper the channel relative to its width (Xu, 1999). Vigorous lateral erosion of unconsolidated stream banks with rates as high as 3 m/min (perpendicular to channel) has been documented at Mount Pinatubo accompanying bed aggradation (Rodolfo et al., 1996). About 40 km downstream from flow source, hundreds of buildings in Angeles City and long sections of flood-control levees were lost due to undercutting and collapse by bank erosion in the first few years following the eruption.

8.7   DISCUSSION

From the variety of experimental and field studies that have been carried out with hyperconcentrated mixtures of sediment and water, hyperconcentrated flow should be considered a distinct flow process on the basis of the following criteria:

1. Concentration of suspended fines is sufficient to impart yield strength to the fluid and maintain high fluid viscosity.

Figure 8.15. Destruction of buildings along a river bank downstream from Mount Pinatubo, Philippines, caused by lateral erosion of channel banks by hyperconcentrated flows.
Photo by Jon Major (USGS).

2. Sand and fine gravel, its settling hindered by the fluid viscosity, is kept in prolonged suspension by turbulence and dynamic grain interactions.
3. Significant bedload transport occurs as a traction carpet.
4. Mean flow velocity is greater and quantity of sediment transported is much greater than for water flow at a similar depth and slope.

These flow characteristics are different from those of either water flow or debris flow and mudflow. Yet many authors use only suspended sediment concentration alone as the defining criterion to identify hyperconcentrated flow, based on the concentration limits set by Beverage and Culbertson (1964). The fatal flaw in this approach is that grain-size distribution and grain density also play extremely important roles in determining the properties of sediment–water suspensions (Cao and Qian, 1990; Wan and Wang, 1994; Xu, 2002b, 2003). The widely used concentration thresholds of 20 vol% and 60 vol% (Beverage and Culbertson, 1964) were somewhat arbitrarily defined and can only be valid for sediment mixtures similar to the ones studied by Beverage and Culbertson.

The importance of grain-size distribution and grain density has been demonstrated experimentally with the use of artificial sediment mixtures and neutrally buoyant particles. For example, fines-free mixtures have Newtonian fluid properties (i.e., develop no yield strength) up to concentrations as high as 35 vol% for poorly sorted mixtures (Fei, 1983) and up to 50 vol% for uniformly sized coarse particles (Howard, 1965). A pure smectite clay suspension, on the other hand, acquires yield strength at concentrations of only about 1 vol% (Hampton, 1975). These values deviate greatly from Beverage and Culbertson's (1964) 20 vol% threshold, as do values for some natural flows. For example, poorly sorted (well-graded), clay-poor sediment mixtures from fresh volcanic terrains will transition from normal streamflow to hyperconcentrated flow when suspended sediment concentrations reach 10–11 vol% and will remain hyperconcentrated up to 53–54 vol%, whereupon they transition to debris flow (Pierson, 1986; Cronin et al., 1997, 1999, 2000; Dinehart, 1999). Flows in the Yellow River of China that have finer, better sorted sediment will transition from water flow to hyperconcentrated flow at a similar total suspended sediment concentration of 8–11 vol%, but they take on mudflow characteristics at much lower concentrations (19–37 vol%) (Qian et al., 1981; Wan and Wang, 1994; Xu, 1999, 2002b). Thus, if total suspended sediment concentration is used to characterize a flow, the sediment size distribution and density of particles in the flow must be stipulated and flow properties must be noted. The ranges in limiting thresholds for experimental and natural hyperconcentrated flows (Figure 8.16) show that the ranges involving mostly fine sediments can be entirely below Beverage and Culbertson's lower limit, while all of them start lower and end lower than the commonly used boundaries.

Hyperconcentrated flow should therefore be defined as a two-phase flow of water and sediment, intermediate in concentration between normal streamflow and debris flow (or mudflow), in which a viscous and yield-strength-maintained suspension of fines in water (the carrier fluid) enables the intermittent, dynamic suspension of large quantities of coarser sediment. Hyperconcentrated flow should not be defined on total suspended sediment concentration alone. One diagnostic test to make in the field would be to take a dip sample at the flow surface (well above the bed) with a bucket of similar container. The flow is hyperconcetrated if sand (and possibly fine gravel) was in suspension at the flow surface but settles out of suspension in the container within seconds. This would indicate that the coarse sediment was in dynamic suspension. The deposit in the bottom of the container should be normally graded, indicating that the coarses grains moved independently of the finer grains and settled out first.

8.8 CONCLUSIONS

Hyperconcentrated flow is a type of two-phase, non-Newtonian flow of sediment and water that operates between normal streamflow (water

flow) and debris flow (or mudflow). It is distinctive in terms of processes acting to transport the sediment. Laboratory and field evidence indicate that the transition to hyperconcentrated flow occurs when the concentration of suspended fines achieves a minimum volumetric concentration of 3–10%, depending on grain-size

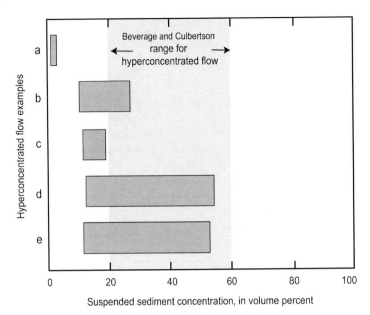

Figure 8.16. Approximate ranges in total suspended sediment concentrations for natural hyperconcentrated flows and laboratory-mixed hyperconcentrated suspensions of varying grain-size distributions and sediment compositions. Limits based on objective criteria discussed below. The shaded area is the widely quoted, arbitrarily defined concentration range assigned to hyperconcentrated flow by Beverage and Culbertson (1964). (a) Sheared experimental mixture of smectite clay and distilled water – concentration range 1–3 vol% (Hampton, 1975), where lower threshold is defined by the onset of yield strength and upper threshold by computation of competence (buoyancy and yield strength) to suspend a 4-mm mineral grain in a mixture. (b) Sheared experimental mixture of kaolinite clay and distilled water – concentration range 10–27 vol% (Hampton, 1975), with the same defining criteria as (a). (c) Clay, silt, and fine sand mixtures from the Yellow River basin, China – concentration range 11–19 vol%. Lower threshold is the concentration value widely accepted by Chinese authors defining the onset of hyperconcentrated flow (criteria not specified). The upper threshold is concentration at which sand grains can no longer settle out and mixture begins moving "as a whole" (Qian et al., 1981). (d) Poorly sorted volcaniclastic sediment (Ruapehu volcano, New Zealand) – concentration range 12–54 vol%, having fractions of 20–50 wt% fines, 50–70 wt% sand, and 0–15 wt% fine gravel (Cronin et al., 1997, 1999, 2000). Threshold criteria not defined. (e) Poorly sorted volcaniclastic sediment (Mount St. Helens, USA) – concentration range 11–53 vol% having fractions of 30–60 wt% fines, 30–70 wt% sand, and 0–10 wt% fine gravel (Pierson and Scott, 1985). Lower threshold based on concentration at which sand starts being suspended in significant quantities (Dinehard, 1999). Upper threshold based on observation of sustained suspension of pebbles and cobbles in moving flows (Pierson, 1986).

distribution, and begins to acquire yield strength. The fines mixture (sometimes referred to as the carrier fluid) then becomes able to transport prodigious quantities of coarser solid particles (sand and some gravel) in suspension. This coarse suspended load is held in prolonged dynamic suspension by turbulence, grain collisions, increased buoyancy, and increased viscosity (decreased fall velocity). However, the coarse suspended load will be selectively deposited during flow as flow velocity decreases.

Both water flow and debris flow/mudflow transport sediment differently from hyperconcentrated flow. In water flow, water has sufficiently low suspended sediment concentrations to behave as a Newtonian fluid, and sand particles are transported primarily as bed load. Large water floods carry some sand in intermittent suspension, but generally the suspended sand is mostly fine-grained, its vertical concentration profile is largely non-uniform, and the sand concentration in the flow is generally less than the fines concentration. Debris flows and mudflows, at the other end of the spectrum, are highly concentrated, relatively homogeneous slurries of sediment and water that behave as non-Newtonian, pseudo-one-phase flows. Because of dense grain packing, debris flows or mudflows cannot selectively deposit transported solid particles by size when velocity decreases or flow stops. This results in massive (non-stratified) and poorly sorted sediment textures that are characteristic of debris-flow and mudflow deposits.

It is not possible to determine whether a flow is hyperconcentrated from concentration values alone. This is because grain-size distributions and grain densities of the transported sediment control the physical properties of sediment–water mixtures, and thus also control the threshold concentrations at which flow transformations occur from water flow to hyperconcentrated flow and from hyperconcentrated flow to debris flow or mudflow. The fundamental characteristic that defines hyperconcentrated flow is the transport of large quantities of coarse sediment (sand and possibly some gravel) at high concentrations *in intermittent dynamic suspension.*

Hyperconcentrated flows generally are not as hazardous as debris flows, because their velocities are usually lower and they tend not to transport the large boulders that are responsible for impact damage in debris flows. However, high-discharge hyperconcentrated flows present a greater hazard to riverside communities than normal floods of similar magnitude, because of their greater potential for doing geomorphic work. In relatively steep channels, hyperconcentrated flows can rapidly incise their channels for tens of metres, undermining bridge piers and other channel structures, such as erosion-control dams. In channels with gentle slopes, deposition of their high sediment loads can (a) cause lateral shifting of active channels and vigorous bank erosion, causing collapse of riverside buildings, bridges, and flood-protection levees, and (b) incrementally fill river valleys with deposits of sand and gravel, burying low-lying areas in sediment and removing channel capacity that can exacerbate later flooding.

8.9 ACKNOWLEDGEMENTS

This paper has benefited greatly from discussions with Shane Cronin and his contributions of photographs and unpublished data. Thoughtful and helpful manuscript reviews were provided at various stages by Nati Bergman, John Costa, Shane Cronin, Matthias Jakob, and Jon Major.

8.10 REFERENCES

Alcoverro, J., Corominas, J., and Gomez, M. (1999) The Barranco de Aras flood of 7 August 1996 (Biescas, Central Pyrenees, Spain). *Engineering Geology*, **51**, 237–255.

Alexandrov, Y., Laronne, J.B., and Reid, I. (2003) Suspended sediment concentration and its variation with water discharge in a dryland ephemeral channel, northern Negev, Israel. *Journal of Arid Environments*, **53**, 73–84.

Aulitzky, H. (1980) Preliminary two-fold classification of debris torrents. In: *Proceedings of "Interpraevent" Conference, Bad Ischl, Austria* (Vol. 4, pp. 285–309, translated from German by G. Eisbacher).

Balog, J.D. (1978) Flooding in Big Thompson River, Colorado, tributaries: Controls on channel erosion and estimates of recurrence interval. *Geology*, **6**, 200–204.

Barnes, H.A. and Walters, K. (1985) The yield stress myth? *Rheological Acta*, **24**, 323–326.

Bates, R.L. and Jackson, J.A. (eds) (1987) *Glossary of Geology* (788 pp.). American Geological Institute, Alexandria, VA.

Batalla, R.J., DeJong, C., Ergenzinger, P., and Sala, M. (1999) Field observations on hyperconcentrated flows in mountain torrents. *Earth Surface Processes and Landforms*, **24**, 247–253.

Best, J., Bennett, S., Bridge, J., and Leeder, M. (1997) Turbulence modulation and particle velocities over flat sand beds at low transport rates. *Journal of Hydraulic Engineering*, **123**, 1118–1129.

Beverage, J.P. and Culbertson, J.K. (1964) Hyperconcentrations of suspended sediment. *Journal of the Hydraulics Division, A.S.C.E.*, **90**, 117–128.

Bondurant, D.C. (1951) Sedimentation studies at Conchas Reservoir in New Mexico. *Transactions of the American Society of Civil Engineers*, **116**, 1292–1295.

Bradley, J.B. (1986) Hydraulics and bed material transport at high fine suspended sediment concentrations (140 pp.). Ph.D. Thesis, Colorado State University, Fort Collins, CO.

Bradley, J.B. and McCutcheon, S.C. (1985) The effects of high sediment concentration on transport processes and flow phenomena. In: *Proceedings of International Symposium on Erosion, Debris Flow, and Disaster Prevention, Tsukuba, Japan, September 3–5, 1985* (pp. 1–7). Erosion-Control Engineering Society, Tokyo.

Bridge, J.S. (2003) Sediment transport by unidirectional water flows. In: G.V. Middleton, M.J. Church, M. Coniglio, L.A. Hardie, and F.J. Longstaffe (eds), *Encyclopedia of Sediments and Sedimentary Rocks* (pp. 609–619). Kluwer Academic Publishers, Dordrecht, The Netherlands.

Cannon, S.H. (2001) Debris-flow generation from recently burned watersheds. *Environmental and Engineering Geoscience*, **7**, 321–341.

Cao, R. and Qian, S. (1990) Sediment transport characteristics of hyperconcentrated flow with suspended load. In: R.H. French (ed), *Proceedings of the International Symposium on Hydraulics/Hydrology of Arid Lands, San Diego, California, July 30–August 2, 1990* (pp. 657–662). American Society of Civil Engineers, New York.

Cao, Z., Egashira, S., and Carling, P.A. (2003) Role of suspended-sediment particle size in modifying velocity profiles in open channel flows. *Water Resources Research*, **39**(ESG 2-1), 15.

Cheng, W., Wang, X., and Zhou, X. (1999) Research on some characteristics of two-phase hyperconcentrated flow. In: A.W. Jayawardena, J.H. Lee, and Z.Y. Wang (eds), *River Sedimentation: Theory and Applications* (pp. 311–317). A.A. Balkema, Rotterdam.

Committee on Methodologies for Predicting Mudflow Areas (1982) *Selecting a Methodology for Delineating Mudslide Hazard Areas for the National Flood Insurance Program* (35 pp.). National Research Council, National Academy Press, Washington, DC.

Costa, J.E. (1978) Colorado Big Thompson flood: Geologic evidence of a rare hydrologic event. *Geology*, **6**, 617–620.

Costa, J.E. (1984) Physical geomorphology of debris flows. In: J.E. Costa and P.J. Fleisher (eds), *Developments and Applications of Geomorphology* (pp. 268–317). Springer Verlag, Berlin.

Costa, J.E. (1988) Rheologic, geomorphic, and sedimentologic differentiation of water floods, hyperconcetrated flows, and debris flows. In: V.R. Baker, R.C. Kochel, and P.C. Patton (eds), *Flood Geomorphology* (pp. 113–122). Wiley-Interscience, New York.

Coussot, P. (1995) Structural similarity and transition from Newtonian to non-Newtonian behavior for clay-water suspensions. *Physical Review Letters*, **74**, 3971–3974.

Coussot, P. and Piau, J.M (1994) On the behavior of fine mud suspensions. *Rheologica Acta*, **33**, 175–184.

Cronin, S.J., Neall, V.E., Lecointre, J.A., and Palmer, A.S. (1997) Changes in Whangaehu River lahar characteristics during the 1995 eruption sequence, Ruapehu volcano, New Zealand. *Journal of Volcanology and Geothermal Research*, **76**, 47–61.

Cronin, S.J., Neall, V.E., Lecointre, J.A., and Palmer, A.S. (1999) Dynamic interactions between lahars and stream flow: A case study from Ruapehu volcano, New Zealand. *Geological Society of America Bulletin*, **111**, 28–38.

Cronin, S.J., LeCointre, J.A., Palmer, A.S., and Neall, V.E. (2000) Transformation, internal stratification, and depositional processes within a channelized, multi-peaked lahar flow. *New Zealand Journal of Geology and Geophysics*, **43**, 117–128.

Dinehart, R.L. (1999) Sediment transport in the hyperconcentrated phase of the March 19, 1982, lahar. In: T.C. Pierson (ed.), *Hydrologic Consequences of Hot-rock/Snowpack Interactions at Mount St. Helens Volcano, Washington, 1982–84* (USGS Professional Paper 1586, pp. 37–52). US Geological Survey, Reston, VA.

Drake, T.G. (1990) Structural features in granular flows. *Journal of Geophysical Research*, **95**(B6), 8681–8696.

Druitt, T.H. (1995) Settling behaviour of concentrated dispersions and some volcanological applications. *Journal of Volcanology and Geothermal Research*, **65**, 27–39.

Durand, R. and Condolios, E. (1952) The hydraulic transport of coal and solid material in pipes. In: *Proceedings of Colloquium on Hydraulic Transportation, France* [cited in Bradley, 1986].

Engelund, F. and Wan, Z. (1984) Instability of hyperconcentrated flow. *Journal of Hydraulic Engineering*, **110**, 219–233.

Fei, X. (1983) Grain composition and flow properties of heavily concentrated suspensions. In: *Proceedings of the 2nd International Symposium on River Sedimentation* (pp. 307–308). Water Resources and Electrical Power Press, Nanjing, China.

Gagoshidze, M.S. (1969) Mud flows and their control. *Soviet Hydrology: Selected Papers*, **4**, 410–422.

Gerson, R. (1977) Sediment transport for desert watersheds in erodible materials. *Earth Surface Processes*, **2**, 343–361.

Glancy, P.A. and Bell, J.W. (2000) *Landside-induced Flooding at Ophir Creek, Washoe County, Western Nevada, May 30, 1983* (USGS Professional Paper 1617, 94 pp.). US Geological Survey, Reston, VA.

Glancy, P.A. and Harmsen, L. (1975) *A Hydrologic Assessment of the September 14, 1974, Flood in Eldorado Canyon, Nevada* (USGS Professional Paper 930, 28 pp.). US Geological Survey, Reston, VA.

Hampton, M.A. (1975) Competence of fine-grained debris flows. *Journal of Sedimentary Petrology*, **45**, 834–844.

Hampton, M.A. (1979) Buoyancy in debris flows. *Journal of Sedimentary Petrology*, **49**, 753–758.

Hanes, D.M. and Bowen, A.J. (1985) A granular-fluid model for steady intense bed-load transport. *Journal of Geophysical Research*, **90**, 9149–9158.

Howard, C.D.D. (1965) Hyperconcentrations of suspended sediment (Discussion). *Journal of the Hydraulics Division, A.S.C.E.*, **91**, 385–388.

Hungr, O., Evans, S.G., Bovis, M.J., and Hutchinson, J.N. (2001) A review of the classification of landslides of the flow type. *Environmental and Engineering Geoscience*, **7**, 221–238.

Iverson, R.M. (1997) The physics of debris flows. *Reviews of Geophysics*, **35**, 245–296.

Iverson, R.M. (2003) The debris-flow rheology myth. In: D. Rickenmann, and C.L. Chen (eds), *Debris-flow Hazards Mitigation: Mechanics, Prediction, and Assessment* (pp. 303–314). Millpress, Rotterdam.

Iverson, R.M. and Vallance, J.W. (2001) New views of granular mass flows. *Geology*, **29**, 115–118.

Janda, R.J., Scott, K.M., Nolan, K.M., and Martinson, H.A. (1981) Lahar movement, effects, and deposits. In: P.W. Lipman, and D.R. Mullineaux (eds), *The 1980 Eruptions of Mount St. Helens, Washington* (USGS Professional Paper 1250, pp. 461–478). US Geological Survey, Reston, VA.

Jarrett, R.D. and Costa, J.E. (1986) *Hydrology, Geomorphology, and Dam-break Modeling of the July 15, 1982 Lawn Lake Dam and Cascade Lake Dam Failures, Larimer County, Colorado* (USGS Professional Paper 1369, 78 pp.). US Geological Survey, Reston, VA.

Johnson, A.M. (1970) *Physical Processes in Geology* (557 pp.). W.H. Freeman, New York.

Julien, P.Y. and Lan, Y. (1991) Rheology of hyperconcentrations. *Journal of Hydraulic Engineering*, **117**, 346–353.

Kang, Z. and Zhang, S. (1980) A preliminary analysis of the characteristics of debris flow. In: *Proceedings of the International Symposium on River Sedimentation* (pp. 225–226). Chinese Society for Hydraulic Engineering, Beijing.

Kikkawa, H. and Fukuoka, S. (1969) The characteristics of flow with wash load. In: *Proceedings 13th Congress, IAHR* (Vol. 2, pp. 233–240). International Association for Hydraulic Research, Ecole Polytechnique Fédérale, Lausanne, Switzerland.

Komar, P.D. (1988) Sediment transport by floods. In: V.R. Baker, R.C. Kochel, and P.C. Patton (eds), *Flood Geomorphology* (pp. 97–111). Wiley-Interscience, New York.

Kostaschuk, R., James, T., and Rishi, R. (2003) Suspended sediment transport during tropical cyclone floods in Fiji. *Hydrological Processes*, **17**, 1149–1164.

Kuang, S.F., Xu, Y.N., Wang, L., and Li, W.W. (1999) Mechanism of ripping up the bottom due to hyperconcentrated flow. In: A.W. Jayawardena, J.H. Lee, and Z.Y. Wang (eds), *River Sedimentation: Theory and Applications* (pp. 283–288). A.A. Balkema, Rotterdam.

Lane, E.W. (1940) Notes on limit of sediment concentration. *Journal of Sedimentary Petrology*, **10**, 95–96.

Laronne, J.B. and Reid, I. (1993) Very high rates of bed load sediment transport by ephemeral desert rivers. *Nature*, **366**, 148–150.

Laronne, J.B., Reid, I., Yitshak, Y., and Frostick, L.E. (1994) The non-layering of gravel streambeds under ephemeral flood regimes. *Journal of Hydrology*, **159**, 353–363.

Lekach, J. and Schick, A.P. (1982) Suspended sediments in desert floods in small catchments. *Israel Journal of Earth Science*, **31**, 144–156.

Maizels, J. (1989) Sedimentology, paleoflow dynamics and flood history of jökulhlaup deposits: Paleohydrology of Holocene sediment sequences in southern Iceland sandur deposits. *Journal of Sedimentary Petrology*, **59**, 204–223.

Major, J.J. (1993) Rheometry of natural sediment slurries. In: *Proceedings of ASCE National Conference on Hydraulic Engineering, July 27–30, 1993, San Francisco* (7 pp.). American Society of Civil Engineers, New York.

Major, J.J. (2003) Hindered settling. In: G.V. Middleton, M.J. Church, M. Coniglio, L.A. Hardie, and F.J. Longstaffe (eds), *Encyclopedia of Sediments and Sedimentary Rocks* (pp. 358–360). Kluwer Academic Publishers, Dordrecht, The Netherlands.

Major, J.J., and Newhall, C.G. (1989) Snow and ice perturbation during historical volcanic eruptions and the formation of lahars and floods: A global review. *Bulletin of Volcanology*, **52**, 1–27.

Major, J.J. and Pierson, T.C. (1992) Debris flow rheology: Experimental analysis of fine-grained slurries. *Water Resources Research*, **28**, 841–857.

Major, J.J., Janda, R.J., and Daag, A.S. (1996) Watershed disturbance and lahars on the east side of Mount Pinatubo during the mid-June 1991 eruptions. In: C.G. Newhall, and R.S. Punungbayan (eds), *Fire and Mud: Eruptions and Lahars of Mount Pinatubo, Philippines* (pp. 895–919). Philippine Institute of Volcanology and Seismology, Quezon City and University of Washington Press, Seattle.

Manville, V., White, J.D., Hodgson, K.A., Cronin, S., Neall, V., Lecointre, J., and Palmer, A. (2000) Dynamic interactions between lahars and stream flow, a case study from Ruapehu Volcano, New Zealand: Discussion and reply. *Geological Society of America Bulletin*, **112**, 1149–1152.

McCutcheon, S.C. and Bradley, J.B. (1984) Effects of high sediment concentrations on velocity and sediment distributions. In: *Water for Resource Development, Proceedings of the Conference, Coeur d'Alene, Idaho* (pp. 43–47). American Society of Civil Engineers, New York.

Meyer, G.A. and Wells, S.G. (1997) Fire-related sedimentation events on alluvial fans, Yellowstone National Park, U.S.A. *Journal of Sedimentary Research*, **67**, 776–791.

Mizuyama, T. (1981) An intermediate phenomenon between debris flow and bed load transport. In: *Erosion and Sediment Transport in Pacific Rim Steeplands* (pp. 212–224). (IAHS Publication No. 132). International Association of Hydrological Sciences, Christchurch, New Zealand.

Montgomery, D.R., Panfil, M.S., and Hayes, S.K. (1999) Channel-bed mobility response to extreme sediment loading at Mount Pinatubo. *Geology*, **27**, 271–274.

Nordin, C.F., Jr. (1963) *A Preliminary Study of Sediment Transport Parameters, Rio Puerco near Bernardo, New Mexico* (USGS Professional Paper 462-C, 21 pp.). US Geological Survey, Reston, VA.

O'Connor, J.E., Hardison, J.H. 3rd, and Costa, J.E. (2002) *Debris Flows from Failures of Neoglacial-age Moraine Dams in the Three Sisters and Mount Jefferson Wilderness Areas, Oregon* (USGS Professional Paper 1606, 93 pp.). US Geological Survey, Reston, VA.

Phillips, C.J., and Davies, T.R.H. (1991) Determining rheological parameters of debris-flow material. *Geomorphology*, **4**, 101–110.

Pierce, R.C. (1917) The measurement of silt-laden streams. In: *Contributions to the Hydrology of the United States, 1916* (USGS Water-Supply Paper 400, pp. 39–51). US Geological Survey, Reston, VA.

Pierson, T.C. (1981) Dominant particle support mechanisms in debris flows at Mt. Thomas, New Zealand, and implications for flow mobility. *Sedimentology*, **28**, 49–60.

Pierson, T.C. (1986) Flow behavior of channelized debris flows, Mount St. Helens, Washington. In: A.D. Abrahams (ed.), *Hillslope Processes* (pp. 269–296). Allen & Unwin, Boston.

Pierson, T.C. (1999) Transformation of water flood to debris flow following the eruption-triggered transient-lake breakout from the crater on March 19, 1982. In: T.C. Pierson (ed.), *Hydrologic Consequences of Hot-rock/Snowpack Interactions at Mount St. Helens Volcano, Washington, 1982–84* (USGS Professional Paper 1586, pp. 19–36). US Geological Survey, Reston, VA.

Pierson, T.C. and Costa, J.E. (1987) A rheologic classification of subaerial sediment-water flows. *Geological Society of America Reviews in Engineering Geology*, **7**, 1–12.

Pierson, T.C. and Scott, K.M. (1985) Downstream dilution of a lahar: Transition from debris flow to hyperconcentrated streamflow. *Water Resources Research*, **21**, 1511–1524.

Pierson, T.C., Daag, A.S., Delos Reyes, P.J., Regalado, M.T.M., Solidum, R., and Tubianosa, B.S. (1996) Flow and deposition of posteruption hot lahars on the east side of Mount Pinatubo, July–October 1991. In: C.G. Newhall and R.S. Punungbayan (eds), *Fire and Mud: Eruptions and Lahars of Mount Pinatubo, Philippines* (pp. 921 950). Philippine Institute of Volcanology and Seismology, Quezon City and University of Washington Press, Seattle.

Pringle, P.T. and Cameron, K.A. (1999) Eruption-triggered lahar on May 14, 1984. In: T.C. Pierson (ed), *Hydrologic Consequences of Hot-rock/Snowpack Interactions at Mount St. Helens Volcano, Washington, 1982–84* (USGS Professional Paper 1586, pp. 81–103). US Geological Survey, Reston, VA.

Qian, Y., Yang, W., Zhao, W., Cheng, X., Zhang, L., and Xu, W. (1981) Basic characteristics of flow with hyperconcentration of sediment. In: *Proceedings of International Symposium on River Sedimentation* (pp. 175–184). Chinese Society of Hydraulic Engineering, Beijing.

Richardson, E.V. and Hanly, T.F. (1965) Hyperconcentrations of suspended sediment (Discussion). *Journal of the Hydraulics Division, A.S.C.E.*, **91**, 215–220.

Rickenmann, D. (1991) Hyperconcentrated flow and sediment transport at steep slopes. *Journal of Hydraulic Engineering*, **117**, 1419–1439.

Rodolfo, K.S., Umbal, J.V., Alonso, R.A., Remotigue, C.T., Paladio-Melosantos, M.L., Salvador, J.H.G., Evangelista, D., and Miller, Y. (1996) Two years of lahars on the western flank of Mount Pinatubo: Initiation, flow processes, deposits, and attendant geomorphic and hydraulic changes. In: C.G. Newhall and R.S. Punungbayan (eds), *Fire and Mud: Eruptions and Lahars of Mount Pinatubo, Philippines* (pp. 989–1013). Philippine Institute of Volcanology and Seismology, Quezon City and University of Washington Press, Seattle.

Scott, K.M. (1988) *Origins, Behavior, and Sedimentology of Lahars and Lahar-runout Flows in the Toutle-Cowlitz River System* (USGS Professional Paper 1447A, 74 pp.). US Geological Survey, Reston, VA.

Scott, K.M. and Gravlee, G.C. (1968) *Flood Surge on the Rubicon River, California: Hydrology, Hydraulics, and Boulder Transport* (USGS Professional Paper 422-M, 40 pp.). US Geological Survey, Reston, VA.

Scott, K.M., Vallance, J.W., and Pringle, P.T. (1995) *Sedimentology, Behavior and Hazards of Debris Flows at Mount Rainier, Washington* (USGS Professional Paper 1547, 56 pp.). US Geological Survey, Reston, VA.

Scott, K.M., Janda, R.J., de la Cruz, E.G., Gabinete, E., Eto, I., Isada, M., Sexon, M., and Hadley, K.C. (1996) Channel and sedimentation responses to large volumes of 1991 volcanic deposits on the east flank of Mount Pinatubo. In: C.G. Newhall and R.S.

Punungbayan (eds), *Fire and Mud: Eruptions and Lahars of Mount Pinatubo, Philippines* (pp. 971–988). Philippine Institute of Volcanology and Seismology, Quezon City and University of Washington Press, Seattle.

Segerstrom, K. (1950) *Erosion Studies at Parícutin, State of Michoacán, Mexico* (USGS Bulletin 965-A, 164 pp.). US Geological Survey, Reston, VA.

Shu, A., Cao, W., and Fei, X. (1999) The resistance laws of two-phase flow with high concentration. In: A.W. Jayawardena, J.H. Lee, and Z.Y. Wang (eds), *River Sedimentation: Theory and Applications* (pp. 303–309). A.A. Balkema, Rotterdam.

Simons, D.B. and Richardson, E.V. (1966) *Resistance to Flow in Alluvial Channels* (USGS Professional Paper 422-J, 61 pp.). US Geological Survey, Reston, VA.

Simons, D.B., Richardson, E.V., and Haushild, W.L. (1963) *Studies of Flow in Alluvial Channels, Some Effects of Fine Sediment on Flow Phenomena* (USGS Water-Supply Paper 1498-G, 46 pp.). US Geological Survey, Reston, VA.

Smith, G.A. and Fritz, W.J. (1989) Volcanic influences on terrestrial sedimentation. *Geology*, **17**, 375–376.

Smith, G.A. and Lowe, D.R. (1991) Lahars: Volcano-hydrologic events and deposition in the debris flow–hyperconcentrated flow continuum. In: R.V. Fisher and G.A. Smith (eds), *Sedimentation in Volcanic Settings* (SEPM Special Publication 45, pp. 59–70). Society for Sedimentary Geology, Tulsa, Oklahoma.

Sohn, Y.K. (1997) On traction-carpet sedimentation. *Journal of Sedimentary Research*, **67**, 502–509.

Todd, O.J., and Eliassen, C.E. (1940) The Yellow River problem. *Transactions of the American Society of Civil Engineers*, **105**, 346–416.

Todd, S.P. (1989) Stream-driven, high-density gravelly traction carpets: Possible deposits in the Trabeg Conglomerate Formation, SW Ireland and some theoretical considerations of their origin. *Sedimentology*, **36**, 513–530.

van Rijn, L.C. (1983) Sediment transportation in heavy sediment-laden flows. In: *Proceedings of the 2nd International Symposium on River Sedimentation, 11–16 October, 1983, Nanjing, China* (pp. 482–491). Water Resources and Electric Power Press, Nanjing, China.

Vanoni, V.A. (1946) Transport of suspended sediment by water. *American Society of Civil Engineers Transactions*, **111**, 67–102.

Vanoni, V.A. and Nomicos, G.N. (1960) Resistance properties of sediment-laden streams. *American Society of Civil Engineers Transactions*, **125**, 1140–1175.

Varnes, D.J. (1978) Slope movement types and processes. In: R.L. Schuster, and R.J. Krizek (eds), *Landslides: Analysis and Control* (Transportation Research Board Special Report 176, pp. 11–33). National Academy of Sciences, Washington, DC.

Waldron, H.H. (1967) *Debris Flow and Erosion Control Problems Caused by the Ash Eruptions of Irazu Volcano, Costa Rica* (USGS Bulletin 1241-I, 37 pp.). US Geological Survey, Reston, VA.

Wan, Z. (1982) *Bed Material Movement in Hyperconcentrated Flow* (Series Paper No. 31). Institute of Hydrodynamics and Hydraulic Engineering, Technical University of Denmark, Lyngby, Denmark.

Wan, Z. (1985) Bed material movement in hyperconcentrated flow. *Journal of Hydraulic Engineering*, **111**, 987–1002.

Wan, Z. and Song, T. (1987) The effect of fine particles on vertical concentration distribution and transport rate of coarse particles. In: W.R. White (ed.), *Topics in Fluvial Hydraulics: Proceedings of the 22nd Congress* (pp. 80–85). International Association for Hydraulic Research, Ecole Polytechnique Federale, Lausanne, Switzerland.

Wan, Z., and Wang, Z. (1994) *Hyperconcentrated Flow* (290 pp.). A.A. Balkema, Rotterdam.

Wannanen, A.O., Harris, D.D., and Williams, R.C. (1970) *Floods of December 1964 and January 1965 in the Far Western States. Part 2: Streamflow and Sediment Data* (USGS Water-Supply Paper 1866B, 861 pp.). US Geological Survey, Reston, VA.

Wang, M., Zhan, Y., Liu, J., Duan, W., and Wu, W. (1983) An experimental study on turbulence characteristics of flow with hyperconcentration of sediment. In: *Proceedings of the 2nd International Symposium on River Sedimentation, 11–16 October, 1983, Nanjing, China* (pp. 36–46). Water Resources and Electric Power Press, Nanjing, China.

Wang, X. and Qian, N. (1989) Turbulence characteristics of sediment-laden flow. *Journal of Hydraulic Engineering*, **115**, 781–800.

Wang, Z. (1990) Limit concentration of suspended sediment. In: R.H. French (ed.), *Proceedings of the International Symposium on Hydraulics/Hydrology of Arid Lands, San Diego, California, July 30–August 2, 1990* (pp. 651–656). American Society of Civil Engineers, New York.

Wang, Z. and Larsen, P. (1994) Turbulent structure of water and clay suspensions with bed load. *Journal of Hydraulic Engineering*, **120**, 577–599.

Wilson, K.C. (1985) Comparison of hyperconcentrated flows in pipes and open channels. In: *Proceedings of International Workshop on Flow at Hyperconcentrations of Sediment* (pp. 115–138). International Research and Training Center on Erosion and Sedimentation, Beijing.

Xu, J. (1999) Erosion caused by hyperconcentrated flow on the Loess Plateau of China. *Catena*, **36**, 1–19.

Xu, J. (2002a) Complex behavior of natural sediment-carrying streamflows and the geomorphological implications. *Earth Surface Processes and Landforms*, **27**, 749–758.

Xu, J. (2002b) Implication of relationships among suspended sediment size, water discharge, and suspended sediment concentration, the Yellow River basin, China. *Catena*, **49**, 289–307.

Xu, J. (2003) Hyperconcentrated flows in the lower Yellow River as influenced by drainage basin factors. *Zeitschrift für Geomorphologie*, **47**, 393–410.

Yang, W. and Zhao, W. (1983) An experimental study of the resistance to flow with hyperconcentration in rough flumes. In: *Proceedings of the 2nd International Symposium on River Sedimentation, 11–16 October, 1983, Nanjing, China* (pp. 47–55). Water Resources and Electric Power Press, Nanjing, China.

Zhang, H. and Ren, Z. (1982) Discussion on law of resistance of hyperconcentration flow in open channel. *Scientia Sinica (Series A)*, **25**, 1332–1342.

Zhou W., Zeng, Q., Fang, Z., Pan, G., and Fan, Z. (1983) Characteristics of fluvial processes for the flow with hyperconcentration in the Yellow River. In: *Proceedings of the 2nd International Symposium on River Sedimentation, 11–16 October, 1983, Nanjing, China* (pp. 618–619). Water Resources and Electric Power Press, Nanjing, China.

9

Subaqueous debris flows

Jacques Locat and Homa J. Lee

9.1 INTRODUCTION

Subaqueous mass movements, including debris flows, have had catastrophic impacts on costal communities and infrastructures. For example, the 1929 Grand Banks earthquake generated a submarine mass movement which initiated a debris flow and a turbidity current that destroyed most of the communication cables over a distance close to 1,000 km away from the epicentre (Piper et al., 1988; Locat and Lee, 2002). The same slide generated a tsunami wave that hit the southern coast of Newfoundland destroying part of a small village and killing 33 people. A similar but less catastrophic event has also been reported in Kitimat (Prior et al., 1982). In that case the trigger was linked to the coupled effect of a high-sedimentation rate at the delta tip and significant tide level changes. Many subaqueous mass movements evolve or include a component which can be regarded as a debris flow which can travel over distances as long as 400 km, as in the case of the Gulf of Mexico which will be presented later (Schwab et al., 1996; Locat et al., 1996). Amongst the largest debris flows are those related to the Storegga slide (3,500 km^3) off the coast of Norway (Bryn et al., 2003), the Big '95 slide off Spain (Lastras et al., 2003), and the Saharan debris flow off the Morocco/Mauritania coast (Masson et al., 1993). A recent example of serious consequences of a submarine mass movement is given by the Papua New Guinea tsunami of 1998 triggered by a slide initiated by a major earthquake. It destroyed various villages along the coast killing about 2,500 people (Tappin et al., 2003).

Submarine mass movements have been classified in various ways (Mulder and Cochonat, 1996; Locat and Lee, 2002). As part of this work we would like to consider subaqueous debris flows as any kind of subaqueous mass movement which can be best described using mostly a fluid mechanics approach, as opposed to soil or rock mechanics. The term mass movement implies here that the moving mass does retain some degree of integrity, the limit being when it transforms into a turbidity current.

M. Jakob and O. Hungr (eds), *Debris-flow Hazards and Related Phenomena*.
© Praxis. Springer Berlin Heidelberg 2005.

This chapter will look at how subaqueous debris flows are formed and deposited, the relevant parameters and properties, their physical aspects, the modelling tools, and two examples. Before concluding, some elements of submarine mass movement hazards will be introduced. An underlying theme of this work is to provide an approach and a framework for undertaking subaqueous debris-flow analysis and evaluating their consequences.

9.2 FORMATION OF SUBAQUEOUS MASS FLOW

What is different between the subaqueous and terrestrial environments in terms of debris flows? The first obvious difference is the fact that the debris flow takes place under water, and this alone may have a significant impact on the onset, evolution, and stoppage of debris flows. For example, Mohrig et al. (1999) have shown that under certain boundary conditions, aquaplaning could be generated with dramatic consequences on the mobility and morphological change of a flowing subaqueous debris flow. In a sense, subaqueous debris flows have more in common with snow avalanches for whom the interstitial fluid is air (Norem et al., 1990). A debris flow, much like a snow avalanche, can also generate a turbidity current (the "cloud" for snow avalanche) that will travel over long distances, almost 1,000 km in the case of the Grand Banks slide (Piper et al., 1988). The second difference is the much greater run out distance. On the seafloor of the Mississippi fan there are numerous channels which are known to have carried debris flows over distances in excess of 400 km (Schwab et al., 1996) on a channel slope less than $0.1°$ for reasons that are presented later.

As the failure takes place, and if the conditions and the configuration of the terrain are appropriate, the failure will transform into a flowing mass via some still poorly understood mechanisms which must, to some extent, imply an increase in the average water content of the flowing mass. So the approach will require the evaluation of the failure, therefore relying on soil or rock mechanics parameters and failure criteria. The post-failure will be mostly governed by fluid mechanics principles. An interesting overall perception of the phenomenon has been proposed by Meunier (1993) (Figure 9.1), which basically considered two types of flows: one phase (mostly clayey material) and two phase (sandy or rocky material). The overall behavior of the failure and post-failure will largely depend on the water-to-solid ratio as a function of time and space. The framework proposed by Meunier (1993) offers a unique way of capturing all the various physics involved in a mass movement and can be used for a simple classification of subaqueous debris flows: granular (i.e., two phase flow) and cohesive (one phase or matrix dominated flow). Under this umbrella definition, we can consider mudflows, debris flows, and debris avalanches. Norem et al. (1990) simplified this even more by separating the flowing mass into two components: dense and suspended, much as for snow avalanches, where the suspension flow represents the turbulent component of the flow (Figure 9.2).

Sec. 9.2] Formation of subaqueous mass flow 205

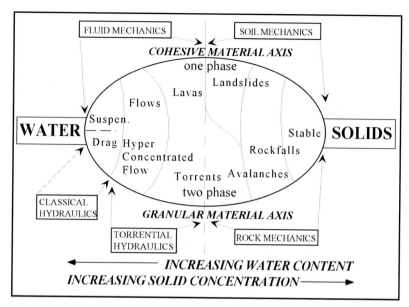

Figure 9.1. Conceptual model of mass movements involving solids and water considering granular and cohesive material.
From Meunier (1993).

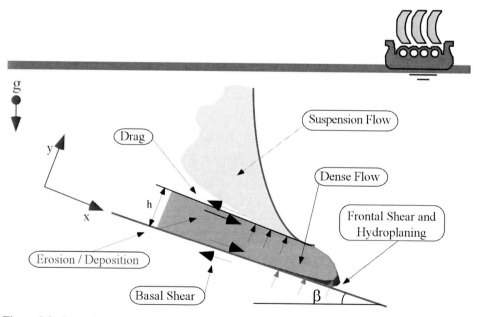

Figure 9.2. Boundary conditions during a subaqueous mass flow event. The suspension flow is created by the drag forces acting on the upper surface of the dense flow.

9.2.1 Triggering mechanism

The initiation of a subaqueous debris flow (as for their land-based counterparts) can originate from a hyperconcentrated flow, a single failure event that can produce a sufficient volume of remoulded material, or from the coalescence of smaller failures into a larger one – in this case the flow is channelled. We will focus here on those debris flows that are generated by an initial failure. The triggering mechanisms of subaqueous mass movements will vary according to each of the causes and the environments in which the mass movements occur. Submarine mass movements are triggered either by an increase in the gravitational (driving) stresses, a decrease in resisting forces (strength), or a combination of the two (Figure 9.3). The following possible triggers show the interplay of these factors.

Earthquakes

Earthquakes are called upon as a cause for many unexplained submarine landslide features (e.g., Hampton et al., 1996; Lee and Edwards, 1986). One of the reasons is that earthquake-induced shear stresses are quite large relative to shear strength in

$$FS = \frac{\Sigma \text{ Resisting Forces}}{\Sigma \text{ Gravitational forces}}$$

Reducing the Strength	Increasing the Stresses
Natural Triggers	
Earthquakes	**Earthquakes**
Wave Loading	**Wave Loading**
Tides	Tides
Sedimentation	**Sedimentation**
Gas and Gas Hydrates	
Groundwater Seepage	
Glaciation	Glaciation
	Erosion
	Diapirs
Man-made Triggers	
Gas-hydrates and reservoir depletion	Excavation Loading Subsidence

Figure 9.3. The factor of safety (FS) and the potential natural and man-made triggers of seabed instability. Triggers in bold are most effective in reducing the FS.
From Locat and Lee (2002).

fully saturated sediment. The seismic shear stress is high because the earthquake must accelerate all of the sediment column including the interstitial water. Earthquakes also generate excess pore pressures through cyclic loading and these can lower the strength more or even induce a state of liquefaction (e.g., Grand Banks slide, Piper et al., 1988). A similar situation has been invoked for the many submarine mass movements in the Saguenay Fjord (Urgeles et al., 2002; Locat et al., 2003; Levesque et al., 2004). The frequently existing layered stratigraphic configuration of subaqueous sediments can also provide a setting such that an earthquake can trigger the development of a weak layer resulting from the formation of a film of water at the interface of layers of contrasting hydraulic conductivity (Kokusho, 1999). This film of water will create a zone of very low strength, thus enhancing a rapid displacement and disintegration of the failing and help transform it into a fluid-like mixture. Such a failure mechanism can also explain why failures can take place on a failure plane dipping at angles of less than 1°.

Waves

As discussed previously, storm-waves can trigger slope failure as illustrated by the failure of offshore drilling rigs during Hurricane Camille in 1969 (Bea et al., 1983). Storm-wave induced failure actually involves several elements. A train of waves subjects the seafloor to alternating increasing and decreasing water pressure as the crests and troughs pass (Poulos et al., 1985). This non-uniform pressure field induces shear stress in the seabed with the greatest shear stresses existing below the neutral point between crest and trough. The sediment can fail after the passage of a series of waves, and it can liquefy if the pore pressures reach a high enough value. Lee and Edwards (1986) showed that there can be a balance in the importance of triggers in environments that are both seismically active and subjected to large storms. In shallower water the largest shear stresses may be induced by storm-waves and these may control offshore stability, whereas in deeper water seismic loading is more important. Clukey et al. (1985) considered another implication of wave loading effects. As the storm-wave-induced pore water pressures build up, the sediment approaches a state of liquefaction. As a result, the current velocity necessary to initiate sediment transport decreases. Accordingly, wave loading, cyclic shear stress development, and pore pressure generation leads not only to slope failure but also to enhanced bottom-current-induced sediment transport (see also Poulos et al., 1985 and Lee et al., 2004 for more details).

Tides

Failures often occur in fjords and other coastal locations during periods of low tides (e.g., Prior et al., 1982). These failures occur because of a phenomenon that engineers term "rapid drawdown" (Lambe and Whitman, 1969, p. 477). When water levels fall rapidly, pore pressures within coastal slopes often cannot adjust quickly enough. This results in an elevated water table directly adjacent to the coast and a resulting accelerated seepage of groundwater. This situation can be modelled as one of "seepage forces", effectively an additional downslope driving stress, or as excess pore pressures reducing the effective stress and the corresponding sediment shear strength. An example of such an event is the Kitimat slide reported by Prior et al. (1982).

Sediment accumulation

Rapid sediment accumulation contributes to failure in several ways. First, when sediment accumulates rapidly, most of the weight of newly added sediment is carried by pore-water pressures. The shear strength probably increases somewhat because some water will always be squeezed out, even if the coefficient of consolidation c_v is low (i.e., low permeability and high compressibility). However, the shear stress acting downslope increases more rapidly. The shear stress increases with the weight of sediment and is not influenced by pore pressure. The shear stress may also increase because more sediment may be deposited at the head of the slope than at the toe. All three of these processes push the slope toward failure: retarded strength development, increased development of shear stress because of thickness of the sediment body, and increased development of shear stress because of increases in the slope steepness. See the papers related to the Fraser River delta for examples of this type of triggering (Chillarige et al., 1997 and Christian et al., 1994).

Erosion

Localized erosion is common in deep-sea channels, submarine canyons, and other active sediment transport systems. When slopes are undercut, this can decrease the stability by increasing shear stress and in some cases decreasing the shear strength. Monterey Canyon offshore California (Greene et al., 2002) shows many examples of erosion-induced slope failures similar to those seen along rivers. Often these failures dam the canyon so that subsequent turbidity current flows are diverted and ultimately erode away part of the dam forming a new meander. These new meanders can lead to further erosion and second generation landslides (Lee et al., 2004).

Volcanoes

The existence of giant submarine landslides on the flanks of the Hawaiian Islands has been the subject of debate for at least 50 years (Normark et al., 1993). Many volcanic islands, including Hawaii, build up over pre-existing pelagic sediment bodies, often clay. This could produce a weak basal layer (Dietrich, 1988) that might contain excess pore-water pressures, although the island basalts are fairly pervious and build over millions of years. Magma pressure in the rift zones has been suggested as a trigger and was evaluated by Iverson (1995) who showed that the zones of enhanced magma pressure are too small to trigger landslides wider than a few kilometers.

Gas and gas hydrates

Gas charging of sediment is not so much a trigger as a means by which shear strength may be altered. Gas charging can affect sediment strength either by decreasing it through the development of excess pore pressures or potentially increasing it by reducing the impact of cyclic loads. In cases where gas charging reduces strength, the actual trigger causing failure is likely some other factor such as an earthquake.

Dissociation of gas hydrates can be considered a trigger because it results from environmental changes. Sea level fall has often been invoked as a means of triggering landslides through destabilization of the base of the gas-hydrate zone. Kayen and Lee (1991) modelled pore pressure generation on the continental slope of the Beaufort Sea during the last eustatic sea level drop. They determined that fluid diffusion properties dominate the process. Sultan et al. (2003) suggested a mechanism by which increases in pressure and temperature associated with the end of the last glacial period can increase the solubility of methane and induce a dissociation of methane hydrate at the top of the hydrate layer which may have contributed to the initiation of the Storegga slide (Bryn et al., 2003).

Groundwater seepage

Sangrey (1977) speculated, based on experience and proprietary information, that underconsolidation and excess pore pressures resulting from artesian reservoir sources are "very common offshore and may be the most significant mechanism" for causing slope failure. Orange and Breen (1992) suggested that pore fluids percolating up from subducted sediment could induce slope failure and lead to the development of headless canyons. Many others (e.g., Saffer and Bekins, 1999) have developed models for the ways in which subducted fluids and the resulting excess pore pressures influence the mechanics of subduction zones.

Groundwater seepage out of coastal aquifers also likely serve as a trigger for landslides. Based on an examination of morphology, Robb (1984) suggested that spring sapping (i.e., erosion of sediment and rock by underwater springs) may have occurred on the lower continental slope off New Jersey during periods of lowered sea level. In support of this suggestion, Robb (1984) observed that nearly fresh interstitial water is found beneath the continental shelf almost 100 km off the New Jersey coast. Hot fluid seeps are known to occur (Hampton et al., 2002) on the continental shelf off the Palos Verdes Peninsula in Southern California near the head of a very large submarine landslide (Bohannon and Gardner, 2004).

Diapirism

Any tectonic or diapiric deformation that results in steepened slopes will lead to a reduction in the FS of the slope and an increase in the likelihood of failure. The northern Gulf of Mexico is an area in which diapiric deformation is one of the major causes of failure on the continental slope (Martin and Bouma, 1982). The vertical growth of these structures causes local steepening of the sea floor and causes the seabed to consist of many hillocks, knolls, and ridges interspersed by topographic depressions and canyon systems. The movement of the salt sheet, or halokenesis, is largely responsible for the surface morphology (Silva et al., 2004).

Human activity

Human-constructed facilities, either along the coastline or on the seafloor have the potential for causing coastal or submarine slope failures. Typically this is because they change the distribution of stresses in the bottom sediment and can increase the downslope component. Often the role of human influence in causing observed landslides is hotly debated because fault must be assigned to damages, injuries, and even death. The debated question is typically whether a natural slope failure event affected human development or whether human development caused the slope to fail. See for example the case of the Nice slide of 1979 (Seed et al., 1988) and at Skagway (Alaska) in 1994 (Rabinovich et al., 1999).

9.2.2 Transformation

Debris-flow formation and behavior will depend on the nature of the material involved, its sensitivity, the triggering mechanism and the amount of available energy. All these components will contribute to the transformation of the initial mass into one which has acquired the necessary flow properties. In general the release and the development of a slide may be divided into three main phases:

- Initial failure and break-up of blocks and slumps.
- Transformation of the released material having a viscoplastic fluid and the development of a debris flow.
- Generation of a turbidity current due to the shear stresses on top of the debris flow producing a "cloud" of suspended matter (see also Figure 9.2).

The initial failure and break-up is illustrated by the Storegga slide (Bryn et al., 2003) as shown in Figure 9.4. It shows a retrogressive mechanism in the escarpment area (Figure 9.4(a, b)) which took place near the end of the event (Bryn et al., 2003) but is still showing a significant disintegration of the mass in the downstream direction. The extreme mobility of large portions of the Storegga sliding mass implies that it disintegrated quite rapidly and at a rate fast enough to maintain instability at the sliding front until it stabilized itself (Figure 9.4(d, e)). For the Storegga slide, the disintegration was such that the debris travelled over 400 km on a slope less than 1° (De Blasio et al., 2003).

During the course of a subaqueous slide event (or also a subaerial slide), there appears to be a process by which there are some changes in the solid-to-water ratio which provides a sufficiently low strength to allow flow to take place (Figure 9.5). If the nature of the material involved and the environmental conditions are favourable, the sliding mass will evolve and transform itself into a flowing mass. An important aspect of debris-flow initiation is the break-up mechanism by which the failing mass is shattered. As this takes place, the whole mass absorbs water with a resulting decrease in the bulk density. This is illustrated in Figure 9.5, which conceptually shows the evolution of the strength, the water content, and the density with shearing and time until it transforms into a flowing material with fluid-like characteristics. It

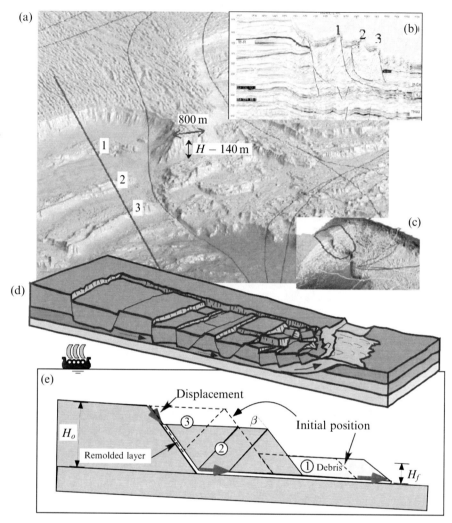

Figure 9.4. Mechanics involved in a lateral spread failure. (a) Example of a spread failure along the headwall of the Storegga slide with a view of the seismic line across several sliding blocks (location shown in (c)). (b) Seismic line along point 1, 2, and 3 in (a). (d) Schematic view of a typical spread failure on land with a well identified remolded layer over which the retrogression has taken place. (e) Cartoon of the mechanics involved in the process of retrogression showing the relative displacement of the various compartments after the lateral movements of blocks 1, 2, and 3.

(a), (b), and (c) are from Norsk-Hydro and Statoil.

is worth pointing out here that for the case of quick clays, the remoulding process reaches the lowest strength at a constant water content.

Whatever the exact nature of the phenomenon, it can be embedded here in the remoulding energy (E_r). Many hypotheses are proposed to explain the development

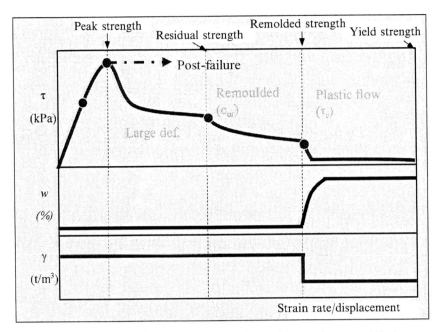

Figure 9.5. Schematic view of strength evolution with strain rate (displacement or deformation) and the corresponding state of the sediment and its transformation into a fluidized sediment by an increase in water content.

of flows, including: (1) they must take place at the time of, or soon after, failure; (2) the transformation of the original mass can result from fragmentation associated with inter-collision in rock masses (Leroueil et al., 1996; Davies et al., 2000); and (3) it may include the effects of impact with the sea floor of the rock mass (e.g., the case of chalk along the coast of England; Hutchinson, 1988) or sediments (Flon, 1982; Tavenas et al., 1983).

9.3 PARAMETERS AND PROPERTIES OF DEBRIS FLOWS

In order to conduct a debris-flow analysis, one has to know, estimate, or back-calculate the various necessary parameters as will be illustrated in the examples at the end of this chapter. The overall approach is shown in Figure 9.6 which has been used for the case of the Palos Verdes debris avalanche (Locat et al., 2004) which is presented in Section 9.6.2. To achieve the final analysis of a debris flow, one may also need to consider the more or less standard failure analysis using, for example, Mohr–Coulomb failure criteria. This analysis may require the knowledge of specific parameters that are listed in Table 9.1 but will not be described here. The focus here is on identifying the relevant parameters for the post-failure analysis of mass movement.

Sec. 9.3] **Parameters and properties of debris flows** 213

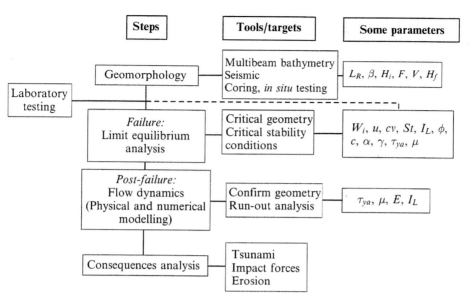

Figure 9.6. Overall approach to the analysis of subaqueous debris flows and their consequences. Symbols are listed in Table 9.1 (except for LR – run-out distance, Hi – initial height, E – energy).

9.3.1 Direct measurements

There is no data on the direct measurement of the flow properties of subaqueous debris flows, although some unsuccessful attempts were made (Syvitski and Schafer, 1989; Locat et al., 1990b; Syvitski, 1992; Couture et al., 1995). The closest to direct measurement can be derived from physical modelling of debris flows as done by Marr et al. (2001) in a large flume. But, as we will see, these are mostly used to understand the behavior and not so much to define the actual mobilized properties of a debris flow.

There is a large collection of laboratory measurements which have been carried out to measure the rheological properties of muddy debris flows (e.g., Coussot and Piau, 1994; Locat, 1997). In most cases, measurements are carried out using various types of viscometer, the Couette type being the most popular (Locat and Demers, 1988). So far, the yield strength calculated using the viscometer is quite close to what can be derived from the geometrical characteristics of the flow. Still, the viscosity measured in the lab is often one or two orders of magnitude lower than the field back-calculated values and this, according to Locat and Demers (1988), may be due to the difference in the shear rate in the lab compared to that in the field (Locat and Lee, 2002). As will be shown below, the extensive existing database on laboratory rheological measurement of muddy debris-flow material has been put together to develop useful empirical relationships.

Table 9.1. List of parameters required in various slope failure and post-failure analysis (X – primary; x – secondary relevance; RO – run out analysis; Tur – turbidity currents; Sub. – subsidence analysis; D – drained case; U – undrained case).

Parameters	Symbol	Sub failure (infinite slope) D	U	Circular failure D	U	Retrogressive failure	RO	Tur
Slope angle (°)	β	X	X	X	X	X	X	x
Slope height (m)	H	X	X	X	X	X	X	x
Geotechnical stratigraphy	–	X	X	X	X	X	x	–
Seafloor morphology	–	x	x	x	x	x	X	X
Water content (%)	w	X	X	X	X	X	X	–
Plastic limit (%)	w_P	X	X	X	X	X	X	–
Liquid limit (%)	w_l	X	X	X	X	X	X	–
Liquidity index	I_L	X	X	X	X	X	X	–
Undrained strength (kPa)	Cu	x	X	x	X	X	x	–
Remolded undrained strength (kPa)	Cu_r	–	X	–	X	X	X	–
Sensitivity	S_t	X	X	X	X	X	X	–
In situ effective stress (kPa)	σ'_v	X	–	X	–	X	–	–
Pre-consolidation pressure (kPa)	σ'_p	X	X	X	X	X	–	–
Overconsolidation ratio	OCR	X	X	X	X	X	–	–
Sensitivity	S_t	X	X	X	X	X	X	–
Friction angle (°)	ϕ'	X	–	X	–	–	x	–
Cohesion (kPa)	c'	X	–	X	–	–	–	–
Pore pressure (kPa)	u	X	–	X	–	X	x	–
Excess pore pressure (kPa)	u_e	X	–	X	–	X	x	–
Gas pressure (hydrates) (kPa)	u_g	X	x	X	x	X	x	–
Pore pressure ratio	r_u	X	–	X	–	–	X	–
Earthquake acceleration	α	X	X	X	X	–	–	–
Viscosity (Pa.s)	μ	–	–	–	–	X	X	X
Yield strength (Pa)	τ_c	–	–	–	–	X	X	X
External loads (kN)	Q	X	X	X	X	X	–	–
Hydraulic conductivity (cm/s)	k	x	–	x	–	–	x	–

9.3.2 Back analyses

Indirect measurements are often obtained from the observation or back analysis of a given event, its local morphology, and the signature left by the event along its path toward a final deposit. A fairly complete example of such an approach can be found in Schwab et al. (1996) and Locat et al. (1996) for debris flows in the Mississippi fan and by Locat et al. (2004) on the Palos Verdes slide (see Section 9.6.2). The basic idea is to derive as much information as possible from knowledge of the geomorphology and stratigraphy in all segments of the area involved in the debris-flow event as indicated in Table 9.2.

Table 9.2. Example of parameters and properties which can be derived from various field and laboratory observations.

Elements	Parameter or property
Flow thickness and slope angle	Yield strength
Run up	Local velocity
Departure zone	Volume
Accumulation zone	Validate energy versus volume
Channel depth	Maximum flow height
Clast size in clay matrix	Yield strength, liquidity index of flow mixture matrix
Core samples	Liquidity index, strength parameters

9.3.3 Empirical relationships

In many cases, we consider the mixture to be a yield stress fluid so that the rheological behavior of the matrix can be represented by a yield strength and a viscosity parameter. It has been proposed that the yield strength and viscosity of cohesive (macro-viscous) mixtures could be related to the liquidity index (I_L; Locat and Demers, 1988; Locat, 1997; Locat and Lee, 2002) for as long as the liquidity index is greater than 0 (i.e., for a water content above the plastic limit). Explanations are provided in Figure 9.7 which will help make the comparison between the terminology used in fluid mechanics and in soil mechanics. The various phases of soils are shown in Figure 9.7(a) while the concept of the liquidity index is introduced in Figure 9.7(b). The liquidity index (I_L) is defined as:

$$I_L = \frac{w - w_P}{w_l - w_P} \tag{9.1}$$

where w is the water content (assuming 100% saturation), w_L and w_P, the liquid and plastic limits, respectively. Figure 9.7(b) shows the laboratory data for three different clay soils in terms of their solid concentration and water contents. It can be readily seen that when the data are portrayed using the liquidity index, all values fall along more or less the same line. This strong normalizing power of the liquidity index is now largely used to estimate many properties including rheological properties of muddy debris flows (Locat, 1997).

Many laboratory experiments reported by Locat (1997) have shown that the rheological measurements can be partly influenced by the floc size and also by salinity in the case of the yield strength τ_c. An interesting observation reported by Locat and Demers (1988) is that the yield strength, for most clayey mixtures, contributes about 1,000 times more than the viscosity (μ) to the resistance to flow of the fluid; this ratio can decrease to 100 for silty mixtures (Jeong et al., 2004). The following relationships between liquidity index and rheological parameters were proposed by Locat (1997) for both fresh and marine water conditions:

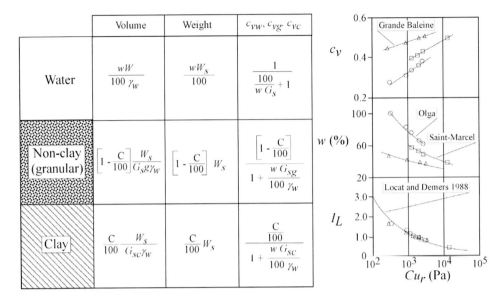

Figure 9.7. Soil phases and the use of the liquidity index and the illustration of the normalisation potential of the liquidity index in comparing the remoulded strength profile for three different clayey soils in terms of their sediment concentration (c_v) and water content (w). Locat (1997).

$$\mu = \left(\frac{9.27}{I_L}\right)^{3.3} \tag{9.2}$$

$$\tau_c = \left(\frac{5.81}{I_L}\right)^{4.55} \text{ for a pore-water salinity of about 0 g/L} \tag{9.3}$$

$$\tau_c = \left(\frac{12.05}{I_L}\right)^{3.13} \text{ for a pore-water salinity of about 30 g/L} \tag{9.4}$$

$$\mu(\text{mPa.s}) = 0.52\tau_c^{1.12} \text{ (Pa)} \tag{9.5}$$

These relationships have successfully been used by Elverhoi et al. (1997) to analyse the behavior of debris flows along the coast of Norway and by Gauer et al. (2004) for the Storegga slide. For mudflow or matrix controlled debris flows, Hampton (1972) has shown that the minimum thickness of the flowing material (H_c, in metres) for a flow to take place can be defined by the following relationship:

$$H_c = \left(\frac{\tau_c}{\gamma' \sin \beta}\right) \tag{9.6}$$

where γ' (in kN/m^3; not to be confused here with the shear rate in fluid mechanics) is the submerged unit weight and β is the slope angle (note that here the unit of τ_c is given in kPa). By modifying Eq. (9.4), for seawater, and considering the units of τ_c in kPa so that:

$$\tau_c = 2.42 I_L^{-3.13} \tag{9.7}$$

we can rewrite Eq. (9.6) partly as a function of the liquidity index:

$$H_c = \frac{2.42 I_L^{-3.13}}{(\gamma - \gamma_w) \sin \beta} \tag{9.8}$$

where γ and γ_w are respectively, the total unit weight of the sediment and the unit weight of water. Equation (9.8) is a generalization of the approach proposed by Schwab et al. (1996) for debris flows in the Gulf of Mexico to estimate the critical flow height from easily determined physical parameters. In Eq. (9.8), all parameters can be obtained easily from intact cores or even remoulded ones.

The liquidity index–yield strength relationship can also be used to back-calculate the yield strength of a debris flow at the time of the event as long as the water content of the clast is greater than that of the matrix (Figure 9.8). This assumes that no consolidation of the clast took place since deposition. Such an approach, based on the work of Hampton (1975), has been used successfully by Schwab et al. (1996) to analyse the mobility of debris flows on the Gulf of Mexico fan. Hampton (1975) considered the mixture to behave like a Bingham fluid so that the largest diameter of the clast (D_{max}) which can be supported by the clay–water slurry is calculated with the following relationship:

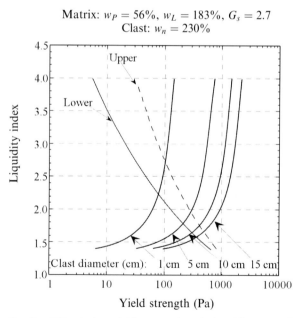

Figure 9.8. Using the liquidity index–yield strength relationship to estimate rheological properties at the time of debris flow formation. Index properties of the soil tested for this computation are given in the figure (w_P – plastic limit; w_L – liquid limit; G_s – grain density, w_n – natural water content of the clast).

$$D_{max} = \frac{8.8\tau_c}{g(\gamma'_c - \gamma'_m)} \qquad (9.9)$$

where g is the acceleration due to gravity, and γ'_c and γ'_m are the submerged unit weight of the clast and the matrix, respectively (adapted from Schwab et al., 1996). The use of Eq. (9.9) could also be another way of estimating the liquidity index, at the time of the mudflow event, using Eqs (9.3) and (9.4).

Coupling many of the above relationships, we analysed muddy sediments containing clay clasts in order to develop a nomogram which can be used to restrain the rheological parameters at the time of the mudflow event. This is illustrated by the results displayed in Figure 9.8, for sediments from the Black Sea with their physical properties as indicated on the figure. We show two extreme curves relating the liquidity index and the yield strength (from Eqs (9.3) and (9.4)), which provide a realistic range of values for both liquidity index and yield strength. Also shown in Figure 9.8 is the computation of Eq. (9.9) for different values of D_{max} (here given in centimetres). For example, if the maximum observed clast diameter is 10 cm (with physical properties as indicated in the figure), the only possible ranges of liquidity index and yield strength values of the matrix would have to fall inside the area bounded by the so-called *upper* and *lower* curves. So, for a maximum clast size of 10 cm, the rheological properties of the matrix were a value of 300–500 Pa for the yield strength and a viscosity between 270 and 370 mPa.s (using Eq. (9.2)). Moreover, if for a given sediment the relationship between I_L and τ_c has been obtained directly using a viscometer, then the potential range of values can be greatly reduced. The end result can be quite useful in trying to determine the rheological conditions under which a mudflow or a debris flow has taken place (provided that the water content of the clast has not changed since deposition, or could be estimated properly).

9.4 PHYSICAL ASPECTS OF SUBAQUEOUS DEBRIS FLOWS

There are many authors who have reviewed the physics of debris flows including Johnson (1970) and more recently Iverson (1997; see also Chapter 6). What follows are basic considerations on boundary conditions as we see it for subaqueous debris flows, and the type of flow to be modelled in order to reproduce or predict the velocity and run out distances of subaqueous mass flows. We will concentrate on aspects that are concerned with boundary conditions, fluid models, and mobility.

9.4.1 Boundary conditions

Possible boundary conditions during a flow event are illustrated in Figure 9.2. As for snow avalanches (Norem et al., 1990), the flowing material is divided into two components: dense and suspension flows. This phenomenon can take place on slopes as shallow as 0.1° (Schwab et al., 1996). Recently, Mohrig et al. (1999) have shown that once a critical velocity is reached, around 5–6 m/s, hydroplaning could also induce added mobility by reducing the shearing resistance at the base of the frontal part of the flowing mass (Figure 9.2). This process of hydroplaning is analogous to what has been observed by Laval et al. (1988) for density surges and turbidity currents. Hydroplaning debris flows can lead to stretching of the flowing mass resulting from higher velocities in the frontal part which could lead to a segmentation (or partitioning) of the debris. Recent modelling using a Computational Fluid Dynamics (CFX[1]) solver were able to reproduce the generation of "out runner" blocks (Gauer et al., 2004). At the upper surface, drag forces may lead to the re-suspension of sediments which can develop into a turbidity current (Norem et al., 1990; Figure 9.2).

During the flow, we should expect some erosion or sedimentation to take place but these phenomena still remain to be described more fully and integrated into numerical models. In some environments (e.g., the Gulf of Mexico), the flow will be channelled and, if the channel is filled and the flow height is in excess of the critical flow height, flow can proceed over long distances (Johnson, 1970; Schwab et al., 1996).

9.4.2 Fluid types

Once a mudflow or a debris flow is generated, the velocity of the flowing mass is considered to be such that the flowing material remains under undrained conditions. In this case, and considering the high rate of movement, the phenomenon is best described by means of fluid mechanics rather than soil mechanics. In the case of mudflows or muddy debris flows, the flow behavior could be represented by three types of fluids (Locat, 1997):

(1) A Bingham fluid (see also Johnson, 1970; Huang and Garcia, 1999):

$$\tau = \tau_c + \mu \gamma^n \quad (9.10)$$

(2) A Herschel–Bulkley fluid (see also Coussot and Piau, 1994):

$$(\tau - \tau_c) = K \gamma^n \quad (9.11)$$

(3) A bilinear fluid (see also O'Brien and Julien, 1988; Locat, 1997; Imran et al., 2001):

$$\tau = \tau_{ya} + \mu_{dh}\gamma - \frac{\tau_{ya}\gamma_o}{\gamma + \gamma_o} \quad (9.12)$$

where τ is the flow resistance, τ_{ya} the yield strength, μ_{dh} the viscosity, γ the shear rate, and γ_o the shear rate at the transition from Newtonian to Bingham behavior (Figure 9.9). K has units equivalent to the viscosity once the mixture is analysed as a non-yield-stress fluid using Eqs (9.11) and (9.12). The exponent n qualifies the state of the mixture, either as pseudo-plastic for $n < 1$, as a dilatant fluid for $n > 1$, or as a Bingham fluid for $n = 1$. The bi-linear model has been successfully tested against

---

[1] A CFX solver is software used in computational fluid dynamics marketed by Ansys Inc. and is oriented towards solving various problems in fluid mechanics including multiphase flow.

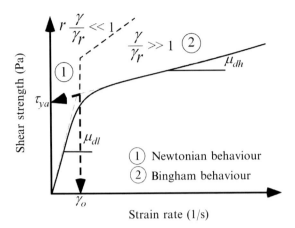

Figure 9.9. Bilinear model of Locat (1997) with the boundary conditions used for BING (1-D numerical model (Imran et al., 2001)), see the text for an explanation of the symbols.

experimental data and provides a good prediction of the movements in the run out zone (Imran et al. 2001).

In addition to the above rheological models, Norem et al. (1990) proposed to analyse the mobility of subaqueous mass movements by using a viscoplastic model described by:

$$\tau = \tau_c + \sigma(1 - r_u)\tan\phi' + \mu\gamma^n \qquad (9.13)$$

where σ is the total stress, r_u the pore-pressure ratio ($u/\gamma'h$) and ϕ' the friction angle. This constitutive equation is a sort of hybrid model, similar to what has been proposed by Suhayada and Prior (1978). The first and third terms of the equation are related to the viscous components of the flow, as in Eqs (9.11), (9.12), and (9.13). The second term is a plasticity term described by the effective stress and the friction angle. An interesting aspect of such an approach is that it can be adjusted to various flow conditions. For example, if we consider a rapidly (undrained) flowing granular flow we would be mostly using the third term of Eq. (9.13) with a value of "n" greater than 1. In the case of a mudflow (undrained), terms one and two of Eq. (9.13) would be used but the value of "n" in Eq. (9.13) would be less than or equal to 1. For flows where the velocity and the material properties are such that excess pore pressures can dissipate, the second term would dominate and the equation would approach the sliding-consolidation model proposed by Hutchinson (1986) who used a soil mechanics numerical model. For rock avalanches, the last two terms of Eq. (9.13) would be considered.

9.4.3 Mobility

In considering the mobility of a mass movement (Figure 9.10), we can distinguish two components: the retrogression (R) and the run-out distance (L). Heim (1932) first proposed looking at the mobility of a given mass involved in a landslide in terms of the geometry of the deposits before and after the slide event and he proposed the

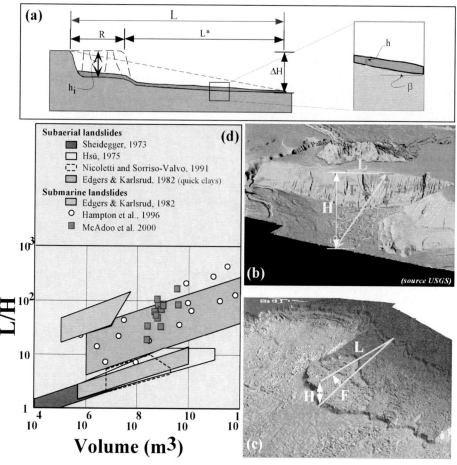

Figure 9.10. The mobility of a mass movement. (a) Basic geometry of a mass movement resulting in a debris flow. Examples for (b) Palos Verdes, and (c) Storegga slides. (d) The compilation of various mobile mass movements.

use of the term Farböschung ($F = \Delta H/L$), which represents the angle of the line joining the escarpment to the maximum distance reached by the debris. The Farböschung is commonly used to characterize the mobility of a mass movement. In such a definition, the term L (see Figure 9.10(b)), would also include R. The R parameter has also been linked to the liquidity index (I_L) by Lebuis et al. (1983). The term R becomes negligible for long travel distances but still remains a critical element for the safe positioning of sea floor structures.

Heim (1932) observed that for subaerial slides, the value of F was inversely proportional to the initial volume (V) of the sliding mass. Edgers and Karlsrud (1982) first reviewed the extent of subaqueous slides and compiled data on values of F and V for subaqueous landslides. This compilation has been updated by

Hampton et al. (1996). Figure 9.10(d) does not distinguish channelled flows, for which cases it would tend to provide much greater run-out distances. In comparison with subaerial slides, subaqueous landslides are much more mobile and tend to involve larger volumes (Figure 9.10(d)). This energy consideration is quite important because it provides a first simple control for the modelling on open slopes: a large run-out distance requires a large volume. This will prove to be important when answering the following question: is the observed mass flow deposit the result of a single event or not?

9.5 MODELLING TOOLS

The modelling of potential or old subaqueous debris flows is mostly done by using numerical models. This is useful for predicting impact pressure and run-out distances, which are the two most critical aspects of subaqueous debris flows.

9.5.1 Physical modelling

In some cases, one must rely on physical modelling in order to get a better idea of the phenomena (i.e., its physics), and also a feeling for some of the rheological parameters. Physical modelling of subaqueous debris flows and mudflows, as for turbidity currents, is typically carried in flumes. Flumes are of various dimensions and shapes. In most instances, they are about 10–20 m long, 1 m wide, and about 1–2 m deep. A recent example of work using such a technique can be found in Mohrig et al. (1998) who illustrated the onset of hydroplaning, and the work of Marr et al. (2001) on the role of a clay matrix on sandy gravity flows (Figure 9.11).

Centrifuge modelling, although quite popular for the initiation of submarine mass movements (Coulter and Phillips, 2003; Phillips et al., 2004), is not yet well designed for run-out simulations.

9.5.2 Numerical modelling

Numerical modelling of debris flows has been initated by Edgers and Karlsrud (1981) who used a simple 1-D Bingham flow model called VIFLOW. Subsequently, significant development has taken place not only on the complexity of the model but also on the types of fluids which can be modelled.

Viscous models

Imran et al. (2001) proposed a pseudo 2-D numerical model called BING which can consider three types of fluids: Bingham, Herschel–Bulkley, and bilinear. The bilinear model described by Locat (1997) (i.e., Eq. (9.12)), has been adapted for numerical modelling by Imran et al. (2001) in the following manner:

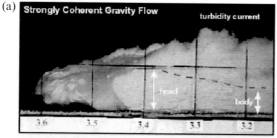
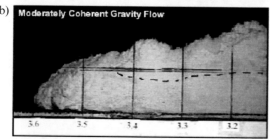
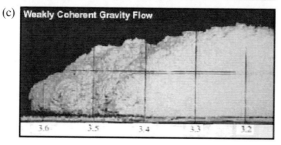

Figure 9.11. Example of flume modelling of debris flows using kaolin and fine sand mixtures (scale is in metres).
Marr et al. (2001).

$$\frac{\tau}{\tau_{ya}} = 1 + \frac{\gamma}{\gamma_r} - \frac{1}{1 + r\frac{\gamma}{\gamma_r}} \qquad (9.14)$$

where γ_r is the strain rate defined as:

$$\gamma_r = \frac{\tau_{ya}}{\mu_{dh}} \qquad (9.15)$$

and r is the ratio of the strain rates written as:

$$r = \frac{\gamma_r}{\gamma_o} \qquad (9.16)$$

One of the main parameters used in this analysis, the yield strength τ_{ya}, can be estimated from field observations of the failed mass in the accumulation zone. The values related to the viscosity (i.e., the strain rate γ_r, and the ratio of strain rates r) are estimated from a parametric analysis to find the best values which can fit the observed geometric characteristics in the runout zone.

VIFLOW, used for the study on the Mississippi fan for debris-flow analysis (Locat et al., 1990a, 1996), is a an earlier 1-D numerical model used to estimate and evaluate the steady-state velocity of debris flow as a function of slope angle, assuming a flow thickness equivalent to the channel depth (or the levee height). The model uses the following constitutive equation (Edgers and Karlsrud, 1981):

$$\tau = \tau_c + \mu_b(dv/dy) + D \tag{9.17}$$

where τ is the resistance to flow; τ_c the yield strength; μ_b the plastic viscosity; D the drag component of the shear resistance; and (dv/dy) the velocity gradient in the y direction. It is analogous to the Bingham portion of BING.

Viscoplastic models

As reported above, submarine debris flows were analysed by Norem et al. (1990) using a viscoplastic model, initially developed for snow avalanches, and who proposed Eq. (9.13) to describe the resistance to flow. The model is named NIS (old name is SKRED, which means "slide" in Norwegian) and works as a 2-D finite-difference model which outputs the shape of the flowing mass and its position in time and space (Norem et al., 1990). It is considered a 2-D model since it provides the evolution of the shape of the moving mass during the event. When applied to submarine mass movements, Norem et al. (1990) concluded that NIS could overestimate the velocity and run-out values of submarine debris flows by as much as 25% because it does not yet take into account the drag resistance from water overlying the debris flow.

Model using two phase fluids with a CFX flow solver

The CFX modelling tool has been successfully adapted recently for post-failure analysis of submarine mass movements (Gauer et al., 2004). It considers the sliding mass and the ambient water as a two-phase flow model, and for both phases the continuity equations are solved. The water is considered to be a turbulent flow and the sliding mass is described as a generalized Newtonian fluid (Balmforth and Craster, 2001) (i.e., the slide is treated as a Bingham fluid with a history-dependent yield strength and viscosity). This is the only numerical model that introduces a decay function that could more or less reproduce the reduction in the yield strength during remolding. Still it does not incorporate a space and time variation in the yield strength due to a water content increase (it assumes a constant density). There are three basic momentum equations used in the model (Gauer et al. 2004): One for the water body ($_w$):

$$\frac{\partial v_w \rho_w V_w}{\partial t} + \nabla \cdot (v_w \rho_w V_w \otimes V_w) = -v_w \nabla p + \nabla \cdot [v_w \mu_w (\nabla V_w)^T] + v_w \rho_w g - D_{WS}$$

(9.18)

one for the sliding mass ($_s$):

$$\frac{\partial v_s \rho_s V_s}{\partial t} + \nabla \cdot (v_s \rho_s V_s \otimes V_s) = -v_s \nabla p + \nabla \cdot [v_s \mu_s (\nabla V_s)^T] + v_s \rho_s g + D_{WS} \quad (9.19)$$

and one for the drag force D_{ws} at the interface between the two phases:

$$D_{ws} = C_D^*(v_s \rho_s + v_w \rho_w) v_s v_w |V_w - V_s|(V_w - V_s) \quad (9.20)$$

where V is the velocity; ρ the intrinsic density; v_w and v_s the volume fraction of the water and slide, respectively; p the pressure; and g the gravitational acceleration. This approach, like NIS, uses a finite volume approach, but provides a full 2-D modelling of the slide. Examples are available in the paper by Gauer et al. (2004).

9.6 EXAMPLES

In order to predict or back-analyse subaqueous debris flows, we propose an overall scenario which has recently been used for an analysis of the Palos Verdes debris avalanche (see Section 9.6.2). This overall approach is presented in Figure 9.6 and illustrates how we can use field (e.g., geomorphology), laboratory, and numerical analyses to get at the various parameters (e.g., Table 9.1) which are useful in the analysis. This approach also shows how the various stages of a mass movement can be used to ascertain parameters which are sometimes more difficult to obtain directly.

9.6.1 Debris flows on the Mississippi fan, Gulf of Mexico

Feature

The Mississippi fan is one of the most studied deep sea fan systems in the world (Bouma et al., 1985; Weimer and Link, 1991). The sedimentary features revealed on the surface and in the shallow subsurface of the fan are the result of various transport and depositional processes, which occurred during much of the Pleistocene until about 11,000 years B.P. (Schwab et al., 1996). Large parts of the sedimentary deposits making up the Mississippi fan originated from deep erosion of the Mississippi slope, delta front failures, and large mass movements on adjacent slopes (Walker and Massingill, 1970; Goodwin and Prior, 1989). Twichell et al. (1991) used GLORIA images and high-resolution acoustic reflection profiles to more precisely define the surface and shallow subsurface of the Mississippi fan. They defined a series of nine lobes (Figure 9.12(a)). The Mississippi canyon is seen as the source of material as revealed by buried scars reaching slope angles as steep as 23° with failure planes at about 0.3°, indicative of a low remolded shear strength (Goodwin and Prior, 1989).

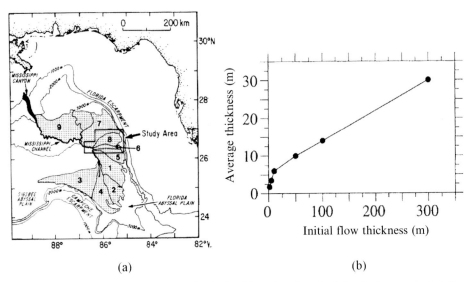

Figure 9.12. (a) Location of the distal Missippi fan showing the lobe that was selected for a study of the far-reaching debris flows. Florida is to the east and the mouth of the Mississippi River is to the north-west. (b) Relationship between the initial flow thickness and the average flow thickness (modelled using SKRED).

From Locat et al. (1996).

For a debris-flow back-analysis, we selected a portion of the youngest channel (Weimer and Link, 1991). This channel extends approximately 600 km from the lower part of Mississippi canyon, at a water depth of about 1,800 m, to the distal part of the fan at a water depth of 3,225 m. The bottom steepness starts at an initial value of 0.3° and ends at 0.06° in the distal part of the fan. The initial slope angle is based on observations of buried failure planes observed in large failures along the canyon by Goodwin and Prior (1989). Levee height information is sparsely distributed but indicates that the height ranges from 174–109 m near the mouth of the canyon to 20–1.5 m in the distal part of the fan (Bouma et al., 1985; Twichell et al., 1991).

Sediment properties

Properties must be inferred from known properties of the sediment in the source and depositional areas (Schwab et al., 1996; Locat et al., 1996) and the results are given in Table 9.3 as an average value for the index properties. Rheological properties were derived from laboratory tests (Schwab et al., 1996) and from existing relationships for soft sediment (Locat and Demers, 1988). The yield strength used for debris-flow analysis was also calibrated by using muddy clasts up to 5 cm in diameter found in the deposit. The coefficient of consolidation (c_v) was not measured but estimated, by considering a remolded sediment with a liquid limit of about 30%, at $1 \times 10^{-5}\,m^2/s$

Table 9.3. Rheological and geotechnical properties of sediment, at flow conditions, used for numerical models of debris flows on the Mississippi fan.

Property	Value
Liquid limit (%)	33
Plastic limit (%)	19
Plasticity index (%)	14
Water content (%)	90
Liquidity index (%)	5.1
Void ratio	2.5
Buoyant unit weight (kN/m^3)	5.0
Sand/mud ratio	0.47
Yield strength (kPa)	0.05
Dynamic viscosity (kPa.s)	0.00002
Coefficient of consolidation c_v (m^2/s)	1×10^{-5}
Coefficient of friction φ (°)	24

(Lambe and Whitman, 1969). The coefficient of friction of 24° is from Hooper and Dunlap (1989).

Analysis of the mobility

The first point to consider is whether or not the failure sizes observed along the canyon wall are of significant thicknesses to produce a debris flow with a flow thickness in the range of the observed levee heights? To approach this question, a parametric analysis (NIS) was carried out with the values indicated in Table 9.4 and for different thicknesses of the initial failing mass. The simulation was carried out for a range of thicknesses between 2 and 300 m, the latter simulating a canyon wall

Table 9.4. Parameters used for modelling the debris flows on the Mississippi fan.

Parameters	Symbol	VIFLOW	SKRED	Consolidation
Friction angle	φ (°)	–	24	–
Apparent friction angle	φ^* (°)[a]	–	0.06	–
Pore-pressure ratio	r_u	–	–	0.9976
Coefficient of consolidation	c_v (m^2/s)	–	–	10^{-5}
Diffusion coefficient	C_d (m^2/s^2)[b]	–	–	10^{-5}
Drag coefficient	D_c	–	–	0.01
Specific weight	γ^* (kN/m^3)	5.0	5.0	5.0
Plastic viscosity	η_b (kPa.s)	0.00002	0.00002	–
Yield strength	τ_c (kPa)	0.005	–	–
Bed friction angle	φ (°)	–	24	–

[a] Apparent friction angle, $\tan \varphi^* = (1 - r_u) \tan \varphi$.
[b] As specified in Locat et al. (1996).

228 Subaqueous debris flows [Ch. 9

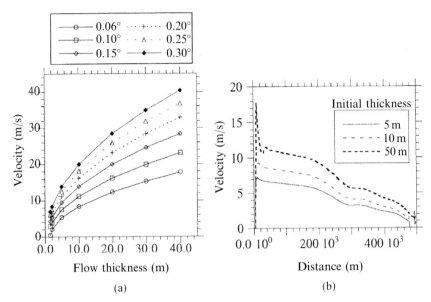

Figure 9.13. (a) Flow thickness and steady-state velocity reached for various slope angles corresponding to those along the slope profile obtained using VIFLOW. (b) Velocity run-out distance simulation using SKRED with an initial thickness varying from 5–50 m.
From Locat et al. (1996).

failure and the former, a pro-delta failure. The total length of the failure plane, in all cases, was set at 5,000 m, which is what has been observed for many large landslide scars along the canyon (Goodwin and Prior, 1989). Results are shown in Figure 9.12(b) for values integrated and averaged over the full run-out distance (maximum of 600 km). For an initial failing mass that is about 100 m thick, the average flow thickness is about 15 m, which is less than the channel depth in most of the proximal part of the fan. The spreading coefficient S ($S = H_f/H_i$; where H_f and H_i are the flow and initial flow thickness respectively) for an initial thickness greater than 10 m can be as low as 0.1. Therefore, a failure in the canyon would have to be about 250 m deep in order to fill the channel at the exit from the canyon and overbank significantly in the proximal part of the fan. Such a large failure has been described within the canyon by Goodwin and Prior (1989) who found displaced blocks about 150 m thick and mapped buried back scarps about 200–300 m high.

Run-out velocities were simulated with VIFLOW (Figure 9.13(a)) and NIS (Figure 9.13(b)). In the latter case, results are given for the frontal velocity. VIFLOW results indicate that an initial flow thickness of 40 m on a slope of 0.3° could reach a steady-state velocity of about 40 m/s. A flow thickness of 10 m, on the same slope, would reach a velocity of about 20 m/s. As most of the profile in the lower fan area is below 0.1°, and levee height is less than 20 m, the velocity would be less than 15 m/s. The flow will cease when the remaining apparent friction angle of the flowing mass is equal to the slope angle. The flow slows when the flow thickness reaches a critical value H_c, estimated using Eq. (9.6), which for this sample is 1.5 m (Schwab et al., 1996).

These results provide a first approximation of the expected flow velocities along the slope profile, assuming that the steady-state velocity can be reached. These velocities are high and could produce some turbulence at the base of the flow underneath the plug zone (Locat et al., 1996).

NIS numerical results are shown in Figure 9.13(b) for a range of flow thickness from 5–50 m. Figure 9.13(b) indicates that the peak velocity varies from 5 to 18 m/s for an initial average flow thickness that varies from 5 to 50 m. For most of the proximal part of the fan, a flow thickness between 5 m and 10 m would provide a range of velocities between 5 and 10 m/s and, for the more distal part of the fan, a velocity less than 5 m/s. Such velocities are within the range measured or estimated for other submarine debris flows (Norem et al., 1990; Hampton et al., 1996). With a flow thickness above 5 m, these velocities would apply to the Mississippi channel and the upper portions of the depositional lobes. This numerical simulation also indicates that, as the distal part of the fan is reached, the average velocity for most initial conditions reaches a value between 2 and 5 m/s. Such values are similar to those estimated for that part of the fan by Schwab et al. (1996).

Locat et al. (1996) have also shown that, considering the fine grained nature of the sediments and the mass flow velocity, the duration (about 0.6 day) was short enough to allow us to neglect any consolidation effect during the flow (i.e., no significant increase in the yield strength or viscosity).

9.6.2 The Palos Verdes debris avalanche, California

Features

The Palos Verdes slide (Bohanon and Gardner, 2004; Lee et al., 2003; Figure 9.14) is located along the San Pedro escarpment just offshore of Los Angeles. The seafloor lying at the base of the escarpment is the San Pedro basin. The slope itself is formed of sedimentary rocks dipping between 10 and 15°. The slope is eroded by a series of gullies which are 2–4 km apart. The base of the slope more or less coincides with the trace of the San Pedro fault (Bohannon and Gardner, 2004). The slide took place along a steep escarpment, mobilized into a debris avalanche and travelled a distance of about 8 km out onto the adjacent basin floor (Figure 9.14(b)). The head scarp is about 600 m high and the slope varies between 10–20°. The debris was dispersed over a wide area shown in Figure 9.14(a). From seismic records the thickness of the debris deposit varies from about 20 m in the lower part of the slope to less than 1 m at a distance of 8 km away from the base of the slope, with an average thickness of 5–10 m (Figure 9.14(b)). The slide occurred about 7,000 years ago (Normark et al., 2004). The coupling of both the seismic survey and the multibeam survey provides a comprehensive picture of the nature of this slide and its extent. An analysis of the run-out distance of the debris, indicates that the initial sliding mass was large enough and had sufficient potential energy to trigger a tsunami and reach the observed run-out distance (Locat et al., 2004).

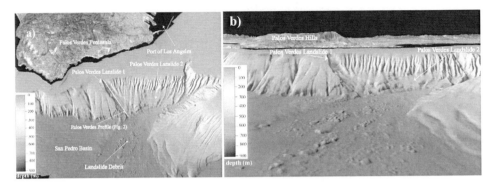

Figure 9.14. (a) Shaded relief multibeam imagery of the Palos Verdes margin near Los Angeles, California, showing landslide features. (b) An oblique view of the slide.
From Lee et al. (2003).

Selecting the parameters

Before using BING, we needed to evaluate the required parameters, in particular, the yield strength (τ_{ya}). This was achieved by observing the morphology in the depositional zone. As proposed by Johnson (1984), and using Eq. (9.6), with a buoyant unit weight of $15\,\text{kN/m}^3$, and assuming that the debris-flow rheology can be approximated by a Bingham fluid, we can use field observations to evaluate the yield strength. Results of such a parametric analysis are shown in Figure 9.15. Here, we have calculated yield strength τ_{ya} for various slope angles β_f and thicknesses H_f. The gray zone in Figure 9.15 delineates the range of the yield strength for the observed field thickness varying between 10 m and 15 m. Based on this analysis, a yield strength of 5,000 Pa is used in most of the following computations. We do not know of published values for the yield strength of debris avalanche materials. However, a value of 5,000 Pa is comparable to measured values obtained for bouldery debris flows in China, where the value of τ_{ya} was measured at 2,500 Pa for a flow thickness of about 3 m.

The next step is to determine terms related to rheology (γ_r and r, Eqs. (9.16) and (9.17)) using the run-out distance as a way of calibrating the computations. This is done using BING with the following conditions: $H_i = 70\,\text{m}$, $\tau_{ya} = 5,000\,\text{Pa}$, and $W_i = 1,000\,\text{m}$ with a slope profile corresponding to $\beta = 10°$ in the starting zone. The shape of the mass involved at the onset of the slide is given a parabolic form, a default geometry in BING. According to Imran et al. (2001), one could build in a rectangular shape, but the overall impact is minimal, considering the length of the sliding mass relative to its average thickness. First, we looked at the strain rate γ_r by assuming a value of 1,000 for r. Results are given in Figure 9.16(a) which shows that if the value of γ_r exceeds 10, the computed run-out distance becomes much larger than the observed range. A value of 1 was selected.

A similar approach was used for estimating the value of r. Assuming a value of γ_r equal to 1.0, we found that a value of r smaller than 1,000 would quickly result in

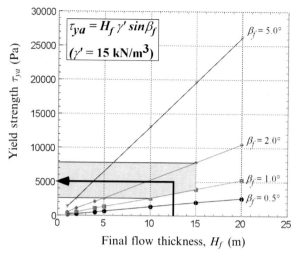

Figure 9.15. Using the flow deposit thickness (H_f) in the run-out zone to estimate the yield strength using the relationship proposed by Johnson (1984) for the Palos Verdes debris avalanche.
From Locat et al. (2004).

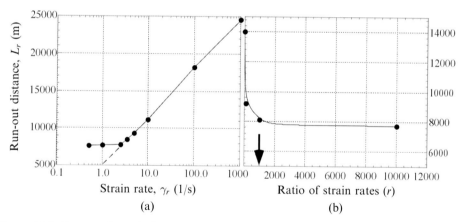

Figure 9.16. Parametric analysis on the values of the strain rate (γ_r) and the ratio of strain rates r, using the run-out distance L_R for calibration for the Palos Verdes debris avalanche.
From Locat et al. (2004).

very high computed values of run-out distances (Figure 9.16(b)). In summary, and for the rest of the analysis, we will use a value for γ_r of 1.0 and a value for r of 1,000.

Before completing the final selection of parameters, we applied the above values for viscosity in an analysis to show the effect of both the slope angle β in the starting zone and the selected yield strength τ_{ya}, on the computed values of the run-out distance L_R.

Results are given in Figure 9.17(a) with the grey zone delimiting the most probable values. From the results shown in Figure 9.17(a), it appears that, for the

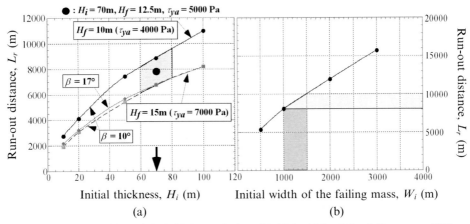

Figure 9.17. Effect of both inital thickness (H_i) and initial width of the failing mass (W_i) on the run-out distance (L_R) for the Palos Verdes debris avalanche.
From Locat et al. (2004).

model used here, the selection of the yield strength has a significant effect on the computed value of the run-out distance. On the other hand, with the same value of yield strength, the value of slope angle β in the starting zone has little influence on the overall run-out distance.

Finally, using the mid-point value, shown as a large dot in Figure 9.17(a), we investigated the effect of the initial size W_i of the failed mass on the computed run-out distance (Figure 9.17(b)). Results indicate that the range of observed run-out distances could be reached for an initial width between 1,000 m and 1,500 m.

From an analysis of Figure 9.17, we make an important observation: in order for the failed mass to reach the observed run-out distances, the initial size has to be equivalent to that deduced in the failure analysis and, moreover, it has to more or less move as a single mass and in a single event.

Mobility analysis

The mobility of the Palos Verdes slide will be investigated with the parameters selected from the above scenario. The conditions used are: $W_i = 1,000$ m, $H_i = 70$ m, $\beta = 10°$, $\tau_{ya} = 5,000$ Pa, $\gamma_r = 1.0$, and $r = 1,000$.

For post-failure analysis of the failed mass, we used BING (Imran et al., 2001). In our analysis, we assumed that the rheological properties of the Palos Verdes debris avalanche could be described by a bilinear rheological model. The bilinear model assumes that the initial phase of the flow is Newtonian (1 in Figure 9.9) and evolves, after reaching a threshold shear rate value (γ_o), into a Bingham-type flow (2 in Figure 9.9).

The velocity of the frontal element is shown in Figure 9.18(a). The peak velocity predicted by BING reaches a peak value of about 45 m/s after only about 1,000 m of

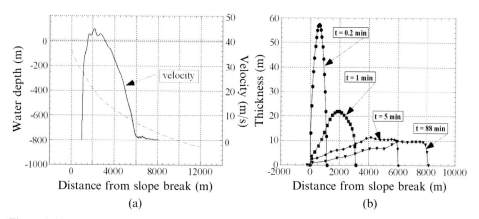

Figure 9.18. Evolution of (a) the velocity of the frontal element and (b) shape of the flowing mass during the flow event for the Palos Verdes debris avalanche.
From Locat et al. (2004).

movement (Figure 9.18(a)). Then it decreases regularly until the flow reaches a distance of about 6,000 m (measured from the crest of the slope). After this point, the velocity diminishes very slowly until the flow stops at a distance of about 8,000 m.

As the flow takes place, the initial mass evolves from a parabolic shape into a stretched mass with an average final thickness (H_f) that is 7–8 times less than the average initial value (Figure 9.18(b)). The final thickness of about 10 m has been more or less pre-built into the computation given that we used the final thickness to estimate yield strength. When using the BING model, the flowing mass must still be anchored to the initial point so that it is not possible for it to split during the flow. The final thickness is more or less reached only 5 minutes after the start of the flow event.

It is interesting to follow the mobility of the frontal and middle elements as a function of time (Figure 9.19). As observed by others (e.g., Norem et al., 1990), the flowing mass stretches. To do so, the acceleration of each individual element must gradually decrease as we move back into the flowing mass. Therefore, the frontal element must move much faster and farther than an element at the middle (Figure 9.19(a)). From a point of view of the acceleration, the same will apply (i.e., the frontal element will experience a much more rapid acceleration than those in the back). Here, the initial peak acceleration of the frontal element is at about $2.5\,\text{m/s}^2$, while it is about $0.75\,\text{m/s}^2$ for the middle element.

In summary, to reach the observed run-out distances, the flowing mass must be large and the mass must move more or less as a single mass. The initial geometry of the sliding mass, as determined from the 2-D limit equilibrium analysis, provides a good estimate of the conditions under which this sliding event took place, and shows that the final debris avalanche deposits resulted from a single event.

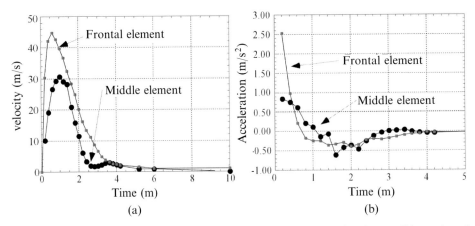

Figure 9.19. Velocity (a) and acceleration (b) of the flowing mass for the conditions given in Figure 9.16 for the Palos Verdes debris avalanche.
From Locat et al. (2004).

9.7 SUBAQUEOUS DEBRIS FLOWS HAZARDS AND CONSEQUENCES

With the recent development of improved sea-floor mapping techniques (Prior and Doyle, 1993; Locat et al., 1999; Locat and Sanfaçon, 2002) numerous subaqueous mass movements and their deposits have been mapped all around the world (Locat and Mienert, 2003; Mienert and Weaver, 2003; McAdoo et al., 2000). For example, on the Atlantic seaboard of the USA, Booth et al. (1993) reported more than 120 mass movements signatures ranging in aerial extents from less than 1 km^2 to more than 1,000 km^2 with more than 60% on slopes inclined at less than 4°. In many areas, such as in the Gulf of Mexico, they pose a threat to offshore facilities (Silva et al., 2004; Orange et al., 2003; Young et al., 2003). What is at risk are coastal populations (e.g., effect of tsunamogenic landslides), coastal infrastructure, offshore facilities related to resources development and exploitation, and transport facilities like pipelines and communication cables.

As shown below, recent events can be used to illustrate the type of risk undergone by coastal infrastructures and communities.

9.7.1 Subaqueous debris-flow hazards

Interest for hazard and risk assessment for submarine mass movements is quite recent, the first known detailed analysis being the work done by Favre et al. (1992). Many of the approaches proposed by Cruden and Fell (1997) for subaerial mass movements would apply here also. The same is true for post-failure behavior (Leroueil et al., 2003; Hungr, 2004). Varnes et al. (1984) defined the total risk R_T as the set of damages resulting from the occurrence of a phenomenon. It can be described by the following equation:

$$R_T = \Sigma H R_i V_i \tag{9.22}$$

in which H is the hazard or the phenomenon occurrence probability within a given area (i.e., that can have an impact on the area, similar to the meaning of "danger" in French) and a given time period; R_i (for $i = 1$ to n) are the elements at risk, potentially damaged by the phenomenon; V_i is the vulnerability of each element represented by a damage degree ranging from 0 (no loss) to 1 (total loss). As indicated by Leroueil et al. (2003) and Hungr (2004) for the case of debris-flow hazard (H_{pf} of Leroueil et al., 2003), the hazard is not only related to the initial slide event (H_f), but also to the encounter probabilities:

$$H = H_f \cdot H_{pf} \qquad (9.23)$$

If the possibility of a tsunami is considered, the probability that a given coast be struck by a tsunami (H_{ts}) of a given amplitude is:

$$H = H_f \cdot H_{pf} \cdot H_{ts} \qquad (9.24)$$

A major issue here relates both to the frequency of a potential slide event in the context of the geological evolution of the area (Locat, 2001), and to the likelihood of evolving into a debris flow. To consider the geological evolution of a given area requires that some knowledge must be acquired that can evaluate in which direction the triggering factors and site morphology are evolving. For example, a canyon which is being infilled should be evolving in an increased stability mode rather than when it is forming.

Resolving the above issues requires the establishment of a database with well documented case histories which, if possible, need to have been dated. In some cases, if the frequency is difficult to evaluate, one can also carry a regional analysis leading to the development of a landslide susceptibility map, as was done by Lee et al. (1999) for the area offshore of Los Angeles.

For ocean margins, the hazard component lacks data on frequencies and magnitude. To achieve this component, one may have to rely on detailed studies of the sediment record (e.g., turbidites, seismites, debrites) which can be dated according to various methods. Still, in most cases, we know only the intensity but little about the timing. Much more effort is required for this aspect.

Leroueil et al. (2003) have proposed an approach to evaluate the hazards and risk associated with submarine mass movements and their consequences which is illustrated in Figure 9.20. Such an approach incorporates knowledge of the geological setting, the environmental forcing, the hazard evaluation, the elements at risk, and the risk assessment itself. Some elements of this approach have been used for the Hudson-Apron post-failure analysis by Desgagnés et al. (2001). For the Hudson-Apron, the debris-flow mechanics was analysed using a Bingham and a bilinear fluid model. The predisposition factors were related to the presence of weak layers and possibly the presence of excess pore pressures in the underlying formation and of gas hydrates which may have helped provide a rapid acceleration of the flowing mass. The revealing factors were the presence of many debris flows. Since these events occurred a few thousand years ago, their consequences were limited to the destruction of marine habitats. In today's situation it could result in the destruction of a

236 Subaqueous debris flows [Ch. 9

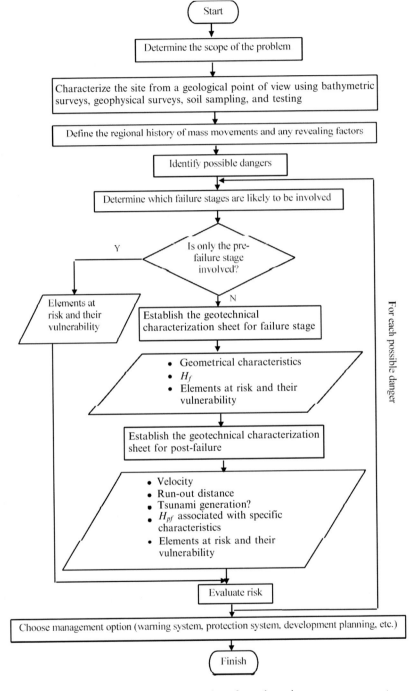

Figure 9.20. Risk management procedure for submarine mass movements.
From Leroueil et al. (2003).

seafloor structure, erosion, and possibly the generation of a tsunami along coastal areas.

9.7.2 Consequences of debris flows

The consequence or risk associated with submarine mass movements and their related debris flows are briefly illustrated by three cases: Nice Airport (1979), Skagway terminal in Alaska (1994), and Papua New Guinea (1998).

Nice, France

This event took place offshore of Nice in France in 1979 and involved a portion of the Var delta and airport runway fill under construction (Seed et al., 1988). The slide contained about 10 million m^3 of fill and native material and occurred over a period of about 4 minutes. The debris flow moved down the sloping face of the delta deposit, along an offshore canyon and finally out onto an abyssal plain, eventually rupturing 2 sets of cables as far as 120 km offshore from Nice. Investigators estimated that several hundred million m^3 of material were involved in the event, much more than the amount that failed near the construction area. A tsunami struck 120 km of coastline with a maximum amplitude of 3 m. Several lives were lost and considerable damage was done to local communities and harbours. Two hypotheses were advanced to explain the failure at Nice: (1) the construction area failed first, perhaps because of the construction activity, the sliding mass transformed into a debris flow which moved downslope, undercutting canyon walls and causing continued failure of considerably more natural material. (2) There was a large natural underwater landslide that caused a tsunami. The tsunami caused a "rapid drawdown" condition, as discussed above, and this produced a failure in the newly constructed fill. Seed et al. (1988) presented evidence supporting the second explanation. Many others supported the first explanation, and considerable debate has followed both in the scientific literature and in court.

Skagway, Alaska

The second case is similar in that it involved a coastal landslide, a death, and legal action. This case occurred in Skagway, Alaska, in 1994, when a dock that was under construction slid into a fjord (Cornforth and Lowell, 1996). The event occurred during a particularly low tide and was accompanied by a series of tsunami waves estimated to be as high as 11 m. Subsequent surveys showed that a submarine landslide had occurred. Again the debate concerned whether the dock construction caused the landslide or if a large natural landslide caused the dock to fail. Rabinovich et al. (1999) developed a model that appeared to show that the tsunami was caused by the dock failure and not by an external submarine landslide. Others (e.g., Mader, 1997) have argued to the contrary. As part of the run-out, little damage was observed, with the potential of destroying or disrupting fishing habitats.

Seattle, USA

A final example of a human-induced submarine landslide occurred in 1985 near Duwamish Head in Seattle, Washington (Kraft et al., 1992). The landslide occurred during dredging operations that were undertaken to extend a sewage effluent pipe about 3 km out into Puget Sound. The slide occurred during low tide, involving 400,000 m^3 of sediment. The slide was attributed to a collapse mechanism in the sediment, causing static liquefaction. The landslide was clearly triggered by the low tide, but the dredging operation was an underlying cause.

Of great importance for risk assessment of debris flows is there potential long run-outs (i.e., a few hundred of kilometres). This alone requires that for some projects, the size of the investigated area has to be quite wide in order to properly assess the risk.

9.8 CONCLUSIONS

Subaqueous debris flows have many similarities with their terrestrial equivalent, mostly in terms of the basic flow mechanics:

- The same types of fluid can be used to model the flow (Bingham, Herchel-Bulkley, and bilinear).
- Empirical relationships using the liquidity index to estimate the rheological properties can be used successfully.

Still, the subaqueous environment provides particular conditions that make their study different on many aspects, including:

- the triggering mechanism can include the role of waves, tides, and gas hydrates;
- the role of water during the transition from failure to post-failure as a mechanism to reduce the strength;
- the presence of strong sediment layering favours the formation of a weak layer resulting from the generation of a film of water at the interface of layers of contrasting hydraulic conductivity;
- the role of water for providing high mobility and long run-out (hydroplaning);
- they can flow on slopes with an angle often less than 1°;
- they can reach very long distances, in some cases in excess of 500 km.
- numerical models must incorporate a two-phase flow or at least take into account the drag forces on the top and buoyancy;
- reflection seismic and multibeam surveys are particularly relevant to the study of architecture and morphology of debris-flow deposits; and
- there has not yet been any real-time observation of subaqueous debris flows in nature and our vision of the process relies largely on the use of flume experiments.

With these above considerations, it is clear that the evaluation of a subaqueous debris-flow hazard and the associated risk is not easy and requires the mobilization of significant equipment often over a very large area. The field of submarine mass movements and their consequences is growing rapidly and this is largely due to an increase in economic pressure to develop the natural resources along the margins, the need for improve communication cable deployment, the need to minimize risk to coastal communities, and the protection of marine habitats.

9.9 ACKNOWLEDGEMENTS

This chapter on subaqueous debris flows is the results of many years of research supported mainly by the following organizations: the US Office of Naval Research, National Science and Engineering Research Council of Canada, and the Fond québécois pour la recherche en sciences naturelles et en technologie (FQRNT). This work has greatly benefited from the contribution of the many colleagues and graduates students involved in our projects over the years.

9.10 REFERENCES

Balmforth, N.J. and Craster, R.V. (2001) Geophysical aspects of non-Newtonian fluid mechanics. In: N.J. Balmforth and A. Provenzale (eds), *Geomorphological Fluid Mechanics* (Lecture Notes in Physics No. 582). Springer-Verlag, New York.

Bea, R.G., Wright, S.G., Sircar, P., and Niedoroda, A.W. (1983) Wave-induced slides in South Pass Block 70, Mississippi Delta. *Journal of Geotechnical Engineering*, **109**, 619–644.

Bohannon, R.G. and Gardner, J.V. (2004) Submarine landslides of San Pedro Escarpment, southwest of Long Beach, California. *Marine Geology*, **203**, 261–268.

Bouma, A.H., Stelting, C.E., and Coleman, J.M. (1985). Mississippi Fan, Gulf of Mexico. In: A.H. Bouma, W.R. Normark and N.E. Barnes (eds), *Submarine Fans and Related Turbidity Systems* (pp. 143–150). Springer-Verlag, New York.

Bryn, P., Solheim, A., Berg, K., Lien, R., Forsberg, C.F., Haflidason, H., Ottensen, D., and Rise, L. (2003) The Storegga slide complex: Repeated large slumping in response to climatic cyclicity. In: J. Locat and J. Mienert (eds), *Submarine Mass Movements and Their Consequences* (Kluwer Advances in Natural and Technological Hazards Research No. 19, pp. 215–222). Kluwer Academic, Boston.

Chillarige, V.A., Morgenstern, N.R., Robertson, P.K., and Christian, H.A. (1997) Seabed instability due to flow liquefaction in the Fraser River delta. *Canadian Geotechnical Journal*, **34**, 520–533.

Christian, H.A., Mulder, T., Courtney, R.C., Mosher, D.C., Barrie, J.V., Currie, R.G., Olynyk, H.W., and Monahan, P.A. (1994) Slope stability on the Fraser River delta foreslope, Vancouver, British Columbia. *Proceedings of the 47th Canadian Geotechnical Conference, Halifax* (pp. 155–165). ByTech Publishing.

Clukey, E., Kulhawy, F.H., Liu, P.L.F., and Tate, G.B. (1985) The impact of wave loads and pore-water pressure generation on initiation of sediment transport. *Geo-Marine Letters*, **5**, 177–183.

Cornforth, D.H. and Lowell, J.A. (1996) The 1994 subaqueous slope failure at Skagway, Alaska. In: K. Senneset (ed.), *Proceedings of the 7th International Symposium on Landslides* (Vol. 1, pp. 527–532). A.A. Balkema, Rotterdam.

Coulter, S.E. and Philipps, R. (2003) Simulating submarine slope instability initiation using centrifuge model testing. In: J. Locat and J. Mienert (eds), *Submarine Mass Movements and Their Consequences* (Kluwer Advances in Natural and Technological Hazards Research No. 19, 29–36). Kluwer Academic, Boston.

Coussot, P. and Piau, J.-M. (1994) On the behavior of fine mud suspension. *Rheologica Acta*, **33**, 175–184.

Couture, R., Konrad, J.-M., and Locat, J. (1995) Analyse de la liquéfaction et du comportement non drainé des sables du delta de Kenamu (projet ADFEX). *Canadian Geotechnical Journal*, **32**, 137–155 [in French].

Cruden, D.M. and Fell, R. (1997) *Landslide Risk Assessment* (371 pp.). A.A. Balkema, Rotterdam.

Davies, T.R., McSaveney, M.J., and Hodgson, K.A. (2000) A fragmentation-spreading model for long run-out rock avalanches. *Canadian Geotechnical Journal*, **36**, 1096–1110.

De Blasio, F.V., Issler, D., Elverhoi, A., Harbitz, C.B., Ilstad, T., Bryb, P., Lien, R., and Lovholt, F. (2003) Dynamics, velocity and run out of the giant Storegga slide. In: J. Locat and J. Mienert (eds), *Submarine Mass Movements and Their Consequences* (Kluwer Advances in Natural and Technological Hazards Research No. 19, pp. 223–230). Kluwer Academic, Boston.

Desgagnés, P., Locat, J., Lee, H., and Leroueil, S. (2001) Le glissement du Hudson Apron: un exemple d'application de la caractérisation géomécanique des mouvements de masse. *Proceedings of the 54th Canadian Geotechnical Conference, Calgary, Alberta* (pp. 816–823).

Dietrich, J.H. (1988) Growth and persistence of Hawaiian rift zones. *Journal of Geophysical Research*, **93**, 4258–4270.

Edgers, L. and Karlsrud, K. (1981) *Viscous Analysis of Submarine Flows* (Report 52207, 21 pp.). Norwegian Geotechnical Institute, Norway.

Edgers, L. and Karlsrud, K. (1982) Soil flows generated by subaqueous slides: Case studies and consequences. In: C. Chryssostomomidis and J.J. Connor (eds), *Proceedings of the 3rd International Conference on the Behavior of Offshore Structures* (pp. 425–437). Hemisphere Publ. Corporation, U.S.A.

Elverhoi, A., Norem, H., Andersen, E.S., Dowdeswell, J.A., Fossen, I., Haflidason, H., Kenyon, N.H., Laberg, J.S., King, E.L., Sejrup, H.P., Solheim, A., and Vorren, T. (1997) On the origin and flow behavior of subaqueous slides on deep-sea fans along the Norwegian–Barents Sea continental margin. *Geo-marine Letters*, **17**, 119–125.

Favre, J.L., Barakat, B., Yoon, S.H., and Franchomme, O. (1992) Modèle de risque de glissements de fonds marins. In: D.H. Bell (ed.), *6th International Symposium on Landslides* (pp. 937–942). A.A. Balkema, Rotterdam [in French].

Flon, P. (1982) Énergie de remaniement et régression des coulées d'argiles. M.Sc. thesis, Department of Civil Engineering, Laval University, Québec, Canada [in French].

Gauer, P., Kvalstad, T.J., Forsberg, C.F., Bryn, P., and Berg, K. (2004) The last phase of the Storegga slide: Simulation of retrogressive slide dynamics and comparison with slice-scar morphology. *Marine Geology* (in press).

Goodwin, R.H., and Prior, D.B. (1989) Geometry and depositional sequences of the Mississippi Canyon, Gulf of Mexico. *Journal of Sedimentary Petrology*, **59**, 318–329.

Greene, H.G., Maher, N.M., and Paull, C.K. (2002) Physiography of the Monterey Bay national Marine Sanctuary and implications about continental margin development. *Marine Geology*, **181**, 55–82.

Hampton, M.A. (1972) The role of subaqueous debris flows in generating turbidity currents. *Journal of Sedimentary Petrology*, **42**, 775–793.

Hampton, M.A. (1975) Competence of fine debris flows. *Journal of Sedimentary Petrology*, **45**, 834–844.

Hampton, M.A., Lee, H.J., and Locat, J. (1996) Subaqueous landslides. *Reviews of Geophysics*, **34**, 33–59.

Hampton, M.A., Karl, H.A., and Murray, C.J. (2002) Acoustic profiles and images of the Palos Verdes margin: Implications concerning deposition from the White's Point outfall. *Continental Shelf Research*, **22**, 841–858.

Heim, A. (1932) *Bergstruz und Menschenleben* (218 pp.). Fretz und Wasmuth Verlag, Zurich, Switzerland [in German].

Hooper, J.R. and Dumlap, W.A. (1989) Modeling soil properties on the continental slope, Gulf of Mexico. *Proceedings, Offshore Technology Conference*, **21**(1), 677–688.

Huang, X. and Garcia, M.H. (1999) Modeling of non-hydroplaning mudflows on continental slopes. *Marine Geology*, **154**, 131–142.

Hungr, O. (2004) Geotechnique and management of landslide hazard. *Proceedings of the 57th Canadian Geotechnical Conference, Québec City* (Paper G33.411, Session 4C, pp. 1–10).

Hutchinson, J.N. (1986) A sliding-consolidation model for flow slides. *Canadian Geotechnical Journal*, **23**, 115–126.

Hutchinson, J.N. (1988) Morphology and geotechnical parameters of landslides in relation to geology and hyrdogeology. *Proceedings of the 5th International Symposium on Landslides, Lausanne* (Vol. 1, pp. 3–35).

Imran, J., Parker, G., Locat, J., and Lee, H.J. (2001) 1-D numerical model of muddy subaqueous and subaerial debris flows. *ASCE Journal of Hydraulic Engineering*, **127**, 959–968.

Iverson, R.M. (1995) Can magma-injection and groundwater forces cause massive landslides on Hawaiian volcanoes? *Journal of Volcanology and Geothermal Research*, **66**, 295–308.

Iverson, R.M. (1997) The physics of debris flows. *Reviews of Geophysics*, **35**, 245–296.

Jeong, S.W., Locat, J., and Leroueil, S. (2004) A preliminary analysis of the rheological transformation due to water infiltration as a mechanism for high mobility of submarine mass movements. *Proceedings of the 57th Canadian Geotechnical Conference, Quebec City* (Paper G36.362, Session 7g, pp. 15–22).

Johnson, A.M. (1970) *Physical Processes in Geology* (557 pp.). W.H. Freeman, New York.

Johnson, A.M. (1984) Debris flows. In: D. Brunsden and D.B. Prior (eds), *Slope Instability* (pp. 257–361). John Wiley & Sons, New York.

Kayen, R.E. and Lee, H.J. (1991) Pleistocene slope instability of gas hydrate-laden sediment on the beaufort sea margin. *Marine Geotechnology*, **10**, 125–141.

Kokusho, F.K. (1999) Water film in liquefied sand and its effect on lateral spread. *Journal of Geotechnical and Geoenvironmental Engineering*, **125**, 817–826.

Kraft, L.M., Gavin, T.M., and Bruton, J.C. (1992) Submarine flow slide in Puget Sound. *ASCE Journal of Geotechnical Engineering*, **118**, 1577–1591.

Lambe, T.W. and Whitman, R.V. (1969) *Soil Mechanics* (553 pp.). John Wiley & Sons, New York.

Lastras, G., Canals, M., and Urgeles, R. (2003) Lessons from sea-floor and subsea-floor imagery of the Big '95 flow scar and deposit. In: J. Locat and J. Mienert (eds),

Submarine Mass Movements and Their Consequences (Kluwer Advances in Natural and Technological Hazards Research No. 19, pp. 425–431). Kluwer Academic, Boston.

Laval, A., Cremer, M., Beghin, P., and Ravenne, C. (1988) Density surges: Two-dimensional experiments. *Sedimentology*, **35**, 73–84.

Lebuis, J., Robert, J.-M., and Rissmann, P. (1983) Regional mapping of landslide hazard in Québec. *Symposium on Slopes in Soft Clays, Linkoping, Sweden* (Report No. 17, pp. 205–262). Swedish Geotechnical Institute, Sweden.

Lee, H.J. and Edwards, B.D. (1986) Regional method to assess offshore slope stability. *Journal of Geotechnical Engineering*, **112**, 489–509.

Lee, H.J., Locat, J., Dartnell, P., Israel, K. and Wong, F. (1999) Regional variability of slope stability: Application to the Eel margin, California. *Marine Geology*, **154**, 305–321.

Lee, H.J., Kayen, R.E., Gardner, J.V., and Locat, J. (2003) Characteristics of several tsunamigenic submarine landslides. In: J. Locat and J. Mienert (eds), *Submarine Mass Movements and Their Consequences* (Kluwer Advances in Natural and Technological Hazards Research No. 19, pp. 357–366). Kluwer Academic, Boston.

Lee, H.J., Locat, J., Desgagnés, P., Parsons, J., McAdoo, B., Orange, D., Puig, P., Wong, F., Dartnell, P., and Boulanger, E. (in preparation) Submarine mass movements. In: C. Nittrouer (ed.), *Continental-Margin Sedimentation: Transport to Sequence* (in preparation).

Leroueil, S., Vaunat, J., Picarelli, L., Locat, J., Lee, H., and Faure, R. (1996) Geotechnical characterization of slope movements. *Proceedings of the 7th International Symposium on Landslides, Trondheim, 1, 63–74*, Balkema, Rotterdam.

Leroueil, S., Locat, J., Levesque, C., and Lee, H.J. (2003) Towards an approach for the assessment of risk associated with submarine mass movements. In: J. Locat and J. Mienert (eds), *Submarine Mass Movements and Their Consequences* (Kluwer Advances in Natural and Technological Hazards Research No. 19, pp. 59–67). Kluwer Academic, Boston.

Levesque, C., Locat, J., Leroueil, S., and Urgeles, R. (2004) Preliminary overview of the morphology in the Saguenay Fjord with a particular look at mass movements. *Proceedings of the 57th Canadian Geotechnical Conference* (Paper G36.350, Session 6G, pp. 23–30).

Locat, J. (1997) Normalized rheological behavior of fine muds and their flow properties in a pseudoplastic regime. *Debris-flow Hazards Mitigation: Mechanics, Prediction, and Assessment* (pp. 260–269). Water Resources Engineering Division, American Society of Civil Engineers, New York.

Locat, J. (2001) Instabilities along ocean margins: A geomorphological and geotechnical perspective. *Marine and Petroleum Geology*, **18**, 503–512.

Locat, J. and Demers, D. (1988) Viscosity, yield stress, remolded shear strength, and liquidity index relationships for sensitive clays. *Canadian Geotechnical Journal*, **25**, 799–806.

Locat, J. and Lee, H.J. (2002) Submarine landslides: Advances and challenges. *Canadian Geotechnical Journal*, **39**, 193–212.

Locat, J. and Mienert, J. (2003) *Submarine Mass Movements and Their Consequences* (Kluwer Advances in Natural and Technological Hazards Research No. 19, 540 pp.). Kluwer Academic, Boston.

Locat, J. and Sanfaçon, R. (2002) Multibeam surveys: A major tool for geosciences. *Sea Technology*, June, 39–47.

Locat, J., Norem, H., and Schieldrup, N. (1990a) Modélisation de la dynamique des glissements sous-marins. *Proceedings of the 6th Congress of the International Association of Engineering Geology, Amsterdam* (pp. 2849–2855). [in French].

Locat, J., Syvitski, J.P.M., Norem, A., Hay, A., Long, B., LeBlond, P., Schafer, C.T., and Brughnot, G. (1990b) Une approche à l'etude de la dynamique des coulées sous-marines: le project ADFEX. *Proceedings of the 43rd Canadian Geotechnical Conference* (Vol. 1, pp. 105–113). [in French].

Locat, J., Lee, H.J., Nelson, H.C., Schwab, W.C., and Twichell, D.C. (1996) Analysis of the mobility of far reaching debris flows on the Mississippi Fan, Gulf of Mexico. In: K. Senneset (ed.), *7th International Symposium on Landslides* (pp. 555–560). A.A. Balkema, Rotterdam.

Locat, J., Gardner, J.V., Lee, H., Mayer, L., Hughes-Clarke, J.E., and Kammerer, E. (1999) Using multibeam surveys for subaqueous landslide investigations. In: N. Yagi (ed.), *Slope Stability Engineering* (pp. 127–134). A.A. Balkema, Rotterdam.

Locat, J., Martin, F., Levesque, C., Locat, P., Leroueil, S., Konrad, J.-M., Urgeles, R., Canals, M., and Duchesne, M. (2003) Submarine mass movements in the Upper Saguenay Fjord, (Québec, Canada), triggered by the 1663 earthquake. In: J. Locat and J. Mienert (eds), *Submarine Mass Movements and Their Consequencs* (Kluwer Advances in Natural and Technological Hazards Research No. 19, pp. 509–519). Kluwer Academic, Boston.

Locat, J., Lee, H.J., Locat, P., and Imran, J. (2004) Analysis of the Palos Verdes Slide, California, and its implication for the generation of tsunamis. *Marine Geology*, **203**, 269–280.

Mader, C.L. (1997) Modeling the 1994 Skagway tsunami. *Scientific Tsunami Hazards*, **15**, 41–48.

Marr, J.D., Harff, P., Shanmugam, G., and Parker, G. (2001) Experiments on subaqueous sandy debris flows: The role of clay and water content in flow dynamics and depositional structures. *Geological Society of America Bulletin*, **113**(10).

Martin, R.G. and Bouma, A.H. (1982) Active diapirism and slope steepening, northern Gulf of Mexico continental slope. *Marine Geotechnology*, **5**, 63–91.

Masson, D.G., Hugget, Q.J., and Brunsden, D. (1993) The surface texture of the Saharan debris flow and some speculations on subaqueous debris-flow processes. *Sedimentology*, **40**, 583–598.

McAdoo, B.G., Pratson, L.F., and Orange, D.L. (2000) Submarine landslide geomorphology, US continental slope. *Marine Geology*, **169**, 103–136.

Meunier, M. (1993) Classification of stream flows. *Proceedings of the Pierre Beghin International Workshop on Rapid Gravitational Mass Movements, CEMAGREF, Grenoble, France* (pp. 231–236). CEMAGNEF, Grenoble.

Mienert, J. and Weaver, P. (2003) *European Margin Sediment Dynamics* (309 pp.). Springer-Verlag, New York.

Mohrig, D., Whipple, K.X., Hondzo, M., Ellis, C., and Parker, G. (1998) Hydroplaning of subaqueous debris flows. *Geological Society of America Bulletin*, **110**, 387–394.

Mohrig, D., Elverhoi, A., and Parker, G. (1999) Experiments on the relative mobility of muddy subaqueous and subaerial debris flows, and their capacity to remobilize antecedent deposits. *Marine Geology*, **154**, 117–129.

Mulder, T. and Cochonat, P. (1996) Classification of offshore mass movements. *Journal of Sedimentary Research*, **66**, 43–57.

Norem, H., Locat, J., and Schieldrop, B. (1990) An approach to the physics and the modelling of subaqueous landslides. *Marine Geotechnology*, **9**, 93–111.

Normark, W.R., Moore, J.G., and Torresan, M.E. (1993) Giant volcano-related landslides and the development of the Hawaiian Islands. In: W.C. Schwab, H.J. Lee, and D.

Twichell (eds), *Submarine Landslides: Selected Studies in the U.S. Exclusive Economic Zone* (USGS Survey Bulletin 2002, pp. 184–196). US Geological Survey, Reston, VA.

Normark, W.R., McGann, M., and Sliter, R. (2004) Age of Palos Verdes submarine debris avalanche, southern California. *Marine Geology*, **203**, 247–259.

O'Brien, J.S. and Julien, P.Y. (1988) Laboratory analysis of mudflow properties. *ASCE Journal of Hydraulics Engineering*, **114**, 877–887.

Orange, D.L. and Breen, N.A. (1992) The effects of fluid escape on accretionary wedges. II: seepage force, slope failure, headless subaqueous canyons and vents. *Journal of Geophysical Research*, **97**, 9277–9295.

Orange, D.L., Saffer, D., Jeanjean, P., Al-Khafaji, Z., Riley, G., and Humphrey, G. (2003) Measurements and modeling of the shallow pore pressure regime at the Sigsbee Escarpment: Successful prediction of overpressure and ground-truthing with borehole measurements. *Proceedings of Offshore Technology Conference* (CD-ROM, 11 pp.).

Phillips, R., Coulter, S.E., and Tu, M. (2004) Earthquake simulator development fot the C-CORE geotechnical centrifuge. *Proceedings of the 57th Canadian Geotechnical Conference, Québec City* (Paper G36.398, Session 7G, pp. 1–7).

Piper, D.J.W., Shor, A.N., and Hughes-Clarke, J.E. (1988) *The 1929 Grand Banks Earthquake, Slump and Turbidity Current* (GSA Special Paper 229, pp. 77–92). Geological Society of America, Boulder, CO.

Poulos, S.G., Castro, G., and France, J.W. (1985) Liquefaction evaluation procedure. *Journal of Geotechnical Engineering*, **111**, 772–791.

Prior, D.B. and Doyle, E.H. (1993) Submarine landslides: The value of high resolution geophysical geohazard surveys for engineering. *Proceedings of the Royal Academy of Engineering Conference on Landslide Hazards Mitigation with Particular Reference to Developing Countries, London*, pp. 67–82.

Prior, D.B., Bornhold, B.D., Coleman, J.M., and Bryant, W.R. (1982) Morphology of a subaqueous slide, Kitimat Arm, British Columbia. *Geology*, **10**, 588–592.

Rabinovich, A.B., Thomson, R.E., Kulikov, E.A., Bornhold, B.D., and Fine, I.V. (1999) The landslide-generated tsunami November 3, 1994, in Skagway. *Geophysical Research Letters*, **26**, 3009–3012.

Robb, J.M. (1984) Spring sapping on the lower continental slope, offshore New Jersey. *Geology*, **12**, 278–282.

Saffer, D.M. and Bekins, B.A. (1999) Fluid budgets at convergent plate margins: Implications for the extent and duration of fault-zone dilation. *Geology*, **29**, 1095–1098.

Sangrey, D., 1977. Marine geotechnology-state of the art. Marine Geotechnology, 2: 45–80.

Schwab, W.C., Lee, H.J., Twichell, D.C., Locat, J., Nelson, H.C., McArthur, M., and Kenyon, N.H. (1996) Sediment mass-flow processes on a depositional lobe, outer Mississippi Fan. *Journal of Sedimentary Research*, **66**, 916–927.

Seed, H.B., Seed, R.B., Schlosser, F., Blondeau, F., and Juran, I. (1988) *The Landslide at the Port of Nice on October 16, 1979* (Report No. UCB/EERC-88/10, 68 pp.). Earthquake Engineering Research Center, University of California Berkeley.

Silva, A.J., Baxter, C.D.P., LaRosa, P.T., and Bryant, W.R. (2004) Investigation of mass wasting on the continental slope and rise. *Marine Geology*, **203**, 355–366.

Suhayada, J.N. and Prior, D.B. (1978) Explanation of subaqueous landslide morphology by stability analysis and rheological models. *Proceedings of the 10th Offshore Technology Conference, Houston* (Vol. 2, pp. 1075–1082).

Sultan, N., Cochonat, P., Foucher, J.P., Mienert, J., Haflidason, H., and Sejrup, H.P. (2003) Effect of gas hydrates dissociation on seafloor slope stability. In: J. Locat and J. Mienert (eds), *Submarine Mass Movements and Their Consequences* (Kluwer Advances in Natural and Technological Hazards Research No. 19, pp. 103–111). Kluwer Academic, Boston.

Syvitski, J.P.M. and Schafer, C.T. (1989) *ADFEX: Environmental Impact Statement (EIS)* (GSC Open-file Report 2312, 81 pp.). Geological Survey of Canada, Vancouver.

Syvitski, J.P.M. (1992) ADFEX: An international effort to generate and monitor the dynamics of mesoscale slides. *Abstracts of the 29th International Geological Congress, Kyoto, Japan, August 24–September 3, 1992* (Vol. 3, p. 936).

Tappin, D.R., Watts, P., and Matsumoto, T. (2003) Architecture and failure mechanism of the offshore slump responsible for the 1998 Papua New Guinea tsunami. In: J. Locat and J. Mienert (eds), *Submarine Mass Movements and Their Consequences* (Kluwer Advances in Natural and Technological Hazards Research No. 19, pp. 383–389). Kluwer Academic, Boston.

Tavenas, F., Flon, P., Leroueil, S., and Lebuis, J. (1983) Remolding energy and risk of slide retrogression in sensitive clays. *Symposium on Slopes in Soft Clays, Linkoping, Sweden* (Report No. 17, pp. 423–454). Swedish Geotechnical Institute, Sweden.

Twichell, D.C., Kenyon, N.H., Parson, L.M., and McGregor, B.A. (1991) Depositional patterns of the Mississippi Fan surface: Evidence from GLORIA II and high-resolution seismic profiles. In: P. Weimer and M.H. Link (eds), *Seismic Facies and Sedimentary Processes of Submarine Fans and Turbidite Systems* (pp. 349–363). Springer-Verlag, New York.

Urgeles, R., Locat, J., Lee, H., and Martin, F. (2002) The Saguenay Fjord, Québec, Canada: Integrating marine geotechnical and geophysical data for spatial seismic slope stability and hazard assessment. *Marine Geology*, **185**, 319–340.

Varnes, D.J. and the IAGE Commission on Landslides and other Mass Movements on Slopes (1984) *Landslide Hazard Zonation: A Review of the Principles and Practice*. UNESCO, Paris.

Walker, J.R. and Massingill, J.V. (1970) Slump features on the Mississippi Fan, Northeastern Gulf of Mexico. *Geological Society of America Bulletin*, **81**, 3101-3108.

Weimer, P., and Link M.H. (1991) *Seismic Facies and Sedimentary Processes of Submarine Fans and Turbidite Systems* (447 pp.). Springer-Verlag, New York.

Young, A.G., Bryant, W.R., Slowey, N.C., Brand, J.R., and Gartner, S. (2003) Age dating of past slope failures of the Sigsbee Escarpment within Atlantis and Mad Dog developments. *Proceedings of the Offshore Technology Conference* (CD-ROM).

10

Volcanic debris flows

James W. Vallance

10.1 INTRODUCTION

Volcanic debris flows, more commonly known as lahars, behave similarly to debris flows in other settings, but can most strikingly differ in origin and size. Whereas debris flows in other settings commonly range in size from about $10^2\,m^3$ to $10^7\,m^3$, volcanic debris flows range in size from $10^4\,m^3$ to more than $10^9\,m^3$. This size difference can be explained by the abundance of loose debris on the steep slopes and aprons of volcanoes; the presence of weakened hydrothermally altered rock within some volcanic edifices; the abundance of water stored in glaciers, crater lakes, and hydrothermal systems; rainfall that washes over denuded slopes after eruptions; and the potential for releasing both water and sediment during and immediately after eruptions.

Origins of debris flows at volcanoes include the direct and indirect effects of eruptions, as well those typical of other environments. Volcanic eruptions can generate avalanches of hot rock and ash (pyroclastic flows) that move across glacial ice or snow, melt it, and generate debris flows. Volcanic eruptions through crater lakes sometimes generate floods of water that subsequently entrain sediment and generate lahars. Eruptions or earthquakes associated with eruptions can cause volcanic edifices to collapse. If the resulting debris avalanches contain sufficient volumes of weakened, altered rock and water, they can evolve to form debris flows. Explosive eruptions coat volcano slopes with voluminous easily erodible deposits of volcanic debris that form lahars during periods of heavy rain.

Because of their large sizes and propensity for long-distance transport, volcanic debris flows show downstream evolutions less commonly observed in debris flows from other environments. Many volcanic debris flows evolve from water floods. Those that move down rivers tend to incorporate river water with distance downstream and evolve over river reaches of tens of kilometres to more dilute flow types (see Chapter 8). Volcanic edifice collapses generate debris avalanches that sometimes

---

M. Jakob and O. Hungr (eds), *Debris-flow Hazards and Related Phenomena*.
© Praxis. Springer Berlin Heidelberg 2005.

evolve partially or completely to debris flows. There is a complete spectrum of behavior between unsaturated debris avalanche and water-saturated debris flow in volcanic settings (see Chapter 27).

Because the precise timing of lahar events is unpredictable and working with active flows can be hazardous, much of our present knowledge of flow behavior is inferred from study of lahar deposits. Nonetheless, a few key observational studies have improved understanding of debris-flow processes. The chief purpose of this chapter is to summarize what is known about the causes and behavior of volcanic debris flows on the basis of observations and careful examination of deposits, and to further describe carefully the nature of hazards that derive from such events.

10.2 DEFINITIONS

Lahar is an Indonesian (Javanese) word used in that country to describe highly concentrated flowing mixtures of rock debris, mud, and water coming from volcanoes, and was introduced into the volcanological literature through the work of Schmidt (1934) and van Bemmelen (1949). The term has wide usage in the volcanological and geological literature as a synonym for "volcanic mudflow or debris flow" (e.g., Crandell, 1971; Fisher and Schmincke, 1984; Pierson and Scott, 1985; Smith, 1986). A rheologically specific definition causes confusion, however, because some sediment-water flows from volcanoes transform from water flood, to hyperconcentrated flow, to debris flow, and back again to more dilute phases during a single event sequence (Pierson and Scott, 1985; see also Chapters 8 and 27) and other sediment-water flows evolve wholly or partly from debris avalanches (e.g., Crandell, 1971; Vallance and Scott, 1997).

A rheologically non-specific definition adopted by the 1988 Geological Society of America Penrose Conference on Volcanic Influences on Terrestrial Sedimentation (Smith and Fritz, 1989) is: *lahar, a general term for a rapidly flowing, gravity-driven mixture of rock debris and water (other than normal streamflow) from a volcano.* A lahar event can vary in character with time and distance downstream. It may comprise one or more flow *phases*, which include a debris-flow phase, transitional or hyperconcentrated-flow phase, and streamflow phase (Vallance, 2000). A *debris-flow phase* is one in which the solid and liquid fractions are approximately equal volumetrically and in which the two fractions in a vertical section move downstream approximately in unison. A *streamflow* phase is one in which fine-grained sediment moves in suspension with the fluid (suspended load) and coarse-grained sediment moves along the bed at discrete intervals (bedload). It is useful to define a *transitional flow phase*, commonly known as *hyperconcentrated flow*, intermediate between that of debris flow and streamflow. Unlike streamflow, this transitional phase carries very high sediment loads, and unlike the debris-flow phase coarse-grained solids tend to separate vertically from the liquid-and-fine-solids mixture. Although the literature distinguishes the hyperconcentrated-flow phase from more dilute and more concentrated phases in terms of solids fraction, transitions are gradational and dependent on other factors like sediment-size distribution and energy of the flow. Thus, flow-phase transitions cannot be defined precisely. Following common usage in the volcanologic literature, the term lahar includes high-sediment-fraction volcanic debris flows, low-sediment-fraction hyperconcentrated flows, and gradational flows in between, but does not include unsaturated parts of debris avalanches or sediment-water floods. Readers should note that this usage contrasts somewhat to that of Chapter 2.

Lahar events can include processes, such as debris avalanches, flood flows, and pyroclastic flows, not normally encompassed by the terms lahar and debris flow. A *debris avalanche* here is a flowing mixture of debris, rock, and moisture that moves downslope under the influence of gravity. Debris avalanches differ from debris flows in that they are not water saturated and in that the load is entirely supported by particle–particle interactions.

In this chapter, the *phase* is the flow type in a lahar event at some time and place. Phases associated with lahars include debris avalanche, debris flow, hyperconcentrated flow, flood flow, and streamflow. The height above the channel bottom of the flowing lahar is the *stage*. Examples of lahar stages include the initial rising or waxing stage, the peak-inundation stage, and the final long-duration falling or waning stage.

A *pyroclastic flow* is a mixture of hot rock, ash, and gas with roughly equal proportions of solids and gas that move rapidly (10s to 100s of m/s) away from source during eruptions. *Pyroclastic surges* are more dilute. The temperature of these flows is invariably greater than $100°C$ and is usually greater than $500°C$ thus they are completely dry. The hot gas and ash mixture is highly fluid, supports sand and gravel-size particles, and makes pyroclastic flows more mobile than normal dry avalanches. The interactions of such flows with ice and snow cause lahars. There is much argument in the literature (c.f. Scott, 1988; Waitt, 1989) about whether pyroclastic flows and more dilute surges transform directly to lahars but no well-documented examples. The passage of these hot energetic flows across glaciers and snow pack causes water floods that generate lahars (Table 10.1).

For certain types of primary lahars, size and frequency are influenced by volcano explosivity, a measure of the power of volcanic eruptions and production of pyroclastic debris at a volcano. The volcanic explosivity index (VEI) is a measure of eruption size and duration as well as explosivity (Newhall and Self, 1982). Values of VEI range from 0 to 8 with each interval indicating an increase of about a factor of ten. A VEI of 0 means non-explosive and gentle, 1 means effusive with small explosions, 2–3 mean explosive and moderate in size, 4–5 mean large and severe to cataclysmic, and 6–8 are paroxysmal and rare. Recent eruptions of Kilauea from the south-west rift zone and that of Manua Loa, USA in 1984 have $VEI = 0$. Eruptions of Stromboli in Italy are commonly $VEI = 2$. Mount Lassen in 1914 produced a VEI 3 eruption and Cotopaxi, Ecuador, in 1877 produced a VEI 4 eruption. Well-known eruptions of Mount St. Helens on 18 May, 1980 and Pinatubo on 15 June, 1991 were VEI 5 and 6 respectively. Although there is some tendency for large explosive eruptions to produce large lahars, in many cases,

Table 10.1. Notable examples of volcanic debris flows (lahars) and their triggering mechanisms. Adapted from Rodolfo (2000), data from Simkin and Siebert (1994).

Triggering mechanism	Volcano, occurrence	Fatalities	VEI[1]	Volumes, areas	Remarks
		Syneruptive (primary)			
Crater lake expulsion	Galunggung, Indonesia, 1822	>400	5	~10^8 m^3	4,011 fatalities, at least 10% owing to lahars
	Kelut, Indonesia, 1826	yes	4		65 villages destroyed
	Kelut, Indonesia, 1848	21	3		11 villages destroyed
	Kelut, Indonesia, 1864	54	2		
	Kelut, Indonesia, 1901	yes	3		
	Kelut, Indonesia, 1919	5,110	4	4×10^7 m^3 H$_2$O, 130 km^2	Expelled 4×10^7 m^3 lake. Beginning 1923 tunneled to drain lake to 2×10^6 m^3. No major lahars. Eruption blocked tunnels, deepened lake to 2.4×10^7 m^3
	Kelut, Indonesia, 1951		3		
	Kelut, Indonesia, 1966	211	4	2×10^7 m^3 H$_2$O, 45 km^2	Drained lake again in 1967
	Ruapehu, New Zealand, 1969, 1975				
	Ruapehu, New Zealand, 1995		4	~10^7 m^3	
Crater lake breakout	Pelée, Martinique, 1902	25	2		
Jökulhlaups	Öraefajökull, Iceland, 1727	4			
	Katla, Iceland, 1918, 1955	4, 1			
	Gjálp, Iceland, 1996		5?	4×10^9 m^3	$Q = 3\text{–}6 \times 10^6$ m^3/s, $L = 330$ km
Pyroclastic flows, hot rock avalanches, and hot blasts melting snow and ice	Cotopaxi, Equador, 4500 BP	yes	3		
	Cotopaxi, Equador, 2400 BP	yes	3		
	Cotopaxi, Equador, 1698	many	3		
	Cotopaxi, Equador, 1742	many	4		
	Cotopaxi, Equador, 1768	many	4		
	Cotopaxi, Equador, 1877	>400	4	~10^8 m^3	
	Nevado del Ruiz, Columbia, 1595	636	3	3×10^8 m^3	
	Nevado del Ruiz, Columbia, 1845	~1,000	3	~10^8 m^3	
	Nevado del Ruiz, Columbia, 1985	23,080		V_{max}, $V_{glacier}$	Longest historical runout, $L = 230$ km
	Individual lahars, Ruiz, 1985			4×10^7, ~9×10^6 m^3	
	Azufrado-Lagunillas	>21,000		3.2×10^7, $5\text{–}8 \times 10^6$ m^3	Bulking ~×4, $Q_{max} = 5 \times 10^5$ m^3/s
	Molinos-Nereidas	>1000		1.6×10^7, $6\text{–}9 \times 10^6$ m^3	Bulking ~×5, $Q_{max} = 2 \times 10^5$ m^3/s
	Guali	yes			Bulking ~×3, $Q_{max} = 2 \times 10^5$ m^3/s

Trigger	Location, Year	Fatalities	Ref.	Volume	Comments
	Rainier, USA, 1100 BP		3	$\sim 10^8$ m^3	
	St. Helens, USA, 1980		5	3×10^7 m^3	
	St. Helens, USA, 1982		2	1.2×10^7 m^3	
	Redoubt, USA, 1989		2		
Lava flows melt ice and ponded water breaks out	Villarica, Chile, 1949	36	2		
	Villarica, Chile, 1963/1964	22	2		
	Villarica, Chile, 1971	15	3		
Pyroclastic flows entering streams	Asama, Japan, 1783	~1,000	4		Pyroclastic flow dammed river, ephemeral lake breakout 1 hr later
Avalanche generated	Rainier, USA, 7200 BP		3	10^8 m^3	Eruption related avalanche liquefied
	Rainier, USA, 5600 BP	yes	3	4×10^9 m^3	$Q = 3$–15×10^6 m^3/s, avalanche liquefied almost completely to form lahar
	Rainier, USA, 2500 BP		2	2×10^8 m^3	Incomplete liquefaction of avalanche
	Rainier, USA, 1100 BP		2	10^7–10^8 m^3	Eruption related avalanche liquefied
	Rainier, USA, 500 BP		?	2.5×10^8 m^3	No clear evidence of an eruption
	St. Helens, USA, 1980		5	1.4×10^8 m^3	Avalanche first, lahar 3 hours later Major et al., this vol.
Rain during eruption	Agung, Indonesia, 1963	200	4		Rain during and after eruption
	Pinatubo, Philippines, 1991	~100	6		Typhoon Yunya coeval with climactic eruption of 15 June
Posteruptive (secondary)					
Crater lake failure	Kelut, Indonesia, 1875	yes			Failure of west crater wall, destroyed 30 villages
	Kelut, Indonesia, 1990	4			Secondary lahar about 1 month after eruption
Heavy rain after eruptions	Ruapehu, New Zealand, 1953	151			$Q = 850$ m^3/s
	Parker, Philippines, 1995	310		2×10^6 m^3	Caused by illegal gold mining activity?
	Sakurajima, Japan, post-1955				
	Irazu, Costa Rica, post-1964				
	Semeru, Indonesia, 1909	221			
	Semeru, Indonesia, 1981	372			
	Mayon, Indonesia, 1875	1,500			
	Unzen, Japan, 1990–1995				Lahars have remobilized 2.5 or 5.5 km^3 of pyroclastic-flow deposits
	Pinatubo, Philippines, post-1991				

continued

Table 10.1 (*cont.*)

Triggering mechanism	Volcano, occurrence	Fatalities	VEI[1]	Volumes, areas	Remarks
Lake breakouts (not crater lakes)	St. Helens, USA, Pine Creek, 2500 BP				Spirit Lake breakout after debris avalanche
	Santa María, Guatemala, post-1902				
	Pinatubo, Philippines, post-1991				
	Unrelated to eruptions				
Earthquake-induced avalanche	White Island, New Zealand, 1914	14	n/a		
	Ontake, Japan, 1984		n/a		Preceding period of rain, debris flows occurred a few hours after earthquake
Heavy-rain-induced slope failure	Agua, Guatemala, 1541	>1300	n/a		Destroyed capital, Cuidad Viejo
	Mombacho, Nicaragua, 1570	400	n/a		Destroyed town of Mombacho
	Mayon, Philippines, 1981	>200	n/a		Typhoon caused lahars, last eruption 1978
	Casita, Nicaragua, 1998	2,500	n/a		Destroyed 2 towns, Hurricane Mitch
Glacial outburst floods	Rainier, USA, 1947		n/a	4×10^7 m^3	
	Rainier, USA, 1960's to present		n/a		Yearly

[1] VEI is volcano explosivity index; explained in text.

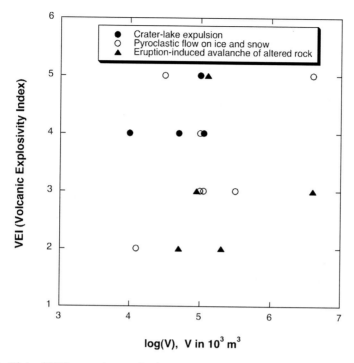

Figure 10.1. Plot of VEI vs. volume of primary lahars having three diverse origins. The plot shows poor correlation between explosivity and lahar volume, thus indicating that small explosive eruptions can potentially trigger devastating primary lahars.

especially at ice-clad volcanoes, small to moderate eruptions can produce huge lahars (Figure 10.1).

10.3 ORIGINS AND DOWNSTREAM BEHAVIOR

At volcanoes lahars may be syneruptive (primary), posteruptive (secondary), or unrelated to eruptions (Table 10.1). Genesis requires: (1) an adequate water source; (2) abundant unconsolidated debris; (3) steep slopes and substantial relief at the source; and (4) a triggering mechanism. Water sources include pore water, rapidly melted snow and ice, subglacially trapped water, crater or other lake water, and rainfall runoff. Triggering mechanisms that derive from volcanic unrest include volcanigenic earthquakes and explosions. The shallow intrusion of hot rock and gases prior to the onset of an eruption drives water of the hydrothermal system toward edifice flanks, and the resultant outwardly directed pore-pressure gradient contributes to edifice instability and debris avalanches that sometimes evolve to lahars. Sudden melting of snow and ice during eruptions by voluminous flows of hot rock and ash that sweep down steep slopes cause water floods that incorporate debris and form lahars. Sudden release of crater lakes, other lakes adjacent to volcanoes, and subglacially stored water can also generate lahars. Lake-breakout-induced lahars occur during eruptions but can also occur months or years after eruptions.

Their temperature and fines content can help to delineate lahar origins. Admixing of hot pyroclastic debris and water results in hot lahars. Such mixtures are necessarily below the boiling point, but can contain individual clasts that are hotter. Hot lahars also derive from the release of water heated during volcanic unrest or eruptions. The hot water can derive from subglacial lakes, crater lakes, and hydrothermal systems. As observed at Mount Rainier (Crandell, 1971), lahar deposits can have clay-rich matrix ($>5\%$ clay/sand + silt + clay) or clay-poor matrix ($<5\%$ clay). An empirical observation is that clay-rich lahars commonly initiate as flank failures of hydrothermally altered, weakened rock, whereas clay-poor lahars commonly originate as floods of water that entrain gravel, sand, and fines. Flood waters leading to the formation of clay-poor lahars have been produced during lake breakout (Pierson, 1999), intense rainfall (Rodolfo and Arguden, 1991; Hodgson and Manville, 1999; Lavigne et al., 2000; Lavigne and Thouret, 2003), and from melting of snow and ice by hot pyroclastic flows (Scott, 1988; Major and Newhall, 1989; Pierson et al., 1990).

10.3.1 Lahars induced by sudden melting of snow and ice, voluminous floods, or torrential rains

Floods of water moving across the easily erodible loose clastic sediment common on the flanks and aprons of volcanoes easily incorporate that debris and may quickly form lahars. Erosion and incorporation of sediment (bulking) is critical to all lahars that begin with sudden water releases. Lahar formation depends on the right mix of readily available sediment and water discharge.

At volcanoes, lahars induced by sudden water release can occur by four principal means. (1) Pyroclastic flows and surges mix with and rapidly melt glacial ice and firn. Generally, such pyroclastic flows come to rest or nearly do, form meltwater which then runs off, coalesces, and erodes the pyroclastic debris to form water-rich lahars. The lahars continue to bulk up with volcaniclastic debris, glacial drift, alluvium, and colluvium as they flow downstream so that within a few to several tens of kilometers they become debris flows. Lahars of this type are considered primary. (2) Eruptions can displace large volumes of crater-lake water that form lahars downstream. Crater lakes, caldera lakes, and volcanically debris-dammed lakes can also break out months to years after eruptions. Such delayed breakouts occur when water levels gradually rise, then overtop and rapidly incise fragile debris dams. (3) Subglacial eruptions can form subglacial lakes that eventually break out when a section of the ice cap becomes buoyant and releases the trapped water. Small-scale outburst floods occur as normal glacial processes during periods of ablation. Although huge eruption-driven outbursts cause sediment-rich, water floods called jökulhlaups, small secondary floods commonly bulk up to form lahars. (4) Lahars owing to intense rainfall often occur after pyroclastic eruptions

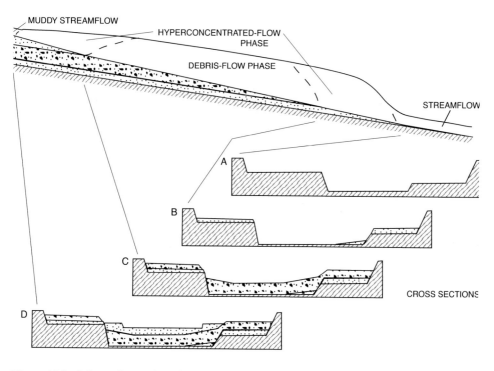

Figure 10.2. Schematic model of a lahar moving down a river undergoing downstream dilution from debris-flow phase to hyperconcentrated-flow phase and deposit facies derived from it. The idealized model shows erosion and deposition of hyperconcentrated-flow facies from A to B; deposition of debris flow facies from B to C; and deposition of hyperconcentrated flow facies and subsequent erosion from C to D.

that deposit abundant loose debris in the form of pyroclastic-flow or fall deposits. Lahars of this type are commonly small but frequent in occurrence during rainy periods. Size and frequency of rain-induced lahars may increase in the months or years following the primary pyroclastic eruption, then slowly decrease as drainage networks and vegetation re-establish themselves (e.g., Mount Pinatubo after its 1991 pyroclastic eruption).

Debris-flow or hyperconcentrated-flow phases that bulk up and evolve rapidly from water floods can deposit rapidly on low gradients or slowly become more dilute downstream as they overrun rivers (Vallance, 2000). Sediment-rich lahars that move down active river channels will push river water in front of them, gradually incorporate that water, become progressively dilute, and undergo transformations to hyperconcentrated-flow and muddy-streamflow phases (Figure 10.2). The flow transformation begins at the flow front and migrates back through the lahar wave as it travels downstream. If lahars of this type occur in arid regions without perennial streamflow, they will bulk up to become debris flows, not undergo significant downstream transformation, and remain debris flows to their termini. Such debris flows will typically develop relatively dry bouldery flow fronts that surround more liquefied interiors (Iverson, 1997).

10.3.2 Edifice or flank-collapse-induced lahars

Although most edifice collapses behave as debris avalanches, those with sufficient, widely dispersed pore and hydrothermal water in the pre-collapse rock liquefy as the material deforms during collapse. A shallow intrusion of magma within an edifice is the likeliest trigger of collapse-induced lahars larger than about $0.2\,km^3$. Smaller collapse-induced lahars may have various triggers, including magmatic or phreatic volcanism and volcanic or tectonic earthquakes.

Hydrothermal alteration, especially at glaciated volcanoes, increases the probability of edifice-collapse lahars. Acid-sulfate leaching in hydrothermal systems removes mobile elements, adds sulfate, and breaks down framework silicates to form silica phases such as cristobalite and opal, and clay minerals such as kaolinite and smectite. This process weakens the rock so that it more readily disintegrates during deformation after collapse. Thus, huge coherent blocks typical of debris avalanches are less common in lahars of this type even though their origin is the same. Abundant alteration minerals, especially clay minerals, increase porosity and decrease permeability of the rock and thus, in combination with a hydrothermal system, trap a widely dispersed reservoir of water within the pre-collapse mass. Because of its water content and its tendency to disintegrate, hydrothermally altered rock, unlike fresh rock, easily liquefies as it deforms. Collapse-induced lahars are typically clay-rich, and most are observed to have greater than 5% clay in the matrix.

Clay-rich, collapse-induced lahars appear to be more common at ice-clad volcanoes than at volcanoes that are ice-free. Glacial erosion tends to expose deeper and potentially more altered parts of volcanoes than does surface-water erosion alone. Furthermore, slopes incised in altered rock are more susceptible to failure not only because the rock is weak but also because it is commonly oversteepened. Lastly, melting glacial ice provides a slow-release source of water that is important to the efficient operation of the acid-sulfate leaching process.

At and near some volcanoes, especially tropical ones, tectonic earthquakes cause multiple slope failures that ultimately coalesce and form collapse-induced lahars. Such lahars are unrelated to volcanism but, owing to characteristics of their deposits (commonly clay-rich, hummocky, and voluminous), can be mistaken for lahars generated by major edifice collapse of hydrothermally altered sectors of volcanoes (Scott et al., 2001).

Collapse-induced lahars are debris flows that transform directly from avalanching debris that contains enough water to be water-saturated upon deformation. Such lahars commonly contain huge fragments composed of relatively fresh volcanic rock from the edifice. As they migrate downstream, they incorporate exotic debris, especially at their flow fronts. Because of their size, avalanche-induced lahars rarely undergo transformation to more dilute phases as they migrate downstream.

10.4 DOWNSTREAM PROCESSES AND BEHAVIOR

Like debris flows in other environments, lahars can change character as they move downstream; however, because of their large sizes and propensity for long-distance transport, some lahars can show much more pronounced changes with distance downstream than other debris flows. Small lahars (those less than $\sim 10^6 \, \text{m}^3$), like many non-volcanic debris flows, commonly coalesce quickly, form bouldery fronts through size-segregation processes (Chapter 6), and come to a halt once they reach low-gradient reaches of the drainage. Lahars larger than $\sim 10^6 \, \text{m}^3$ commonly don't fit the mold of typical debris flows. In the case of huge lahars that travel tens of kilometers, size-segregation processes can become relatively less important and downstream erosion and dilution processes become more important (Scott, 1988; see also Chapter 27). In medial or distal reaches, such large lahars can also transform from a sediment-rich debris-flow phase to water-rich hyperconcentrated-flow or streamflow phases. Another class of lahars rare in other environments comprises those that begin as huge (10^8 to $>10^9 \, \text{m}^3$) debris avalanches proximal to volcanoes and which transform to debris flows as they move downstream (Carrasco-Núñez et al., 1994; Vallance and Scott, 1997; see also Section 10.5.4).

10.4.1 Erosion and bulking

Lahars cause erosion by undercutting steep slopes and terrace scarps and by scouring their beds. Erosion is strongest along steep reaches underlain by loose clastic sediment and weakest either along reaches underlain by highly resistant bedrock or reaches with very gentle gradients. Along any particular reach, water-rich, hyperconcentrated-flow phases are typically more erosive than sediment-rich debris-flow phases (Figure 10.2), but local erosion can occur during any flow phase. Water-rich phases are more turbulent and agitated than equivalent-stage debris-flow phases and thus more effectively erode sediment (Dolan, 2004). In particular, turbulence along the boundary layer of water-rich phases allows more effective sediment entrainment. Fully bulked debris-flow phases entrain additional sediment less easily because they less readily undercut banks and less readily pluck particles from their beds.

Erosion at the base of a lahar occurs by piecemeal dislodgment of particles, by undercutting at upstream migrating knick points, and by rip up of sediment owing to root heave of falling trees. Piecemeal erosion occurs chiefly during more water-rich hyperconcentrated phases and is probably not an important process outside of active channels. If channel beds comprise erodible sediment, sequences of upstream migrating knick points are common during hyperconcentrated-flow phases (Dolan, 2004). Typically knick-point steps are several tens of centimetres to a metre high and spaced on the order of tens to hundreds of metres apart. During more sediment-rich flows, knick points become higher and more widely spaced. Knick points may disappear during debris-flow phases, which commonly aggrade rather than erode their beds. Lahars voluminous enough to escape channels knock trees down and incorporate them. Root balls of falling trees drag considerable sediment into the active flow and loosen even more sediment that is then available for erosion.

Voluminous lahars that inundate large areas of forested terrain can incorporate considerable quantities of sediment and huge amounts of wood in this way.

Undercutting of steep slopes, fluvial terrace scarps, and active stream banks is probably the most important way in which lahars erode and incorporate sediment. Undercutting is active during both debris-flow and hyperconcentrated-flow phases. Large lahars are capable of incorporating huge blocks (>10 m in dimension) of unconsolidated sediment, and sometimes even of bedrock, in this way. These huge clastic blocks may move tens of kilometers downstream before they deform and ultimately break up into individual fragments.

Progressive bulking imposes downstream changes of character on lahars. Bulking transforms flood flows and hyperconcentrated flows to more sediment-rich phases (Figure 10.2). If the process continues, both waxing and waning stages of flow ultimately become debris-flow phases. With continued entrainment, lahars become richer in exotic sediment, like alluvium, colluvium, and glacial drift. Because the waxing flow-front and following peak stages of the flow are the most erosive, these stages most readily incorporate exotic sediment. The sediment-rich falling flow that follows peak flow is less erosive and commonly actively deposits sediment rather than erodes it. The final waning-stage flow is typically more watery, and more erosive, but less voluminous in discharge than the preceding stage. Final waning stages of lahars incise previously deposited laharic sediment.

10.4.2 Particle-size and particle-density segregation processes

Particles in lahars can effectively segregate by density or size, but the most important segregation processes are mediated by solids fraction, proportion of coarse particles, and fluid density, the latter being determined by the proportion of fine-grained particles in suspension (Vallance, 2000). More dilute flow favours particle settling or buoyant rise because there are fewer particles to hinder these processes (Dolan, 2004). Greater solids fractions hinder these processes and may favour percolation, a process that counteracts preferential settling of large particles (Vallance, 1994). Percolation occurs when particles, under the influence of gravitational body forces, move downward into gaps that open beneath them (Savage and Lun, 1988). The process occurs chiefly when individual particles influenced by body forces like gravity rub or collide in vibrating or shearing mixtures (Savage and Lun, 1988).

The net result of percolation is that small particles migrate downward and displace large ones, and large ones gradually migrate upward. The large particles then migrate forward toward the margins of the flow because velocities are greatest near the surface (Figure 10.3). Such a kinetic-sieve process can not only cause the inverse vertical grading that is common in moving debris flows, but it can also cause the accretion of large particles at flow perimeters (Figure 10.3).

Low-density particles commonly collect at the surface of lahars and form rafts of material that appear to move *en masse* (Dolan, 2004). Particles that are less dense than water, like pumice, rise buoyantly. In addition, particles slightly more dense than water rise to the surfaces of lahars whose fluid phases suspend sufficient

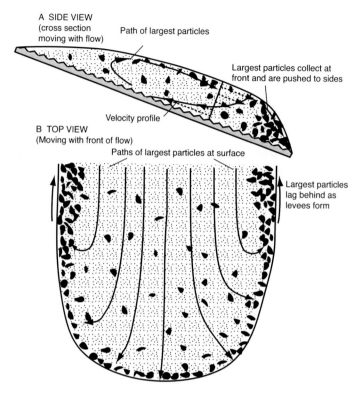

Figure 10.3. Schematic diagram showing how segregation of coarse particles to the surface of a debris flow results in segregation of the coarse particles to the front (a) and margins (b) of the flow. Note that the frame of reference is shown moving at the average speed of the flow.

quantities of fines (generally clay and silt but in some cases fine grained sand) to increase the effective fluid density. Friction at the bed retards lahars and causes vertical profiles in which velocities are smallest at the base and gradually increase upward. Low-density particles will migrate upward toward the flow surface to form rafts. Velocities are greater than average at the surface and these rafts (commonly of pumice) will then migrate toward flow margins.

10.4.3 Downstream dilution and transformation

The gradual incorporation of water at the front of a lahar flowing down an active stream channel causes progressive loss of carrying capacity and an eventual change in the nature of the flow with distance downstream (Figure 10.2) (Pierson and Scott, 1985; Scott, 1988). This process is important only in lahars that follow active rivers or other bodies of water. Though dilution of large lahars can occur, the process has little effect on the behavior of lahars so huge that their volumes are significantly greater than that of the water in the river being overrun (Vallance and Scott, 1997; Mothes et al., 1998).

Once off the flanks of the volcano and confined in river channels, lahars, which typically move faster than normal streamflow, push river water ahead of them and gradually, with distance downstream, begin to mix with that water (Cronin et al., 1997; Scott, 1988). As the flow front becomes progressively more watery, it loses its capacity to carry large gravel particles, and these progressively lag behind the flow front. With time and distance downstream, a dilution front progresses from the front of the lahar to its middle, and eventually the entire lahar becomes more dilute. In lahars that occurred at Mount St. Helens in 1980 and 1982, downstream dilution occurred over the course of tens of kilometres and caused a complete transformation from debris-flow phase to hyperconcentrated-flow phase. In medial reaches, the hyperconcentrated-flow phase preceded the debris-flow phase of the lahar because the dilution process began at the flow front then gradually worked backward toward the tail as the flow migrated downstream (see Chapter 27).

10.4.4 Depositional processes

Emplacement of lahar deposits can occur rapidly, by steady incremental accretion or, most likely, by some intermediate process, in which accretion begins then accelerates, wanes, or does both alternately. Debris-flow-phase deposits are poorly sorted, generally massive, and unstratified. It is common to infer that such massive deposits are emplaced *en masse* and that they represent a frozen portion of the moving debris flow. Debris-flowdeposits may accrete very rapidly; however, they do not freeze from the top down as Bingham models of debris-flow behavior suggest. In contrast, hyperconcentrated-flow deposits are better sorted, commonly show faint stratification, and hence are assumed to accrete during significant time intervals.

An increasing body of evidence suggests that deposits of both hyperconcentrated flows and debris flows accrete incrementally (Vallance and Scott, 1997; Vallance, 2000). Such evidence includes: (1) strong alignment of elongate clasts parallel to flow directions or imbrication of such clasts in upstream directions; (2) strong changes in composition of particles with vertical position in outcrops, especially those which are graded; (3) marks of peak-flow levels in upland valleys that indicate flow depths 5–10 times greater than deposit thicknesses; and (4) stratification in deposits of transitional or hyperconcentrated flows. Stratification in deposits indicates piecemeal accretion but many debris flows have a massive appearance. Nonetheless, careful measurements show that most debris-flow deposits exhibit alignment and imbrication of elongate clasts. This evidence strongly suggests piecemeal emplacement of particles and thus incremental accretion. Vertical compositional change of particles in outcrop suggests incremental accretion from a flow with compositional changes from peak-flow through waning-flow stage. For example, the peak-flow deposits of the Osceola Mudflow had proportions of exotic particles of up 90%, whereas waning-flow deposits had exotic contents of 20% to 50%. In thick vertical sections proportions of exotic particles at the base were similar to those of peak flow and those at the top were similar to those of waning flow. Vallance and Scott

(1997) concluded that these observations were not consistent with sudden emplacement and thus implied accretion during significant time intervals. They also suggested that deposit thicknesses significantly less than those of inundation depths were a common indicator of incremental accretion.

Although rapid deposition of a vertically size-segregated flow can generate normally and inversely graded deposits, incremental accretion of longitudinally size-segregated flow can also be responsible. Because lateral and longitudinal variations in both sizes and compositions of particles commonly occur in moving lahars, accretion from such laterally graded systems during significant time intervals can cause vertically graded deposits. Deposits in positions higher in the valley may also be graded, but often less obviously so.

Debris flows that do not undergo downstream dilution commonly form cobble-boulder-rich perimeters owing to the size segregation process described above (Figure 10.3). The surface of the flow with its concentration of large particles moves faster than the rest of the flow so that cobbles and boulders migrate to the flow front. Once the flow peak has passed at any particular cross section, boulders begin to accrete at flow margins to form coarse levees. Flow fronts, with concentrations of large angular particles, that move onto gentle slopes become progressively drier because water can more easily escape from more permeable coarse-grained flow perimeters than from fines-rich interior flows. The net result is a dry, frictional, resistant flow perimeter that surrounds a liquefied interior (*cf.*, Iverson, 1997). When flows of this type reach sufficiently gentle slopes, the frictional perimeter slows to a stop and leaves behind a steep-fronted fines-poor margin and a partly liquefied fines-rich interior. Because resistance to flow is greater at the margin than the interior of the flow, bifurcation of the flow into fingers can result as the more mobile fines-rich interior diverts around the more static perimeter.

Avulsion can occur in lowland areas with gentle slopes where sediment-rich portions of lahars accrete and fill the active channel so that its cross-sectional capacity is diminished. When channel capacity is sufficiently diminished, the flow overtops this channel and cuts a new one. In the strict sense, the term applies only to the breaking away and establishment of the new channel, although a pre-condition for avulsion is accretion during debris-flow phases of lahars.

10.5 SELECTED CASE STUDIES

10.5.1 Lahars related to crater lakes

Explosive eruptions at volcanoes with crater lakes commonly expel the water or breach the outlet dam and cause floods of water that generate lahars. Crater lakes commonly contain sulfurous, highly acid water and mud that can reach temperatures of more than 40°C during the unrest that presages eruptions. When explosive eruptions or outlet collapses release huge volumes of this nasty brew, it rushes downstream, picks up abundant loose debris along the way, and becomes a particularly unpleasant variety of lahar comprising scalding hot acidic water, mud, and debris.

Kelut volcano, on the island of Java in Indonesia, is notorious for producing lahars in this fashion, and its lahars are notorious for killing people who live on its fertile slopes and for destroying villages within a 40-km radius. Of more than 30 eruptions in the past 1000 years at Kelut, half or more are thought to have produced disastrous lahars (Simkin and Siebert, 1994; Bourdier et al., 1997). Since 1848, the volcano has erupted 10 times. Of these eruptions, seven have blown out the crater lake, and five have produced devastating primary lahars. The most devastating of these lahars occurred when the largest volumes of water were expelled from the lake (Table 10.1). In 1919, an eruption blew $4 \times 10^7 \, \text{m}^3$ of water out of Kelut's lake and produced lahars that swept downstream as far as 37 km. These lahars covered more than $130 \, \text{km}^2$ and killed more than 5,000 people (Scrivenor, 1929).

Kelut volcano offers a spectacular example of lahar hazard mitigation because of tunnels designed to lower its crater-lake level and check-dam installations designed to reduce downstream hazards. In 1905, an earthen sediment-retention dam was built about 10 km downstream in an initial effort to protect the town of Bitur, about 20 km south of Kelut. This effort proved futile, however, when the 1919 lahars overwhelmed the dam and destroyed the town (Zen and Hadikusumo, 1965). The dam was rebuilt but was nearly full by 1925 (Blong, 1986). The 1919 tragedy prompted the Dutch colonial government to build a sequence of tunnels that lowered lake level by about 50 m and reduced the volume of the lake to about $2 \times 10^6 \, \text{m}^3$ (Scrivenor, 1929; Zen and Hadikusumo, 1965). When the volcano erupted again in 1951, it produced only small lahars and few lives were lost. Unfortunately, the 1951 eruption damaged the drainage tunnels, reamed out the crater, and lowered the lake bottom by 50 meters. When the old drainage tunnels were repaired the lake still contained $2 \times 10^7 \, \text{m}^3$ of water. In 1965, two Indonesian geologists recognized the threat posed by the inadequate drainage system and predicted another eruption within 5 years (Zen and Hadikusumo, 1965). Less than two years later, the volcano erupted again and lahars caused by ejection of lake water killed hundreds of people. This needless disaster prompted authorities to dig new drainage tunnels, and to build more diversion and sediment retention structures. The new drainage tunnels and a fortuitous shallowing of the lake bottom reduced lake volume to less than $2 \times 10^6 \, \text{m}^3$, and no primary lahars occurred during the 1990 eruption of Kelut (Bourdier et al., 1997). Sediment retention and diversion structures have been less effective. The sedimentation following the 1966 and 1990 eruptions have filled most of these structures to or near capacity and resulted in the threats of flooding to farms and towns from rivers flowing in channels at levels of 1–2 m above their surroundings (Blong, 1986: Rodolfo, 2000).

Crater-lake dams can fail during dormant periods owing to overflow and rapid incision, sapping and undermining, or collapse caused by weakening or earthquakes. At Kelut in 1875, a failure of the west crater wall during a dormant interval caused lahars that destroyed 30 villages and cost numerous lives (Escher, 1922). An infamous example is the Tangiwai disaster that occurred in 1953 when sapping and undercutting caused the crater-lake dam at Ruapehu, New Zealand to fail (Neall, 1976). Water gushed out of the lake, lowering it by 6 m, and caused a lahar downstream. Tragically, the resultant lahar destroyed a railroad bridge just minutes before the arrival of a passenger train, and the train could not be stopped in time to prevent it and its passengers from plunging into the valley below; 151 fatalities resulted (Neall, 1976). A lahar detection system that detects acidic high water upstream was implemented after 1953. The system now automatically triggers warning signals for passing trains.

10.5.2 Lahars caused by melting of ice and snow

Explosive eruptions of hot debris at ice and snow-clad volcanoes have caused some of the largest known and most catastrophic primary lahars. The most voluminous of these occur when rapidly moving (10–200 m/s) hot (typically >500°C) pyroclastic flows and surges sweep across steep glaciated slopes. The hot flows are energetic, erosive and very efficiently mix hot debris with ice and snow. The pyroclastic flows typically come to rest or nearly do before floods of water pour out of the deposits and adjacent snow pack, picking up sediment from the pyroclastic-flow deposits themselves and from other loose debris downstream. Water floods formed in this way bulk up rapidly by several fold near source and form debris-flow phases downstream (Pierson et al., 1990). The steep incised valleys that typically radiate away from large glaciated volcanoes serve as efficient conduits for lahars to travel tens or even hundreds of kilometres before spreading widely over hundreds of square kilometres in lowlands (Crandell, 1971). Moderate and large lahars can thus inundate unsuspecting populations with lethal effect in distal areas far from the source volcano.

Historically, Cotopaxi volcano, at 5,897 m and clad by 20 km^2 of glacier ice, has been the one of the world's most prolific producers of lahars that derive from pyroclastic flows sweeping across glaciers; furthermore, in prehistoric time, it produced one of the largest known lahars (Mothes, 1992). From 1532 until 1902, lahars had occurred on 27 occasions during explosive eruptions (Simkin and Siebert, 1994). The most widespread and destructive lahars occurred in 1744, 1768, and 1877 during explosive eruptions with VEI = 4 (Table 10.1). In 1877, lahars extended more than 200 km down valleys, destroyed towns, and killed hundreds of people and thousands of animals (Wolf, 1878). About 4,500 years ago, a huge pyroclastic flow caused a lahar much larger than those of the historical time (Chillos valley lahar). The Chillos valley lahar had a volume of almost 4 km^3, covered 440 km^2 and flowed 330 km to the Pacific Ocean (Mothes et al., 1998). On the basis of past lahar inundations at Cotopaxi, Hall and von Hillebrant (1988) have mapped a smaller higher lahar hazard zone and larger lower lahar hazard zone (Figure 10.4, see color section). Almost 50,000 people live in the higher hazard zone and almost 90,000 people live in the more extensive lower hazard zone (Figure 10.4).

The most tragic historical lahar event occurred in 1985 when pyroclastic flows swept across the ice cap of Nevado del Ruiz (5,321 m), Columbia, and produced deadly lahars. Shortly after 9 p.m. on 13 November a rather small (5×10^6 m^3) but explosive (VEI 3) pyroclastic eruption produced 2×10^7 m^3 of melt water (Pierson et al., 1990; Thouret, 1990). The resultant floods, augmented by eroded debris and overrun streamflow produced lahars with a total volume of nearly 10^8 m^3 in four valleys (Pierson et al., 1990). Within four hours these lahars had descended 5,100 m, travelled 100 km, and inundated more than 50 km^2. Two and half to three hours after the explosive eruption, the Lagunillas and Azufrado River lahars had flowed about 70 km and debouched onto the flood plain near the town of Armero. The lahars destroyed Armero (Figure 10.5, see color section) and killed more than 23,000 of its residents.

The tragedy at Nevado del Ruiz provided both scientific and socio-political lessons. First, relatively small pyroclastic eruptions can generate large, destructive lahars, especially at ice-clad volcanoes (Figure 10.1). Second, steep narrow valleys can focus lahars so that they flow tens of kilometres with little attenuation. In fact such lahars are apt to grow within 10–20 km of the volcano. Third, land-use planning could have mitigated the disaster: the previous eruptions of Nevado del Ruiz in 1595 and 1845 had inundated the same valleys and produced lahars that were larger than those of 1985 (Thouret et al., 1990). Armero was built on a lahar plain constructed of deposits from these past two eruptions. Furthermore, fatalities were recorded in the same places during the previous eruptions (Simkin and Siebert, 1994). A land-use plan preventing construction in areas at risk could have saved many lives. Fourth, information about hazardous phenomena has little value if it is not shared in timely fashion with those in harm's way. A better understanding of the risk from lahars and an evacuation plan could have helped many to escape. The eruption sequence began at 15:06; by 17:00 ash began to fall in Armero; at 19:00 the Red Cross in Ibagué ordered evacuation of Armero, but this information was not communicated to the residents; at 21:08 the climactic eruption began; at 22:40 lahars arrived in Chinchiná, sweeping away hundreds of homes and perhaps 1,000 people; finally, at 22:35 lahars arrived in Armero. Residents of Armero were not told to evacuate, had little idea where to go, and did not understand the risk that confronted them (Hall, 1990). Fifth, even a rudimentary alarm system could have provided warning (Rodolfo, 2000). Lahars did not arrive until 4.5 hours after ash began to fall in Armero, 2.5 hours after the onset of the climactic eruption, and 55 minutes after the disaster at Chinchiná.

10.5.3 Lahars derived from rain after eruptions

Large explosive eruptions emplace voluminous unconsolidated debris deposits (pyroclastics) that, because they are loose and unvegetated, are easily remobilized as lahars during heavy rains. Explosive eruptions can produce voluminous fills of pyroclastic flow and fall deposits. Initially, these deposits have no established drainage network and are easily eroded to form lahars. During rainy periods after eruptions, such lahars are typically small ($<10^6$ m^3) but common, in some cases occurring for many years after the eruption. Furthermore, pyroclastic and subsequent lahar deposits commonly dam tributary drainages to form lakes that later break out to form lahars that can be moderate or even large in size ($\sim 10^7$, 10^8 m^3).

Since 1991, when the cataclysmic (VEI 5–6) 9 June to 15 June eruptions of Mount Pinatubo, Philippines, generated about 5.5 km^3 of unconsolidated pyroclastic-flow deposits (Scott et al., 1996b), monsoon rains on loose unconsolidated debris

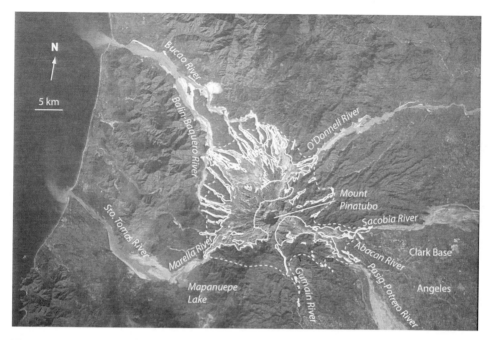

Figure 10.6. Shuttle image of Pinatubo (NASA, 1996). The white lines approximately delineate the extent of the 1991 pyroclastic flows. The dashed line indicates the approximate extent of pyroclastic surges. The light gray in the drainages downstream are largely lahar deposits.

have remobilized 2.5 km^3 of that debris and generated hundreds of lahars (Umbal, 1997; written commun., Phivolcs staff, 2004) (Figure 10.6). Those lahars inundated villages, towns, and once fertile farmland, killed tens of people and displaced thousands of others, and caused millions of dollars in damage to the local infrastructure (Bautista, 1996). Lahars began 12 June, 1991 when rains eroded and mixed with tephra-fall deposits and also remobilized pre-existing channel material. Large numbers of lahars occurred during the cataclysmic eruptions of 14–15 June when rainfall runoff from coeval Typhoon Yunya incorporated tephra and eroded fresh pyroclastic deposits. These syneruptive lahars inundated all of the major drainages, destroyed bridges, eroded channels, and buried houses. Heavy rains during the 1991 monsoon season initiated all subsequent large lahars, and lahars had inundated 360 km^2 of residential arable land by the end of 1991 (Rudolfo, 2000). Since the cataclysmic eruptions of June 1991, nearly all lahars at Pinatubo have resulted from torrential rains on loose unconsolidated pyroclastic debris (Scott et al., 1996a). Water torrents running across huge fans of pyroclastic-flow deposits and through narrow proximal barrancas bulk up in very short distances to become debris flows. Floods during typhoons have breached ephemeral lakes in a few cases. The resulting water floods have also bulked up and caused a small number of post-1991 lahars, which have been up to an order of magnitude larger than rain-induced lahars

266 Volcanic debris flows

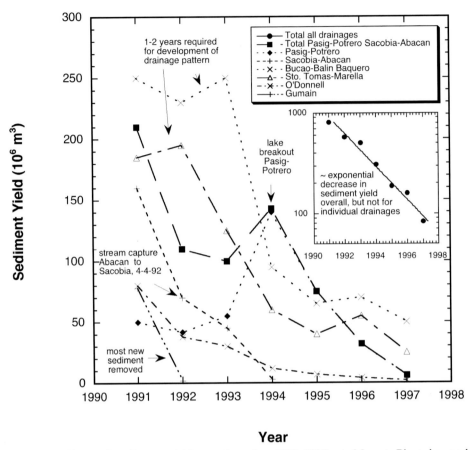

Figure 10.7. Plots of sediment yield vs. time for 1991–1997 at Mount Pinatubo and surrounding drainages. During the period from 1991–1997 the bulk of sediment transport was by lahars.

(Rudolfo, 2000). Lake breakouts from the lahar-dammed Mapanuepe Lake severely affected the Santo Tomas River downstream (Rodolfo et al., 1996; Umbal, 1997). A lake breakout along the Pasig-Potrero River in 1994 generated lahars that overwhelmed an earthen dike system and buried the town of Bacolor under several meters of sediment.

Overall, the sedimentary response to the huge influx of pyroclastic debris was large in 1991 and diminished exponentially through 1997 (Figure 10.7), but the response in individual drainages was more complex owing to factors such as re-establishment of new drainage networks and stream capture. Most of this sedimentary transport was by lahars (Montgomery et al., 1999; Hayes et al., 2002). In drainages with the largest proportions of pyroclastic deposits, the sedimentary transport increased in the first year or two as the drainage network developed (Figure 10.7). In other watersheds, the drainage pattern established itself within the first monsoon season. In February 1992, secondary explosions blocked the Abacan River and rerouted its lahars into the Sacobia drainage. Owing to a similar event in October 1993, the Pasig-Potrero River, which until then had been relatively unaffected by lahars, captured a large portion of the Sacobia River watershed as well as most of its lahar transport (Umbal, 1997; Hayes et al., 2002).

In the first few years after 1991, the immense scale of the sedimentary response overwhelmed efforts to protect property and communities downstream. Dikes, typically constructed of laharic debris eroded easily (Rudolfo, 2000). Sediment retention structures were ineffective because they rapidly filled up. Efforts to remove material from sediment-choked drainages simply could not keep pace. Diversion dams typically rerouted the threat of lahars from one populated area to another. It was nearly impossible to protect communities from large lake-breakout events or rerouting of sediment owing to stream capture.

Torrential rain can also cause lahars at dormant volcanoes as happened at Casita, Nicaragua, in 1998 (Vallance et al., 2004). On 30 October, 1998, five days of heavy rain from Hurricane Mitch triggered a slope failure and lahar high on the south slopes of Casita volcano. Between 10:30 and 11:00 a.m., survivors who lived downslope on the apron of the volcano reported a roaring noise like helicopters or thunder (Scott et al., 2004). Shortly thereafter, a debris-flow wave several metres deep and more than 1 km wide swept off the volcano and obliterated all traces of two towns in its path. Those who saw the flow on the volcano's slopes described it as water, but those who saw it near the town or below described it as mud – "an infernal wave of mud, rocks, and trees" (Barreto, 1998).

Examination of deposits in lahar pathways at Casita and other Central American volcanoes show that such events occur periodically (Scott et al., 2004). In 1541, several days of heavy rain toward the end of the rainy season, when soils were saturated, caused a lahar that swept down the slopes of Agua volcano and destroyed the capital of Guatemala, Cuidad Vieja (Schilling et al., 2001). Original accounts (Feldman, 1993) show that this was a rain-induced lahar not a lake breakout as is commonly believed.

10.5.4 Lahars that derive from debris avalanches

Some debris avalanches contain sufficient water and weakened rock that they transform to water-saturated flows as they deform and move downslope. Lahars of this type are most common at glaciated volcanoes whose rocks have been severely weakened by hydrothermal alteration and seem to be uncommon elsewhere. Such lahars are among the largest documented. Even at glaciated volcanoes, lahars having this origin do not occur often, and our knowledge of such events is based almost entirely on examination of prehistoric deposits.

Mount Rainier in Washington State has produced seven such lahars in the past 10,000 years, one of which, the Osceola Mudflow, is one of the largest lahars yet discovered worldwide (Figure 10.8, see color section). The $\sim$4-km^3 Osceola Mudflow began as a water-saturated avalanche during eruptions at the summit of Mount Rainier about 5,600 years ago. It filled valleys of the White River system north and north-east of Mount Rainier to depths of 80–150 m, flowed northward and westward more than 120 km, covered more than 240 km^2 of the Puget Sound lowland, covered another 160 km^2 under the water of Puget Sound, and extended up to 20 km under the water (Dragovich et al., 1994; Vallance and Scott, 1997). About 10 km from its source the lahar had a velocity of about 160 km/hr and a peak discharge of 10–20 million m^3/s; about 50 km from source the velocity was about 70 km/hr and peak discharge was 2–3 million m^3/s (Vallance and Scott, 1997).

The Osceola Mudflow transformed completely from debris avalanche to clay-rich lahar within 2 km of its source because of the presence within the pre-avalanche mass of large volumes of pore water and abundant weak hydrothermally altered rock (Vallance and Scott, 1997). A survey of clay-rich lahars suggests that the amount of hydrothermally altered rock in the pre-avalanche mass determines whether a debris avalanche will transform into a clay-rich debris flow or remain a largely unsaturated debris avalanche. The potential for transformation has important implications for hazards because lahars are more mobile than debris avalanches, and therefore spread more widely and travel much greater distances down-valley. Furthermore, rather small eruptions can produce very large lahars of this type (Figure 10.1).

10.6 LAHAR-HAZARD ASSESSMENTS

The paradigm most commonly applied to assess lahar hazards at an individual volcano is that past events are the best indicators of what might potentially happen in the future. Lahar-hazard zones are readily delineated at volcanoes for which long historical records exist. Because many volcanoes lack such a detailed historical record, field mapping and reconstructions of relations between lahars and primary volcanic deposits such as ash fallout, pyroclastic flows, and lava flows must serve to delineate temporal and spatial distributions of lahars (a detailed hazard and risk analysis for non-volcanic debris flows is given in Chapter 17).

10.6.1 Rationale for lahar-hazard assessments

A lahar-hazard appraisal includes a forecast of the kinds of events that cause lahars at a volcano and of lahar genesis; anticipation of the timing of such lahar events and the resulting deposits that might endanger people living downstream; and the location of areas that might be inundated by lahars. The information necessary to achieve these objectives derives from four closely related kinds of studies, characterized here as genetic, stratigraphic, chronologic, and cartographic (Crandell and Mullineaux, 1975). Studies of lahar genesis and order of succession (stratigraphy) are generally undertaken together because correlation of individual lahar deposits to eruptive products like pyroclastic flows and tephra-fall deposits aids in the interpretation of lahar origin. Not all lahars at active volcanoes are related to volcanism, and many small to moderate sized lahars have causes similar to debris flows in other environments (Table 10.1).

Both relative and absolute dating methods can help to establish a chronology at a given volcano. Stratigraphic relation of lahars to one another, weathering profiles between deposits, and geomorphic relations where deposits form successive terraces within a valley provide relative age determinations. Stratigraphic units can also be dated absolutely through radiometric dating, especially radiocarbon dating. Accurate chronologies are essential for evaluating frequency–magnitude relationships and statistical patterns of lahar production.

Distribution of deposits is critical for delineating lahar-hazard zones. Future lahars may not inundate exactly the same areas, but a map of areas inundated in the past can nonetheless be used in a general way as a rough guide. Inferences about lahar distribution at a volcano lead to the preparation of a map on which zones of relative risk are defined.

At Mount Rainier, a high-profile volcano because of its large nearby population, more than 60 Holocene lahars have been identified and mapped (Crandell, 1971; Scott et al. (1995), and the lahar chronology and distribution allow excellent delineation of lahar hazards related to volcanism (Scott and Vallance, 1995) (Figure 10.9, see color section). Debris flows unrelated to volcanism occur on almost a yearly basis at Mount Rainier. Common causes of these generic debris flows at volcanoes might include glacial outburst floods, heavy precipitation events, and tectonic earthquakes. Because non-volcanic debris flows are much more common and much smaller than those related to volcanism a separate debris-flow hazard map has been prepared for Mount Rainier (Vallance et al., 2003; *http://vulcan.wr.usgs.gov/Volcanoes/Rainier/Publications/OFR03-368/framework.html*).

10.6.2 An automated method of lahar-hazard delineation

Because the time-consuming and expensive investigations that allow lahar-hazard assessments are not done at many volcanoes, alternative methods are needed. Iverson et al. (1998) have developed a method of delineating lahar-hazard zones based on inundation and size data for well-documented lahars worldwide. Their method depends on empirical relationships:

$$A = 0.05 V^{2/3} \quad \text{and} \quad B = 200 V^{2/3}$$

where A is the cross-sectional area of the valley that is inundated, B is the planimetric area that is inundated, and V is the volume of the lahar. Using these relationships and GIS-based topography, computer-generated lahar-inundation zones can be calculated for given lahar volumes using a routine called LAHARZ (Figure 10.10, see color section). Iverson et al. (1998) suggest using a geometric progression of volumes to delineate hazard zones in potentially affected valleys. The smallest volume simulations delineate the areas that lahars are most likely to inundate and represent areas having the greatest risk. In contrast, largest volume simulations delineate the broadest areas that might be inundated but represent areas least at risk. When combined with reconnaissance geology, the method allows expeditious generation of lahar-hazards maps. At Mount Rainier, for example, a reasonable range of potential lahar volumes would be from 10^7–10^9 m^3, progressing geometrically as a factor of $\sqrt{10}$ (1×10^7, 3×10^7, 1×10^8, 3×10^8, and 1×10^9 m^3) (Figure 10.11, see color section). Given information about size and recurrence these hazard simulations yield a reasonable approximation of lahar-hazard zones (Figures 10.9 and 10.11, see color section).

10.6.3 Mitigation of lahar hazards

Lahar hazards can be mitigated through land-use planning, engineering, or evacuation in time of crisis. Engineering strategies include sediment-retention dams, diversion dams and dikes, and draining of dangerous crater lakes as at Kelut. Draining crater lakes and other dangerous dammed lakes as at Mount St. Helens (Chapter 27) has proved to be effective in mitigating lahar hazards. Dams, diversions and dikes work well for lahars of modest volume. Unfortunately, structures of these types cannot always cope with the huge volume of volcanic debris that is transported downstream during and after eruptions. In such cases, lahar-warning systems can give residents downstream time to escape; however, such systems do not protect property and work only when there is sufficient distance and time between source areas and population centres. Warning systems include those in which an observer is posted upstream with a radio, those in which a connection or series of connections are broken, and those in which a seismic sensor sends a signal to emergency response personnel (Chapter 12).

The surest way to avoid risk from lahars is through strict land-use regulation. Once lahar-hazard maps are prepared, federal and local agencies can prohibit residential and commercial building in areas potentially at risk. Unfortunately, dense populations that are already established in some threatened areas make land-use planning impractical from a political point of view. In many countries, people will choose to tolerate moderately high risk from lahars and other volcanic processes because the most fertile land and most healthy climates are located on or near volcano flanks.

In many cases, evacuation is the only mitigation option. At volcanoes, lahars are considerably more likely during or immediately following eruptions. Furthermore, the most voluminous lahars are associated with explosive eruptions. Fortunately, volcanoes most commonly give advanced warning of a few days to a few weeks prior to eruption and allow time for orderly evacuation of residents at risk. In the past 25 years, tens of thousands of people have successfully been evacuated from areas at risk owing to lahars and other volcanic hazards (Simkin and Siebert, 1994).

Volcanic debris flows pose a grave risk to people and property downstream because eruptions can suddenly release water and debris on steep slopes that coalesce to flow tens of kilometres downstream and because they disturb landscapes thereby facilitating post-eruption debris-flow production. Debris flows related to eruptions can occur suddenly and can be much larger in size than those occurring in other environments. Volcanic debris flows not directly related to eruptions are smaller but can occur more often and less predictably than primary volcanic debris flows. Such secondary debris flows can occur during rainstorms for years after a major volcanic disturbance. Abundant liquid and huge quantities of available sediment allow volcanic debris flows to flow over gentle gradients and inundate areas far away from their sources. The flows can severely damage or destroy structures in their paths and can potentially injure or kill unwary people. It is therefore critical to assess hazards owing to such debris flows and to adopt mitigation strategies for dealing with the hazards that derive from them.

10.7 REFERENCES

Baretto, P.E. (1998) Testimonios relacionados con los ruidos y imagines fatales segundos antes de la tragedia, llegada repentinamente desde el volcán Casita a las comunidades Rolando Rodriguez y El Porvenir. Unpublished report, Managua, Nicaragua.

Bautista, C.B. (1996) The Mount Pinatubo disaster and the people of central Luzon. In: C.G. Newhall and R.S. Punongbayan (eds), *Fire and Mud: Eruptions and Lahars of Mount Pinatubo, Philippines* (pp. 151–164). University of Washington Press, Seattle.

Bourdier, J-L., Pratomo, I., Thouret, J-C., Boudon, G., and Vincent, P.M. (1997) Observations, stratigraphy and eruptive processes of the 1990 eruption of Kelut volcano, Indonesia. *Journal of Volcanology and Geothermal Research*, **79**, 181–203.

Carrasco-Núñez, G., Vallance, J.W., and Rose, W.I. (1994) A voluminous avalanche-induced lahar from Citlaltépetl volcano, Mexico: Implications for hazard assessment. *Journal of Volcanology and Geothermal Research*, **59**, 35–46.

Crandell, D.R. (1971) *Postglacial Lahars from Mount Rainier Volcano, Washington* (USGS Professional Paper 677, 75 pp.). US Geological Survey, Reston, VA.

Crandell, D.R. and Mullineaux, D.R. (1975) Technique and rationale of volcanic-hazards appraisals in the Cascade Range, Northwestern United States. *Environmental Geology*, **1**, 23–32.

Cronin, S.J., Neall, V.E., Palmer, A.S., and Lecointre, J.A. (1997) Changes in Whangaehu River lahar characteristics during the 1995 eruption sequence, Ruapehu volcano, New Zealand. *Journal of Volcanology and Geothermal Research*, **76**, 47–61.

Blong, R. (1986) *Volcanic Hazards: A Sourcebook on the Effects of Eruptions* (400 pp.). Academic Press, Sydney.

Dolan, M.T. (2004) Observations of lahars along the Sacobia-Bamban River systems, Mount Pinatubo, Philippines (48 pp.). MS thesis, Michigan Technological University, Houghton, Michigan.

Dragovich, J.D., Pringle, P.T., and Walsh, T.J. (1994) Extent and geometry of the mid-Holocene Osceola mudflow in the Puget Lowland: Implications for Holocene sedimentation and paleogeography. *Washington Geology*, **22**(3), 3–26.

Ewert, J.W. and Harpel, C.J. (2004) In harm's way: Population and volcanic risk. *Geotimes*, **14**(4), 4–9.

Escher, B.G. (1922) On the hot lahar of the Valley of Ten Thousand Smokes (Alaska). *Proceedings Koninklijke Akademie van Wetenschappen, Amsterdam*, **24**, 282–293.

Feldman, L. (1993) *Mountains of Fire Lands that Shake. Labyrinthos* (295 pp.). Culver City, CA.

Fisher, R.V. and Schmincke, H.-U. (1984) *Pyroclastic Rocks* (472 pp.). Springer-Verlag, Berlin.

Hall, M.L. (1990) Chronology of the principal scientific and governmental actions leading up to the November 13, 1985 eruption of Nevado del Ruiz, Colombia. *Journal of Volcanology and Geothermal Research*, **42**, 101–115.

Hall, M.L. and von Hillebrant, C. (1988) *Mapa de los Peligros Volcánicos Potenciales Asociados con el Volcán Cotopaxi*. Instituto Geofísico, Escuela Politécnica Nacional, Quito.

Hayes, S.K, Montgomery, D.R., and Newhall, C.G. (2002) Fluvial sediment transport and deposition following the 1001 eruption of Mount Pinatubo. *Geomorphology*, **45**, 211–224.

Hoblitt, R.P., Walder, J.S., Driedger, C.L., Scott, K.M., Pringle, P.T., and Vallance, J.W. (1998) *Volcano Hazards from Mount Rainier, Washington, Revised 1998* (USGS Open-file Report 98-428, 11 pp., 2 plates). US Geological Survey, Reston, VA.

Hodgson, K.A. and Manville, V.R. (1999) Sedimentology and flow behavior of a rain triggered lahar, Mangatoetoenui Stream, Ruapehu volcano, New Zealand. *Geological Society of America Bulletin*, **111**(5), 743–754.

Iverson, R.M. (1997) The physics of debris flows. *Reviews of Geophysics*, **35**, 245–296.

Iverson, R.M., Schilling, S., and Vallance, J.W. (1998) Objective delineation of lahar-inundation on hazard zones. *Geological Society of America Bulletin*, **110**, 972–984.

Janda, R.J., Scott, K.M., Nolan, K.M., and Martinson, H.A. (1981) *Lahar Movement, Effects and Deposits* (USGS Professional Paper 1250, pp. 461–478). US Geological Survey, Reston, VA.

Lavigne, F., Thouret, J.C., Voight, B., Suwa, H., and Sumaryono, A. (2000) Lahars at Merapi volcano, Central Java: An overview. *Journal of Volcanology and Geothermal Research*, **100**, 423–456.

Lavigne, F. and Thouret, J.C. (2003) Sediment transportation and deposition by rain-triggered lahars at Merapi Volcano, Central Java, Indonesia. *Geomorphology*, **49**(1/2), 45–69.

Major, J.J. and Newhall, C.G. (1989) Snow and ice perturbation during historical volcanic eruptions and the formation of lahars and floods: A global review. *Bulletin of Volcanology*, **52**, 1–27.

Montgomery, D.R., Panfil, M.S., and Hayes, S.K. (1999) Channel-bed mobility response to extreme sediment loading at Mount Pinatubo. *Geology*, **27**, 271–274.

Mothes, P.A. (1992) Lahars of Cotopaxi volcano, Ecuador: Hazard and risk evaluation. *Geohazards, Natural and Man-made* (pp. 53–63). Chapman & Hall, London.

Mothes, P.A., Hall, M.L., and Janda, R.J. (1998) The enormous Chillos Valley Lahar: An ash-flow-generated debris flow from Cotopaxi volcano, Ecuador. *Bulletin of Volcanology*, **59**, 233–244.

Neall, V.E. (1976) Lahars as major geological hazards. *Bulletin of the International Association of Engineering Geologists*, **14**, 233–240.

Newhall, C.G. and Self, S. (1982) The volcanic explosivity index (VEI): An estimate of explosive magnitude for historical volcanism. *Journal of Geophysical Research*, **87**, 1231–1238.

Pierson, T.C. (1999) Transformation of a water flood to debris flood to debris flow following the eruption-triggered transient-lake breakout from the crater on March 19, 1982. In: T.C. Pierson (ed.), *Hydrologic Consequences of Hot-rock/Snowpack Interactions at Mount St. Helens Volcano, 1982–84* (pp. 19–36). US Geological Survey, Reston, VA.

Pierson, T.C. and Scott, K.M. (1985) Downstream dilution of a lahar: Transition from debris flow to hyperconcentrated streamflow. *Water Resources Research*, **21**, 1511–1524.

Pierson, T.C., Janda, R.J., Thouret, J.C., and Borerro, C.A. (1990) Perturbation and melting of snow and ice by the 13 November 1985 eruption of Nevado del Ruiz, Colombia, and consequent mobilization, flow and deposition of lahars. *Journal of Volcanology and Geothermal Research*, **41**, 17–66.

Rodolfo, K.S. (2000) The hazards from lahars and jökulhlaups. In: H. Sigurdsson, B.F. Houghton, S.R. McNutt, H. Rymer, and J. Stix (eds), *Encyclopedia of Volcanoes* (pp. 973–995). Academic Press, London.

Rodolfo, K.S. and Arguden, A.T. (1991) Rain-lahar generation and sediment-delivery systems at Mayon volcano, Phillipines. In: R.V. Fisher and G.A. Smith (eds), *Sedimentation in Volcanic Settings* (SEPM Special Publication No. 45, pp. 71–87). SEPM, Tulsa, Oklahoma.

Rodolfo, K.S., Umbal, J.V., Alonso, R.A., Remotigue, C.T., Paladio-Melosantos, M., Salvador, J.H.G., Evangelista, D., and Miller, Y. (1996) Two years of lahars on the western flank of Mount Pinatubo: Initiation, flow processes, deposits, and attendant geomorphic and hydraulic changes. In: C.G. Newhall and R.S. Punongbayan (eds), *Fire and Mud: Eruptions and Lahars of Mount Pinatubo, Philippines* (pp. 989–1014). University of Washington Press, Seattle.

Savage, S.B. and Lun, C.K.K. (1988) Particle size segregation in inclined chute flow of dry cohesionless granular solids. *Journal of Fluid Mechanics*, **189**, 311–335.

Schilling, S.P., Vallance, J.W., Matías, O., and Howell, M.M. (2001) *Lahar Hazards at Agua Volcano, Guatemala* (USGS Open-file Report 01-432, 9 pp., 1 plate). US Geological Survey, Reston, VA.

Schmidt, K.G. (1934) Die Schuttströme am Merapi auf Java nach dem Ausbruch von 1930. *De Ingenieur in Nederlandsch-Indië*, **1**(7), 91–120.

Scott, K.M. (1988) *Origins, Behavior, and Sedimentology of Lahars and Lahar-runout Flows in the Toutle-Cowlitz River System* (USGS Professional Paper 1447-A, 75 pp.). US Geological Survey, Reston, VA.

Scott, K.M. and Vallance, J.M. (1995) *Debris Flow, Debris Avalanche, and Flood Hazards at and Downstream from Mount Rainier, Washington* (USGS Hydrologic Investigations Atlas HA-729, 9 pp., 2 plates). US Geological Survey, Reston, VA.

Scott, K.M., Vallance, J.W., Pringle, P.T. (1995) *Sedimentology, Behavior, and Hazards of Debris Flows at Mount Rainier, Washington* (USGS Professional Paper 1547, 106 pp., 1 plate). US Geological Survey, Reston, VA.

Scott, K.M., Janda, R.J., de la Cruz, E.G., Gabinete, E., Eto, I., Isada, M., Sexon M., and Hadley, K.C. (1996a) Channel and sedimentation responses to large volumes of 1991 volcanic deposits on the east flank of Mount Pinatubo. In: C.G. Newhall and R.S. Punongbayan (eds), *Fire and Mud: Eruptions and Lahars of Mount Pinatubo, Philippines* (pp. 971–988). University of Washington Press, Seattle.

Scott, W.E, Hoblitt, R.P, Torres, R.C, Self, L., Marinez, M., and Nillos, T. Jr (1996b) Pyroclastic flows of the June 15, 1991, climactic eruption of Mount Pinatubo. In: C.G. Newhall and R.S. Punongbayan (eds), *Fire and Mud: Eruptions and Lahars of Mount Pinatubo, Philippines* (pp. 151–164). University of Washington Press, Seattle.

Scott, K.M., Macías, J.L., Naranjo, J.A., Rodríguez, S., and McGeehin, J.P. (2001) *Catastrophic Debris Flows Transformed from Landslides in Volcanic Terrains: Mobility, Hazard Assessment, and Mitigation Strategies* (USGS Professional Paper 1630, 59 pp.). US Geological Survey, Reston, VA.

Scott, K.M., Vallance, J.W., Kerle, N., Macías, J.L., Strauch, W., and Devoli, G. (In press) Catastrophic precipitation-triggered lahar at Casita volcano, Nicaragua: Occurrence, bulking, and transformation. *Earth Surface Processes and Landforms*.

Scrivenor, J.B. (1929) The mudstreams (lahars) of Gunoug Keleot in Java. *Geology Magazine*, **66**, 433–434.

Simkin, T. and Siebert, L. (1994) *Volcanoes of the World* (349 pp.). Geoscience Press, Tucson, AZ.

Smith, G.A. (1986) Coarse-grained nonmarine volcaniclastic sediment: Terminology and depositional process. *Geological Society of America Bulletin*, **97**, 1–10.

Smith, G.A. and Fritz, W.J. (1989) Volcanic influences on terrestrial sedimentation. *Geology*, **17**, 375–376.

Thouret, J.C. (1990) Effect of the November 13, 1985 eruption in the snow pack and ice cap of Nevado del Ruiz, Colombia. *Journal of Volcanology and Geothermal Research*, **41**, 177–202.

Thouret, J.C., Cantagrel, J.M., Salinas, R, and Murcia, A. (1990) Quaternary eruptive history of Nevado del Ruiz, Colombia. *Journal of Volcanology and Geothermal Research*, **42**, 225–252.

Vallance, J.W. (1994) Experimental and field studies related to the behavior of granular mass flows and the characteristics of their deposits (197 pp.). PhD. Thesis, Michigan Technological University, Houghton, Michigan.

Vallance, J.W. (2000) Lahars. In: H. Sigurdsson, B.F. Houghton, S.R. McNutt, H. Rymer, and J. Stix (eds), *Encyclopedia of Volcanoes* (pp. 601–616). Academic Press, London.

Vallance, J.W. and Scott, K.M. (1997) The Osceola Mudflow from Mount Rainier: Sedimentology and hazard implications of a huge clay-rich debris flow. *Geological Society of America Bulletin*, **109**(2), 143–163.

Vallance, J.W., Cunico, M.L., and Schilling, S.P. (2003) *Debris-flow Hazards Caused by Hydrologic Events at Mount Rainier, Washington* (USGS Open-file Report 03-368, 4 pp., 2 plates). US Geological Survey, Reston, VA.

Vallance, J.W., Schilling, S.P., Devoli, G., Reid, M.E., Howell, M.M., and Brien, D.L. (2004) *Lahar Hazards at Casita and San Cristóbal volcanoes, Nicaragua* (USGS Open-file Report 01-467, 12 pp., 3 plates). US Geological Survey, Reston, VA.

van Bemmelen, R.W. (1949) *The Geology of Indonesia* (Vol. 1A, 732 pp.). Government Printing Office, The Hague.

Umbal, J.V. (1997) Five years of lahars at Pinatubo volcano: Declining but still potentially lethal hazards. *Journal of the Geological Society of the Philippines*, **52**(1), 1–19.

Waitt, R. (1989) Swift snowmelt and floods (lahars) caused by great pyroclastic surge at Mount St. Helens volcano, Washington, 18 May 1980. *Bulletin of Volcanology*, **52**, 138–157.

Wolf, T. (1878) *Memoria Sobre El Cotopaxi y su Ultima Erupcion Acaecida el 26 de Junio de 1877 (Memoir of Cotopaxi and its Final Eruption of 26 June, 1877)*. Imprenta del Comercio, Guayaquil.

Wolf, T. (1879) Der Cotopaxi und seine letzte eruption am 26 Juni 1877. *Neues Jahrbuch für Mineralogie, Geologie und Paleontologie*, 113–167.

Zen, M.T. and Hadikusumo, D. (1965) The future danger of Mt. Kelut. *Bulletin of Volcanology*, **28**, 275–282.

11

Application of airborne and spaceborne remote sensing methods
Robert T. Pack

11.1 INTRODUCTION

This chapter focuses on the use of electromagnetic remote sensing methods for the quantification of debris flow and debris avalanche hazard and risk. First, various remote sensing methods in common use will be introduced. These include passive systems (i.e., systems that collect light energy reflected from the sun or some other natural source of energy). Examples of passive systems include color film photography, satellite imagery such as IKONOS, and digital infrared imagers that sense thermal energy. Active systems transmit electromagnetic energy that interacts with the environment and then returns back to the instrument. These include synthetic aperture radars and laser radars (lidars) where either radio or light radiation is emitted then received back into the aperture.

11.2 DEFINITIONS

In this chapter, precise definitions of what is meant by *spatial resolution* and *spectral range* will first be reviewed as these terms are used throughout the chapter.

Spatial resolution is a quality measure that affects the utility of imagery for landslide evaluation. Common measures of spatial resolution include the resolving power (RP), modulation transfer function (MTF), or the optical transfer function (OTF). The most intuitive of the measures is the RP. It is determined by projecting a standard three-bar square-wave test pattern, with a range of spatial frequencies, onto a film specimen and exposing it (ISO, 1982; see Figure 11.1). The RP of a camera system is a measure of the sensors ability to transfer the test image to a photograph or image and is influenced by the lens, film, filter, and processing (film or digital). It is

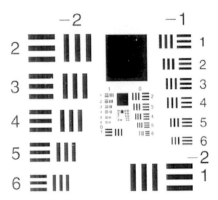

Figure 11.1. US Air Force 1951 lens test chart illustrating black and white line pairs used in defining the resolving power of an imager.

often expressed as line pairs per millimeter (lp/mm). In this chapter, the spatial resolution is defined (in microns) as:

$$SR = \frac{1}{\left(\dfrac{RP}{1000}\right) \times \sqrt{2}} \qquad (11.1)$$

This is roughly equivalent to a spread function that is the Fourier transform of the MTF (Kolbl et al., 1996). Thus an RP of 500 lp/mm, a common specification for aerial film, is equivalent to a spatial resolution of approximately 1.4 μm on the film. Given a resolution i on the focal plane of a sensor, the resolvable object size can be calculated by the formula $o = i/S$ where o is the resolvable object size and S is the scale of the photography (Wolf and Dewitt, 2000). For example, given aerial photography with a scale of 1 : 20,000 and a film resolution of 1.4 μm, the resolvable object size on the ground would theoretically be 28 mm (gravel size particles).

Another important remote sensing concept is *spectral range*. Suites of spectral ranges sensed by an instrument are described by terms such as color, panchromatic, multispectral, hyperspectral, etc. Figure 11.2 (see color section) shows the full span of spectral ranges, from cosmic rays through radio waves, that can be sensed. The most common ranges used in landslide studies include: (1) visible (VIS) 400–700 μm, (2) near-infrared (NIR) 700–3,000 μm, (3) mid-infrared (MIR) 3,000 nm–30 μm, (4) far-infrared (FIR) 30 μm–0.1 mm, millimetre wave (MMW) 0.1–10 mm, and radio frequency (RF) >10 mm. Other terms for spectral range in common usage include visible/near-infrared (VNIR) 400–1,000 nm, color infrared (CIR) also in the 400–1,000 nm range, and short-wave infrared (SWIR) 1,000–2,500 nm. The MIR spectral range spans thermally emitted energy that peaks at 10 μm (Clark, 1999). Panchromatic (sometimes called monochromatic or black and white imagers) collect data for only one spectral range in VIS through IR. Color imagers collect three fairly broad ranges in VIS, usually red, green, and blue. Multispectral imagers collect two or more narrow spectral ranges with VIS and IR. Hyperspectral imagers or imaging spectrometers collect 10s to 100s of narrow spectral ranges in VIS and IR.

Radar systems used in landslide studies use radio waves in three common wavelength bands: X-band (3 cm), C-band (6 cm), and L-band (24 cm) (Henderson and Lewis, 1998).

11.3 REMOTE SENSING METHODS

11.3.1 Aerial cameras

Instruments and film

The highest spatial resolution in imagery for landslide inventories comes from panchromatic (black and white) film taken by a quality aerial camera. This is the product that has been most commonly used for qualitative or interpretive evaluation. This film typically has a spatial resolution of about 1–2 µm on the focal plane of the camera which means that, at a photography scale of 1:20,000, a high-contrast object (e.g., gravel-size rocks) would theoretically be visible. This assumes that the camera is of high enough quality (including a large enough aperture) to project an image of that detail onto the film. It also assumes one uses a zoom stereoscope to view the µm-level detail of the film. Color film typically has a resolution of 4 µm or more, which would resolve a cobble size larger than 8 cm given a photo scale of 1:20,000. This lower resolution is compensated by the fact that photo interpreters can distinguish more variations of color on color photography than shades of gray on panchromatic photography. If photos are to be used for *measuring* landslide geometry or morphology using photogrammetry, panchromatic photography is usually the better choice.

Panchromatic or color film is often scanned into digital format for use in stereo photogrammetry or orthorectification. In this case, the resolution of the image is limited by the resolution of the scanner, which is typically limited to 7–20 µm (Kolbl et al., 1996). For example, resolvable objects in scanned 1:20,000 scale photography would be approximately 14–40 cm in size on the ground. For this reason, photo interpreters often prefer the use of original film prints or diapositives. However, digital imagery is preferred for photogrammetry where image orthorectification and incorporation into GIS databases is necessary. Figure 11.3 shows a scanned panchromatic photograph of a debris avalanche and debris-flow terrain near Lowman Idaho (Pack, 2003). This data set comes from the commonly available US Geological Survey Digital Orthophoto Quadrangle (DOQ) product that covers most of the USA. Note that at a ground resolution of only 3 m, the 3–4 m wide debris-flow path is barely discernable. Debris levees and other morphologic features that distinguish debris flow from hyperconcentrated flow or flood erosion would be difficult to identify with this photography. Figures 11.3 and 11.4 show an orthophoto with a 1-m ground resolution. Note that debris levees are recognizable in this case.

As illustrated above, an important rule when choosing film photography for landslide studies is to choose a spatial *resolution greater than half the size of the*

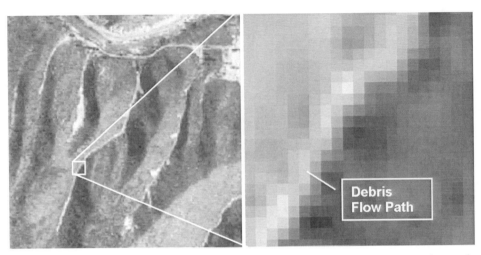

Figure 11.3. Trace of debris flow near Lowman, Idaho, shown on a panchromatic photograph. The image on the right is a blow-up of a short portion of the debris path. Note that the DOQ spatial resolution of approximately 3 m on the ground just enables tracing of the debris flow path.

Data is from the US Geological Survey.

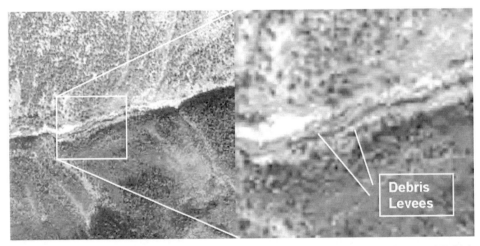

Figure 11.4. Trace of debris flow showing remnant levees in the Chetwynd area of British Columbia. Note that the spatial resolution of 1 m on the ground allows the recognition of debris flow levees.

Data is from Canadian Forest Products.

landform to be observed. Thus, if the expected minimum dimension of the surface process (in this case a debris-flow levee) is expected to be 1 m, at least 0.5 m should be resolved in the photography. This becomes especially important when considering the use of satellite imagery and orthorectified aerial imagery.

Infrared film used in landslide studies includes black and white IR and false-CIR. These films are sensitive to the green, red, and NIR wavelengths. IR and CIR is especially useful in differentiating vegetation species and soil moisture conditions that might be useful in identifying potential debris-flow initiation areas. With IR, moist soil appears darker gray than in a panchromatic image. Also, young deciduous vegetation that often grows in landslide scars will appear light gray in contrast to dark gray coniferous vegetation. Similarly with CIR, damp ground where seepage prevails is distinctly darker compared to dry ground and coniferous vegetation is bright red compared with the brownish red associated with conifers. Figure 11.5 (see color section) shows a CIR photograph of debris slide scar showing seepage areas as dark grayish-blue patches within the scar.

The recognition and mapping of debris avalanches and debris flows requires the interpretation of geomorphic expression and vegetative patterns. Difficulties in doing this are amplified in heavily forested terrain where not only photo interpretation is adversely affected but where the topography derived from photogrammetry can be in error. Some studies have been completed that analyse the relative influence of forest obscuration on the statistics of landslide inventories based on photos. When comparing aerial photo-based reconnaissance mapping with detailed field mapping in the Santa Cruz Mountains, California, Wills and McCrink (2002) found that reconnaissance mapping consistently underestimates the percentage of shallow debris slides and debris flows. This is important in studies that aim to develop cumulative frequency–magnitude relationships and base management decisions on a power function fitted to the statistical distribution, because the distribution may not reflect the true occurrence of landslides in an area (Pelletier et al., 1997; Hovius et al., 2000; Dai and Lee, 2003; Stark and Hovius, 2001). Identification of landslide morphology from the air can be improved through the use of lidar technology that has the ability to penetrate through holes in the vegetation. This relatively new technology is discussed in Section 11.2.3.

Landslide studies sometimes require the acquisition of static and quasi-dynamic variables that enable the prediction of the spatiotemporal probabilities. In combination with field studies, many of these data can be collected from aerial photographs. Multi-temporal aerial photographs have been particularly useful in estimating temporal probabilities in many studies. For example, Dai and Lee (2003) completed a study of landslides in North Lantau, Hong Kong, where data for spatial variables including geology, topography, and geomorphology were qualitatively combined with data for temporal variables including rainfall records and landslide occurrence periods. A predictive logistic regression model was developed that achieved an overall accuracy of 87.2%. The model was then applied to rainfalls of a variety of return periods to predict the spatiotemporal probability of landsliding on natural slopes.

Multi-temporal digital aerial imagery has been used to automatically map and monitor landslide recurrence intervals. Hervas et al. (2003) have used digital VNIR (see Section 11.1) aerial imagery acquired at different dates and at high spatial resolution to perform automatic change detection. The method requires that the imagery first be geometrically registered and radiometrically normalized. This is followed by the thresholding of multiple images to create landslide-related change pixels. Subsequent filtering based on the degree of rectangularity of regions was also used to eliminate pixel clusters corresponding to man-made land use changes. This method was used in the eastern Italian Alps with scanned aerial diapositives with a 1-m pixel resolution.

11.3.2 Digital imagers

Instruments

Airborne and spaceborne digital imagers use an array of solid state detectors, either charge-coupled devices (CCDs) or complementary metal oxide semiconductors (CMOSs). The array is arranged in a matrix and is placed on the image focal plane. They have the advantage of being able to provide rich multispectral or hyperspectral data sets that may assist in the interpretation of landslide hazards. Multispectral imagers usually sense 3–8 spectral bands. Hyperspectral imagers can have up to several hundred spectral bands. Operators of a digital imager can adjust the radiometry to fit changing conditions thus ensuring maximum dynamic range and contrast to reveal more details in images.

When compared to aerial film, the added spectral information comes at the expense of spatial detail. To duplicate the spatial resolution of a $9'' \times 9''$ aerial camera having film with a $2\,\mu m$ spatial resolution, an array of $115,000 \times 115,000$ pixels (13.2 gigapixels) is necessary. Such cameras do not exist. Alternatively, to match film subsequently scanned to a 15-μm resolution, a $15,240 \times 15,240$ pixel (0.232 gigapixel) digital camera would be required. Recently developed aerial digital aerial cameras such as the Leica ADS40 and Zeiss DMC have pixel dimensions of 12,000 and 13,000 respectively, and are close to achieving this standard (Heier and Alexander, 1999; Fricker et al., 2001). Smaller format cameras used from aircraft typically have 16 megapixels or less and are obviously less comparable to aerial film in spatial resolution.

Spaceborne multispectral digital imagers applied to landslide studies include Landsat TM (30 m),[1] SPOT (20 m), IRS (23 m), and IKONOS (4 m) (Jadkowski, 1987; Rosenqvist et al., 1990; Kasa et al., 1992; Hervas, 2000). As imagers with spatial resolutions of only a few metres have recently been launched, satellite data has become increasingly useful for debris avalanche and debris-flow delineation. Airborne multispectral imagers including the Daedelus Airborne Thematic Mapper (2 m) are also being used in landslide studies (Whitworth et al., 2002).

Analysis techniques

Spectral bands associated with digital cameras can be used for image rectification and restoration, image enhancement, image classification, data merging, and Geographic Information System (GIS) integration. Restored, enhanced, or classified images facilitate landslide identification and characterization in hazard studies. An

---
[1] The number in parentheses is the associated spatial resolution of the sensor.

image analyst can use these products, in combination with field studies, to assist in the landslide interpretive process as with film imagery.

Various statistical methods are used to choose the spectral band combination containing the most information for landslide studies. Methods might include the optimum index factor, maximum variance–covariance determinant, and principal component analysis. In order to evaluate the applicability of these methods to landslide studies, one study used 11-band Airborne Thematic Mapper (ATM) imagery of the inland slopes of Stonebarrow Hill in West Dorset, England (Ramli et al., 2002). These slopes are extensively mantled with relict landslide features. The best band combination results obtained from these methods are evaluated against the visually checked imagery, where all possible bands are generated and classified in terms of texture and color. It was found that, for this area, two combinations of three bands produced the best result. A similar study by Whitworth et al. (2002) in Worcestershire, England, found that the most useful bands are the NIR and MIR, possibly owing to the relatively higher moisture contents of the landslides.

11.3.3 Passive microwave sensors

Passive microwave sensors operate in the same spectral range as radar (30 cm), except that they sense the naturally available microwave energy within the field of view rather than emitted energy. All objects in the natural environment emit microwave radiation, albeit faintly (e.g., Lillesand et al., 2004). For a given object, a microwave signal would include: (1) an emitted component related to surface temperature, moisture and other attributes; (2) an emitted component coming from the atmosphere; (3) a surface-reflected component from sunlight and skylight; and (4) a transmitted component having a subsurface origin. By appropriate selection of wavelength, passive microwave systems can either look *through* or look *at* the atmosphere. An important capability of importance in landslide studies is the ability to measure soil temperature and soil moisture, properties often indicative of potentially unstable areas. Another capability yet to be fully exploited is the ability of multispectral microwave radiometry to peer through overburden to detect geologic structure, material type, and subsurface voids.

Pelletier et al. (1997) investigated the use of passive microwave remote sensing in studying the cumulative frequency–size distributions of shallow landslides induced by precipitation in Japan and Bolivia as well as landslides triggered by the 1994 Northridge California earthquake. The soil moisture field was used with slope stability analysis to model the frequency–size distribution. The study depended on the fact that soil shear strength, as governed by soil cohesion, soil pore water pressure, and internal friction angle, is primarily affected by soil moisture measurable by a microwave sensor. An interesting result of the study was the observation that strong ground motion from earthquakes lowers the shear stress necessary for failure, but does not change the frequency–size distribution of failed areas.

11.3.4 Lidar sensors

The term *lidar* is an acronym for light detection and ranging and is the optical equivalent to radar. Lidars designed for sensing terrain are typically of the *energy detection* type and transmit a laser pulse that travels to the ground, reflects off rocks and bushes, then returns to the sensor. The time it takes for the light to make the round trip is precisely timed, then converted into the distance or range to the objects. Short laser pulses in the order of a few nanoseconds (ns) enable measurement of the surface geometry of an object with an accuracy of a few centimetres. The resulting lidar data is displayed as either a cloud of 3-D plotted points or as a range image. A range image is similar to a digital image except that each pixel represents the exact distance or range to the target from the camera. The spatial resolution is determined by the beam divergence of the laser. Typical divergence values range from 5 mrad to 100 μrad, which, at a range of 1,000 m is a pixel size of between 5 m and 10 cm respectively.

Figure 11.6 (see color section) shows a lidar range image of a hillside subject to debris slides and ravelling (Pack, 2002). Figure 11.7 (see color section) shows the same hillside rendered in 3-D on a computer graphics system. Color digital imagery from a separate camera has been draped onto the surface generated by the lidar. Even though it was acquired from a tripod, the 3-D image is displayed from the perspective of a helicopter. These data enable the direct evaluation of the surface attributes and morphology of potential debris source areas. These data also are of sufficient precision to form the basis for engineering design of stabilization measures, protection berms, or other types of rehabilitation.

Lidar has proven useful for the monitoring of ground movements. A ground-based lidar system with a 2-km range has been used to monitor debris/rock slide movement in the Alps (Paar et al., 2000). This system has a beam divergence of 0.5 mrad, a range accuracy of 2 cm, and a maximum range of 2 km and was used to map the surface of the scarp face of the Schwaz Landslide, Austria (see Figure 11.8, color section). A measurement interval of every hour for regions of interest and twice a day for the entire face was used to detect slope movements. Isoline plots were then made and overlaid with imagery to visualize the slide dynamics. Figure 11.9 (see color section) shows movement isolines plotted on top of an image. It shows several zones where boulder movements were detected suggesting that the scarp face is still active. Similar work was completed in Austria by Scheikl et al. (2000) and in England by Rowlands et al. (2003).

An advantage of lidar is its ability to detect the ground surface under a fairly dense tree canopy. A series of tests conducted by Carter et al. (2001) have demonstrated the ability of airborne lidar to detect landslide scarps beneath a tree canopy. An area near the Oahe Dam, South Dakota (USA) was chosen for a comparative study because it contains landslides that have been mapped in detail by classical ground surveying and photogrammetric techniques. Three lidar shot densities were obtained by an airborne lidar flying at 500 m above ground level and at a speed of 60 m/s. Distances between lidar shots for swath mapping varied from 0.7–2.0 m. Also, a profiling mode was used with a linear spacing of less than 1 cm. As one would expect, the greatest penetration to ground was achieved by the profiling mode. This work verified the ability of lidar to closely match ground measurements and detect detailed landslide features that would otherwise remain undetected.

11.3.5 Synthetic aperture radar (SAR) sensors

Radar imagers such as the SAR scatter millimetre wave or radio wave energy off objects in the scene, then sense the amplitude and phase of return energy to determine the object reflectivity and range. Unlike lidar, SAR does not directly observe the power of each individual echo. Many echoes are received simultaneously from which both the geometric and radiometric response are calculated through post-mission processing. A SAR image can be thought of as a panchromatic photographic print with an oblique illumination, but where the brightness is a measure of the scene reflectivity. Because of constructive and destructive interference between members of a signal ensemble, speckle is seen in most SAR imagery (e.g., Raney, 1998). Many SAR images are 2-D and are used for qualitative interpretation. The 3-D geometry of objects can be determined by the use of radargrammetry with multiple SAR images (Liberl, 1998) or interferometric SAR, sometimes referred to as InSAR, IfSAR, or ISAR. Delta-R interferometry uses two antennae on the same platform and allows automatic computation of the third dimension. Two-pass delta-R interferometry (D-InSAR or Delta-t InSAR) enables small soil or rock displacements, of the order of a few centimetres, to be resolved over a time interval, even from a satellite platform (Raney, 1998; Leva et al., 2003; Dzurisin, 2004).

The utility of the use of D-InSAR techniques with the ERS-1 and the JERS-1 L-band satellites in the study of landslide dynamics has been demonstrated (Fruneau et al., 1996; Rott et al., 1999; Kimura and Yamaguchi, 2000; Rizo and Tesauro, 2000). However, relatively large displacements are needed for this approach to work. Rott (2000) suggests that for ERS-1 SAR to detect slope movements, the slide should be at least of the order of several pixels in size which in this case is about 200 m.

Alternatively, airborne or ground-based SAR provides much higher resolution and a capability of measuring fine movement. Some of the earliest ground-based work on landslides was completed by Tarchi et al. (2000). More recently, Leva et al. (2003) used a ground-based D-InSAR to study the dynamics of the same debris avalanche headscarp studied by Paar et al. (2000). This work enabled the production of a velocity field map that is in agreement with field GPS measurements. Figure 11.10 (see color section) shows their ground deformation image where movements of between 0.5 and 5 cm are detected. The central zone is characterized by fast slope deformation of the order of a few centimetres per week while the scarp areas experience slow deformations of a few millimetres per week.

11.4 REMOTE SENSING METHODS IN HAZARD/RISK ASSESSMENT

The various remote sensing methods can contribute to the primary goal of the landslide investigator to understand and mitigate landslide risks. The means to achieve this usually first requires event inventories and landslide characterization. This step is essential to the understanding of the phenomena and is the first step to avoidance or mitigation. The next step typically involves terrain mapping that helps in the understanding of the natural setting and geomorphic history that provides a context for past events. The final step combines the previous two studies to predict spatial and temporal probabilities of occurrence and the expected consequences. The final product is a hazard and/or risk map that provides useful information to the land manager, engineer, or other persons responsible for hazard mitigation.

11.4.1 Event inventories and characterization

The visual interpretation of stereo aerial photographs is probably the most common tool for inventories of debris avalanches and debris flows. Through the use of a magnifying stereoscope, superior 3-D observations can be made at this highest of spatial resolutions. The other remote sensing methods reviewed in this chapter universally suffer from inferior spatial resolution. Table 11.1 is a summary of terrain features frequently associated with slope movements, the relationship of these features to debris avalanches and debris flows, and their characterization on aerial photographs. These elements can then be used to develop an interpretation and classification of debris avalanches and debris flows according to the characteristics in Table 11.2 (adapted from Soeters and Westen, 1996).

Multi-temporal aerial photographs enable estimates of landslide recurrence intervals. In the absence of historical information, estimates of landslide dates are limited to intervals between photographic missions. However, in some cases it is possible to correlate debris flows in a given time period if a significant hydroclimatic event falls within the period of aerial photography.

Multispectral digital imagery has a place in facilitating the identification of bare soil and seepage associated with landslides in heavily vegetated areas. The advantages of the spectral information in this product must be carefully weighed against the inferior spatial resolution compared to film photography.

It has been shown that lidar is a helpful technology in situations where is it known that forest vegetation is obscuring significant portions of smaller debris avalanches and debris flows. Depending on the thickness, type, and density of cover, the ability of lidar to penetrate vegetation and discover debris avalanche morphology could help avoid the inventory bias described by Wills and McCrink (2002) and others. A side benefit of lidar is the automatic production of contour maps of use in terrain mapping.

D-InSAR techniques have been shown to enable automatic landslide change detection over time intervals between InSAR missions. However, the greatest strength of this remote sensing technology has been associated with the characterization of the complex movements of relatively slow-moving earth flows.

The use of passive microwave sensors for landslide inventory work is relatively rare due to poor instrument availability and the difficulties in processing this data.

Table 11.1. Morphologic, vegetation, and drainage features characteristic of debris avalanches and debris flows, and their photographic characteristics.

Terrain features	Relation to slope stability	Photographic characteristics
Morphology		
Semicircular backscarp and steps	Head part of slide with outcrop of failure plane	Light-toned scarp, associated with small slightly curved lineaments
Hummocky and irregular slope morphology	Microrelief associated with shallow movements or small retrogressive slide blocks	Coarse surface texture, contrasting with smooth surroundings
Berms or levees parallel to a stream channel in a gully or canyon	Microrelief associated with the deposition of debris during a debris flow	Raised ridges immediately adjacent to and on one or both sides of a stream (see Figure 11.4)
Concave/convex slope features	Landslide scar and associated deposit	Concave/convex anomalies in stereo model
Step-like morphology	Retrogressive sliding	Step-like appearance of slope
Vegetation		
Lack of vegetation immediately below breaks in slope	Removal of vegetation by translational sliding at debris avalanche headscarps	Light-toned elongated areas at the head of gullies or just below breaks in slope
Irregular linear swaths of denuded vegetation or new regrowth	Slip surface of debris avalanches and the path of debris flows	Light-toned bare soil tracing a path down the fall-line of the slope
Drainage		
Areas of stagnated drainage	Landslide hollow, back-tilting landslide blocks, and hummocky landslide topography	Tonal differences and darker tones associated with ponds or wet areas
Seepage and springs in hillslope hollows	Naturally wet areas on slopes sometimes naturally occur at debris slide headscarps	Dark patches in hollows sometimes enhanced by differential vegetation
Interruption in drainage lines	Drainage anomaly caused by a headscarp	Drainage line abruptly broken by a break in slope
Anomalous drainage pattern	Streams curving around the lobe of a debris deposit	Stream disruption by a debris fan deposit

However, their ability to sense soil moisture and subsurface structures is intriguing and will likely entice more pilot studies in the future.

11.4.2 Terrain mapping

Spatiotemporal landslide analysis requires the use of natural terrain features for the prediction of landslides in areas where they have not yet occurred. Geomorphologic

Table 11.2. Photo interpretable characteristics of debris avalanches and debris flows.

Type of movement	Characterization	
Debris avalanche	Morphology:	Small shallow denuded spoon-shaped scars on steep slopes (>30°) with a clear linear path downslope; debris frequently absent due to erosion
	Vegetation:	Head and path are denuded or covered by secondary vegetation
	Drainage:	Seepage can appear at headscarp, channel or gully head frequently starts at the headscarp
Debris flow	Morphology:	Linear denuded paths that follow gullies or streams; evidence of aggressive scouring and vegetation removal; sometimes marked by depositional levees; debris distribution area (debris fan) showing subtle flow paths and structure
	Vegetation:	Recent flows show complete absence of vegetation; older flows are covered by secondary vegetation
	Drainage:	Stream flow along debris path common, sometimes originating at the head; bifurcation evidence on debris fan; evidence of temporary stream blockage by fans deposited at the junction with tributary channels.

mapping of terrain features typically involves manual delineation of homogeneous areas or polygons on aerial photographs using a stereoscope. The delineations are then transferred to a base map using photogrammetric techniques.

If digital photography is available, stereo viewing on a 3-D-enabled computer workstation is possible. In the 3-D stereo environment of the computer (using stereo glasses), geomorphic interpretations can be directly input into a georeferenced database. Moreover, because the photos are set up photogrammetrically, 3-D surface mapping is possible within this environment. Thus contour and slope maps can be created.

The creation of slope maps is an important aspect of terrain mapping as this is a required parameter in landslide prediction. Such maps have traditionally been prepared using analogue or analytic stereoplotters that capture elevation points in a specific pattern and along breaklines in the map area. These data are then input into a computer mapping system that uses these points to interpolate a digital elevation model (DEM) surface. Depending on the quality of the stereocompilation, the quality of the resulting slope map can drastically vary; especially considering that surface differentiation inherently magnifies errors. This is particularly true in heavily forested areas.

As an alternative to DEM production using aerial photography, lidar sensors can produce densely spaced elevation points, can penetrate trees, and can therefore produce high-quality DEMs. However, the ability to filter out the effects of vegetation to create a "bare earth model" is critical to the resulting data quality. Otherwise bushes or half-removed trees could be mistaken for landslide hummocks, large boulders, or other geomorphic features.

11.4.3 Hazard/risk assessment

Once the landslide inventory and terrain mapping is complete (with sufficient amount of ground truthing) the data are then used with a chosen hazard/risk prediction model which produces hazard/risk maps that are of use to decision makers. Where the rate of soil deformation at a headscarp requires monitoring for hazard warning purposes, it has been shown that lidar or InSAR methods can be quite helpful. These methods are capable of detecting centimetre-level deformations and can replace or nicely supplement ground survey methods including GPS. Their use in year-round continuous slope hazard monitoring has not yet been reported. However, as these instruments become cheaper to operate, such use in the future should be anticipated.

11.5 REFERENCES

Carter, W., Shrestha, R., Tuell, G., Bloomquist, D., and Sartori, M. (2001) Airborne laser swath mapping shines new light on earth's topography. *EOS Transactions, American Geophysical Union*, **82**(46), 549–555.

Clark, R.N. (1999) Spectroscopy of rocks and minerals, and principals of spectroscopy. In: A. N. Rencz (ed.), *Remote Sensing for the Earth Sciences* (3rd edn, Volume 3). John Wiley & Sons, New York.

Dai, F.C. and Lee, C.F. (2003) A spatiotemporal probabilistic modeling of storm-induced shallow landsliding using aerial photographs and logistic regression. *Earth Surface Processes and Landforms*, **28**(5), 527–545.

Dzurisin, D. (2004) *Volcano Geodesy*. Springer-Praxis, Chichester, UK.

Fricker, P., Schaeppi, R., and Walker, S. (2001) Results from test flights of the LH Systems ADS40 airborne digital sensor. *22nd Asian Conference on Remote Sensing, Singapore, CRISP*, 5–9 November.

Fruneau, B,, Achache, J. and Delancourt, C. (1996) Observation and modeling of the Saint-Etienne-de-Tinée landslide using SAR interferometry. *Technophysics*, **265**, 181–190.

Heier, H. and Alexander, H. (1999) A digital airborne camera system for photogrammetry and thematic applications. *ISPRS Joint Workshop on Sensors and Mapping from Space, September 27–30, Hanover, Germany*. International Society for Photogrammetry and Remote Sensing/Elsevier, Amsterdam.

Herderson, F.M. and Lewis, A.J. (eds) (1998) Principles and applications of imaging radar. In: *Manual of Remote Sensing* (Vol. 2). American Society for Photogrammetry and Remote Sensing and John Wiley & Sons, New York.

Hervas, J. (2000) Optical remote sensing for landslide investigations. *Natural Hazards Workshop, Igls Austria*, 5–7 June.

Hervas, J., Barredo, J.I., Rosin, P.L., Pasuto, A., Mantovani, F., and Silvano, S. (2003) Monitoring landslides from optical remotely sensed imagery: The case history of the Tessina landslide, Italy. *Geomorphology*, **54**, 63–75.

Hovius, N., Stark, C.P., Chu, H.-T., and Lin, J.-C. (2000) Supply and removal of sediment in a landslide-dominated mountain belt: Central Range, Taiwan. *Journal of Geology*, **108**, 73–89.

ISO (1982) *Photography—Photographic materials—Determination of ISO Resolving Power* (ISO 6328). International Organization for Standardization, Geneva.

Jadkowski, M.A. (1987) Multispectral remote sensing of landslide susceptible areas. Ph.D. dissertation, Utah State University, Logan, UT.

Kasa, H., Kurodai, M., Obayashi, S., and Kojima, H. (1992) On the applicability of remote sensing data for landslide prediction model. *Journal of the Remote Sensing Society of Japan*, **12**(1), 5.

Kimura, H. and Yamaguchi, Y. (2000) Detection of landslide areas using satellite radar interferometry. *Photogrammetric Engineering and Remote Sensing*, **66**(3), 337–344.

Kolbl, O., Best, M.P., Dam, A., Douglass, J.W., Mayr, W., Philbrik, R.H., Seitz, P., and Wehrli, H. (1996) Scanning and state-of-the-art scanners. In: *Digital Photogrammetry, An Addendum to the Manual of Photogrammetry*. American Society for Photogrammetry and Remote Sensing, Bethesda, MD.

Leva, D., Nico, G., Tarchi, D., Fortuny-Guasch, J., and Sieber, A.J. (2003) Temporal analysis of a landslide by means of a ground-based SAR Interferometer. *IEEE Transactions on Geoscience and Remote Sensing*, **41**(4), 745–752.

Liberl, F.W. (1998) Radargrammetry. In: F.M. Henderson and A.J. Lewis (eds), *Principles and Applications of Imaging Radar*. American Society of Photogrammetry and Remote Sensing and John Wiley & Sons, New York.

Lillesand, T.M., Kiefer R.W., and Chipman, J.W. (2004) *Remote Sensing and Image Interpretation* (5th edn, 763 pp.). John Wiley & Sons, New York.

Paar, G., Nauschnegg, B., and Ullrich, A. (2000) Laser scanning monitoring: Technical concepts, possibilities and limits. *Natural Hazards Workshop, Igls, Austria, 5–7 June*.

Pack, R.T. (2002) Engineering geologic mapping using 3D imaging technology. *37th Annual Symposium on Engineering Geology and Geotechnical Engineering, Boise, Idaho*.

Pack, R.T. (2003) Data from research project entitled "Quantifying the Exposure of Streams to Sediment Inputs from Managed Forests, A Risk-Based Approach" conducted at Utah State University, funded by U.S. Department of Agriculture contract 9901085, 1999–2002.

Pelletier, J.D., Malamud, B.D., Blodgett, T.A., and Turcotte, D.L. (1997) Scale-invariance of soil moisture variability and its implications for the frequency-size distribution of landslides. *Engineering Geology*, **48**, 254–268.

Ramli, M.F., Petley, D.N., Murphy, W., and Inkpen, R. (2002) Integration of high resolution thematic mapper and digital elevation model for landslide mapping. *Proceedings of the Regional Symposium on Environmental and Natural Resources, Kuala Lumpur* (pp. 647–653).

Raney, R.K. (1998) Radar fundamentals: Technical perspectives. In: F.M. Henderson and A.J. Lewis (eds), *Principles and Applications of Imaging Radar*. American Society of Photogrammetry and Remote Sensing and John Wiley & Sons, New York.

Rizo, V. and Tesauro, M. (2000) SAR interferometry and field data of Randazzo landslide (Eastern Sicily, Italy). *Physics and Chemistry of the Earth. Part B: Hydrology, Oceans, and Atmosphere*, **25**(9), 771–780.

Rosenqvist, A., Murai, S., Ochi, S., and Vibulsresth, S. (1990) A study on the causes of debris flow damage '88 in southern Thailand. *Journal of the Japan Society of Photogrammetry and Remote Sensing*, **29**(4), 16.

Rott, H. (2000) MUSCL: A European project on monitoring urban subsidence, cavities and landslides by means of remote sensing. *Natural Hazards Workshop, 5–7 June 2000, Igls, Austria.*

Rott, H., Scheuchl, B., Siegel A., and Grasemann, B. (1999) Monitoring very slow slope movements by means of SAR interferometry: A case study from a mass waste above a reservoir in the Otztal Alps, Austria. *Geophysical Research Letters*, **26**, 1629–1632.

Rowens, K.A., Jones L.D., and Whitworth, M. (2003) Landslide laser scanning: Anew look at an old problem. *Quarterly Journal of Engineering Geology and Hydrogeology*, **36**, 155–157.

Rowlands, K.A., Jones, L.D., and Whitworth, M. (2003) Landslide laser scanning: A new look at an old problem. *Quarterly Journal of Engineering Geology and Hydrogeology*, **36**(2), 155–157.

Scheikl, M., Poscher, G., and Grafinger, H. (2000) Application of the new Automatic Laser Remote Monitoring System (ALARM) for the continuous observation of the mass movement at the Eiblschrofen rockfall area, Tyrol, Austria. *Natural Hazards Workshop, Igls, Austria, 5–7 June.*

Soeters, R. and vanWeston, C.J. (1996) Slope instability recognition, analysis, and zonation. *Landslides Investigation and Mitigation* (Chapter 8, Special Report 247). Transportation Research Board, National Research Council, National Academy Press, Washington, DC.

Stark, C.P. and Hovis, N. (2001) The characterization of landslide size distributions. *Geophysical Research Letters*, **28**(6), 1091–1094.

Tarchi, D., Leva, D., and Sieber, A. (2000) Monitoring of landslides in mountain areas by radar interferometry. *Natural Hazards Workshop, 5–7 June 2000, Igls, Austria.*

Whitworth, M.C.Z., Giles, D.P., and Murphy, W. (2002) Identification of landslides in clay terrains using Airborne Thematic Mapper (ATM) multispectral imagery. *Proceedings of SPIE*, **4545**, 216–224.

Wills, C.J. and McCrink, T.P. (2002) Comparing landslide inventories: The map depends on the method. *Environmental Engineering and Geoscience*, **4**, 279–293.

Wolf, P.R. and Dewitt, B.A. (2000) *Elements of Photogrammetry* (3rd edn, p. 127). McGraw-Hill, Boston.

12

Debris-flow instrumentation

Richard LaHusen

Risk to people and property from debris flows may be mitigated by active measures such as deflection structures, debris basins, or passive measures such as land-use zoning. Where such measures are impractical, it may be necessary to use field instrumentation to detect and warn of debris flows in progress or conditions that might initiate debris flows. In response to such warnings, roads can be closed, people can be evacuated and other avoidance measures can be taken. Importantly, instrumentation systems to detect flows and warn people at risk are only effective when used as one part of a more comprehensive civil defense programme.

Instrumentation systems are also applied to scientific studies of debris-flow mechanics and behaviour. Scientific study of debris flow is undertaken in field settings as well as small-scale and large-scale experimental flumes. Although immediate short-term goals may differ, the ultimate justification for any instrumentation application is the reduction of hazards from debris flows.

12.1 SENSORS

Although many characteristics of debris flows can be instrumented, the essential parameters to be measured to meet specific objectives must be identified before appropriate sensors and other instruments are specified. In addition, project budget and duration, site access, and available expertise in the use and installation of monitoring equipment influence the design of a suitable instrumentation scheme. For example, a long-term warning system intended to reliably detect and communicate the occurrence of debris flows in real time has very different design criteria than a short-term study to test hypotheses of debris-flow dynamics. Whatever the application, specific sensors must not only meet scientific objectives, they must be compatible with the data acquisition system (DAS) to which they will interface. Ongoing technological development of sensors, data acquisition systems, and communications equipment for consumer markets enable the design and application of powerful, yet cost-effective instrumentation systems for debris-flow investigation.

12.1.1 Sensing debris-flow antecedents

Most debris flows are triggered by intense rainfall or rapid snowmelt (Couture and Evans, 2000) therefore monitoring antecedent hydrologic conditions can provide advance knowledge of hazardous conditions so advisory watches and warnings can be given. Precursory hydrologic monitoring may be limited to simple observations of rainfall or may include more intensive instrumentation systems to measure subsurface responses to rain or snowmelt.

Precipitation monitoring

Precipitation is often an essential parameter of interest for both research and warning applications. Rainfall is most often directly measured with tipping bucket rain gauges that have resolutions of 0.25 mm per tip. Such gauges can be equipped with electric or propane-fueled heating systems that can accommodate limited amounts of snowfall. More elaborate gauges filled with antifreeze solution can be installed to collect precipitation that falls as rain or snow. Accumulated precipitation in these gauges may be sensed by directly measuring the level of accumulation in the gauge or by measuring overflow with a tipping bucket mechanism.

Indirect estimates of rainfall rates can be made from radar reflectivity measurements. Meteorological radar data can be obtained in near real time from public agencies and commercial services or complete stand-alone systems can be purchased. Estimates of rainfall rates made from radar reflectivity measurements have variable accuracy and can be greatly affected by the presence of frozen or partially frozen precipitation. Accuracy can be improved with calibration using telemetered rain gauges. Meteorological radar data is particularly useful for tracking high-intensity rainfall over large areas with a single instrument. This capability allows greater lead-time for warning in areas where debris flows are triggered by short-duration high-intensity rainfall. Debris-flow warning systems based radar data and telemetered rain gauges were successfully implemented in the San Francisco area (Keefer et al., 1987).

Precipitation can be measured using rain gauges and changes in snow-pack water equivalent with snow pillows and calculate the resultant water release to the soil surface. However, it is simpler to directly measure the water delivered to the soil surface with a snowmelt lysimeter. Details of lysimeter construction vary but they generally consist of a large (1–100 m^2) shallow pan on the soil surface with a drain that routes meltwater to an underground tipping bucket. The tipping bucket mechanism should be buried to a depth sufficient to avoid freezing and adequate drainage should be insured. Relatively large lysimeter areas minimize effects of non-uniform water flow through the snow pack.

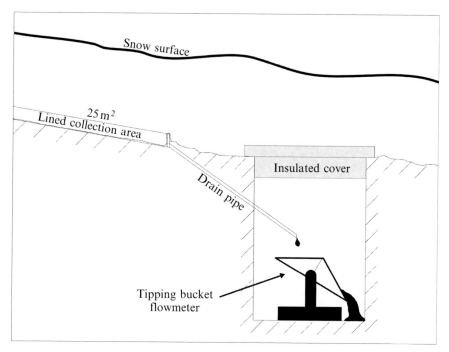

Figure 12.1. Cross-section diagram of a snowmelt lysimeter showing collection pan and tipping bucket flowmeter that is buried in a well-drained pit to avoid freezing.

Pore pressure, soil moisture, and water levels

After precipitation infiltrates the soil surface, it moves through the soil as both unsaturated and saturated subsurface flow. A variety of instrumentation may be used to monitor subsurface hydrologic conditions that are antecedent to debris-flow initiation. However, because the ultimate trigger for many debris flows is elevated pore-water pressure gradients, nests of piezometers are an obvious first choice for instrumentation in source areas. Most soils that are susceptible to mobilization into debris flows have relatively high permeability and therefore allow the use of standpipe or Casagrande piezometers with small diameters, which have sufficiently rapid responses to changes in pore pressures. Such piezometers are economical and have excellent long-term reliability.

Open standpipe piezometers can be installed by directly driving a small diameter metal pipe with a screened drive point into the ground with a fence post driver or with mechanized equipment such as a percussion hammer. Alternatively, piezometers can be installed in boreholes, typically as a small-diameter PVC pipe with a screened tip section. The tip should be placed in a pocket of clean sand at the bottom of the hole and then backfilled with bentonite granules down the space between the pipe and borehole to create an expansive seal above the sand pocket. Bentonite granules are dense enough to sink through standing water and small enough to effectively flow into the annulus. Hydraulic potential gradients can be determined by installing piezometers at multiple depths. After installation, fill the standpipe and observe the water level drop to confirm an acceptable response time for each piezometer. Water levels can be measured manually with a sounding weight or electrical continuity sensor or automated measurements can be made with an electronic pressure transducer installed near the bottom of the standpipe. An advantage of using open standpipes with pressure transducers is the ability to easily replace a failed sensor.

Alternatively, pressure transducers can be directly buried in boreholes without a standpipe. Nested piezometric sensors can be situated in a single borehole by placing each transducer in its own separate sand pocket at successive depths then sealing intervening sections of the hole with granular bentonite. Although sensors buried in this manner are not easily replaceable, the scheme has the advantages of easy installation in a single hole and buried sensors have a rapid response to pore pressure changes in soils with relatively lower permeability.

Some hillslope soils may intermittently approach saturated conditions, but only rarely become saturated and experience positive pore pressures that can be measured with piezometers. Consequently, soil moisture tensiometers may be installed to complement piezometers as soils approach saturation because they sense hydraulic potential under both saturated and unsaturated conditions. Tensiometers are installed as closed, rigid water-filled tubes with a porous ceramic bottom cup in direct contact with the soil. They require protection from freezing and periodic maintenance to remove accumulated air, thus maintaining a rapid response. Hydraulic potential gradients can also be determined from measurements of multiple tensiometers placed at different depths. Manual measurements are made with vacuum gauges or pressure transducers can be installed for automated measurements (Richards, 1965).

Selecting a suitable pressure transducer for field use from the vast array of commercially available sensors can be a daunting task. Minimum performance requirements for an application should be defined including fluid compatibility, range, resolution, and accuracy because costs increase with increasing resolution and accuracy. Absolute pressure transducers have an internal vacuum reference so changes in barometric pressure must also be monitored, whereas, differential transducers have two open ports so one port is vented to the atmosphere to automatically compensate for barometric pressure changes. Level and type of electrical output should be compatible with the capability of the DAS being used. Where long cable runs are needed, current, frequency, or differential signals are superior to single-ended voltage outputs.

Time domain reflectometry (TDR) has recently been used in field studies to measure soil moisture and earth deformation. TDR uses sophisticated electronics to accurately measure the travel time of an electromagnetic pulse as it travels down a transmission line and reflects back from discontinuities in the line. Pulse velocity, hence travel time, varies with the dielectric constant of the path. When part of the transmission path is a waveguide consisting of parallel conductors placed in soil, presence of pore water in the soil increases the travel time of the pulse. Cable testing instruments were the first instruments used to measure volumetric soil moisture using this technique, but dedicated soil moisture instruments are now available. Although the electronics are expensive, many inexpensive probes can be connected to a single instrument. TDR can rapidly perform accurate measurements of soil moisture ranging from completely dry to fully saturated, however, no measure of hydraulic potential or gradients can be made from these instruments (Topp et al., 1988).

Sensing complementary parameters

For robust debris-flow warning systems, it is not enough to only detect debris-flow occurrence but it is prudent to sense complementary parameters that can confirm hazardous conditions and help prevent false alarms. Some of the world's largest debris flows have occurred during or following volcanic eruptions (Chapter 10). Also known as lahars, debris flows originating on volcanoes are formed in several ways. Primary lahars form during eruptions as hot ash and eruptive products scour and melt snow and ice. Secondary lahars that follow eruptions are often generated by rainfall and surface runoff mobilizing loose ash or pyroclastic deposits. Some extremely large lahars have occurred when lakes dammed by volcanic deposits catastrophically failed. Complementary parameters in a lahar-warning program would be those that confirm these triggering events. For example, confirmation of volcanic eruptive activity is typically accomplished with seismic monitoring networks, whereas conditions that trigger secondary lahars may be measured with precipitation or lake level gauges as appropriate (Anonymous, 2004).

Some slow-moving landslides spawn debris flows, either when a portion of a slide breaks loose or when an entire landslide mass accelerates catastrophically and mobilizes into a debris flow (Reid et al., 2003). Therefore, monitoring landslide movements can be an integral part of some debris-flow studies. In addition to the hydrologic monitoring previously discussed, ground deformation measurements are often desired. Cable extensometers can be used to measure extension or contraction across lateral or transverse scarps, tension cracks, or advancing lobes. They can also be mounted with a cable extended down a borehole and through the failure zone to stable ground beneath a slide. Subsequent downslope movement of the slide extends the cable as the borehole is distended.

Inclinometers can be used to make intermittent or manual measurements of a minor deformation of a plastic or aluminum casing that is installed in a borehole. Alternatively, tilt meters can be permanently installed to make continuous measurements when interfaced to a DAS.

Ground deformation can also be intermittently or continuously measured using geospatial positioning systems (GPS). GPS receivers collect precisely timed messages that are broadcast from a constellation of orbiting satellites. To obtain high-precision locations, raw GPS data are logged in the receiver for subsequent retrieval or telemetered in real time to a personal computer (PC) for processing. Short-term (<20 minute) GPS position solutions are inherently noisy so high-accuracy positions require extended periods to accumulate data and filter results. Nevertheless, measurements with sub-centimeter accuracy can be made without long-term drift that plagues some other deformation sensors. On the Mission Peak Landslide near Fremont, California, a system using inexpensive consumer-grade receivers, telemetry, and near-real-time data processing on a PC has tracked seasonal landslide movements with sub-centimeter accuracy for more than four years. High accuracy differential measurements between receivers located up to 10 km apart using static post-processing methods is a technique well suited to landslide investigations where a clear view of the sky is possible.

12.1.2 Measuring debris-flow dynamics

Sensing ground vibrations

Installation and maintenance of sensors that require contact or close proximity to unstable channels where debris flows occur is problematic. Alternatively, sensors that can detect debris-flow occurrence and do not require close proximity to the channel are preferred for long-term reliable operation. Ground vibration sensors have proven to be such an option. Eyewitness accounts of debris flows have described the roaring noise and strong ground vibrations that accompany debris flow. Seismic systems intended to monitor volcanic eruptions have incidentally recorded the strong ground vibrations of passing lahars. Consequently, some seismic networks for volcano hazard monitoring now include seismometers deliberately located near channels to detect the distinctive sustained seismic signal that is characteristic of debris flow.

Debris-flow researchers in China and Japan installed broad-spectrum accelerometers to investigate the character of ground vibrations caused by debris flows (Suwa and Okuda, 1985). Although the channel and flow characteristics in the two studies differed greatly, both studies showed that debris flows generate broad-spectrum vibrations with the strongest vibrations occurring in the range 20–50 Hz. This finding has important significance to debris-flow vibration monitoring because geophones sensitive to this dominant frequency range are readily available and are much more rugged and cost one-tenth as much as lower frequency seismometers used for earthquake monitoring. Furthermore, because higher frequency ground vibrations rapidly attenuate with distance, high-frequency geophones are most sensitive to vibration sources near a channel where they are installed and are not as responsive to low-frequency signals such as earthquakes and eruptive explosions from more distant sources. The Acoustic Flow Monitor (AFM) is a debris-flow warning system responding to frequencies at the low end of the acoustic range (20–200 Hz) developed by the US Geological Survey. It contrasts sub-acoustic systems used for earthquake seismology (Brantley, 1990; LaHusen, 1996; Arattano, 1999; Berti et al., 2000).

Vibration sensors output signals proportional either to ground velocity or acceleration. Both parameters have proven useful for debris-flow detection in both vertical and horizontal orientations. Moving-coil geophones generate millivolt level AC voltages proportional to ground velocity and are sensitive to a somewhat limited frequency range. In contrast, accelerometers commonly use a piezoelectric

Figure 12.2. Loowit AFM station (Acoustic Flow Monitor, in housing centre) and precipitation gauge (on right), looking upstream towards Mount St. Helens' crater. Levee from a 1997 lahar is visible along the channel. Mount St. Helens' east crater wall is visible in the background.

USGS photograph taken on 30 August, 2004 by Kurt Spicer.

element that responds to a broad range of frequencies and requires electronic signal-conditioning circuitry to produce a signal that can connect to a DAS. Regardless of the sensor selected, interference from surface environmental noise can be avoided by burial of the sensor as deep as is practical, away from trees and poles that cause ground vibrations in response to wind.

Flow height sensing with ultrasonic range finders

Where debris flow is channelized and it is possible to install an overhead boom or cable, distance between a range finder and the flow surface can be sensed. Ultrasonic range finders, also known as distance meters, can measure distances up to 20 m or more by emitting ultrasonic or microwave bursts and measuring the round-trip travel time of the emissions that reflect back off the flow surface. The sensors should be oriented vertically so the beams meet the flow surface perpendicularly and obtain a strong reflection. High-speed, high-precision range finders that are useful for distances up to 2 m use optical triangulation of an infrared laser beam.

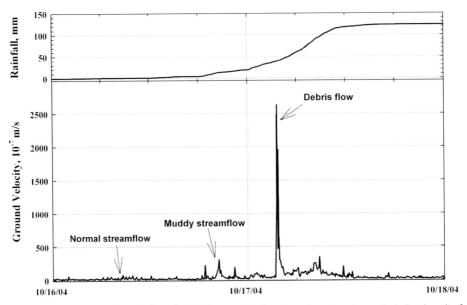

Figure 12.3. Graph of rainfall and AFM ground vibration showing characteristic signal of a debris flow with a rapid rise to more than 10 times background storm flow levels and a more gradual cessation of flow. Subsequent sediment-laden flow is evident as channel was flushed by additional rain.

These short-range sensors are best suited for experimental studies where flow depths will not be unexpectedly large and damage the instrument (Berti et al., 2000). New developments in laser range finders offer additional opportunities for flow height measurement (Iverson et al., 1992).

Surface velocity sensing with Doppler radar

Velocity measuring systems similar to those used for traffic law enforcement or sporting events have been used to measure the surface velocity of flows. The systems work by emitting microwaves at a fixed frequency towards an advancing or retreating flow. The Doppler shift in the frequency of the reflected waves is calculated and translated into a velocity measurement. Sensors may have analogue (voltage) or serial digital outputs that can be interfaced to a DAS.

Video imaging techniques

Video cameras and recorders have been used to record debris flows and later extract measurements of flow velocity, depth, superelevation, and even track the motion of individual clasts. Portable video recording systems can be activated by other sensors to conserve power and allocate storage media (Berti et al., 2000). Video telemetry requires considerable power and specialized equipment so it is not practical for most applications. Automated imaging analysis has been experimentally applied in a system to detect debris flows and trigger traffic-warning devices (Chang, 2000). Commercial equipment designed for real-time motion detection from video images is available for security systems and traffic-control systems. The utility of video and other optical imaging techniques is limited during periods of low visibility such as darkness, fog, heavy rain, or snowfall. Because of these limitations, optical monitoring systems have chiefly been used in scientific studies of flow behaviour and mechanics.

Force/load/pressure; contacting sensors

Knowledge of impact forces and loads are needed to design engineered structures strong enough to withstand the forces of debris flows. The greatest challenge in instrumentation of debris flows is measurement of forces that require direct sensor contact with a flow. Not only do such sensors need to survive extreme forces experienced during debris flow, they must be installed in such a manner that insures they measure only the forces they are intended to sense. For example, measurement of fluid pressure at the base or interior of a debris flow has been accomplished with pressure transducers that have rigid porous plates to isolate the sensors from the loads carried by solids in the flow (Iverson and LaHusen, 1989). In contrast, pressure transducers that are connected to a fluid-filled bladder or earth pressure cell measure the total load exerted by both solid and fluid phases of a flow. Total basal load has also been sensed with rigid plates and load cells mounted in a

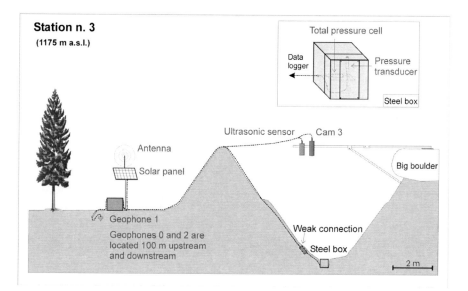

Figure 12.4. Illustration of a multiparameter debris-flow monitoring station at Acquabona channel, Italy. The system includes multiple geophones for debris-flow detection and velocity determination, over head video camera, ultrasonic depth sensor, basal pressure and load cells, as well as telemetry and on-site high-speed data logging.

rigid channel bed (Berti et al., 2000). Channel sections with rock or concrete beds are preferred sites for installation of basal force or pressure sensors because it is easier to install a sensor that will not be scoured away. Forces in a debris flow can fluctuate rapidly and the frequency response of a sensor can be severely reduced by the physical manner by which it is installed. Therefore, *in situ* measurement of the frequency response of sensors should be performed when possible to determine the resultant sensitivity and frequency response of the installed sensing assemblies.

12.2 DATA ACQUISITION SYSTEMS

12.2.1 Data loggers and other devices

After identifying parameters to measure and select corresponding sensors, a suitable DAS is needed to record or transmit measurements. An easy-to-use data logger that simply records measurements at regular intervals may suffice for many investigations, whereas more complex or power consumptive instruments may require a more versatile data logger for instrument control and power management. For example, a programmable DAS with both inputs and outputs would be needed to start a video recorder whenever a specific geophone level is exceeded. Anywhere adequate power and shelter are available close to a monitoring site, a PC with a data acquisition card or other device may be used as a complete and powerful DAS. Finally, where real-time telemetry is desired, a remote terminal unit (RTU) designed for supervisory control and data acquisition (SCADA) can be selected to control, measure, and transmit data. An Internet search for these devices will result in numerous suitable DAS devices and their availability and capabilities may be overwhelming. Evaluation of the following critical specifications will help select a suitable instrument for application to a debris-flow study.

The *number and type of inputs* of a DAS must be compatible with the number and type of sensors selected. Common types of inputs are analogue (voltage), digital, and serial (RS232). Analogue inputs are most prevalent, but the analogue range of the sensors and DAS should correspond. Analogue inputs can be of single-ended or differential type. A single-ended input is measured with reference to a common system ground whereas a differential signal measurement is a comparison of one signal to the other. Differential measurements are useful to minimize the offsets caused by long cable distances between the DAS and sensors. Another useful technique to handle long cable runs is the use of sensors with current outputs. The current can be sensed on an analogue input with the use of a quality resistor. Be sure that a current output transducer can operate with the systems supply voltage because many such sensors need higher supply voltages for operation.

Power consumption can be a critical specification for remote sites where long-term installation is needed to monitor infrequent flows. Memory size of a data logger should be evaluated in terms of the *number of measurements* that can be recorded so the interval between site visits can be considered. *Resolution* of signal magnitude is often specified in units of volts, % of the full-scale measurement range, or bits.

Low-resolution devices offer 8-bit resolution, equivalent to 0.4% of full scale range (F.S.). More commonly, DAS systems have 12-bit resolution (0.025% F.S.), but 16-bit (0.00015% F.S.) and 24-bit devices are increasingly available. Notably, typical noise levels and sensor inaccuracies limit the advantage of high-resolution systems. Only the most carefully engineered instrumentation systems benefit from device resolutions over 12–16 bits without over sampling and signal averaging. Temporal resolution is typically expressed as *sampling frequency* and can be very important for dynamic measurements of debris-flow processes.

Basic signal conditioning such as power supply regulation and amplification may be accomplished as part of a transducer electronics package or the DAS may need to supply regulated excitation voltage and amplify the low-level outputs of a sensor. Therefore, data acquisition hardware and software should be capable of capturing signal frequencies of interest such as impact forces or ground velocity. Anti-aliasing filters are typically needed when making AC measurements to prevent erroneous frequency analysis of recorded data. In order to simplify data acquisition and reduce telemetry data-rate requirements, simple electronic circuits have been used to rectify, integrate, capture, and hold transient signals. Such signal conditioning techniques can be used for measurements of mean and/or peak ground vibration or impact forces without needing high-powered and complex DASs.

Finally, one of the most important factors to consider when selecting a DAS is its ease of programming compared to the abilities of personnel involved in the project. Although many systems are quite versatile and powerful, their use may require an accomplished programmer and/or electronics expert to successfully connect, program, and maintain the equipment.

12.2.2 Real-time monitoring from remote sites

Real-time data acquisition from remote sites is a requisite for some debris-flow warning schemes to provide ample time to evacuate people or close access to hazardous roads or other sites. However, real-time or near-real-time monitoring is also useful for other reasons. It can help maximize data continuity through early detection of instrument malfunctions without the expense and inconvenience of unnecessary site visits. Another benefit is enhanced personnel safety resulting from fewer visits to hazardous field sites to retrieve data. Incorporating real-time capabilities in an instrumentation system requires additional system components including telemetry equipment and real-time data-management software.

Telemetry

Proliferation of wireless technology in the consumer market has produced numerous options for digital data telemetry from remote sites. Although most radio frequencies require a government issued license, there are several frequency bands now available for unlicensed (but restricted) use in industrial, scientific, and medical applications. These so-called ISM bands avoid interference on the shared frequencies through spread-spectrum techniques. Depending on the country of installation, telemetry may be authorized in 900, 2,400, and 5,800-MHz bands. Allowable frequencies, transmit power levels, and antenna gain restrictions vary, so consult regulations for the country of interest. Importantly, radio waves at these frequencies propagate in a straight line with minimal refraction or reflection; therefore, a clear line-of-sight is needed between telemetry links. Fortunately, implementing data repeaters is a straightforward task with many of these ISM spread-spectrum radios. Most radios of this type easily connect to DAS systems with a standard RS232 serial data connection. Telemetry distances up to 100 km can be achieved between elevated high-gain antennas.

Satellite telemetry is now a viable technique for data transmission from remote sites. Consumer oriented Internet service and equipment is offered by many satellite television providers for modest monthly charges. These very small aperture terminals (VSAT) consist of a dish and a satellite modem with an ethernet port that connects to the Internet. Serial device servers, also called serial-to-ethernet converters, are useful in such applications to convert data streams from a DAS or digital radio into an ethernet format for direct connection to a VSAT terminal.

Real-time data management software

Custom software expressly developed for specific real-time monitoring systems is a considerable undertaking. Alternatively, proprietary software is offered by manufacturers of SCADA equipment, but features may be quite limited. In 1996, an international industrial consortium was formed to define and support an open standard for interoperability of sensors and real-time plant operation software. The OPC Foundation (*http://opcfoundation.org*) now constitutes over 300 member companies. As a result of this effort, numerous compatible software packages that feature real-time data acquisition, graphing, logging, alarming, and web services are all available that conform to the standard. OPC data servers and clients are being deployed by the US Geological Survey Volcano Disaster Assistance Program as part of real-time debris-flow warning systems for international crisis responses.

12.2.3 Power systems

Where AC line power is available at a monitoring site, power consumption of instrumentation is not an important issue. However, lightning surge protection should be installed and an uninterruptible power supply is needed to insure data continuity during power outages. More likely, line power is not available at debris-flow monitoring sites so system designs should minimize power consumption so that battery powered operation is possible. A common power configuration for remote instruments is a deep-cycle lead-acid battery bank that is recharged by photovoltaic panels. Panel size can be estimated by summing the average current consumption of the system components then multiplying by 10–20 depending on the availability of solar energy at the site. For example, a system that consumes an average of 2 watts will require 20–40 watts of photovoltaic panel output. Battery capacity is determined by selecting the number of days the system might need to operate with insufficient solar input for battery charging. Wind generators are an effective option in some areas and should be considered, especially for higher power systems. Thermoelectric generators that burn liquefied petroleum gas and produce 10–100 watts with no noise or moving parts are also useful as a power source.

In special situations where excessive snow or ash precludes use of solar panels or where an instrument needs to be hidden to reduce risk of vandalism, non-rechargeable or primary batteries are useful. Small power demands can be met with lithium batteries but for larger systems, air-alkaline batteries are preferred because they have exceptionally high energy density and very low self-discharge rates so they excel in extended operation at remote sites. Air-alkaline batteries are typically used in railroad signalling or marine navigation buoys, with outputs of 1.5 volts per cell and capacities in excess of 1,000 amp hours.

12.3 SUMMARY

Design of debris-flow instrumentation for research or warning purposes involves specification and application of sensors, data acquisition electronics, telemetry, power supplies, and software for acquisition and management of data. Some sensing instruments are well known such as rain gauges, however, newer technologies such as meteorological radar greatly enhances our abilities to monitor hazardous conditions. In fact, some of the most promising system components might be borrowed from other disciplines. Geophones for geophysical exploration have proven to be rugged, inexpensive sensors for detection of debris flows for warning and monitoring. Products mass-produced for consumer markets such as automotive pressure sensors or digital telemetry equipment are readily available and surprisingly affordable. Even software products developed for industrial plant operations and monitoring are useful out-of-the-box in debris-flow research or warning systems. Application of traditional instruments and using technologies from other sources offers the greatest potential to creating affordable and effective debris-flow monitoring systems.

12.4 REFERENCES

Anonymous (2004) *Mt. Ruapehu Crater Lake Lahar Threat Response* (Fact Sheet, May). New Zealand Department of Conservation, Turangi, New Zealand.

Arattano, M. (1999) On the use of seismic detectors as monitoring and warning systems for debris flows. *Natural Hazards*, **20**(2), 197–213.

Berti, M., Genevois, R., LaHusen, R., Simoni, A., and Tecca, P. (2000) Debris flow monitoring in the Acquabona watershed on the Dolomites (Italian Alps). *Physics and Chemistry of the Earth, Part B: Hydrology, Oceans and Atmosphere*, **25**(9), 707–715.

Brantley, S.R. (ed.) (1990) *The Eruption of Redoubt Volcano, Alaska, December 14, 1989– August 31, 1990* (USGS Circular 1061, p. 30). US Geological Survey, Reston, VA.

Chang, S.Y. (2000) Evaluation of a system for detecting debris flows and warning road traffic at bridges susceptible to debris-flow hazard. *Debris-flow Hazards Mitigation, Mechanics, Prediction and Assessment.*

Couture, R. and Evans, S.G. (2000) *The East Gate Landslide, Beaver Valley, Glacier National Park, Columbia Mountains, British Columbia* (GSC Open File 3877, 26 pp.). Geological Survey of Canada, Vancouver.

Hadley, K.C. and LaHusen, R.G. (1995) *Technical Manual for an Experimental Acoustic Flow Monitor* (USGS Open-file Report 95-114, pp. 1–25). US Geological Survey, Reston, VA.

Iverson, R.M., Costa, J., and LaHusen, R. (1992) *Debris-flow Flume at H.J. Andrews Experimental Forest, Oregon* (USGS Open-file Report 92-483). US Geological Survey, Reston, VA.

Iverson, R., and LaHusen, R. (1989) Dynamic pore-pressure fluctuations in rapidly shearing materials. *Science*, **246**, 796–799.

Keefer, D.K., Wilson, R.C., Mark, R.K., Brabb, E.E., Brown, W.M., Ellen, S.D., Harp, E.L., Wieczorek, G.F., Alger, C.S., and Zatkin, R.S. (1987) Real-time landslide warning during heavy rainfall. *Science*, **238**, 921–925.

LaHusen, R.G. (1996) *Detecting Debris Flows Using Ground Vibrations* (USGS Fact Sheet FS-236-96, pp. 1–2). US Geological Survey, Reston, VA.

Lavigne, F., Thouret, J.-C., Voight, B., Young, K., LaHusen, R., Marso, J., Sumaryono, A., Dejean, M., and Sayudi, D.S. (2000) Instrumental lahar monitoring at Merapi volcano, Central Java, Indonesia. *Journal of Volcanology and Geothermal Research*, **100**, 457–478.

Reid, M.E., Brien, D.L., LaHusen, R.G., Roering, J.J., de la Fuente, J., and Ellen, S.D. (2003) Debris-flow initiation from large, slow-moving landslides. In: D. Rickenmann and C-L. Chen (eds), *Debris-flow Hazards Mitigation: Mechanics, Prediction and Assessment.*

Richards, S.J. (1965) Soil suction measurements with tensiometers. In: C.A. Black (ed.), *Methods of Soil Analysis* (Part 1, pp. 153–163). American Society of Agronomy, Madison, WI.

Stein, J. and Kane, D.L. (1983) Monitoring the unfrozen water content of soil and snow Using time domain reflectometry. *Water Resources Research*, **19**(6), December, 1573–1584.

Suwa, H. and Okuda, S. (1985) Measurement of debris flows in Japan. *Proceedings IV International Conference and Field Workshop on Landslides, Tokyo* (pp. 391–400).

Topp, G.C., Yanuka, M., Zebchuk, W.D., and Zegelin, S. (1988) Determination of electrical conductivity using time domain reflectometry: Soil and water experiments in coaxial lines. *Water Resources Research*, **24**(7), July, 945–952.

Zhang, S. (1993) A comprehensive approach to the observation and prevention of debris flows in China. *Natural Hazards*, **7**, 1–23.

13

Runout prediction methods

Dieter Rickenmann

13.1 INTRODUCTION

Runout prediction of natural hazards is important for determining affected areas and flow intensity parameters, which are essential elements for producing hazard maps (Petrascheck and Kienholz, 2003). A discussion of methods for runout assessment of debris flows is given, for example, in Hungr et al. (1984), Cannon (1989), Bathurst et al. (1997), Fannin and Wise (2001), and McDougall and Hungr (2003). Deposition and runout of debris flows are governed by several factors, especially a decrease in slope and a lack of flow confinement in the runout zone (e.g., Hungr et al., 1984; Corominas, 1996). For example, a Japanese rule of thumb claims that for debris fan slopes less than 10°, deposition occurs if the slope angle is reduced by a factor of two or if the flow widens out two or three times the flow width (Ikeya, 1979). Other influences on runout distances include momentum loss in bends (Benda and Cundy, 1990; Fannin and Wise, 2001; Lancaster et al., 2003), the presence of large organic debris in the debris-flow front (Lancaster et al., 2003), and the presence of trees or other obstacles in the runout zone (Corominas, 1996).

For specific debris-flow types, typical slope ranges can be defined where erosion and entrainment, transport, or deposition of sediments predominates (e.g., VanDine, 1996; Fannin and Wise, 2001). Considering the large variability of material composition and water content of debris flows, these ranges may differ considerably. Very coarse-grained debris flows, typically originating in small watersheds, may start to deposit at gradients as high as 27° (Rickenmann and Zimmermann, 1993), while lahars may travel several tens of kilometres, arresting at gradients of only a few degrees (Pierson, 1995).

Prediction of runout distance may be divided into empirical–statistical and dynamic methods. Though empirical–statistical approaches are often easy to use, they should only be applied to conditions similar to those on which their development is based. Dynamic models are physically based and consider the momentum or

Table 13.1. Overview of runout prediction methods discussed in the text.

General approach	Keywords to characterize method	Main references
Total travel distance (entire path length)		
Travel distance and event magnitude	Travel angle	Corominas (1996)
	Volume and descent height	Rickenmann (1999)
Volume balance	Without entrainment	Cannon (1993)
	With entrainment	Fannin and Wise (2001)
Mass point models	Voellmy approach	Zimmermann et al. (1997)
	Iverson approach	Lancaster et al. (2003)
Limiting criteria	Critical slope and junction angle	Benda and Cundy (1990)
Runout length (depositional part of flow)		
Critical slope and deposition on fan	Several empirical methods	VanDine (1996); Bathurst et al. (1997)
Volume balance	Deposition area and flow cross section	Iverson et al. (1988); Crosta et al. (2003)
Analytical approaches	Mass point models	Körner (1980); Perla et al. (1980)
	Constant discharge model	Hungr et al. (1984); Takahashi (1991)
Runout distance (entire path or depositional part only)		
Continuum based simulation models	Various constitutive equations	Iverson (1997); McDougall and Hungr (2003)

energy conservation of the flow. Such approaches for predicting runout of mass movements have been developed either for mass point or lumped mass models, or for continuum based models which also simulate the deformation of the moving mass along the flow path. A major difficulty in developing dynamic models for debris-flow runout prediction is the choice of appropriate friction parameters or material rheologies (e.g., Hungr, 1995; Iverson, 1997). Similar approaches have been originally developed for snow avalanches (Voellmy, 1955; Salm, 1966) and for landslides or rock avalanches (Cannon, 1989; Hungr, 1992).

In this chapter, *travel distance* refers to the entire (horizontal) path length, *runout length* is used for the depositional part or terminal flow path downstream of a defined point, and *runout distance* is used generally and may refer to either of the first two cases. An overview of the methods discussed below is given in Table 13.1 which also reflects the structure of the chapter.

13.2 TOTAL TRAVEL DISTANCE

13.2.1 Travel distance and event magnitude

The total travel distance of a debris flow is useful for rough delineations of potential hazard areas. For rock avalanches or landslides, the "travel angle" or

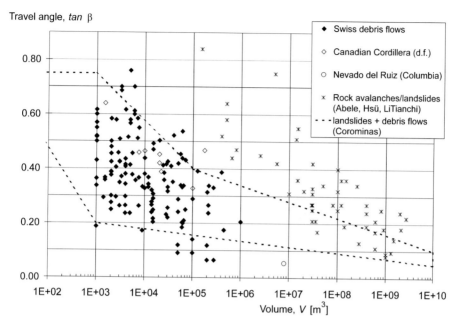

Figure 13.1. Travel angle vs. volume of mass movement. Data include 140 debris flows from the Swiss Alps and 51 large landslides/rock avalanches (for data sources see Rickenmann 1999). Also shown is the approximate range for 101 data points of unobstructed landslides plus unobstructed and channelized debris flows discussed in Corominas (1996).

"Fahrböschung" depends to some extent on the volume V of the mass movement (e.g., Scheidegger, 1973; Corominas, 1996; Legros, 2002). The travel angle β is defined here as $\tan \beta = H/L$, where H is the elevation difference between the starting point and the lowest point of deposition of the mass movement and L is the corresponding horizontal distance. In a physically based approach of runout analysis, the centers of mass both of the failing mass and the deposits should be considered. However, L and H are easily measured parameters, and the relative error to the horizontal and vertical distances related to the center of mass is estimated to be less than 20% in most cases (Legros, 2002). For many (larger) debris flows, this error is much smaller.

For a given volume, debris flows usually show a larger mobility, or lower travel angles, than landslides and rock falls (Figure 13.1; see also Corominas, 1996; Iverson, 1997; Rickenmann, 1999; Legros, 2002). In a study of the mobility of 204 landslides, Corominas (1996) concluded that the scattering in the plots of $\tan \beta$ v. V is partly due to effects of obstacles and topographic constraints, which generally tend to reduce the landslide mobility. For a subset of 71 debris flows, including debris slides and debris avalanches but excluding mudflows and mudslides, he developed the following empirical relationship:

$$\log(H/L) = -1.05 \log V - 0.012 \qquad (13.1a)$$

where H and L are in m and V varies between 10^2 and $10^{10}\,\mathrm{m}^3$. Equation (13.1a) has a correlation coefficient $r^2 = 0.76$ for this data set, and can be transformed into:

$$L = 1.03 V^{0.105} H \qquad (13.1\mathrm{b})$$

Both unobstructed and channelized debris flows show a similar mobility, although the latter ones are reported to be slightly more mobile. Unobstructed flows refer to mass displacements over a regular topographic surface without obstacles or lateral restrictions. For volumes smaller than $10^4\,\mathrm{m}^3$, debris flows occurring in dense forests appear to be clearly less mobile than channelized or unobstructed flows.

A regression analysis using mainly debris-flow data from the Swiss Alps produced the following equation (Rickenmann, 1999):

$$L = 1.9 V^{0.16} H^{0.83} \qquad (13.2)$$

where L and H are in m and V is in m^3, Equation (13.2) has a correlation coefficient $r^2 = 0.75$ for the dataset which includes 160 debris-flow events with L ranging from 300 to 12,600 m, M from 7×10^2 to $10^6\,\mathrm{m}^3$, and H from 110 to 1,820 m.

A comparison of (13.1b) and (13.2) with debris flows data (Figures 13.2(a) and 13.2(b)) show similar scatter of the data in relation to the two equations. In these figures, additional data on rock avalanches follow a similar trend but have comparatively shorter travel distances than the debris flows, which may be explained by lower water concentrations in the flowing masses. The Mount St. Helens lahars and the Nevado del Ruiz mudflow, on the other hand, have comparatively high L values, possibly reflecting either larger water concentrations in the flowing mixtures or higher clay and silt contents compared to the majority of the mostly alpine-type debris-flow data.

When applying (13.1) or (13.2) for predictive purposes, a debris-flow starting point and the longitudinal profile of the expected flow path must also be defined. For a given estimate of V, a solution for L can then be determined either graphically or mathematically. A similar procedure has been proposed by Iverson et al. (1998) to predict the depositional extent of lahars (see Section 13.3.2).

Based on almost the same data set as used by Rickenmann (1999), including 144 debris flows from the Swiss Alps, Zimmermann et al. (1997) defined a lower envelope of $\tan \beta$ as a function of the catchment area A_c [km^2]:

$$\tan \beta_{min} = 0.20 A_c^{-0.26} \qquad (13.3)$$

Equation (13.3) may be used to determine a probable maximum travel distance (Figure 13.3). Including eight additional observations from the Canadian Cordillera, the lowest travel angles (lowest value $\tan \beta_{min} = 0.07$) observed for the Swiss data are associated with debris flows with a high proportion of fine material (data points labeled "matrix supported"). For coarser-grained debris flows (data points labeled "clast supported" or "sand and gravel"), the minimum observed travel angle is $\tan \beta_{min} = 0.19$ (Rickenmann and Zimmermann, 1993).

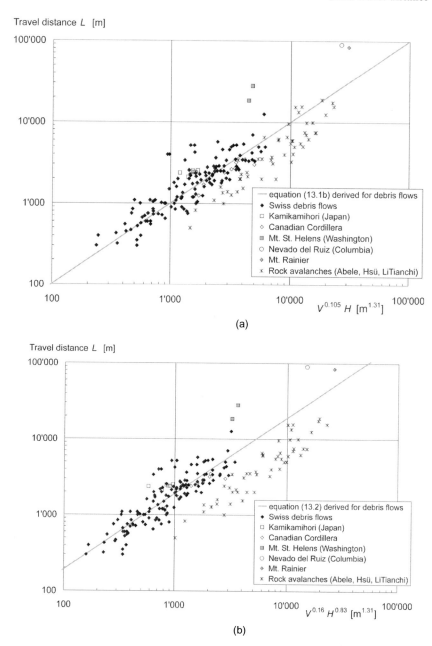

Figure 13.2. (a) Observed runout distance of debris flows and rock avalanches compared to predictions with (13.1b), which was derived for debris flows. Data include 140 debris flows from the Swiss Alps and 51 large landslides/rock avalanches. (b) Observed runout distance of debris flows and rock avalanches compared to predictions with (13.2), which was derived for debris flows. Data include 140 debris flows from the Swiss Alps and 51 large landslides/rock avalanches (for data sources see Rickenmann, 1999).

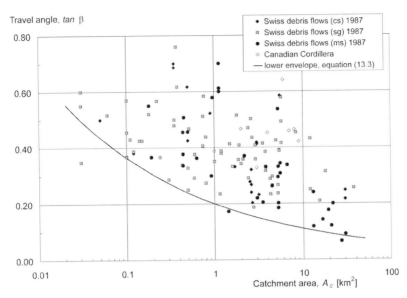

Figure 13.3. Travel angle of debris flows vs. watershed area. Data include 144 debris flows from the Swiss Alps, where the deposits are characterized as "cs" = clast supported, "sg" = sand and gravel, and "ms" = matrix supported.
Data from Zimmermann et al. (1997).

13.2.2 Volume balance approach

Debris flows can erode and entrain material in the steep channel reaches and typically start to deposit sediment below some critical gradient. If a sediment budget is established along the flow path, the terminal deposition location can be estimated. Hungr et al. (1984) introduced the concept of a debris yield rate which depends on the characteristics of the channel and ranges from zero to about $50\,\mathrm{m}^3$ per metre of channel length. An estimate of the debris-flow magnitude can be made by multiplying the mean yield rate with the length of the channel. Then the deposit area may be derived by the dividing the expected volume by a typical average thickness (Hungr et al., 1987).

Cannon (1993) developed an empirical model for the volume-change behavior of debris flows, based on 26 debris flows in Hawaii. The initial debris-flow volume must be known or assumed, and no material entrainment is considered. The average volume change is assumed to be a function of the slope angle θ along the flow path and the degree of flow confinement by the channel. Consideration of vegetation type through which the flow travelled did not significantly improve the model. Flow confinement is defined by the radius of a circle R in the vertical plane, perpendicular to the flow direction. Initial volumes ranged from 25–$938\,\mathrm{m}^3$, travel distances L from 19–$220\,\mathrm{m}$, slope angles θ from 47–$26°$, and radii R from 14–$2{,}000\,\mathrm{m}$ (as upper limit also for planar and convex hillsides). The regression analysis resulted in the equation:

$$\log[(V_i - V_f)/D] = 0.14 \log R - 1.40 \log \theta + 2.16 \qquad (13.4)$$

where V_i and V_f are the volumes [m^3] entering and leaving a segment, and D [m] is the length of the segment. As expected, volume loss is promoted by lack of confinement (large R, in m) and small gradients (θ, in degrees).

A similar but more comprehensive approach used 449 landslide/debris-flow events from Queen Charlotte Island, Canada, on clear-cut, glaciated hillslopes (Fannin and Wise, 2001). Both entrainment and deposition are considered for the volume balance. Most initial failures volumes are less than 500 m^3, with peak volumes less than 4,000 m^3, and travel distances less than 600 m (Wise, 1997). Model development is based on 131 debris-flow events comprising morphological information on 533 reaches. Five different regression equations describing the volume change were developed, considering separately three typical flow modes (unconfined, confined, and transition flow). Entrainment and/or deposition functions are specified for different slope ranges, and depend on several of the following geometric and derived parameters for each reach: slope angle θ, reach length D, width W of entrainment or deposition, entering flow volume V_i, and a bend-angle function (BAF) (which depends on vertical and horizontal change in flow direction and on V_i). As an example, the regression equation describing deposition $(-\Delta V_i)$ is given for transition-flow reaches (defined as open slope reaches immediately downstream of a gully channel):

$$\ln(-\Delta V_i) = -1.54 \ln W - 0.90 \ln D + 0.123 \, \text{BAF} \qquad (13.5)$$

where ΔV_i is in m^3 and W and D are in m. Equation (13.5) is based on 39 reaches and has a correlation coefficient r^2 = 0.95 for this data set. A further model development includes a probabilistic formulation where both entrainment and deposition may occur for unconfined flows in a critical slope range of 19–24° (Fannin and Wise, 2001).

13.2.3 Mass point models (dynamic models)

In snow avalanche modelling, the two-parameter model by Voellmy (1955) is often used. This model has a turbulent and a sliding friction component. The basic equation for a mass point formulation is given as (Körner, 1980; Perla et al., 1980):

$$\frac{dv}{dt} = g(\sin \theta - \mu_m \cos \theta) - (1/k)v^2 \qquad (13.6)$$

where v is the flow velocity, t is the time, g is the gravitational acceleration, μ_m is the sliding friction coefficient, and k is the turbulent friction coefficient. An analytical solution for velocity is possible if the path is approximated by discrete segments. The two friction parameters can be varied from segment to segment. Körner (1980) shows that (13.6) may also be related to a unit element of the flow with a given thickness h, and that k is then related to the flow depth h and a Chezy-like friction coefficient C. If k becomes very large, (13.6) becomes identical with the simple sliding block or sled model (see also (13.14) below). The runout distance can only be modelled if the μ_m value is larger than the actual terrain slope $\tan\theta$ of the depositional reach. If μ_m approaches this slope, the computed runout distance is very sensitive to small changes in μ_m.

This model has also been applied to rock avalanches at Huascaran in Peru (Körner, 1980), Mount Ontake (Moriwaki et al., 1985; Voight and Sousa, 1994), and Mount Tombi (Ouchi and Mizuyama, 1989) in Japan, and to debris flows in Switzerland (Rickenmann, 1990). This latter application was extended by Zimmermann et al. (1997) and Gamma (2000) to 54 Swiss debris flows, with event volumes ranging from 3,000 to 450,000 m^3, and by Genolet (2002) to another 21 Swiss debris flows. In these studies, back-calibrations of the two model parameters were made based on travel distance information and in some cases also on flow velocity (estimated from superelevation). In principle, there are many possible combinations of μ_m and k that satisfy the relatively poor field data. Therefore, an additional restriction has been introduced by limiting the maximum velocity, for example, to 15 m/s for slope reaches not steeper than 25° (Gamma, 2000). For all 75 Swiss events, the optimized parameters were found to be in the following range: $\mu_m = 0.02$–0.32, $k = 20$–120 m. There is a tendency for fine-grained debris flows to have smaller μ_m values for given k values than coarser grained (or "clast supported") debris flows. Genolet (2002) showed that μ_m is strongly correlated with the terrain slope at terminal deposition of the debris flow. According to the simulation results, μ_m values tend to decrease with increasing watershed area (Gamma, 2000; Genolet, 2002), which may be related to the observation that larger basins generally have larger fans with flatter slopes.

Iverson (1997) developed a two-phase mixture model that generalized the Savage-Hutter (1989) granular avalanche model, and Iverson and Denlinger (2001) extended the 1-D Iverson (1997) model to multidimensional flow (see also Section 13.4). In the context of assessing the effect of organic debris on debris-flow runout at the watershed scale, Lancaster et al. (2003) use a simplified form of the Iverson and Denlinger (2001) equation:

$$\frac{d}{dt}\rho h v = -\operatorname{sgn} v(\rho h g \cos\theta - p_b)f_a \tan\phi_b + \rho h g \sin\theta \qquad (13.7)$$

where h is the slope normal debris-flow depth, v is the slope-parallel velocity, p_b is pore pressure at the bed, f_a is a velocity-dependent expression to modify the normal stress by centripetal acceleration due to changes in slope angle, ϕ_b is the bed friction angle, and $-\operatorname{sgn} v$ indicates the direction opposite to that of the debris-flow velocity. The debris flow is treated here as a 1-D point process, and the velocity and depth are functions only of time. Changes in density and depth are prescribed by entrainment of sediment, wood, and water and by changes in channel and valley geometry. Then (13.7) is solved for the change in velocity. The effect of wood on debris-flow runout is considered by accounting for the momentum needed to accelerate entrained wood and its effect on bulk density and flow depth.

The Lancaster model was calibrated by adjusting the pore-pressure term p_b such that the simulated debris-flow travel distance distribution best matched the observed distribution. A pore pressure of 1.8 times the hydrostatic pressure (a similar value as used by Denlinger and Iverson, 2001) resulted in a satisfactory calibration. The calibration procedure included 29 debris flows mapped in a study area in the Oregon Coast Range, which have travel distances of 30–680 m and known deposit volumes of 35–2,500 m^3. The average wood fraction in the deposits was 60% by volume. Disregarding wood effects substantially increased modelled runout distances for flow conditions associated with larger travel distances including wood effects (Lancaster et al., 2003).

13.2.4 Limiting criteria method

Another empirical approach was developed based on 14 channel-confined debris flows in the Pacific North-west, USA (Benda and Cundy, 1990). The travel distance is determined by two criteria which postulate that deposition occurs if the channel gradient drops below 0.06 or the tributary junction angle is greater than 70°. The main implicit assumptions are momentum loss at bends and on flat slopes, where the relative loss of material is probably also significant. The model was tested using 65 debris-flow events from the same basin where the frontal part of the deposits, which had volumes up to several 1,000 m^3, typically contained a considerable amount of woody debris.

13.3 RUNOUT LENGTH IN THE DEPOSITIONAL ZONE

13.3.1 Critical slope and deposition on the fan

For a more detailed delineation of potentially endangered areas, the runout length on the debris-flow fan, or more generally, the runout length downstream of the point where major deposition occurs should be known.

Observations from different catchments indicate that for many (larger) debris flows deposition starts once the channel gradient becomes smaller than 6–12° (Ikeya, 1989; Rickenmann and Zimmermann, 1993; Bathurst et al., 1997). However, for smaller and unconfined debris flows the critical deposition slope may be as high as 27° (Rickenmann and Zimmermann, 1993) or about 35° (Fannin and Wise, 2001). The critical slope also depends on the characteristics of the debris flow (Scott et al., 1992; Jordan, 1994).

Several empirical approaches to estimate debris-flow and landslide runout distance are discussed by Bathurst et al. (1997). It is often assumed that debris-flow deposition begins at slopes of 10°. These approaches estimate runout distance using one or a combination of the following factors: flow volume, mean slope of the transportation zone, elevation difference from the starting point to the point where deposition begins, or travel angle. Bathurst et al. (1997) tested various approaches with landslide and hillslope debris-flow data from Idaho and found that none of the approaches are accurate over the full range of their field data. To estimate sediment delivery into streams at a watershed scale, they recommended using at least two approaches in parallel, thus providing information on the uncertainty of the prediction.

From geometric considerations the runout length on the fan L_f depends to some extent on the volume of the mass movement. This is partly supported by data on debris flows and rock avalanches covering several orders of magnitude (Rickenmann, 1999). However, for any tested empirical relationship the scatter is quite large between predicted and observed values of L_f for the debris-flow field data from different sites. Ikeya (1981, 1989) reports empirical equations from Japan where L_f depends on the event volume and the average gradient of the transportation zone, without providing information on the background and performance of the equations.

13.3.2 Volume balance approach

The empirical–statistical LAHARZ model (Iverson et al., 1998) predicts runout lengths and areas affected by lahars, and is based on a scaling analysis of key parameters and a statistical analysis using data from 27 lahars at 9 volcanoes. The following two semi-empirical relationships are used in the model:

$$A = 0.05 V^{2/3} \qquad (13.8a)$$

and

$$B = 200 V^{2/3} \qquad (13.8b)$$

where $A[m^2]$ is the cross-sectional area, $B[m^2]$ is the planimetric area of the inundated valley, and $V[m^3]$ is the lahar volume. The model calculations are most easily accomplished using a digital elevation model (DEM) implemented in a geographic information system (GIS) but may also conducted manually. Input parameters include lahar volume, starting position and an appropriate travel angle defining the upstream limit of the deposits. For the given lahar volume V, the inundation cross-sectional area A is given by (13.8a) and assumed constant for any location along the depositional reach. The calculation proceeds downstream for a defined segment length, determining innundation width at each location and summing the planimetric area of the inundated valley until it equals the final value B given by (13.8b).

In the analysis leading to (13.8a) and (13.8b), a limited number of data on non-volcanic debris flows were included by Iverson et al. (1998). These show a slight systematic deviation from the trends defined by the lahar data, indicating that non-volcanic debris flows move less fluidly and form appropriately thicker deposits than most lahars. A similar conclusion is made by Crosta et al. (2003) who derived the following relationship based on 116 granular debris flows in the Italian Alps:

$$B = 6.2 V^{2/3} \qquad (13.9)$$

where the units are as in (13.8b), the volumes range between about 10 and 100,000 m^3, and (13.9) has a correlation coefficient r^2 = 0.97. The approach of Iverson et al. (1998) has been implemented by Hofmeister and Miller (2003) in a

GIS-based model of debris-flow deposition zones for regional hazard assessments in Oregon. Iverson and colleagues have also expanded their methodology successfully to smaller debris flows in non-volcanic environments (written commun., R.M. Iverson, 2003).

13.3.3 Analytical approches

For a given torrent channel and fan topography, larger flows tend to result in longer runout lengths than smaller flows having similar material properties. Empirical evidence shows that larger debris-flow volumes also result in higher peak discharges (Mizuyama et al., 1992; Jakob and Bovis, 1996; Rickenmann, 1999) which in turn are associated with higher flow velocities and/or larger flow cross sections and flow depths.

Based on a momentum consideration for a flow travelling over a surface with a constant slope, the runout length s_T can be described by the following theoretical equation (Hungr et al., 1984; Takahashi, 1991):

$$s_T = A_V^2/G \tag{13.10}$$

$$A_V = v_u \cos(\theta_u - \theta)[1 + gh_u \cos\theta_u)/(2v_u^2)] \tag{13.11}$$

$$G = g(S_f \cos\theta - \sin\theta) \tag{13.12}$$

where θ = runout slope angle, θ_u = entry channel slope angle, v_u = entry velocity, h_u = entry flow depth, and S_f = friction slope which is assumed to be constant along the runout path and accounts only for sliding friction. The model assumes a constant discharge from upstream and that there is no change in flow width after the break in slope.

For given values of slopes and entry flow depth, (13.10) can be differentiated for v_u, to find an expression for v_u for which s_T assumes a minimum value:

$$v_u = [(1/2)(gh_u \cos\theta_u)]^{1/2} \tag{13.13}$$

For typical runout conditions on alpine torrential fans (i.e., $h_u = 1\text{–}3\,\text{m}$ and $\theta = 3\text{–}10°$), the runout length calculated by (13.10–13.12) assumes a minimum value for flow velocities v_u ranging between about $2\,\text{m/s}$ and $4\,\text{m/s}$, according to (13.13).

Hungr et al. (1984) assumed $S_f = 10°$ and report good agreement between observed values of s_T and those predicted by (13.10) for five debris flows in western Canada. However, when (13.10) is applied to 14 debris flows at the Kamikamihori valley in Japan using measured flow parameters (Okuda and Suwa, 1984), better predictions of s_T are obtained if $S_f = 1.12 \tan\theta$ is used instead of $S_f = 10°$ (Figure 13.4). Application of (13.10) to 12 Swiss debris flows of 1987 (VAW, 1992) also results in reasonable predictions of runout length, by using $S_f = 1.08 \tan\theta$, when observed flow depths are used to estimate the entry velocity v_u (Figure 13.5). For the analysis of both the Japanese and the Swiss data, it has been assumed that the main surge travelled in the existing channel on the fan to the lowest point of deposition with essentially no change in flow width.

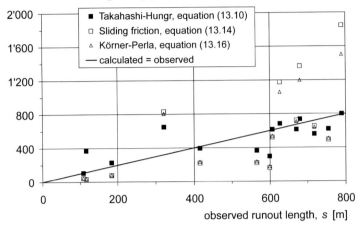

Figure 13.4. Comparison between observed and calculated runout lengths for 14 debris flows at the fan of the Kamikamihori valley in Japan, using measured flow depth and velocity (data from Okuda and Suwa, 1984). Calculations were made with (13.10) and $S_f = 1.12 \tan\theta$, with (13.14) and $\mu_s = 1.012 \tan\theta$, and with (13.16) and $\mu_m = 1.01 \tan\theta$, $C = 100\,\mathrm{m}^{0.5}/\mathrm{s}$ (C is a turbulent friction coefficient).

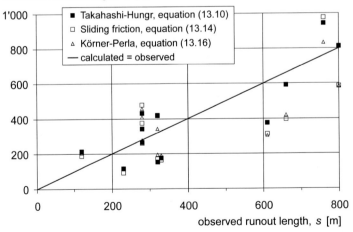

Figure 13.5. Comparison between observed and calculated runout lengths for 12 debris flows in Switzerland, using observed flow depths h_u (data from VAW, 1992) and entry flow velocities estimated with $v_u = (1/0.16) h_u^{0.57} (\sin\theta_u)^{0.5}$ (Rickenmann and Weber, 2000). Calculations were made with (13.10) and $S_f = 1.09 \tan\theta$, with (13.14) and $\mu_s = 1.028 \tan\theta$, and with (13.16) and $\mu_m = 1.01 \tan\theta$, $C = 40\,\mathrm{m}^{0.5}/\mathrm{s}$ (C is a turbulent friction coefficient).

A similar model as given by (13.10–13.12) had earlier been developed for dynamically similar snow avalanches (Salm, 1966). It is worthwhile considering other analytical expressions proposed for the estimation of the runout length s of avalanches, landslides, or debris flows for comparison:

$$s_s = v_u^2/[2g(\mu_s \cos\theta - \sin\theta)] \qquad (13.14)$$

$$s_v = v_u^2/[2g(\mu_v \cos\theta - \sin\theta + (v_u^2/2C^2h))] \qquad (13.15)$$

$$s_m = (C^2h/2g)\ln[1 + v_u^2/(C^2h(\mu_m \cos\theta - \sin\theta))] \qquad (13.16)$$

where (13.14) refers to the sliding block model (e.g., Sassa, 1988; Van Gassen and Cruden, 1989), and (13.15) and (13.16) additionally include a turbulent friction term. Equation (13.15) is given by Voellmy (1955) for a block element, and (13.16) represents the mass point model given by (13.6) and a solution for the last segment (Körner, 1980; Perla et al., 1980), where $k = (C^2h)/g$. Here, $C[\mathrm{m}^{0.5}/\mathrm{s}]$ is a Chezy-like turbulent friction coefficient, and $h[\mathrm{m}]$ is the flow depth in the runout zone. Similar parameters appear in (13.10) and (13.14–13.16), but the sliding friction coefficients are defined differently.

A comparison of the Takahashi-Hungr model (equation 13.10), the sliding friction model (equation 13.14) and the Körner-Perla model (equation 13.16) applied to the Kamikamihori data with a relatively flat deposition slope ($\theta = 5.4°$) suggests that the two latter models produce a systematic deviation from observed values of runout length for debris flows with higher entry velocities (Figure 13.4). When applied to the limited Swiss data with generally steeper deposition slopes ($\theta = 7$–$19°$), all models show a similar scatter (Figure 13.5). If, for the same data set, v_u is estimated via the surge volume and maximum discharge according to Rickenmann (1999), the back-calculated "best" friction parameters are $S_f = 1.12\tan\theta$, $\mu_s = 1.04\tan\theta$, $\mu_m = 1.01\tan\theta$, and $C = 35\,\mathrm{m}^{0.5}/\mathrm{s}$, to obtain similar predictions and scatter as in Figure 13.5. The Körner-Perla model (equation 13.16) requires μ_m values very close to $\tan\theta$, otherwise calculated runout lengths are far too small for the two data sets. Chezy C values back-calculated here with (13.16) are unrealistically large when compared to the Körner-Perla model (equation 13.6) applied over the entire flow path to a subset of the Swiss events used in Figure 13.5 (Rickenmann, 1990), or when compared to simulations of the continuum based Voellmy model in the Kamikamihori valley (Rickenmann and Koch, 1997).

In the idealized case of a mass point displacement described by the sliding block model (equation 13.14), the friction slope μ_s is equal to the slope of the energy line defined by the travel angle β if the velocity height $v_u^2/2g$ is equal to the vertical distance between the energy line and the ground surface at that point (Körner, 1980).

Equations (13.15) and (13.16) represent essentially the same model, and for typical parameter values for debris flows at the fan, differences in the calculated runout lengths are relatively small; differences increase if $\mu_v \approx \mu_m$ becomes very close to $\tan\theta$. Equations (13.10), (13.14), and (13.16) may be compared by assuming $\cos(\theta_u - \theta) \approx 1$ (which is true for $(\theta_u - \theta)$ smaller than about 10°), and by first setting $S_f = \mu_s = \mu_m$, which yields:

$$s_T/s_m = (B_T s_s)/(B_m s_s) = [2 + gh_u \cos\theta_u/v_u^2]/[(A_m/v_u^2)\ln(1 + v_u^2/A_m)] \quad (13.17)$$

with

$$A_m = C^2 h(\mu_m \cos\theta - \sin\theta) \quad (13.18)$$

It can be shown that B_T is always greater than 2 and B_m is always smaller than 1. For typical Alpine-type debris flows, B_T is about one order of magnitude larger than B_m. If the three models are applied to the same case, this implies that the friction slopes satisfy the condition: $S_f > \mu_s > \mu_m$. This is an indirect confirmation of the empirical observation made here that for the mass point model (i.e., equations 13.6 and 13.16) μ_m values are very close to the depositional terrain slope $\tan\theta$, whereas a limited number of applications of equation (13.10) suggests to select values of S_f which are somewhat larger than $\tan\theta$.

For the examined cases, the friction slope S_f in (13.10) appears to be close to $S_f \approx 1.1 \tan\theta$. The friction slope reflects the characteristics of the flowing material. A torrential fan has been constructed by similar flows in the past, and therefore the fan slope at the expected location of deposition ($\tan\theta$) may be be used to estimate the friction slope S_f.

13.4 CONTINUUM BASED DYNAMIC SIMULATION MODELS

Continuum based simulation models allow determination of the flow parameters and deformation of the mass along the entire track, including deposition. In many numerical debris-flow simulation models, the solid–fluid mixture is considered as a quasi-homogeneous fluid. A number of models are partly or fully based on a rheologic formulation for a Bingham or viscoplastic fluid (e.g., Fraccarollo and Papa, 2000; Imran et al., 2001; Laigle et al., 2003). Several applications to natural debris flows modified the pure Bingham model by adding a friction term accounting for channel roughness and turbulence (O Brien et al., 1993; Han and Wang, 1996; Jin and Fread, 1999). Others assume that inertial grain flow determines the flow behavior. Most of these models closely follow the approach of Takahashi et al. (Takahashi, 1991; Takahashi et al., 1992; Nakagawa et al., 2000; Ghilardi et al., 2003). In general, these two-phase models allow for a variation in sediment concentration and can also calculate bed level changes.

Iverson and Denlinger (2001) extended the 1-D Iverson (1997) model to multi-dimensional flow (see also Section 13.2.3). Mixture theory is used to develop the two-phase flow equations. The approach assumes that the apparent debris rheology can evolve as debris-flow motion evolves. Constitutive equations (or rheologic "laws") are specified for the solid and fluid phases independently of one another, and the solid and fluid phases interact through coupling stresses. Iverson and Denlinger (2001) assume that the coupling stresses result from Darcian flow of fluid relative to the solid grains, and that this flow evolves with time and position. In their approach pore pressure may vary within the solid–fluid mixture. The model is reported to require little or no calibration because model parameters like basal and internal friction angles of solid grains, fluid viscosity, and mixture diffusivity can be constrained by independent measurements. The model appears to perform quite well in predicting runout distance (written commun., R. M. Iverson, 2003).

Hungr (1995) proposed a one-phase model for rapid flow slides and debris flows that better represents some soil mechanics properties of the fluid–solid mixture. Different rheologic laws can be selected. The model has been applied to several field cases (Ayotte and Hungr, 2000), and an extended version is under development (McDougall and Hungr, 2003). A similar one-phase model was applied by Rickenmann and Koch (1997) to compare different flow resistance equations for debris flows at the Kamikamihori valley in Japan.

A comparison of different modelling approaches with field observations in the European Alps (McArdell et al., 2003; Rickenmann et al., 2003) used the following flow laws (or constitutive equations) in the simulation models: turbulent laws (e.g., Manning, Chézy), some having also a stop term (e.g., Voellmy), laminar (Bingham, Newtonian laminar, Generalized Viscoplatic Fluid), and inertial formulations (dilatant/grain shearing), as well as combinations of flow laws when appropriate. If the correct stopping location was simulated in the 1-D applications, it was possible to approximately match the velocity observations in most cases. However, in this case there is a general trend of under-prediction of flow depth or flow area. A general conclusion from the 2-D simulations is that the topography (for given input conditions) plays a key role with regard to the deposition pattern. A more detailed spatial resolution of the channel and fan topography strongly improves the model results in many cases. In comparison, the choice of a particular constitutive equation appears to be relatively less important.

For several 1-D and 2-D simulations, a comparison was also made of the relative importance of the three friction terms of the hybrid FLO-2D flow law approach (O'Brien et al., 1993). In this approach, the total friction is determined as a combination of yield, viscous, collisional, and turbulent stress components where the latter two are represented by one single coefficient. The comparison indicates that in many cases the turbulent-collisional friction term dominates for faster moving flow in channelized reaches; at slower velocities and in particular for depositional flows on the fan outside the channel, a yield strength or stopping term becomes more important as well (Rickenmann et al., 2003). As a relatively simple model, the Voellmy fluid approach was found to be rather robust in terms of numerical stability of the simulations, and the back-calculated Voellmy parameters are consistent between the 1-D and the 2-D simulations.

For many of these continuum based models, application to a real debris-flow event produced reasonable agreement with the observed deposition pattern. However, for reliable runout prediction, the models should be first tested rigorously against several field events. The simple friction law approaches in many of these models require either estimation of the magnitudes of the friction parameters or calibration of the models to match previous events.

13.5 SUMMARY AND CONCLUSIONS

A variety of approaches are presented for estimating the total travel distance L or runout length s of debris flows. The majority of the simpler approaches are empirical–statistical methods which should only be applied for predictive purposes to similar conditions as those on which their development is based. In general, these models appear to be more useful for prediction of L than of s.

Topography and volume of mass movement control runout distance to a considerable extent. If an analysis similar to the LAHARZ method by Iverson et al. (1988) can be made for debris flows based on detailed data from the runout zone topography, this method can potentially improve the success of empirical–statistical methods for the prediction of the runout length. However, the assumption of a constant inundation cross-sectional area along the depositional reach may have to be modified for example on the fan.

The two parameter mass point model incorporating a Voellmy fluid approach has been applied to 75 debris-flow events in Switzerland to back-calibrate model parameters, and has been used for preliminary hazard assessments. A mass point model based on the approach of Iverson and Denlinger (2001) has been successfully applied to a study area in Oregon. These dynamic models are more physically based but they also require some prior parameter calibration.

Among the analytical approaches, the comparison of the Takahashi-Hungr model (equation 13.10), the sliding friction model (equation 13.14) and the Körner-Perla model (equation 13.16) with two limited data sets suggests that (13.10) is possibly more suitable for runout length predictions of debris flows than the other two models. According to the limited number of applications, suitable values of the friction slope S_f in (13.10) may be around $S_f \approx 1.1 \tan\theta$, possibly reflecting the characteristics of the flowing material that form the fan.

The most complete description of debris-flow behavior can be provided by continuum based dynamic simulation models. Only for some model applications were exact values of model parameters known or determined a priori. Typically, appropriate values for the rheologic or friction parameters were assumed or back-estimated from field observations. However, most of these models have not yet undergone rigorous testing with field events, which is necessary before they can be reliably used for predictive purposes. When comparing model simulation results with observations of natural debris flows, some general debris-flow characteristics needed for hazard assessment can be reasonably well simulated with simple dynamic models if prior calibration of the model parameters is possible.

For the empirical–statistical methods, experience applying one method to different settings is still limited; however, potential exists for applying them in preliminary hazard assessments.

Analytic approaches may be better suited for reliable runout predictions for detailed hazard assessments. In general, more systematic applications using comprehensive data sets are necessary to determine appropriate model parameters which often cannot be assessed independently of a previous model calibration for similar conditions.

13.6 ACKNOWLEDGEMENT

I thank the reviewers and editors for constructive suggestions and Brian McArdell and Melissa Swartz for comments on an earlier version of the manuscript.

13.7 REFERENCES

Ayotte, D. and Hungr, O. (2000) Calibration of a runout prediction model for debris flows and avalanches. In: G.F. Wieczorek and N.D. Naeser (eds), *Debris-flow Hazards Mitigation: Mechanics, Prediction, and Assessment: Proceedings 2nd International Conference, Taipei, Taiwan* (pp. 505–514). A.A. Balkema, Rotterdam.

Bathurst, J.C., Burton A., and Ward, T.J. (1997) Debris flow run-out and landslide sediment delivery model tests. *Journal of Hydraulic Engineering*, **123**(5), 410–419.

Benda, L.E. and Cundy, T.W. (1990) Predicting deposition of debris flows in mountain channels. *Canadian Geotechnical Journal*, **27**, 409–417.

Cannon, S.H. (1989) An approach for estimating debris flow runout distances. In: *Proceedings Conference XX, International Erosion Control Association, Vancouver, British Columbia* (pp. 457–468).

Cannon, S.H. (1993) An empirical model for the volume-change behavior of debris flows. In: H.W. Shen, S.T. Su, and F. Wen (eds), *Hydraulic Engineering '93* (Vol. 2, pp. 1768–1773). American Society of Civil Engineers, New York.

Corominas, J. (1996) The angle of reach as a mobility index for small and large landslides. *Canadian Geotechnical Journal*, **33**, 260–271.

Crosta, G.B., Cucchiaro, S., and Frattini, P. (2003) Validation of semi-empirical relationships for the definition of debris-flow behavior in granular materials. In: D. Rickenmann and C-L. Chen (eds), *Debris-flow Hazards Mitigation: Mechanics, Prediction, and Assessment: Proceedings 3rd International DFHM Conference, Davos, Switzerland* (pp. 821–831). Millpress, Rotterdam.

Denlinger, R.P. and Iverson, R.M. (2001) Flow of variably fluidized granular masses across three-dimensional terrain. 2: Numerical predictions and experimental tests. *Journal of Geophysical Research*, **106**(B1), 537–552.

Fannin, R.J. and Wise, M.P. (2001) An empirical-statistical model for debris flow travel distance. *Canadian Geotechnical Journal*, **38**, 982–994.

Fraccarollo, L. and Papa, M. (2000) Numerical simulation of real debris-flow events. *Physics and Chemistry of the Earth, B*, **25**(9), 757–763.

Gamma, P. (2000) *dfwalk—Ein Murgangsimulationprogramm zur Gefahrenzonierung* (Geographica Bernensia, G66, 144 pp.). Geographisches Institut der Universität Bern [in German].

Genolet, F. (2002) Modélisation de laves torrentielles: Contribution à la paramétrisation du modèle Voellmy-Perla (70 pp. + annexes). Postgraduate thesis, Ecole Polytechnique Fédérale de Lausanne, Switzerland [in French].

Ghilardi, P., Natale, L., and Savi, F. (2003) Experimental investigation and mathematical simulation of debris-flow runout distance and deposition area. In: D. Rickenmann and C-L. Chen (eds), *Debris-flow Hazards Mitigation: Mechanics, Prediction, and Assessment: Proceedings 3rd International DFHM Conference, Davos, Switzerland* (pp. 601–610). Millpress, Rotterdam.

Han, G. and Wang, D. (1996) Numerical modeling of Anhui debris flow. *Journal of Hydraulic Engineering*, **122**(5), 262–265.

Hofmeister, R.J. and Miller, D.J. (2003) GIS-based modeling of debris-flow initiation, transport and deposition zones for regional hazard assessments in western Oregon, USA. In: D. Rickenmann and C-L. Chen (eds), *Debris-flow Hazards Mitigation: Mechanics, Prediction, and Assessment: Proceedings 3rd International DFHM Conference, Davos, Switzerland* (pp. 1141–1149). Millpress, Rotterdam.

Hungr, O. (1992) Runout prediction for flow-slides and avalanches: Analytical methods. In: *Proceedings of the Geotechnical and Natural Hazards Symposium, Vancouver, British Columbia* (pp. 139–144). Vancouver Geotechnical Society/Canadian Geotechnical Society and Bitech Publishers, Richmond, Canada.

Hungr, O. (1995) A model for the runout analysis of rapid flow slides, debris flows, and avalanches. *Canadian Geotechnical Journal*, **32**, 610–623.

Hungr, O., Morgan, G.C. and Kellerhals, R. (1984) Quantitative analysis of debris torrent hazards for design of remedial measures. *Canadian Geotechnical Journal*, **21**, 663–677.

Hungr, O., Morgan, G.C., VanDine, D.F., and Lister, D.R. (1987) Debris flow defences in British Columbia. In: J.E. Costa and G.F. Wieczorek (eds), *Debris Flow: Process, Description and Mitigation* (GSA Reviews in Engineering Geology, Vol. 7, pp. 201–222). Geological Society of America, Boulder, CO.

Ikeya, H. (1979) *Introduction to Sabo Works: The Preservation of Land against Sediment Disaster* (first English edn, 168 pp.). The Japan Sabo Association, Tokyo.

Ikeya, H. (1981) A method of designation for area in danger of debris flow. *Erosion and Sediment Transport in Pacific Rim Steeplands* (IAHS Publ. No. 132). International Association of Hydrological Sciences, Christchurch, New Zealand.

Ikeya, H. (1989) Debris flow and its countermeasures in Japan. *Bulletin International Association of Engineering Geologists*, **40**, 15–33.

Imran, J., Parker, G., Locat, J., and Lee, H. (2001) 1D Numerical model of muddy subaqueous and subaerial debris flows. *Journal of Hydraulic Engineering*, **127**(11), 959–967.

Iverson, R.M. (1997) The physics of debris flows. *Review of Geophysics*, **35**(3), 245–296.

Iverson, R.M. and Denlinger, R.P. (2001) Flow of variably fluidized granular masses across three-dimensional terrain. 1: Coulomb mixture theory. *Journal of Geophysical Research*, **106**(B1), 537–552.

Iverson, R.M., Schilling, S.P., and Vallance, J.W. (1998) Objective delineation of lahar-inundation zones. *Geological Society of America Bulletin*, **110**(8), 972–984.

Jakob, M. and Bovis, M.J. (1996) Morphometric and geotechnical controls of debris flow activity, southern Coast Mountains, British Columbia. *Zeitschrift für Geomorphologie*, **Supplement Band 104**, 13–26.

Jin, M. and Fread, D.L. (1999) 1D modeling of mud/debris unsteady flows. *Journal of Hydraulic Engineering*, **125**(8), 827–834.

Jordan, R.P. (1994) Debris flows in the southern Coast Mountains, British Columbia: Dynamic behaviour and physical properties. Ph.D. thesis, University of British Columbia, Vancouver.

Körner, H.J. (1980) Modelle zur Berechnung der Bergsturz- und Lawinenberechnung. *Internationales Symposium "Interpraevent", Bad Ischl, Austria* (Tagungspublikation, Band 2, pp. 15–55). Internationale Forschungsgesellschaft Interpraevent, Klagenfurt, Austria [in German].

Laigle, D., Hector, A.F., Hübl, J., and Rickenmann, D. (2003) Comparison of numerical simulation of muddy debris flow spreading to records of real events. In: D. Rickenmann and C-L. Chen (eds), *Debris-flow Hazards Mitigation: Mechanics, Prediction, and Assess-*

ment: Proceedings 3rd International DFHM Conference, Davos, Switzerland (pp. 635–646). Millpress, Rotterdam.

Lancaster, S.T., Hayes, S.K., and Grant, G.E. (2003) Effects of wood on debris flow runout in small mountain watersheds. *Water Resources Research*, **39**(6), 1168, doi:10.1029/2001WR001227, 21 pp.

Legros, F. (2002) The mobility of long-runout landslides. *Engineering Geology*, **63**, 301–331.

McArdell, B.W., Zanuttigh, B., Lamberti, A., and Rickenmann, D. (2003) Systematic comparison of debris flow laws at the Illgraben torrent, Switzerland. In: D. Rickenmann and C-L. Chen (eds), *Debris-flow Hazards Mitigation: Mechanics, Prediction, and Assessment: Proceedings 3rd International DFHM Conference, Davos, Switzerland* (pp. 647–657). Millpress, Rotterdam.

McDougall, S.D. and Hungr, O. (2003) Objectives for the development of an integrated three-dimensional continuum model for the analysis of landslide runout. In: D. Rickenmann and C-L. Chen (eds), *Debris-flow Hazards Mitigation: Mechanics, Prediction, and Assessment: Proceedings 3rd International DFHM Conference, Davos, Switzerland* (pp. 481–490). Millpress, Rotterdam.

Mizuyama, T., Kobashi, S., and Ou, G. (1992) Prediction of debris flow peak discharge. *Internationales Symposium* (Tagungspublikation, Band 4, pp. 99–108). Interpraevent, Bern.

Moriwaki, H., Yazaki, S., and Oyagi, N. (1985) A gigantic debris avalanche and its dynamics at Mount Ontake caused by the Naganoken-Seibu earthquake, 1984. In: *Proceedings 4th International Conference and Field Workshop on Landslides, 1985, Tokyo* (pp. 359–362).

Nakagawa, H., Takahashi, T., and Satofuka, Y. (2000) A debris-flow disaster on the fan of the Harihara River, Japan. In: G.F. Wieczorek and N.D. Naeser (eds), *Debris-flow Hazards Mitigation: Mechanics, Prediction, and Assessment: Proceedings 2nd International Conference, Taipei, Taiwan* (pp. 193–201). A.A. Balkema, Rotterdam.

O'Brien, J.S., Julien, P.Y., and Fullerton, W.T. (1993) Two-dimensional water flood and mudflow simulation. *Journal of Hydraulic Engineering*, **119**(2), 244–261.

Okuda, S. and Suwa, H. (1984) Some relationships between debris flow motion and microtopography for the Kamikamihori fan, North Japan Alps. In: T.P. Burt and D.E. Walling (eds), *Catchment Experiments in Fluvial Geomorphology* (pp. 447–464). GeoBooks, Norwich, UK.

Ouchi, S. and Mizuyama, T. (1989) Volume and movement of Tombi landslide in 1858, Japan. *Transactions of the Japanese Geomorphological Union*, **10**(1), 27–51.

Perla, R., Cheng, T.T., and McClung, D.M. (1980) A two parameter model of snow avalanche motion. *Journal of Glaciology*, **26**(94), 197–208.

Petrascheck, A. and Kienholz, H. (2003) Hazard assessment and mapping of mountain risks in Switzerland. In: D. Rickenmann and C-L. Chen (eds), *Debris-flow Hazards Mitigation: Mechanics, Prediction, and Assessment: Proceedings 3rd International DFHM Conference, Davos, Switzerland* (pp. 25–38). Millpress, Rotterdam.

Pierson, T.C. (1995) Flow characteristics of large eruption-triggered debris flows at snow-clad volcanoes: Constraints for debris-flow models. *Journal of Volcanology and Geothermal Research*, **66**, 283–294.

Rickenmann, D. (1990) *Debris Flows 1987 in Switzerland: Modelling and Sediment Transport* (IAHS Publ. No. 194, pp. 371–378). International Association of Hydrological Sciences, Christchurch, New Zealand.

Rickenmann, D. (1999) Empirical relationships for debris flows. *Natural Hazards*, **19**, 47–77.

Rickenmann, D. and Koch, T. (1997) Comparison of debris flow modelling approaches. In: C-L. Chen (ed.), *Debris-flow Hazards Mitigation: Mechanics, Prediction, and Assessment:*

Proceedings 1st International DFHM Conference, San Francisco, CA (pp. 576–585). American Society of Civil Engineers, New York.

Rickenmann, D. and Weber, D. (2000) Flow resistance of natural and experimental debris flows in torrent channels. In: G.F. Wieczorek and N.D. Naeser (eds), *Debris-flow Hazards Mitigation: Mechanics, Prediction, and Assessment: Proceedings 2nd International Conference, Taipei, Taiwan* (pp. 245–254). A.A. Balkema, Rotterdam.

Rickenmann, D. and Zimmermann, M. (1993) The 1987 debris flows in Switzerland: Documentation and analysis. *Geomorphology*, **8**, 175–189.

Rickenmann, D., Laigle, D., Lamberti, A., Zanuttigh, B., Armanini, A., Fraccarollo, L., Giuliani, M., Rosati, G., McArdell, B.W., Ng, D., Swartz, M., and Graf, Ch. (2003) *Evaluation of Existing Numerical Simulation Models for Debris Flows* (Report on work package 3 of the research project THARMIT of the European Union, EU Contract EVG1-CT-1999-00012). EU, Brussels.

Sassa, K. (1988) Geotechnical model for the motion of landslides (Special lecture). In: C. Bonnard (ed.), *Proceedings 5th International Symposium on Landslides* (Vol. 1, pp. 37–55). A.A. Balkema, Rotterdam.

Salm, B. (1966) Contribution to avalanche dynamics. *Proceedings International Symposium on Scientific Aspects of Snow and Ice Avalanches, Christchurch, New Zealand* (IAHS Publ. No. 69, pp. 199–214). International Association of Hydrological Sciences, Christchurch, New Zealand.

Savage, S.B. and Hutter, K. (1989) The motion of a finite mass of granular material down a rough incline. *Journal of Fluid Mechanics*, **199**, 177–215.

Scheidegger, A.E. (1973) On the prediction of the reach and velocity of catastrophic landslides. *Rock Mechanics*, **5**, 231–236.

Scott, K.M., Pringle, P.T., and Vallance, J.W. (1992) *Sedimentology, Behavior, and Hazards of Debris Flows at Mount Rainier, Washington* (USGS Open-file Report 90-385, 106 pp.). US Geological Survey, Reston, VA.

Takahashi, T. (1991) *Debris Flow* (IAHR Monograph Series, 165 pp.). International Association for Hydraulic Research, Ecole Polytechnique Fédérale, Lausanne, Switzerland and A.A. Balkema, Rotterdam.

Takahashi, T., Nakagawa, H., Harada, T., and Yamashiki, Y. (1992) Routing debris flows with particle segregation. *Journal of Hydraulic Engineering*, **118**(11), 1490–1507.

VanDine, D.F. (1996) *Debris Flow Control Structures for Forest Engineering* (Ministry of Forests Research Program, Working Paper 22/1996, 75 pp.). Government of the Province of British Columbia, Vancouver.

Van Gassen, W. and Cruden, D.M. (1989) Momentum transfer and friction in the debris of rock avalanches. *Canadian Geotechnical Journal*, **26**, 623–628.

VAW (1992) *Murgänge 1987: Dokumentation und Analyse* (unpublished report, No. 97.6, 620 pp.). Versuchsanstalt für Wasserbau, Hydrologie und Glaziologie, ETH, Zurich [in German].

Voellmy, A. (1955) Über die Zerstörungskraft von Lawinen. *Schweizerische Bauzeitung*, **73**(12), 159–162, (15), 212–217, (17), 246–249, (19), 280–285 [in German].

Voight, B. and Sousa, J. (1994) Lessons from Ontake-san: A comparative analysis of debris avalanche dynamics. *Engineering Geology*, **38**, 261–297.

Wise, M.P. (1997) Probabilistic modelling of debris flow travel distance using empirical volumetric relationships. M.Sc. thesis, University of British Columbia, Vancouver.

Zimmermann, M., Mani, P., Gamma, P., Gsteiger, P., Heiniger, O., and Hunziker, G. (1997) *Murganggefahr und Klimaänderung: ein GIS-basierter Ansatz*. (Schlussbericht NFP 31, 161 pp.). ETH, Zurich [in German].

14

Climatic factors influencing occurrence of debris flows
Gerald F. Wieczorek and Thomas Glade

14.1 INTRODUCTION

A wide variety of climatic factors influence the occurrence of debris flows, which include mudflows and lahars (see Chapter 2). Climatic factors are an important subject for a better understanding of hydrologic response of soils and of how global climate change can influence debris-flow activity. In addition, climatic factors are essential for developing debris-flow warning systems. Climatic factors have extreme spatial and temporal variability. Rapid infiltration of prolonged intense rainfall, causing soil saturation and a temporary increase in pore-water pressure, is generally believed to be the mechanism by which most shallow landslides, and more specifically debris flows, are generated during rainstorms (Iverson, 2000).

This chapter focuses on the primary and secondary climatic factors that influence the occurrence of debris flows. Primary climatic influences are those that directly trigger debris flows, such as intense rainstorms or rapid snowmelt. Secondary climatic influences are those, such as antecedent rainfall or antecedent snowmelt, that influence whether debris flows are triggered during an earthquake, volcanic event, or intense rainstorm. Thresholds for the triggering of debris flows will also be discussed, including application of rainfall thresholds for hazards assessment and mitigation such as warning systems. In addition to short-term climatic influences, the possible effects of longer term climatic changes are examined. This chapter does not attempt to present new scientific material on this subject, but does try to summarize the extent of worldwide research to gain a better understanding of this complex subject.

14.2 PRIMARY CLIMATIC FACTORS

Intense rainstorms and rapid snowmelt are two primary climatic factors that are recognized as directly associated with near-immediate triggering of debris flows.

14.2.1 Intense rainstorms

The relationship of high-intensity rainfall in the triggering of shallow landslides that transform into debris flows was first noted by Campbell (1975) in the Santa Monica Mountains of southern California. He postulated that after sufficient antecedent rainfall, infiltration of intense storm rainfall created temporary perched aquifers with positive pore-water pressures that reduced the effective strength of surface soils and initiated shallow landslides (Figure 14.1). Starkel (1979) conceived a critical rainfall threshold as a combination of rainfall intensity and duration. Storms of very high intensity, but relatively short duration, such as less than one hour, may cause high surface runoff, but generally insufficient infiltration for high pore-water pressures for triggering shallow landslides. Conversely, low-intensity, lengthy storms, lasting a few days, may increase deep groundwater levels, but often result in insufficient pore pressure within near surface soils for triggering shallow landslides. Although the instability effects of rainfall intensity and duration depend on thickness, porosity, and permeability of the local regolith, the measure of the combination of rainfall intensity-duration is useful for comparing regional triggering of debris flows. Caine (1980) assembled worldwide rainfall data to further support the concept of thresholds of rainfall intensity and duration for the

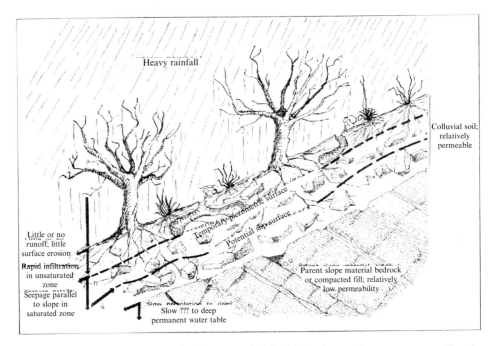

Figure 14.1. Conceptual model for intense rainfall, infiltration and temporary aquifers in shallow hillside soils.

Campbell (1975).

triggering of debris flows. Subsequently, studies from many parts of the world have documented different intensity and duration of rainfall for the triggering of debris flows in different regions.

Worldwide observations have identified the minimum and maximum rainfall over various periods of time critical for triggering debris flows. A minimum rainfall threshold defines the lowest amount of rainfall capable of initiating a landslide (i.e., below which no landslide has occurred). What we here call a maximum threshold defines the amount of rainfall that has always triggered landslides historically (Crozier, 1997). The probability of debris-flow occurrence increases as rainfall progresses from the minimum threshold towards the maximum threshold.

Caine (1980) developed a minimum rainfall intensity–duration threshold using published rainfall data from 73 worldwide storm events, with durations up to 10 days associated with the triggering of different types of shallow landslides. Only landslide triggering events were used to obtain the threshold; rainfall conditions which did not trigger landslides were not considered, and so the definition of the threshold incorporates only part of the evidence. Subsequent studies, however, have included both triggering and non-triggering events in the analysis, and thresholds obtained in this manner have been applied for a single drainage in the Italian Alps (Marchi et al., 2002), in New Zealand (Crozier, 1997), and in southern British Columbia (Jakob and Weatherly, 2003).

Govi and Sorzana (1980) used the timing of mudflow and debris flows in combination with hourly rainfall data to characterize rainfall thresholds in a variety of geologic settings in north-western Italy. They found that the spatial density of debris flows within a 30-year period did not correlate to bedrock and soil type. However, they discovered that the rainfall threshold varied from one region to another mainly as a function of the mean annual precipitation.

During January 1982, an intense storm lasting for about 32 hours triggered more than 18,000 mainly shallow landslides (mostly debris flows) in soil and weathered rock in the San Francisco Bay region (Ellen and Wieczorek, 1988) (Figure 14.2). Documentation of initiation times of debris flows, coupled with continuous measurements of rainfall, permitted the identification of rainfall thresholds (Wieczorek and Sarmiento, 1983; Cannon and Ellen, 1985; Wieczorek, 1987; Cannon, 1988; Wieczorek and Sarmiento, 1988).

Subsequent to the 1982 storm in the San Francisco Bay region, a real-time landslide warning system was established. It was operated by the US Geological Survey and the National Weather Service (NWS) between 1986 and 1995 (Wilson et al., 1993; Wilson, 2005). Using a real-time rainfall monitoring system and NWS satellite-based quantitative rainfall forecasts, regional landslide warnings were issued using the rainfall thresholds during storms in 1986, 1991, 1993, and 1995. The times of landslide warnings in the storms of February 1986, which were based on the thresholds of Cannon and Ellen (1985) were found to correspond with documented times of shallow landslides (Keefer et al., 1987) (Figure 14.3). Organizational changes and decreases in funding and staffing forced the termination of the USGS/NWS debris-flow warning system in December 1995 (Wilson, 2005).

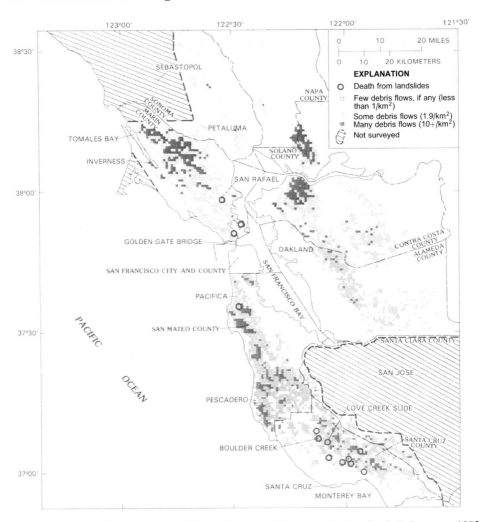

Figure 14.2. Distribution and spatial density of debris flows during the 3–5 January, 1982 storm in the San Francisco Bay region.
Ellen and Wieczorek (1988).

On 9–11 March, 1988 Tropical Cyclone Bola hit the east coast of the North Island of New Zealand dropping 753 mm of rain within a four-day period. Thousands of landslides were triggered, mostly earth flows, debris flows, and debris slides. Similar triggering events hit New Zealand frequently (e.g., near Wairarapa during 1977 and in Gisborne during 2002) (Figure 14.4, see also color section for 14.4(b)). Within an area of approximately 50 km^2 more than 19,000 landslides were mapped (Glade, 1997), resulting in an average spatial density of 380 landslides per square kilometer. A comparison of the landslide history with climate measurements in this region showed that a rainfall threshold could be estab-

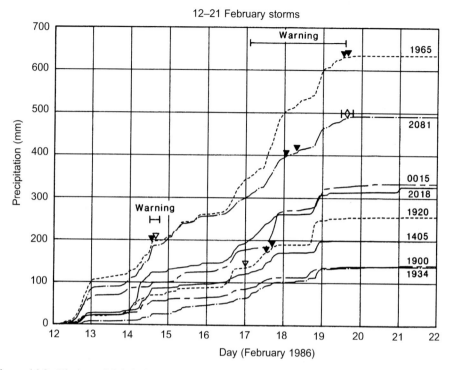

Figure 14.3. Timing of debris-flow warnings in San Francisco Bay region during the storms of 12–21 February, 1986 and documented debris flows (filled triangles), slump (diamond), and undetermined types of landslides (open triangles) at locations near specific numbered rain gauges.
Keefer et al. (1987).

lished (Page et al., 1994). Of added interest, Page et al. also demonstrated that a smaller rainstorm in 1938 (692 mm of rain within a four-day period) had produced more sediment from landslides than the larger 1988 storm, resulting in much larger environmental impact. They concluded that conditions in the drainage basin had been changed by the removal of a large amount of surficial materials, so that by 1988 the slopes had still not recovered from previous events (Preston, 1999).

In some areas, rainfall from intense storms infiltrates and percolates quickly into the regolith. This often results in the saturation of soils, development of perched aquifers, and rapid rise in groundwater levels. The temporary creation of perched aquifers and resulting high pore-water pressures can trigger debris flows on steep hillsides by effectively reducing soil strength. However, positive pore-water pressure can reach high levels without triggering debris flows. In the San Francisco Bay region, California, Johnson and Sitar (1990) used tensiometers to measure both negative and positive pore-water pressures through several intense rainstorms with associated debris flows. These measurements substantiated the importance of antecedent rainfall, which reduced negative pore pressure and thereby facilitated

(a)

Figure 14.4. Spatial distribution of debris flows and other landslides following extreme rainstorm events in New Zealand. (a) Rainstorm in 1977, Kiwi Valley, Wairoa. See color section for 14.4(b).

Photograph of Hawke's Bay Catchment Board.

development of transient positive pore-water pressures during subsequent high-intensity rainfall. Pore pressures rose and fell quickly over a period of less than 24 hours, and the pore-pressure pulses slowly progressed downhill (Johnson and Sitar, 1990). Theoretically, during initial infiltration during a rainstorm, the dissipation of negative pore-water pressure, or capillary soil suction, reduces the soil strength and could result in shallow landslides which may transform into debris flows (De Campos et al., 1994). However, this concept, which has been proposed primarily in tropical environments, such as Brazil, has not been verified by field measurements of the dissipation of negative pore pressure temporally associated with the triggering of debris flows.

14.2.2 Rapid snowmelt

Snowmelt enhanced by rainfall or sudden temperature increase can also lead to increased water infiltration. Horton (1938) monitored and studied the infiltration

Sec. 14.2] Primary climatic factors 331

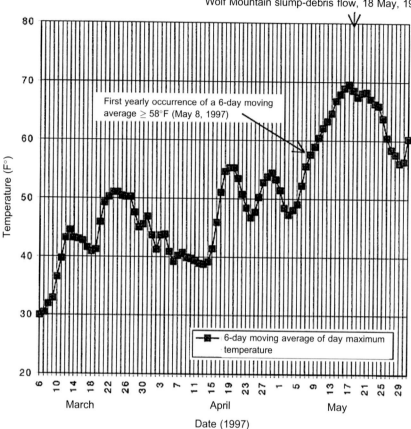

Figure 14.5. Temperature patterns for snowmelt triggering of a debris flow at Wolf Mountain, near Jackson Hole, Wyoming.
Chleborad (1998).

of melting snow into soil, including the case of the effects of rainfall on a snowpack. Horton found that the process of snowmelt provides a more continuous supply of water over longer time periods than infiltration from rain. Matthewson et al. (1990) found that snowmelt may also recharge shallow fractured bedrock and raise pore-water pressures beneath shallow soils, thus triggering debris flows. Spatial variability from infiltration of snowmelt is not as high as that from intense rainstorm cells. The rate of snowmelt depends on air temperature, which in turn relates to the timing of debris flows (Chleborad, 1997, 1998). Chleborad et al. (1997) used a 6-day moving average of daily maximum temperature and 14.4°C as an optimum threshold of this average for anticipating the onset of snowmelt-generated landslides in the central Rocky Mountains of the USA (Figure 14.5). This threshold was used successfully to time the deployment of instrumentation prior to the occurrence of debris-flow activity at a site near Aspen, Colorado (Chleborad et al., 1997). Debris flows have also been found in south-western British Columbia to be associated with moderate rainfall with snowmelt, low intensity rainfall and heavy snowmelt, and heavy rainfall onto deeply frozen, thawing ground (Church and Miles, 1987).

14.3 SECONDARY CLIMATIC FACTORS

14.3.1 Antecedent rainfall

Campbell (1975) identified a seasonal antecedent rainfall threshold of 267 mm as necessary for debris flows to be produced by intense storms in the Santa Monica Mountains of southern California. This value of antecedent rainfall is an empirical representation of the minimum field moisture capacity required of slope materials for intense storms to trigger debris flows. A seasonal characterization of antecedent rainfall applies to regions, such as southern California, and the Pacific Northwest of the USA and Canada, where wet and dry seasons are distinct. For California, the wet season falls between late fall and early spring, generally during the months of November through to April.

The importance of antecedent, or pre-storm, seasonal rainfall for the triggering of debris flows was also demonstrated in northern California (Wieczorek and Sarmiento, 1988). Seasonal rainfall is defined as the total amount of rain beginning after the dry summer period, which in northern California usually ends in late October. A set of 22 storms between 1975 and 1984 were divided into three groups, based on the pre-storm seasonal rainfall and the ability of the storms to trigger debris flows in a study area near La Honda, California (Wieczorek, 1987, table 1). No storm, however intense, produced debris flows unless the seasonal antecedent rainfall exceeded 280 mm. The 10 intense storms in the first group, which occurred after a seasonal rainfall exceeding 280 mm, triggered debris flows; the intense storms in the second group, which occurred before 280 mm of seasonal rainfall, triggered no debris flows; and the storms in the third group, which occurred after a seasonal rainfall of 280 mm, but were of low intensity and short duration, triggered no debris flows.

The significance of pre-storm seasonal rainfall becomes clear when one examines the rainfall characteristics and the effects of the first and second group of storms. Even though the storms in the second group had storm rainfall totals, 24-h maximums, 1-hr maximums, and high intensity for durations of 2–8 hours exceeding most of the storms in the first group, the storms in the second group did not trigger debris flows, apparently because of insufficient pre-storm seasonal rainfall. Antecedent rainfall was evaluated for 2, 7, 15, and 30-day periods (Wieczorek and Sarmiento, 1988). Rainfall during the 7 and 15-day periods before the three storms which caused the greatest number of debris flows was significantly higher than for other storms, an observation suggesting that rainfall during the 1 or 2-week period preceding an intense storm may be more significant than the earlier pre-storm seasonal rainfall.

In regional contrast, comparison of storms in the Moscardo basin of the Italian Alps provides an example of the lack of importance of antecedent rainfall for the triggering of debris flows (Deganutti et al., 2000). In an examination of 73 storms, 15 of which triggered debris flows between late June and late September, antecedent rainfall of 24 hours, 5, 10, and 15 days showed no significant statistical correlation to debris-flow or non-debris-flow storm events. During the summer the melting of snow contributes to the high moisture content of soils and along with the presence of springs can account for the lack of significance of antecedent rainfall.

Other investigations have found that antecedent rainfall conditions contribute differently to debris-flow occurrence and distribution depending upon regional climate. Research in Italy (Wasowski, 1998) and New Zealand (Crozier and Eyles, 1980; Crozier, 1989, 1997) has demonstrated that antecedent climatic conditions play a vital role in determining debris-flow occurrence. In Korea, Kim et al. (1992) demonstrated that antecedent rainfall is significant in some climatic-terrain regions and not so important in others. In Hong Kong, Brand et al. (1984) and Brand (1989) found that antecedent climatic conditions are not important for the occurrence of debris flows and that rainfall intensity is the only critical triggering factor, despite the earlier contrary findings of Lumb (1975).

Although pre-storm rainfall is widely recognized as an important factor in the rainfall conditions that trigger debris flows, there is little agreement on the time period significant for the build-up of antecedent soil moisture (Cannon and Ellen, 1988). Lumb (1975), Eyles (1979), and Govi and Sorzana (1980) reported rainfall totals for time periods ranging from 2 to 45 days before a storm as contributing to the soil-moisture conditions that lead to debris flows. Other authors have defined the critical period of antecedent climatic conditions on soil saturation as 15 days in Italy (Pasuto and Silvano, 1998), 25 days in Colombia (Terlien, 1997, 1998), and 4 weeks in Seattle, Washington (Chleborad, 2000), and North Vancouver, British Columbia, Canada (Jakob and Weatherly, 2003). Seasonal variations of rainfall and temperature, affecting evapotranspiration could be significant to the importance of antecedent rainfall. For example, in the San Francisco Bay region, most intense storms occur during the cool fall and winter seasons. Evapotranspiration is minimal and soils would remain partly saturated for long periods of time. In contrast, in central Virginia, intense convective storms occur most often during the warm summer period when evapotranspiration could remove much of the soil moisture within days or a few weeks preceding another storm. Consequently, the significance of antecedent rainfall may vary depending upon the regional climate.

In November of 1998, intense rainfall from Hurricane Mitch triggered two catastrophic debris flows from the slopes of Casita volcano in Nicaragua. The first (larger) debris flow began as a landslide representing the collapse of a small flank ($200,000 \, m^3$) near the summit of an inactive volcanic edifice. The resulting large debris flow increased its volume by a factor of nine as it travelled 4 km, destroying several towns and killing more than 2,500 people (Scott, 2000). Intense rainfall of 750 mm over 83 hours between 26 October and the time of the Casita flank collapse at 10:30–11:00 a.m. on 30 October, probably increased pore pressure in the highly fractured, but only slightly altered, bedrock. In addition to the intense rainfall of

Mitch, antecedent rainfall was considered significant (Kerle et al., 2003) because of comparison of Hurricane Mitch (1,538 mm total rainfall) to 1982 Tropical Storm Alleta of similar magnitude (1457 mm). Whereas more than 1,900 mm of rain fell during the 6 months prior to Hurricane Mitch, the 1982 rainfall event, which occurred at the beginning of the rainy season with only 164 mm of antecedent rainfall over the same time interval, resulted in limited shallow debris flows causing only two fatalities (Kerle et al., 2003).

The influence of different lengths of antecedent conditions on landslide initiation has been investigated in Portugal by Zêzere et al. (1999) and in New Zealand by Glade (2000a). Whereas Zêzere et al. (1999) used cumulative rainfall over periods from 1–120 days, Glade (2000a) considered the loss from evapotranspiration and soil water drainage. Four antecedent periods (2, 3, 5, and 10 days) were applied in Glade's *Soil Water Status* model. Results show no significant changes in landslide occurrence with durations of antecedent rainfall beyond 2 days (Glade, 2000a). Thus, the length of record, in this case, is not very important. This result makes sense because the New Zealand study area consists of pasture lands that have shallow-rooted vegetation and coarse-grained soils developed on volcanic ash. In these conditions, rapid drainage and evapotranspiration can be expected to reduce the importance of long-term antecedent conditions. This example shows how local conditions affect the duration of significant antecedent conditions. In these examples, soil conditions, such as rainfall infiltration and soil water percolation, are generally not considered explicitly. For many regions, these data are not available. If values can be obtained, however, it is important to include them in modeling approaches.

Secondary climatic factors, such as antecedent rainfall or the melting of large snowpacks, increase soil moisture influencing the triggering of debris flows by earthquakes and volcanic events (Waldron, 1967; Waitt et al., 1983; Pierson et al., 1990; Schuster, 1991; Pierson, 1999; Scott et al., 2001; see also Chapter 10). Two earthquakes (M 6.1 and 6.9) on 5 March, 1987, which occurred shortly after a period of extended rainfall east of Quito, Ecuador, triggered thousands of earth slides, debris avalanches, and earth and debris flows that destroyed nearly 70 km of the Trans-Ecuadorian oil pipeline (Schuster, 1991). After a period of rainfall that saturated residual soils on steep slopes, a M 6.4 earthquake on 6 June, 1994 in the Rio Paez basin of Colombia triggered earth and debris flows that destroyed homes, a school, and other buildings (Martinez et al., 1999). A comparison of different earthquakes occurring in the same region can demonstrate the influence of antecedent rainfall on the earthquake triggering of debris flows. For example, the 16 April, 1906 San Francisco, California, earthquake (M ~ 8.2) triggered many large, deep-seated landslides throughout northern California, as well as several debris flows near Half Moon Bay (Youd and Hoose, 1978). In comparison, the 17 October, 1989 Loma Prieta earthquake (M 7.1), which affected a large portion of the same region, triggered mostly shallow landslides, but not any debris flows (Schuster et al., 1998). Although it had not rained for 17 days before the 18 April, 1906 earthquake (Youd and Hoose, 1978), the month of March 1906 was exceptionally wet. Cumulative rainfall for the region for the 1, 3, and 6-month periods before the 1906 earthquake was about 150–200% of normal. In comparison, regional rainfall in the month before the 17 October, 1989 earthquake was minimal; the earthquake occurred near the end of the dry summer season and since 1 June, only 30 mm of rain had fallen (Keefer and Harp, 1998). Thus, the timing of earthquakes in relation to seasonal rainfall can play a critical role in the triggering of debris flows. The timing of earthquakes with respect to climatic influence on groundwater levels and soil saturation is of importance to its impact on triggering landslides of any type.

On 13 November, 1985 a sequence of pyroclastic flows and surges interacted with snow and ice on the summit ice cap atop Nevado del Ruiz volcano in Colombia, triggering lahars that killed about 23,000 people living at the base of the volcano (Pierson et al., 1990). Combined with seismic shaking, the hot eruptive materials quickly melted about 10 km^2 of the snowpack and produced large volumes of meltwater that combined with new volcanic deposits to generate avalanches of snow, ice, and rock debris into the upper reaches of river valleys. Rapid incorporation of valley fill materials transformed the dilute flows and avalanches into debris flows. A total of about 9×10^7 m^3 of debris was transported up to 104 km from the source area before deposition. Key lessons related to climatic influences in this event included: (1) catastrophic lahars can be generated on ice and snow-capped volcanoes by relatively small eruptions; (2) the surface area of snow on an ice cap can be more critical than total ice volume; and (3) the mechanical mixing of hot rock debris with snow increases the rate of heat transfer and provides an efficient mechanism for generating lahars.

14.4 USE OF CLIMATIC DATA FOR FORECASTS AND WARNINGS

14.4.1 Rainfall thresholds

During the past two decades the concept of rainfall thresholds for the triggering of debris flows as presented by Caine (1980) has been widely applied worldwide. Within the USA, rainfall thresholds have been developed for many regions, including the San Francisco Bay region (Cannon and Ellen, 1985), Honolulu, Hawaii (Wilson et al., 1992), Puerto Rico (Jibson, 1989; Larsen and Simon, 1993), Seattle, Washington (Chleborad, 2000, 2003), and the Appalachian and Blue Ridge Mountains of the eastern USA (Neary and Swift, 1987; Wieczorek et al., 2000) (Figure 14.6). Rainfall thresholds for triggering shallow landslides or debris flows have also been developed in many other countries (Table 14.1). Some of these thresholds are based on abundant data and have been incorporated into early warning systems, for example in California (Keefer et al., 1987; Wilson, 2005), Italy (Iritano et al., 1998), Brazil (Ortigao et al., 2003), and Hong Kong (Brand 1985; Hansen et al., 1995). These examples show that debris-flow initiation follows some general trends in climatic conditions.

Rainfall thresholds are valuable for prediction and warning of landslide events, particularly for debris flows, which are high-velocity and high-hazard events. Such thresholds can be either derived for single catchments (e.g. Marchi et al., 2002) or on a regional scale (e.g. Cannon and Ellen, 1985; Glade, 1998; Jakob and Weatherly,

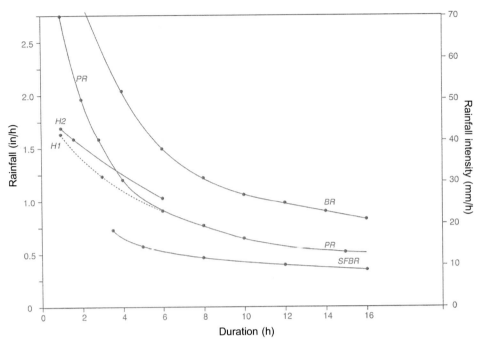

Figure 14.6. Comparison of rainfall thresholds for triggering of debris flows in Hawaii (H1, H2), Puerto Rico (PR), Blue Ridge Mountains of central Virginia (BR), and San Francisco Bay region (SFBR).

Wieczorek et al. (2000).

2003). As noted by Wilson (2000), successful regional rainfall thresholds are based on local conditions of geology, slope geometry, and climate.

Threshold models can be developed with a range of complexity. The simplest models compare only the rainfall amounts within a given period with landslide occurrence. For example, Glade (1998) compares daily rainfall magnitude with landslide occurrence (Figure 14.7). Other periods for which rainfall totals have been applied to determine rainfall thresholds include a single storm event (Corominas and Moya, 1999), a month (Flageollet et al., 1999), a season (Jäger and Dikau, 1994), and a year (Slosson and Larson, 1995; Cuesta et al., 1999).

Rainfall intensity is important in influencing the spatial distribution of debris flows as well as their timing. For several regions of Italy, Crosta and Frattini (2000) examined the correlation between the intensity vs. duration plots for different events and the number and spatial density of debris flows (Figure 14.8). These authors also compared worldwide thresholds using very detailed rainfall intensity information (Figure 14.9).

Regional variation in rainfall thresholds for the triggering of debris flows depends upon many factors, such as morphology, geology, hydrology, and vegetation. Most of the threshold models apply to debris flows triggered on slopes, and different models apply to debris-flow initiation in channel beds (Tognacca et al., 2000).

In examining the effects of climatic variation on rainfall thresholds along the Pacific Coast of the USA, Wilson (2000) noted the importance of the frequency of rainfall as well as mean annual precipitation (MAP). Although MAP is higher in the northern Pacific states of Oregon and Washington than in southern California, for example, the rainfall frequency is also higher in the north, so that the average daily rainfall is less than in southern California. Consequently, the rainfall threshold required to trigger debris flows is greater in the southern region than in the north. This result casts doubt on any simple relation between rainfall thresholds in different climatic regions and MAP.

The *Antecedent Daily Rainfall* model (Crozier and Eyles, 1980) provides a method to calculate the relation between daily rainfall and antecedent rainfall. In this model, triggering rainfall conditions are represented by a combination of antecedent rainfall and rainfall on the day of the event. A study applying this model was conducted in three regions susceptible to landslides on the North Island of New Zealand (Glade et al., 2000). A decay coefficient, derived for each region from the recessional behaviour of storm discharge hydrographs, was used to produce an index for antecedent rainfall. Statistical techniques were used to obtain the thresholds which best separate rainfall conditions associated with landslide occurrence from those associated with non-occurrence or a given probability of occurrence (Figure 14.10).

Modeling techniques have been developed in combination with physical measurements to relate antecedent rainfall to soil saturation and its influence on pore-water pressure. Wilson (1989) developed a numerical model, based on the simple physical concept of a "leaky barrel", which receives additional water from above at one rate, while losing water through leakage below at another rate. The model is used to represent the accumulation of infiltrated rainfall to form a zone of saturation. This *Leaky Barrel* model was tested using rainfall and piezometric data (Figure 14.11) collected at La Honda, California (Wilson and Wieczorek, 1995).

The *Leaky Barrel* model was subsequently used to develop a new threshold for the triggering of debris flows at La Honda based on cumulative rainfall and accounting for soil drainage and rainfall duration (Wilson and Wieczorek, 1995). This threshold identifies conditions for the triggering of single or isolated debris flows within this small area of high landslide susceptibility. In an area of historic debris-flow events in Campania, Italy, Chirico et al. (2000) developed a similar conceptual model of hydrological debris-flow initiation based on buildup of water levels within fractured bedrock and pyroclastic mantle incorporating rainfall, infiltration, surface runoff, evapotranspiration, and subsurface outflow.

Another method involves the weighting of antecedent climatic and soil moisture conditions preceding a rainstorm event (Glade, 2000a). Regional averages of soil depth, porosity, texture, and soil moisture capacity, which provide information on physical properties of soils for the specific regions, were established from the literature. In addition to rainfall inputs and water loss through drainage, the loss of water to the atmosphere through evapotranspiration and the ability of the regolith to retain water have to be taken into account. These factors are incorporated into the *Soil Water Status* model originally developed by Crozier and Eyles (1980) and

Table 14.1. A selection of worldwide criteria for debris-flow triggering threshold.

Continent	Country, location	Type(s) of threshold criteria	Reference(s)
North America	Canada, North Shore Mountains of Vancouver, British Columbia	Antecedent rainfall, rainfall intensity–duration, stream discharge	Jakob and Weatherly (2003)
	USA, San Francisco Bay region, California	Antecedent rainfall, rainfall intensity–duration	Cannon and Ellen (1985)
	USA, Santa Monica Mountains, southern California	Antecedent rainfall, rainfall intensity	Campbell (1975)
	USA, Blue Ridge Mountains, central Virginia	Rainfall intensity–duration	Wieczorek et al. (2000)
	USA, Puerto Rico	Rainfall intensity–duration	Jibson (1989), Larsen and Simon (1993)
	USA, Rocky Mountains, Colorado	Air temperature related to rate of snowmelt	Chleborad et al. (1997)
	USA, Seattle, Washington	Air temperature related to rate of snowmelt, antecedent rainfall, rainfall storm total	Chleborad (2000, 2003)
	USA, Honolulu District, Oahu, Hawaii	Rainfall intensity–duration	Wilson et al. (1992)
South America	Colombia, Manizales	Daily rainfall total	Terlien (1997, 1998)
	Brazil, Rio de Janeiro	Hourly rainfall	Ortigao et al. (2003)
Europe	Germany, Rheinhessen	Cumulative rainfall in specific time interval	Dikau and Jäger (1995), Glade et al. (2001)
	Germany, Bonn Area	Prolonged rainfall, groundwater, no thresholds have been derived	Hardenbicker and Grunert (2001)
	Iceland	Cumulative rainfall in specific time interval	Saemundsson et al. (2003)
	Italy, Piedmont region	Cumulative rainfall in specific time interval	Govi et al. (1985)
	Italy, Tiber River basin	Stream discharge	Reichenbach et al. (1998)

Region	Location	Rainfall measure	References
	Italy, Moscardo Torrent, Italian Alps	Total storm rainfall, maximum 1-hr rainfall intensity	Marchi et al. (2002)
	Italy, Cancia area, Dolomites	Rainfall intensity–duration	Bacchini and Zannoni (2003)
	Italy, Cordevola River basin, Dolomites	Antecedent rainfall, cumulative rainfall in specific time interval	Pasuto and Silvano (1998)
	Italy, Valtellina area, northern Italy	Rainfall intensity–duration	Polloni et al. (1992), Crosta (1998), Crosta and Frattini (2000)
	Italy, Calabria	Daily cumulative rainfall	Petrucci and Polemio (2000, 2002)
	Norway	Cumulative rainfall in specific time interval	Sandersen (1997)
	Portugal, north of Lisbon	Antecedent rainfall, rainfall intensity–duration	Zêzere et al. (1999), Zêzere and Rodrigues (2002)
	Poland, Carpathians	Rainfall intensity	Starkel (1996)
	Scotland	Antecedent rainfall, daily rainfall total	Ballantyne (2002)
	Spain, Llobregat River, Eastern Pyrenees	Daily rainfall total	Corominas and Moya (1999)
Asia	China, Hong Kong	Rainfall intensity	Brand et al. (1984), Brand (1989), Premchitt et al. (1994), Chan et al. (2003), Pun et al. (2003)
	India and Bhutan, Himalayas	Daily rainfall total, total storm rainfall	Starkel and Sarkar (2002)
	Japan, Higashi-Hiroshima	Antecedent rainfall, hourly rainfall	Kaibori et al. (2003)
	Japan, Ibi River, Gifu Prefecture	Stream discharge	Onda et al. (2003)
	Korea	Antecedent rainfall, rainfall intensity–duration	Kim et al. (1992)
	Malaysia, Kuala Lumpur, Karak	Rainfall intensity, antecedent precipitation	Lloyd et al. (2003)
	Nepal	Rainfall intensity, hourly, daily, monthly, and annual rainfall totals	Gerrard and Gardner (2000)
	Taiwan, Nan-Tou County	Antecedent rainfall	Fan et al. (2003)
Australia	Australia, Illawarra area of New South Wales	Antecedent rainfall, daily rainfall total	Chowdhurry and Flentje (2002)
	New Zealand	Antecedent rainfall, rainfall intensity, rainfall storm total	Crozier and Eyles (1980), Crozier (1989, 1997), Glade (2000b)

340 Climatic factors influencing occurrence of debris flows [Ch. 14

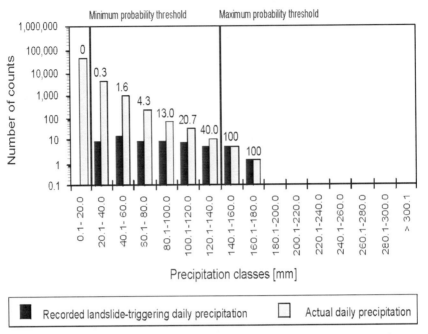

Figure 14.7. Rainfall probability thresholds established by applying the "*Daily Rainfall*" model (Glade, 1998) for the period from 1862 to 1995 for Wellington, New Zealand. The number of counts refers to the total number of rain days used in the analysis. An upper probability rainfall threshold of 140 mm and a lower threshold of 20 mm were established based on all past rainfall events. Any rainfall event greater than 140 mm triggered landslides, whereas no historical landslides were recorded for rainfall less than 20 mm.

further refined by Crozier (1999) and Glade (2000a). Glade determined a decay function for loss of water through drainage using recession curves from hydrographs of different streams with various catchment sizes within the study area. This method requires some assumptions. First, it is assumed that landslides are triggered by the maximum daily rainfall in the region. Second, the location of the triggered landslides is assumed similar to the position where the maximum daily rainfall was recorded. Further, it is assumed that the lowest values of potential evapotranspiration in the region are concurrent with maximum rainfall, and that a landslide is most likely to occur at the location where field moisture capacity is reached. Although the underlying assumptions seem restrictive, previous research in the two New Zealand regions of Wellington (Eyles et al., 1978) and Wairoa (Eyles and Eyles, 1981) support their validity. Crozier (1999) applied this model to the Wellington region and compared rainfall with landslide data from files provided by the Wellington City Council. The model provided a daily update of the soil water status and hence the amount of rainfall necessary to trigger landslides the following day. The probability of rainfall for the following day was calculated using frequency/magnitude statistics. The model results give a satisfactory level of landslide prediction, particularly during periods of intense landslide activity (Crozier, 1999). Despite its capability as an early warning

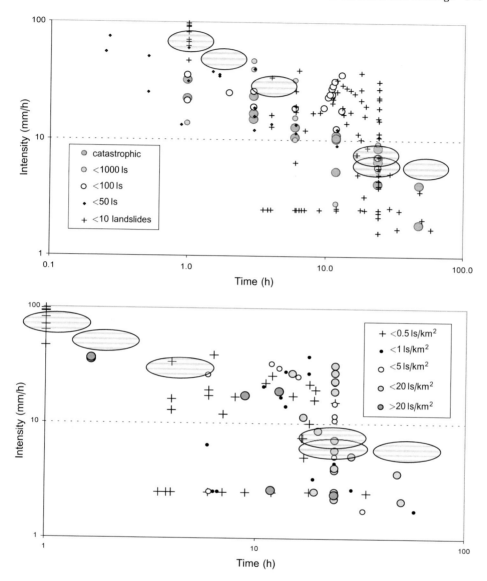

Figure 14.8. Number and spatial density of debris flows for rainstorms of different intensity and duration in the Alps, Prealps, and the Sarno regions of Italy. The gray ellipses indicate the position of the cluster of data points for major events in the Alps, Prealps, and the Sarno areas. From Crosta and Frattini (2000).

system on a daily basis and its successful application for an eight month period, the model has not yet been implemented as a hazard alert system.

The result of applying the *Soil Water Status* model is shown for the Wellington region (Figure 14.12). The graph shows that every landslide occurred with positive soil water status indices, indicating that the field capacity and consequent positive

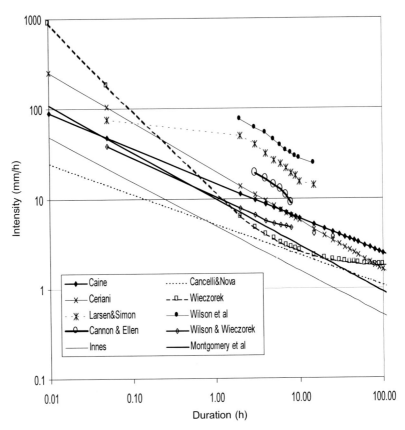

Figure 14.9. Worldwide rainfall thresholds from the literature. The thresholds by Caine (1980), Innes (1983), Cancelli and Nova (1985), and Ceriani et al. (1992) are of global type (i.e., they were prepared using all worldwide data available at the time) (Caine, 1980, Innes, 1983) or for large areas with different soil, morphologic and rainfall characteristics (Ceriani et al., 1992). Local or regional thresholds are from Larsen and Simon (1993), Cannon and Ellen (1985), Wieczorek (1987), Wilson and Wieczorek (1995), and Montgomery et al. (2000).
From Crosta and Frattini (2000).

pore-water pressures were reached (Glade 2000a). Rectilinear thresholds give the probability of landslide occurrence for a given combination of daily rainfall (horizontal line) and soil water status index (vertical line) (Figure 14.12). A similar decay function for soil drainage was used by Jakob and Weatherly (2003) for an application on the North Shore Mountains of Vancouver, Canada.

A worldwide model predicting the principal probabilistic periods of debris-flow danger depending upon the temporal variation of climatic factors in different climatic zones was developed by Belaya (2003). The systematic classification of regions with similar climatic conditions favourable to triggering debris flows has been conducted. The model uses monthly temperatures and precipitation data as well as documented times of rainfall-initiated debris flows occurring in the broad

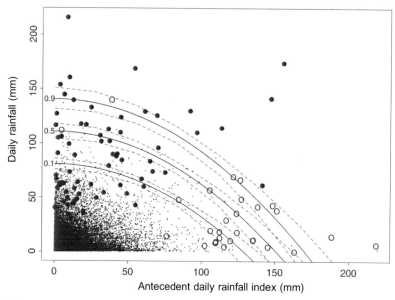

Figure 14.10. Rainfall probability thresholds established by applying the "*Antecedent Daily Rainfall*" model for the period 1862–1995 for Wellington, New Zealand (Glade et al., 2000). The antecedent daily rainfall index compromises the length of the antecedent rainfall period, including water loss to the atmosphere through evapotranspiration, and a decay factor representing the rate of soil moisture decrease in a specific period of time. Large dots relate to rainfall which triggered landslides, open circles relate to rainfall with probable landslide occurrence, and small dots relate to rainfall which did not trigger landslides. Confidence intervals are indicated for each probability curve by dashed lines.

climatic regions of permafrost areas, middle climate regions, and tropical areas. The Climate Research Unit (CRU) Global Climate Dataset, which consists of a mean monthly climatic database, with 0.5° latitude by 0.5° longitude resolution for global land areas, excluding Antarctica, was used for the period 1961–1990. The model characterizes three principal periods of debris-flow hazard: a debris-flow danger period (DFDP) as the part of the calendar year during which 100% of all debris flows occur; the main debris-flow danger period (MDFDP) within the DFDP and accounts for 90% of debris-flow events; and the extreme debris-flow danger period (EDFDP) accounting for 50% of debris-flow events (Figure 14.13). The number of debris flows per unit time increases progressively with each of these periods, from DFDP to MDFDP to EDFDP. The model was applied to all debris-flow regions in all continents to identify DFDP, MDFDP, and EDFDP months and the number of days of the DFDP for the present-day climate.

14.4.2 Forecasts and early warning

For practical applications, debris-flow forecasts, preferably coupled with early warning systems, are a major component of debris-flow risk management. The

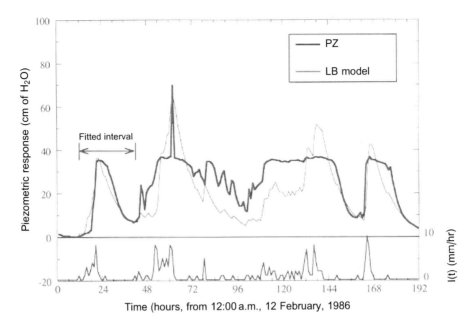

Figure 14.11. Piezometric (PZ) response (dark line) and "*Leaky Barrel*" (LB) modeling (gray line) vs. rainfall intensity for storm events starting at 12:00 a.m., 12 February, 1986 near La Honda, California.
Wilson and Wieczorek (1995).

various approaches include both regional forecasts and local early warnings. A regional forecast might issue a warning such as "*Tomorrow there will be a 50% probability of debris flows somewhere in the XY region*". No specific location can be identified. In contrast, local or site-specific warnings might be given through closure of railways, bridges, or roads to prevent damage from debris-flow impact.

In parts of the western USA, telemetered data on snowpack water equivalent, total precipitation, and air temperature are collected with the SNOTEL acquisition system (Crook, 1983). These data provide an indication of regional slope stability, which is useful for hazard evaluation and warning. Following the numerous debris-flow events of 1983 along the Wasatch Front in Utah (Wieczorek et al., 1989), several potential landslide sites were instrumented in 1984 to measure temperature, precipitation, and slope movement and to relay the data by telemetry to local officials (McCarter and Kaliser, 1985). After temperatures began rising and snowmelt was almost complete by late-May of 1984 in Rudd Canyon, near Salt Lake City, Utah, alarms generated by slow landslide movement gave advance warning of debris flows to local officials (McCarter and Kaliser, 1985).

Rainfall thresholds have been used for regional real-time landslide warning in the San Francisco Bay region, California (Keefer et al., 1987), Hong Kong, China (Hansen et al., 1995), and Rio de Janeiro, Brazil (Ortigao et al., 2003). These warning systems, which rely on nearly continuous ground-based rainfall measure-

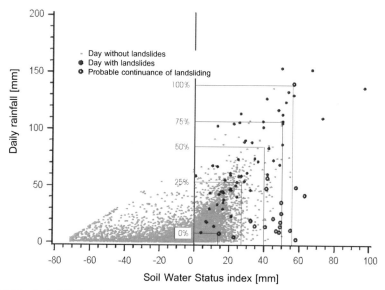

Figure 14.12. Rainfall thresholds established by applying the "*Soil Water Status*" model for the Wellington region, New Zealand (Glade, 2000a). The "*Soil Water Status*" index represents the soil water content critical for slope failure (Crozier and Eyles, 1980; Crozier, 1997).

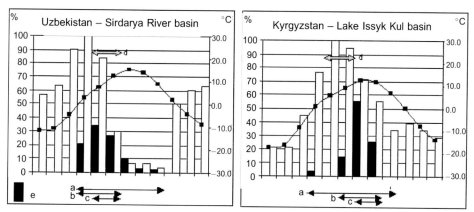

Figure 14.13. Distribution of debris-flow events among calendar months in percent of total (e). Periods of debris-flow danger: (a) DFDP, (b) MDFDP, (c) EDFDP, and (d) snow melt period. Solid lines with squares are monthly air temperature (°C); white bars are percent of maximum monthly precipitation.
Belaya (2003).

ments, can be very accurate temporally, but they may have serious spatial limitations related to the limited number and regional distribution of monitoring stations. Frequent temporal remote sensing of rainfall data, such as from Doppler radar, can provide the information necessary for assessing regional slope stability on a detailed spatial basis during near real time. Post-event comparisons of ground-based rainfall measurements with remote-sensed data for major storms are needed, particularly in areas with major topographic relief, to improve understanding of storm processes and to improve rainfall-estimating techniques based on remote sensing. Improved assessment of localized tropical storms is needed for providing accurate estimates of rainfall totals. Improved use of remote-sensing rainfall data for landslide hazard assessment and warning will depend upon shorter time intervals between measurements to allow sufficient time for analysis, communication of warning, and public response to warning (Wieczorek et al., 2003).

The regional landslide (debris-flow) warning system operated by the US Geological Survey and National Weather Service (NWS) from 1986 to 1995 in the San Francisco Bay region issued public advisories when rainfall conditions reached intensities likely to trigger debris flows from susceptible hillsides (Wilson, 2005). The warning system was based on a set of rainfall thresholds for triggering significant debris-flow activity developed by Cannon and Ellen (1985), forecasting of severe storms by the NWS, and real-time monitoring of rainfall by both agencies. The first public warnings were issued during a severe storm sequence in February 1986 which triggered debris flows that corresponded well with the time intervals of issued advisories (Keefer et al., 1987). Subsequent public advisories were issued during or just before severe storms in 1991, 1993, and 1995. The rapid distribution of a landslide warning and the proper response by the public were problems beyond the measurement of rainfall related to the thresholds for triggering debris flows. Although relatively few people listened directly to the landslide warnings broadcast on the NWS Weather Radio, the warnings would be picked up and rebroadcast by commercial radio and television stations; consequently, these warnings were widely distributed to the public. In addition, many local fire and law enforcement agencies regularly monitored the NWS Weather Radio broadcasts, so these warnings were heard by those responsible for public safety (Wilson, 2005). Although attempts were made to educate the public to respond wisely to landslide warnings, there was no documentation of how the public responded to the issued warnings and how many lives and property values were saved.

In Hong Kong, risk of debris flows from torrential rainstorms with the passage of typhoons, tropical depressions, and severe thunderstorms (Brand, 1985, 1988) has inspired the development of the most comprehensive landslide warning system in the world (Hanson et al., 1995; Wilson, 2005). In May 1982, over 1,000 landslides occurred during four days of severe rain, followed by another 500 during another intense storm in August of that year. The warning system, operated jointly by the Geotechnical Engineering Office of the Hong Kong Government and the Hong Kong Observatory, began issuing warnings in 1984. The warnings are issued when weather forecasts and rainfall data suggest that numerous (>10) debris flows are expected within the city. The rainfall thresholds for numerous debris flows were 175 mm during a 24-h period or 70 mm within one hour. On 5 November, 1993 Lantau Island, west of Hong Kong, experienced an intense storm with a peak 1-hr intensity of 94 mm, peak 6-hr intensity of 423 mm, and 24-hour intensity of 742 mm, resulting in about 600 natural slope failures (Hansen et al., 1995).

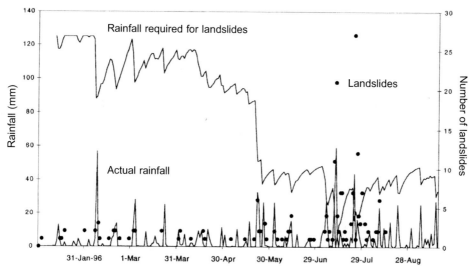

Figure 14.14. Landslide occurrence, daily rainfall, and rainfall required to trigger landslides based on calculations applying the "*Antecedent Soil Water Status*" model (Crozier, 1999). Bold dots refer to the number of landslides, the lower line is the actual rainfall, and the top line indicates the rainfall required to trigger landslides. Note that landslides are also initiated by low-magnitude rainstorm events.

As previously mentioned, Crozier (1999) coupled the previously introduced *Soil Water Status* model with climatic conditions to develop a landslide-triggering threshold for Wellington, New Zealand. Using a water-balance routine with input of daily evaporation and daily rainfall, and constants for soil water storage capacity and a drainage function, daily soil moisture was evaluated using the *Antecedent Soil Water Status model* (Figure 14.14). Based on the evaluation of daily soil moisture a predictive evaluation of the probability of rainfall sufficient for the triggering of landslides was developed (Glade, 1997; Crozier and Glade, 1999; Glade et al., 2000).

Hydrologic models have been developed for forecasting activation of various types of landslides (Reid, 1994; van Asch and Buma, 1997; Bonomi and Cavallin, 1999; see also Chapter 4). A hydrologic model of rainfall infiltration into shallow soils (Iverson, 2000) has been used to represent temporal changes in pore pressure and their effects on regional slope stability for the triggering of debris flows (Baum et al., 2002; Crosta and Frattini, 2003). Continuous remotely-sensed rainfall data from Doppler radar has been utilized with four different hydrologic models to compare predicted slope instability with the timing, number, and distribution of documented debris flows following intense rainstorms (Morrissey et al., 2005; Crosta and Frattini, 2003) (Figure 14.15).

In combination with rainfall thresholds and real-time remotely-sensed rainfall data, regional slope stability models could be used to provide a means of near-real-time prediction and warning of debris-flow hazards. For local sites, detection of debris flows, combined with instrumentation of climatic factors, could be used for

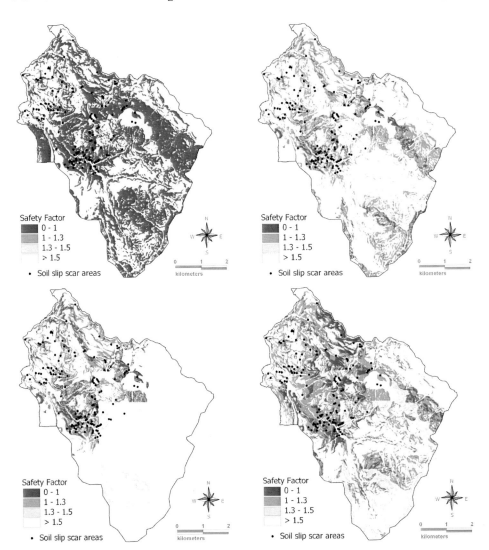

Figure 14.15. Slope stability maps from simulation with different hydrologic models: (a) steady-state model; (b) piston flow model with uniform precipitation; (c) piston flow model with distributed precipitation; and (d) diffusive model with distributed precipitation.

Crosta and Frattini (2003).

hazard warning (Chang, 2003). For example, Arattano et al. (1997) used a monitoring system, including rain gauges, ultrasonic sensors, and a video camera in the upper part of the Moscardo basin, Italy, to detect twelve debris flows between 1989 and 1995. For local hazard warning, monitoring of climatic conditions could serve as a preliminary indicator of potential debris-flow initiation, which could then be further verified by detection of debris-flow movement.

14.5 EFFECTS OF CLIMATE CHANGE ON DEBRIS-FLOW ACTIVITY

The time frame of significant variations in climatic factors relevant to the triggering of debris flows ranges from hourly for rainfall intensity, to daily or weekly for antecedent rainfall, to yearly or multi-yearly for seasonal patterns such as El Niño Southern Oscillation (ENSO), to many thousands of years for fluctuations during the Holocene and Pleistocene. Long-term climatic variations can significantly alter vegetation, topography, and geologic and hydrologic factors, all of which may influence debris-flow susceptibility.

14.5.1 El Niño-Southern Oscillation

The North and South American continents have historically been affected by an irregularity in seasonal patterns of precipitation, known as ENSO. In North America, ENSO generally causes increased rainfall in the southern part of the continent (south of 40° latitude) and dryer conditions in the northern part. The increased storm intensity and seasonal rainfall have resulted in an increase in many types of landslides, including debris flows. Severe El Niño-related storms in 1982–1983 and 1983–1984 triggered thousands of landslides, ranging from debris flows to deep-seated slumps and slides, in Nevada, Utah, and western Colorado (Schuster and Wieczorek, 2002).

In southern coastal Peru, flood and debris-flow deposits have been correlated with previous El Niño events (Keefer et al., 2003). An El Niño seasonal event in this region causing more severe effects than any in recent history has been dated within the period of 1607–1608 AD. Older deposits dominated by flood and debris-flow deposits of similar scale indicate that severe El Niño events occurred throughout the late Pleistocene and Holocene. The period of greatest debris-flow frequency in this part of Peru began about 12,000 years ago and lasted for about 3,600 years during the early Holocene when at least 6 debris-flow events occurred at one site. No severe debris-flow events were detected during the Middle Holocene between about 8,400 and 5,300 years ago, when other evidence indicates that the ENSO pattern was particularly weak (Keefer et al., 2003).

14.5.2 Warmer and dryer climates

Warming trends can be a longer term, secondary climatic influence that causes retreat of glaciers and consequent slope instability. Evans and Clague (1994) demonstrated an increase in slope instability in deglaciated mountainous regions worldwide during climatic warming during the last 100–150 years. Recent retreat of glacial ice has resulted in widespread destabilization of many mountainous areas, resulting in large floods, debris flows, and other types of landslides. In the Swiss Alps, numerous large debris flows were triggered by intense rainfall during the summer of 1987 (Zimmerman and Haeberli, 1992). In this case, glacial melt had uncovered areas of steep, unconsolidated materials which were now prone to mass movement activity. Similarly, the deterioration of alpine permafrost led to the failure of previously frozen talus slopes. In many cases, the source of debris is Little Ice Age glacial deposits left uncovered and oversteepened following recent glacial retreat (Evans and Clague, 1994). Outbursts from moraine-dammed and glacier-dammed lakes also generate floods and debris flows. Recent slope instability due to warming in glacial regions during the past 150 years are probably at least an order of magnitude less than those associated with late Pleistocene deglaciation around 15,000 to 10,000 years ago (Evans and Clague, 1994; Wieczorek and Jäger, 1996).

Examination of prehistoric evidence has shown that in some areas warmer and dryer climate periods affect vegetation, resulting in increased fire frequency and consequently more fire-related debris flows (see Chapter 15). Meyer and Pierce (2003) used ^{14}C-dated geologic records to examine spatial and temporal variations in climate and sedimentation. They found that in Yellowstone National Park, USA episodes of fire-induced debris flows occurred at 300–450-year intervals during the past 3,500 years. Debris-flow deposition decreased during cooler episodes during the Little Ice Age ~1,200–1,900 AD because wetter conditions prevented most fires from spreading. However, the warmer period between 900 and 1,200 AD caused many large fire-related debris flows.

In the Holocene, humans converted large areas from native forest and brush land into agricultural production in various areas throughout Europe. The linkage between climate impact, human impact, and landslide initiation during the Holocene has been shown for Germany (Grunert and Hardenbicker, 1997), Scotland (Innes, 1997), England (Ibsen and Brunsden, 1997), and Italy (Rodolfi, 1997; Wasowski, 1998). Common to these examples is the uncertainty involved in trying to correlate past landslide occurrence with former climate regimes. Dating techniques applicable to mass movements are reviewed by Lang et al. (1999). Examples of applications of these methods are given for the Carpathians in Poland and for Europe by Starkel (1997) and for northern Spain by Gonzalez-Diéz et al. (1999). The complexity of these analyses is increased by human interference with nature and the resulting changes in frequency and magnitude of geomorphic processes, including debris flows.

14.5.3 Forecasting effects of climate change

Speculation regarding global climate change can raise questions regarding the potential for changes in frequency of debris flows in the mid to long-term future. Although forecasts of climate change are uncertain, a tendency for future debris flows can be estimated. Belaya (2003) predicted the likelihood of debris-flow hazards for a scenario of climate change by the year 2050 for a territory within the former USSR. An example from the Caucasus identifying the number of days of the DFDP for the present-day climate, as well as for a scenario of the climate change by the year 2050 is shown in Figure 14.16.

The following examples give applications of General Circulation Models (GCMs) to (1) a complex landslide involving both earth flow and debris flow, and (2) the previously introduced regional landslide-triggering rainfall thresholds based on the *Soil Water Status* model.

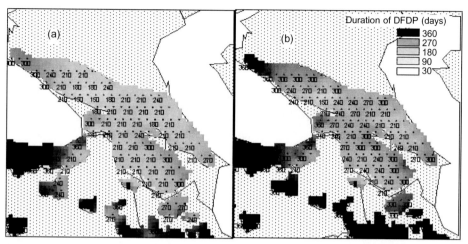

Figure 14.16. Grid points (0.5° latitude by 0.5° longitude spacing) of the CRU Global Climate Dataset and duration of DFDP (in days) in the Caucasus region: (a) for present-day climate and (b) for the year 2050, according to a scenario of the Hadley Center for Climate Prediction and Research (HadCM2).
Belaya (2003).

Dehn and Buma (1999) applied an analogue statistical downscaling technique to predict landslide activity depending on climatic variables in the Barcelonnette basin of the French Alps for the periods of 2020–2049 and 2070–2099. Mean monthly precipitation and temperature were used to calculate potential evapotranspiration for three different GCMs. Climatic data were coupled with a simple hydrological model which simulates groundwater levels in the slope and leads to predictions of landslide movement. The results give no consistent picture of future landslide activity. Some models give decreased activity, some increased. Consequently, improved GCMs and an optimization of the approach are necessary to obtain better information on landslide movement.

Schmidt and Glade (2003) took a similar downscaling approach, and applied GCMs to the regional *Soil Water Status* model for Hawke Bay and Wellington, both located on the North Island of New Zealand. The results show good agreement between observations and the control run for the period 1950–1979 (Figure 14.17(a, b), see color section). In contrast, the predicted values of soil water status for 2070-2099 are shifted towards the y-axis, indicating a decrease in soil water (Figure 14.17(c)). Thus, the probability of landslide events in the period 2070–2099 appears to decrease compared to current conditions (Figure 14.18, see color section).

These examples illustrate the uncertainty associated with such predictions. However, such calculations provide a preliminary approximation of what might happen in the future. Climate change will exert spatially and temporally different responses to the surface and subsurface hydrology of hillslopes. Having established that the interaction of antecedent conditions and rainfall intensity is important to predict debris-flow occurrence, any predictions of the impacts of climate change on debris flows will require knowledge of the climatic effects on each variable. For example, coastal British Columbia has become approximately 10% wetter during the past 100 years (a 10% increase in the total annual precipitation). However, a detailed study of rainfall intensities has not shown any long-term trends, but has identified decadal variations as determined by the Pacific Decadal Oscillation (PDO) (Jakob et al., 2004). This means that any predictions of future debris-flow activity will have to take into account changes in total rainfall, changes in storm frequency, and changes in storm intensity which seem to undergo decadal-scale cycles. It is therefore safe to say that predictions of debris-flow activity, as a consequence of climate change, are in their infancy and much more work has to be done before concrete predictions can be made.

14.6 CONCLUSIONS

Many methods and concepts are available to investigate the relation between climatic factors and debris-flow occurrence. These range in complexity and depend on the scale of investigation. Both site-specific and regional models have demonstrated their validity, and offer a set of choices that can be evaluated for applicability to a given region and civil authority. Further research is necessary to:

- evaluate existing methods;
- transfer them to other regions by calibrating input parameters; and
- develop improved methods by introducing new observational techniques for both monitoring of debris flows and measurement of climatic parameters.

14.7 ACKNOWLEDGEMENTS

We are indebted to many people, especially Matthias Jakob, David Keefer, Susan Cannon, Giovanni Crosta, and Nina Belaya, who provided useful information on climatic influences on debris flows. We are very grateful to Steve Ellen, Ray Wilson, and Matthias Jakob for many constructive review comments on a preliminary version of this paper.

14.8 REFERENCES

Arattano, M., Deganutti, A.M., and Marchi, L. (1997) Debris flow monitoring activities in an instrumented watershed on the Italian Alps. In: C-L. Chen. (ed.), *Proceedings of the 1st International Conference on Debris-flow Hazards Mitigation: Mechanics, Prediction, and Assessment, San Francisco, CA* (pp. 506–515). American Society of Civil Engineers, New York.

Bacchini, M. and Zannoni, A. (2003) Relations between rainfall and triggering of debris-flow: Case study of Cancia (Dolomites, Northeastern Italy). *Natural Hazards and Earth System Sciences*, **3**(1/2), 71–79.

Ballantyne, C.K. (2002) Debris flow activity in the Scottish Highlands: Temporal trends and wider implications for dating. *Studia Geomorphologica Carpatho-Balcanica*, **XXXVI**, 7–28.

Baum, R.L., Savage, W.Z., and Godt, J.W. (2002) *TRIGRS – A Fortran Program for Transient Rainfall Infiltration and Grid-based Regional Slope-Stability Analysis* (USGS Open-File Report 02-424). US Geological Survey, Reston, VA. Available at *http://pubs.usgs.gov/of/2002/ofr-02-424/*

Belaya, N.L. (2003) Distribution model for periods of debris-flow danger: In: D. Rickenmann and C-L. Chen (eds), *Debris-flow Hazards Mitigation: Mechanics, Prediction, and Assessment: Proceedings of the 3rd International Conference, Davos, Switzerland, September 10–12* (pp. 59–70). Millpress, Rotterdam.

Bonomi, T. and Cavallin, A. (1999) Three-dimensional hydrogeological modelling application to the Alvera mudslide (Cortina d'Ampezzo, Italy). *Geomorphology*, **30**(1/2), 189–199.

Brand, E.W. (1985) Predicting the performance of residual soil slopes (Theme lecture). *Proceedings of the 11th International Conference on Soil Mechanics and Foundation Engineering, San Francisco* (Vol. 5, pp. 2541–2578). Norges Geotekniske Institutt, Oslo.

Brand, E.W. (1988) Landslides risk assessment in Hong Kong. In: C. Bonnard (ed.), *Proceedings 5th International Symposium on Landslides, Lausanne* (Vol. 1, pp. 1059–1074). A.A. Balkema, Rotterdam.

Brand, E.W. (1989) Occurrence and significance of landslides in Southeast Asia. In: E.E. Brabb and B.L. Harrod (eds), *Landslides: Extent and Economic Significance* (pp. 303–324). A.A. Balkema, Rotterdam.

Brand, E.W., Premchitt, J., and Phillipson, H.B. (1984) Relationship between rainfall and landslides in Hong Kong. In: Canadian Geotechnical Society (ed.), *Proceedings of the 4th International Symposium on Landslides* (Vol. 1, pp. 377–384). University of Toronto Press.

Caine, N. (1980) The rainfall intensity-duration control of shallow landslides and debris flows. *Geografiska Annaler*, **62A**, 23–27.

Campbell, R.H. (1975) *Soil Slips, Debris Flows, and Rainstorms in the Santa Monica Mountains and Vicinity, Southern California* (USGS Professional Paper 851, 51 pp.). US Geological Survey, Reston, VA.

Cancelli, A. and Nova, R. (1985) Landslides in soil debris cover triggered by rainstorms in Valtellina (central Alps, Italy). *Proceedings of 4th International Conference on Landslides* (pp. 267–272). Japanese Landslide Society, Tokyo.

Cannon, S.H. (1988) Regional rainfall-threshold conditions for abundant debris-flow activity. In: S.D. Ellen and G.F. Wieczorek (eds), *Landslides, Floods, and Marine Effects of the January 3–5, 1982, Storm in the San Francisco Bay Region, California* (USGS Professional Paper 1434, pp. 35–42). US Geological Survey, Reston, VA.

Cannon, S.H. and Ellen, S. (1985) Rainfall conditions for abundant debris avalanches in the San Francisco Bay region, California. *California Geology*, **38**(12), 267–272.

Cannon, S.H. and Ellen, S. (1988) Rainfall that resulted in abundant debris-flow activity during the storm: In: S.D. Ellen and G.F. Wieczorek (eds), *Landslides, Floods, and Marine Effects of the January 3–5, 1982, Storm in the San Francisco Bay Region, California* (USGS Professional Paper 1434, pp. 27–34). US Geological Survey, Reston, VA.

Ceriani, M., Lauzi S., and Padovan, N. (1992) Rainfalls and landslides in the Alpine area of Lombardia Region, Central Alps Italy. *Proceedings of the International Symposium*

"Interpraevent", Bern (Vol. 2, pp. 9–20). Internationale Forschungsgesellschaft Interpraevent, Klagenfurt, Austria.

Chan, R.K.S., Pang, P.L.R., and Pun, W.K. (2003) Recent developments in the Landslip Warning System in Hong Kong. In: K.K.S. Ho and K.S. Li (eds), *Geotechnical Engineering – Meeting Society's Needs: Proceedings of the 14th Southeast Asian Geotechnical Conference, December 10–14, 2001, Hong Kong* (pp. 219–224). A.A. Balkema, Rotterdam.

Chang, S.Y. (2003) Evaluation of a system for detecting debris flows and warning road traffic at bridges susceptible to debris-flow hazard. In: D. Rickenmann and C-L. Chen (eds), *Debris-flow Hazards Mitigation: Mechanics, Prediction, and Assessment: Proceedings of the 3rd International Conference, Davos, Switzerland, September 10–12* (pp. 731–742). Millpress, Rotterdam.

Chirico, G.B., Claps, P., Rossi, F., and Villani, P. (2000) Hydrologic conditions leading to debris-flow initiation in the Campanian volcanoclastic soils. In: P. Claps and F. Siccardi (eds), *Mediterranean Storms: Proceedings of the EGS Plinius Conference, Maratea, Italy, October 14–16, 1999* (pp. 473–484). Editoriale Bios, Cosenza, Italy.

Chleborad, A.F. (1997) *Temperature, Snowmelt, and the Onset of Spring Season Landslides in the Central Rocky Mountains* (USGS Open-File Report 97-27, 35 pp.). US Geological Survey, Reston, VA.

Chleborad, A.F. (1998) *Use of Air Temperature Data to Anticipate the Onset of Snowmelt-season Landslides* (USGS Open-File Report 98-0124). US Geological Survey, Reston, VA. Available at *http://pubs.usgs.gov/of/1998/ofr-98-0124/*

Chleborad, A.F. (2000) *Preliminary Method for Anticipating the Occurrence of Precipitation-induced Landslides in Seattle, Washington* (USGS Open-File Report 00-0469). US Geological Survey, Reston, VA. Available at *http://pubs.usgs.gov/of/2000/ofr-00-0469/*

Chleborad, A.F. (2003) *Preliminary Evaluation of a Precipitation Threshold for Anticipating the Occurrence of Landslides in the Seattle, Washington, Area* (USGS Open-File Report 03-463). US Geological Survey, Reston, VA. Available at *http://pubs.usgs.gov/of/2003/ofr-03-463/*

Chleborad, A.F., Ellis, W.L., and Kibler, D. (1997) *Results of Site Investigation and Instrumentation of the Keno Gulch Landslide/Debris-flow Source Area, Aspen, Colorado* (USGS Open-File Report 97-717, 17 pp.). US Geological Survey, Reston, VA.

Chowdhury, R. and Flentje, P. (2002) Uncertainties in rainfall-induced landslide hazard. *Quarterly Journal of Engineering Geology and Hydrogeology*, **35**, 61–69.

Church, M., and Miles, M.J. (1987) Meteorological antecedents to debris flow in southwestern British Columbia: Some case studies. In: J.E. Costa and G.F. Wieczorek (eds), *Debris Flows/Avalanches: Process, Recognition and Mitigation* (Reviews in Engineering Geology, Vol. 7, pp. 63–80). Geological Society of America, Boulder, CO.

Corominas, J. and Moya, J. (1999) Reconstructing recent landslide activity in relation to rainfall in the Llobregat River basin, Eastern Pyrenees, Spain. *Geomorphology*, **30**(1/2), 79–93.

Crook, A.G. (1983) The SNOTEL data acquisition system, and its use in mitigation of natural hazards. *Proceedings of the International Technical Conference on Mitigation of Natural Hazards through Real-time Data Collection Systems and Hydrologic Forecasting* (83 pp.). World Meteorological Organization, US National Oceanic and Atmospheric Adminstration, and California Department of Water Resources.

Crosta, G. (1998) Regionalization of rainfall thresholds: An aid to landslide hazard evaluation. *Environmental Geology*, **35**(2/3), 131–145.

Crosta, G.B. and Frattini, P. (2000) Rainfall thresholds for triggering soil slips and debris flow. In: A. Mugnai, F. Guzzetti, and G. Roth (eds), *Mediterranean Storms: Proceedings* of the EGS 2nd Plinius Conference, Siena, Italy, October 16–18 (pp. 463–487). Tipolitografia Grifo, Perugia, Italy.

Crosta, G.B. and Frattini, P. (2003) Distributed modeling of shallow landslides triggered by intense rainfall. *Natural Hazards and Earth System Sciences*, 3(1/2), 81–93.

Crozier, M.J. (1989) *Landslides: Causes, Consequences and Environment*. Routledge, London.

Crozier, M.J. (1997) The climate-landslide couple: A southern hemisphere perspective. In: J.A. Matthews, D. Brunsden, B. Frenzel, B. Gläser, and M.M. Weiß (eds), *Rapid Mass Movement as a Source of Climatic Evidence for the Holocene: Palaeoclimate Research* (Vol. 19, pp. 333–354). Gustav Fischer Verlag, Stuttgart.

Crozier, M.J. (1999) Prediction of rainfall-triggered landslides: A test of the antecedent water status model. *Earth Surface Processes and Landforms*, 24(9), 825–833.

Crozier, M.J. and Eyles, R.J. (1980) Assessing the probability of rapid mass movement. In: Technical Groups (eds), *Proceedings of 3rd Australia–New Zealand Conference on Geomechanics, Wellington* (Vol. 6, pp. 2.47–2.51). New Zealand Institution of Engineers, Wellington.

Crozier, M.J. and Glade, T. (1999) Frequency and magnitude of landsliding: Fundamental research issues. *Zeitschrift für Geomorphologie N.F., Suppl.*, 115, 141–155.

Cuesta, M.J.D., Sanchez, M.J., and Garcia, A.R. (1999) Press archives as temporal records of landslides in the North of Spain: Relationships between rainfall and instability slope events. *Geomorphology*, 30(1/2), 125–132.

De Campos, T.M.P., De N. Andrade, M.H., Gescovich, D.M.S., and Jargas, E.A., Jr (1994) Analysis of the failure of an unsaturated gneissic residual soil slope in Rio de Janeiro, Brazil. *Proceedings of the 1st Panamerican Symposium on Landslides, Guayaquil, Ecuador* (Vol. 1, pp. 201–213). Sociedad Ecuatoriana de Mecánica de Suelos y Rocas, Guayaquil, Ecuador.

Deganutti, A.M., Marchi, L., and Arattano, M. (2000) Rainfall and debris-flow occurrence in the Moscardo basin (Italian Alps). In: G.F. Wieczorek and N.D. Naeser (eds), *Debris-flow Hazards Mitigation: Mechanics, Prediction, and Assessment: Proceedings of the 2nd International Conference, Taipei, Taiwan, August 16–18* (pp. 67–72). A.A. Balkema, Rotterdam.

Dehn, M. and Buma, J. (1999) Modelling future landslide activity based on general circulation models. *Geomorphology*, 30(1/2), 175–187.

Dikau, R. and Jäger, S. (1995) Landslide hazard modelling in New Mexico and Germany. In: D.F.M. McGregor and D.A. Thompson (eds), *Geomorphology and Land Management in a Changing Environment* (pp. 51–68). John Wiley & Sons, Chichester, UK.

Ellen, S.D. and Wieczorek, G.F. (eds) (1988) *Landslides, Floods, and Marine Effects of the Storm of January 3–5, 1982, in the San Francisco Bay Region, California* (USGS Professional Paper 1434, 310 pp.). US Geological Survey, Reston, VA.

Evans, S.G. and Clague, J.J. (1994) Recent climatic change and catastrophic geomorphic processes in mountain environments. *Geomorphology*, 10, 107–128.

Eyles, R.J. (1979) Slip-triggering rainfalls in Wellington City, New Zealand. *New Zealand Journal of Science*, 22(2), 117–121.

Eyles, R.J. and Eyles, G.O. (1981) Recognition of storm damage events. *Proceedings of 11th New Zealand Geography Conference, Wellington* (pp. 118–123). New Zealand Geographical Society, Wellington.

Eyles, R.J., Crozier, M.J., and Wheeler, R.H. (1978) Landslips in Wellington City. *New Zealand Geographer*, 34(2), 58–74.

Fan, J.C., Liu, C.H., Wu, M.J., and Yu, S.K. (2003) Determination of critical rainfall thresholds for debris flow occurrence in central Taiwan and their revision after the 1999 Chi-Chi great earthquake. In: D. Rickenmann and C-L. Chen (eds), *Debris-flow Hazards Mitigation: Mechanics, Prediction, and Assessment: Proceedings of the 3rd International Conference, Davos, Switzerland, September 10–12* (pp. 103–114). Millpress, Rotterdam.

Flageollet, J.C., Maquaire, O., Martin, B., and Weber, D. (1999) Landslides and climatic conditions in the Barcelonnette and Vars basins (Southern French Alps, France). *Geomorphology*, **30**(1/2), 65–78.

Gerrard, A.J. and Gardner, R.A.M. (2000) Relationships between rainfall and landsliding in the Middle Hills, Nepal. *Norsk Geografisk Tidsskrift*, **54**, 74–81.

Glade, T. (1997) The temporal and spatial occurrence of rainstorm-triggered landslide events in New Zealand (380 pp.). PhD thesis, Victoria University of Wellington.

Glade, T. (1998) Establishing the frequency and magnitude of landslide-triggering rainstorm events in New Zealand. *Environmental Geology*, **35**(2/3), 160–174.

Glade, T. (2000a) Modeling landslide-triggering rainfalls in different regions in New Zealand: The soil water status model. *Zeitschrift für Geomorphologie*, **122**, 63–84.

Glade, T. (2000b) Modeling landslide triggering rainfall thresholds at a range of complexities. In: E. Bromhead, N. Dixon and M.-L. Ibsen (eds), *Landslides in Research, Theory and Practice* (Vol. 2, pp. 633–640). Thomas Telford, London.

Glade, T., Crozier, M.J. and Smith, P. (2000) Applying probability determination to refine landslide-triggering rainfall thresholds using an empirical "Antecedent Daily Rainfall Model". *Pure and Applied Geophysics*, **157**(6/8), 1059–1079.

Glade, T., Kadereit, A., and Dikau, R. (2001) Landslides at the Tertiary escarpment of Rheinhessen, southwest Germany. *Zeitschrift für Geomorphologie, Suppl.*, **125**, 65–92.

Gonzalez-Diéz, A., Remondo, J., de Teran, J.R.D. and Cendrero, A. (1999) A methodological approach for the analysis of the temporal occurrence and triggering factors of landslides. *Geomorphology*, **30**(1/2), 95–113.

Govi, M. and Sorzana, P.F. (1980) Landslide susceptibility as a function of critical rainfall amount in Piedmont Basin (north-western Italy). *Studia Geomorphologica Carpatho-Balcanica, Krakow*, **14**, 43–61.

Govi, M., Mortara, G., and Sorzana, P. (1985) Eventi idrologici e frane. *Geologia Applicata e Idrogeologia, Università Bari*, **20**, 359–375.

Grunert, J. and Hardenbicker, U. (1997) The frequency of landsliding in the north Rhine area and possible climatic implications. In: J.A. Matthews, D. Brunsden, B. Frenzel, B. Gläser and M.M. Weiß (eds), *Rapid Mass Movement as a Source of Climatic Evidence for the Holocene: Palaeoclimate Research* (Vol. 12, pp. 159–170). Gustav Fischer Verlag, Stuttgart.

Hanson, A., Brimicombe, A.J., Franks, C.A.M., Kirk, P.A., and Tung, F. (1995) Application of GIS to hazard assessment, with particular reference to landslides in Hong Kong. In: A. Carrara and F. Guzzetti (eds), *Geographical Information Systems in Assessing Natural Hazards* (pp. 273–298). Kluwer Academic, Dordrecht, The Netherlands.

Hardenbicker, U. and Grunert, J. (2001) Temporal occurrence of mass movements in the Bonn area. *Zeitschrift für Geomorphologie, Suppl.*, **125**, 13–24.

Horton, R.E. (1938) Phenomena of the contact zone between the ground surface and a layer of melting snow. *Association Internationale d'Hydrologie Scientifique, Paris*, **244**, 545–561.

Ibsen, M.L. and Brunsden, D. (1997) Mass movement and climatic variation on the south coast of Great Britain. In: J.A. Matthews, D. Brunsden, B. Frenzel, B. Gläser and M.M. Weiß (eds), *Rapid Mass Movement as a Source of Climatic Evidence for the Holocene: Palaeoclimate Research* (Vol. 12, pp. 171–182). Gustav Fischer Verlag, Stuttgart.

Iiritano, G., Versace, P. and Sirangelo, B. (1998) Real-time estimation of hazard for landslides triggered by rainfall. *Environmental Geology*, **35**(2/3), 175–183.

Innes, J.L. (1983) Debris flows. *Progress in Physical Geography*, **7**, 469–501.
Innes, J.L. (1997) Historical debris-flow activity and climate in Scotland. In: J.A. Matthews, D. Brunsden, B. Frenzel, B. Gläser and M.M. Weiß (eds), *Rapid Mass Movement as a Source of Climatic Evidence for the Holocene: Palaeoclimate Research* (Vol. 12, pp. 233–240). Gustav Fischer Verlag, Stuttgart.
Iverson, R.M. (2000) Landslide triggering by rain infiltration. *Water Resources Research*, **36**(7), 1897–1910.
Jäger, S. and Dikau, R. (1994) The temporal occurrence of landslides in South Germany. In: R. Casale, R. Fantechi, and J.C. Flageollet (eds), *Temporal Occurrence and Forecasting of Landslides in the European Community* (Vol. 1, pp. 509–564). EU, Brussels.
Jakob, M. and Weatherly, H. (2003) A hydroclimatic threshold for landslide initiation on the North Shore Mountains of Vancouver, British Columbia. *Geomorphology*, **54**(3–4), 137–156.
Jakob, M., McKendry, I., and Lee, R. (2004) Changes in rainfall intensity in the Greater Vancouver Regional District, British Columbia. *Canadian Water Resources Journal*, **28**(4), 587–604.
Jibson, R.W. (1989) Debris flows in southern Puerto Rico. In: A.P. Schultz and R.W. Jibson (eds), *Landslide Processes of the Eastern United States and Puerto Rico* (Special Paper 236, pp. 29–55). Geological Society of America, Boulder, CO.
Johnson, K.A. and Sitar, N. (1990) Hydrologic conditions leading to debris flow initiation. *Canadian Geotechnical Journal*, **27**(6), 789–801.
Kaibori, M., Kuwada, S., and Umeki, K. (2003) Some features of debris flow movements from the view point of disaster prevention. In: K.K.S. Ho and K.S. Li (eds), *Geotechnical Engineering – Meeting Society's Needs: Proceedings of the 14th Southeast Asian Geotechnical Conference, December 10–14, 2001, Hong Kong* (pp. 251–256). A.A. Balkema, Rotterdam.
Keefer, D.K. and Harp, E.L. (1998) Large landslides near the San Andreas fault in the Summit Ridge area, Santa Cruz Mountains, California. In: D.K. Keefer (ed.), *The October 17, 1989, Loma Prieta, California, Earthquake: Landslides and Stream Channel Change* (USGS Professional Paper 1551-C, 2, C71C127). US Geological Survey, Reston, VA.
Keefer, D.K., Wilson, R.C., Mark, R.K., Brabb, E.E., Brown, W.M. III, Ellen, S.D., Harp, E.L., Wieczorek, G.F., Alger, C.S., and Zatkin, R.S. (1987) Real-time landslide warning during heavy rainfall. *Science*, **238**, 921–925.
Keefer, D.K., Moseley, M.E., and deFrance, S.D. (2003) A 38000-year record of floods and debris flows in the Ilo region of southern Peru and its relation to El Niño events and great earthquakes. *Palaeogeography, Palaeoclimatology, Palaeoecology*, **194**, 41–77.
Kerle, N., van Wyk de Vries, B., and Oppenheimer, C. (2003) New insight into the factors leading to the 1988 flank collapse and lahar disaster at Casita volcano, Nicaragua. *Bulletin of Volcanology*, **65**, 331–345.
Kim, S.K., Hong, W.P., and Kim, Y.M. (1992) Prediction of rainfall-triggered landslides in Korea. In: D.H. Bell (ed.), *Proceedings of the 6th International Symposium on Landslides, 10–14 February, Christchurch, New Zealand* (Vol. 2, pp. 989–994).
Lang, A., Moya, J., Corominas, J., Schrott, L., and Dikau, R. (1999) Classic and new dating methods for assessing the temporal occurrence of mass movements. *Geomorphology*, **30**(1/2), 33–52.
Larsen, M.C. and Simon, A. (1993) A rainfall-intensity-duration threshold for landslides in a humid-tropical environment, Puerto Rico. *Geografiska Annaler*, **75A**(1/2), 13–23.

Lloyd, D.M., Wilkinson, P.L., Othmann, M.A. and Anderson, M.G. (2003) Predicting landslides: Assessment of an automated rainfall based landslide warning system. In: K.K.S. Ho and K.S. Li (eds), *Geotechnical Engineering – Meeting Society's Needs: Proceedings of the 14th Southeast Asian Geotechnical Conference, December 10–14, 2001, Hong Kong* (pp. 135–139). A.A. Balkema, Rotterdam.

Lumb, P. (1975) Slope failures in Hong Kong. *Quarterly Journal of Engineering Geology*, 8(1), 31–65.

Marchi, L., Arattano, M., and Deganutti, A.M. (2002) Ten years of debris-flow monitoring in the Moscardo Torrent (Italian Alps). *Geomorphology*, 46(1/2), 1–17.

Martinez, J., Avila, G., Agudelo, A., Schuster, R.L., Casadevall, T.J., and Scott, K.M. (1999) Landslides and debris flows triggered by the 6 June 1994 Paez earthquake, southwestern Colombia. In: K. Sassa (ed.), *Landslides of the World* (pp. 227–230). Japan Landslide Society, Kyoto.

Mathewson, C.C., Keaton, J.R., and Santi, P.M. (1990) Role of bedrock ground water in the initiation of debris flows and sustained post-storm stream discharge. *Bulletin of Association of Engineering Geologists*, 27(1), 73–78.

McCarter, M.K. and Kaliser, B.N. (1985) Prototype instrumentation and monitoring programs for measuring surface deformation associated with landslide processes. In: D.S. Bowles (ed.), *Proceedings of a Specialty Conference on Delineation of Landslide, Flash Flood, and Debris Flow Hazards in Utah* (pp. 30–49). Utah State University, Logan, UT.

Meyer, G.A. and Pierce, J.L. (2003) Climatic controls on fire-induced sediment pulses in Yellowstone National Park and central Idaho: A long-term perspective. *Forest Ecology and Management*, 178, 89–104.

Montgomery, D.R., Schmidt, K.M., Greenberg, H.M., and Dietrich, W.E. (2000) Forest clearing and regional landsliding. *Geology*, 28(4), 311–314.

Morrissey, M.M., Wieczorek, G.F., and Morgan, B.A. (2005) Transient hazard model using radar for predicting debris flows in Madison County, Virginia. *Environmental and Engineering Geoscience*, X(4), 285–296.

Neary, D.G. and Swift, L.W., Jr (1987) Rainfall thresholds for triggering a debris-avalanching event in the southern Appalachian Mountains. In: J.E. Costa and G.F. Wieczorek (eds), *Debris Flows/Avalanches: Process, Recognition and Mitigation* (Reviews in Engineering Geology No. 7, pp. 81–92). Geological Society of America, Boulder, CO.

Onda, Y., Mizuyama, T., and Kato, Y. (2003) Judging the timing of peak rainfall and the initiation of debris flow by monitoring runoff. In: D. Rickenmann and C-L. Chen (eds), *Debris-flow Hazards Mitigation: Mechanics, Prediction, and Assessment: Proceedings of the 3rd International Conference, Davos, Switzerland, September 10–12* (pp. 147–153). Millpress, Rotterdam.

Ortigao, J.A.R., Justi, M.G., D'Orsi, R. and Brito, H. (2003) Rio-Watch 2001: The Rio de Janeiro landslide alarm system. In: K.K.S. Ho and K.S. Li (eds), *Geotechnical Engineering – Meeting Society's Needs: Proceedings of the 14th Southeast Asian Geotechnical Conference, Hong Kong, December 10–14, 2001* (pp. 237–241). A.A. Balkema, Rotterdam.

Page, M.J., Trustrum, N.A., and Dymond, J.R. (1994) Sediment budget to assess the geomorphic effect of a cyclonic storm, New Zealand. *Geomorphology*, 9, 169–188.

Pasuto, A. and Silvano, S. (1998) Rainfall as a trigger of shallow mass movements. A case study in the Dolomites, Italy. *Environmental Geology*, 35(2/3), 184–189.

Petrucci, O. and Polemio, M. (2000) Catastrophic geomorphological events and the role of rainfalls in South-Eastern Calabria (Southern Italy). In: P. Claps and F. Siccardi (eds),

Mediterranean Storms: Proceedings of the EGS Plinius Conference, Maratea, Italy, October 14–16, 1999 (pp. 449–459). Editoriale Bios, Cosenza, Italy.

Petrucci, O. and Polemio, M. (2002) Hydrogeological multiple hazard: A characterisation based on the use of historical data. In: J. Rybár, J. Stemberk and P. Wagner (eds), *Landslides: Proceedings of 1st European Conference on Landslides, Prague, June 24–26, 2002* (pp. 269–274). A.A. Balkema, Rotterdam.

Pierson, T.C. (1999) Rainfall-triggered lahars at Mt. Pinatubo, Philipppines, following the June 1991 eruption. In: K. Sassa (ed.), *Landslides of the World* (pp. 284–289). Japanese Landslide Society, Kyoto University Press.

Pierson, T.C., Janda, R.J., Thouret, J.C., and Borrero, C.A. (1990) Perturbation and melting of snow and ice by the 13 November 1985 eruption of Nevado del Ruiz, Colombia, and consequent mobilization, flow, and deposition of lahars. *Journal of Volcanology and Geothermal Research*, **41**, 17–66.

Pollini, G., Ceriani, M., Lauzi, S., Padovan, N., and Crosta, G. (1992) Rainfall and soil slipping in Valtellina. In: D.H. Bell (ed.), *Proceedings of the 6th International Symposium on Landslides, 10–14 February, Christchurch, New Zealand* (Vol. 1, pp. 183–188). A.A. Balkema, Rotterdam.

Premchitt, J., Brand, E.W., and Chen, P.Y.M. (1994) Rain-induced landslides in Hong Kong. *Asia Engineer*, June, 43–51.

Preston, N.J. (1999) Event-induced changes in landsurface condition: Implications for subsequent slope stability. *Zeitschrift für Geomorphologie, Suppl.*, **115**, 157–173.

Pun, W.K., Wong, A.C.W., and Pang, P.L.R. (2003) A review of the relationship between rainfall and landslides in Hong Kong. In: K.K.S. Ho and K.S. Li (eds), *Geotechnical Engineering – Meeting Society's Needs: Proceedings of the 14th Southeast Asian Geotechnical Conference, December 10–14, 2001, Hong Kong* (pp. 211–216). A.A. Balkema, Rotterdam.

Reichenbach, P., Cardinali, M., De Vita, P., and Guzzetti, F. (1998) Regional hydrological thresholds for landslides and floods in the Tiber River Basin (central Italy). *Environmental Geology*, **35**(2), 146–159.

Reid, M.E. (1994) A pore-pressure diffusion model for estimating landslide-inducing rainfall. *Journal of Geology*, **102**, 709–717.

Rodolfi, G. (1997) Holocene mass movement activity in the Tosco-Romagnolo Apennines (Italy). In: J.A. Matthews, D. Brunsden, B. Frenzel, B. Gläser, and M.M. Weiß (eds), *Rapid Mass Movement as a Source of Climatic Evidence for the Holocene: Palaeoclimate Research* (Vol. 12, pp. 33–46). Gustav Fischer Verlag, Stuttgart.

Saemundsson, T., Petursson, H.G., and Decaulne, A. (2003) Triggering factors for rapid mass movements in Iceland. In: D. Rickenmann and C-L. Chen (eds), *Debris-flow Hazards Mitigation: Mechanics, Prediction, and Assessment: Proceedings of the 3rd International Conference, Davos, Switzerland, September 10–12* (pp. 167–178). Millpress, Rotterdam.

Sandersen, F. (1997) The influence of meteorological factors on the initiation of debris flows in Norway. In: J.A. Matthews, D. Brunsden, B. Frenzel, B. Gläser, and M.M. Weiß (eds), *Rapid Mass Movement as a Source of Climatic Evidence for the Holocene: Palaeoclimate Research* (Vol. 19, pp. 321–332). Gustav Fischer Verlag, Stuttgart.

Schmidt, M. and Glade, T. (2003) Linking global circulation model outputs to regional geomorphic models: A case study of landslide activity in New Zealand (Case studies from New Zealand). *Climate Research*, **25**(2), 135–150.

Schuster, R.L. (1991) Introduction. In: R.L. Schuster (ed.), *The March 5, 1987, Ecuador Earthquakes: Mass Wasting and Socioeconomic Effects* (Natural Disaster Studies No. 5, pp. 11–22). National Research Council, Washington, DC.

Schuster, R.L. and Wieczorek, G.F. (2002) Landslide triggers and types. In: J. Rybár, J. Stemberk, and P. Wagner (eds), *Proceedings of 1st European Conference on Landslides, Prague, June 24–26* (pp. 59–78). A.A. Balkema, Rotterdam.

Schuster, R.L., Wieczorek, G.F., and Hope, D.G., II (1998) Landslide dams in Santa Cruz County, California, resulting from the October 17, 1989, Loma Prieta earthquake. In: D.K. Keefer (ed.), *The October 17, 1989, Loma Prieta, California, earthquake: Landslides and Stream Channel Change* (USGS Professional Paper 1551-C, 2, C51C70). US Geological Survey, Reston, VA.

Scott, K.M. (2000) Precipitation-triggered debris flow at Casita Volcano, Nicaragua: Implications for mitigation strategies in volcanic and tectonically active steeplands. In: G.F. Wieczorek and N.D. Naeser (eds), *Debris-flow Hazards Mitigation: Mechanics, Prediction, and Assessment: Proceedings of the 2nd International Conference, Taipei, Taiwan, August 16–18* (pp. 3–13). A.A. Balkema, Rotterdam.

Scott, K.M., Macias, J.L., Naranjo, J.A., Rodriquez, S., and McGeehin, J.P. (2001) *Catastrophic Debris Flows Transformed from Landslides in Volcanic Terrains: Mobility, Hazard Assessment, and Mitigation Strategy* (USGS Professional Paper 1630, 59 pp.). US Geological Survey, Reston, VA.

Slosson, J.E. and Larson, R.A. (1995) Slope failures in Southern California: Rainfall thresholds, prediction, and human causes. *Environmental and Engineering Geoscience*, **1**(4), 393–401.

Starkel, L. (1979) The role of extreme meteorological events in the shaping of mountain relief. *Geographia Polonica*, **41**, 13–20.

Starkel, L. (1996) Geomorphic role of extreme rainfalls in the Polish Carpathians. *Studia Geomorphologica Carpatho-Balcanica*, **XXX**, 21–38.

Starkel, L. (1997) Mass movements during the Holocene: A Carpathian example and the European perspective. In: J.A. Matthews, D. Brunsden, B. Frenzel, B. Gläser, and M.M. Weiß (eds), *Rapid Mass Movement as a Source of Climatic Evidence for the Holocene: Palaeoclimate Research* (Vol. 12, pp. 385–400). Gustav Fischer Verlag, Stuttgart.

Starkel, L. and Sarkar, S. (2002) Different frequency of threshold rainfalls transforming the margin of Sikkimese and Bhutanese Himalayas. *Studia Geomorphologica Carpatho-Balcanica*, **XXXVI**, 51–68.

Terlien, M.T.J. (1997) Hydrological landslide triggering in ash-covered slopes of Manizales (Colombia). *Geomorphology*, **20**(1/2), 165–175.

Terlien, M.T.J. (1998) The determination of statistical and deterministic hydrological landslide-triggering thresholds. *Environmental Geology*, **35**(2/3), 124–130.

Terwilliger, V.J. and Waldron, L.J. (1991) Effects of root reinforcement on soil-slip patterns in the Transverse Ranges of southern California. *Geological Society of America Bulletin*, **103**, 775–785.

Tognacca, C., Bezzola, G.R., and Minor, H.-E. (2000) Threshold criterion for debris-flow initiation due to channel-bed failure. In: G.F. Wieczorek and N.D. Naeser (eds), *Debris-flow Hazards Mitigation: Mechanics, Prediction, and Assessment: Proceedings of the 2nd International Conference, Taipei, Taiwan, August 16–18* (pp. 89–97). A.A. Balkema, Rotterdam.

van Asch, T.W.J. and Buma, J.T. (1997) Modelling groundwater fluctuations and the frequency of movement of a landslide in the Terres Noires Region of Barcelonnette (France). *Earth Surface Processes and Landforms*, **22**(2), 131–142.

Waitt, R.B., Jr, Pierson, T.C., MacLeod, N.S., Janda, R.J., Voight, B., and Holcomb, R.T. (1983) Eruption-triggered avalanche, flood, and lahar at Mount St. Helens: Effects of winter snowpack. *Science*, **221**(4618), 1394–1397.

Waldron, H.H. (1967) *Debris Flow and Erosion Control Problems Caused by the Ash Eruptions of Irazú Volcano, Costa Rica* (USGS Bulletin 1241-I, 37 pp.). US Geological Survey, Reston, VA.

Wasowski, J. (1998) Understanding rainfall-landslide relationships in man-modified environments: A case-history from Caramanico terme, Italy. *Environmental Geology*, **35**(2/3), 197–209.

Wieczorek, G.F. (1987) Effect of rainfall intensity and duration on debris flows in central Santa Cruz Mountains, California. In: J.E. Costa and G.F. Wieczorek (eds), *Debris flows/Avalanches: Process, Recognition and Mitigation* (Reviews in Engineering Geology No. 7, pp. 93–104). Geological Society of America, Boulder, CO.

Wieczorek, G.F., and Jäger, S. (1996) Triggering mechanisms and depositional rates of postglacial slope-movement processes in the Yosemite Valley, California. *Geomorphology*, **15**, 17–31.

Wieczorek, G.F. and Sarmiento, J. (1983) Significance of storm intensity-duration for triggering debris flows near La Honda, California (Abstracts with Programs No. 15, pp. 5, 289). Geological Society of America, Boulder, CO.

Wieczorek, G.F. and Sarmiento, J. (1988) Rainfall, piezometric levels and debris flows near La Honda, California in the January 3–5, 1982, and other storms between 1975 and 1983. In: S.D. Ellen and G.F. Wieczorek (eds), *Landslides, Floods, and Marine Effects of the January 3–5, 1982, Storm in the San Francisco Bay Region, California* (USGS Professional Paper 1434, pp. 43–62). US Geological Survey, Reston, VA.

Wieczorek, G.F., Lips, E.W., and Ellen, S.D. (1989) Debris flows and hyperconcentrated floods along the Wasatch Front, Utah, 1983 and 1984. *Association of Engineering Geologists Bulletin*, **26**(2), 191–208.

Wieczorek, G.F., Morgan, B.A., and Campbell, R.H. (2000) Debris-flow hazards in the Blue Ridge of central Virginia. *Environmental and Engineering Geoscience*, **VI**(1), 3–23.

Wieczorek, G.F., Coe, J.A., and Godt, J.W. (2003) Debris-flow hazard assessment using remote sensing of rainfall. In: D. Rickenmann, and C-L. Chen (eds), *Debris-flow Hazards Mitigation: Mechanics, Prediction, and Assessment: Proceedings of the 3rd International Conference, Davos, Switzerland, September 10–12* (pp. 1257–1268). Millpress, Rotterdam.

Wilson, R.C. (1989) Rainstorms, pore pressures, and debris flows: A theoretical framework. In: P.M. Sadler and D.M. Morton (eds), *Landslides in Semi-Arid Environment* (Publication No. 2, pp. 101–117). Inland Geological Society, Riverside, CA.

Wilson, R.C. (2000) Climatic variations in rainfall thresholds for debris-flow activity. In: P. Claps and F. Siccardi (eds), *Proceedings of the 1st Plinius Conference on Mediterranean Storms, Maratea, Italy, October 14–16, 1999* (pp. 415–424). European Geophysical Union and Editoriale Bios.

Wilson, R.C. (2005) The rise and fall of a debris flow warning system for the San Francisco Bay region, California. In: T. Glade, M. Anderson, and M. Crozier (eds), *Landslide Hazard and Risk* (pp. 493–514). John Wiley & Sons, New York.

Wilson, R.C. and Wieczorek, G.F. (1995) Rainfall thresholds for the initiation of debris flows at La Honda, California. *Environmental and Engineering Geoscience*, **1**(1), 11–27.

Wilson, R.C., Torikai, J.D., and Ellen, S.D. (1992) *Development of Rainfall Warning Thresholds for Debris Flows in the Honolulu District, Oahu* (USGS Open-file Report 92-521, 45 pp.). US Geological Survey, Reston, VA.

Wilson, R.C., Mark, R.K., and Barbato, G.E. (1993) Operation of a real-time warning system for debris flows in the San Francisco Bay area, California. In: H.W. Shen, S.T. Su, and F. Wen (eds), *Hydraulic Engineering '93: Proceedings of the 1993 Conference, Hydraulics Division, American Society of Civil Engineers, San Francisco, CA, July 25–30* (Vol. 2, pp. 1908–1913).

Youd, T.L. and Hoose, S.N. (1978) *Historic Ground Failures in Northern California Associated with Earthquakes* (USGS Professional Paper 993, 177 pp.). US Geological Survey, Reston, VA.

Zêzere, J.L., Ferreira, A.D., and Rodrigues, M.L. (1999) The role of conditioning and triggering factors in the occurrence of landslides: A case study in the area north of Lisbon (Portugal). *Geomorphology*, **30**(1/2), 133–146.

Zêzere, J.L. and Rodrigues, M.L. (2002) Rainfall thresholds for landsliding in the Lisbon Area (Portugal). In: J. Rybár, J. Stemberk, and P. Wagner (eds), *Proceedings of 1st European Conference on Landslides, Prague, June 24–26* (pp. 333–340). A.A. Balkema, Rotterdam.

Zimmermann, M. and Haeberli, W. (1992) Climatic change and debris flow activity in high mountain areas: A case study in the Swiss Alps. *Catena, Suppl.*, **22**, 59–72.

15

Wildfire-related debris flow from a hazards perspective

Susan H. Cannon and Joseph E. Gartner

15.1 INTRODUCTION

Wildland fire can have profound effects on the hydrologic response of a watershed. Consumption of the rainfall-intercepting canopy and of the soil-mantling litter and duff, intensive drying of the soil, combustion of soil-binding organic matter, and the enhancement or formation of water-repellent soils can change the infiltration characteristics and erodibility of the soil, leading to decreased rainfall infiltration, subsequent significantly increased overland flow and runoff in channels, and movement of soil (e.g., Swanson, 1981; Spittler, 1995; Doerr et al., 2000; Martin and Moody, 2001; Moody and Martin, 2001b; Wondzell and King, 2003). Unit–area peak discharges measured following wildfire have shown between 1.45 and 870-fold increases over pre-fire rates (Moody and Martin, 2001a). Removal of obstructions by wildfire through consumption of vegetation can also enhance the erosive power of overland flow, resulting in accelerated erosion of material from hillslopes (Meyer, 2002). Increased runoff can erode significant volumes of material from channels, either by bank failure or channel bed erosion. Over longer time periods, decreased rates of evapotranspiration caused by vegetation mortality and decay of root structure may result in increased soil moisture and the loss of soil cohesion (Klock and Helvey, 1976; Swanson, 1981; Schmidt et al., 2001). Rainfall on burned watersheds thus has a high potential to transport and deposit large volumes of sediment both within and down-channel from the burned area.

Debris flows can be one of the most hazardous consequences of rainfall on burned hillslopes (e.g., Parrett, 1987; Morton, 1989; Meyer and Wells, 1997; Cannon, 2001) (Figure 15.1a, b, see color section). They pose a hazard distinct from other sediment-laden flows because of their unique destructive power. Debris flows can occur with little warning, can exert great impulsive loads on objects in their paths, and even small debris flows can strip vegetation, block drainage ways, damage structures by impact and erosion, and endanger human life. The deaths of sixteen

---

M. Jakob and O. Hungr (eds), *Debris-flow Hazards and Related Phenomena.*
© Praxis. Springer Berlin Heidelberg 2005.

people during the Christmas Day 2003 storm that impacted recently-burned hillslopes in southern California highlight the most drastic consequences of post-wildfire debris flows (Chong et al., 2004). In addition to the lives lost, US$9.5 million were spent to remove the 4.1 million cubic meters of material deposited in debris retention basins following this event. Understanding the processes that result in fire-related debris flows, the conditions under which they occur, and their size and frequency of occurrence are critical elements in effective post-fire hazard assessments.

The objective of this chapter is to provide an overview of the current understanding of post-wildfire debris-flow processes and their occurrence. We examine the physical processes by which post-wildfire debris flows initiate in different settings, over varying timescales, and in response to variable storm rainfall conditions. We describe the lithologic, soil, basin-configuration, and burn-severity conditions known to have produced debris flows following wildfire, and examine relations between the magnitude of debris-flow response and storm-rainfall conditions, lithology, basin gradient, and burn severity conditions. This chapter is intended as a review and synthesis of the published literature; in-depth examination of specific issues is available from the literature citations in this chapter.

In addition to specific findings reviewed in this chapter, we use here data from two compilations: Gartner et al. (2004) and Gartner et al. (in press). Gartner et al. (2004) assembled published data on the magnitude of post-fire debris-flow events from 95 basins throughout the western USA, and from our own monitoring efforts (Figure 15.2, see color section). This database includes estimates of debris-flow volume or peak discharge; the area, relief ratio, and the percentage burned of each debris-flow producing basin; the lithology underlying the basin; and the reported storm rainfall conditions that triggered the event. Further, Gartner et al. (in press) compiled data from the literature and from our own monitoring efforts on debris-flow initiation process; basin area and average gradient; area of basin burned at high, moderate, and low severities; underlying rock type; grain-size distribution of burned hillslope-mantling materials; triggering storm rainfall total, duration, average intensity, and, where available, the peak intensities of different durations. Although this compilation consists of data for 210 debris-flow producing basins located throughout the western USA (Figure 15.2, see color section), because of differences in data reporting, not every parameter is characterized for every basin.

These data compilations allow for the unique opportunity to examine issues related to the generation of post-wildfire debris flows across a variety of environments and under a variety of conditions, and to move from a qualitative conception of the controls on post-fire debris-flow generation to the definition of specific conditions that result in their occurrence.

15.2 FIRE-RELATED DEBRIS-FLOW INITIATION PROCESSES

Two primary processes for the initiation of fire-related debris flows have been identified in the literature: runoff-dominated erosion by surface overland flow, and infiltration-triggered failure and mobilization of a discrete landslide mass.

15.2.1 Runoff-dominated erosion by surface overland flow

The process

Debris-flow initiation in recently burned areas is most frequently attributed to significantly increased rates of rainfall runoff. Johnson (1984), Wells (1987), Spittler (1995), and Cannon et al. (2001a, 2003b) traced debris-flow deposits upslope through small gullies into a series of rills, and concluded that the debris flows initiated high on the hillslopes from material eroded by surface runoff. Wells (1987) described debris flows initiating as miniature soil slips in a saturated layer of soil a few millimetres thick above a subsurface water-repellent zone, and Gabet (2003) continued this characterization by modeling shallow failure along the lengths of hillslopes. Cannon et al. (2001a, 2003b) did not observe the presence of miniature soil slips at the rill heads as the source of the debris flows; rather, they described a progressive entrainment of hillslope materials into runoff that resulted in debris-flow conditions with travel downslope. Note that these small, hillslope-generated debris flows do not necessarily evolve into more destructive debris flows once they travel into channels (Cannon et al., 2003b).

Meyer and Wells (1997) and Cannon et al. (2001b) describe a somewhat different process focused within channels. These workers also observed that runoff high on hillslopes resulted in the generation of rills. However, convergence and concentration of flow within hollows[1] and in low-order channels resulted in considerable erosion, often to bedrock, and the transport of material downslope. Debris-flow deposits in the form of levees and lobes consisting of poorly-sorted, unstratified deposits with a fine-grained, consolidated matrix, and muddy veneers, first occurred well down the drainage network (Figure 15.3, see color section). Meyer and Wells (1997) concluded that debris flows initiated through progressive bulking of surface runoff with sediment entrained by rill erosion in steep, upper-basin slopes and from deep incision as the flows progressed down the channels. Parrett (1987) and Cannon et al. (2001b) also noted the lack of landslide scars in a burned area that experienced debris flows, and suggested a similar mechanism. Scott (1971), Parrett (1987), and Meyer and Wells (1997) emphasize that sediment input from hillslope rilling, as well as material entrained by extensive channel incision, are important in the bulking process that led to the formation of debris flows. The sheer volume of material that is often excavated from the drainage network, however, suggests that a large proportion of the material within the debris flows originated from this source.

Cannon et al. (2001, 2003b) further examined the process of progressive sediment bulking in burned areas using detailed field mapping of transitions from debris floods (see Chapter 2) to debris flow within channels. This work identified a threshold location within channels where sufficient eroded material is incorporated, relative to the volume of surface runoff, to generate debris flows that persist down the length of the channel. Above this location in a given basin, the attainment of

---

[1] As described by Hack (1965) and Reneau and Dietrich (1987), unchannelized areas on hillslopes in which the contours are concave outward away from the ridge, and that occur in the valley axis upslope from the stream head.

debris-flow conditions can be transitory; variations in sorting and grain-size distributions in the deposits indicate that the flow fluctuates between debris flow and more dilute flows before persistent debris-flow conditions are achieved. Cannon et al. (2003b) attributes the fluctuation in deposit character to an episodic sediment input to the flows, and demonstrates that these episodic fluxes increase in volume with travel down the channels. The episodic sediment contributions appear to be necessary in order to entrain sufficient material, relative to the amount of runoff, to impart debris-flow characteristics to the flow. Although Tognacca and Bezzola (1997), Tognacca et al. (2000), and Istanbulluoglu et al. (2003), describe possible theoretical frameworks and experimental and field work that characterize the erosive processes that can result in channel erosion on steep hillslopes, they do not specifically address the critical transition from sediment-laden water flow to debris flows. This is an important consideration in burned areas because given sufficient rainfall, although channel incision can occur in nearly every basin, debris flows are not necessarily generated from all incised channels.

As a variation on the process described above, debris flows have also been generated from burned basins in response to increased runoff by water cascading over a steep, bedrock cliff and incorporating material from a readily erodible bed (Larsen, 2003). Johnson and Rodine (1984) described this process in unburned terrain as the "firehose" effect.

Note that although this paper focuses exclusively on debris flows generated from burned basins, debris flows generated through what appear to be a similar process have also been described on unburned, yet primarily unvegetated, hillslopes (e.g., Scott and Williams, 1978; Johnson and Rodine, 1984; Davies et al., 1992; Coe et al., 2003). Gostner et al. (2003) outline a methodology for debris-flow hazard assessment that integrates the transition from clear-water flow to debris flow in steep, unburned channels. See also Chapter 7.

Settings and abundances

The runoff-dominated process for generating wildfire-related debris flows described above is the most frequently reported in the literature; 76% of the 210 basins documented as having produced debris flows showed this process (Gartner et al., in press) (Figure 15.4). Runoff-initiated debris flows have been observed in a variety of environments, including the northern Rocky Mountains of the US (Klock and Helvey, 1976; Meyer and Wells, 1997; Parrett et al., 2003), and British Columbia (written commun., Timothy Smith, Westrek Geotechnical Services, 2004), the central and southern Rocky Mountains (Cannon et al., 2001a, b; Meyer et al., 2001), the Uinta Mountains (Larsen, 2003), the Wasatch Front of Utah (McDonald and Giraud, 2002), north-eastern Oregon (written commun., William Russell, Oregon State University, 2003), the Huachuca Mountains of south-eastern Arizona (Wohl and Pearthree, 1991), the Mazatzal Mountains of south-central Arizona (written commun., Anne Yoberg, Arizona Geological Survey, 2004), southern California (Doehring, 1969; Wells, 1981, 1987; Morton, 1989; Booker, 1998; Cannon, 2001), and the Big Sur coast of California (Cleveland, 1973; Johnson, 1984) (Figure 15.2,

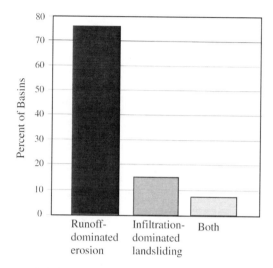

Figure 15.4. Frequency of debris-flow initiation processes for 210 debris-flow-producing basins from data in Gartner et al. (in press).

see color section). Debris flows generated through this process have also been described in the Alps of Switzerland (Conedera et al., 2003).

Post-wildfire debris flows generated through runoff bulking have not been reported in the Coastal and Cascade Mountains of the Pacific North-west (Wondzell and King, 2003). Swanson (1981) and Beschta (1990) suggest that these areas are characterized by long-duration, low-intensity rainfall, so that infiltration rates are seldom exceeded, even after intense wildfires. Wondzell and King (2003) further propose that rapid recovery of fire-caused reductions in infiltration rates, high antecedent soil moisture, and rapid rates of vegetation regrowth after fires in these regions might explain the absence of overland flow.

Rainfall conditions and time frames

Debris flows generated through runoff-dominated processes are most frequently produced in response to high-intensity, short-duration storms, and all events reported in the literature occurred within two years of the fire (Gartner et al., in press). Both winter frontal storms and summer monsoon thunderstorms have triggered debris flows by this process. Debris flows have occurred in response to storms with durations as short as 18 min (Cannon et al., 2003a), and during 10-day-long, greater-than-100-year recurrence interval storms (Scott, 1971) (Figure 15.5, filled circles). However, most debris-flow-producing storms are between about 30 min and 24 hr in duration (Figure 15.5). Although debris flows have been reported in response to average storm rainfall intensities of more than 100 mm/hr (Klock and Helvey, 1976), most debris-flow-producing storm rainfall intensities are greater than about 4 mm/hr.

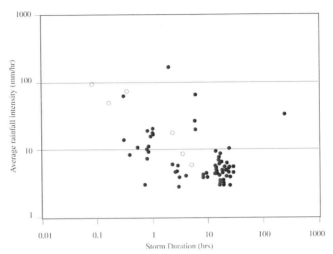

Figure 15.5. Average storm rainfall intensity and duration for debris-flow producing storms (filled circles). Open circles show storm rainfall intensities and durations leading up to debris-flow events in Colorado and southern California where time of occurrence is known.
From Gartner et al. (in press).

The rainfall conditions leading up to the time of debris-flow generation within a storm are more specific measures of the triggering event than the total storm rainfall. For the few cases where the times of debris-flow occurrence within a storm are known, we can see that debris flows have occurred after as little as 6 min of rainfall at intensities of 95 mm/hr (Cannon et al., 2003a), and up to 5 hr of rainfall at intensities of 6 mm/hr (Gartner et al., in press) (Figure 15.5, open circles).

By comparing the rainfall conditions in storms that produced debris flows from recently burned basins with those that produced sediment-laden floods or showed no response, and defining those conditions that are unique to debris-flow-producing storms, Cannon et al. (2003a) defined threshold rainfall intensity–duration conditions for runoff-dominated debris flows in the form:

$$I = 7.0 D^{-0.6} \qquad (15.1)$$

where I = rainfall intensity (in mm/hr) and D = duration of that intensity (in hours) (Figure 15.6). This threshold is a useful tool for issuing warnings and planning for emergency response in mid-latitude, temperate climate settings with steep, recently burned basins underlain by sedimentary rocks and with thin colluvial covers that experience convective thunderstorms. Additional thresholds are necessary for different environments and storm conditions.

Note that the rainfall conditions that trigger fire-related debris flows defined here are attained at durations at least an order of magnitude less than those described for the generation of debris flows in unburned settings, and at significantly lower intensities (e.g., Campbell, 1974; Caine, 1980; Larsen and Simon, 1993). This

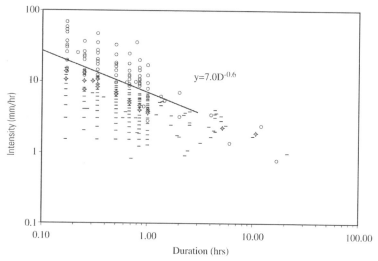

Figure 15.6. Rainfall intensity–duration threshold for the generation of fire-related debris flows from recently burned, steep basins underlain by sedimentary rocks from Cannon et al. (2003a). Open circles represent measures of storm rainfall from gauges within 1 km of basins that produced debris flows; diamonds represent measures of storm rainfall from gauges within 1 km of basins that produced sediment-laden flows; and dashes represent measures of storm rainfall from gauges within 1 km of basins that showed no response.

difference can likely be attributed to the difference between infiltration and runoff-initiation mechanisms.

How frequently will debris flows be produced from burned basins? Work by Cannon (2001) and Cannon et al. (2003a) demonstrated that basins with thin colluvial covers and minimal channel-fill deposits produced debris flows only in response to the first significant rainfall of the season, reflecting a supply-limited setting (e.g., Bovis and Jakob, 1999). In contrast, basins with thick accumulations of colluvium in the channels, and those mantled with thick talus deposits, produced repeated debris-flow events throughout the two summer monsoon seasons following the fire.

15.2.2 Infiltration-dominated landslide failure and mobilization

The process

Debris-flow initiation by failure of discrete landslide masses on hillslopes has been documented in wildfire-affected watersheds. The landslide failures observed in burned areas can range in thickness from a few tens of centimetres to more than 6 meters, and generally involve the soil and colluvium-mantled hillslopes; the failures mobilize into the muddy slurries characteristic of debris flow (Figure 15.7, see color section).

Landslides occur if shear stresses (driving forces) equal or exceed the shear strength (resisting forces) (see Chapter 4). Shear stresses are imparted by the mass of the soil and gravity. Shear strength consists of the combined resistance to movement provided by friction on the shear surface and the cohesion of the mineral soil and plant roots. The balance between driving and resisting forces depends primarily on slope steepness, pore-water pressure, and the thickness and physical characteristics of the hillslope-mantling materials. In order to attribute landslide activity to wildfires, it is necessary to consider the possible changes imparted by the fire to this balance of forces. Although three possible wildfire-related landslide-triggering effects have been proposed in the literature, there is little well-controlled data available on this subject. Increases in soil moisture after fires were measured by Klock and Helvey (1976) and Helvey (1980), who attributed these to reduced interception and transpiration rates caused by vegetation consumption and mortality. These workers postulated that the increases in soil moisture (and attendant increases in pore pressures) could promote shallow landslide failure. Although Klock and Helvey (1976) and Helvey (1980) documented the occurrence of landslides that mobilized into debris flows at their study site, the role of the increased soil moisture in the landslide failure was postulated. Wildfire-induced tree mortality can also lead to the decay of regolith-anchoring roots, and Swanson (1981) proposed that this could result in decreased soil cohesion and the increased probability of landsliding. Although the impact of logging and the associated increased soil moisture and root decay rates on slope stability has been extensively studied (see Chapter 16), a comparable body of definitive work on the impact of wildfire does not exist. As an example, Megahan (1983) measured increased pore pressures and attendant rates of subsurface flow in a watershed that had experienced both clear-cutting and wildfire, but did not detect these responses in a watershed that was burned by wildfire alone. Some physical evidence specific to wildfires that supports root-decay promoted failure is provided by DeGraff (1997), who demonstrated that the character of the sheared roots exposed along the slip plane in a shallow landslide scar within an area burned by wildfire 10 years previously demonstrated failure of decaying, rather than live, material. Finally, Wondzell and King (2003) suggest that increased peak flows that occur after fire can contribute to accelerated bank erosion, with a concurrent increase in rate of bank-side failure. Booker (1998) also describes such a process following fires in southern California.

Settings and abundances

The process of generation of debris flows exclusively by failure of discrete landslides on hillslopes has been documented in burned areas in a number of different settings, including southern California (Scott, 1971; Morton, 1989; Menitove, 1999; Cannon 2001), the Sierra Nevada of California (DeGraff, 1997), central Idaho (Megahan, 1983; Meyer et al., 2001; Shaub, 2001), and north central Washington (Klock and Helvey, 1976; Helvey, 1980). Debris-flow deposits assumed to have originated from landslides related to forest-fire activity have been described in the Oregon Coast Range by May and Gresswell (2003) and in east-central British Columbia by Sanborn et al. (2002).

Of the 210 basins included in Gartner et al. (in press), only 33 (16%) were characterized by observations of debris flows originating exclusively from landslide failures (Figure 15.4). These data came from reports of just four events: two in southern California following winter storms that impacted basins burned by fires the previous summer (Morton, 1989; Cannon, 2001), one in the Sierra Nevada of California 10 years after the fire (DeGraff, 1997), and one in response to a greater than 100-year recurrence prolonged rainstorm that culminated in a rain-on-snow event and triggered widespread flooding and landsliding throughout the north-western USA (Meyer et al., 2001; Shaub, 2001).

Note that Scott (1971), Klock and Helvey (1976), Cannon (2001), and Cannon et al. (2001b, 2003a) described evidence of both runoff and infiltration-triggered debris-flow initiation processes within individual burned basins; these events account for 8% of the sample of 210 burned basins (Figure 15.4). Importantly, Cannon et al. (2001b) found that when this was the case, considerably more material was contributed to the debris flows from hillslope runoff and channel erosion than from the landslide scars.

Rainfall conditions and timeframes

Debris flows generated from landslide failure in burned areas most frequently occur in response to prolonged periods of storm rainfall, usually of a day or more in duration, or prolonged rainfall in combination with rapid snowmelt or rain-on-snow events. Landslides have been documented as occurring during the first rainy season immediately after the fire (e.g., Morton, 1989; Cannon, 2001; Cannon et al., 2001a, b; 2003a), one to two years after the fire (Scott, 1971; Klock and Helvey, 1976; Meyer et al., 2001), and up to 10 years (DeGraff, 1997), or even 30 years (May and Gresswell, 2003) after the fire. Note that it is important to establish if landslide failure can indeed be attributed to fire, and not simply to extreme meteorological events which would have triggered failures even without the effect of the fire. This is particularly important for failures that occur after significant time periods.

Although we found no reported instances of fire-related debris-flow activity triggered by spring snowmelt alone, Gray and Megahan (1981), Meyer et al. (2001), and Shaub (2001), described debris flows generated from burned watersheds in response to rain-on-snowmelt events.

15.3 FIRE-RELATED DEBRIS-FLOW SUSCEPTIBILITY

Numerous studies have documented an increased occurrence of debris flows following wildfire (e.g., Parrett, 1987; Morton, 1989; Meyer and Wells, 1997; Cannon et al., 2003a). However, in a study of the response of 95 recently burned basins to storm rainfall, Cannon (2001) found that not all basins that experience heavy rainfall produced debris flows; only about 40% showed evidence of debris flow. The remainder of the basins showed either a sediment-laden flood response, or no response. The fact that not all burned basins produce debris flows suggests the existence of a set of geologic, geomorphic, and rainfall conditions that may indicate a susceptibility specifically to debris-flow activity following wildfires.

In the following section we describe conditions in basins known to have produced fire-related debris flows, with the expectation that debris flows can be produced in the future from basins with similar conditions. The parameters we consider include bedrock lithology, surficial materials, basin area, average basin gradient, burn extent and severity, and water repellent soils. Other conditions may certainly affect debris-flow occurrence from burned basins; here we consider these parameters as possible first-order effects that can be readily characterized within the relatively short time frames necessary for hazard assessments.

15.3.1 Bedrock lithology and surficial materials

Basins underlain by some rock types appear to be more likely to produce debris flows than others following wildfires. More than 70% of the 160 basins that generated debris flows through runoff-dominated surface erosion included in Gartner et al. (2004) are underlain by metamorphic and sedimentary rock types, in nearly equal proportions (Figure 15.8). Significantly smaller numbers of debris flows have been generated from basins underlain by volcanic, granitic, and mixed lithologies through this process. More than 90% of the basins that produced debris flows through the mobilization of discrete landslides were underlain by decomposed granite, although some landslide-triggered debris flows have also been documented in sedimentary terrains (Figure 15.8). The basins that produced debris flows through the combina-

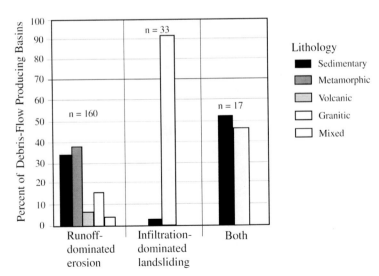

Figure 15.8. Frequency of lithologies identified by initiation process underlying 210 debris-flow producing basins included in Gartner et al. (in press).

tion of mobilization of infiltration-triggered landslides and runoff-dominated progressive sediment bulking were underlain in nearly equal proportions by granitic and sedimentary materials. Post-fire landslide activity has not yet been documented in metamorphic or volcanic terrains. Note that Spittler (1995) further suggests that the presence of highly-fractured, hard bedrock may affect debris-flow generation.

An abundance of loose, unconsolidated materials lining channels and mantling hillslopes is thought to play a role in post-fire debris-flow generation. Doehring (1969), Wells (1987), Spittler (1995), Booker (1998), Menitove (1999), and Cannon (2001b) emphasized the importance of dry-ravel[2] deposits that are stored in channels and incorporated into the passing debris flows. Further, Cannon et al. (2003a) documented fire-related debris-flow occurrence in a setting with extensive glacial deposits mantling hillslopes and infilling channels. The observation that debris flows have been generated from hillslopes and channels that do not show extensive dry-ravel deposits indicates that the presence of dry-ravel material is not a prerequisite for debris-flow generation in all settings, and suggests perhaps that its presence may affect the magnitude of the event, rather than susceptibility. At any rate, characterization of the availability of readily eroded material is a necessary element in a comprehensive hazard assessment.

The physical properties of surficial materials may certainly influence debris-flow generation from burned basins. Cannon et al. (2004) found that the sorting of the grain-size distribution of samples of burned surficial soils, soil permeability, and percentage organic matter of unburned soils were significant factors in distinguishing debris-flow-producing basins from those that showed a different response. However, Cannon (2001), found no significant differences in the proportions of fine materials or dispersion ratios of samples collected from debris-flow-producing basins and those that did not. The necessity of additional work to determine the soil properties that best indicate a propensity toward debris-flow production is clear, as is the need to put this understanding in a physical context.

15.3.2 Basin area and average gradient

Fire-related debris flows can be produced from basins with broad ranges in area and average gradient (Gartner et al., in press). Debris-flow-producing basins range in area from $0.02\,\text{km}^2$ up to $25\,\text{km}^2$; the average basin area of $2.5\,\text{km}^2$ and median of $0.65\,\text{km}^2$ indicates that most debris-flow-producing basins are the lower-order tributaries (Figure 15.9). The basins ranged in average gradient between 14 and 42 degrees (Figure 15.9). Importantly, debris flows were not observed at the outlets of basins larger than about $25\,\text{km}^2$; although debris flows may have been generated in the lower order drainages of such basins, they were not of sufficient size or energy to travel the entire length of the basin.

---

[2] Dry ravel is the process of rapid, downhill movement of individual regolith and organic particles solely under the influence of gravity, and without the effect of water (Swanson, 1981). This process occurs in response to drying of the soil and combustion of soil-binding organisms and has been observed occurring both during and after the passage of the fire.

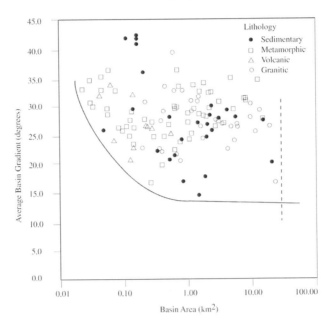

Figure 15.9. Relations between basin area and average basin gradient, relative to lithology, of basins reported to have produced fire-related debris flows. Heavy line indicates threshold conditions, above which, runoff-initiated debris flows can be expected. Dashed vertical line emphasizes that debris flows were not observed beyond the outlets of basins larger than about 25 km^2.

Threshold conditions for basin area and average gradient combinations that are most likely to produce runoff-initiated debris flows are shown as the heavy line in Figure 15.9. Basins with areas and average gradients that fall above this line are those most likely to produce this type of debris flow, given, of course, sufficient rainfall and readily erodible material. The data in Gartner et al. (in press) are not sufficient to delineate such thresholds for landslide-initiated debris flows.

The basin area and gradient characteristics of fire-related debris-flow-producing basins do not appear to vary significantly with rock type (Figure 15.9), although debris-flow generating basins underlain by volcanic rock types might be considered to be somewhat smaller than those underlain by granites. Debris-flow producing basins underlain by metamorphic and sedimentary materials show wide variations in both area and average gradient.

15.3.3 Burn extent and severity

Burn severity is a qualitative description of the effects of fire on soil hydrologic function (Miller, 1994). Areas classified as high burn severity generally exhibit complete consumption of the forest litter and duff, and combustion of all fine fuels in the canopy. A deep ash layer may be present, and the top layer of the mineral soil may be changed in color due to significant soil heating where large diameter fuels were consumed. The layer below may be blackened from charring of organic matter in the soil. Areas burned at moderate severity can be characterized by the consumption of litter and duff in discontinuous patches, and leaves or needles, although scorched, may remain on trees. Foliage and twigs are consumed, and some heating of the mineral soils may occur if the soil organic layer was thin. Areas of low burn severity may show charring of the relatively intact litter and duff, consumption of small diameter wood debris, intact fine roots within the soils, and very little effect of fire on the canopy. Essentially no soil heating occurs in this case.

Although debris flows have been produced from basins that have experienced very little, or even no, high-severity fire (Figure 15.10(a)), the area of a basin burned at a combination of high and moderate severities strongly influences debris-flow occurrence (Figure 15.10(b)); most (91%) of the fire-related debris flows included in Gartner et al. (in press) were produced from basins with more than 65% of their areas burned at a combination of high and moderate severities (Figure 15.10(b)). Further, using logistic regression statistical analyses, Cannon et al. (2004) found that the area of a basin burned at a combination of both high and moderate severity best separated debris-flow-producing basins from those that showed a different response. Work by Agee (1973) and Neary et al. (1999) demonstrates that the removal of the soil-mantling litter and duff by fire is the indicator of burn severity that most closely affects rates of post-fire runoff and erosion; the fact that the litter and duff consumption is part of both the high and moderate burn severity classification provides a physical basis for this finding.

15.3.4 Water repellent soils

Although increased rates of runoff and erosion after wildfires are frequently attributed to the enhancement or development of water-repellent soils with the passage of the fire (e.g., Doerr et al., 2000; Shakesby et al., 2000; Letey, 2001), and this increased runoff and erosion is generally assumed to influence debris-flow generation (e.g., Spittler, 1995), the role of this phenomenon relative to debris-flow occurrence has not been extensively examined. Wells (1981) and Gabet (2003) described debris-flow initiation as failure of a few millimetre thick saturated layers of soil above a subsurface water-repellent zone. However, in an evaluation of the response of 95 recently burned basins, Cannon (2001) found that debris flows that originated within low-order channels were more likely to be generated from basins without water-repellent soils than from basins with water repellency. The lack of detectable water repellency in debris-flow-producing basins led Meyer and Wells (1997) and Cannon (2001) to conclude that the properties of the bare, burned soils alone are sufficient to result in runoff sufficient to generate debris flows. These results indicate the necessity of additional work to evaluate the physical effects of this phenomenon on the generation of debris flows.

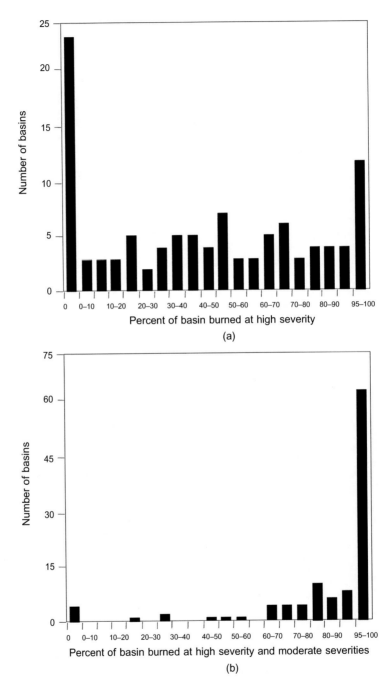

Figure 15.10. (a) Frequency distributions of percentage of basin area burned at high severity, and (b) at moderate and high severities for 108 debris-flow producing basins included in Gartner et al. (in press).

15.4 MAGNITUDE OF DEBRIS-FLOW RESPONSE

Obtaining measurements of the magnitude of the debris-flow response to fire is one of the most challenging efforts in characterizing the effects of rainfall on a basin, and the most important parameter in a hazard assessment. Debris-flow magnitude is generally characterized as either peak discharge or volume. The 61 peak discharge estimates reported from runoff-dominated debris flows included in Gartner et al. (2004) vary from 2–240 m^3/s (Figure 15.11). The 34 reported volumes range from as little as 600 m^3 (Wells, 1981) to 300,000 m^3 (Wells, 1987).

The dominance of runoff processes over infiltration processes in recently burned basins indicates that methodologies developed for unburned basins to map landslide potential (see Chapter 4) may be appropriate only in limited settings. As an alternative, relations traditionally defined between peak discharges of floods, basin characteristics, and storm rainfall can be useful in predicting the magnitude of potential debris-flow response from burned basins. The relations described below can be used to prioritize mitigation efforts in burned basins, to aid in the design of mitigation structures, and to guide decisions for warning, evacuation, shelter, and escape routes.

15.4.1 Relations between peak discharge, area of basin burned, and lithology

Figure 15.11 shows the estimates of peak discharge compiled by Gartner et al. (2004) as a function of the area of the basins burned at high and moderate severities. When

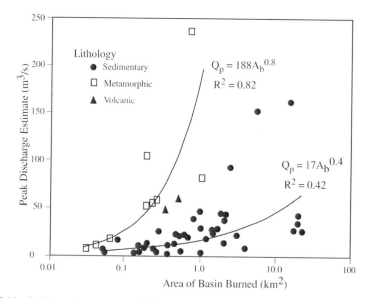

Figure 15.11. Relation between peak debris-flow discharge estimates (Q_p) and area of basins burned at high and moderate severities (A_b), identified by lithology.
From Gartner et al. (2004).

the data points are distinguished by lithology, peak discharge and burned area of basins underlain by metamorphic rock types are strongly related by the power law:

$$Q_p = 188 A_b^{0.8} \qquad (15.2)$$

where Q_p is the estimate of peak discharge in m³/s and A_b is the area burned. Due to the asymptotic form of the relation, this is probably most appropriate for basins with areas less than about 1.0 km². The relation is not as robust for basins underlain by sedimentary rock types, where:

$$Q_p = 17 A_b^{0.4} \qquad (15.3)$$

The data are not sufficient to define such a relation for basins underlain by volcanic or granitic lithologies, or using the measured volumes (a preferred measure of debris-flow magnitude) as the dependent variable. Note that these relations are based on the assumption that the bedrock lithology reflects some unknown quality of the erodibility, and thus propensity for debris-flow production, of the surficial materials.

15.4.2 Relations between peak discharge, area of basin burned, average basin gradient, and storm rainfall

Using data from Gartner et al. (2004) and multi-variate statistical analyses, Cannon et al. (2004) found that the peak discharge of debris flows issuing from the outlet of recently-burned basins could be estimated by the following relation:

$$Q_p = 171 + 0.552\theta + 2.84 \log A_b + 3.6 I \qquad (15.4)$$

where Q_p is the debris-flow peak discharge (in m³/s), θ is the average basin gradient (in percent), A_b is the area of the basin burned at all severities (in m²), and I is the average storm rainfall intensity (in mm/hr). This relation has an adjusted R^2 of 0.67 for this data set. Although volume may be a preferred measure of debris-flow magnitude, it was not possible to develop similar relations for this measure with the available data set (Cannon et al., 2004).

15.5 SUMMARY AND CONCLUSIONS

Data compiled from studies of debris-flow processes following wildfires throughout the western USA can answer some of the questions fundamental to post-fire hazard assessments – what, where, why, when, how big, and how often? Not all elements of all questions have satisfactory answers, but what follows is what can be gleaned from the preceding pages.

What and why? Fire-related debris flows have been found to initiate through two primary processes: runoff-dominated erosion by surface overland flow, and infiltration-triggered failure of a discrete landslide mass. Runoff-dominated processes are by far the most prevalent (76% of a sample of 210 basins), and occur in response to decreased infiltration and attendant increased runoff and erosion brought about by the immediate effects of the fires. Infiltration-triggered landslide activity is frequently attributed to both increased soil moisture brought about by vegetation-mortality-induced reduced transpiration rates, and root decay associated with decreases in soil cohesion.

Where? Debris flows that initiate through runoff-dominated erosion have been documented throughout the intermountain west and southern California. Basins underlain with sedimentary and metamorphic rock types with more than about 65% of their areas burned at a combination of high and moderate severities, and with areas and average gradients that fall above the threshold shown in Figure 15.9, are those most likely to produce this type of debris flow.

Debris flows generated through mobilization of infiltration-triggered landslides have been documented in southern California, the Sierra Nevada of California, Washington, Idaho, and Colorado, and in basins underlain by sedimentary and granitic rock types.

When? Runoff-initiated debris flows are produced in response to storms that occur up to two years after the fire, and often in response to the first significant rainfall of the storm season. They occur most frequently in response to storms with average intensities greater than about 4 mm/hr and between 30 minutes and 24 hours in duration. However, debris flows have occurred within a storm after as little as 6 minutes of rainfall at intensities of 95 mm/hr.

Threshold rainfall intensity–duration conditions in the form $I = 7.0D^{-0.6}$, where I = rainfall intensity (in mm/hr) and D = duration of that intensity (in hours), can be used to determine the conditions under which to expect runoff-initiated debris flows in steep, recently burned basins underlain by sedimentary rock types and with thin colluvial covers that experience convective storms and are located in temperate climate mid-latitudes.

Debris flows generated through mobilization of landslides can occur during the first rainy season immediately after the fire, and up to about 10 years after the fire. These events generally occur in response to prolonged rainfall events, and in some cases, considerably more material is contributed to the debris flows from hillslope runoff and channel erosion than from the landslide scars. The most extensive landslide events have occurred in response to week long, or multi-week storms, or prolonged rainfall in combination with rapid snowmelt or rain-on-snow events. Although these events might be among the most destructive, they occur in response to infrequent meteorologic events.

How big? Reported peak discharge estimates for runoff-initiated debris-flow events vary between 2 and 240 m^3/s and reported volumes range from as little as 600 m^3 to 300,000 m^3. Relations between peak discharge estimates and area of the basin burned at high and moderate severities for basins underlain by metamorphic rock types are defined by $Q_p = 188 A_b^{0.8}$, and in sedimentary rock types by $Q_p = 17 A_b^{0.48}$, where Q_p is the estimate of peak discharge in m^3/s and A_b is the area burned in m^2. Peak discharge for a given storm event can be estimated using the relation: $Q_p = 171 + 0.552\theta + 2.84 \log A_b + 3.6I$, where θ is the average basin gradient (in percent), A_b is the area of the basin burned at all severities (in m^2), and I is the average storm rainfall intensity (in mm/hr). It was not possible to develop similar relations for volume with the available data set.

How often? Basins with thin colluvial covers and minimal channel-fill deposits generally produce debris flows only in response to the first significant rainfall of the season. Basins with thick channel-fill deposits, and those mantled with thick accumulations of talus, frequently produce numerous debris flows throughout the rainy season.

In the absence of similar data in other settings throughout the world, the relations developed here may be appropriate for preliminary hazard assessments. However, we would expect that local conditions strongly affect debris-flow occurrence, and collection and analysis of site-specific data can only help but to improve such assessments.

15.5.1 Future research needs

Neither the progressive sediment bulking or shallow landsliding process for debris-flow generation described in this chapter is well understood in the context of burned areas. The mechanics of generation of debris flow through progressive sediment bulking, with a focus on the transition from water flow to debris flow, can benefit from examination through a combination of theoretical, experimental, and field work. Although possible wildfire-related landslide-triggering effects have been proposed in the literature, there is little well-controlled data available on this subject. Quantitative examination of the effects of decreased transpiration, root decay, and revegetation following fires is necessary to define the role of post-fire landsliding in the generation of debris flows. It will be necessary to determine if landslide activity over long time frames can indeed be attributed to wildfires, or simply extreme meteorologic events.

The discussion of post-fire debris-flow susceptibility presented above has focused on examining the univariate effects of a limited number of parameters, while susceptibility is a multi-variate issue, and factors other than those examined here may well affect debris-flow occurrence. Further work is necessary to identify those conditions that best separate debris-flow-producing basins from those that do not produce debris flows, and to develop relations that characterize the combined effects of these variables on debris-flow susceptibility. Variations in storm-rainfall patterns, material properties, and other effects will most certainly require that these relations be region specific. For example, the fire and debris-flow response history of basins that burn frequently may be an important factor in debris-flow susceptibility in southern California, where fires are a frequent occurrence.

To develop relations that better characterize post-fire debris-flow susceptibility, the physical role of water-repellency in the generation of post-fire debris flows needs to be better understood. Further, most fires show an extremely patchy mosaic of burn severity, and little attention has focused on examining the effects of variations within the spatial distribution of the fire within a basin on debris-flow generation. Lastly, there is a need to identify specifically those soil properties that indicate propensity to debris-flow production, and put this understanding in a physical framework.

Because measures of debris-flow volume are preferable to those of peak discharge for quantifying debris-flow magnitude, we suggest that such measures be systematically collected and catalogued so that relations between volume and burned area, basin gradient, rainfall intensity, and other controlling variables can be developed.

Definition of rainfall conditions that can potentially lead to post-wildfire debris flows is a critical element in any hazard assessment, and there is a need for definition of such conditions in settings other than those described here. Collection and analysis of information of the times of debris-flow occurrences within a storm will provide an invaluable addition to such an analysis.

Finally, definition of the locations and potential volumes of sediment sources within burned areas is a critical element in hazard assessments. This requires the development of rigorous methodologies for characterizing the amount of material stored in channels and available from hillslopes in order to define the magnitude of the potential hazards posed by post-wildfire debris flow.

We are just in the first stages of understanding the effects of wildfire on debris-flow processes. The extensive recent fires in western North America can provide significant opportunities to move beyond the empirical evaluations presented here to develop an improved understanding of the physical controls on this hazardous phenomenon, and to develop useful and appropriate tools and methodologies for characterizing the hazards.

15.6 REFERENCES

Agee, J.K. (1973) *Prescribed Fire Effects on Physical and Hydrologic Properties of Mixed-conifer Forest Floor and Soil* (Contribution Report No. 143, 57 pp.). University of California Resources Center, Davis, CA.

Beschta, R.L. (1990) Effects of fire on water quantity and quality. In: J.S. Walsad, S.R. Radosevich, and D.V. Sandberg (eds), *Natural and Prescribed Fire in the Pacific Northwest Forests* (pp. 219–231). Oregon State University Press, Corvallis, OR.

Booker, F.A. (1998) Landscape and management response to wildfires in California. MSc thesis, University of California, Berkeley.

Bovis, M.J. and Jakob, M. (1999) The role of debris supply conditions in predicting debris flow activity. *Earth Surface Processes and Landforms*, **24**, 1039–1054.

Caine, N. (1980) The rainfall intensity-duration control of shallow landslides and debris flows. *Geografisk Annaler*, **62A**, 23–37.

Campbell, R.H. (1974) *Soil Slips, Debris Flows, and Rainstorms in the Santa Monica Mountains and Vicinity, Southern California* (USGS Professional Paper 851). US Geological Survey, Reston, VA.

Cannon, S.H. (2001) Debris-flow generation from recently burned watersheds. *Environmental and Engineering Geoscience*, **7**, 321–341.

Cannon, S.H., Bigio, E.R., and Mine, E. (2001a) A process for fire-related debris-flow initiation, Cerro Grande Fire, New Mexico. *Hydrological Processes*, **15**, 3011–3023.

Cannon, S.H., Kirkham, R.M., and Parise, M. (2001b) Wildfire-related debris-flow initiation processes, Storm King Mountain, Colorado. *Geomorphology*, **39**, 171–188.

Cannon, S.H., Gartner, J.E., Holland-Sears, A., Thurston, B.M. and Gleason, J.A. (2003a) Debris-flow response of basins burned by the 2002 Coal Seam and Missionary Ridge fires, Colorado. In: D.D. Boyer, P.M. Santi, and W.P. Rogers (eds), *Engineering Geology in Colorado: Contributions, Trends, and Case Histories* (AEG Special Publication 14, on CD-ROM). Association of Engineering Geologists.

Cannon, S.H., Gartner, J.E., Parrett, C., and Parise, M. (2003b) Wildfire-related debris flow generation through episodic progressive sediment bulking processes, western U.S.A. In: D. Rickenmann and C-L. Chen (eds), *Proceedings of 3rd International Conference on Debris-flow Hazards Mitigation: Mechanics, Prediction, and Assessment, September 10–12, Davos, Switzerland* (pp. 71–82). Millpress, Rotterdam.

Cannon, S.H., Gartner, J.E., Rupert, M.G., and Michael, J.A. (2004) *Emergency Assessment of Debris-flow Hazards from Basins Burned by the Cedar and Paradise Fires of 2003, Southern California* (USGS Open-File Report 04-1011). US Geological Survey, Reston, VA.

Chong, J., Renaud, J., and Ailsworth, E. (2004) Flash floods wash away lives, dreams. *Los Angeles Times* (January 3, 2004, p. B.1).

Cleveland, G.B. (1973) Fire + rain = mudflows, Big Sur. *California Geology*, **26**, 127–135.

Coe, J.A., Godt, J.W., Parise, M., and Moscariello, A. (2003) Estimating debris-flow probability using fan stratigraphy, historic records, and drainage-basin morphology, Interstate 70 highway corridor, central Colorado, USA. In: D. Rickenmann and C-L. Chen (eds), *Proceedings of 3rd International Conference on Debris-flow Hazards Mitigation: Mechanics, Prediction, and Assessment, September 10–12, Davos, Switzerland* (pp. 1085–1096). Millpress, Rotterdam.

Conedera, M., Peter, L., Marxer, P., Forster, F., Rickenmann, D., and Re, L. (2003) Consequences of forest fires on the hydrogeological response of mountain catchments: A case study of the Riale Buffaga, Ticino, Switzerland. *Earth Surface Processes and Landforms*, **28**, 117–129.

Davies, T.R., Phillips, C.J., Pearce, A.J., and Zhang, X.B. (1992) Debris flow behavior: An integrated overview. In: *Erosion, Debris Flows, and Environment in Mountain Regions: Proceedings of the Chengdu Symposium, July* (IAHS Publ. No. 209). International Association of Hydrological Sciences, Christchurch, New Zealand.

DeGraff, J.V. (1997) *Geologic Investigation of the Pilot Ridge Debris Flow, Groveland Range District, Stanislaus National Forest* (USDA Forest Service, 20 pp.). US Department of Agriculture, Washington, DC.

Doehring, D.O. (1969) The effect of fire on geomorphic processes in the San Gabriel Mountains, California. *Contributions to Geology*, **7**, 43–65.

Doerr, S.H., Shakesby, R.A., and Walsh, R.P.D. (2000) Soil water repellency: Its causes, characteristics and hydro-geomorphological significance. *Earth-Science Reviews*, **51**, 33–65.

Gabet, E.J. (2003) Post-fire thin debris flows: Sediment transport and numerical modeling. *Earth Surface Processes and Landforms*, **28**, 1341–1348.

Gartner, J.E., Bigio, E.R., and Cannon, S.H. (2004) *Compilation of Post-wildfire Runoff-event Data from the Western United States* (USGS Open-File Report 2004-1085). US Geological Survey, Reston, VA.

Gartner, J.E., Cannon, S.H., Bigio, E.R., Davis, N.K., McDonald, C., Pierce, K.L., Rupert, M.G. (in press) *Compilation of Basin Morphology, Burn Severity, Soils and Rock Type, Erosive Response, Debris-flow Initiation Process, and Event-triggering Rainfall for 599*

Recently Burned Basins in the Western U.S. (USGS Open-File Report). US Geological Survey, Reston, VA.

Gostner, W., Bezzola, G.R., Schatzmann, M., and Minor, H.E. (2003) Integral analysis of debris flow in Alpine torrent: The case study of Tschengls. In: D. Rickenmann and C-L. Chen (eds), *Proceedings of 3rd International Conference on Debris-flow Hazards Mitigation: Mechanics, Prediction, and Assessment, September 10–12, Davos, Switzerland* (pp. 1129–1140). Millpress, Rotterdam.

Gray, D.H. and Megahan, W.F. (1981) *Forest Vegetation Removal and Slope Stability in the Idaho Batholith* (USDA Forest Service Research Paper INT-271, 23 pp.). US Department of Agriculture, Washington, DC.

Hack, J.T. (1965) *Geomorphology of the Shenandoah Valley, Virginia and West Virginia, and Origin of the Residual Ore Deposits* (USGS Professional Paper 484). US Geological Survey, Reston, VA.

Helvey, J.A. (1980) Effects of a north central Washington wildfire on runoff and sediment production. *Water Resources Bulletin*, **16**, 627–634.

Istanbulluoglu, E.D.G., Tarbotton, R.T., Pack, R.T., and Luce, C.H. (2003) A sediment transport model for incision of gullies on steep topography. *Water Resources Research*, **39**, 1103.

Johnson, A.M. (1984) with contributions by J.R. Rodine. Debris flow. In: D. Brunsden and D.B. Prior (eds), *Slope Instability* (pp. 257–361). John Wiley & Sons, New York.

Klock, G.O. and Helvey, J.D. (1976) Debris flows following wildfire in North Central Washington. *Proceedings of the 3rd Federal Inter-Agency Sedimentation Conference, March 22–25, Denver, Colorado* (pp. 91–98). Water Resources Council, Denver, CO.

Larsen, I.J. (2003) From the rim to the river: The geomorphology of debris flows in the Green River Canyons of Dinosaur National Monument, Colorado and Utah (196 pp.). MSc thesis, Utah State University, Logan, UT.

Larson, M.C. and Simon, A. (1993) A rainfall intensity-duration threshold for landslides in a humid-tropical environment, Puerto Rico. *Geografiska Annaler*, **75A**, 13–23.

Letey, J. (2001) Causes and consequences of fire-induced soil water repellency. *Hydrological Processes*, **15**, 2867–2875.

Martin, D.A. and Moody, J.A. (2001) Comparison of soil infiltration rates in burned and unburned mountainous watersheds. *Hydrological Processes*, **15**, 2893–2903.

May, C.L. and Gresswell, R.E. (2003) Processes and rates of sediment and wood accumulation in headwater streams of the Oregon Coast Range, USA. *Earth Surface Processes and Landforms*, **28**, 409–424.

McDonald, G.N. and Giraud, R.E. (2002) *September 12, 2002, Fire-related Debris Flows East of Santiaquin and Spring Lake, Utah County, Utah* (Technical Report 02-09, 15 pp.). Utah Geological Survey, Salt Lake City, UT.

Megahan, W.F. (1983) Hydrologic effects of clearcutting and wildfire on steep granitic slopes in Idaho. *Water Resources Research*, **19**, 811–819.

Menitove, A. (1999) Wildfire related debris-flow susceptibility in the Santa Monica Mountains, Los Angeles and Ventura Counties, California. MSc thesis, Colorado School of Mines, Golden, CO.

Meyer, G.A. (2002) *Fire in Western Conifer Forests: Geomorphic and Ecologic Processes and Climatic Drivers* (Abstracts with Programs 34, p. 46). Geological Society of America, Boulder, CO.

Meyer, G.A. and Wells, S.G. (1997) Fire-related sedimentation events on alluvial fans, Yellowstone National Park, U.S.A. *Journal of Sedimentary Research*, **67**, 776–791.

Meyer, G.A., Pierce, J.L., Wood, S.L., and Jull, A.J.T. (2001) Fire, storms and erosional events in the Idaho Batholith. *Hydrological Processes*, **15**, 3025–3038.

Miller, M. (1994) Fire behavior and characteristics. In: M. Miller (ed.), *Fire Effects Guide*. National Wildfire Coordinating Group, Boise, ID.

Moody, J.A. and Martin, D.A. (2001a) Post-fire, rainfall intensity–peak discharge relations for three mountainous watersheds in the western USA. *Hydrological Processes*, **15**, 2981–2993.

Moody, J.A. and Martin, D.A. (2001b) Initial hydrologic and geomorphic response following a wildfire in the Colorado Front Range. *Earth Surface Processes and Landforms*, **26**, 1049–1070.

Morton, D.M. (1989) Distribution and frequency of storm-generated soil slips on burned and unburned slopes, San Timoteo Badlands, Southern California. In: P.M. Sadler and D.M. Morton (eds), *Landslides in a Semi-arid Environment with Emphasis on the Inland Valleys of Southern California* (Publication No. 2, pp. 279–284). Inland Geological Society, Riverside, CA.

Neary, D.G., Klopatek, C.C., DeBano, L.F., and Ffolliott, P.F. (1999) Fire effects on below-ground sustainability: A review and synthesis. *Forest Ecology and Management*, **122**, 51–71.

O'Loughlin, C.L. (1974) The effect of timber removal on the stability of forest soils. *New Zealand Journal of Hydrology*, **13**, 121–123.

Parrett, C. (1987) *Fire-related Debris Flows in the Beaver Creek Drainage, Lewis and Clark County, Montana* (USGS Water-Supply Paper 2330, pp. 57–67). US Geological Survey, Reston, VA.

Parrett, C., Cannon, S.H., and Pierce, K.L. (2003) *Wildfire-related Floods and Debris Flows in Montana in 2001* (USGS Water-Resources Investigations Report 03-4319). US Geological Survey, Reston, VA.

Reneau, S.L. and Dietrich, W.E. (1987) The importance of hollows in debris flow studies: Examples from Marin County. In: J.E. Costa and G.F. Wieczorek (eds), *Debris Flows/ Avalanches: Process, Recognition, and Mitigation* (Reviews in Engineering Geology No. VII, pp. 165–180). Geological Society of America, Boulder, CO.

Sanborn, P. Jull, T.J., and Hawkes, B. (2002) *Holocene Fire History and Slope Processes in an Inland Temperate Rainforest, East-central British Columbia, Canada* (Abstracts with Programs No. 34, p. 319). Geological Society of America, Boulder, CO.

Schmidt, K.M., Roering, J.J., Stock, J.D., Dietrich, W.E., Montgomery, D.R. and Schuab, T. (2001) The variability of root cohesion as an influence on shallow landslide susceptibility in the Oregon Coast Range. *Canadian Geotechnical Journal*, **38**, 995–1024.

Scott, K.M. (1971) *Origin and Sedimentology of 1969 Debris Flows near Glendora, California* (USGS Professional Paper 750, pp. C242–C247). US Geological Survey, Reston, VA.

Scott, K.M. and Williams, R.P. (1978) *Erosion and Sediment Yields in the Transverse Ranges, Southern California* (USGS Professional Paper 1030). US Geological Survey, Reston, VA.

Shakesby, R.A., Doerr, S.H., and Walsh, R.P.D. (2000) The erosional impact of soil hydrophobicity: Current problems and future research directions. *Journal of Hydrology*, **231/ 232**, 178–191.

Shaub, S. (2001) Landslides and wildfire: An example from the Boise National Forest. MSc thesis, Boise State University, ID.

Spittler, T.E. (1995) Fire and the debris flow potential of winter storms. In: J.E. Keeley and T. Scott (eds), *Brushfires in California Wildlands: Ecology and Resource Management* (pp. 113–120). International Association of Wildland Fire, Fairfield, WA.

Swanson, F.J. (1981) Fire and geomorphic processes. In: H.A. Mooney, T.H. Bonniksen, N.L. Christensen, J.E. Lotan, and W.A. Reiners (eds), *Fire Regimes and Ecosystem Properties* (USDA Forest Service General Technical Report WO-26, pp. 401–420). US Department of Agriculture, Washington, DC.

Swanston, D.N. (1975) *Slope Stability Problems Associated with Timber Harvesting in Mountainous Regions of the Western United States* (USDA Forest Service General Technical Report PNW-21, 14 pp.). US Department of Agriculture, Washington, DC.

Tognacca, C. and Bezzola, G.R. (1997) Debris-flow initiation by channel-bed failure. In: C-L. Chen (ed), *Debris-flow Hazards Mitigation: Mechanics, Prediction, and Assessment: Proceedings of 1st International Conference, San Francisco, California, August 7–9* (pp. 44–53). American Society of Civil Engineers, New York.

Tognacca, C., Bezzola, G.R., and Minor, H.E. (2000) Threshold criterion for debris-flow initiation due to channel bed failure. In: G.F. Wieczorek and N.D. Nasser (eds), *Debris-flow Hazards Mitigation: Mechanics, Prediction, and Assessment* (pp. 89–97). A.A. Balkema, Rotterdam.

Wells, W.G., II (1981) Some effects of brushfires on erosion processes in coastal Southern California. *Erosion and Sediment Transport in Pacific Rim Steeplands, Christchurch, New Zealand* (Vol. 132, pp. 305–342). International Association of Hydrological Sciences, Christchurch, New Zealand.

Wells, W.G., II (1987) The effects of fire on the generation of debris flows in southern California. In: J.E. Costa and G.F. Wieczorek (eds), *Debris Flows/Avalanches: Process, Recognition, and Mitigation* (Reviews in Engineering Geology No. VII, pp. 105–113).

Wohl, E.E. and Pearthree, P.P. (1991) Debris flow geomorphic agents in the Huachuca Mountains of southeastern Arizona. *Geomorphology*, **4**, 273–292.

Wondzell, S.M. and King, J.G. (2003) Post-fire erosional processes in the Pacific Northwest and Rocky Mountain Regions. *Forest Ecology and Management*, **178**, 75–87.

Ziemer, R.R. (1981) *Roots and the Stability of Forested Slopes* (IAHS No. 132, pp. 343–357). International Association of Hydrological Sciences, Christchurch, New Zealand.

16

Influence of forest harvesting activities on debris avalanches and flows

Roy C. Sidle

16.1 INTRODUCTION

16.1.1 Environmental damage

Debris flows and debris avalanches are natural geomorphic phenomena in steep forested terrain. However, their timing, magnitude, and spatial extent can be altered by forest logging practices (e.g., Jakob, 2000; Sidle, 2000; Guthrie, 2002). While the consequences of these mass soil movements in remote forest environments are often not seriously considered, particularly in the developing world, the encroachment of urban centres and residential settlements into potentially hazardous impact zones has posed recent concerns in many regions (e.g., Johnson et al., 1982; Smyth and Royle, 2000; Sidle and Chigira, 2004). Furthermore, debris flows inflict damage to valuable aquatic habitats and streams (e.g., Lamberti et al., 1991; Gomi et al., 2002). Exposed debris avalanche and debris-flow tracks are generally subject to recurring surface erosion for many years after the mass failure, which further reduces forest productivity (e.g., Crozier et al., 1980). In the Pacific Northwest of North America, alder (*Alunus rubra*) corridors along headwater channels are common legacies of debris-flow occurrence more than 50 years ago (Figure 16.1). The rate of site recovery on mass movement scars is influenced by the area of the scour zone, influx rates of soil and organic matter (into the scar), weathering of parent material or exposed substrate, availability of natural seed sources, nature of the substrate, colonization by nitrogen-fixing vegetation, and available soil water (Harris, 1967; Adams and Sidle, 1987; Shimokawa et al., 1989; Aust et al., 1997). Studies on the recovery of landslide sites in the tropics and subtropics are lacking, but high rates of rainfall and weathering appear to promote more rapid recovery than in temperate zones; thus landslides may reoccur at the same site in periods ranging from less than one-hundred years to several hundred years (Shimokawa et al., 1989).

M. Jakob and O. Hungr (eds), *Debris-flow Hazards and Related Phenomena.*
© Praxis. Springer Berlin Heidelberg 2005.

Figure 16.1. Alder (*Alunus rubra*) corridors along headwater channels on Prince of Wales Island, southeast Alaska, mark disturbances caused by debris flows that occurred approximately 50 years prior to the photograph.

The interactions between forest harvesting and mass wasting have received particular attention due to the potential for forest removal to reduce rooting strength and, in some cases, change the hydrologic regime. Additionally, roads and skid trails needed for forest management are frequent sites of landslides and debris flows. Especially throughout the Pacific Rim, widespread forest conversion and clear-cutting in the past two centuries coupled with high and intense rainfall, mountainous terrain, active tectonics and volcanism, and vulnerable lithologies has accelerated mass wasting activity (Sidle et al., 1985, 2004). As a result, sediment transport in streams and rivers increases, changing channel form and aquatic habitat.

16.1.2 Triggering mechanisms

In many forested areas, rainfall induced landslides prevail; however, seismically active regions such as Papua New Guinea, Taiwan, China, and Japan have experienced widespread mass movements during major earthquakes, especially during the rainy season (Garwood et al., 1979; Fukuda and Ochiai, 1993; Cheng et al., 2000). Taiwan has particularly unstable terrain due to large earthquakes, extensive tectonic uplift, past conversion of forest land, cultivation of steep slopes, high rainfall, and active bank erosion by rivers and streams. Worldwide, debris avalanches and debris flows triggered by rainfall produce more widespread, albeit less spectacular environmental damage than those triggered by earthquakes (Sidle et al., 1985; Sidle and

Dhakal, 2002). Even in Japan, where the government has made extensive investments in countermeasures against landslides and debris flows, casualties attributed to such sediment disasters are still around 100 per year (on average), with the majority of these caused by rainfall initiated landslides and debris flows (Nunamoto et al., 1999). The recent devastating landslides and debris flows in southern Kyushu are an example (Sidle and Chigira, 2004).

This chapter focuses on: (1) how forests enhance slope stability; (2) the interaction of forest ecosystem processes with debris avalanche and debris-flow dynamics; and (3) the effects of timber harvesting and land conversion activities. Practical measures for ameliorating adverse effects of forest management practices are addressed. Forest roads and skid trails, while a major concern related to mass wasting, are discussed only briefly in this chapter. Readers are referred to the following papers that deal with forest road and skid trail effects on the stability of forest terrain: Burroughs et al. (1976); Megahan (1977, 1987); Sidle et al. (1985); Coker and Fahey (1993); Larsen and Parks (1997); Wemple et al. (2001).

16.2 HOW DO FORESTS ENHANCE SLOPE STABILITY?

Forest vegetation provides two major benefits to the stability of slopes: (1) modifies the soil moisture regime through evapotranspiration processes; and (2) provides root cohesion to the soil mantle (Gray and Megahan, 1981; Greenway, 1987; Phillips and Watson, 1994). The first factor is not particularly important for shallow debris avalanches and debris flows that occur during an extended rainy season, except possibly in the tropics and sub-tropics where evapotranspiration is high throughout the year. In temperate regions, soils are near saturation and evapotranspiration is low during autumn and winter rainstorms when slope failures frequently occur, thus, soil water content is only minimally affected by forest vegetation (Sidle et al., 1985). However, evapotranspiration, or the lack thereof, could alter the "window of susceptibility" for shallow debris avalanches and debris flows if a large storm occurred near the beginning or end of the rainy season (Megahan, 1983; Sidle et al., 1985). Recent findings in the Pacific Northwest indicate that reduced evapotranspiration following logging may increase pore-water pressures during winter rainstorms, especially for moderate storms (Dhakal and Sidle, 2004). Also, when large and intense storms occur during drier conditions, evapotranspiration can be a factor related to shallow debris avalanche and debris-flow initiation. Forests in wet tropical regions likely play a more significant role in altering soil moisture content since evapotranspiration is high throughout the year (Greenway, 1987). Vegetation rooting depth in deep unstable soils is an important control on soil water depletion. Deeper-rooted vegetation species can sustain maximum transpiration rates for greater durations, thus drying the soil at greater depths compared to shallow-rooted vegetation (McNaughton and Jarvis, 1983). Such mechanisms would only be important in cases where debris avalanches and flows initiate within deep soil mantles.

Trees and shrubs influence soil hydrologic properties by providing organic matter to the forest floor. The physical role of ground cover is important to slope stability only because it affects the infiltration capacity of the soil and the interception and ultimate evaporation of rainwater and snow. Forest fires consume surface organic matter and may create a shallow hydrophobic layer that promotes overland flow; however, recent studies have noted patchy patterns of fire-induced runoff may be significant at the local scale, but not necessarily at the catchment scale (Prosser and Williams, 1998; Cannon and Reneau, 2000). At forest sites where the balance between organic matter supply to the forest floor and outputs (decomposition and export) are in a delicate balance, forest floor disturbances can alter hydrologic flow paths. Because accumulated organic litter on the forest floor enhances infiltration capacity, the maintenance of high rates of recharge would appear to decrease the stability of hillslopes. In spite of such enhanced infiltration, many unstable sites have well-developed tension cracks that facilitate rapid and deep percolation, thus maintaining high recharge rates over the entire slope will generally be beneficial as peak water inputs through the soil are dispersed (Sidle et al., 1985).

A potentially more important effect of trees is related to interception and subsequent evaporation of rainfall and snow. Water intercepted by canopies will reduce inputs to the soil and potentially reduce the potential for debris avalanches and debris flows. During rainstorms any combination of wet forest canopies, high rainfall intensities, long storm durations, and high wind velocities all tend to minimize interception losses. In the tropics and subtropics, groundwater levels may be more augmented by forest clearance because of higher evaporation losses. Conversion of forest cover to agriculture crops and plantations will undoubtedly have longer term consequences for slope stability compared to timber harvesting (with subsequent regeneration) due to permanent changes in canopy structure and related interception.

In areas with high snowfall, snow intercepted by and evaporated from canopies of dense conifer stands will reduce the recharge of meltwater into the soil mantle (e.g., Golding and Swanson, 1986). While this canopy–snow interaction is important in controlling runoff from high-elevation forested catchments (e.g., Troendle and King, 1985), the effects of snow evaporation from canopies on debris avalanche and debris-flow initiation has only been speculated (Sidle et al., 1985). The removal of extensive areas of forest canopy by clear-cutting or forest conversion allows for more snow to directly reach the forest floor without first being subject to evaporation from the canopy (i.e., interception losses) (Berris and Harr, 1987). Subsequent accumulations of deeper snowpacks can potentially raise shallow groundwater levels during snowmelt periods, thereby increasing the probability of debris avalanches and debris flows.

The contribution of vegetation roots to soil shear strength is generally recognized as more important in stabilizing hillslopes (Wu et al., 1979; Gray and Megahan, 1981; Phillips and Watson, 1994) compared to effects of vegetation on site hydrology. Field investigations in steep forested terrain worldwide have noted a two- to greater than tenfold increase in rates of mass erosion 3–15 years after timber harvesting (Bishop and Stevens, 1964; Endo and Tsuruta, 1969; O'Loughlin and

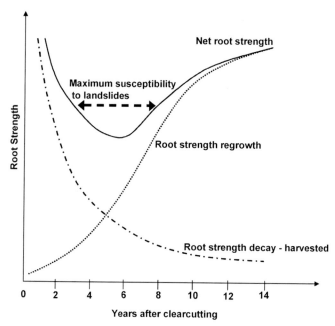

Figure 16.2. An example of typical changes in forest vegetation rooting strength after timber harvesting (clear-cutting). Root decay and recovery curves are based on numerous data worldwide compiled by Sidle (1991, 1992). Net rooting strength is the sum of the decay and recovery curves.

Pearce, 1976; Megahan et al., 1978; Wu and Sidle, 1995; Jakob, 2000). This increase in landslide frequency and volume is related to the period of minimum rooting strength after clear-cut harvesting and prior to substantial regeneration (Figure 16.2). Tests of the effects of timber harvest on rooting strength based on mechanical straining of roots (Burroughs and Thomas, 1977; Ziemer and Swanston, 1977; Wu et al., 1979; Abe and Ziemer, 1991) have confirmed these empirical observations. During storms or snowmelt periods when hillslope soils are in a tenuous state of equilibrium, reinforcement from tree roots may provide the critical difference between stable and unstable sites, especially when soils are partly or completely saturated (Sidle, 1992). Shallow landslides occurred during the first major storm after clear-cutting in stands of shallow-rooted Sitka spruce (*Picea sitchensis*) – western hemlock (*Tsuga heterophylla*) in coastal Alaska (Sidle, 1984). Here, cable yarding extensively disturbed the root-reinforced organic horizon exposing weaker mineral soil.

The stability of shallow hillslope soils is much more influenced by vegetation rooting strength than deeper soil mantles (Sidle et al., 1985). In shallow soils, roots may penetrate the entire soil mantle and provide vertical anchors into the more stable substrate (Wu et al., 1979; Gray and Megahan, 1981; Greenway, 1987). Dense lateral root systems in the upper soil horizons provide a membrane of strength that stabilizes the soil (Sidle et al., 1985). This membrane is much more significant in protecting against shallow landslides compared to deep-seated mass movements (Swanston and Swanson, 1976). Tree roots may lend some stability to deeper soils by lateral reinforcement across planes of weakness (Swanson and Swanston, 1977; Schroeder, 1985); however, this beneficial effect diminishes with increasing size (depth and area) of the potential failure site.

16.3 INTERACTION OF FOREST ECOSYSTEMS WITH GEOMORPHIC PROCESSES

Complex linkages exist between shallow debris avalanches and debris slides on hillslopes and debris flows that initiate in forested headwater channels (Gomi et al., 2002). Both the mechanics and processes controlling these interrelated mass movements have been investigated separately, thus the linkages have not been satisfactorily clarified, particularly for forested sites where inputs of wood and the structure of riparian vegetation play important roles. Debris avalanches and debris flows can be distinguished on the basis of the higher fluidity and pore water content of the latter (e.g., Iverson et al., 1997). Nonetheless, linkages between landslides and debris flows in steep, forested terrain encompass a more complex set of site characteristics, including terrain and vegetation roughness, woody debris inputs and dams in channels, and channel tributary junctions (e.g., Benda and Cundy, 1990; Gomi et al., 2001, 2002) that cannot be attributed to rheology alone. Such geomorphic and biological heterogeneities in headwater ecosystems complicate the magnitude–frequency relationships of in-channel debris flows in steep forested terrain. For example, small woody debris dams, typical of second-growth and recently clear-cut forest streams, may store considerable quantities of bedload sediment; however, these structures may collapse during debris flows releasing additional sediment downstream (Gomi and Sidle, 2003). In contrast, debris dams composed of larger wood (such as in old-growth forest streams) may survive small debris flows and hyperconcentrated sediment flows (Gomi and Sidle, 2003), thus affecting the temporal release of sediment to downstream environments. The long-term supply of wood to streams can be greatly influenced by timber harvesting strategies, particularly in riparian corridors (Bilby and Ward, 1991; Gomi et al., 2002).

Debris flows that occur in forested headwaters can be categorized into two types based on triggering conditions: Type 1, those that occur in concert with debris slides and debris avalanches; and Type 2, those that occur independently of such landslides (or at least partially independently). In the first category, the debris slides and debris avalanches mobilize rapidly into debris flows and continue downslope to a channel (e.g., Rapp and Nyberg, 1981; Benda, 1990). While such phenomena occur in steep forested terrain, due to the surface roughness caused by forest vegetation, these long-travelling failures are uncommon unless the displaced mass is quite large and moves on continuously steep slopes (Figure 16.3). Similarly, debris slides and avalanches may enter a headwater channel where, due to the increased water supply, a debris flow is immediately triggered (e.g., Costa, 1991; Palacios et al., 2003; Sidle and

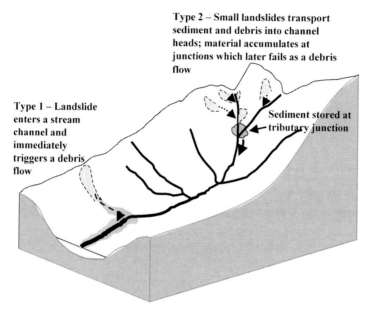

Figure 16.3. Examples of two types of debris-flow conditions that predominate in steep headwaters: Type 1, landslides immediately or rapidly mobilize into debris flows and travel downstream; Type 2, sediment from smaller hillslope landslides accumulates in channel heads or at headwater tributary junctions, later to be mobilized into a debris flow once sufficient material has accumulated and in conjunction with a storm event.

Chigira, 2004) (Figure 16.3). For this simpler type of combination debris avalanche–debris flow, the geomorphic threshold is related to the return interval of the triggering storm or snowmelt event (e.g., Caine, 1980; Sidle et al., 1985; Larsen and Simon, 1993) as well as hydrological properties of the substrate (Chigira, 2001; Sidle and Chigira, 2004). Debris flows resulting from such a rapid chain of processes can cause widespread destruction due to their velocity, mass, and runout distance (e.g., Sidle and Chigira, 2004). With respect to the second category, small debris slides and debris flows may transport sediment and organic debris into channel heads and channels, but a debris flow does not immediately occur (e.g., Jonasson et al., 1991; Bovis and Jakob, 1999; Gomi et al., 2002). Such headwater sites gradually fill with debris and reach a critical mass that is destabilized by a large (but not necessarily episodic) runoff event, resulting in a debris flow (VanDine, 1985; Gomi et al., 2002; Palacios et al., 2003). The rates of headwater channel recharge are a major factor influencing the repetition of debris flows. Similarly, small landslides may deposit debris on the hillslope that can potentially be transported to the headwater channel during future storms (e.g., Wieczorek et al., 1989; Fannin and Rollerson, 1993) (Figure 16.3). Many field observations attest to the importance of such sporadic and chronic infilling of headwater channels and gullies in steep terrain (e.g., Shimokawa, 1984; Benda, 1990; Bovis and Jakob, 1999), but process-based investigations of these types of failures are rare.

The complex nature of infilling of channel heads in forested sites may involve chronic processes such as surface wash, soil creep, bedrock weathering, freeze–thaw processes, dry ravel (both sediment and organic debris), as well as episodic processes such as small debris slides, rockfall, blowdown of trees, earthflows, and small slumps (Shimokawa, 1984; Trustrum and DeRose, 1988; Reneau and Dietrich, 1991; Gomi et al., 2001; Palacios et al., 2003). Because of the wide array of infilling and soil accretion processes, debris flows in such headwater channels (Type 2) may be more frequent compared to hillslope landslides that mobilize directly into debris flows (Type 1). This more regular occurrence of in-channel debris flows is supported by studies in British Columbia (Bovis and Jakob, 1999), coastal Alaska (Gomi and Sidle, 2003), Wasatch Mountains in Utah (Wieczorek et al., 1989), Lapland (Rapp and Nyberg, 1988), Scotland (Innes, 1985), the French Alps (van Steijn, 1996), Japan (Okunishi and Suwa, 1985), Loess Plateau of China (Xu, 1999) and the Buzău Subcarpathians of Romania (Balteanu, 1976), among other areas. In tropical forests, wood may be a less significant component of the overall infilling process due to higher rates of decomposition, but weathering rates of substrate may be higher. The role of small landslides that do not directly enter channels is poorly understood, but these failures certainly contribute to the long-term sediment redistribution and recharge of forest headwater systems (Dietrich et al., 1982; Swanson et al., 1982; Sidle, 2000; Gomi et al., 2002).

For cases where debris avalanches do not immediately mobilize into debris flows (Type 2), the timing of debris avalanche initiation is separated, but not independent from the timing of the debris-flow initiation processes. Factors dictating the timing of hillslope landslide initiation (e.g., rainfall intensity, storm duration, antecedent soil moisture) may be more stochastic (e.g., Iida, 1993; Sidle and Wu, 1999; Dhakal and Sidle, 2003) compared to the partly deterministic factors that control in-channel debris flows (e.g., sediment and woody debris accretion, channel sediment storage sites, topographic and channel roughness) (e.g., Benda and Cundy, 1990; Montgomery and Buffington, 1997; Gomi and Sidle, 2003). Sediment deposited in the vicinity of channel heads by hillslope landslides reaches some critical level of accumulation after which time the material fails as a debris flow during a large (but not uncommon) rainfall/flood event (Figure 16.3). Thus, it may be possible to predict such debris flows from a more deterministic perspective based on infilling rates. Similar phenomenon would likely apply to landslides and resultant debris flows triggered by snowmelt or rain-on-snow events (e.g., Megahan, 1983; Wieczorek et al., 1989).

16.4 EFFECTS OF MANAGEMENT PRACTICES IN FORESTS

16.4.1 Silvicultural practices and forest stand management

As discussed previously, timber harvesting affects the occurrence of debris avalanches and flows by modifying the soil moisture regime and reducing rooting strength. Long-term rates of mass erosion from steep clear-cut forests in North

America summarized by Sidle et al. (1985) range from about 0.3–4.0 t/ha/yr; corresponding rates from nearby undisturbed forests ranged from 0.1–1.1 t/ha/yr. In New Zealand, where beech-podocarp-hardwood forests were clear-cut and replanted with weaker-rooted radiata pine, mass erosion was approximately 10 times greater (15–50 t/ha/yr) compared to the North American sites (O'Loughlin and Pearce, 1976). It must be recognized that these rates are long-term averages (typically based on records of 15–35 yr) and since landslides are episodic processes, mass erosion will be much higher in some years and much less (or even negligible) in other years. Also, the effects of timber harvesting on landslide initiation decline significantly after 15–25 yr; thus, increased mass erosion rates should not persist throughout the entire rotation of regenerating forest, except when stands are harvested at a very young age. Many of the studies cited in these summaries and investigations were based on aerial photograph interpretation. This methodology has been shown to underestimate the number of small landslides occurring in forested stands and may miss as much as 30% of the landslide volume (Brardinoni et al., 2003). Thus, some of the previously reported increases in landslides due to clear-cutting could be overestimates.

In addition to the many field investigations in managed forest terrain cited earlier, the period of maximum landslide susceptibility following timber harvesting (3–15 yr after logging) has been confirmed in several catchment scale simulations of slope stability (Sidle and Wu, 1999; Dhakal and Sidle, 2003). The reported 2–10-fold increase in landslide erosion following clear-felling is strongly influenced by the timing of an episodic triggering event (e.g., rainfall or snowmelt) as well as the topographic and geologic attributes of the forested site.

Few studies have assessed the effects of partial cutting and stand tending on landslide occurrence. A recent survey conducted in the northern California Coast Range found that on slopes >60%, landslide density was about 5–9-fold higher in clear-cut compared to thinned and unthinned second-growth forests (Rollerson, pers. commun., 2003). Unthinned stands had slightly higher landslide densities compared to thinned stands, which may have been attributed to foresters selecting more stable terrain for intensive management (personal commun., Rollerson, 2003).

A series of modelling studies have shed some light on the long-term consequences of various silvicultural and stand tending practices on landslide erosion (Sidle, 1991, 1992; Sidle and Wu, 1999; Dhakal and Sidle, 2003; Sakals and Sidle, 2004). Simulations of failure probability indicate that alternate thinnings and clear-cuts as well as clear-cuts alone produce less stable conditions compared to shelterwood harvesting systems and partial cuts (Sidle, 1991, 1992). Partial cutting reduced landslide volume 1.4–1.6-fold compared to clear-cutting in forest management simulations in Vancouver Island, British Columbia (Dhakal and Sidle, 2003) (Figure 16.4). Sakals and Sidle (2004) modelled small-scale changes in patterns of Douglas-fir (*Pseudotsuga mensiezii*) root strength related to different silvicultural practices and found that a random selection cutting (25% of the trees older than 60 yr harvested every 20 yr) caused the smallest decrease in rooting strength (81% of pre-harvest strength), followed by strip cutting (58% of pre-harvest root strength), and clear-cutting (an average of 47% of pre-harvest root strength). Repeated har-

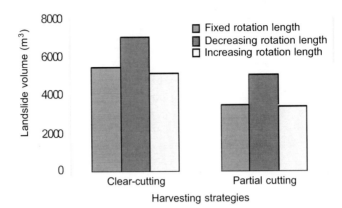

Figure 16.4. Simulated effects of clear-cutting vs. partial cutting on landslide volume for different rotation lengths and conditions in a 57-ha subcatchment at Carnation Creek, Vancouver Island, British Columbia.
From Dhakal and Sidle (2003).

vesting cycles with progressively shorter rotations increased landslide probability for sites modelled in the Oregon Coast Ranges (Sidle and Wu, 1999) and Vancouver Island (Dhakal and Sidle, 2003). Such findings may be of concern to current forest practices in areas such as New Zealand where rotation lengths of plantations continue to be shortened. When geomorphic hollows and sites steeper than 40° were protected from harvesting during slope stability simulations, landslide volume decreased by 1.6-fold in a catchment in coastal Oregon (Sidle and Wu, 1999) and by 1.8–2.9-fold in Carnation Creek, Vancouver Island, British Columbia (Dhakal and Sidle, 2003). Retaining vigorous understory vegetation has also been demonstrated to be beneficial in preventing landslides following harvesting (Dhakal and Sidle, 2003; Sakals and Sidle, 2004). Logging methods that destroy understory vegetation or reduce the regeneration potential of new trees will conversely decrease slope stability. Longer intervals between initial and final shelterwood cuttings promote greater rooting strength than short intervals (Sidle, 1991).

16.4.2 Effects of logging systems

The type of logging system employed probably has a minor impact on slope stability compared to the more significant effects due to silvicultural practices (e.g., clear-cutting, partial cutting) and road and skid trail systems associated with different harvesting practices. Exceptions include cases where ground-based (e.g., tractor) logging extensively reduces the understory vegetation and the previously noted situation where disturbance of shallow rooted organic horizons during cable logging resulted in small failures in the underlying mineral soil (Sidle, 1984). The stability of geomorphic hollows and small V-shaped headwater channels (referred to as gullies in glaciated terrain of North America) can be affected by the selection of logging systems because ground disturbances influence the rates of infilling and eventual unloading (failure) of these sites (Dietrich and Dunne, 1978; Sidle et al., 1985; Swanson et al., 1988). Thus, tractor logging and certain types of ground-based cable logging practices that compact soils and destroy surficial preferential flow paths in soils, will augment surface runoff and surface wash. These increased surface processes together with increases in loading of organic debris (due to logging) may increase the frequency of landslide initiation in hollows and gullies (Dietrich and Dunne, 1978; Shimokawa, 1984; Iida, 1993; Bovis and Pellerin, 1998). Timber harvesting methods that underutilize small diameter wood may contribute to stability problems in these depressions as woody debris accumulates and traps sediment from surface sources (Sidle, 1980).

16.4.3 Harvesting – soil moisture interactions affecting slope stability

Increases in soil moisture due to logging that may increase debris avalanche and debris-flow activity are poorly documented, but such cases would appear more probable in the tropics where evapotranspiration from rain forest canopies is high year-round. An example of how timber harvesting can affect the water balance in relatively shallow (1 m) tropical soils and thus influence the potential occurrence of shallow landslides is illustrated for the Bukit Tarek forest catchment in Peninsular Malaysia. Rainfall data were synthesized for one year based on seasonal distributions and daily rainfall amounts at Bukit Tarek (Noguchi et al., 1996). Steady-state evapotranspiration rates of 3.5 mm/day and 1.5 mm/day are assumed for forested and clear-cut conditions, respectively (Bruijnzeel, 1990; Jones, 1997). Interception losses were assumed to be 30% of total daily rainfall when storms were small ($\leq$10 mm) and 10% when rainfall was >10 mm (based on data in Dingman, 1994); these losses were reduced to 10% and 3%, respectively, after logging. Subsurface storm runoff was assumed to be 10% of rainfall for these shallow soils based on data from Bukit Tarek (Noguchi et al., 1997). After storms, an exponential decay of soil moisture was used to simulate deep seepage. For moisture levels >250 mm in shallow soils, 90% of the drainage (down to a minimum of 250 mm) occurred in 5 days (including the day of the storm). Moisture content for saturated conditions was set at 600 mm.

It is assumed that a rapid landslide would initiate only when the soil was nearly saturated (i.e., >500 mm of stored moisture). Such conditions did not occur for either the case with or without trees (Figure 16.5). Nevertheless, this example shows that forest harvesting can increase soil moisture seasonally. The three largest storms of the year (100 mm, 75 mm, and 75 mm) were all preceded by rather wet periods. During these events soil moisture for forested conditions was only 9–13 mm lower than without the influence of trees (i.e., simulated clear-cut) (Figure 16.5). Thus, during such wet antecedent periods it appears logging has only a minor influence on landslide potential. A notable exception to this trend is the smaller storm (60 mm on day 174) that occurred during dry antecedent conditions. Soil moisture increased by about the same amount for both scenarios (with and without trees) during the storm, but because the moisture content at the site with

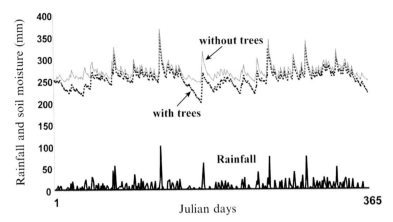

Figure 16.5. Simulations of the effects of timber harvesting on soil moisture in a Peninsular Malaysia rain forest.

trees was initially much lower due to evapotranspiration, the final soil moisture was 48 mm higher after the storm in the site without trees (Figure 16.5). Since this daily water budget cannot capture the dynamic response to intense storm bursts that often trigger landslides (e.g., Sidle, 1984), the actual peak pore-water pressure could have been much higher. Nevertheless, this example illustrates that the effects of timber harvesting in the tropics on pore water pressure response (compared with unharvested forests) are realized during moderate storms that occur following periods of low rainfall.

In temperate forests, timber harvesting may exert some influence on pore-water pressure response in steep slopes, but such impacts have not been well documented and it is difficult to disentangle such effects from other environmental factors. In one of the few studies that has assessed the effects of timber harvesting on pore pressure response, Dhakal and Sidle (2004) noted increases in maximum pressure heads (varying from 90–280 mm) during storms following timber harvesting (all measurements were taken after road construction) in seven out of nine piezometers in a subbasin of Carnation Creek, Vancouver Island, British Columbia. Moderate storms caused the largest relative increases in pressure head after harvesting; for large storms such increases were small, lending support to the concept that timber harvesting in temperate forests enhances hydrologic response only during small and moderate storms because soils are already near saturation and the evapotranspiration is very low during the winter rainy season (Thomas and Megahan, 1998). Increases in peak pore pressure due to decreases in evapotranspiration after clear-felling appear smaller than pore pressure increases related to soil disturbance after harvesting and road construction for these wet, winter storms. It is especially interesting that pressure head changes during large storms that typically trigger landslides in the area were less affected by timber harvesting (Dhakal and Sidle, 2004).

16.4.4 Vegetation conversion

Although many examples of permanent forest vegetation conversion on steep hillslopes can be observed in developing countries of Southeast and East Asia, Africa, and Latin America, practices such as permanent conversion of forests to grazing and cropland, as well as to recreational areas, still persist in North America and Europe to some extent. Most erosion research has focused on surface erosion losses from converted forest sites, but cases of increases in debris avalanches and debris flows in steep terrain have been documented in a few cases. Conversion of forest vegetation in steep terrain to grassland, cropland, monoculture plantations, agroforestry, or other vegetation cover has the potential to accelerate mass-wasting processes, largely by reducing rooting strength, but also by modifying the soil moisture regime. In southern California, converted grasslands had five times higher mass-erosion rates compared to native brushlands (Corbett and Rice, 1966). Widespread conversion of mixed evergreen forests of North Island, New Zealand, during development of pastoral hill country between 1860 and 1920 caused severe mass erosion and reduced productivity of the land (MWD, 1970; Garrett, 1980; Trustrum et al., 1983; Sidle et al., 1985). A total of 9,280 km^2 of converted lands on North Island were affected by moderate to extreme mass movement erosion – 78.5% of all North Island areas (inclusive of surface erosion) eroded to this degree (Sidle et al., 1985). No studies have quantitatively addressed the impacts of forest land conversion to agriculture on slope stability in Southeast Asia. However, progressive forest clearing, conversion to plantations and agricultural cropland, and deterioration of the land base in developing countries of Africa, Asia, and Latin America has been associated with general increases in mass wasting (Haldemann, 1956; Harwood, 1996; Fischer and Vasseur, 2000).

Rapid conversion of tropical forests to agricultural lands during the past few decades has prompted an increased interest in agroforestry (Steiner, 1988). Although various types of agroforestry are used in steep terrain, one of the oldest methods is the much-maligned practice of shifting cultivation (commonly called slash and burn). Shifting cultivation involves leaving intensively cropped land fallow to restore soil fertility; however, it may increase the probability of landslide erosion if practiced on steep hillsides. Increases in landslide erosion following progressive deforestation in Tanzania from the mid to late nineteenth century are partly attributed to loss of rooting strength after clearing and burning (Haldemann, 1956). Because shifting cultivation clears forested hillsides for long periods and because reforestation on abandoned, nutrient-depleted agricultural plots is very slow, the impact of slash and burn agriculture is longer-lived compared to timber harvesting.

Modern, sustainable agroforestry systems have the potential of reducing the probability of landslide erosion in developing countries in areas where the agriculture frontier is increasingly expanding into forested hillslopes. The forestry component of agroforesty systems offers benefits for rooting strength albeit less than the typical rooting strength of mature forests (Figure 16.6). The agricultural component of these integrated systems may provide soil cover, but sometimes at the expense of long-term physical and chemical soil properties depending on the amount

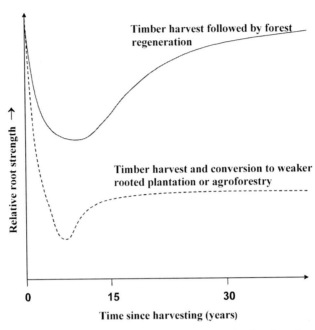

Figure 16.6. Hypothetical comparison of tree root strength deterioration and recovery following timber harvesting with regeneration, as well as for forest conversion to weaker rooted plantation or agroforestry cover.

of cultivation required. Terraces incorporated into agroforestry systems can destabilize slopes if they impound excessive water. Landsides have been observed on large terraces constructed to control surface erosion in the Loess Plateau of China (Billard et al., 1993) (Figure 16.7); often the steep faces of these terraces are unstable, in addition to the increased opportunity to impound water. Essentially no attention has been paid to off-site consequences or cumulative effects of agroforestry practices at the catchment scale nor have any systematic studies evaluated the effects of this management system on landslide initiation.

16.4.5 Roads

Forest roads affect slope stability by: (1) altering natural hydrologic pathways and concentrating water onto unstable portions of the hillslope; (2) undercutting unstable hillslopes, thereby removing support; and (3) overloading and oversteepening hillslopes (via fillslope material) (Sidle et al., 1985). The relative importance of these destabilizing factors depends on the design and construction standards (if any) of the road and associated drainage system, as well as the natural instability of the terrain.

A thorough discussion of the impact of forest roads on slope stability is beyond the scope of this chapter; however, studies worldwide have shown that roads increase

Figure 16.7. A landslide initiating on a constructed agricultural terrace in the Loess Plateau, Shanxi Province, China.

landslide erosion by approximately two orders of magnitude compared with undisturbed forest land (e.g., O'Loughlin, 1972; Morrison, 1975; O'Loughlin and Pearce, 1976; Gray and Megahan, 1981; Amaranthus et al., 1985). While roads are a necessary part of most forest land uses, the critical concerns related to slope stability are: the length of roads in steep terrain; cutting roads at mid-slope locations (including the width of the road); interception and removal of water in the road right-of-way (including drainage design); recognition of highly unstable landscape features (e.g., geomorphic hollows, old slump blocks); overall road design, layout, and construction considerations; maintenance; and ultimate life and use of the road, including deactivation strategies. Because mechanical slope stabilization is generally not economically feasible along most low-volume roads and trails, landslide prevention can be partly achieved by prudent road location and construction methods (Megahan, 1977; Sidle et al., 1985; Hinch, 1986). Such management measures have been shown to reduce road-related landslides (e.g., Duncan et al., 1987; Skaugset et al., 1996); however, landslide problems persist along forest roads in developing nations where little planning and engineering design are utilized (e.g., Haigh, 1984; Thakur, 1996).

While government agencies have expended many funds on forest road rehabilitation and deactivation in steep, unstable terrain, especially in the Pacific Northwest (e.g., Slaney and Zaldokas, 1997), little emphasis has been placed on assessing the potential success of such efforts. A recent study in coastal British Columbia suggests that proper road reclamation techniques on only 10% of the existing forest road network would yield 98% of the cumulative expected net benefits (related to landslides and debris-flow protection) and would incur only 18% of cost of restoring the entire hillslope road system once logging was completed (Allison et al., 2004). Thus, better geotechnical investigations conducted prior to forest road reclamation may be highly beneficial.

16.5 SUMMARY

Forest vegetation affords protection against landslides in unstable terrain by imparting root cohesion into the soil mantle and, to a lesser extent, modifying the soil moisture regime through evapotranspiration. The benefits of root cohesion are most important in shallow soils where tree roots can anchor directly into underlying bedrock or other stable substrate. Timber harvesting in steep terrain can increase landslide erosion rates two- to greater than tenfold, and impacts are greatest in sites with shallow and wet soils. The window of susceptibility for increased landslide erosion after clear-felling and prior to significant regeneration ranges from about 3–15 years. A clear distinction must be made among various sustainable forest land uses and conversion of forests to other vegetation types. Rates of mass movement are generally much lower for sustainable forestry practices compared to converted lands. Once forests are converted to grasslands, monoculture plantations, or even agroforestry, the lower rooting strength of the converted vegetation persists indefinitely.

Little research has focused on understanding the interactions of vegetation with slope stability at large scales and over long periods of time. To better understand this important interaction amongst geomorphic, hydrologic, topographic, and biological processes, it is necessary to evaluate slope stability in relation to both the temporal and spatial distributed attributes of catchments. One approach is to apply distributed, physically-based slope stability models at the catchment scale (Wu and Sidle, 1995; Sidle and Wu, 1999; Dhakal and Sidle, 2003). Such models can assess the effects of various timber harvesting strategies on slope stability in both the temporal and spatial dimensions; however, such models require substantial data inputs to characterize variable soil and site conditions. Alternative approaches to landslide hazard assessment in more remote areas where such distributed data are difficult to collect or assess rely on empirical comparisons of unstable site indicators. Using a Geographic Information System (GIS) framework, instability indicators derived from digital terrain data, geological information, climatic data, land use and vegetation maps, and other remotely sensed data can be overlain and evaluated quantitatively related to relative slope stability (Gupta and Joshi, 1990; Dhakal et al., 2000). Successful application of such GIS-based models is contingent upon using terrain indicators that are closely linked with processes that control slope stability.

16.6 REFERENCES

Abe, K. and Ziemer, R.R. (1991) Effect of tree roots on a shear zone: Modeling reinforced shear stress. *Canadian Journal of Forest Research*, **21**, 1012–1019.

Adams, P.W. and Sidle, R.C. (1987) Soil conditions in three recent landslides in southeast Alaska. *Forest Ecology and Management*, **18**, 93–102.

Allison, C., Sidle, R.C., and Tait, D. (2004) Application of decision analysis to forest road deactivation in unstable terrain. *Environmental Management*, **33**, 173–185.

Amaranthus, M.P., Rice, R.M., Barr, N.R., and Ziemer, R.R. (1985) Logging and forest roads related to increased debris slides in southwestern Oregon. *Journal of Forestry*, **83**, 229–233.

Aust, W.M., Schoenholtz, S.H., Zaebst, T.W., and Szabo, B.A. (1997) Recovery status of a tupelo-cypress wetland seven years after disturbance: Silvicultural implications. *Forest Ecology and Management*, **90**, 161–169.

Balteanu, D. (1976) Two case studies of mudflows in the Buzau Subcarpathians. *Geografiska Annaler*, **58A**, 165–171.

Benda, L. (1990) The influence of debris flows on channels and valley floors in the Oregon Coast range, U.S.A. *Earth Surface Processes and Landforms*, **15**, 457–466.

Benda, L.E. and Cundy, T.W. (1990) Predicting depositions of debris flows in mountain channels. *Canadian Geotechnical Journal*, **27**, 409–417.

Berris, S.N. and Harr, R.D. (1987) Comparative snow accumulation and melt during rainfall in forested and clear-cut plots in the western Cascades of Oregon. *Water Resources Research*, **23**, 135–142.

Bilby, R.E. and Ward, J.W. (1991) Characteristics and function of large woody debris in streams draining old-growth, clear-cut, and second-growth forests in Southwestern Washington. *Canadian Journal of Fisheries and Aquatic Science*, **48**, 2499–2508.

Billard, A., Muxart, T., Derbyshire, E., Wang, J.T., and Dijkstra, T.A. (1993) Landsliding and land use in the loess of Gansu Province, China. *Zeitschrift für Geomorphologie, N.F. Suppl.*, **87**, 117–131.

Bishop, D.M. and Stevens, M.E. (1964) *Landslides on Logged Areas, Southeast Alaska* (USDA Forest Service Research Report NOR-1, 18 pp.). US Department of Agriculture, Juneau, AK.

Bovis, M.J. and Jakob, M. (1999) The role of debris supply conditions in predicting debris flow activity. *Earth Surface Processes and Landforms*, **24**, 1039–1054.

Bovis, M.J. and Pellerin, D. (1998) Rates of revegetation of gullies in coastal British Columbia: Implications for fine-sediment production (Unpublished annual report). Scientific Council of British Columbia, Victoria, Canada.

Brardinoni, F., Slaymaker, O., and Hassan, M.A. (2003) Landslide inventory in a rugged forested watershed: A comparison between air-photo and field survey data. *Geomorphology*, **54**, 179–196.

Bruijnzeel, L.A. (1990) *Hydrology of Moist Tropical Forests and Effects of Conversion* (A state of knowledge review). UNESCO International Hydrological Programme, Humid Tropics Programme, Paris.

Burroughs, E.R. and Thomas, B.R. (1977) *Declining Root Strength in Douglas-fir after Felling as a Factor in Slope Stability* (USDA Forest Service Research Paper INT-190, 27 pp.). US Department of Agriculture, Intermountain Forest and Range Experiment Station, Ogden, UT.

Burroughs, E.R., Chalfant, G.R., and Townsend, M.A. (1976) *Slope Stability in Road Construction* (102 pp.). Bureau of Land Management, Portland, OR.

Caine, N. (1980) Rainfall intensity-duration control of shallow landslides and debris flows. *Geografiska Annaler*, **62A**, 23–27.

Cannon, S.H. and Reneau, S.L. (2000) Conditions for generation of fire-related debris flows, Capulin Canyon, New Mexico. *Earth Surface Proceedings and Landforms*, **25**, 1103–1121.

Cheng, C.T., Lee, C.T., and Tsai, Y.B. (2000) Fault rupture plane and attenuation model associated with the 1999 Chi-Chi earthquake. *Proceedings of the 2000 Annual Meeting of the Geological Society of China* (pp. 21–23). Geological Society of China, [in Chinese].

Chigira, M. (2001) Micro-sheeting of granite and its relationship with landsliding specifically after the heavy rainstorm in June 1999, Hiroshima Prefecture, Japan. *Engineering Geology*, **59**, 219–231.

Coker, R.J. and Fahey, B.D. (1993) Road-related mass movement in weathered granite, Golden Downs and Motueka Forests, New Zealand (A note). *Journal of Hydrology (N.Z.)*, **31**, 65–69.

Corbett, E.S. and Rice, R.M. (1966) *Soil Slippage Increased by Brush Conversion* (USDA Forest Service Research Note PSW-128, 8 pp.). US Department of Agriculture, Pacific Southwest Research Station, Berkeley, CA.

Costa, J.E. (1991) Nature, mechanics, and mitigation of the Val Pola landslide, Valtellina, Italy, 1987–1988. *Zeitschrift für Geomorphologie, N.F. Suppl.*, **35**, 15–38.

Crozier, M.J., Eyles, R.J., Marx, S.L., McConchie, J.A., and Owen, R.C. (1980) Distribution of landslips in the Wairarapa hill country. *New Zealand Journal of Geology and Geophysics.*, 23, 575–586.

Dhakal, A.S. and Sidle, R.C. (2003) Long-term modeling of landslides for different forest management practices. *Earth Surface Processes and Landforms*, **28**, 853–868.

Dhakal, A.S. and Sidle, R.C. (2004) Pore water pressure assessment in a forest watershed: Simulations and distributed field measurements related to forest practices. *Water Resources Research*, **40**, W02405, doi:1029/2003WR002017.

Dhakal, A.S., Amada, T., and Aniya, M. (2000) Landslide hazard mapping and its evaluation using GIS: An investigation of sampling scheme for grid-cell based quantitative method. *Photogrammetric Engineering and Remote Sensing*, **66**, 981–989.

Dietrich, W.E. and Dunne, T. (1978) Sediment budget for a small catchment in mountainous terrain. *Zeitschrift für Geomorphologie, N.F.*, **29**, 191–206.

Dietrich, W.E., Dunne, T., Humphrey, N.F., and Reid, L.M. (1982) Construction of sediment budgets for drainage basins. *Sediment Budgets and Routing in Forested Drainage Basins* (USDA Forest Service General Technical Report PNW-141, pp. 5–23). US Department of Agriculture, Pacific Northwest Forest and Range Experimental Station, Portland, OR.

Dingman, S.L. (1994) *Physical Hydrology* (575 pp.). Macmillan, New York.

Duncan, S.H., Ward J.W., and Anderson, R.J. (1987) A method for assessing landslide potential as an aid in forest road placement. *Northwest Science*, **61**, 152–159.

Endo, T. and Tsuruta, T. (1969) *The Effect of the Tree's Roots on the Shear Strength of Soil* (Annual Report, 1968, pp. 167–182). Hokkaido Branch, Forestry Experimental Station, Sapporo, Japan.

Fannin, R.J. and Rollerson, T.P. (1993) Debris flows: some physical characteristics and behaviour. *Canadian Geotechnical Journal*, **30**, 71–81.

Fischer, A. and Vasseur, L. (2000) The crisis in shifting cultivation practices and the promise of agroforestry: A review of the Panamanian experience. *Biodiversity and Conservation*, **9**, 739–756.

Fukuda, F. and Ochiai, H. (1993) Landslides caused by the 1993, Hokkaido Nansei-oki earthquake. *Shin Sabo Journal*, **46**, 62–63 [in Japanese].

Garrett, J. (1980) *Catchment Authority Work in the Rangitikei Area* (Aokautere Scientific Centre Internal Report 21, pp. 23–26). New Zealand Ministry of Works and Development, Wellington.

Garwood, N.C., Janos, D.P., and Brokaw, N. (1979) Earthquake induced landslides: A major disturbance to tropical forests. *Science*, **205**, 997–999.

Golding, D.L. and Swanson, R.H. (1986) Snow distribution patterns in clearings and adjacent forest. *Water Resources Research*, **22**, 1931–1940.

Gomi, T. and Sidle, R.C. (2003) Bedload transport in managed steep-gradient headwater streams of southeast Alaska. *Water Resources Research*, **39**, 1336, doi:10.1029/2003 WR002440.

Gomi, T., Sidle, R.C., Bryant, M.D., and Woodsmith, R.D. (2001) Characteristics of woody debris and sediment distribution in headwater streams, southeast Alaska. *Canadian Journal of Forest Research*, **31**, 1386–1399.

Gomi, T., Sidle, R.C., and Richardson, J.S. (2002) Understanding processes and downstream linkages of headwater systems. *BioScience*, **52**, 905–916.

Gray, D.H. and Megahan, W.F. (1981) *Forest Vegetation Removal and Slope Stability in the Idaho Batholith* (USDA Forest Service Research Paper INT-271, 23 pp.). US Department of Agriculture, Ogden, UT.

Greenway, D.R. (1987) Vegetation and slope stability. In: M.G. Anderson and K.S. Richards (eds.), *Slope Stability* (pp. 187–230). John Wiley & Sons, Chichester, UK.

Gupta, R.P. and Joshi B.C. (1990) Landslide hazard zoning using the GIS approach: A case study from the Ramganga catchment, Himalayas. *Engineering Geology*, **28**, 119–131.

Guthrie, R.H. (2002) The effects of logging on frequency and distribution of landslides in three watersheds on Vancouver Island, British Columbia. *Geomorphology*, **43**, 273–292.

Haigh, M.J. (1984) Landslide prediction and highway maintenance in the Lesser Himalaya, India. *Zeitschrift für Geomorphologie, N.F. Suppl.*, **51**, 17–37.

Haldemann, E.G. (1956) Recent landslide phenomena in the Rungwe volcanic area, Tanganyika. *Tanganyika Notes Record*, **45**, 3–14.

Harris, A.S. (1967) *Natural Reforestation on a Mile-square Clearcut in Southeast Alaska* (USDA Forest Service Research Paper PNW-52, 11 pp.). US Department of Agriculture, Pacific Northwest Forest and Range Experimental Station, Portland, OR.

Harwood, R.R. (1996) Development pathways toward sustainable systems following slash-and-burn. *Agriculture, Ecosystems and Environment*, **58**, 75–86.

Hinch, L.W. (1986) The location and geotechnical design of roads in mountainous terrain. *Proceedings of the Sino-British Highways and Urban Traffic Conference, Beijing, November 17–22, 1986* (pp. 87–96). Transportation Research Laboratory, Crowthorne, UK.

Iida, T. (1993) A probability model of slope failure and hillslope development. *Transactions of Japanese Geomorphological Union*, **14**, 17–31 [in Japanese].

Innes, J.L. (1985) Magnitude-frequency relations of debris flows in northwest Europe. *Geografiska Annaler*, **67A**, 23–32.

Iverson, R.M., Reid, M.E., and LaHusen, R.G. (1997) Debris-flow mobilization from landslides. *Annual Review of Earth and Planetary Sciences*, **25**, 85–138.

Jakob, M. (2000) The impacts of logging on landslide activity at Clayoquot Sound, British Columbia. *Catena*, **38**, 279–300.

Johnson, K., Olson, E.A., and Manandhar, S. (1982) Environmental knowledge and response to natural hazards in mountainous Nepal. *Mountain Research and Development*, **2**, 175–188.

Jonasson, C., Kot, M., and Kotarba, A. (1991) Lichenometrical studies and dating of debris flow deposits in the High Tatra Mountains, Poland. *Geografiska Annaler*, **73A**, 141–146.

Jones, J.A.A. (1997) *Global Hydrology: Processes, Resources and Environmental Management* (399 pp.). Longman, Harlow, UK.

Lambereti, G.A., Gregory, S.V., Ashkenas, L.R., Wildman, R.C., and Moore, K.M.S. (1991) Stream ecosystem recovery following a catastrophic debris flow. *Canadian Journal of Fisheries and Aquatic Science*, **48**, 196–208.

Larsen, M.C. and Parks, J.E. (1997) How wide is a road? The association of roads and masswasting in a forested montane environment. *Earth Surface Processes and Landforms*, **22**, 835–848.

Larsen, M.C. and Simon, A. (1993) A rainfall intensity-duration threshold for landslides in a humid-tropical environment, Puerto Rico. *Geografiska Annaler*, **75A**, 13–23.

McNaughton, K.G. and Jarvis, P.G. (1983) Predicting effects of vegetation changes on transpiration and evaporation. In: T.T. Kozlowski (ed.), *Water Deficits and Plant Growth* (Vol. 7, pp. 1–47). Academic Press, New York.

Megahan, W.F. (1977) Reducing erosional impacts of roads. *Guidelines for Watershed Management* (FAO Conservation Guide, pp. 237–251). UN Food and Agriculture Organization, Rome.

Megahan, W.F. (1983) Hydroligic effects of clearcutting and wildfire on steep granitic slopes in Idaho. *Water Resources Research*, **19**, 811–819.

Megahan, W.F. (1987) Effects of forest roads on watershed function in mountainous areas. In: A.S. Balasubramaniam et al. (eds), *Environmental Geotechnics and Problematic Soils and Rocks* (pp. 335–348). A.A. Balkema, Rotterdam.

Megahan, W.F., Day, N.F., and Bliss, T.M. (1978) Landslide occurrence in the western and central northern Rocky Mountain physiographic province in Idaho. *Proceedings of the 5th North American Forest Soils Conference* (pp. 116–139). Colorado State University, Fort Collins, CO.

Montgomery, D.R. and Buffington, J.M. (1997) Channel-reach morphology in mountain drainage basins. *Geological Society of America Bulletin*, **109**, 596–611.

Morrison, P.H. (1975) Ecological and geomorphological consequences of mass movements in the Alder Creek watershed and implications for forest land management (102 pp.). B.A. thesis, University of Oregon, Eugene, OR.

MWD (1970) *Wise Land Use and Community Development* (Report of technical committee of inquiry into the problems of the Poverty Bay–East Cape District of New Zealand, 119 pp.). Ministry of Works and Development, Wellington.

Noguchi, S., Abdul Rahim, N., Sammori, T., Tani, M., and Tsuboyama, Y. (1996) Rainfall characteristics of tropical rain forest and temperate forest: Comparison between Bukit Tarek in Peninsular Malaysia and Hitachi Ohta in Japan. *Journal of Tropical Forest Science*, **9**, 206–220.

Noguchi, S., Abdul Rahim, N., Zulkifli, Y., Tani, M., and Sammori, T. (1997) Rainfall-runoff responses and roles of soil moisture variations to the response in tropical rain forest, Bukit Tarek, Malaysia. *Journal of Forest Research*, **2**, 125–132.

Nunamoto, S., Suzuki, M., and Ohta, T. (1999) Decreasing trend of deaths and missing persons caused by sediment disasters in the last fifty years in Japan. *Sabo Gakkai Journal*, **51**, 3–12 [in Japanese].

Okunishi, K. and Suwa, H. (1985) Hydrological approach to debris flow. *Proceedings of International Symposium on Erosion, Debris Flow and Disaster Prevention, September 3–5, Tsukuba, Japan* (pp. 243–247). Public Works Research Institute, Tsukuba, Japan.

O'Loughlin, C.L. (1972) The stability of steepland forest soils in the Coast Mountains, southwest British Columbia. Ph.D. thesis, University of British Columbia, Vancouver.

O'Loughlin, C.L. and Pearce, A.J. (1976) Influence of Cenozoic geology on mass movement and sediment yield response to forest removal, North Westland, New Zealand. *Bulletin of International Association of Engineers and Geologists*, **14**, 41–46.

Onda, Y. (1992) Influence of water storage capacity in the regolith zone on hydrologic characteristics, slope processes, and slope form. *Zeitschrift für Geomorphologie, N.F. Suppl.*, **36**, 165–178.

Palacios, D., García, R., Rubio, V., and Vigil, R. (2003) Debris flows in a weathered granitic massif: Sierra de Gredos, Spain. *Catena*, **51**, 115–140.

Phillips, C.J. and Watson, A.J. (1994) *Structural Tree Root Research in New Zealand: A Review* (Landcare Research Science Series No. 7, 71 pp.). Landcare, Lincoln, New Zealand.

Prosser, I.P. and Williams, L. (1998) The effect of wildfire on runoff and erosion in native Eucalyptus forest. *Hydrological Processes*, **12**, 251–265.

Rapp, A. and Nyberg, R. (1981) Alpine debris flows in northern Scandinavia: Morphology and dating by lichenometry. *Geografiska Annaler*, **63**, 183–196.

Rapp, A. and Nyberg, R. (1988) Mass movements, nivation processes and climatic fluctuations in northern Scandinavian mountains. *Norsk Geografiska Tidsskrift*, **42**, 245–253.

Reneau, S.L. and Dietrich, W.E. (1991) Erosion rates in the southern Oregon Coast Range: Evidence for an equilibrium between hillslope erosion and sediment yield. *Earth Surface Proceedings and Landforms*, **16**, 307–322.

Sakals, M.E. and Sidle, R.C. (2004) A spatial and temporal model of root cohesion in forest soils. *Canadian Journal of Forest Research*, **34**, 950–958.

Schroeder, W.L. (1985) The engineering approach to landslide risk analysis. In: D.N. Swanston (ed.), *Proceedings of a Workshop on Slope Stability: Problems and Solutions in Forest Management, Seattle, Washington, 6–8 February 1984* (USDA Forest Service General Technical Report PNW-180, pp. 43–50). US Department of Agriculture, Pacific Northwest Forest and Range Experimental Station, Portland, OR.

Shimokawa, E. (1984) A natural recovery process of vegetation on landslide scars and landslide periodicity in forested drainage basins. *Proceedings of a Symposium on Effects of Forest Land Use on Erosion and Slope Stability* (pp. 99–107). East-West Center, Honolulu, HI.

Shimokawa, E., Jitousono, T., and Takano, S. (1989) Periodicity of shallow landslide on Shirasu (Ito pyroclastic flow deposits) steep slopes and prediction of potential landslide sites. *Transactions of Japanese Geomorphological Union*, **10**, 267–284 [in Japanese].

Sidle, R.C. (1980) *Impacts of Forest Practices on Surface Erosion* (USDA Extension Publication No. PNW-195, 15 pp.). US Department of Agriculture, Pacific Northwest Forest and Range Experimental Station, Corvallis, OR.

Sidle, R.C. (1984) Shallow groundwater fluctuations in unstable hillslopes of coastal Alaska. *Zeitschrift Gletscherkunde und Glazialgeologie*, **20**, 79–95.

Sidle, R.C. (1991) A conceptual model of changes in root cohesion in response to vegetation management. *Journal of Environmental Quality*, **20**, 43–52.

Sidle, R.C. (1992) A theoretical model of the effects of timber harvesting on slope stability. *Water Resources Research*, **28**, 1897–1910.

Sidle, R.C. (2000) Watershed challenges for the 21st Century: A global perspective for mountainous terrain. *Proceedings of Land Stewardship in the 21st Century: The Contributions of Watershed Management* (USDA Forest Service Proceedings RMRS-P-13, pp. 45–56). US Department of Agriculture, Rocky Mountain Research Station, Fort Collins, CO.

Sidle, R.C. and Chigira, M. (2004) Landslides and debris flows strike Kyushu, Japan. *Eos, Transactions of the American Geophysical Union*, **85**(15), 145–151.

Sidle, R.C. and Dhakal, A.S. (2002) Potential effects of environmental change on landslide hazards in forest environments. In: R.C. Sidle (ed.), *Environmental Change and Geomorphic Hazards in Forests* (International Union of Forestry Research Organizations Research Series No. 9, pp. 123–165). CAB International Press, Oxford, UK.

Sidle, R.C. and Wu, W. (1999) Simulating effects of timber harvesting on the temporal and spatial distribution of shallow landslides. *Zeitschrift für Geomorphologie, N.F. Suppl.*, **43**, 185–201.

Sidle, R.C., Pearce, A.J., and O'Loughlin, C.L. (1985) *Hillslope Stability and Land Use* (Water Resources Monograph Vol. 11, 140 pp.). American Geophysical Union, Washington, DC.

Sidle, R.C., Taylor, D., Lu, X.X., Adger, W.N., Lowe, D.J., deLange, W.P., Newnham, R.N., and Dodson, J.R. (2004) Interactions of natural hazards and humans: Evidence in historical and recent records. *Quaternary International*, **118/119**, 181–203.

Skaugset, A., Swall, S., and Martin, K. (1996) The effect of forest road location, construction, and drainage standards on road-related landslides in western Oregon associated with the February 1996 storm. *Proceedings of the Pacific Northwest Floods of February 1996 Water Issues Conference, October 7–8, Portland, Oregon* (pp. 201–206). American Institute of Hydrology, St Paul, MN.

Slaney, P.A. and Zaldokas, D. (eds) (1997) *Fish Habitat Rehabilitation Procedures* (Watershed Restoration Technical Circular No. 9, Watershed Restoration Program). British Columbia Ministry of Environment, Vancouver.

Smyth, C.G. and Royle, S.A. (2000) Urban landslide hazards: Incidence and causative factors in Niterói, Rio de Janeiro State, Brazil. *Applied Geography*, **20**, 95–117.

Steiner, F. (1988) Agroforestry's coming of age. *Journal of Soil and Water Conservation*, **43**, 157–158.

Swanson, F.J. and Swanston, D.N. (1977) *Complex Mass Movement Terrains in the Western Cascade Range, Oregon* (Reviews in Engineering Geology Vol. 3, Landslides, pp. 113–124). Geological Society of America, Boulder, CO.

Swanson, F.J., Fredriksen, R.L., and McCorison, F.M. (1982) Material transfer in a western Oregon forested watershed. In: R.L. Edmonds (ed.), *Analysis of Coniferous Forest Ecosystems in the Western United States* (pp. 233–266). Hutchison Ross, Stroudsburg, PA.

Swanson, F.J., Kratz, T.K., Caine, N., and Woodmansee, R.G. (1988) Landform effects on ecosystem patterns and processes. *BioScience*, **38**, 92–98.

Swanston, D.N. and Swanson, F.J. (1976) Timber harvesting, mass erosion, and steepland forest geomorphology in the Pacific Northwest. In: D.R. Coats (ed.), *Geomorphology and Engineering* (pp. 199–221). Dowden, Hutchison & Ross, Stroudsburg, PA.

Thakur, V.C. (1996) *Landslide Hazard Management and Control in India* (51 pp.). International Center for Integrated Mountain Development, Kathmandu, Nepal.

Thomas, R.B. and Megahan, W.F. (1998) Peakflow responses to clear cutting and roads in small and large basins, western Cascades, Oregon: A second opinion. *Water Resources Research*, **34**, 3393–3403.

Troendle, C.A. and King, R.M. (1985) The effect of timber harvest on the Fool Creek watershed, 30 years later. *Water Resources Research*, **21**, 1915–1922.

Trustrum, N.A. and DeRose, R.C. (1988) Soil depth-age relationship of landslides on deforested hillslopes, Taranaki, New Zealand. *Geomorphology*, **1**, 143–160.

Trustrum, N.A., Lambert, M.G., and Thomas, V.J. (1983) The impact of soil slip erosion on hill country pasture production in New Zealand. *Proceedings of the 2nd International Conference on Soil Erosion and Conservation, University of Hawaii, Honolulu.*

van Dine, D.F. (1985) Debris flows and debris torrents in the Southern Canadian Cordillera. *Canadian Geotechnical Journal*, **22**, 44–68.

van Steijn, H. (1996) Debris-flow magnitude–frequency relationships for mountainous regions of Central and Northeast Europe. *Geomorphology*, **15**, 259–273.

Wasson, R.J. (1978) A debris flow at Reshkn, Pakistan Hindu Kush. *Geografiska Annaler*, **60A**, 151–159.

Wemple, B.C., Swanson, F.J., and Jones, J.A. (2001) Forest roads and geomorphic process interactions, Cascade Range, Oregon. *Earth Surface Processes and Landforms*, **26**, 191–204.

Wieczorek, G.F., Lips, E.W., and Ellen, S.D. (1989) Debris flows and hyperconcentrated floods along the Wasatch Front, Utah, 1983 and 1984. *Bulletin of Association of Engineers and Geologists*, **26**, 191–208.

Wu, T.H., McKinnel, W.P., and Swanston, D.N. (1979) Strength of tree roots and landslides on Prince of Wales Island, Alaska. *Canadian Geotechnical Journal*, **16**, 19–33.

Wu, W. and Sidle, R.C. (1995) A distributed slope stability model for steep forested hillslopes. *Water Resources Research*, **31**, 2097–2110.

Xu, J. (1999) Erosion caused by hyperconcentrated flow on the Loess Plateau of China. *Catena*, **36**, 1–19.

Ziemer, R.R. and Swanston, D.N. (1977) *Root Strength Changes after Logging in Southeast Alaska* (USDA Forest Service Research Note PNW-306, 10 pp.). US Department of Agriculture, Pacific Northwest Forest and Range Experimental Station, Portland, OR.

17

Debris-flow hazard analysis

Matthias Jakob

17.1 INTRODUCTION

Much of the research presented earlier in this book was motivated by the potential for loss or damage by debris flows and the need to assess and mitigate the hazard. An increasing number of government agencies, infrastructure and utility owners, and insurance companies are realizing that debris flows are a serious hazard that, if not recognized and addressed, can lead to substantial loss of revenue and life.

Landslide hazard and risk management are developing fields and a number of national and local governments and agencies have provided guidelines (e.g., Cave, 1992; Hong Kong Government Planning Department, 1994; IUGS Working Group on Landslides, 1997; AGS, 2000; British Columbia Ministry of Forest, 2004).

Methods of debris-flow hazard analyses vary widely from country to country, from region to region and from practitioner to practitioner. Few countries or jurisdictions have developed guidelines on how to quantify and map debris-flow hazards. Notable exceptions are Austria (Fiebiger, 1997), which first developed and legislated a comprehensive method for debris-flow hazard analysis, and Switzerland (Petracheck and Kienholz, 2003; see also Chapter 24), which modified the Austrian system. In Austria, debris-flow hazard management was legislated in the forestry law in 1975. Since then, 1,200 of the 2,300 Austrian municipalities subject to debris flows, avalanches, and other ground hazards, have received detailed hazard zoning plans (Fiebiger, 1997). Over the past 30 years, the guidelines for debris-flow hazard mapping have undergone rigorous scrutiny by the affected municipalities, by experts from the Ministry of Agriculture and Forestry and the Federal Service for Torrent and Avalanche Control, and by representatives of the provincial authorities. Finally, hazard maps are approved by the Austrian government (Scheuringer, 1998).

In Japan, 70,000 debris-flow channels have been identified with at least five buildings along the channel or in the runout zone. This situation necessitated Guidelines for Zoning Debris Flow Vulnerable Areas to quantify debris-flow

Table 17.1. Debris flow hazard analysis.

Step	Action
1	Debris flow hazard recognition
2	Estimation of debris flow probability
3	Estimation of debris flow magnitude and intensity
4	Production of debris flow frequency–magnitude relationships
5	Estimation design of debris flow magnitude and intensity
6	Presentation of debris flow hazards on maps and writing a debris flow hazard analysis report

hazard and risk (Japan Ministry of Construction, 1979). The Japanese system is based on a combination of methods but, unlike in Europe or North America, does not attribute a specific return period to the hazard zones.

In North America, debris-flow recognition, hazard assessment, and mitigation design are, in comparison, still in their infancy with studies reaching back only some 25 years. No country or state/province-wide system for debris-flow hazard and risk analysis exists in North America. A few states, such as Washington, have a documented process for overview debris-flow hazard analyses, which lead to detailed fans-specific studies for the highest risk creeks (e.g., personal commun., Whatcom County (Washington) Flood Control District, 2003).

The objective of this chapter is to summarize and evaluate methods of, and to provide guidance for, debris-flow hazard analyses (Table 17.1). It synthesizes some information and methods discussed in previous chapters and provides guidelines to quantify debris-flow hazards in a variety of settings. Emphasis is placed on debris-flow hazards on individual fans rather than on a regional scale, which has been discussed in Chapters 4 and 11.

Hazard analyses and assessments are sometimes extended to risk analyses, which include the identification of elements at risk and their vulnerabilities, and the calculation of specific and total risks. These analyses may be followed by an evaluation of hazard or risk acceptability or tolerance, and hazard or risk mitigation (Fell, 1994). Risk evaluation and mitigation are covered in Chapter 18.

17.2 DEBRIS-FLOW HAZARD RECOGNITION

Debris-flow hazards are not always easily recognized, particularly on fans that are subject to high magnitude, low-frequency events. Time after time, debris-flow professionals hear the phrase "I have been living here for so many years and the creek has never come over its banks". This statement has often been proven incorrect after detailed investigations have revealed evidence of destructive debris flows. Debris-flow hazard recognition is therefore the first, and possibly most important, step in any debris-flow hazard analysis.

17.2.1 Significance of fans

The vast majority of loss or damage occurs in the depositional zone of debris flows, which is referred to as the creek fan. Fans are a preferred location for urban development because they are well drained, gently sloping, and often provide good aquifers. In many of the densely developed mountainous regions of the world, fans provide the only readily developable land. Unfortunately, many fans are prone to extreme floods, hyperconcentrated flows and/or debris flows. Fans are classified as *alluvial fans* if formed by fluvial processes, and *colluvial fans* if formed by landside processes including debris flows. Often colluvial fans, formed predominantly by debris flows or hyperconcentrated flows, are incorrectly classified as alluvial fans. The majority of fans at the mouth of steep creeks are formed by a number of processes that vary in space and time. Consequently, if the exact fan-forming processes are unknown they should be referred to as composite fans, or simply as stream fans.

Several researchers have attempted to correlate fan gradient to the dominant geomorphic fan-forming process. For example, Jackson et al. (1987), in a study of 103 fans in the Canadian Rocky Mountains, noted that those fans at least partially formed by debris flows have a fan gradient $>4°$. Regional fan gradient thresholds, however, cannot readily be transferred to other regions. For example, volcanic debris flows (lahars) may form fans with slopes as low as 1–2°, while debris-flow fans formed in granitic terrain with a sandy matrix may show average fan gradients several degrees higher than the 4° quoted for the Canadian Rockies. Furthermore, low-frequency debris-flow deposits may be overprinted by more frequent fluvial processes and, in aggrading valleys, may loose their fan morphology over time. Therefore, to recognize debris-flow potential, stratigraphic evidence needs to be analysed in conjunction with age control.

17.2.2 Geomorphic evidence

Geomorphic evidence is the first step to determine if a debris-flow hazard exists. As a general rule, watersheds smaller than about $5 \, \text{km}^2$ with gradients steeper than approximately 15° are likely to have debris-flow potential. Clearly, there are numerous exceptions.

In the field, possible signs of debris-flow activity in the transport zone include:

- well-defined boulder trains and levees (Figure 17.1);
- scour marks, mudlines and debris impact scars on trees well above the flood limit (Figure 17.3); and
- boulders much larger than could be moved by flood flow, including megaclasts rolled in large debris flows (Figure 17.2).

Signs of debris-flow activity on the fan can be recognized by:

- lack of sorting or imbrication (Figure 27.9);
- angularity of boulders (Figure 27.12B);

Figure 17.1. Well-defined debris lobes and levees at Gunbarrel 1 Gully, Lillooet, British Columbia. At this location subtle changes in gradient or flow confinement encourage a debris flow to stall in mid channel.

- inverse grading of fan or levee deposits (sometimes); and
- buried logs with frayed ends.

The use of several geomorphic pointers is recommended to recognize debris-flow hazard potential. It is also important to realize that evidence of other geomorphic processes does not necessarily preclude debris-flow occurrence. For example,

Figure 17.2. The Hinkelstein Boulder rolled at least 200 m by a debris flows at Hummingbird Creek, British Columbia. Boulders transported by a 200-year return period flood at this creek likely do not exceed 40 cm in diameter.

abundant well-rounded boulders embedded in a matrix of unsorted fines could be a result of debris-flow entrainment of fluvial sediments. Similarly, a hummocky deposit may initially suggest deposition from a debris avalanche or rock avalanche. However, this deposit may only be the proximal phase of a debris or

416 Debris-flow hazard analysis   [Ch. 17

Figure 17.3. Scour marks and impact scars in channels well above the flood limit at Canyon Creek, Washington State. The 100-year return period flood would reach approximately the upper limit of bedrock.

Photo: Doug Goldthorp.

rock avalanche that has transformed into a debris flow farther downstream (see Chapter 27.

17.2.3 Aerial photographs, satellite imagery, and topography

Aerial photographs and satellite imagery are indispensable tools in the recognition and analysis of debris-flow hazards. Most countries have at least some aerial photographs available, with the earliest aerial photographs dating back to the 1930s. In populated parts of Canada, large-scale aerial photography (1 : 20,000 and larger) is flown on average every 5 years.

Specifics of these and other remote sensing techniques are provided in Chapter 11. Of note is that despite recent advances in remote sensing techniques for landslide hazard recognition such as InSAR and LIDAR, the resolution of black and white aerial photography has not yet been surpassed.

Figure 17.4 shows an aerial photograph stereopair of a watershed and colluvial fan in British Columbia displaying numerous features typical of an active debris-flow watershed.

Figure 17.4. Aerial photograph stereopair of a typical debris flow prone watershed in south-western British Columbia displaying the typical signs such as actively eroding areas, a steep channel, and an active fan.

Vegetation sometimes prohibits or impedes identification of past debris flows using remote sensing. Depending on the flow velocity, flow depth, and boulder size of the debris flow, it is possible that the vegetation on the fan survived. For example, debris flows up to several thousand cubic meters in volume have been observed in the south-western British Columbia rainforest that did not have sufficient impact force to destroy several 100-year-old cedars and Douglas firs, thus avoiding creation of a canopy opening (Brardinoni et al., 2003).

Topographic information can also be used to differentiate if a specific watershed is prone to debris flows or other processes. Jackson et al. (1987) used the Melton ratio to differentiate floods and debris-flow-prone watersheds in the Canadian Rocky Mountains. More recently, Wilford et al. (2004) extended the model by adding watershed length to the Melton ratio. This model was tested on 65 alluvial and colluvial fans in British Columbia and correctly identified 92% of the debris-flow-prone watersheds.

17.2.4 Historical accounts and records

Carrara et al. (2003) found that historical records such as interviews with residents (anecdotal evidence), newspaper articles, and hospital records are used by public administrators without supplementary geomorphic information because they do not require any training in earth sciences for interpretation. These records lack spatial completeness, resolution, and precision, and are biased towards events that caused damage to structures or loss of life, and under-sample debris flows in unpopulated areas. An additional bias is introduced when interviewing residents because human memory is highly selective, and the record will be biased towards more frequent events in the recent past. This approach therefore should be supplemented with other techniques.

17.3 DEBRIS-FLOW PROBABILITY AND MAGNITUDE

A debris-flow hazard can be defined as a combination of debris-flow probability of occurrence and magnitude. *Probability* is usually expressed as an annual probability of occurrence or as a long-term probability (e.g., 10% probability in 50 years). Probability is the likelihood of debris flows to occur in the future, while *frequency*, the inverse of the return period, defines how often an event occurs. Debris-flow *magnitude* is usually expressed as either total debris-flow volume or peak discharge. As was pointed out in Chapter 2, the concept of hazard *intensity* is a useful tool since the debris-flow effects vary with location along the transportation zone and on the fan. Hazard intensity parameters include velocity, runout distance, flow depth, and maximum deposit thickness as well as impact forces and run-up on structures in the debris-flow path.

17.3.1 Debris-flow occurrence probability

Debris-flow occurrence probability is a function of the availability of erodible sediments, the occurrence probability of debris slides, and the frequency at which hydroclimatic thresholds are exceeded. To determine the availability of erodible sediments it is helpful to differentiate basins into *supply-limited* and *supply-unlimited*.

Basin differentiation

Debris flows occur at vastly different return periods. More frequent debris flows usually occur in gullies eroded into unconsolidated Quaternary deposits, loess or volcanic materials that continually provide a supply of sediment to the creek channel. Figure 17.5 shows such a watershed in Japan. A creek of this type is capable of producing a debris flow every time a critical hydroclimatic threshold is exceeded independent of the debris flow being triggered by a landslide or triggered in-channel. This type of basin is termed supply-unlimited or transport-limited; a concept that was originally applied to debris flows by Stiny (1910).

Figure 17.5. Example of a supply-unlimited watershed in Japan (Kamikamihori Gully) in which debris flows likely occur whenever a critical hydroclimatic threshold is exceeded. The blocks in the foreground are up to 3 m in diameter.
Photo: H. Suwa.

Less frequent debris flows occur on streams filled with coarse rockfall debris with high hydraulic conductivity. Figure 17.6 shows such a basin in the Coast Mountains, British Columbia. Debris flows in such basins are only triggered during exceptional climatic events, often combined with unusually high antecedent

Figure 17.6. Example of a channel with slow recharge rates in the southern Coast Mountains of British Columbia where competent rock and infrequent mass movements hinder rapid debris recharge.

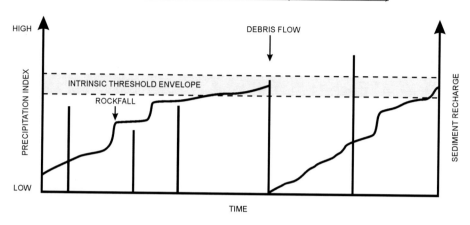

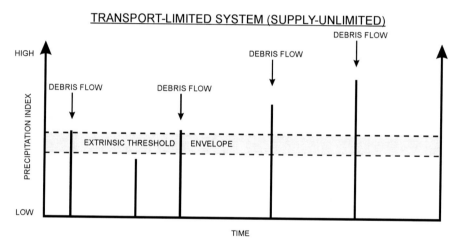

Figure 17.7. Conceptual sketch illustrating the difference between supply-limited and supply-unlimited watersheds with respect to debris flow initiation.

moisture conditions. In-channel mobilization of debris in these basins is atypical. After a debris flow, the channel is often completely scoured. Future occurrence depends on the time needed to recharge the channel with sediment, appropriate antecedent moisture conditions, and the occurrence of another low-frequency climatic event (see Chapter 14). This type of basin is termed supply-limited or weathering-limited.

Figure 17.7 conceptually shows the difference between the two basin types. In supply-limited basins, debris flows are often difficult to recognize as a hazard because of the low frequency.

If debris flows are considered as discrete, rare, independent events (where multiple trials of "similar" events have equally possible outcomes), the probability

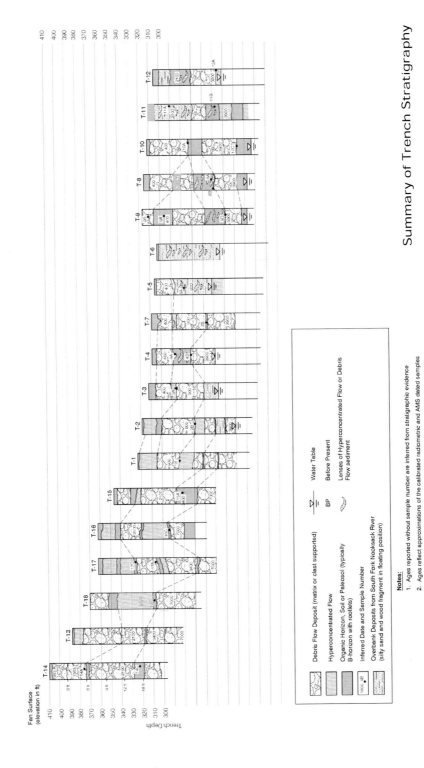

Figure 17.8. Summary of stratigraphic evidence of debris flows during the past 7,000 years at Jones Creek, Whatcom County, Washington State, USA. KWL Ltd (2004).

of occurrence of a debris flow during a period of n years can be estimated using the binomial formula:

$$E_p = 1 - (1 - 1/T)^n \qquad (17.1)$$

where T is the return period of debris flows.

While the first two assumptions mentioned are fulfilled in most debris-flow cases, the condition of independence is only given in cases where E_p is not a function of debris recharge. This condition is not met in most supply-limited basins because the probability of occurrence after a debris flow will be much lower for the same time period. Research by Jakob et al. (in press) found that debris recharge rates decline exponentially as the channel sidewalls and channel bottom stabilize over time.

Dating methods

Debris flows can be dated by relative or absolute methods. The first estimates the relative age of debris-flow deposits (e.g., deposit A is older than deposit B), while the latter estimates a fixed date or age range. Relative dating methods include lichenometry, soil development, and weathering rinds on surface boulders (e.g., Birkeland, 1999; Decaulne and Saemundsson, 2003). Absolute dating methods applicable to debris flows include repeat photography (e.g., Griffith et al., 1996), radiocarbon dating (e.g., Jakob et al. 2003), tephrochronology (e.g., Jackson, 1987), and dendrochronology (e.g., Hupp, 1984). Aside from radiocarbon dating, other methods exist (e.g., ^{3}He, ^{137}Cs) but are rarely used because they are imprecise and their resolution over the time periods between most debris flows is low (Melis et al., 1994; Hereford et al., 1996; Webb et al., 1996).

Particularly the science of dendrogeomorphology has been very well developed over the years with over 100 papers being published on this subject. Descriptions of the methodology and its application to debris-flow dating can be found, for example, in Sigafoss (1964), Jackson (1977), Shroeder (1978), Butler (1979), Hupp (1984), Hupp et al. (1987), and Strunk (1995).

The type of debris-flow dating method is governed by the study's objective. For most hazard analyses the time period over which a frequency–magnitude relationship of a debris flow is required spans several hundred years. A combination of methods is usually the best approach to determine reliable debris-flow frequencies. The following points should be considered when making decisions as to which methods should be used:

- Irrespective of which dating method was used, a bias toward a higher frequency in the more recent past will likely exist, due to erosion of older morphologic evidence used for dating, limiting tree age, lichen age, etc.
- Debris-flow frequency cannot be assumed a constant. The statistical requirement for data stationarity (no long-term trends) for frequency analyses must be examined in each case. For example, in many areas of the world subject to glaciation, erosion rates, and thus debris-flow activity was much greater in the early part of the Holocene when large amounts of unconsolidated and unvegetated materials were readily available for erosion (see Chapter 26). Alternatively,

Table 17.2. Suggested semi-quantitative probability scale for debris flows.
From Hungr (1997).

Probability term	Range of annual probability	Significance
Very high	>1/20	The hazard is imminent, and very likely to occur within the lifetime of a person or structure
High	1/20 to 1/100	The hazard is likely to happen within the lifetime of a person or structure
Moderate	1/100 to 1/500	The hazard occurrence within a given lifetime is possible. A probability of 10% in 50 years (1/475 annual probability) is used by the British Columbia Ministry of Transportation as a design standard for natural hazards
Low	>1/500	The hazard lies outside the probability generally used for debris flow hazard management and decision making. Lower probabilities may be considered for very high risk areas (e.g., in lahar runout areas)

wetter conditions due to climate change may increase debris-flow frequency in supply-unlimited watersheds. Finally, forest fires or land use change in the basin can have a profound effect on the frequency (and magnitude) of debris flows as is demonstrated in some examples in Chapters 15 and 16.

Once debris-flow frequency has been established, a scale can be used to qualify probability and return periods (Table 17.2). In Austria, for example, debris-flow hazard zones are defined by the 150-year return period event, while in British Columbia the 500-year return period has begun to be a standard for land use zoning and construction of debris-flow mitigation measures. In Japan, no such return period is specified. The four-fold scale of debris-flow probabilities shown in Table 17.2 can be integrated into a qualitative hazard analysis.

17.3.2 Debris-flow magnitude

To estimate the debris-flow hazard and to compile hazard maps, debris-flow probability of occurrence must be complemented by an estimate of debris-flow magnitude. Debris-flow magnitude estimates can be expressed as debris-flow volume, peak discharge, or area inundated, all of which are related. The method of estimating and expressing magnitude depends on the primary objective of the hazard analysis:

- debris-flow volume must be estimated if the hazard analysis is to lead to the design of a containment structure;

- peak discharge must be estimated to size culverts, bridge and pipeline crossings, as well as to size the necessary conveyance for the design of deflection berms and channelization works; and
- area inundated is the key variable for planning development in the runout area of a debris flow.

Debris-flow volume and peak discharge also relates to the likelihood of a larger-order stream being impounded at the confluence of the debris-flow stream. Once established, estimates of debris-flow magnitude can be combined to establish a *debris-flow magnitude classification* that may be valuable for overview studies and comparison of debris-flow hazards in different study areas (Table 17.3).

Debris-flow volume

Debris-flow volume is defined as the total amount of inorganic sediment, organic material, and water transported past a specific point of reference (usually the fan apex). Debris-flow volume also needs to be estimated to calibrate debris-flow models (e.g., Hungr, 1995; Griswold, 2004; O'Brien et al., 1993).

Debris-flow volume is a function of three elements:

- the volume of the initiating failure or failures;
- the volumes entrained along the transport reach; and
- the volumes deposited along the transport reach.

This simple model can apply to the majority of small debris-flow creeks, but can be more complicated for events in complex terrain with variable channel. Furthermore, debris-flow material deposited in previous surges can be remobilized in subsequent surges. The equation to determine total debris-flow volume (V_t) reaching the fan apex can be mathematically expressed as:

$$V_t = \Sigma V_i + \Sigma V_e - \Sigma V_d \tag{17.2}$$

with V_i being the initiating failure volume, V_e the entrainment volume, and V_d the deposition volume. When the total debris-flow volume downstream from the fan apex is required, and where the potential for fan scour exists, (17.2) is extended to include the amount of sediment entrained (V_{fe}) and deposited on the fan (V_{fd}) between the fan and the site of interest:

$$V_t = \Sigma V_i + \Sigma V_e - \Sigma V_d + \Sigma V_{fe} - \Sigma V_{fd} \tag{17.3}$$

In practice, all these terms can be very difficult to quantify, and rely heavily on past observations.

Point source volume estimates

Point sources responsible for debris-flow initiation can be divided into direct triggers such as debris avalanches, debris slides, rock avalanches, slumps, rock fall, and wet

Table 17.3. Suggested debris-flow magnitude classification.
Jakob (in press).

Size class	V, range (m^3)	Q_b, range (m^3/s)	Q_v, range (m^3/s)	B_b (m^2)	B_v (m^2)	Potential consequences
1	$<10^2$	<5	<1	$<4 \times 10^2$	$<4 \times 10^3$	Very localized damage, known to have killed forestry workers in small gullies, damage small buildings
2	10^2-10^3	5–30	1–3	$4 \times 10^2 - 2 \times 10^3$	$4 \times 10^3 - 2 \times 10^4$	Could bury cars, destroy a small wooden building, break trees, block culverts, derail trains
3	10^3-10^4	30–200	3–30	$2 \times 10^3 - 9 \times 10^3$	$2 \times 10^4 - 9 \times 10^4$	Could destroy larger buildings, damage concrete bridge piers, block or damage highways and pipelines
4	10^4-10^5	200–1,500	30–300	$9 \times 10^3 - 4 \times 10^4$	$9 \times 10^4 - 4 \times 10^5$	Could destroy parts of villages, destroy sections of infrastructure corridors, bridges, could block creeks
5	10^5-10^6	1,500–12,000	$300-3 \times 10^3$	$4 \times 10^4 - 2 \times 10^5$	$4 \times 10^5 - 2 \times 10^6$	Could destroy parts of towns, destroy forests of 2 km^2 in size, block creeks and small rivers
6	10^5-10^6	N/A	$3 \times 10^3 - 3 \times 10^4$	$>2 \times 10^5$	$2 \times 10^6 - 3 \times 10^7$	Could destroy towns, obliterate valleys or fans up to several tens of km^2 in size, dam rivers
7	10^6-10^7	N/A	$3 \times 10^4 - 3 \times 10^5$	N/A	$3 \times 10^7 - 3 \times 10^8$	Could destroy parts of cities, obliterate valleys or fans up to several tens of km^2 in size, dam large rivers
8	10^7-10^8	N/A	$3 \times 10^5 - 3 \times 10^6$	N/A	$3 \times 10^8 - 3 \times 10^9$	Could destroy cities, inundate large valleys up to one hundred km^2 in size, dam large rivers
9	10^8-10^9	N/A	$3 \times 10^6 - 3 \times 10^7$	N/A	$3 \times 10^9 - 3 \times 10^{10}$	Vast and complete destruction over hundreds of km^2
10	$>10^9$	N/A	$3 \times 10^7 - 3 \times 10^8$	N/A	$>3 \times 10^{10}$	Vast and complete destruction over hundreds of km^2

V is the total volume, Q_b and Q_v are the peak discharge for bouldery and volcanic debris flows, respectively, B_b and B_v are the areas inundated by bouldery and volcanic debris flows, respectively. N/A signifies that bouldery debris flows of this magnitude have not been observed. The constant in (17.2) was rounded so that B by non-volcanic debris flows is 10 times smaller than that of volcanic debris flows.

snow avalanches, and indirect triggers such as outbreaks from landslide dams, moraine dams, mine tailings dams, glacial lakes, log jams, and beaver dams.

Estimating the point source volume for discrete landslides is difficult, but can be accomplished by detailed remote sensing techniques or in the field. Landslide scar areas can be measured photogrammetrically or in the field, and an average depth can be estimated. One problem associated with the use of aerial photos is the inability to capture smaller failures hidden under a dense tree canopy (Pyles and Froehlich, 1987; Brardinoni et al., 2003) because numerous small failures can contribute greatly to sediment recharge in steep watersheds.

A challenge in estimating the volume of individual landslides is to determine which failures occurred during a specific event. Multiple simultaneous failures in a watershed are common in tropical and subtropical environments, but have also been observed in mid-latitude mountains, particularly in clear-cut watersheds. Historical data and very detailed dendrochronological investigations of old landslide scars can help in determining how many failures contributed to a specific debris flow. Another approach is to estimate the magnitude of multiple slope failures in a watershed by multiplying the predicted area of landslide scars by the average erosion depth (see Chapter 7).

Entrainment volume estimates

Debris entrainment (bulking) along channels is comprehensively treated in Chapter 7. This chapter introduces some of the key considerations that can be included in hazard analyses.

To estimate the volume of material that can be entrained in a debris flow along the transport reach it can be simply (and conservatively) assumed that in continuously steep and confined bedrock channels, all stored material will be mobilized and deposited on the fan. This approach, however, can result in significant error.

An example where simple entrainment estimates would not have resulted in reliable volume estimates is the debris flow from Casita volcano in Nicaragua in 1998 described by Scott (2000). A debris slide had transformed into a debris avalanche, then into a hyperconcentrated flow, before transforming into a debris flow and travelling some 6 km before becoming diluted and once again transformed into a hyperconcentrated flow. Each slide/flow phase had a different rheologic character and traversed terrain with variable geotechnical characteristics. Debris entrainment alternated with debris deposition several times during the complex event. The Casita volcano disaster, which killed some 2,000 people, had an originating failure volume 9 times smaller than the total debris-flow volume deposited. Similarly, but on a smaller scale, a debris flow in south-western British Columbia entrained more than 90% of its total volume from a steep colluvial channel which showed no signs of ever having undergone such a deep scour (Jakob et al., 1997).

Volume estimates from fan deposits

Volumes of past debris flows can be estimated by detailed mapping of surficial deposits and from subsurface evidence from test pits and trenches of fan deposits.

Volumes of surficial deposits can be estimated photogrammetrically or by ground-based surveying techniques such as plane tables or electronic distance metering. In densely vegetated terrain, a measuring tape, hip chain, and compass may be more suitable.

Photogrammetry is particularly useful when the fan deposits are visible from the air, where there is little or no vegetation, or the vegetation has been removed by the debris flow. Chronosequential aerial photographs also allow for the approximate dating, or date bracketing, of debris-flow deposits if the area has been re-photographed with respect to the occurrence of debris flows. Usually, fresh debris-flow deposits appear bright on both black and white and colored aerial photographs.

Depth estimates represent the more difficult aspects of volume estimates. They usually involve the interpretation of limited data since, in the case of older deposits, natural cross sections are rare and can yield ambiguous depths. However, it is often possible to estimate at least the depth of the most recent debris flow. Volume–area relationships can then be used to determine the approximate volumes of older deposits with known areas (e.g., Jakob and Podor, 1995).

Test pits, test trenches, and boreholes can provide information of the stratigraphy of the fan (see Chapter 26). For example, a 7,000-year chronology of large debris flows was determined at Jones creek fan in Washington State by excavating 18 test trenches up to 6 m deep, logging the stratigraphy, and radiocarbon dating the individual deposits (Jakob, 2003). Figure 17.8 provides an example of the stratigraphic evidence collected from this study.

Ground penetrating radar (Ekes and Friele, 2003), electronic sensitivity soundings, or seismic methods can also yield valuable information on the fan architecture but should be calibrated with some observational data for better interpretation of the geophysical output.

Debris-flow volume–peak discharge relationships

Another approach to estimate debris-flow volume is to correlate it with peak discharge. Mizuyama et al. (1992) collected data worldwide in which both debris-flow volume and peak discharge were known. Linear regression analysis yielded an equation that allowed the estimation of debris-flow volumes from peak discharge estimates. Dividing Mizuyama et al.'s sample set into muddy debris flows and bouldery debris flows improved the precision of this simple method. Jakob and Bovis (1996) used the same method to predict debris-flow volumes at numerous sites in south-western British Columbia. Rickenmann (1999) summarized approaches used by a number of researchers and compared the volume–peak discharge relationships. He found a large regional variety of predictive equations and emphasizes the need for regional calibration. Table 17.4 summarizes the debris-flow volume–peak discharge relationships determined by a number of researchers.

Volume estimations from watershed characteristics

Many researchers have correlated debris-flow volume with watershed characteristics (Hampel, 1977; Ikeya, 1981; Okubo and Mizuyama, 1981; Takahashi, 1981;

Table 17.4. Equations for indirect determination of debris-flow velocities.

Equations	Author	Equation number
$v = (gr_c \cos \Theta \tan \alpha)^{0.5}$	Chow (1959)	(17.4)
$v = (2g\Delta h)^{0.5}$	Chow (1959)	(17.5)
$v = (1.21g\Delta h)$	Wigmosta (1983)	(17.6)
$v = (\gamma S/K\mu)H^2$	Hungr et al. (1984)	(17.7)
$v = 2.1Q^{0.33} S^{0.33}$	Rickenmann (1999)	(17.8)
$v = (\gamma S/K\mu_B)H^2 F$	Jordan (1994)	(17.9)

v is debris-flow velocity, r is radius of curvature of the channel bend, α is the channel gradient, Θ is the superelevation gradient, Δh is the runup height, g is the mass acceleration constant, μ is the dynamic viscosity of the debris flow, μ_B is the Bingham viscosity, S is the channel slope, γ is the unit weight, H is the flow thickness, and K is a shape factor for various channel forms.

Watanabe, 1981; Ikeya and Mizuyama, 1982; Mizuyama, 1982; Kronfellner-Kraus, 1983; Hungr et al., 1984; Takei, 1984; VanDine, 1985; Johnson et al., 1991; Jakob and Bovis, 1996; D'Agostino et al., 1996; Marchi and Tecca, 1996; Bottino and Crivellari, 1998; Bianco and Franzi, 2000). Most have attempted to correlate debris-flow volume with either basin area (A_B) or channel length (L_C).

Several authors have recognized that a univariate approach oversimplifies the correlation and have used variables describing basin geometry and ruggedness as well as the geotechnical properties of the materials. Some of those variables include basin hypsometry (H_B), relief ratio (R_r), bifurcation ratio (B_r), basin elongation ratio (B_{er}), drainage density (D_d), total basin relief (Z_T), basin ruggedness (N_M) (Johnson et al., 1991; Jakob and Bovis, 1996). Further differentiation between supply-limited and supply-unlimited basins improves the predictive capabilities to an extent that the multiple regression equations can be used for cursory volume and peak discharge estimates in similar climatic and geologic environments (Bovis and Jakob, 1999).

One shortcoming with many of the debris-flow volume–watershed characteristic correlations that use individual parameters is the introduction of multiplicative effects and spurious correlations when interdependent variables, such as creek gradient and basin area, are correlated. Some researchers have created specific indices to better relate geologic information and debris-flow volume (D'Agostino et al., 1996; Bottino and Crivellari, 1998; Bovis and Jakob, 1999; Bianco and Franzi, 2000).

Multivariate approaches suggested by Johnson et al. (1991), D'Agostino et al. (1996), Jakob and Bovis (1996), Bovis and Jakob (1999), and Bianco and Franzi (2000) usually yield better correlations, but cannot be transferred beyond the area or region for which they have been developed. Even within the calibration region, the error in volume estimates is usually too large for individual fan studies that require accurate volume estimates necessary for land-use decisions and the design of debris-flow mitigation measures (Rickenmann, 1999). For this reason the application of volume estimates from watershed characteristics should be reserved for overview

Debris-flow peak discharge

Debris-flow discharge is an important variable when designing debris-flow mitigation structures such as deflection berms, culverts, flumes, bridges, debris-flow barriers, and check dams. Debris-flow peak discharge is defined as:

$$Q_{max} = A_{max} v_t \qquad (17.10)$$

where A_{max} is the maximum cross-sectional area of a debris-flow and v_t is the velocity during time t when the maximum cross-section flow occurs.

Debris-flow peak discharge can rarely be measured directly. In those cases where it is measured directly (usually along instrumented creeks), the design event may not occur within the time frame in which instrumentation is set up. At least decades of observations are necessary to estimate a frequency–magnitude relationship that would allow for the estimation of a frequency–peak discharge relationship that can be applied to 100-year or 500-year return periods. For very active volcanoes with eruption periods within the design return period, this may be even more difficult since the eruptive cycles may constitute the primary variable in determining debris-flow magnitude.

For the above reasons, debris-flow peak discharge is usually estimated by indirect methods. These methods can be classified into (i) field observations (ii) empirical methods, and (iii) numerical methods.

Field observations – cross-sectional area

Field observations include cross-sectional measurements along preferably bedrock channel sections where unambiguous physical evidence of debris flows is still visible. As outlined in Section 17.3.1, typical physical evidence includes mudlines or scour marks on the channel sidewalls well above the flood discharge level. Relatively small cross-sectional areas can be measured by measuring tape, inclinometer, and stadia rod. For larger cross-sectional areas, detailed topographic maps and hand-held laser range finders are useful.

Velocity

Debris-flow velocity can be indirectly estimated from well-known fluid mechanics equations that use flow superelevation, runup against obstacles, or channel characteristics (e.g., Chow, 1959). Some of these equations are summarized in Table 17.5.

The superelevation equation (Equation (17.4) in Table 17.4) is based on the assumption that the square of the velocity based on an average cross-sectional area can be substituted for the mean velocity, that the channel gradient is constant, and that the channel width is much smaller than the radius of curvature.

Wigmosta (1983) found that these assumptions produced compensating errors, though the first assumption may lead to a net underestimate of the velocity. Using large-scale flume experiments, Iverson et al. (1994) showed that the superelevation

Table 17.5. Equations for indirect determination of debris-flow peak discharge, Q_p.

Equation	Author	Equation number
$Q_p = 0.135 V^{0.78}$ (bouldery debris flows)	Mizuyama et al. (1992)	(17.11)
$Q_p = 0.019 V^{0.79}$ (muddy debris flows)	Mizuyama et al. (1992)	(17.12)
$Q_p = 0.006 V^{0.83}$ (volcanic debris flows)	Jitousono et al. (1996)	(17.13)
$Q_p = 0.04 V^{0.90}$ (bouldery debris flows)	Bovis and Jakob (1999)	(17.14)
$Q_p = 0.003 V^{1.01}$ (volcanic debris flows)	Bovis and Jakob (1999)	(17.15)
$Q_p = 0.293 V_w^{0.56}$	Costa (1988)	(17.16)
$Q_p = 0.016 V_w^{0.64}$	Costa (1988)	(17.17)
$Q_p = 0.1 V^{0.83}$	Rickenmann (1999)	(17.18)

V is debris-flow volume and V_w is the water volume behind the natural dam.

formula provides reasonable velocity estimates. The second assumption may be violated when cross sections for peak discharge estimates are obtained in channel bends (Webb et al., 1989; Jakob et al., 1996). This is due to the tendency of particularly fast and fine-grained debris flows to display strong concavities in their flow surface that may exaggerate the peak discharge estimates (Scott and Yuyi, 2003).

The runup or velocity head equation can be applied to debris flows impacting obstructions oriented perpendicular to the flow direction. The equation assumes that all kinetic energy of the moving object is converted to potential energy. The runup equation was found to yield velocities up to 30% lower than those observed in large-scale flume experiments (Iverson et al., 1994). Wigmosta's (1983) equation (Equation (17.6) in Table 17.4) is based on a combination of empirical measurements and theoretical analysis of laminar viscous flow around a cylinder. Wigmosta found errors less than 15% for Reynolds numbers greater than 20.

Hungr et al.'s (1984) formula (Equation (17.7) in Table 17.4) is based on a Newtonian model and uses a shape factor of 3 for infinitely wide rectangular channels and 8 for semicircular channels based on resistance experiments by Straub et al. (1958) based on earlier work by Boussinesq (1868).

Ideally, the results from several indirect observation and equations should be obtained. Estimated velocities should be rounded to one significant number since debris-flow velocities are highly variable and for most applications, more precise measurements of debris-flow discharge is unnecessary.

Empirical correlations – peak discharge and volume

Debris-flow discharge can be correlated to (i) debris-flow volume or (ii) watershed characteristics. The first approach was pioneered by Mizuyama et al. (1992) and was followed by others (Jakob and Bovis, 1996; Rickenmann, 1999) due to its ease of use. Rickenmann (1999) provides a detailed summary of methods to determine debris-flow discharge as well as volume. Table 17.5 summarizes the different empirical equations used by various researchers for regional and global data sets. Figure 17.9 shows the curves for the data sets in Table 17.5 and shows the large variability that

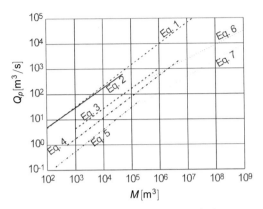

Figure 17.9. Summary of peak discharge–total volume correlations compiled by Rickenmann (1999).

Reproduced with permission

can be explained by variable debris-flow rheology (muddy vs. bouldery flows), initiation mechanism, and/or channel morphology. Figure 17.9 also shows that empirical correlations need to be verified regionally. Where a conservative hazard analysis is required, it is prudent to use the 95% or 99% confidence interval as the design event, rather than the best fit as indicated by linear regression analysis. Given the variability of the equations presented in Table 17.4, it is recommended to only use empirical correlations for regional studies in which a rough estimate of debris-flow peak discharge is needed.

The large difference in discharge–volume correlations between muddy and bouldery debris flows can be explained by studying the flow characteristics of each. Muddy debris flows are often initiated in fine-grained volcanic soils or loess, travel along low-gradient streams, and usually occur as many individual surges. In contrast, bouldery flows are generally coarse-grained, often follow steep mountain channels with very high friction slopes, and result in few slower moving coarse and steep surge fronts, often followed by a tail of finer grained, faster moving debris.

Empirical correlations – watershed charcteristics

Peak discharge can also be correlated to watershed characteristics (Jakob, 1996). Using a larger number of morphometric variables and indices, Bovis and Jakob (1999) developed empirical equations for the estimation of Q_p:

$$\log Q_p = -0.77 + 1.66 Z_T + 0.30 A_\% \quad \text{(supply-limited basins)} \quad (17.19)$$

$$Q_p = 40 A_I^{0.72} N_S^{0.40} \quad \text{(supply-unlimited basins)} \quad (17.20)$$

The adjusted coefficient of variance (r^2) is 0.74 for both equations, showing a satisfactorily degree of estimation at least in the geographical region (the southwestern Coast Mountains of British Columbia) in which they were calibrated.

Discharge–volume relationships (Equations 9–17) cannot be used to determine peak discharge of the design event. While empirical correlations that have been calibrated for a specific geographic area may be useful for regional overview studies, they are often too imprecise for detailed creek-specific debris-flow hazard analyses.

Correlations with water discharge

Attempts have been made to correlate flood discharge with debris-flow discharge. At present, this correlation forms the basis of calculating debris-flow peak discharge in Taiwan (Jan et al., 2003). This work was pioneered by Takahashi (1978, 1991) stating that:

$$Q_p = C^*/(C^* - C)Q_w \tag{17.21}$$

where C^* is the maximum sediment concentration and C is the average sediment concentration. Jan et al. (2003) rewrote (17.21) as $Q_t = Q_w/(1 - C_v)$ where Q_t is the debris-flow discharge at any time t, Q_w is the water peak discharge, and C_v is the volumetric sediment concentration of the debris flow. Using a probabilistic approach, Jan et al. then computed the design debris-flow peak discharge (Q_d) as:

$$Q_d = Q_t + \beta S(G) \tag{17.22}$$

where β represents a safety index based on the normal probability distribution and $S(G)$ is the standard deviation of $G(S(G) = Q_d - Q_t)$. Jan et al. (2003) noted that (17.21) is only valid for in-channel debris-flow initiation since it does not account for point source failure volumes. A similar approach is used in Japan to determine the discharge of debris flow triggered by in-channel mixing of water and sediment to form a debris flow (Japan Ministry of Construction, 1979).

It is important to realize that debris flow peak discharge is largely dependent on the initiation mechanism (discrete landslide point source vs. in-channel mobilization), the amount of debris entrained and deposited in the channel, and channel morphology. None of these three variables are included in (17.20) and (17.21), which suggests the equations may be unreliable in accurately estimating debris flow peak discharge even if sediment concentrations are known.

Area inundated

The area inundated by debris flow will influence land-use decisions and the selection and design of mitigation measures. Iverson et al. (1998) and Griswold (2004) have found a correlation between the area inundated by a debris flow (B) and the volume of a debris flow (V):

$$B = 200 V^{2/3} \quad \text{(volcanic debris flows, lahars)} \tag{17.23}$$

$$B = 20 V^{2/3} \quad \text{(non-volcanic debris flows)} \tag{17.24}$$

Regression equations for lahars (Equation 17.23) and non-volcanic debris flows (Equation 17.24), in conjunction with Geographic Information Systems (GISs) can provide a relatively simple and objective alternative to traditional methods of determining the area inundated or to the computer modeling of debris flows.

The order-of-magnitude difference between the two equations can be explained by the higher mobility of volcanic debris flows that allows debris to spread at much lower slopes than coarser grained non-volcanic flows (see Chapter 10). In particular (17.24) should not be applied without comparing the calibration data set with local site conditions. For example, debris flows in clay-rich sedimentary or altered metamorphic rocks may result in a relationship intermediate between (17.23) and (17.24).

17.3.3 Debris flow intensity

Debris flow intensity includes parameters such as maximum flow velocity, maximum flow depth, and the deposit thickness (e.g., Hungr, 1997). These parameters are discussed below:

- Maximum flow velocity influences impact forces and runout distance. Indirect methods of estimating maximum flow velocity have been discussed in Section 17.3.2.
- Maximum flow depth is a variable needed to assess the vulnerability of structures to debris flow impact and to design mitigation measures.
- Deposit thickness must be known to determine the effectiveness of debris flow deflection berms and debris basins.

Runout distance will determine the type and scale of structural debris flow mitigation measures. At the same time it will be instrumental in delineating land-use zones and dictate restrictions and covenants for development. Methods of determining runout distance have been discussed in detail in Chapter 13.

17.4 FREQUENCY–MAGNITUDE RELATIONSHIPS

Frequency–magnitude relationships are a necessary input for debris flow hazard analysis, evaluation, and the decision for debris flow mitigation measures because they allow estimation of the debris flow magnitude for any given return period. The following sections provide general information on frequency–magnitude relationships of debris flows, followed by two approaches by which frequency–magnitude relationships have been used for land-use planning decisions and the design of debris flow mitigation structures. This chapter deals with site-specific applications and therefore does not consider small-scale overview studies focusing on area-wide frequency–magnitude relationships (e.g., Innes, 1985; Pelletier et al., 1997; Hovius et al., 2000; Stark and Hovius, 2001; Brardinoni and Church, 2004).

17.4.1 Process recognition

In single watersheds, debris flow frequency–magnitude relationships can be established using the methods outlined in this chapter. It is important to differentiate the frequency and magnitude relationships of the hazardous processes (e.g., debris flows vs. hyperconcentrated flows vs. floods) and to identify the originating failure mechanisms. For example, rock avalanches in volcanic terrain that transform into debris flows will likely have distinctly different frequency–magnitude relationships than debris flows triggered by smaller, relatively frequently occurring debris avalanches or triggered in-channel by a critical discharge threshold. The significance of variable frequency–magnitude relationships is illustrated in Figure 17.10 that shows a frequency–magnitude curve for debris flows at Cheekye River in south-western British Columbia (see Chapter 26). In this case, for a 500-year return period debris flow, a simple power law (straight line between the two end points) would

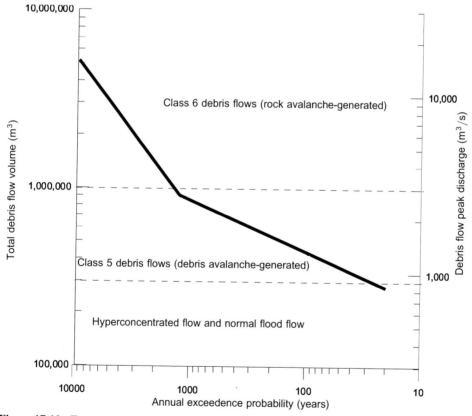

Figure 17.10. Frequency–magnitude graph of debris flows at Mount Garibaldi separating debris flows of different initiation mechanisms.
Jakob and Friele (in press).

have resulted in a volume estimate of approximately 1.6 million m^3 while process differentiation suggests a design volume of only 800,000 million m^3.

17.4.2 The design magnitude approach

The previous sections have demonstrated a number of methods to determine both parameters of hazard – probability and magnitude. For any land-use decision (zonation, property acquisitions, etc.) or for the design of mitigative structures, a particular debris flow magnitude must be estimated. Using the concept of frequency–magnitude, a particular debris flow magnitude will have a corresponding return period. In many countries floodplain maps and floodplain legislation define the *design flood* as the 100-year or 200-year return period flood. There is no such well-defined return period for debris flows. This is likely because debris flow frequency–magnitude relationships are more difficult to estimate and because many insurance companies do not cover landslide (including debris flow) loss or damage. Some local governments, though, are beginning to use return periods to define design hazardous events, including debris flows. In some jurisdictions in British Columbia and the USA, a return period of 500 years has been chosen to define the design debris flow (personal commun., Whatcom County, Washington State, 2003).

Unlike for floods, the episodic nature of debris flows does not allow the production of frequency–magnitude curves from gauged streams. Therefore, the estimation of the magnitude of debris flows with known, or time-bracketed, dates is necessary. Once a number of debris flows have been dated and their respective magnitudes (volume, peak discharge or area inundated) have been determined, a frequency–magnitude curve can be produced (e.g., Figure 17.10). From Figure 17.10, the 500-year, or any other chosen return period, can be read. In some cases, frequency–magnitude curves for floods can be added if required (Jakob and Jordan, 2001).

17.5 DEBRIS FLOW HAZARD MAPPING

Once debris flow frequency–magnitude curves have been produced, hazard can be expressed on topographic maps of suitable scale. Debris flow hazard maps can be defined as the topographic expression of zones of equal debris flow intensity (maximum flow depth and velocity, runout distance and area, and final deposition depth). Methods of how to arrive at intensity variables have been discussed in Chapter 13 and earlier in this chapter. In cases where the area inundated is the only variable illustrated it is also possible to combine flows of different return periods. An example of this approach is shown in Figure 10.9 (see color section) that synthesizes more than a decade of forensic evidence collected by researchers of the US Geological Survey. They mapped the likely area affected from lahars around Mount Rainier for return periods <100 years, 100–500 years, and 500–1,000 years (Vallance and Scott, 1997). For this mapping project, the estimates of debris flow intensity need to be transferred to specific locations (e.g., channel section to be spanned by a bridge or culvert, pipeline crossing, or debris barrier) or to the potential runout (usually the debris fan or confluence of the debris flow stream with a higher-order stream). In the case of Figure 10.9, the hazard is expressed by runout distance and area inundated, rather than classified into high, medium, and low.

A different approach to that used at Mount Rainier uses a specific return period and then plots ranges of hazards for this specific debris flow. The use of several colors facilitates the delineation of ranges of hazards on creek fans. Figure 17.11 (see color section) shows an example of this kind of plotting for a small debris flow fan near Vancouver, British Columbia in which red is designated the highest hazard, orange a moderate hazard, and yellow a low hazard (see caption for definitions of high, moderate, and low hazards). Areas of no hazard are left white. Isohazard zones (zones with an equal level of estimated hazard) are defined by deposition depth, flow velocity, and largest boulder size transported as these variables determine impact forces and thus consequences of debris flow impact.

Drawing lines of equal hazard on a map, though science-based, can be an artful exercise. Because of possible legal implications and far-reaching land-use decisions, the exact location of each line must be scrutinized carefully. Ideally, the delineation of hazard zones should be replicable by others. This, however, is difficult to achieve given the use of different methods and the different opinions of the practitioners. In Figure 17.11 (see color section), the boundaries between hazard zones were purposely graded from one to another to avoid the illusion of precision, which is rarely achievable in debris flow hazard zoning (KWL, 2003).

A fan-specific debris flow hazard map is a snapshot in time. Channel scour by fluvial processes, debris flows with less-than design magnitude, landscaping, and construction of roads, bridges and buildings may alter the fan surface topography and change one or several of the debris flow intensity parameters, thus changing the hazard delineation on the fan. In this case, local regulators have to decide if and when to re-draw a hazard map. Budget allowances should be made to allow for such work. Similarly, the construction of debris flow mitigation structures will require a post-construction hazard map on which regulators can base landuse decisions.

17.6 CONCLUSIONS

In geoscience and engineering practice in many parts of the world there is an increasing demand for quantitative debris flow hazard analysis. Debris flow hazard analyses are no trivial exercise. Depending on the scale of the investigation, costs associated with a single fan study can range between tens of thousands to hundreds of thousands of US dollars. Debris flow hazard analyses require knowledge in the fields of Quaternary stratigraphy, sedimentology, absolute and relative dating methods, geomorphology, hillslope processes, soil and rock mechanics, fluid mechanics, hydrology, and sometimes volcanology or botany, skills that are rarely mastered by a single practitioner. Furthermore, computer models that facilitate debris flow modeling and hazard estimates, on which mitigation decisions can be based, should become part of the hazard analysis (see Chapter 18). These skills should be coupled with some basic understanding of the regulatory framework in a given jurisdiction.

Debris flow hazard analyses will always require a combination of fieldwork, numerical methods, and some degree of judgment. An understanding and evaluation of uncertainty is not only necessary but also strengthens the credibility in geotechnical engineering in general and specifically in debris flow hazard analyses (Whitman, 2000; Einstein and Karam, 2001).

The evolution of empirical methods and multi-dimensional computer models for debris flow runout will provide the practitioner with an increasing number of tools to assess debris flow hazard. However, it is stressed that empirical equations and computer modeling are only as good as their input. Fieldwork is still crucial to a good understanding of the debris flow hazard.

During the hazard analysis, tradeoffs must be made as to the data necessary to complete an assignment, desirable and affordable (NRC, 1994), and it is important to remember that "the things we would like to know may be unknowable" (Southwood, 1985).

This chapter has summarized and evaluated tools and methods for debris flow hazard analyses, and has provided some guidelines of how to carry out such analyses. Chapter 18 builds upon this knowledge by addressing different debris flow mitigation strategies.

17.7 ACKNOWLEDGEMENTS

This chapter has been improved through a review by Doug VanDine and Oldrich Hungr. Kris Holm helped with the literature search.

17.8 REFERENCES

AGS (2000). Landslide risk management concepts and guidelines. *Australian Geomechanics*, **35**(1), 51–92.

Bianco, G. and Franzi, L. (2000) Estimation of debris flow volumes from storm events. In: G.F. Wieczorek and N.D. Naeser (eds), *Debris-flow Hazards Mitigation: Mechanics, Prediction, and Assessment: Proceedings of 2nd International DFHM Conference, Taipei, Taiwan, August 16–18* (pp. 441–448). A.A.Balkema, Rotterdam.

Birkeland, P. (1999) *Soils and Geomorphology* (448 pp.). Oxford University Press, New York.

Bottino, G and Crivellari, R. (1998) Analisi di collate detritiche connesse con l'evento alluvionale del 5-6 Novembre 1994 nell'anfiteatro morenico de Ivrea. *Hydrogeological Risk, Countermeasures and Use of the Canavese Territory: Proceedings National Conference, Ivrea, Italy* (pp. 36–46) [in Italian].

Boussinesq, J. (1868) Mémoire sur l'influence des frottements dans les mouvements réguliers des fluides. *Journal de Mathématiques Pures et Appliquées, Series 2*, **13**, 377–424 [in French].

Bovis, M.J. and Jakob, M. (1999) The role of debris supply conditions in predicting debris flow activity. *Earth Surface Processes and Landforms*, **24**, 1039–1054.

Brardinoni, F. and Church, M. (2004) Representing the landslide magnitude-frequency relation: Capilano River basin, British Columbia. *Earth Surface Processes and Landforms*, **29**(1), 115–124.

Brardinoni, F., Hassan, M.A., and Slaymaker, O. (2003) Landslide inventory in a rugged forested watershed: A comparison between air-photo and field survey data. *Geomorphology*, **54**, 179–196.

Butler, D. (1979) Snow avalanche path terrain and vegetation, Glacier National Park, Montana. *Arctic and Alpine Research*, **11**(1), 17–32.

Carrara, A., Crosta, G., and Frattini, P. (2003) Geomorphological and historical data in assessing landslide hazard. *Earth Surface Processes and Landforms*, **28**, 1125–1142.

Cave, P.W. (1992) Natural hazards, risk assessment and landuse planning in British Columbia: Progress and problems. *First Canadian Symposium on Geotechnique and Natural Hazards, Vancouver, BC* (pp. 1–12). Bitech Publishers.

Chow, V.T. (1959) *Open Channel Hydraulics* (680 pp.). McGraw-Hill, New York.

Costa (1988) Floods from dam failures. In: V.R. Baker and P.C. Patton (eds), *Flood Geomorphology* (pp. 439–463). John Wiley & Sons, New York.

D'Agostino, V., Cerato, M., and Coali, R. (1996) Il trasporto solido di eventi estremi nei torrenti del trentino orientale. *Proceedings International Symposium "Interpraevent", Garmisch-Partenkirchen, Germany* (Vol. I, pp. 377–386). Internationale Forschungsgesellschaft Interpraevent, Klagenfurt, Austria.

Dai, F.C. and Lee, C.F. (2001) Frequency-volume relation and prediction of rainfall-induced landslides. *Engineering Geology*, **59**, 253–266.

Decaulne, A. and Saemundsson, T. (2003) Debris-flow characteristics in the Gleidarhjalli area, northwestern Iceland. *Debris-flow Hazards Mitigation: Mechanics, Prediction, and Assessment: Proceedings of 3rd International DFHM conference, Davos, Switzerland* (pp. 1107–1128). Millpress, Rotterdam.

Einstein, H.H. and Karam, K.S. (2001) Risk assessment and uncertainties. In: M. Kühne, H.H. Einstein, H. Krauter, H. Klapperich, and R. Pöttler (eds), *Proceedings of the International Conference on Landslides, Davos* (pp. 457–488). Verlag Glückauf, Essen, Germany.

Ekes, C. and Friele, P.A. (2003) Sedimentary architecture and post-glacial evolution of Cheekye fan, southwestern British Columbia, Canada. In: C.S. Bristow and H.M. Jol (eds), *Ground Penetrating Radar in Sediments* (Special Publication 211, pp. 87–98). Geological Society of London.

Fell, R. (1994) Landslide risk assessment and acceptable risk. *Canadian Geotechnical Journal*, **31**, 261–272.

Fiebiger, G. (1997) Zonage des risques naturels en Autriche [Natural hazard risk zoning in Austria]. *France–Autriche conférence en restauration du terrain en montagne Grenoble, France*. Translated in *Journal of Torrent, Avalanche, Landslide and Rockfall Engineering*, **134**(61).

Griffith, P.G., Webb, R.H., and Melis, T.S. (1996) *Initiation and Frequency of Debris Flows in Grand Canyon, Arizona* (USGS Open-file report 96-491). US Geological Survey, Reston, VA.

Griswold, J.P. (2004) Mobility statistics and hazard mapping for non-volcanic debris flows and rock avalanches. Master thesis, Portland State University, OR.

Hovius, N., Stark, C.P., Chu, H.-T., and Lin, J.-C. (2000) Supply and removal of sediment in a landslide-dominated mountain belt: Central Range, Taiwan. *Journal of Geology*, **108**, 73–89.

Hungr, O. (1997) Some methods of landslide hazard intensity mapping (Invited paper). In: R. Fell and D.M. Cruden (eds), *Proceedings of Landslide Risk Workshop* (pp. 215–226). A.A. Balkema, Rotterdam.

Hungr, O. (2002) Hazard and risk assessment in the runout zone of rapid landslides (Keynote paper). *Natural Terrain Hazards, a Constraint to Development? Proceedings, Annual Meeting of the Institution of Mining and Metallurgy, Hong Kong* (pp. 10–25). Institution of Mining and Metallurgy, Hong Kong.

Hungr, O., Morgan, G.C., and Kellerhals, R. (1984) Quantitative analysis of debris torrent hazards for design of remedial measures. *Canadian Geotechnical Journal*, **21**, 663–677.

Hampel, R. (1977) Geschiebewirtschaft in Wildbächen. *Wildbach- und Lawinenverbau*, **41**(1), 3–34.

Hereford, R., Thomson, K.S., Burke, K.J., and Fairley, H.C. (1996) Tributary debris fans and the late Holocene alluvial chronology of the Colorado River, eastern Grand Canyon, Arizona. *Geological Society of America Bulletin*, **108**, 3–19.

Hupp, C.R. (1984) Dendrogeomorphic evidence of debris flow frequency and magnitude at Mount Shasta, California. *Environment Geology and Water Sciences*, **6**(2), 121–128.

Hupp, C.R., Ostercamp, W.R., and Thornton, J.L. (1987) *Dendrogeomorphic Evidence and Dating of Recent Debris Flows on Mount Shasta, Northern California* (USGS Professional Paper 1396-B, 39 pp.). US Geological Survey, Reston, VA.

Ikeya, H. (1981) A method of designation for area in danger of debris flow. In: *Erosion and Sediment Transport in Pacific Rim Steeplands* (pp. 576–587). International Association of Hydrological Sciences, Christchurch, New Zealand.

Ikeya, H. and Mizuyama, T. (1982) *Flow and Deposit Properties of Debris Flows* (Report 157-2, pp. 88–153). Public Works Research Institute, Tsukuba, Japan [in Japanese].

Innes, J.L. (1985) Magnitude-frequency relations of debris flows in northwest Europe. *Geografiska Annaler*, **67A**(1/2), 23–32.

IUGS Working Group on Landslides, Committee on Risk Assessment (1997) Quantitative risk assessment for slopes and landslides: The state of the art. In: D.M. Cruden and R. Fell (eds), *Proceedings of the International Workshop on Landslide Risk Assessment, Honolulu* (pp. 3–12). A.A. Balkema, Rotterdam.

Iverson, R.M., LaHusen, R.G., Major, J.J., and Zimmerman, C.L. (1994) Debris flow against obstacles and bends: Dynamics and deposits (Abstract). *Eos*, **75**(44), 274.

Iverson, R.M., Schilling, S.P., and Vallance, J.W. (1998) Objective delineation of lahar-inundation hazard zones. *Geological Society of America Bulletin*, **110**(8), 972–984.

Jackson, L.E. (1977) *Dating and Recurrence Frequency of Prehistoric Mudflows near Big Sur, Monterey County, California* (USGS Professional Paper 1250, pp. 461–478). US Geological Survey, Reston, VA.

Jackson, L.E., Kostachuk, R.A., and MacDonald, G.M. (1987) Identification of debris flow hazard on alluvial fans in the Canadian Rocky Mountains. In: J.E. Costa and G.F. Wieczorek (eds), *Debris Flows/Avalanches: Process, Recognition, and Mitigation* (Reviews in Engineering Geology No. VII). Geological Society of America, Boulder, CO.

Jakob, M. (1996) Morphometric and geotechnical controls of debris flow frequency and magnitude in southwestern British Columbia. Ph.D. thesis, University of British Columbia.

Jakob, M. (in press) A debris flow size classification. *Engineering Geology*.

Jakob, M. and Bovis, M.J. (1996) Morphometric and geotechnical controls of debris flow activity, southern Coast Mountains, British Columbia, Canada. *Zeitschrift für Geomorphologie, Suppl.*, **104**, 13–26.

Jakob, M. and Friele, P. (in press) A 200-year history of debris flows at Cheekye River, British Columbia. *Canadian Journal of Earth Sciences*.

Jakob, M. and Jordan, P. (2001) Design flood estimates in mountain streams. *Canadian Journal of Civil Engineering*, **28**(3), 425–439.

Jakob, M. and Podor, A. (1995) Frequency and magnitude of debris flows. *48th Canadian Geotechnical Conference, September 25–27, Vancouver, British Columbia* (pp. 491–498). Canadian Geotechnical Society, Vancouver.

Jakob, M., Hungr, O. and Thomson, B. (1997) Two debris flows with anomalously high magnitude. In: C-L. Chen (ed.), *Debris-flow Hazards Mitigation: Mechanics, Prediction and Assessment: Proceedings of the 1st International Conference, American Society of Civil Engineers* (pp. 382–394). American Society of Civil Engineers, New York.

Jakob, M., Weatherly, H., and Pittman, P. (2003) 7000 years of debris flow history at Jones Creek, Whatcom County, USA. *Geological Society of America Annual Conference, Seattle, November 2–5* (Program and Abstracts). Geological Society of America, Boulder, CO.

Jakob, M., Bovis, M.J., and Oden, M. (in press) Estimating debris flow magnitude and frequency from channel recharge rates. *Earth Surface Processes and Landforms*.

Jan, C.D., Lee, M.H., and Chen, J.C. (2003) Reliability analysis of design discharge of debris flow. In: D. Rickenmann and C-L. Chen (eds), *Debris-flow Hazards Mitigation: Mechanics, Prediction and Assessment* (pp. 1163–1171). Millpress, Rotterdam.

Jitousono, T., Shimokawa, E., and Tsuchiqa, S. (1996). Debris flow following the 1994 eruption with pyroclastic flows in Merapi volcano, Indonesia. *Journal of the Japanese Society of Erosion Control Engineering*, **48**, 109–116.

Johnson, P.A., McCuen, R.H., and Hromadka, T.V. (1991) Magnitude and frequency of debris flows. *Journal of Hydrology*, **123**, 69–82.

Jordan, P.R. (1994) Debris flows in the southern Coast Mountains, British Columbia: Dynamic behaviour and physical properties (258 pp.). Ph.D. thesis, University of British Columbia, Vancouver.

Kronfellner-Kraus, G. (1983) Torrect erosion and its control in Europe and some research activities in this field in Austria. *SABO – The Erosion Control Engineering Society, Japan*, **3**(126), 33–44.

KWL Ltd (2003) *Debris Flow Study and Risk Mitigation Alternatives for Percy Creek and Vapour Creek* (Final report, December). District of North Vancouver.

Marchi, L. and Tecca, P.R. (1996) Magnitudo delle collate detritiche nelle Alpi Orientali Italiane. *Geoingegneria Ambientale e Mineraria*, **33**(2/3), 79–86.

Melis, T.S., Webb, R.H., Griffith, P.G., and Wise, T.J. (1994) *Magnitude and Frequency Data for Historic Debris Flows in Grand Canyon National Park and Vicinity, Arizona* (USGS Water Resources Investigations Report 94-4214, 285 pp.). US Geological Survey, Reston, VA.

Mizuyama, T. (1982) Analysis of sediment yield and transport data for erosion control works. In: *Recent Developments in the Explanation and Prediction of Erosion and Sediment Yield: Proceedings of the Exeter Symposium* (No. 137, pp. 177–182). International Association of Hydraulics Research.

Mizuyama, T., Kobashi, S., and Ou, G. (1992) Prediction of debris flow peak discharge, *Proceedings of the International Symposium "Interpraevent", Bern, Switzerland* (Vol. 4, 99–108). Internationale Forschungsgesellschaft Interpraevent, Klagenfurt, Austria.

NRC (1994) *Science and Judgment in Risk Assessment*. National Research Council and National Academy Press, Washington, DC.

O'Brien, J.S., Julien, P.Y., and Fullerton, W.T. (1993) Two-dimensional water flood and mudflow simulation. *Journal of Hydraulic Engineering*, **119**(2), 244–261.

Okubo, S. and Mizuyama, T. (1981) Planning countermeasures against debris flow. *Civil Engineering Journal*, **23**(9) [in Japanese].

Pelletier, J.D., Malamud, B.D., Blodgett, T.A., and Turcotte, D.L. (1997) Scale-invariance of soil moisture variability and its implications for the frequency-size distribution of landslides. *Engineering Geology*, **48**, 254–268.

Petracheck, A. and Kienholz, H. (2003) Hazard assessment and mapping of mountain risks in Siwtzerland. In: D. Rickenmann and C-L. Chen (eds), *Debris-flow Hazards Mitigation: Mechanics, Prediction, and Assessment* (pp. 23–38). Millpress, Rotterdam.

Pyles, M.R. and Froehlich, H.A. (1987) Discussion of "Rates of landsliding as impacted by timber management activities in north-western California" by M. Wolfe and J. Williams. *Bulletin of the Association of Engineering Geologists*, **24**(3), 425–431.

Rickenmann, D. (1999) Empirical relationships for debris flows. *Natural Hazards*, **19**, 47–77.

Scheuringer, E. (1998) Grundlagen und Grundsätze der Gefahrenzonenausweisung der Wildbach- und Lawinenverbauung. *Der Alm- und Bergbauer*, **48**, 58–61.

Scott, K.M. (2000) Precipitation-triggered debris flow at Casita Volcano, Nicaragua: Implications for mitigation strategies in volcanic and tectonically active steeplands. In: G.F. Wieczorek and N.D. Naeser(eds), *Debris-flow Hazards Mitigation: Mechanics, Prediction, and Assessment* (pp. 3–14). A.A. Balkema, Rotterdam.

Scott, K.M. and Vallance, J.W. (1995) *Debris Flow, Debris Avalanche, and Flood Hazards at and Downstream from Mount Rainier, Washington* (USGS Hydrologic Investigations Atlas 729: 2 sheets and accompanying pamphlet). US Geological Survey, Reston, VA.

Scott, K. and Yuyi, W. (2003) *Debris Flows – Geological Process and Hazard – Illustrated by a Surge Sequence at Jiangjia Ravine, Yunnan, China* (USGS Professional Paper 1671). US Geological Survey, Reston, VA.

Shroeder, J.F. (1978) Dendrogeomorphological analysis of movement of Table Cliffs Plateau, Utah. *Quaternary Research*, **9**, 168–185.

Sigafoss, R.S. (1964) *Botanical Evidence of Floods and Flood-plain Position* (USGS Professional Paper 485-A). US Geological Survey, Reston, VA.

Southwood, T.R.E. (1985) The roles of proof and concern in the work of the Royal Commission on Environmental Pollution. *Marine Pollution Bulletin*, **16**, 346–350.

Stark, C.P. and Hovis, N. (2001) The characterization of landslide size distributions. *Geophysical Research Letters*, **28**(6), 1091–1094.

Straub, L.G., Silberman, E., and Nelson, H.C. (1958) Open channel flow at small Reynolds numbers. *Transactions of the American Society of Civil Engineers*, **123**, 685–706.

Stiny, J. (1910) *Die Muren* (106 pp.). Verlag der Wagner'schen Buchhandlung.

Strunk, H. (1995) *Dendrogeomorphologische Methoden zur Ermittlung der Murfrequenz und Beispiele ihrer Anwendung* (195 pp.). Roderer, Regensburg, Germany.

Takahashi, T. (1978) Mechanical characteristics of debris flow. *Journal of the Hydraulics Division ASCE*, **HY8**, 1153–1169.

Takahashi, T. (1981) Estimation of potential debris flows and their hazardous zones: Soft countermeasures for a disaster. *Journal of Natural Disaster Science*, **3**, 57–89.

Takahashi, T. (1991) *Debris Flow* (IAHR Monograph Series). International Association for Hydraulic Research, Ecole Polytechnique Fédérale, Lausanne, Switzerland and A.A. Balkema, Rotterdam.

Takei, A. (1984) Interdependence of sediment budget between individual torrents and a river-system. *Proceedings International Symposium "Interpraevent", Villach, Austria* (Vol. I, pp. 35–48). Internationale Forschungsgesellschaft Interpraevent, Klagenfurt, Austria.

Vallance, J.W. and Scott, K.M. (1997) The Osceola Mudflow from Mount Rainier: Sedimentology and hazard implications of a huge clay-rich debris flow. *Geological Society of America Bulletin*, **109**, 143–163.

VanDine, D.F. (1985) Debris flow and debris torrents in the Southern Canadian Cordillera. *Canadian Geotechnical Journal*, **22**, 44–68.

Watanabe, M. (1981) Debris flows and associated disasters. *Civil Engineering Journal*, **23**(6) [in Japanese].

Webb, R.H., Pringle, P.T., and Rink, G.R. (1989) *Debris Flows from Tributaries of the Colorado River, Grand Canyon National Park, Arizona* (USGS Professional Paper 1492, 39 pp.). US Geological Survey, Reston, VA.

Webb, R.H., Melis, T.S., Wise, T.W., and Elliott, J.G. (1996) *The Great Cataract: The Effects of Late Holocene Debris Flows on Lava Falls Rapid, Grand Canyon National Park, Arizona* (USGS Open-File Report 96-460, 96 pp.). US Geological Survey, Reston, VA.

Whitman, R.V. (2000) Organizing and evaluating uncertainty in geotechnical engineering. *Journal of Geotechnical and Geoenvironmental Engineering, ASCE*, **126**(7), 583–593.

Wigmosta, M.S. (1983) Rheology and flow dynamics of the Toutle debris flows from Mt. St. Helens (184 pp.). M.Sc. thesis, University of Washington, Seattle.

Wilford, D.J., Sakals, M.E., Innes, J.L., Sidle, R.C., and Bergerud, W.A. (2004) Recognition of debris flow, debris flood and flood hazard through watershed morphometrics. *Landslides*, **1**, 61–66.

18

Debris-flow mitigation measures

Johannes Huebl and Gernot Fiebiger

18.1 STRATEGY OF PROTECTION

Integrated risk management is a tool to prevent, intervene, and avoid natural hazards (Amman, 2001). This includes a combination of land use planning and technical and bioengineering measures to guarantee an optimal cost-benefit ratio.

An essential aspect of risk management is the design of mitigation measures which reduce the existing risk to an accepted level of residual risk. Two types of mitigation measures can be distinguished (Zollinger, 1985): active measures and passive measures.

Active measures focus on the hazard, while passive measures focus on the potential damage (Huebl and Steinwendtner, 2000; Kienholz, 2003). It is of fundamental importance to risk management to clearly define the spatial and temporal objectives of the desired degree of protection, with an understanding of acceptable residual risk.

The strategy of protection describes the best combination of protection measures (Figure 18.1). Once protection objectives have been established from the risk assessment, the protection concept describes the strategy selected to reach those targets. Therefore, special management tasks are assigned for unique elements of the watershed. Each task defines a desired modification of the debris-flow system and the performance of the elements with regard to the hazard. Ultimately, the sum of each element's functions must lead to the fulfilment of the overall protection objective (Huebl, 2001).

In the next step appropriate measures are chosen to meet the tasks derived from the protection concept. This conceptual plan of measures is called the "safety system" (Kettl, 1984, 1998; Wehrmann, 2000). It includes measures that guarantee the effective performance of debris-flow mitigation. These measures must be evaluated with respect to their technical, economical, ecological, and political feasibility, and should be combined to fulfil their functional demands.

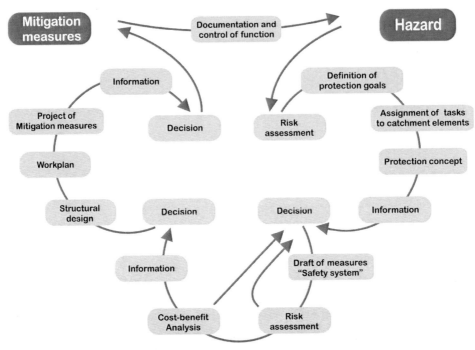

Figure 18.1. Strategy of protection.
Huebl et al. (2004).

The final step within the planning process includes the detailed structural design of the mitigation measures and the development of a work plan of all projected works.

Once all mitigation measures are established in the debris-flow catchment the utility of these measures is documented to allow a better understanding of the interrelationship between debris flows and mitigation measures.

18.2 MITIGATION MEASURES

18.2.1 Active mitigation measures

Active debris-flow mitigation measures (Table 18.1) may affect the initiation, transport, or deposition of debris flows and can therefore change its magnitude and frequency characteristics. This can be achieved either by changing the probability of occurrence of a debris flow (disposition management) or by manipulating the debris flow itself (event management).

Table 18.1. Active measures.

Objective	Task	Measure
Disposition management		
Decrease runoff	Decrease peak discharge	• Forestry measures
		• Watershed management
		• Diversion of runoff to other catchments
Decrease erosion	Decrease surficial erosion due to overland flow	• Forestry measures and soil bioengineering
		• Watershed management
		• Drainage
	Increase slope stability	• Forestry measures and soil bioengineering
		• Terrain alteration (grading, scaling)
		• Drainage control
		• Stabilization of the toe slope (e.g., consolidation, rock buttresses)
	Decrease vertical and lateral erosion in the channel bed	• Channel enlargement
		• Channel-bed stabilization
		• Transverse structure (sill, ramp, check dam)
		• Longitudinal construction
		• Groyne
		• Soil bioengineering
	Decrease water discharge at high erodible channel-reach	• Diversion of runoff to other catchments
		• Bypass
Event management		
Discharge control	Decrease peak discharge to prevent damage	• Water storage
		• Channel enlargement
		• Enlargement of the cross section at channel crossings (e.g., bridges)
Debris control	Transformation process	• Debris flow breaker
	Deposition debris under controlled conditions	• Permanent debris deposition
		• Temporary debris deposition
	Debris flow deflection to adjacent areas	• Deflection to area of low consequence
	Organic debris filtration	• Organic debris rake

Especially in the alpine countries (France, Switzerland, Austria) several books about torrent control works were published during the 19th and 20th centuries (Aretin, 1808; Duile, 1826; Breton, 1875; Pestalozzi, 1882; Salis, 1883; Demontzey, 1889; Thiery, 1891; Seckendorff, 1886; Wang, 1901, 1903; Stiny, 1931; Strele, 1950). During the last three decades many more contributions to all types of active measures have been published worldwide.

In the following, a number of case studies are presented with a description of active debris-flow mitigation measures.

Forest management activities

In the past century, Schmittenbach was amongst the most dreaded mountain torrents in the Salzburg region in Austria. Zell am See (about 60 km south of the city of Salzburg), a town located on the fan of Schmittenbach, had been affected by several severe debris flows. In the past century the upper catchment area was characterized by alpine pasture lands in which cattle grazing had severly damaged its sparse forest cover. In 1885 a project was initiated to decrease debris-flow activity. Its main feature was afforestation of a 158-ha area to increase the soil water absorption capacity and thus limit erosion due to high runoff events.

The afforestation was carried out by specifying the appropriate tree species for the respective elevation belt. Below 1,750 m, a mix of 80% spruce (*Picea abies*) and 20% larch (*Larix deciduas*) were used; while up to 1,780 m, a 50% mix of *Picea abies* and *Larix decidua* were planted. Above 1,780 m only Swiss stone pine (*Pinus cembra*) was used. Planting was performed in clusters. Areas with extended snowcover were left bare, because of biotic endangerment. The plants were produced from native seeds, at the corresponding altitude.

From 1950–1999 an additional afforestation campaign was carried out, this time using spruce, larch, Swiss stone pine, Scots pine (*Pinus mugo*), European mountain ash (*Fraxinus excelsior*), red beech (*Fagus sylvatica*), birch (*Betula pendula*), willow (*Larix sp.*), and alder (*Alnus sp.*) (Hartwagner, 2002).

The afforestation (Figures 18.2 and 18.3) resulted in a decrease of surface runoff of about 42% and a significant reduction of bedload. Continuous maintenance of the forest cover is necessary to minimize biotic and abiotic damage to the stands. In addition, in order to raise the elevation of timberline, measures against snow avalanches were implemented.

Soil bioengineering and terrain alteration

Soil bioengineering addresses the technologies and applications of dead and live plants for erosion control. Soil bioengineering structures start or accelerate phytosociological successions and processes, minimize erosion, and govern the groundwater supply.

Soil bioengineering measures can be applied to:

- channels, gullies, rivers, and streams (Figure 18.4);
- slope stabilization and bank redevelopment (Figure 18.5); and
- road ditch stabilization.

Most effective in soil bioengineering is the combination of surface protection constructions like seeding with stabilizing constructions. The range of longitudinal structures extends from tree spurs (rough coniferous trees), branch layering in gullies, vegetated channels, live brush mattresses, living slope grids, different

Sec. 18.2] Mitigation measures 449

Figure 18.2. Schmittenbach in 1887, Salzburg, Austria.
Photo: courtesy of WLV.

Figure 18.3. Schmittenbach in 1976 after afforestation, Salzburg, Austria.
Photo: courtesy of WLV.

Figure 18.4. Soil bioengineering at Fendlermure in 1995, Tyrol, Austria.

fascines, vegetated revetments of different materials, log brush barrier construction, live pole construction, brunch and brush packing, and double-row palisades. At the transverse structures there are living groynes, live siltation construction, living combs, brushes and palisade constructions, brush sills, fascine sills, log cribwalls with brushlayers, as well as planted gabions and wooden crib dams.

The success of soil bioengineering measures depends on the effect of the previously implemented technical control measures and the stability thus gained. Therefore, most engineers prefer to combine soil bioengineering with hard engineering structures.

Integrated watershed management

The purpose of integrated watershed management is to minimize the need for costly protective measures along the channel and in the runout zone. Vegetation should be managed under the principles of conservation and sustainability, while protecting against natural hazards.

Figure 18.5. Slope stabilization at Filprittertobel in 1898, Vorarlberg, Austria.
Photo: courtesy of WLV.

Several options are available to the engineer:

- afforestation, sub-alpine forestation, protection forest rehabilitation, and stand conversion;
- bioengineering measures, such as slope and erosion protection, and stream bank protection;
- agricultural measures such as grazing management, replacement of forest pasture, management of alpine grazing; and
- hard engineering measures such as barriers, deflection berms, or debris basins.

In Austria's Zillertal (Tyrol) the forest of a 200-km^2 study area was heavily damaged due to forest grazing, forest litter utilization, and using branches as green fodder (pollarding). Further damage was caused by an excessive game stock. The unhealthy state of the forest, combined with natural erosion processes, caused frequent debris flows resulting in severe damage to the villages and agricultural areas at the fans. To control this situation, a number of technical measures were applied, such as the installation of debris basins, bedload control at potential sources of erosion, and the installation of a drainage system on slopes susceptible to erosion. Moreover, suspending all forest cattle grazing and litter utilization, as well as reducing the game stock, improved the forest condition. Furthermore, 1,200 ha of new forests were planted above the timberline between 1,700 and 2,100 m. Improvement of land development, rationalization of alpine farming, and a change in land ownership, were also implemented (Stauder, 1975).

While technical measures may result in immediate debris-flow protection, forest management has proven to be very successful. It succeeded in decreasing the consequences of natural hazards and also produced positive economic effects on the local forestry and agricultural industries and increased the region's touristic values. These days, for example, pasture access roads are being used as scenic routes for thousands of tourists.

Drainage

Drainage systems drain water from wet areas and/or hillsides to stabilize the ecosystems. Water drainage stabilizes unstable areas and prevents the build-up of high pore-water pressure along potential shear surfaces. The principles of drainage stabilization are based on the following:

- Prevention of superficial and subsurface runoff from the area above the slide as well as collecting wells by a horseshoe-like diversion drainage (horseshoe drainage).
- To drain subsurface water to prevent the formation of shear surfaces.

The drainage channels are constructed as ditches with a sealed base (loam layer or seal foils) on which suckers (drainage pipes with an impermeable bottom and perforated top surface to carry off superficial and subsurface water within the landslide mass) or other draining material is installed. The surface of the drain is covered by permeable layers of coarse material like small boulders. After an overall length of 30–50 m, these drains are then guided together in wells. The collected drained water is delivered to a receiving creek. The drains are either linear, branched out, triple-pole, or herring bone type.

Toe slope stabilization

Bed or bank erosion can cause a destabilization of banks along channels. Depending on the process and magnitude of this destabilization, either a transverse (Figure 18.6) or longitudinal (Figure 18.7) structure can be constructed. Transverse structures are mainly used to prevent stream bed erosion, while longitudinal structures are used for bank erosion.

Transverse structures are designed to raise channel beds and reduce stream gradients. The original channel slope and the slope created by the installation of a check dam determine the height of, and distance between, successive structures.

During the construction of these transverse structures, it is important to consider the foundations of the abutments, scour depth or depositional grade, and the design of the outlet structure. In Europe, until the 1920s, these structures were mainly

Figure 18.6. Transverse toe slope stabilization, Bretterwandbach, Tyrol, Austria.
Photo: courtesy of WLV.

Figure 18.7. Longitudinal toe slope stabilization, Eugenbach, Austria.

constructed of dry stone. Later, cemented stone was used, until being replaced by concrete and reinforced concrete during the 1950s (Figures 18.8 and 18.9).

Special check dam types have been developed for channel sections subject to high lateral pressures. For these check dams, the wing walls are positioned in relation to the central structure and can consequently be moved perpendicular to the stream centre line. The first check dams of this type were built with wing walls arranged upstream of the central section (Figure 18.10); however, in modern structures, they are situated downstream (Figure 18.11).

Transverse structures are damaged or destroyed mainly by:

- The impact of a debris flow, when the body of the structure is not adequately supported by earth fill from behind (Figure 18.12).
- The scouring of the lateral abutment.
- Scouring downstream of the check dam.
- Lateral bypassing, caused either by the absence of a sloping wing wall top or when discharge is blocked by bedload, as a consequence of sediment accumulation downstream of the dam (Figure 18.13).

Sec. 18.2] Mitigation measures 455

Figure 18.8. Niedernsiller Muehlbach, Salzburg, Austria.
Photo: courtesy of WLV.

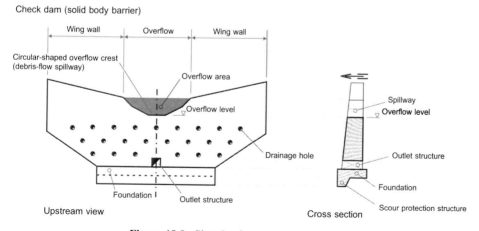

Figure 18.9. Sketch of a typical check dam.

Figure 18.10. Upstream wing walls, Duernbach, Salzburg, Austria.

Therefore, it is important to consider that:

- The spillway cross sectional design is equal to a debris-flow surge profile.
- The wing wall top equals the expected depositional grade (usually higher than 10%).
- The wing walls are laterally consolidated.
- No lateral erosion occurs upstream of the dam.
- The dam foundation reaches below the expected scour depth.

Sec. 18.2] **Mitigation measures** 457

Figure 18.11. Downstream wing walls, Wagrainer Ache, Salzburg, Austria.

Figure 18.12. Niedernsiller Muehlbach in 1970 after a debris-flow event, Salzburg, Austria.
Photo courtesy of WLV.

Figure 18.13. Gimbach in 2002 after a debris-flow event, Upper Austria, Austria.

Longitudinal structures prevent the widening of channels but can also serve as slope toe stabilization. In debris-flow-prone creeks, they should be constructed from stone or concrete. If its effect as a gravity wall is insufficient, additional anchors can be placed (Figure 18.14).

Bypass

One possible preventive measure is a debris-flow bypass to avoid excessive sediment recruitment along specific channel reaches. For example, in the "St. Julien" torrent in France, a tributary of the Rhône River, a 202 m long channel with a cross section of 44 m^2 was designed to bypass the "Mont Denis" landslide (Figure 18.15). Debris and water was directed into the artificial channel by transverse structures. After exiting the artificial channel, the torrent drops 83 m into the original stream bed (Mougin, 1900; cited by Wang, 1903). Similar structures were also constructed in Switzerland and Austria.

After the devastating debris flows in Hollersbach (Salzburg) of 1501 and 1538, numerous protective structures have been built. An array of check dams was

Figure 18.14. Anchored longitudinal structure, Kirchbachgraben, Carinthia, Austria.

constructed to stabilize the bedload source ("Grosse Blaike") on the Buergerbach tributary. However, the check-dam array was eventually abandoned due to high maintenance costs. As a result, debris flows occurred almost every year. In 1989 all surface and subsurface runoff was collected into a tunnel (Figure 18.16).

Figure 18.15. Bypass tunnel at St. Julien, France.

Water retention

Many debris flows have occurred in Wartschenbach (Eastern Tyrol) during the past centuries. This has led to a destabilization of the middle reaches of the creek. However, due to geological conditions (highly fractured gneiss and schist) and a channel gradient of approximately 60%, no measures were taken to stabilize the channel or the side slopes.

Three tributaries drain the watershed, which is intensively utilized by farming and the ski industry. Therefore, three flood-control reservoirs (one is shown in Figure 18.17), were constructed in the upper catchment area to avoid debris-flow initiation in the torrent s middle reach. Before these reservoirs could be constructed, a geological and hydrological study was carried out to determine the location and the ideal retention capacity of the reservoirs. Today the reservoirs retain approximately 40,000 m^3 of water, thereby reducing the 100-year return period discharge in the middle reach by 50%.

The reservoirs were built as earth-fill dikes with concrete outlet structures and a sluice gate. To control the amount of water being retained by the dam, ultrasonic sensors monitor water levels, and alarm the facility operator when the water level reaches a critical stage.

Figure 18.16. Bypass at Buergerbach, Salzburg, Austria.
Photo courtesy of H. Wehrmann.

Permanent debris deposition

Until the end of the 1960s, solid dams (Figure 18.18) were built at the exits of depositional areas to prevent dangerous debris flows from reaching high-consequence areas. These dams are effective until completely full. Later designs

Figure 18.17. Flood-control reservoir in the middle reach of Wartschenbach, Tyrol, Austria.

Figure 18.18. Solid body barrier, Einachgraben, Austria.

Figure 18.19. Small slot barrier, Koednitzbach, Tyrol, Austria.

included small drains that were embedded in the dam body (Figure 18.19) to minimize static water pressures (Kronfellner-Kraus, 1970). The development of solid dams with large slots or slits to regulate sediment transport began with the use of concrete and reinforced concrete. Slots or slits are designed to pass medium sized floods through the opening without producing backwater effects which necessitates their extension to the channel bed (Hoffmann, 1955).

Once debris basins have completely filled, they completely interrupt bedload transport, and can lead to excessive erosion downstream. Some fish species require a renewed supply of spawning gravels which may be interrupted with debris basins.

Debris basins can also cause the water flow to cease completely with detrimental effects to aquatic life and may therefore be rejected for ecological reasons.

In other reaches of the creek, which suffer from a sediment deficit, the retention of debris in tributaries can lead to severe erosion in the receiving streams and can cause a drop in the groundwater levels.

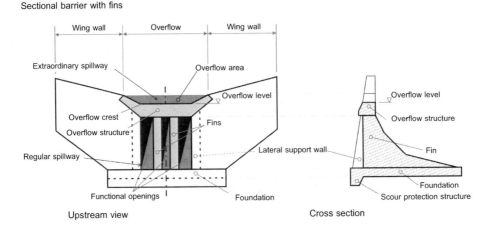

Figure 18.20. Sketch of an open barrier.

Temporary debris deposition

As a result of the apparent problems with debris retention dams, attempts have been made to use temporary debris retention for bedload management downstream. This concept allows the passing of some smaller sized particles while large boulders with high destructive potential are retained (Aulitzky, 1986). The intermediate storage of the accumulated material is designed to balance hazard mitigation and a healthy riverine environment.

During the 1970s, the concepts of "sorting" and "dosing" were introduced (Ueblagger, 1973). For both, it is indispensable to have a sufficient capacity for sediment storage in the dam area. Sediment "sorting" allows particle segregation by grain size to allow only a given grain size to pass the structure. During "dosing" on the other hand, an unsorted sedimentation is encouraged by creating backwater in a bottleneck. In both cases, bedload transport is only influenced by the occurrence of an event. The surplus of solids produced by a debris flow would be stored, and later released if the tailwater was deficient of area sediments. After small and medium-sized events, the deposition should be regularly cleared and drains should be unclogged.

In debris dams (Figure 18.20), sorting and dosing could be installed in the form of large slots or slits (open barriers). The first works of this type were erected at the Maerzenbach and Riedbach in the years 1968/1969 as beam barriers (Stauder, 1972).

Subsequently different types of construction were developed, differing in drainage works and opening covers (Leys, 1965a, b, c, 1967, 1971a, b, 1973, 1976; Stauder, 1973; Hampel, 1974; Leys, 1976; Kettl, 1984; Riccabona, 1988; Eckersdorfer, 1998; Krimpelstaetter, 1998). This Austrian style of barrier (Huebl et al., 2003: Table 18.2) is nowadays used in many European countries as well as Japan, Taiwan, and Canada (Figures 18.21–18.41).

Table 18.2. Types of open barriers.

Slot barriers

Large slot barrier — Small slot barrier

Slit barriers

 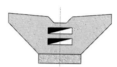 

Slit barrier with vertical slits — Slit barrier with horizontal slits — Gap-crested slit barrier with vertical slits

Compound barriers

 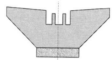

Compound barrier with openings — Compound barrier with teeth

Sectional barriers

Sectional barrier with fins — Sectional barrier with piles — Sectional barrier with braces

Lattice barriers

  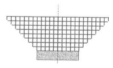

Rake barrier — Beam barrier — Grill barrier

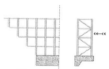

Frame barrier

Net barriers

Figure 18.21. Large slot barrier, Nieschenbach, Tyrol, Austria.
Photo: courtesy of M. Holub.

Figure 18.22. Small slot barrier, Truebenbach, Carinthia, Austria.
Photo: courtesy of M. Holub.

Sec. 18.2] Mitigation measures 467

Figure 18.23. Slit barrier with vertical slits, Zinkenbach, Salzburg, Austria.

Figure 18.24. Slit barrier with horizontal slits, Schnannerbach, Tyrol, Austria.

Figure 18.25. Sectional barrier with fins and beams, Maerzenbach, Tyrol, Austria.

Figure 18.26. Sectional barrier with fins, Sallabach, Styria, Austria.
Photo: courtesy of M. Holub.

Sec. 18.2]　　　　　　　　　　　　　　　　　　　　　　　　　**Mitigation measures**　469

Figure 18.27. Sectional barrier with piles, Waldbach, Styria, Austria.
Photo: courtesy of M. Holub.

Figure 18.28. Beam barrier, Truebenbach, Carinthia, Austria.
Photo: courtesy of M. Holub.

Figure 18.29. Sectional rake barrier, Loehnersbach, Salzburg, Austria.

Figure 18.30. Beam barrier, Istalanzbach, Tyrol, Austria.

Sec. 18.2] Mitigation measures 471

Figure 18.31. Cross-slit barrier, Luggauerbach, Salzburg, Austria.

Figure 18.32. Cross-slit barrier in 2000 after a debris flow, Luggauerbach, Salzburg, Austria.
Photo: courtesy of M. Leitgeb.

Figure 18.33. Sectional barrier with fins, Fong-Chiu, Nan-Tou County, Taiwan.

Figure 18.34. Frame barrier, Ashiya River, Japan.
Photo courtesy of Comm. Steel Sabo Structures.

Figure 18.35. Net barrier, Gleiersbach, Tyrol, Austria.
Photo courtesy of WLV.

Figure 18.36. Sectional barrier with fins, Luggauerbach, Salzburg, Austria.

Figure 18.37. Sectional barrier with fins in 2000 after a debris-flow event, Luggauerbach, Salzburg, Austria.

Photo: courtesy of M. Leitgeb.

Figure 18.38. Sectional barrier with fins, Rastelzenbach, Salzburg, Austria.

Figure 18.39. Sectional barrier with fins, Ellmaubach, Salzburg, Austria.

Figure 18.40. Debris-flow grill, Furano River, Japan.
Photo: courtesy of Comm. Steel Sabo Structures.

Figure 18.41. Debris-flow grill, Dorfbach Randa, Switzerland.

When designing an open barrier, the following considerations ought to be made:

- Assure a sufficient drain width.
- Drainage must reach the channel bed to permit automatic sediment removal during normal streamflow, and maintenance of a minimum streamflow to assure survival of aquatic organisms.
- Prevention of barrier clogging by woody debris (possibly by using a primary barrier or wood rake upstream of the main open barrier).
- Determination of variable beam, rake, or grill widths for efficient debris-flow retention.
- Protection of the wing walls with riprap or earth fill.
- Placement of the maintainance access against the flow direction and at variable elevations (dam opening must be within the reach of a backhoe or bulldozer).
- Construction of small, narrow basins with high gradients (Zollinger, 1985).

Experience shows that sediment removal by normal streamflow after a debris flow rarely works (drain rates of 50% are considered successful). This is because big boulders and woody debris interlock at the opening. In most cases, the filling-up of the debris basin upstream of an open barrier takes approximately 3–10 times longer than in a permanent debris retention dam. This fact allows bedload transport downstream and helps to maintain the processes of dosing and sorting.

In modern debris-flow mitigation, debris barriers are located downstream of debris-flow breakers, followed by an array of check dams aimed at stabilizing the channel. This assembly of measures is referred to as a "torrent training system" or a "functional chain" (Kettl, 1984).

Debris-flow breaker

Debris-flow breakers are designed to reduce debris-flow energy (Kettl, 1984; Fiebiger, 1997). By slowing and depositing the surge front of the debris flow, downstream reaches of the stream channel and settlement areas are exposed to considerably lower dynamic impact.

In an array of different debris-flow mitigation structures, debris-flow breakers are always in the most upstream position. A debris-flow breaker should retain at least the volume of the surge wave. Numerous functional structures with modern sediment management systems could be installed downstream of the breakers.

Debris-flow breakers can be designed as independent structures, but they are often combined with dosing and sorting barriers (bifunctional barrier), using "fins" to divide debris flow. Independent breakers consist of several fins, designed to withstand 7–11 times the hydrostatic water pressure (Lichtenhahn, 1973; Armanini, 1997; Huebl and Holzinger, 2003). For the design of such structures the flow behavior is an important criterion, because liquid mudflows and stony debris flows show distinct flow dynamics.

The upstream side of the fins are typically armoured with a steel sheeting to protect the concrete from abrasion and impact.

The spacing between fins depends on the predicted maximum and average sediment size and the amount of expected woody debris. Each fin may have several knick points and features a vertical segment designed to stop the bouldery front of the debris-flow surge wave.

Recently, check dams are being fitted with small debris-flow breakers so that, in case of an unusually large debris flow, the intermittent retention capacity can be used.

Breakers have been tested in numerous debris flows and have proven to be an essential element in any torrent training system. The deposition area upstream of the breakers must be excavated after each debris flow which necessitates access to the structure.

Deflection

Deflection structures are constructed to direct debris flow toward areas of low consequences. This requires the existence of areas with low economic value in which debris flows are allowed to deposit. Deflection structures include dikes, groynes, and deflection walls (Figures 18.42 and 18.43) constructed of concrete, reinforced concrete, boulder revetments, gabions, and other construction materials. Deflection structures are commonly constructed as the last element of a systematic torrent training control to diminish any remaining risk.

Woody debris rake

Woody debris catching facilities are used to separate large woody debris from mineral debris. These structures consist of steel, reinforced concrete, or massif-concrete. Open barriers can grant the function of filtering woody debris, too.

Figure 18.42. Deflection walls at Luggauerbach, Salzburg, Austria.
Photo: courtesy of WLV.

Woody debris rakes are always planned and constructed in combination with other bedload managing structures.

18.2.2 Passive mitigation measures

Passive mitigation measures are used to reduce the potential loss by, for example, altering the spatial and temporal character of either the damage produced by debris

Figure 18.43. Deflection wall at Niedernsiller Muehlbach, Salzburg, Austria.

flows or the associated vulnerability. Vulnerability of a disaster can be changed either with land-use planning like hazard mapping, or through immediate disaster response (Aulitzky, 1972). Different types of passive mitigation measures are summarized in Table 18.3.

Hazard mapping

The mapping of debris-flow hazards is an important instrument for:

- Disaster prevention.
- Disaster management.

Usually hazard mapping is legalized by law (e.g., in Austria by the Forest Law of 1975). The combination of intensity and frequency of the process is symbolized in the maps by different colors and have to be referenced in land-use planning.

The delineation of debris-flow-affected areas normally is accompanied with restrictions to property rights. It is therefore important that hazard mapping has to be comprehensible and readily comparable to other regions. The result must

Table 18.3. Passive mitigation measures.

Objective	Task/function	Measure
Preventive Reduction of potential loss	Debris-flow transport and deposition without damage Local protection of an object (e.g., house, person, traffic route)	• Land-use planning (local, regional) • Information, education, and disaster management • Specification of construction rules
Event Response Reduction of potential loss	Debris-flow transport and deposition without damage Upkeep of protective measures	• Closing of traffic route • Information • Warning and evacuating of hazardous areas • Immediate technical assistance

match high-quality criteria and should include strategies of how to address residual risk. For further details see Chapter 17.

Land-use zoning

Building regulations in hazardous areas can help to reduce damage to buildings and infrastructures. Based on different legal and administrative procedures, expert opinions lead to decisions being made by authorities and agencies.

Technical self-protection for buildings in minor dangerous locations is easy to apply and includes, for example, deflection walls or dams, reinforced concrete foundations, reinforced concrete walls, jacketing, heightened entrance levels, and downstream adjustment of doors.

In areas with a high impact load of debris flows, development of further settlements and infrastructure should be prohibited.

Warning systems

An important tool in risk management is an early warning system (Yano and Senoo, 1985).

Debris-flow early warning systems are capable of detecting debris flows and automatically triggering an alarm. They allow evacuation or road closures to prevent or mitigate potential damage or loss of life.

Such warning systems can be used to protect settlement areas, traffic routes, and construction sites. They can also be used to control and monitor safety measures, to support active measures, and to aid scientific studies. The most important elements of an early warning system are:

• Data collection.

Figure 18.44. Contactless warning by ultrasonic sensors at Lattenbach, Tyrol, Austria.

- Data transfer.
- Data management.
- Distribution of information.
- A decision hierarchy structure.
- Response planning and organization.

The combination of these components is optimised based on several requirements (e.g., site, transmission rate, power supply, response time) and possible sources of error (e.g., destruction). Malfunction or breakdown of one component can lead to the collapse of the whole system.

In Japan and China, early warning systems have been used since the 1980s. Similar systems have since been installed in the USA and Europe (Italy, Switzerland, France, Austria (Figure 18.44), and Slovenia). Today, debris-flow warning systems

Figure 18.45. Contact warning by DLT (détecteur lave torrentielle) sensors at Ravoire de Pontamafrey, France.

are used principally in connection with traffic routes (e.g., Ravoire de Pontamafrey in France (Figure 18.45) and Log pod Mangrtom in Slovenia (Figures 18.46 and 18.47)). For further details on warning systems see Chapter 12.

Immediate technical assistance

The urgent technical measures following a disaster serve above all the restoration of the infrastructure and the avoidance of consequential loss and damage from following events. These include excavation of buried objects, reconstruction of the infrastructure, and cleaning of the inundated areas.

Figure 18.46. Log pod Mangrtom, Slovenia.

18.2.3 Documentation and control

Established mitigation measures must be monitored either regularly (e.g., yearly) or following a debris-flow event. Inspection of a debris-flow-prone creek is necessary to evaluate the condition of the catchment and the status of the existing mitigation measures (Table 18.4). The effectiveness of these measures should then be evaluated subsequent to an event. As a result, weak elements in the mitigation concept or safety system can be identified and additional measures can be planned accordingly.

18.3 CONCLUSION

In this chapter only a few examples of debris-flow mitigation measures are described. There are a large variety of measures that can be combined in different ways. For

Figure 18.47. Log pod Mangrtom, Slovenia.

selecting the best adjusted arrangement great importance should be given to the knowledge of all ongoing geomorphic processes and their possible interaction with the mitigation measures. This means a multidisciplinary approach has to be applied, including specialized skills in applied geology, geomorphology, hydrology, fluid dynamics, forestry, and structural engineering.

Although there is a large pool of experience gained by practitioners working in this field of activity, a lot of scientific gaps still exist. The rare occurrence of debris flows of design magnitude obliges the engineers to visualize the effect of the mitigation measures on the initiation, transportation, and deposition of the debris flow. Therefore, it is most important to collect and exchange the experience of existing mitigation measures worldwide.

Table 18.4. Documentation and control.

Objective	Task/function	Measure
Inspection of measures	Documentation of the effect of the established debris-flow counter measures	• Control measurements
Maintenance of protective measures	Control of the channel and the established debris-flow counter measures	• Inspection of the catchment and documentation of the existing condition (e.g., channel, slopes, forest, mitigation measures)

Beside the assignment of technical structures, hazard mapping in combination with land-use planning seems to be a proper and cost-effective tool to reduce future losses.

18.4 REFERENCES

Amman, W.J. (2001) Integrales Risikomanagement von Naturgefahren. In: *Forum für Wissen, Tagungsband Risiko + Dialog Naturgefahren vom 16.11.2001, Birmensdorf* (pp. 27–31) [in German].

Aretin, F. von (1808) Ueber Bergfaelle und die Mittel, denselben vorzubeugen oder wenigstens ihre Schaedlichkeit zu vermindern, Innsbruck [in German].

Armanini, A. (1997) On the dynamic impact of debris flows. In: A. Armanini and M. Michiue (eds), *Recent Developments on Debris Flows* (Lecture Notes in Earth Sciences No. 64, pp. 208–225). Springer-Verlag, New York.

Aulitzky, H. (1972) *Gefahrenzonenplaene im Bereich der Wildbach- und Lawinenverbauung. 10: Flussbautagung* (Bericht, pp. 95–113). Bundesministerium fuer Land- und Forstwirtschaft, Sektion IV, Vienna [in German].

Aulitzky, H. (1986): Die Wildbaeche und ihre Verbauung. Unpublished manuscript, Vienna [in German].

Breton, P. (1875) *Etudes sur le système général de défence contre les torrents*. Dunod, Paris [in French].

Demontzey, P. (1889) *La restauration des terrains en montagne, au Pavillion des forêts*. Imprimerie Nationale, Paris [in French].

Duile, J. (1826) *Ueber die Verbauung der Wildbaeche in Gebirgs-Laendern, vorzueglich in der Provinz Tirol und Vorarlberg*. Rauch-Verlag, Innsbruck [in German].

Eckersdorfer, Th. (1998) Type of a dam with combination of dosing and sizing. *Wildbach- und Lawinenverbau*, 62. Jahrgang, **136**, 127–132.

Fiebiger, G. (1997): Structures of debris flow countermeasures. In: C-L. Chen (ed.), *Debris-flow Hazards Mitigation: Mechanics, Prediction, and Assessment: Proceedings of 1st International Conference, San Francisco* (pp. 596–605). American Society of Civil Engineers, New York.

Hampel, R. (1974) Die Wirkungsweise von Wildbachsperren. *Wildbach- und Lawinenverbau*, **1**, 82 [in German].

Hartwagner, W. (2002) Entwicklung Projekt Schmidten-Wildbaeche 1889. *Zukunft braucht Vergangenheit: 150 Jahre Oesterreichischer Forstverein* (Berichte zu den Lehrwanderungen). Forsttagung, Salzburg, Austria [in German].

Hoffmann, L. (1955): Geschiebestausperren mit selbsttaetiger Entleerung von Feingeschiebe. *Wildbach- und Lawinenverbau*, **6**, 49–60 [in German].

Huebl, J. (2001) Strategy of protection. In: R. Didier and F. Zanolini (eds), *Risques torrentiels*. Université Européenne d'Eté sur les Risques Naturels, Grenoble, France.

Huebl, J. and Holzinger, G. (2003) *Kleinmassstaebliche Modellversuche zur Wirkung von Murbrechern* (WLS Report 50, Vol. 3). Institut für Alpine Naturgefahren, Universität für Bodenkultur-Wien [in German].

Huebl, J. and Steinwendtner, H. (2000) Debris flow hazard assessment and risk mitigation. *Felsbau, Rock and Soil Engineering* (Vol. 1, pp. 17–23). Verlag Glueckauf [in German].

Huebl, J., Holzinger, G., and Wehrmann, H. (2003) *Klassifikation von Wildbachsperren* (WLS Report 50, Vol. 2). Institut für Alpine Naturgefahren, Universität für Bodenkultur-Wien [in German].

Huebl, J., Ganahl, E., Gruber, H., Holub, M., Holzinger, G., Moser, M., and Pichler, A. (2004) Grundlagenerhebung für das Schutzkonzept Lattenbach (Catchrisk). *Grundlagen fuer eine Murenprognose und darauf aufbauend die Entwicklung eines Warn- und Alarmsystems* (WLS Report 95, Vol. 1). Institut für Alpine Naturgefahren, Universität für Bodenkultur-Wien [in German].

Kettl, W. (1984) Vom Verbauungsziel zur Bautypenentwicklung: Wildbachverbauung im Umbruch. *Wildbach- und Lawinenverbau*, 48. Jahrgang, Sonderheft, 61–98 [in German].

Kettl, W. (1998) Vom Gefahrenpotential zum Verbauungssystem. *Wildbach- und Lawinenverbau*, 62. Jahrgang, **136**, 9–13 [in German].

Kienholz, H (2003) Early warning systems related to mountain hazards. In: J. Zschau and A. Kueppers (eds), *Early Warning Systems for Natural Disaster Reduction: 3rd International IDNDR Conference on Early Warning Systems for the Reduction of Natural Disasters. Potsdam, 1998* (pp. 555–564). Springer-Verlag, Berlin.

Krimpelstaetter, L. (1998) Ausgestaltung von Rostauflagen bei Sortierwerken. *Wildbach- und Lawinenverbau*, **136**, 107–111 [in German].

Kronfellner-Kraus, G. (1970) Ueber offene Wildbachsperren. *Mitteilungen der forstlichen Bundesversuchsanstalt-Wien*, **88**, 7–77 [in German].

Leys, E. (1965a) Beispiele fuer Wildholzfaenge bei Sperrendolen und bei Abflusssektionen. *Wildbach- und Lawinenverbau*, **23**, 45–48 [in German].

Leys, E. (1965b) Wann sind Murverteiler vorteilhaft. *Wildbach- und Lawinenverbau*, **23**, 49–56 [in German].

Leys, E. (1965c) Schlitzdolen und Eisenrechen bei geraden Auslaufsperren von Geschiebeablagerungsbecken. *Wildbach- und Lawinenverbau*, **23**, 57–65 [in German]..

Leys, E. (1967) Rechen- und Balkenbauten in der Wildbachverbauung zur Regulierung des Geschiebetriebes. *Wildbach- und Lawinenverbau*, **3**, 44–51 [in German].

Leys, E. (1971a) Erkenntnisse und Folgerungen aus den Unwetterkatastrophen des Jahres 1965 in den Bezirken Imst und Landeck in Tirol. *"Interpraevent" 1971* (Vol. 3, pp. 431–440). Internationale Forschungsgesellschaft Interpraevent, Klagenfurt, Austria [in German].

Leys, E. (1971b) Die Bedeutung der grossdoligen und der kronenoffenen Bauweise in der Wildbachverbauung zur Vorbeugung von Hochwasser- und Murschaeden. *"Interpraevent" 1971* (Vol. 3, pp. 441–449). Internationale Forschungsgesellschaft Interpraevent, Klagenfurt, Austria [in German].

Leys, E. (1973) Das Geschiebe und das Wildholz als Bemessungswert fuer die Oeffnungsweite bei den Entleerungsbauwerken in der Wildbachverbauung. *Mitteilungen der forstlichen Bundesversuchsanstalt-Wien*, **102**, 317–333 [in German].

Leys, E. (1976) Die technischen und wirtschaftlichen Grundlagen in der Wildbachverbauung der grossdoligen und der kronenoffenen Bauweise. PhD-These an der Universität für Bodenkultur, Wien [in German].

Lichtenhahn, C. (1973) Die Berechnung von Sperren in Beton und Eisenbeton. *Mitteilungen der forstlichen Bundesversuchsanstalt-Wien*, **102**, 91–118 [in German].

Mougin, M. (1900) *Consolidation des berges par dérivation d'un torrent (Torrent de Saint Julien)*. Imprimerie Nationale, Paris [in French].

Pestalozzi, K. (1882) Verbauung der Wildbaeche. In: *Handbuch der Ingenieurwissenschaften* (III. Band). Der Wasserbau, Leipzig [in German].

Riccabona, B. (1988) Bisherige Erfahrungen mit Entleerungssperren und ihre Wirkungsweise bei Murstoessen und Hochwaessern in den letzten 20 Jahren. *"Interpraevent" 1988* (Band 3, pp. 119–129). Internationale Forschungsgesellschaft Interpraevent, Klagenfurt, Austria [in German].

Salis, A. (1883) *Das schweizerische Wasserbauwesen*. Staempfli'sche Buchdruckerei, Bern [in German].

Seckendorff, A.F. von (1886) Das forstliche System der Wildbachverbauung. Paper presented at *Oesterreichischer Ingenieur- und Architektenverein, 27.3.1886*, Wien [in German].

Stauder, S. (1972): Balkensperren im Zillertal: Eine neue Verbauungstype der Wildbachverbauung. *Wildbach- und Lawinenverbau*, 36. Jahrgang, **1**, 1–45 [in German].

Stauder, S. (1973) Verschiedene Konstruktionen von Balkensperren. *Mitteilungen der forstlichen Bundesversuchsanstalt-Wien*, **102**, 307–316 [in German].

Stauder, S. (1975) Integralmelioration im Zillertal nach dem forsttechnischen System der Wildbach- und Lawinenverbauung. In: *Hochwasser- und Lawinenschutz in Tirol*. Land Tirol, Innsbruck [in German].

Stiny, J. (1931) *Die geologischen Grundlagen der Verbauung der Geschiebeherde*. Springer-Verlag, Vienna [in German].

Strele, G. (1950) *Grundriss der Wildbachverbauung* (2. Auflage). Springer-Verlag, Vienna [in German].

Thiery E. (1891) Restauration des montagnes, correction des torrents, reboisement. *Encyclopédie des Travaux Publics*. Librairie Polytechnique, Paris [in French].

Ueblagger, G. (1973) Retendieren, Dosieren und Sortieren. *Mitteilungen der forstlichen Bundesversuchsanstalt-Wien*, **102**, 335–372 [in German].

Wang, F. (1901) *Grundriss der Wildbachverbauung* (Teil 1). Verlag Hirzel, Leipzig [in German].

Wang, F. (1903) *Grundriss der Wildbachverbauung* (Teil 2). Verlag Hirzel, Leipzig [in German].

Wehrmann, H. (2000): Vergleichende Betrachtung von Wildbachverbauungssystemen im Pinzgau: Vom Gefahrenpotential zum Sicherungssystem. MSc-These an der Universität für Bodenkultur, Wien [in German].

Yano, K. and Senoo, K. (1985) How to set standard rainfall for debris flow warning and evacuation. In: A. Takei (ed.), *International Symposium on Erosion, Debris Flow and Disaster Prevention, Tsukuba, Japan* (pp. 451–455). Tsukuba, Japan.

Zollinger, F. (1985) Debris detention basins in the European Alps. In: A. Takei (ed.), *International Symposium on Erosion, Debris Flow and Disaster Prevention, Tsukuba, Japan* (pp. 433–438). Tsukuba, Japan.

19

Debris avalanches and debris flows of the Campania Region (southern Italy)

Francesco M. Guadagno and Paola Revellino

19.1 INTRODUCTION

On 5–6 May, 1998, prolonged rainfall triggered a large number of slope failures in the Sarno, Quindici, Siano, and Bracigliano area of the Campania Region (southern Italy) (Figure 19.1). The casualties (161 people lost their lives), the huge economic damage, and the severe destruction of this landslide event attracted great attention on the part of the Italian authorities and of the scientific community, leading to careful scrutiny of the landslide-type mechanism and the associated risk. Historical and geological analyses have shown that similar landslide events have previously affected nearby towns and infrastructure, mainly involving the airfall products of the Vesuvius and Phlegrean Field eruptions. Geologically, these landslides should be considered as secondary and delayed effects of volcanic activity.

Despite many points of disagreement and opinion still existing among different research groups and technicians due to the complexity of the phenomenon and the different research aims, a correct comprehension of the landslide mechanism is required to provide parameters for measures of mitigation. Moreover, the area of the Campania Region that can be considered subject to hazard is of about $2,000\,km^2$, a part of which is devoted to tourist activity (e.g., the Sorrento and Amalfi areas). The towns considered at risk (about 100) and the demands of new urbanization that could increase the relevant risks, make correct land planning necessary, given that landslides constitute one of the major problems. Even if single slope instabilities are generally characterized by a small volume of materials involved ($<10,000\,m^3$), clusters of quasi-simultaneous instabilities can involve wide areas, inducing catastrophic situations.

In this chapter the geological setting, the landslide characteristics, and the triggering features are described as well as the failure mechanism and runout.

Figure 19.1. Debris avalanche paths reaching the town of Sarno and Episcopio (aerial view).
Courtesy of the Fire Department of Campania.

19.2 HISTORIC AND RECENT PHENOMENA: AN OVERVIEW

Historical analyses of landslide occurrence show that flow-type instabilities have affected a widespread sector of the Campanian slopes. Figure 19.2 summarizes the sites and dates of past destructive landslides as recorded in the literature. Landslides have been distinguished as single, multiple, or areal, depending on the number of instabilities triggered by single storms.

Flow-type landslides also occurred during some eruptions, as reported in historic documents (Alfano and Friedlander, 1929). For example, the area of Lauro and Palma Campania was hit by debris flows several days after the catastrophic Vesuvian eruption of 1631, induced by heavy rainfalls, involving the fresh and unconsolidated pyroclastic deposits on the slopes.

The first systematic reports on the Campania landslides are those of Montella (1841) and Ranieri (1841) describing landslides involving areas in the vicinity of the Gragnano valley, covered by 1 m thick pyroclastic deposits from the 79 AD Vesuvian eruption that destroyed Pompeii and Herculaneum. The detailed descriptions of the events constitute one of the most significant documents on flow-type landslides in historic times. In that year, approximately 100 people died as a consequence of widespread landsliding and Gragnano village suffered considerable destruction. Ranieri (1841) describes the mass movements as having triangular shapes, a recurrent geometrical characteristic of this landslide type (Figure 19.3). The same

Sec. 19.2] Historic and recent phenomena: an overview 491

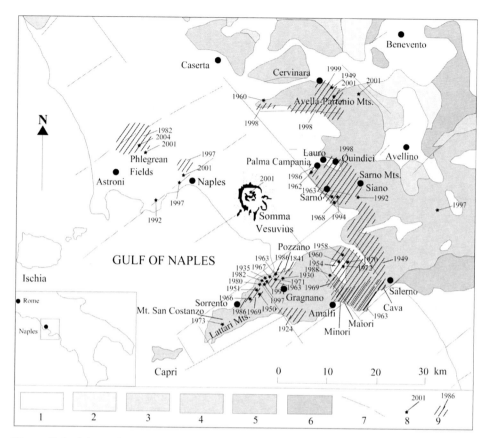

Figure 19.2. Schematic geo-structural map, showing the location of the destructive debris flows and avalanches in the Campania Region. (1) Plio-Pleistocene deposits. (2) Altavilla units (Tortonian Middle Pleistocene). (3) Irpinia units (Langhian–Tortonian). (4) Carbonate Matese unit (Trias-Paleocene). (5) Carbonate unit of the Alburno-Cervate unit (Trias-Paleocene). (6) Sicilide units (Cretaceous–Eocene). (7) Principal faults. (8) Single landslide event. (9) Multiple landslide event.

area was affected by debris flows in August 1935, February 1963, January 1971, and January 1997, emphasizing the very high hazard status of the area.

Debris flows have deeply affected the Amalfi–Salerno coast and the Sorrento peninsula. In 1910, 1924, and above all in 1954, after prolonged rainfalls, a large number of landslides occurred, causing casualties and destruction of houses and infrastructure. In 1954, debris flows and hyperconcentrated flows (Figure 19.4) involved millions of cubic meters of slope materials, that destroyed parts of Maiori, Minori, and Cava dei Tirreni, and created large fans at the river mouth (Figure 19.5(a)). Photos taken at that time testify that the initial movements took place in the steepest parts of the slopes (Figure 19.5(b)), seeming to have been controlled by the presence of natural scarps, whose role in landsliding will be

Figure 19.3. The 1841 Gragnano landslides, illustrated by Ranieri (1841).

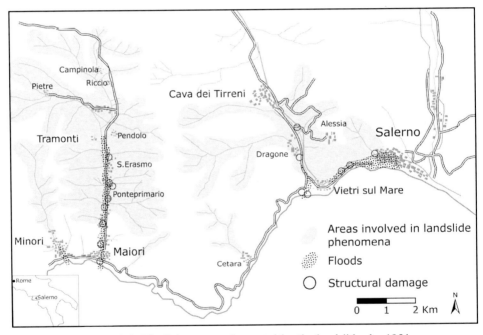

Figure 19.4. The Salerno area impacted by the landslides by 1954.

analysed below. As already noted by Ranieri (1841) one century earlier, Lazzari (1954) describes the triangular shape of the 1954 landslide as a constant character of this type of slope failure (Figure 19.5(b) and (c)).

In 1973, a debris flow, on the Mount San Costanzo slope in the Sorrento area, was triggered by the fall of a limestone block from a vertical scarp during a heavy rainstorm (Figure 19.5(d)). In February 1986, the town of Palma Campania was struck by a debris flow that destroyed two buildings killing eight people (Figure 19.5(e)).

Figure 19.5. (a) The extension of the debris fan of Vietri following the 1954 landslide event. (b) Some source areas of 1954 landslides. (c) Spatially diffused slope instabilities (1954 landslide event). (d) The area of Mount San Costanzo involved in debris flows triggered by a rock fall. (e) Palma Campania landslide. (f) Pozzano landslides.

(b) After Lazzari (1954). (c) After Lazzari (1954). (e) After Guadagno et al. (1988). (f) Courtesy of Gerardo Lombardi.

Celico et al. (1986) proposed at that time that the initial detachment of material along a slip surface could have been favoured by water under pressure at the contact between the fractured and karstic limestone and the pyroclastic rocks. Now, based on experience of studying the Sarno–Quindici landslide type, the triggering mechanism can be assumed to be related to a soil slide at the edge of a natural scarp, as will be explained in Section 19.5.

In January 1997, a high mobility debris flow occurred on the Sorrento coast impacting the town of Pozzano (Figure 19.5(f)). Despite the small initial volume, the movement travelled down the slopes for a distance of about 2 km, leaving a track similar to a narrow chute. The violent impact of the high-velocity mass caused, also in this case, deaths and damage, although the mass had the possibility of spreading out on the bed of a quarry.

On 5–6 May, 1998, as already mentioned in Section 19.1, a large number of debris avalanches and debris flows occurred impacting the towns of Sarno, Quindici, Siano, and Bracigliano (Figure 19.6). In this case, several million cubic metres of slope materials were displaced. The field surveys showed that the landsliding process initiated as shallow debris slides that expanded by incorporating material and water from the slope to become fluid flows (Figure 19.7). In particular, hundreds of small initial failures (Guadagno, 2000) combined to develop about twenty-five long debris flows sufficiently mobile to reach built-up areas at the base of the slopes.

One year later, in December, triggered by heavy rainfall, debris flows and debris avalanches occurred in the Mount Partenio area. The most damaging flow reached the village of Cervinara causing the death of six people. As can be seen in Figure 19.8, subsequently to the initial failure, the flow was forced to make a 90° turn after reaching the base of the slope where it became channelled into a gully, thus being transformed into a high-velocity debris flow.

As described above, flows in pyroclastic soils can affect a wide sector of the Campanian Apennines, where the convergence of different factors, discussed below, permits the development of catastrophic slope failures, characterized by high energy and absence of premonitory signs.

Finally, it must be noted that similar phenomena can also affect the pyroclastics deposited on the slopes in the city of Naples and those of the Phlegrean Fields. Figure 19.9 shows landslides that impacted the San Martino hill in the centre of Naples, where the bedrock is formed of tufaceous rocks. Even if only structural damage was recorded on that occasion, these examples indicate that geomorphological conditions favouring the development of Sarno-type flows may be found in these highly populated areas. In addition, the occurrence of this type of event, numbering more than twenty-five in the last ten years, demonstrates the low importance of the nature of the bedrock, the surficial morphological characteristics being the determining factor for the development of the landslides.

19.3 A FRAGILE GEOLOGICAL ENVIRONMENT

The western side of the Campanian Apennines (Figure 19.2) is characterized by a unique geological setting with an extraordinary convergence of geological and

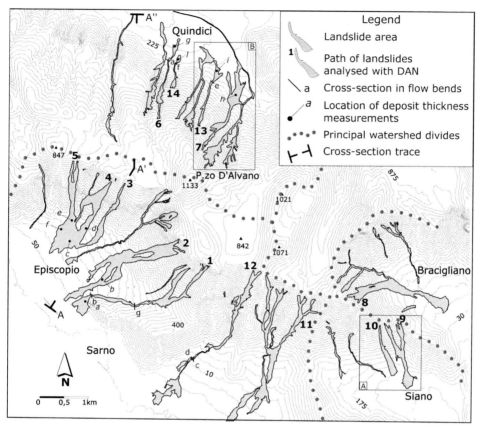

Figure 19.6. Patterns of the 1998 flows along the slopes of the Pizzo d'Alvano ridges. For analysis results with DAN program (as explained in Section 19.5) and their comparison with real measurements, see Table 19.3 (p. 507). Cross section (A–A′–A″) is shown in Figure 19.10. Locations of Figure 19.9 (a and b) are also shown.

morphological factors and of volcanic activity. Three of these factors are important (Figures 19.2 and 19.10):

1. The first factor is the slope morphology and hydrogeology, strongly influenced rock structure. Some of the limestone slopes are controlled by normal faults (e.g., Sarno) and these range from 20° (basal zone) to 50–90° (top). Others are dip slopes belonging to monocline structures formed by limestone sequences (Ippolito et al., 1975). These have angles ranging from 30–45°. The fault slopes are characterized by irregular surfaces where one or more natural scarps can be present in the form of strata outcrops. On the other hand, the lower angled slopes have a more uniform morphology, even though fault-controlled gullies interrupt their morphological continuity with many localized dips. Karst morphology, including a number of dolines is present, as well as a

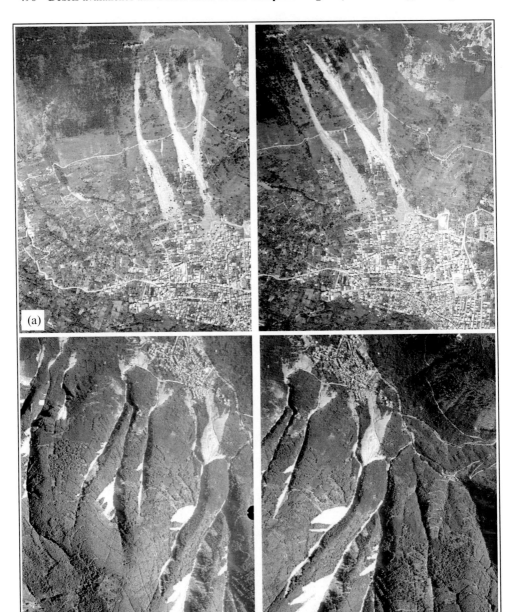

Figure 19.7. Two couples of vertical aerial photographs of the Siano and Quindici landslides respectively (for locations, see Figure 19.6).

Published by licence no. 2004.0961987-031December/2004 of Regione Campania Grant.

Sec. 19.3] A fragile geological environment 497

Figure 19.8. The major landslide event in the Cervinara area. (a) Debris avalanche path. (b) Lateral deposit zone. (c) Debris-flow path. (d) Main deposit area.
After Fiorillo et al. (2001).

Figure 19.9. The landslide of the San Martino area in the city of Naples.

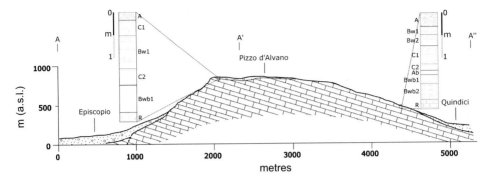

Figure 19.10. Cross section of Pizzo d'Alvano (for location, see Figure 19.6). Soils are named following the *soil taxonomy* (Soil Survey Staff, 1992) as follows: surficial horizon (A); weathered deep horizon (Bw1–Bw2); deep horizon less weathered (C1–C2); buried surficial horizon (Ab); buried weathered deep horizon (Bwb1–Bwb2); suffixes (1) and (2) in the soil description mark similar horizons belonging to different eruptions.

tectonic–karstic water circulation as demonstrated by a number of perennial springs. The most important springs occur at the foot and along the slopes in Sarno.

2. The second factor is the stratigraphy of the pyroclastic mantles. The limestone relief has been mantled by airfall and pyroclastic flow deposits as a result of intermittent and cyclic volcanic activity of the Phlegrean Fields and Somma–Vesuvius volcanoes. The pyroclastic sequence covers the bedrock with no native regolith and is formed by ashy and pumiceous layers alternating with buried horizons of soil modified by pedogenic evolution. They dip at angles similar to those of the bedrock surface. It should be noted that the slope angles are very close to the friction angles of the material, making the geomorphological setting peculiar. The thickness of the pyroclastic sequences increases from the top of the slope (0.5–2 m) to the foot of the hill (over 10 m) as a consequence of the eruption type and prevailing wind directions, the morphologic characteristics of the slopes and their exposure, as well as the effects of erosion and colluvial processes.

3. The third factor is the presence of numerous artificial roads constructed in recent decades for agricultural work on the slopes.

The block diagram of Figure 19.11 schematically depicts the typical geomorphological setting of the slopes. Even if the pyroclastic mantle covers surface irregularities and creates a smooth topography, scarps and cornices are visible. It can be compared to snow mantles in mountain areas. The snow is deposited in successive layers as the winter progresses, its layers characterized by variable density and mechanical properties. Similarly, the cyclic deposition of volcanic airfall, separated by periods of weathering, formed a complex sequence of soils (Figure 19.12(a)). The lithology, grain-size composition, and thickness of each layer are different, as too are their hydrogeological and mechanical properties. Even if horizons can be observed,

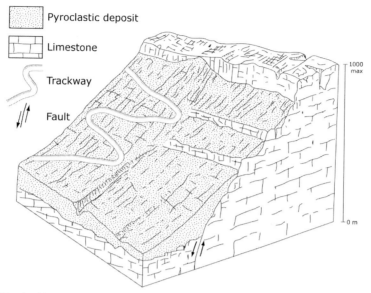

Figure 19.11. A block diagram showing the peculiar geomorphological setting in the Campanian Apennines.

layering is not well defined at the edges of scarps and at the flanks of gullies due to erosion and creep phenomena. Similar to snow avalanches, the described layered structure is fundamental in the development of landsliding processes.

This description of the landscape of the Campanian Apennines should be completed by some notes on land use. Since Roman times the population has developed chestnut and hazelnut plantations on the thin pyroclastic mantle that is very fertile. In previous times, access to the plantations was on foot and hence only limited pathways were created. Moreover, in the past, regulations controlled the use of the plantations with the purpose of wood conservation, preventing slope modifications. The Bourbon regulation (Manuale Forestale, 1838) is probably one of the first legal documents that recognized the fragility of a particular environment. Consequently a wide program of improvements such as debris control structures and channels were developed.

Despite the long-standing knowledge of the delicate equilibrium of the slopes, trackways have been developed during the last twenty-five years for vehicular access. The roadways have been cut into the pyroclastic mantles rising in a zigzag manner up the slope being almost parallel to the contours. Since they are for service access only, trackway construction has generally been carried out without any concern for drainage or compaction works (Figure 19.12(b)).

The fragility of the slope could have been increased by fires. Effects resulting from the destruction of vegetation and the increase of soil temperature are well known, for example the potential formation of hydrophobic surficial layers and the change in water retention capacity because of loss or alteration of organic matter. In the case of Campania these effects seem to be amplified due to the

Figure 19.12. (a) A detail of a typical pyroclastic sequence. (b) A track during the winter. (c) Landslide deposits and a large boulder transported by the flow. (d) Aerial view of Quindici depositional fans. (e) Another crown of a typical debris avalanche scar: here also the initial failure initiated a short distance above the cut slope of a trackway. (f) The crown of a typical debris avalanche scar: the initial failure initiated a short distance above the natural scar.

presence of allophane minerals[1] in weathered volcanic soils, which are very sensitive to temperature increase (Rao, 1995; Terribile et al., 1999). Grain size analyses carried out by Guadagno and Magaldi (2000) have shown a decrease in the fine fractions in samples heated up to 750°C with respect to those left in normal condition. This can be explained by an aggregation effect of the clay particles. The same phenomenon influences the consistence limits too, as proved by decreases in the liquid limit by about 30–35%.

19.4 SPECIAL GEOTECHNICAL PROPERTIES AND BEHAVIOUR OF PYROCLASTIC SOILS

The sequences of pyroclastic layers and soil horizons forming the cover of the carbonate ridges consist of discontinuous layers with varying geotechnical properties. These have been characterized through analyses of representative samples and *in situ* tests. Table 19.1 shows the variation range of selected geotechnical parameters from different sites in the area.

The contrast in basic properties between pumiceus layers and horizons of buried soils is quite evident. Pumice should be considered as a graded granular material (sand with gravel and silt with a small clay fraction) while soil horizons should be considered as typical cohesive materials. If we consider these characteristics, it is possible to infer the complex behaviour of the sequences forming the cover of the slopes. But some special characteristics of the materials induce anomalous behaviour. In the pumice layers, the presence of interconnected capillary-sized voids within the grains influences water flow circulation, causing a suction phenomena and complex water diffusion (Whitam and Sparks, 1986). As shown by Esposito and Guadagno (1998), owing to the retention capacity of the pumice, a greater volume of water is needed to obtain fully saturated conditions than that necessary for soils made of non-porous clasts.

The presence of allophane minerals (Maeda et al., 1977) with some organic matter in the horizons determines specific geotechnical characteristics in the materials. There are problems connected with soil testing, where a special methodology must be used to avoid the effects of temperature and chemical agents on the allophane particles (Guadagno and Magaldi, 2000). Generally, these soils exhibit high liquid limits at relatively low clay contents. As already described by Mitchell (1976) the residual friction angle is generally high (close to 30°) and comparable to the peak angle as defined for some volcanic soils of Campania (Guadagno and Magaldi, 2000).

As expected, layers and horizons of soil exhibit extremely variable permeability (Table 19.1). The contrast in hydraulic conductivity (k) of the single components of the pyroclastic sequences and the specific suction characteristics are therefore the origin of a complex system of water circulation, where the pumice layers can act as drain layers, while the clayey horizons are quasi-impervious. In such a

---

[1] Allophane is a low-crystallinity clay mineral.

Table 19.1. Key geotechnical parameter ranges of a typical pyroclastic sequence.

Thickness of range (cm)	Horizons	G_s	γ_d (kN/m³)	W_n (%)	S (%)	CF (%)	Finer <60 μm (%)	W_L (%)	PI (%)	OM (%)
0–30	A	2.69–2.74	8.25–11.15	20.6–40.3	25.5–70.4	4–9	30–35	35.6–63.2	8.5–13.5	6.6–15.4
0–80	Bw	2.65–2.75	6.88–9.49	36.3–58.1	40.4–82.0	0–16	9–38	49.8–57.1	11.1–15.1	6.6–11.2
0–100	Cl	2.46–2.51	—	24.8–33.2	—	—	4–8	—	—	3.1–3.9
0–20	Ab	2.65	6.7	51.8	46.4	—	—	—	—	10.0
0–45	Bwb	2.67–2.68	6.6–8	67–64.8	62.4–74.0	16–14	54	75.7–62.2	15.1–16.8	7.6–8.5
0–250	Bt	2.71–2.77	6.57–8.74	64.4–86.7	55.6–94.9	15–31	54–68	55.2–93.0	15.5–19.8	7.6–11.8
	Bedrock									

G_S = specific gravity; γ_d = dry unit weight; W_n = water content; S = degree of saturation; CF = clay fraction; Finer <60 μm = % size finer than <60 μm; W_L = liquid limit; PI = plasticity index; OM = organic matter.

hydrogeological setting, geometrical changes due to the cutting of a roadway can provoke important changes in the groundwater flow pattern.

Finally, the granulometric characteristics of the matrix of debris-flow deposits range from *silty sand with gravel* and *sand with silt and gravel*. A large amount of limestone clasts is also present. Calcareous boulders (up to 2 m in diameter) as well as trees or anthropogenic elements were observed in the depositional area (Figure 19.12(c)).

19.5 CHARACTERISTICS OF THE INSTABILITIES

One of the points presently under discussion in the Italian scientific community in connection with the landslides involving the pyroclastic soils of Campania, concerns the characteristics of the initial instability and, consequently, the classification of the flows. As is well known, the determination of the flow type is important for hazard analysis. Therefore, geomorphological location of the initial instabilities, volume and discharge material, velocity, and runout are key parameters for flow prediction.

Based on studies carried out in different sectors of the Campania area by Civita et al. (1975), Guadagno et al. (1988), Calcaterra et al. (1997), Del Prete et al. (1998), and Fiorillo et al. (2001) it was recognized that the initial stage of the movements is usually located in the highest part of the slopes involved. In these typical source areas, translational slides of portions of the pyroclastic cover can turn into channelled or non-channelled flows through acceleration and incorporation of materials present along the slopes or gullies. As seen in the case of the Mount San Costanzo landslide (Figure 19.5(d)), in some cases the initial movement can be connected to the fall of a calcareous block from a scarp.

Following the 1998 phenomena some authors hypothesized that static liquefaction was the triggering mechanism of the source volume of the landslides (Picarelli and Olivares, 2001; Olivares et al., 2003). Such behaviour is characteristic in flow slides as defined by Hutchinson (1986) and more recently by Hungr et al. (2001). Fully-saturated conditions of the material and the total absence of high-permeability pumiceus layers and cohesion have been hypothesized, and in some cases the trigger zone was placed in the middle part of the slope.

In contrast, detailed analyses carried out on the Sarno area instabilities and comparisons with those of the more recent past (Civita et al., 1975; Guadagno, 1991; Calcaterra, 1997; Fiorillo et al., 2001), showed that the initial movement consists of the failure of relatively coherent slabs of pyroclastics in the highest parts of the slope. The slabs slide on failure surfaces, usually located within the pyroclastic sequence, often at the bottom of a pumiceous horizon. The plastic behaviour of the initial sliding mass is proven by tensional cracks present on some sites where the landsliding process was aborted and by the fact that many initial slides occur in the back-slope of trackways and natural scarps.

The mobilized mass of the initial failure generally impacts on the downslope pyroclastic blanket where liquefaction phenomena, by means of rapid undrained loading (Johnson, 1984; Sassa, 1984), could be triggered. In nearly all cases the initial slides transformed into extremely rapid flows, growing substantially in volume by incorporating materials and surface water.

These landslide mechanisms can be compared to those described by Hungr et al. (2001) as *debris avalanches*, phenomena that involve open slopes and therefore are typically not confined in gullies, at least in their initial stages.

In the case of the Campania landslides, many debris avalanches became confined in gullies in middle or lower portions of the slope, transforming into debris flows, still eroding the pyroclastic and colluvial cover from the slope.

At the base of the slopes, the flows spread out in thin depositional fans, continuing to move at relatively high velocities, destroying vegetation and structures. Deposition generally occurs in the basal plain where the velocities decrease (Figure 19.12(d)). If large amounts of water are present in the area or in the gullies, the landslide material can be reworked by hyperconcentrated flows.

In such a reconstruction of the phenomena, the analysis of the morphological characteristics of the initial slides and debris avalanches appears crucial for hazard evaluation. For the large cluster (numbering 176) of avalanches from the Sarno area, Guadagno and Perriello (2000), Guadagno (2000), and Guadagno et al. (2005) showed that morphological discontinuities, associated with natural scarps and with road tracks, appear to be the controlling factors. Figure 19.12(e) shows a typical example of an initial slide that occurred at a short distance above a cut for trackways, whereas Figure 19.12(f) shows a slide at the edge of a scarp. In both cases, the pyroclastic masses possess the kinematic freedom necessary for failure. Possible initial morphological settings resulting from different cases are summarized in Figure 19.13.

Statistical analyses of initial instabilities in the Pizzo d'Alvano slopes (Table 19.2) show that most slides were connected with artificial tracks, even in different morphostructural settings (Guadagno et al., 2005). In the entire area, 61% of the initial instabilities were generated by the mechanism illustrated in Figure 19.13(b).

Guadagno et al. (2005) considered the development of the avalanches as failure phenomena of the pyroclastic slab. There is clearly a recurrent triangular shape to the avalanches, always recognizable in the highest parts of the failures. Similar development is observed in snow avalanches (McClung and Schaerer, 1993) and can be connected with the thrusting action of the moving mass. Taking advantage of the morphological and lithostratigraphic similarity of the landslides, analyses of significant morphometric parameters were performed.

Figure 19.14 indicates the principal morphometric parameters of the analyses of the Sarno area avalanches. In particular, it appears that the width of the apex angle of the triangular-shaped avalanche scars is systematically controlled by the slope angle and the height of the natural or man-made scarps (Figure 19.15).

It is interesting to observe that the initial sliding instability can involve small masses of a few cubic metres. In the presence of favourable morphological conditions along an open slope, the final debris avalanche volume can be up to 20,000 times the initial slide volume (Guadagno et al., 2005). In particular, final avalanche volumes may be correlated to the slope angles in the failure zone and to the length of the slope, as the trend of the plot in Figure 19.16 would seem to show.

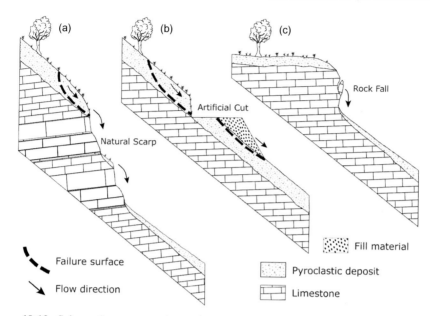

Figure 19.13. Schematic cross sections along a pyroclastic-mantled slope. (a) Instability mechanism at the location of natural scarps. (b) Instability mechanism at a location of man-made cuts along a trackway. (c) Instability mechanism due to a rock fall.

Table 19.2. Recurrence of the morphological settings recognized in the failure areas. After Guadagno et al. (2005).

Morphological conditions of initial failures	Location				
	Sarno (57)*	Quindici (88)*	Siano (11)*	Bracigliano (20)*	Total (176)*
Above natural scarps	21 (37%)	17 (20%)	8 (73%)	5 (25%)	51 (29%)
Below natural scarps	2 (3%)	0 (0%)	1 (9%)	0 (0%)	3 (2%)
Above man-made cut	18 (31%)	57 (65%)	0 (0%)	11 (55%)	86 (49%)
Involving fills	8 (14%)	11 (12%)	2 (18%)	0 (0%)	21 (12%)
Without morphological control	8 (14%)	3 (3%)	0 (0%)	4 (20%)	15 (8%)

*Number of initial failures.

Flow velocities were estimated, for recent phenomena, by measuring the superelevation of the landslide mass in channel bends. As can be observed in Table 19.3, velocities may be estimated up to 12.8 m/s near the toe of the slopes. These values are confirmed by video clips taken during the landslide events and by eyewitnesses.

The use of the *DAN* program (Hungr, 1995), based on an explicit Lagrangian solution to the equations of unsteady non-uniform flow in a shallow open channel, has permitted the debris avalanches and flows to be modelled. The methodological

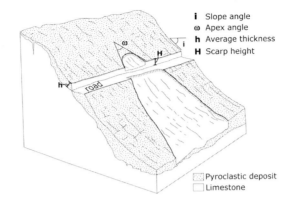

Figure 19.14. The morphometrical parameters used in the analysis of debris avalanches.
After Guadagno et al. (2005).

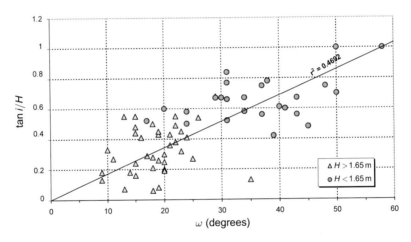

Figure 19.15. Apex angle (ω) vs. the ratio between slope angle tangent (tan i) and scarp height (H).
After Guadagno et al. (2005).

basis of the application of the model to the Campania flows can be found in Revellino et al. (2004). The analyses assumed that the source volumes consisted of a slab of constant depth (1.5 m) and that, downslope of the source area, the debris avalanches were eroding the same constant thickness of material.

Through a systematic program of back-analysis of real cases and by using a trial-and-error procedure, runout distances, velocity at points on the path where it was estimated by measuring superelevation in path bends, and the thickness and distribution of the debris deposits were obtained. An example of the analyses is shown in Figure 19.17, while Table 19.3 presents data of comparisons between some real measurements of the flows and those derived from DAN analyses.

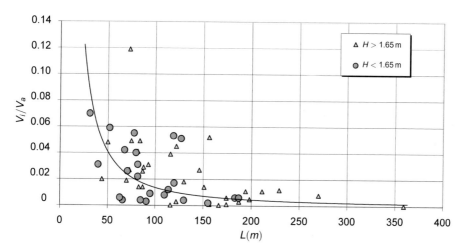

Figure 19.16. Ratio between initial volume (V_i) and avalanche volume (V_a) vs. slope length (L). H is the scarp height.

Table 19.3. Summary of the analysis of measured cross sections, velocity, runout, and deposit thickness and their comparison with DAN model analysis. (Slide numbers correspond to landslides in Figure 19.6. For location of cross sections in flow bends and deposit thickness measurements see Figure 19.6.)
After Revellino et al. (2004).

Slide	Cross section	Velocity Superelevation (m/s)	Velocity Model (m/s)	Runout Actual (m)	Runout Model (m)	Site	Deposit thickness Actual (m) (min/max)	Deposit thickness Model (m)
1	b	5.5	6.1	3,397	3,280	1	1.5/3.0	2.8
	g	12.8	10.2					
2				2,560	2,591	b	0.7/1.4	1.5
3				1,895	1,995			
4				1,860	1,890	c	0.8/1.0	0.6
						d	0.4/0.5	0.2
5				2,051	2,074	e	0.4/0.7	0.4
						f	0.6/0.8	0.5
6	a	10.3	9.4	1,535	1,589	g	0.4/1.0	2.3
7				1,955	2,069	h	0.2/0.4	0.2
8				2,028	2,077			
9				1,122	1,145			
10				1,052	1,170			
11				1,965	2,058			
12	c	5.9	6.2	3,210	2,990			
	d	6.1	5.9					
13	e	10.9	13.9	1,234	1,250	i	0.4/0.8	0.4
14	f	10.3	14.2	736	760	l	0.6/1.0	1.5

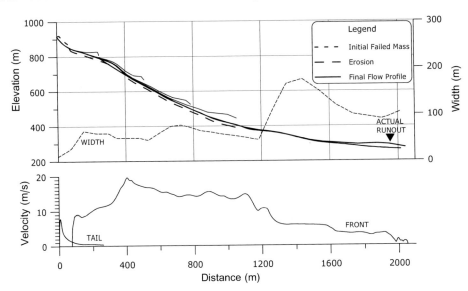

Figure 19.17. Example of a DAN back-analysis using the Voellmy model and showing flow and velocity profiles. The flow profiles are plotted at 20-second intervals. All normal depths (flow depths and erosion depths) are exaggerated 10 times. The landslide is Number 7 in Figure 19.6.

After Revellino et al. (2004).

The results described demonstrate that the Voellmy model (Voellmy, 1955), selected in the program DAN to simulate the reological behaviour of the moving masses, is capable of realistic representation of the dynamic behaviour of the Campania debris flows and avalanches, using a narrowly constrained set of resistance parameters. This must be considered a useful result in view of the need for hazard analysis and the planning of mitigation measures.

19.6 INSTABILITY MECHANISMS AND TRIGGERING CONDITION

Since the flows occur during periods of rainfall, surficial water and groundwater seepage play a fundamental role in triggering the initial failures. As the detailed analyses show, instabilities generally occur where specific hydrogeological conditions can be recognized on the slopes. In particular, natural and anthropogenic morphological settings can create concentrated runoff to specific points that generally correspond to the source points of the avalanches. Hydraulic conditions can be induced in such a zone, in terms of full saturation, pore pressure, and infiltration, necessary for the triggering of the instabilities. Moreover, the saturation of soil masses should be considered, as already stated in Section 19.4, to be a key factor in the development of undrained loading mechanisms within the pyroclastic multilayer as well as in the transformation of plastic sliding mechanisms into more or less viscous flows.

Sec. 19.6] **Instability mechanisms and triggering condition** 509

Figure 19.18. A typical source area on the Sarno slopes.

A typical natural setting favouring initial instability in the Sarno area is shown in Figure 19.18. As can be seen, the movement initiates in a natural infiltration zone where water can concentrate.

Cuts for trackways interrupt the normal downslope flow of surface water. Water flowing downslope through the more permeable deposits can be temporarily confined by the relatively impermeable mixed overlying soils resulting from track construction. In this condition hydrostatic pressures could be created, inducing failures.

Figure 19.19 shows the case of a Cervinara landslide where the initial instability is connected to the presence of two trackways. In this case, erosion traces indicate that the water flows followed the road along the longitudinal section where it could seep through the more permeable pumice layer. Such major water infiltration would explain the development of the pore pressure necessary to induce instability (Fiorillo et al., 2001).

Slope stability analyses have also been performed by means of numerical models, such as the finite difference codes, and in different geomorphological and hydrogeological situations, to verify the described initial failure mechanisms (Crosta

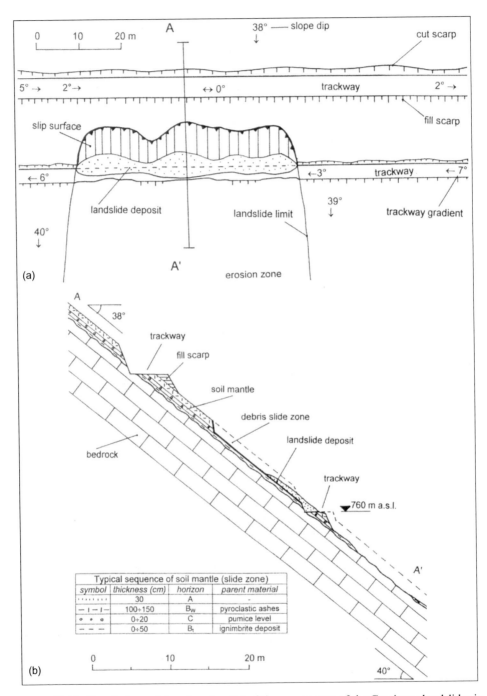

Figure 19.19. Detailed map (a) and section (b) of the source area of the Cervinara landslides in Figure 19.8.
After Fiorillo et al. (2001).

and Dal Negro, 2003; Guadagno et al., 2003). The results show that the continuous and uninterrupted multilayered pyroclastic covers can be considered at equilibrium. On the contrary, human-induced cuts along the slopes cause significant modification of the equilibrium conditions, leading to the development of local yield zones typically localized in the pyroclastic sequence. The analysis shows that the geometry of the cuts can have a significant impact on the shape of the slip surfaces.

The results of the numerical modelling simulations are compatible with the field observation and the statistical analyses on selected morphological parameters characterizing the initial failure.

Regarding the instability of masses upslope of the cuts, an important role is played by the fertility of the pyroclastic material. As schematized in Figure 19.20, following variable periods of time, weathering processes favour the formation of new soil along the steep upslope of the cuts. This condition can induce the development of a partially impervious horizon obstructing the drainage of the permeable layers. Therefore, the analysis of the geomorphological and pedological evolution of the slopes can help in understanding why the instability does not affect relatively recent cuts. In the absence of weathering, the pumiceus layers could have better drainage.

Another fundamental effect can be associated with the channelling of rainfall along the trackways. The roads, with a mean gradient of about 3–4%, induce channelling toward specific points, in particular to the curving segments of the zigzag pathways. As a result, such localized concentrations of water can induce pore pressures triggering the failure of the downslope side of the trackway, where uncompacted materials form fills whose angle is generally higher than that assumed by the pyroclastic deposits in a natural condition. The aerial view of the Quindici side of the Pizzo d'Alvano (Figure 19.21) is an excellent example of this condition.

The Campanian Apennines have a typical Mediterranean climate with hot dry summers and warm wet winters. Although there can be many days in winter without a measurable rainfall, 60–80 mm of precipitation in a day is not atypical. Higher values should be considered exceptional.

The analysis of pluviometric data for some historic and recent phenomena shows that long-duration storms appear to be more important in triggering Campanian shallow landslides than short-duration high-intensity storms. Due to the physiographic situation (mountains near the sea), long periods of rainfall can occur localized in typical cellular areas where the cells are embedded in stationary fronts on the Tyrrhenian side of the Apennines (Mazzarella et al., 2000). Such conditions have occurred several times during the last few centuries, inducing landsliding.

The role of antecedent rainfall in triggering the avalanches of Campania was evidenced by Celico et al. (1986), Guadagno (1991), Del Prete et al. (1998), Rossi and Chirico (1998), Onorati et al. (1999), Calcaterra et al. (2000), and Fiorillo et al. (2001).

Esposito and Guadagno (1998) connected the effects of long-duration rainfall to the peculiar characteristics of the pumice elements, characterized by capillary internal voids. The high retention capacity and suction of the pyroclastic material can induce a delay in water accumulation and in the consequent development of pore

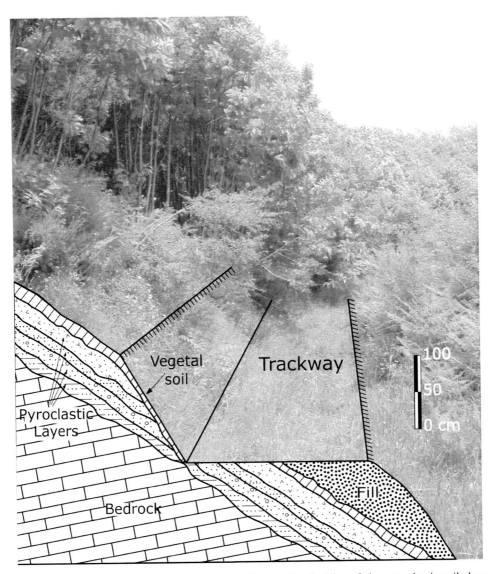

Figure 19.20. A schematic section along a trackway. The fertility of the pyroclastic soils has permitted the rapid growth of vegetation and soil formation along the free face of the cut.
After Guadagno et al. (2005).

pressure. This behaviour is anomalous for shallow landslides, which are commonly controlled by short-duration high-intensity rainfall.

The complexity of the geomorphological settings and of the hydrogeological behaviour of the soils and the lack of appropriate data on the rainfalls triggering landslides, at present make thorough analyses of the rainfall trigger difficult.

Figure 19.21. The slope involved by debris avalanches in the area of San Francisco gullies: note the remarkable link between initial failure and man-made cut. It is possible to observe the same area through photo-interpretation of the coupled images of Figure 19.7(b).

Tentative thresholds were suggested by Guadagno (1991), Onorati et al. (1999), Rossi and Chirico (1998), and De Vita (2000).

This chapter has stressed the crucial role played by the morphological conditions, both natural and anthropogenic, that make local concentrated runoff possible. These control the occurrence of individual landslides, even if the general conditions of the area are the same from a geological and meteorological point of view. These conditions should be considered as the main determinant in the development of the landslide process. Hydrological models should take into account the peculiar properties of the layered pyroclastic sequences.

In the case of the 1998 events, it should be noted that the debris avalanches occurred in a period of about 12 hr. Landslides took place in a discontinuous manner, without a temporal and spatial order. In the Quindici sector the time range was concentrated in about 4 hr. The discontinuous time–spatial distribution of landslide initiation also demonstrates that the local setting has a fundamental importance in triggering the debris avalanches of Campania.

19.7 HAZARD EVALUATION: SOME NOTES ON THE PRESENT SITUATION AND FUTURE PROSPECTIVE

Risks connected to debris avalanches and flows in Campania have rapidly increased in the last few decades because of the intense urbanization and of the changes in landuse.

At the present time, local authorities, following national laws and regulations, are attempting to define the hazard in the different areas and the consequent levels of risk. Even if the criteria utilized for this purpose have not been strictly the same, inventory and landslide susceptibility maps are being used as the basis for hazard evaluations. The criterion of landslide susceptibility has been principally based on the presence of pyroclastic cover along the slopes and slope angles greater than threshold values. Different weights have been assigned to the presence of morphological characteristics such as discontinuities, natural scarps, or cuts. The evaluation of the possible volumes and runout of the flows has been generally determined by using morphological criteria.

It should be noted that, as already stated in Section 19.1, the classification of the phenomena, and the triggering mechanisms, have not been universally agreed on. This situation produces significant uncertainties for decision-makers and local technicians.

In this paper, based on the available field data, the local morphological conditions have been stressed as the principal factor controlling the failures in the source areas, while the dynamic evolution of the phenomena has been considered as typical of debris avalanches and flows.

The landslide inventory and susceptibility maps available at present should be considered as the first step and prerequisite to producing more detailed maps, and specifically, probabilistic hazard maps.

Detailed maps showing the local morphological setting and characteristics should be provided on an adequate scale, to define the possible location of the initial instabilities through deterministic evaluations and the volume of material involved in the avalanche process. Where channelling is possible, characteristics of the paths and deposits filling the gullies should be evaluated.

The dynamic model used in the simulation (DAN) can be used to produce quite realistic first-order predictions of the runout of potential slides. Such predictions could be used to map maximum potential runouts from specific landslide sources, as well as velocities and discharges required for the planning of protective measures.

Another intriguing question related to hazard evaluation is connected to the return period of failures at the same site after a period of time. In those zones where the unstable material is a result of the weathering of a bedrock, the minimum time interval is linked to the time necessary to form a slope cover. In the case of the pyroclastic materials, as allochthonous deposits, the probability of a new event dramatically decreases once the pyroclastic cover is removed. To restore geological conditions similar to those before the landslide, it is necessary to wait for the deposition of new eruptions.

It should be noted that in areas impacted by the debris avalanches in the past, thin unlayered slope mantles have been slowly reconstituted (Vallario, 1992) as a result of deposition of materials eroded from upslope, by creep process, and by the growth of vegetation. This condition is not comparable with the pre-slide stratigraphy, but is typically chaotic. Gullies traversed by the flows have been "cleaned" of the infill deposits. Also in this case it is necessary to reconstitute the filling materials over a period of time. Zones and gullies recently involved in debris avalanches and debris flows can be considered to have hazard levels much lower than others which have never experienced instabilities.

In the probabilistic approach to hazard zoning, crucial evaluation is connected to the return time of the triggering causes. In the Campania avalanches the primary triggering cause is linked to rainfall. As seen, these effects are different according to the geomorphologic setting but also to anthropogenic changes and the elapsed time.

19.8 CONCLUSIONS

The calcareous slopes of the Campania Region, mantled by recent pyroclastic deposits, have to be considered as a singular and fragile geological environment.

Analyses of past and recent debris avalanches and flows in Campania, that caused extensive damage, allowed the authors to define the main predisposing factors. These are connected with the geomorphological and lithostratigraphical setting, and with the geotechnical properties of soils and horizons. In particular, the main determinants of the initiation of the debris avalanches are:

- the peculiar nature and setting of the cover deposits, composed of flow and fall-type pyroclastic materials instead of soils resulting from weathering of the outcropping bedrock; and
- the presence of natural scarps and man-made cuts within the limestone bedrock and the pyroclastic covers producing kinematic freedom for the masses along the edges and at the same time modifying groundwater patterns.

Any changes in slope geometry, and therefore in water seepage both superficial and underground, can be a determining factor in causing instabilities in the surficial deposits. In the last few decades, mechanized forest management practices have favoured the formation of a dense trackway network, which has significantly altered the already sensitive original setting.

Correct land-use practices, in particular management of woods without making inappropriate changes to the local topography, seem to be a very important factor in avoiding the increase of hazard that, as shown in the case of Sarno–Quindici, has increased dramatically.

19.9 REFERENCES

Alfano, G.B. and Friedlander, I. (1929) *La storia del Vesuvio illustrata dai documenti coevi.* Verlag Dr K. Hohn, Ulm, Germany [in Italian].

Calcaterra D., Santo, A., De Riso, R., Budetta, P., Di Crescenzo, G., Franco, I., Galietta, G., Iovinelli, R., Napolitano, P., and Palma, B. (1997) Fenomeni franosi connessi all'evento pluviometrico del gennaio 1997 in Penisola Sorrentina. *Atti del IX Congresso Nazionale Geologi, Roma, 17–20 Aprile* (pp. 223–231) [in Italian].

Calcaterra, D., Parise, M., Palma, B., and Palella, L. (2000) The influence of meteoric events in triggering shallow landslides in pyroclastic deposits of Campania, Italy. *Proceedings of the 8th International Symposium on Landslides, Cardiff* (Vol. 1, pp. 209–214). Thomas Telford, London.

Celico, P., Guadagno, F.M., and Vallario, A. (1986) Proposta di un modello interpretativo per lo studio delle frane nei terreni piroclastici. *Geologia Applicata e Idrogeologia*, **21** [in Italian].

Civita, M., de Riso, R., Lucini, P., and Nota d'Elogio, E. (1975) Sulle condizioni di stabilità dei terreni della Penisola Sorrentina (Campania). *Geologia Applicata e Idrogeologia*, **10**, 129–188 [in Italian].

Crosta, G.B. and Dal Negro, P. (2003) Observations and modelling of soil slip-debris-flow initiation process in pyroclastic deposits: The Sarno 1998 event. *Natural Hazard and Earth System Sciences*, **3**, 53–69.

Del Prete, M., Guadagno, F.M., and Hawkins, B. (1998) Preliminary report on the landslides of 5 May 1998, Campania, southern Italy. *Bulletin of Engineering Geology and Environment*, **57**, 113–129.

De Vita, P. (2000) Fenomeni di instabilità delle coperture piroclastiche dei Monti Lattari, di Sarno e di Salerno (Campania) ed analisi degli eventi pluviometrici determinanti. *Quaderni di Geologia Applicata*, **7**(2), 213–235 [in Italian].

Esposito, L. and Guadagno, F.M. (1998) Some special geotechnical properties of pumice deposits. *Bulletin of Engineering Geology and Environment*, **57**, 41–50.

Ippolito, F., D'Argenio, B., Pescatore, T.S., and Scandone, P. (1975) Structural-stratigraphic inits and tectonic framework of Southern Apennines. *Geology of Italy (Tripoli)*, 317–328.

Johnson, A.M. (1984) Debris flows. In: D. Brudsen and D.E. Prior (eds), *Slope Instability* (pp. 257–361). John Wiley & Sons, London.

Fiorillo, F., Guadagno, F.M., Aquino, S., and De Blasio, A. (2001) The December 1999 Cervinara landslides: Further debris flows in the pyroclastic deposits of Campania (southern Italy). *Bulletin of Engineering Geology and Environment*, **60**(3), 171–184.

Guadagno, F.M. (1991) Debris flows in the Campanian volcaniclastic soil. *Slope Stability Engineering* (pp. 109–114). Thomas Telford, London.

Guadagno, F.M. (2000) The landslides of 5th May 1998 in Campania, Southern Italy: Natural disasters or also man-induced phenomena? *Journal of Nepal Geological Society*, **22**, 463–470.

Guadagno, F.M. and Magaldi, S. (2000) Considerazioni sulle proprietà geotecniche dei suoli allofanici di copertura delle dorsali carbonatiche campane. *Quaderni di Geologia Applicata*, **7**(2), 143–155 [in Italian].

Guadagno, F.M. and Perriello Zampelli, S. (2000) Triggering mechanisms of the landslides that inundated Sarno, Quindici, Siano and Bracigliano (S. Italy) on May 5–6, 1998. *Proceedings of the 8th International Symposium on Landslides, Cardiff* (Vol. 22, pp. 671–676).

Guadagno, F.M., Palmieri, M., Siviero, V., and Vallario, A. (1988) La frana di Palma Campania del 22 Febbraio 1986. *Geologia Tecnica*, **4**, Roma [in Italian].

Guadagno, F.M., Martino, S., and Scarascia Mugnozza, G. (2003) Influence of man-made cuts on the stability of pyroclastic covers (Campania – Southern Italy): A numerical modelling approach. *Environmental Geology*, **43**, 371–384.

Guadagno, F.M., Forte, R., Revellino, P., Fiorillo, F., and Focareta, M. (2005) Geomorphology of the source areas of the flows involving the pyroclastic soils of Campania (Southern Italy). *Geomorphology* (available online).

Hungr, O. (1995) A model for the runout analysis of rapid flow slides, debris flows, and avalanches. *Canadian Geotechnical Journal*, **32**, 610–623.

Hungr, O., Evans, S.G., Bovis, M., and Hutchinson, J.N. (2001) Review of the classification of landslides of the flow type. *Environmental and Engineering Geoscience*, **7**(3), 1–18.

Hutchinson, J.N. (1986) A sliding-consolidation model for flow slides. *Canadian Geotechnical Journal*, **23**, 115–126.

Lazzari, A. (1954) Aspetti geologici dei fenomeni verificatisi nel Salernitano in conseguenza del nubifragio del 25–26 ottobre 1954. *Bollettino della Società dei Naturalisti*, **238**, 921–925 [in Italian].

Maeda, T., Takenake, H., and Warkentin, B.P. (1977) Properties of allophane soils. *Advances in Agronomy*, **29**, 229–264.

Manuale Forestale (1838) Tipografia Flautina, Napoli [in Italian].

Mazzarella, A., Martone, M., and Tranfaglia, G. (2000) Il recente evento alluvionale del 4–5 maggio 1998 nel sarnese e il deficit risolutivo della rete pluviometrica. *Quaderni di Geologia Applicata*, **7**(2) [in Italian].

McClung, D.M. and Schaerer, P. (1993) *The Avalanche Handbook*. The Mountaineers, Seattle, WA.

Mitchell, J.K. (1976) *Fundamentals of Soil Behavior*. John Wiley & Sons, New York.

Montella, N. (1841) *Sposizione del disastro avvenuto in Gragnano diretta ad allontanare il timore di ulteriori pericoli* (pp. 5–25). Tipografia Del Petrarca, Napoli [in Italian].

Olivares, L., Damiano, E., and Picarelli, L. (2003) Wetting and flume tests on a volcanic ash. *International Conference on Fast Slope Movement: Prediction and Prevention for Risk Mitigation, Naples*.

Onorati, G., Braca, G., and Iritano, G. (1999) Evento idrologico del 4, 5 e 6 Maggio 1998 in Campania. Monitoraggio ed analisi idrologica. *Atti dei convegni dei Lincei 154: Il rischio idrogeologico e la difesa del suolo, Roma, 1–2 Ottobre 1998* (pp. 103–108) [in Italian].

Picarelli, L. and Olivares, L. (2001) Innesco e formazione di colate di fango in terreni sciolti di origine piroclastica. *Forum per il Rischio Idrogeologico in Campania, Napoli, Giugno 2001* [in Italian].

Ranieri, C. (1841) *Sul funesto avvenimento della notte dal 21 al 22 gennaio 1841 nel comune di Gragnano*. Boerio, Napoli [in Italian].

Rao, S.M. (1995) Mechanistic approach to the shear strength behaviour of allophanic soils. *Engineering Geology*, **40**, 215–221.

Revellino, P., Hungr, O., Guadagno, F.M., and Evans, S.G. (2004) Velocity and runout simulation of destructive debris flows and debris avalanches in pyroclastic deposits, Campania Region, Italy. *Environmental Geology*, **45**, 295–311.

Rossi, F. and Chirico, G.B. (1998) *Definizione delle soglie pluviometriche di allarme* (Rapporto Unità Operativa 2.38). C.N.R.-G.N.D.C.I., Salerno, Italy [in Italian].

Sassa, K. (1984) The mechanism starting liquefied landslides and debris flows. *Proceedings IV International Symposium on Landslides, Toronto, June* (Vol. 2, pp. 349–354).

Soil Survey Staff (1992) *Keys to Soil Taxonomy* (5th edn, 306 pp.). Pocahontas Press, Blacksburg, VA and Department of Agriculture, VA.

Terribile, A., Basile, A., di Gennaro, A., Aronne, A., Buonanno, M., De Mascellis, R., Vingiani, S., and Mulacelli, F. (1999) The soil of the landslide of Sarno and Quindici. *Proceedings Symposium on Degradation Processes in Volcanic Soils* (pp. 48–64). University of Napoli.

Vallario, A. (1992) La frana di Palma Campania. In: *Frane e territorio*. Ed. Liguori, Naples [in Italian].

Voellmy, A, (1955) Über die Zerstorungskraft von Laimen. *Schwerzerische Bauzertung*, **73**, 212–285 [in German].

Whitam, A.G. and Sparks, R.S.J. (1986) Pumice. *Bulletin of Engineering Geology*, **48**, 209–223.

20

Debris flows of December 1999 in Venezuela

Reinaldo García-Martínez and José Luis López

20.1 INTRODUCTION

The extreme rainfall that occurred in the north-central Venezuelan coast in December 1999 caused the worst disaster from natural causes in Venezuela and perhaps amongst the largest in Latin America (Wieczorek et al., 2001). On 16 December, 1999 simultaneous debris flows occurred in twenty-four streams along 50 km of a narrow coastal strip in the State of Vargas, causing extensive damage in the urban developments located on the alluvial fans that border the Caribbean Sea. Over the course of a few hours large parts of towns were destroyed by the debris avalanches and debris. About 15,000 people were killed, 23,000 houses were destroyed and 65,000 houses suffered severe damage. Economical losses have been estimated to more than US$ two billion (López et al., 2003).

From its occurrence we started a research programme on several aspects of debris flows, including the revision and adaptation to local conditions of methods to define hazard zones, development and application of numerical models to simulate debris flooding in urbanized areas, and the establishment of an experimental watershed in the Vargas region to study correlation between extreme storm precipitation distribution, runoff, and sediment generation, and transport.

This chapter presents some of the most important characteristics of the Vargas disaster including a review of the precipitation that triggered the event, the general geology of the region, and some geomorphologic features of the watersheds. We also present the proposed methodology to produce hazard maps based on the application of the FLO-2D model in the alluvial fans and estimations from water and sediment runoff resulting from a hydrological model and analysis of aerial photographs and field work. An example of the hazard maps produced using this methodology is also provided.

20.2 DEBRIS-FLOW EVENTS IN LATIN AMERICA

Floods and debris flows have a frequent occurrence in Latin America, particularly in the cities of the Andes Mountains. Recent events include the landslide disaster in the Paute River in 1993, near the city of Cuenca, Ecuador, better known as the disaster of La Josefina (Zevallos et al., 1993). Ayala et al. (1994) report four large events related to debris flows in Chile in a period between 1987 and 1993, which caused hundreds of fatalities and losses of US$ tens of millions. The most dramatic event was caused by the eruption of Nevado Ruiz volcano, which on 14 November, 1985 produced mudflows that destroyed the town of Armero, Colombia, killing 25,000 people. The most recent event occurred when Hurricane Mitch passed through Central America, generating debris flows that destroyed the town of Casita Volcano in Nicaragua, with 2,000 victims (Must, 1999). Other minor-scale mudflows have also been reported recently in some states of Mexico in 1999 (Caldino and Bonola, 2000). In Venezuela some debris-flow events have been documented. Oral accounts tell that when the Spaniards founded the port of Caraballeda, they invited the local aborigines to move to the lower coastal lands in order to be closer to port. They received a negative response from the aborigines because according to their experience "that mountain spits rocks from time to time". The oldest written testimony refers to the storm of February 1798 in the region of La Guaira, reported by Alexander von Humboldt (1804) who described the 70-hour duration rainfall that produced a large flood in the Osorio River, carrying wood trunks and boulders of considerable size. Another catastrophic storm occurred in the same region (north-central coast) in February 1951, affecting a similar area as the 1999 storm, including the capital city of Caracas (Sardi, 1959). The town of La Guaira was reported to have been buried to a depth of up to 4 m. The most dramatic event in Venezuela occurred near the city of Maracay in September 1987, where heavy rainfalls produced flash floods and debris flows in the Rio Limon, causing a death toll of 300 people. However, geological evidence suggests that large-scale debris flows have occurred in the Rio Limon basin, and also in the valley of Caracas, in pre-Colombian times between 1,100 and 1,500 AD (Singer et al., 1983).

20.3 GEOGRAPHIC CHARACTERISTICS OF THE DISASTER AREA

Venezuela is located in the northern part of South America between 1° and 12°N (Figure 20.1). The north coastal range of Venezuela runs parallel to the Caribbean Sea attaining elevations of up to 2,800 m above sea level. Mountains are very steep descending to sea level in a horizontal distance between 6 and 10 km (Figure 20.2). The southern slopes descend to a valley at 1,000 m above sea level, where the capital city of Caracas is located 15 km south of Maiquetia International Airport. In the coastal areas, urban and tourist developments have taken place within the State of Vargas, a very narrow land strip whose width varies between 200 m and 2,000 m. Alluvial fans, canyons and steep slopes at the foot of the mountain range have been

Sec. 20.3] Geographic characteristics of the disaster area 521

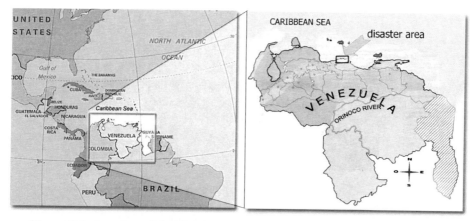

Figure 20.1. Location of the disaster area in the north coastal range of Venezuela.

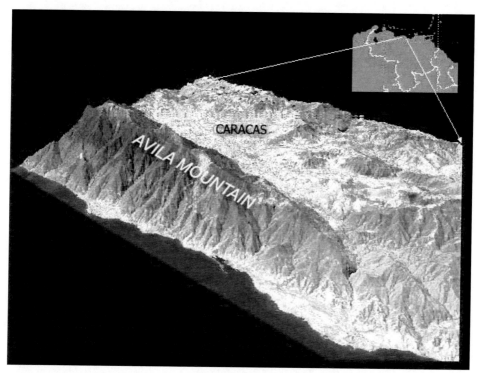

Figure 20.2. Aerial view of Avila Mountain and the north coastal range of Venezuela (Ikonos satellite image). Longitudinal distance is approximately 30 km.

urbanized over the past 50 years (Figure 20.2). The population was estimated at 300,000 before the December 1999 disaster. Most of the upper parts of the catchments are well protected by vegetation and belong to the Avila National Park. The slopes support tropical rain forests at higher elevations including large trees of more

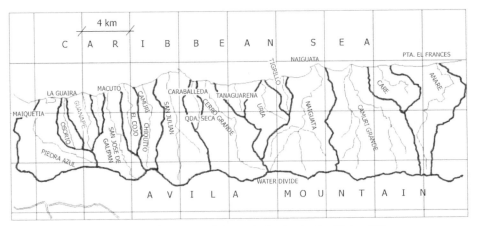

Figure 20.3. Location map of the main basins in the State of Vargas.

than 30 m height, while regions near the sea level are arid with scarce shrubby vegetation. Catchment areas range from 2–120 km^2 with stream bed slopes of up to 90% in the upper reaches and 3–6% in the lower reaches (Figure 20.3). Most streams are ephemeral. The mean annual rainfall is about 500 mm near sea level and rises to 1,200 mm at elevations higher than 1,500 m. The mean annual temperature in the coastal zone is 26°C.

20.4 THE DECEMBER 1999 STORM

The rainy season in Venezuela extends typically between May and September. However, some of the large storms, flash floods, and inundations in the State of Vargas have occurred between November and February, due to cold fronts originated in the North Atlantic Ocean. In the period prior to the 1999 storm, large humid air masses originating over Colombia and the Pacific Ocean were blocked by a low-pressure trough which remained stationary over the Caribbean Sea for about 20 days. By the middle of December, satellite images indicated that the frontal system in the Caribbean region was covering an area of 20,000 km^2 and extended for about 900 km along the north-central coastline of Venezuela. Rainfall began by the end of November 1999. Unfortunately, only two weather stations were operating in the State of Vargas during the storm. These stations are located in the eastern part of the disaster area. Figure 20.4 shows rainfall data recorded in the stations of Maiquetia (43 m above sea level) and Mamo (81 m above sea level) for December 1999. Low intensity but continuous rainfall occurred between 1–13 December, amounting to 293 mm in Maiquetia, followed by heavy rainfall during the following three days. Maiquetia station reported 911 mm from 14–16 December and a total precipitation of 1,207 mm for the first 17 days of the month. This station measured 72 mm of rain between 6 a.m. and 7 a.m. of the morning of 16 December. In Mamo, however, the cumulative value for the same period was

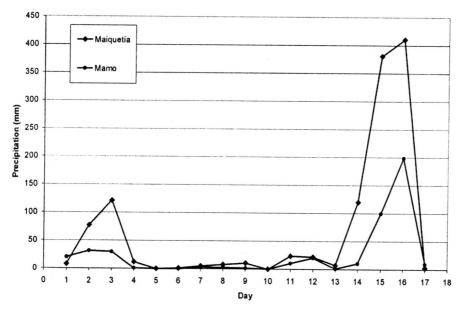

Figure 20.4. Rainfall data during December 1999 in the disaster area.

438 mm. Other stations located in the upper part of the basins were not operating. However, past observations have shown that the amount of rainfall at higher elevations could be twice the amount at sea level. For instance, for the 1951 storm (60 h) one station (Infiernito) located at 1,000 m above sea level reported a total precipitation of 529 mm, compared to 282 mm in Maiquetia.

Over a period of 52 years (excluding 1999), Maiquetia station shows an annual average of 523 mm; annual maximum of 961 mm (1951); and annual minimum of 205 mm in 1959. If we compare these values with the 1999 storm, an evident conclusion is that an extraordinary amount of precipitation fell in the mountains of the State of Vargas in December 1999. The long period of nearly continuous rainfall, between 1–13 December, created conditions of near saturation prior to the high-intensity rainfalls of 16 December. In the morning of 16 December debris flows were generated almost simultaneously in about 20 streams of the State of Vargas (Figure 20.3). The unusual amount of available water suggest that the events could be classified as multiphase flows where, depending on the sediment concentration, the mass movement altered between debris floods and debris flows capable of carrying large amounts of sediments including boulders of up to 10 m in size.

An estimation of the return period for the 1999 storm was based on the 52-year record data at Maiquetia Station. A frequency analysis for the daily annual maximum values shows that the Log Pearson III curve provides a better fit than the Extreme Value Distribution. When the 1999 storm total rainfall (911 mm) is included in the analysis, the Log Pearson III frequency distribution indicates that the return period of the 1999 storm is approximately 270 years.

20.5 GEOLOGICAL ASPECTS

The Avila Mountain is a part of the northern Coast Range of Venezuela which covers an area of 30,000 km^2 and extends east–west along the northern fringe of the South American continent. The greatest elevations are reached north of Caracas at 2,765 m (Pico Naiguata) and 2,640 m (La Silla de Caracas). Alluvial fills, such as encountered in the Caracas valley, are found in the south of the Coast Range. The rocks of the coastal mountains are composed mainly of metamorphic and sedimentary rocks of Precambrian and Mesozoic age respectively.

20.5.1 Lithological units

Table 20.1 and Figure 20.5 show the most important lithological units within the area affected by the December 1999 disaster. Rock outcrops are primarily from Precambrian and Mesozoic ages. Topographic relief differences in Avila Mountain are controlled by differences in rock strength. The oldest Precambric unit is represented by San Julian schists (gneisses and feldspathic schists) and Augen-gneiss of Peña de Mora (quartz micaceous gneiss). Most of the large blocks and boulders found in the alluvial valleys belong to the upper lithologic units (San Julian schists and Peña de Mora gneisses). However some of them seem to have been transported by past events and moved again by the 1999 flood, as suggested by signs of subaerial weathering observed in the surface of some rocks indicating long-time exposure to the atmosphere. Large proportions of fine material transported by the debris flows originated from the Tacagua Phase (marbles, calcareous and epidotic schists), which can be found in the lower elevations of the basins and is characterized by a reddish color. This variety of geological characteristics was responsible for the different types of material supply to the valley bottoms and sediment transported through the channels.

20.5.2 Erosion and weathering

The most common mass movement processes occurring during the storm were debris avalanches which formed long scars from the top of the ridges to the toes of natural slopes (Figure 20.6), exposing the bedrock underneath. Some scars are up to 500 m long and 200 m wide, in the El Cojo basin. The evolution of some slide scars in San

Table 20.1. Lithological units in the Avila Mountain.

Phase	Lithology
Tacagua Phase	Marbles, calcareous and epidotic schists
Nirgua Phase	Amphibolites and amphibolitic schists
Las Brisas Formation	Quartz and feldspathic schists, and meta-sandstones
San Julian Formation	Gneisses and feldspathic schists
Peña de Mora	Quartz micaceous gneiss and augen-gneiss

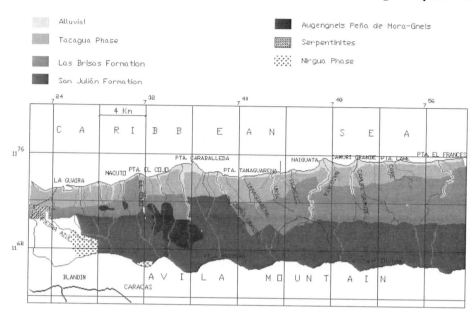

Figure 20.5. Lithological units in the State of Vargas.

Jose de Galipan basin indicates that two years after the event, the upstream migration of knick points had almost reached the top of the basin divide (Figure 20.6). Similar behaviour has been observed for the scars located on the north flank of the Avila Mountain. The material involved in the landslides originates from residual or colluvial soils saturated by the intense rainfall, and include the entire vegetation cover. Observation of aerial photographs shows that in the zones most heavily damaged by landslides, the area of denudation reaches 30% of the total basin area. However, the density of the scars is much greater in the lower parts of the basins than in the upper ones (Figure 20.6), which could be related to the geology and type of vegetation. In the lower basins the soil and weathered rock with a lower vegetation cover was eroded more easily than the hard rocks (schists) protected by large trees. Considering the basins as a whole, estimates of the total areas affected by denudation and landslides range between 10–20% of the total area exposed to rainstorms.

The thickness of the soil layer in the Avila Mountain varies between 1 and 5 m depending of the slope gradient and elevation. Residual soils have developed by weathering of the rock mass lying below the plant cover. The weathered zone, with thicknesses estimated between 80 and 100 m, is due to mechanical and chemical alteration of the rock in the presence of high moisture, organic acids, and temperature effects. The maximum depths are found along the river courses that coincide with geologic faults, where the chemical alteration is greater due to water percolation transporting minerals and organic acids through joints and rock fractures.

Figure 20.6. (*top*) Massive land slides in Cerro Grande and Quebrada Seca basins. (*bottom*) Debris avalanches and debris flow scars of the 1999 storm in the upper part of San Jose de Galipan river basin (photograph taken in January, 2001).

20.5.3 Geologic fault system

Two geologic fault systems are present in the area affected by the debris flows. One of them is parallel to the coastal range and the second is transverse. Some of the streams, like San Julian, Quebrada Seca, and Cerro Grande, have developed their channels along transverse faults. Blocks that fall and slip down to the gullies are important mass sources for debris flows. Slope instability is enhanced by the high earthquake hazard due to the vicinity of a seismic zone where intensive earthquakes have often taken place in the past. Regionally, some of the faults are important from a seismic point of view. The most important one is the San Sebastian Fault, which is considered responsible for the linear trace of the Central Coast. This fault is considered part of the meridian limit of the Caribbean Plate.

20.6 TERRAIN CHANGES

20.6.1 Longitudinal profiles

One of the most striking aspects of the Avila Mountain is its abrupt relief and steep profiles of valley sides and channels. River courses descend steeply from near 3,000-m elevations into sea level in a horizontal distance not greater than 10 km. Channel slopes up to 90% are common in the upper part of the basins. Table 20.2 summarizes some channel characteristics for the main basins. Fifteen basins are listed from west to east (the first three basins, located to the west of La Guaira Port, are not shown in the map of Figure 20.3). The longitudinal profiles of these streams are shown in Figure 20.7. The greatest channel lengths are measured at the Mamo and Tacagua Rivers, two streams located to the west of Maiquetia. Significant changes of slopes are observed above the 400-m elevation, in the transition between the gorge and the upper basin. The maximum average slope corresponds to Guanape stream with a value of 32.7%. The alluvial fans show slopes between 1.6% and 6.7%. Not many boulders were found in the alluvial fans of streams of the Mamo, Tacagua, Naiguata, and Camuri Rivers, where the slopes are less than 2.5% and sand and gravel predominated in the sediment deposits. However, for most of the other streams, with fan slopes greater than 4%, large cobbles and boulders reached the alluvial fans.

20.6.2 Changes in bed profiles

Closer views of the longitudinal profile in the lower reach of Cerro Grande and Uria streams are presented in Figures 20.8 and 20.9, showing conditions before and after December 1999. Sediment deposition occurred in the canyon and alluvial fan of the Cerro Grande River. The slope of the channel bed did not significantly change after the storm, keeping an average value of 4% in the canyon and a value of 2.5% in the fan. The alluvial fan was extended about 120 m into the sea and sediment deposition along the river course amounted to 4–5 m in the valley and 2–3 m in the canyon. The

Table 20.2. Geometric and physiographic characteristics of main basins in the disaster area.

Stream	Area (km²)	Max. basin elev. (m)	Total length (km)	Average slope (%)	Alluvial channel length (m)	Alluvial fan slope (%)	Canyon slope (%)	Channel (aggraded (A) or scoured (S)
Piedra Azul	24.8	1950	9.0	17.3	1500	4.8	6.0	A
Osorio	4.6	1700	5.0	31.2	500	4.4	11.0	A
Guanape	5.7	1960	5.5	28.4	450	6.7	10.0	A
San Jose de Galipán	14.0	2300	8.0	19.5	700	6.3	6.5	A
El Cojo	6.8	1550	5.5	28.4	600	4.0	7.0	A
Camuri Chico	11.2	2350	7.5	20.8	600	5.7	8.5	A
San Julián	23.6	2490	9.8	15.9	2000	5.0	8.0	A
Seca	5.3	1240	3.8	41.1	800	6.2	10.0	A
Cerro Grande	26.6	2750	10.5	14.9	950	2.5	6.0	A
Uria	11.6	2150	7.5	20.8	750	6.3	7.0	S
Naiguatá	33.4	2000	12.0	13.0	1200	1.6	4.0	A
Camuri Grande	42.9	2190	10.2	15.3	1100	2.5	5.0	A
Mamo	141.0	2100	38.5	5.5	3.4	1.5	2.0	A
La Zorra	6.2	900	6.1	15.0	1.0	3.0	6.0	A
Tacagua	93.5	1900	22.4	7.6	3.1	1.3	2.0	A

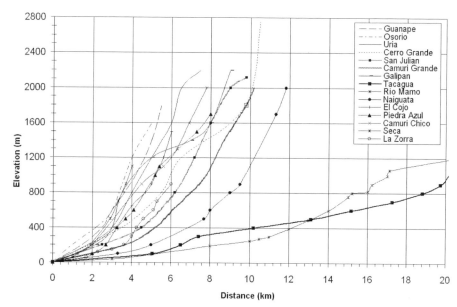

Figure 20.7. Longitudinal profiles of main streams in the disaster area.

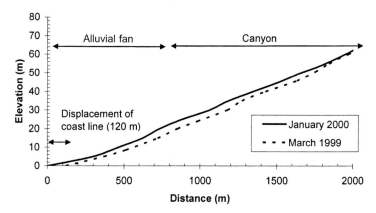

Figure 20.8. Changes in the bed profiles of Cerro Grande River.

majority of streams showed pronounced aggradation. In contrast, at Rio San Julian, approximately 500,000 m^3 of sediment were eroded from the canyon (average slope 8%) and transported downstream to be deposited on the alluvial fan (average slope 5%). On the contrary, the alluvial fan of Uria stream was subjected to extreme erosion that opened a channel 30 m wide and 7 m deep along a heavy urbanized area, destroying and dragging many houses down to the sea (Figure 20.9). The abandoned channel and the new channel for the Uria River can be seen in Figure 20.10. Actually, Uria is the only alluvial fan of the region where erosion occurred.

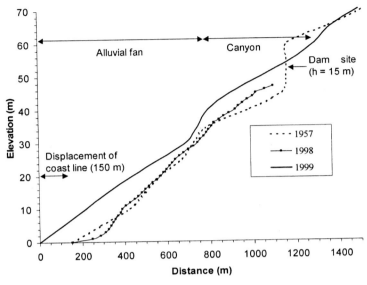

Figure 20.9. Changes in the bed profiles of Uria River.

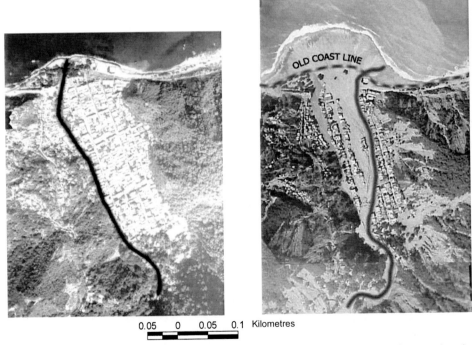

Figure 20.10. Modification of the river channel and coastline in Uria. (*left*) Picture taken in March 1999. (*right*) Picture taken in December 1999 after the disaster (solid line indicates river channel).

20.7 SEDIMENT CHARACTERISTICS

20.7.1 Sediment accumulation

Based on field observations and aerial photographs taken on 21 December 1999, the amount of sediment accumulation on the alluvial fans has been estimated for some of the basins shown in Figure 20.6 and Table 20.2. The volumes of sediment deposition for the 12 basins listed in Table 20.3 refers only to the accumulation of sediment on the alluvial fans and does not include subaqueous deposition. The new land created by fan migrating into the ocean is also included in Table 20.3.

A total volume of sediment deposition for all basins in the study area has been estimated to 20 million m3. However, a significant portion of the sediment transported by the debris flow was washed into the ocean, and is not included in these deposition estimates. The deposition of debris flows changed the existing coast configuration with new beaches and areas that could be used for recreational purposes (Figure 20.10). The amount of land gained to the sea by the 1999 event is estimated to 150 ha.

20.7.2 Characteristics of the bed material

Samples of bed material have been collected in the alluvial fans of different streams in the area of interest. Grain sizes smaller than gravel (6.4 cm) were analysed in the laboratory and distribution curves for greater fractions were obtained directly in the field by Wolman's technique (Wolman, 1954). The finer fractions show quite

Table 20.3. Volume of sediment deposition in alluvial fans of main streams.

Stream	Area of sediment deposition (ha)	Volume of sediment deposition (10^6 m^3)	New land created (ha)
San Julián	127.0	2.60	24.5
Camurí Grande	79.0	1.90	22.4
Cerro Grande	36.6	1.60	13.6
Uria	37.2	1.50	10.7
Seca	74.1	1.48	0.0
Piedra Azul	25.6	1.37	2.7
El Cojo	16.4	1.10	1.7
Naiguatá	66.7	1.05	12.1
Camuri Chico	82.7	0.75	12.8
Galipan	25.4	0.64	5.4
Osorio	17.8	0.45	5.6
Guanape	18.1	0.40	2.1

uniform distribution with particle mean diameters between 0.1 and 10 mm. The maximum percentage of clays and silts was obtained in Cerro Grande (12%), suggesting that enough fine material was available to generate mudflows. A matrix of cobbles, gravel, and fine material was observed under the base of large boulders in many streams, indicating the presence of the so-called "inverse grading". For the coarser fractions analysed, the mean diameters varies between 1 and 2 m. Proportions of sediment sizes in the alluvial fan of the San Julian stream was estimated to be 50% boulders, 20% cobbles, 20% gravel, and 10% sand and finer material. Similar proportions for the Cerro Grande stream are 40% boulders, 15% cobbles, 25% gravel, and 20% sand and finer material. One of the largest blocks (8.2 m × 7.5 m × 4.4 m) in the alluvial fan was found in Cerro Grande with a weight of 318 metric tons.

20.8 ESTIMATED DEBRIS-FLOW DISCHARGES

Table 20.4 summarizes the maximum flood discharges estimated in Uria and Cerro Grande using different methods. The historical data for measured extreme floods given by Creager et al. (1945), as a function of the basin area, is used to obtain an upper boundary for the flow discharge. A second estimate is obtained for a return period of 500 years from a rainfall-runoff model, similar to HEC-1, and historic precipitation data for different stations in Vargas State (Gonzalez and Cordova, 2002). Based on flood marks left by the flow just upstream of a rectangular weir in the canyon of Cerro Grande, an estimate of 1,230 m³/s was obtained from the classical weir equation. Similarly, from measured flood marks on a bend of Uria canyon, the equation of bend superelevation has been used to obtain a value of 1,670 m³/s for the flow peak in Uria. It can be seen that these two values are much greater than the values estimated above.

Other estimates can be made for the discharge produced by dam breaks formed by landslides that temporarily blocked the channel or by the accumulation of woody debris, tree trunks, sediment, and rocks. Figure 20.11 shows two small natural dams formed in a street of Los Corales town in the San Julian alluvial fan (left picture) and in the river channel of Galipan (Macuto) during the 1999 storm.

For discussion purposes we can estimate the flow discharge produced by a hypothetical dam 15 m high subjected to a sudden break. If the classical equation is used (Stoker, 1957):

$$Q = \Psi B g^{0.5} H^{1.5} \tag{20.1}$$

Table 20.4. Comparison of maximum flow discharges estimated by different methods.

River	Area of basin (km²)	Maximum flood discharge (Creager, 1945) (m³/s)	Water discharge for 500-year storm (m³/s)	Estimated peak discharge from flood marks (m³/s)	Dam-break discharge for $H = 15$ m (m³/s)	Discharge with 50% sediment concentration (m³/s)
Cerro Grande	26.6	850	635	1,230	1,078	1,270
Uria	11.6	400	316	1,670	1,725	632

Figure 20.11. Natural dams formed by debris (tree trunks) in Los Corales (*top*) and Macuto (*bottom*). See man for scale.

where B is the channel width; H is the upstream flow depth; g is the acceleration due to gravity; Q is the peak-flow discharge; and $\Psi = 0.21$. The results are quite similar to the measured values.

Table 20.4 also shows the total discharge assuming the bulk corresponds to a sediment concentration of 50% by volume. For the case of Cerro Grande River, this value agrees well with the one estimated from flooding marks, but for the case of Uria River, it is much smaller than the discharge estimated with the curve super-elevation formula.

This is not surprising for the Uria stream, where an artificial concrete dam 15 m high collapsed during the 1999 flood, perhaps the most plausible explanation for such a high discharge.

The above discussion illustrates the limitations of flood flow calculations for determining peak-flow discharges for debris flows. In practical applications such as the present case, detailed field work and comparison of several approaches may be the only way to arrive at accurate estimates of peak discharges.

20.9 STRATEGIES FOR MITIGATION

Mitigation strategies against debris flows in urban areas may require both structural and non-structural measures. Structural measures include debris basins, debris barriers, and deflection dikes. Chapter 18 provides a summary of various methods used worldwide. Non-structural measures include: monitoring meteorological, hydrological, and geomorphic variables in the catchments; development of hazard and risk mapping; implementation of warning systems; land-use planning; and design of contingency plans. Some of these measures are presently being implemented in the north coastal range of Venezuela, and are briefly described herein.

20.9.1 Methodology for hazard maps

The Instituto de Mecánica de Fluidos has developed hazard maps to show the hazard level on alluvial fans. A methodology has been proposed to delineate hazard maps due to mud and debris-flow events, based on the application of mathematical models (FLO-2D) combined with Geographic Information Systems (GIS) (Garcia et al., 2003). The methodology includes criteria to define potential flood hazard zones depending on the event frequency and intensity. Several processor programs complement the use of the FLO-2D model and automate the process of generating the hazard maps. The methodology was tested in 23 sites in the Caracas and Vargas State region in Venezuela. The hazard maps for the Vargas region are now used by planners of the Venezuelan Ministry of Environment and Natural Resources and other agencies to design emergency plans and new land-use policies. The methodology is also being expanded to other flood hazard regions in Venezuela.

The hazard delineation map criteria used in this work were first proposed in the PREVENE 2001 project (PREVENE, 2001), where it was applied to two alluvial

Table 20.5. Definition of hazard level.

Hazard level	Map color	Description
High	Red	Persons are in danger both inside and outside their houses. Buildings are in danger of being destroyed.
Medium	Orange	Persons are in danger outside their houses. Buildings may suffer damage and possible destruction depending on construction characteristics.
Low	Yellow	Danger to persons is low or non-existent. Buildings may suffer little damage, but flooding or sedimentation may affect house interiors.

fans in Caracas. Later, this methodology was extended and applied to other fans in Caracas and the Vargas region. This methodology is based on Swiss and Austrian standards (OFEE et al., 1997; Fiebiger, 1997). The delineation method involves establishing three zones to identify the hazard level in a particular location.

A location may suffer flooding of different intensities and probabilities. Stronger events are generally less frequent. The intensity of the event is a function of the flow depth and velocity. The flood hazard level is then defined as a discrete combined function of the event intensity (severity of the event) and return period (frequency) as shown in Figure 20.12 (see color section).

The limits between the probability regions are defined for return periods of 10, 100, and 500 years, considering that the event of 10 years is highly probable, the event of 100 has an intermediate probability of occurrence and the event of 500 years has a low probability of occurrence. The color over a map translates into specific hazard levels as shown in Table 20.5.

To define the event intensity most existing methods use a combination of flow depth and velocity. The Austrian method (Fiebiger, 1997) uses the total energy defined as $h + v^2/2g$, where h is the flow depth, v is the velocity, and g is the gravitational acceleration. The Swiss method (OFEE et al., 1997) defines the intensity in terms of a combination of h and the product of h and v. This allows assigning high intensities to high depths independently of the velocities.

Following the Swiss method, the criteria used in this work make the distinction between water flooding and mud or debris flows. Intensities are defined in terms of the maximum water depth generated throughout the event and the product of the maximum velocity multiplied by the maximum depth. For example, in the case of water flooding, the intensities are defined in Table 20.6.

The mud or debris-flow events are more destructive than water floods, therefore, the intensity criteria are more conservative as shown in Table 20.7.

The hazard level criteria also encompass the probability of occurrence of a particular mudflow event for three return periods of 10, 100, and 500 years. The general procedure begins with the application of the FLO-2D model (O'Brien, 2003) for events with the above return periods. The model predicts the maximum depths and velocities on the alluvial fan. The water hydrograph at the fan entrance is

Table 20.6. Event intensities for water flooding.

Water flood event intensity	Maximum depth h (m)		Product of maximum depth h times maximum velocity v (m^2/s)
High	$h > 1.5$ m	OR	$vh > 1.5$ m^2/s
Medium	0.5 m $< h < 1.5$ m	OR	0.5 m^2/s $< vh < 1.5$ m^2/s
Low	0.1 m $< h < 0.5$ m	AND	0.1 m^2/s $< vh < 0.5$ m^2/s

Table 20.7. Event intensities for mud and debris flow.

Mud or debris-flow event intensity	Maximum depth h (m)		Product of maximum depth h times maximum velocity v (m^{2}s)
High	$h > 1.0$ m	OR	$vh > 1.0$ m^2/s
Medium	0.2 m $< h < 1.0$ m	AND	$0.2 < vh < 1.0$ m^2/s
Low	0.2 m $< h < 1.0$ m	AND	$vh < 0.2$ m^2/s

obtained applying a hydrological model in the watershed. The mud or debris-flow yield stress and viscosities depend on the fluid concentration by volume through correlations developed by O'Brien (2003). For each scenario the model is run with the corresponding liquid hydrograph and with an initially assumed concentration distribution. The total sediment volume calculated by the model for the complete hydrograph is then compared with the estimated available sediment volume in the watershed that could be mobilized. The concentration distribution is then adjusted accordingly.

For a given location on the alluvial fan the event intensity for a return period determines the hazard level according to Figure 20.12 (see color section).

The described methodology has been accepted by the Venezuela's Simon Bolivar Geographical Institute of the Ministry of Environment and Natural Resources for general application through the country. Figure 20.13 (see color section) shows the hazard map for the San Julian alluvial fan.

20.9.2 Structural measures

Government agencies through CORPOVARGAS with local and foreign funds have initiated an intensive program to canalize the water courses in the alluvial fans, and to build sediment control dams in the canyons of the streams affected by the torrential flows of 1999. The original plans and design projects recommended hydraulic structures made of concrete, but authorities decided to construct them using gabions. As of the time of writing, 18 dams have been completed. The uncertainties regarding the effectiveness of these gabion dams to stop debris flows is a matter of debate between experts and local authorities.

20.10 OTHER RESEARCH EFFORTS

An experimental basin has been established in the Avila Mountain and is now being used for water and sediment observation. The San Jose de Galipan Observation System consists of a network of rain gauges distributed over the basin, a network of water level gauges in the streams, and use of sediment dams for sediment surveys and sampling of bed material. The general objective is to collect water and sediment data to study and investigate the mechanism of debris-flow formation and sediment transport in mountain areas. Specific objectives are to obtain relationships between rainfall and sediment yield, calibrate mathematical models for flow and sediment transport, and improve the design of mitigation measures against debris flows.

20.11 CONCLUSIONS

Twenty basins generated debris flows of extreme magnitude in the morning of 16 December 1999 in the coastal range of northern Venezuela. Approximately 15,000 people lost their lives in one of South America's worst disasters. The volume of deposited material on several alluvial fans are amongst the largest on record from rainfall-induced debris flows. The main causes for the formation of debris flows are associated to: (a) the presence of very steep slopes in the channels and valley sides of the Avila Mountain; (b) the occurrence of a long rainstorm (293 mm in 13 days) that saturated the soils prior to heavy rainfall; (c) the occurrence of high-intensity rainfall (911 mm in 72, and 72 mm in 1 hr) during the event; and (d) the presence of abundant sediment, sufficient thickness of soil, large plants, and weathered bedrock. Collapse of natural dams within the channels is believed to have been responsible for the occurrence of large debris-flow discharges that cannot be explained by the usual response of rainfall-runoff processes in the basins. The high toll of victims is associated to the indiscriminate occupation of the alluvial fans with no control works to regulate and transport the water and sediment flows. Volumes of sediment deposition on the alluvial fans have been estimated in the order of 20 million m^3 for the towns located along 40 km of the coastal area of the State of Vargas. This accumulation of sediment reached into the sea modifying the coastline and creating new lands whose aerial extent has been estimated to 150 ha.

A methodology was applied to delineate hazard maps for 23 alluvial fans of the Avila Mountain watershed. The results show that large areas of the urbanized alluvial fans, both in Caracas and in Vargas State, are located in high flood-hazard areas and require urgent mitigation measures.

20.12 ACKNOWLEDGEMENTS

This research was partially supported by FONACIT through Millennium Scientific Initiative Project and FONACIT Project G-2000001528.

20.13 REFERENCES

Ayala, L., López, A., Tamburrino, A., and Vera, G. (1994) Aspectos hidrometeorológicos e hidrodinámicos de algunos eventos aluvionales recientes en Chile. *Memorias del XVI Congreso Latinoamericano de Hidráulica, Santiago, Chile, noviembre* [in Spanish].

Caldino, I. and Bonola, I. (2000) Viscosidad de mezclas agua-arcilla en relación con los flujos de lodos y debris. *XIX Congreso Latinoamericano de Hidráulica, Cordoba, Argentina, octubre*.

Creager, W.P., Justin, J.D., and Hinds, J. (1945) *Engineering for Dams* (Vol. 1). John Wiley & Sons. New York.

Fiebiger, G. (1997) Hazard mapping in Austria. *Journal of Torrent, Avalanche, Landslide and Rockfall Engineering*, **61**(134), 121–133.

Garcia, R., Lopez, J.L., Noya, M., Bello, M.E., Bello, M.T., Gonzalez, N., Paredes, G., Vivas, M.I., and O'Brien, J.S. (2003) Hazard mapping for debris-flow events in the alluvial fans of northern Venezuela. *3rd International Conference on Debris-flow Hazards Mitigation: Mechanics, Prediction, and Assessment, Davos, Switzerland, September 13–15*.

Gonzalez, M. and Cordova, J.R. (2002) Estimation of extreme water discharge hydrographs in basins of coastal range after the 1999 torrential flows of December 1999 in Venezuela. *Acta Científica Venezolana* [in Spanish].

Humboldt, A. de (1804) *Viaje a las Regiones Equinocciales del Nuevo Continente en los años de 1799 a 1804* (1985 edn). Monte Avila Editores, Caracas.

Lopez, J.L., Perez, D., and Garcia, R. (2003) Hydrologic and geomorphologic evaluation of the 1999 debris-flow event in Venezuela. *3rd International Conference on Debris-flow Hazards Mitigation: Mechanics, Prediction, and Assessment, Davos, Switzerland, September 13–15*.

Must, V. (1999) Hurricane Mitch. *National Geographic Magazine*, November.

O'Brien, J.S. (2003) *FLO-2D User's Manual* (Version 2003.06). FLO-2D, Nutrioso, AZ.

OFEE, OFAT, OFEFP (1997) *Prise en compte des dangers dûs aux crues dans le cadre des activités de l'aménagement du territoire*. Office fédéral de l'économie des eaux (OFEE), Office fédéral de l'aménagement du territoire (OFAT), Office fédéral de l'environnement, des forêts et du paysage (OFEFP), Bienne [in French].

PREVENE (2001) *Contribution to "Natural" Disaster Prevention in Venezuela* (Technical report Project VEN/00/005). Cooperation: Venezuela–Switzerland–PNUD.

Sardi, V. (1959) Gastos máximos de los ríos y las quebradas del litoral central. *Revista del Colegio de Ingenieros de Venezuela*, **275**, febrero.

Singer, A., Rojas, C., and Lugo, M. (1983) *Inventario Nacional de Riesgos Geológicos*. FUNVISIS, Caracas.

Stoker, J.J. (1957) *Water Waves*. Interscience, New York.

Wieczorek, G.F., Larsen, M.C., Eaton, L.S., Morgan, B.A., and Blair, J.L. (2001) *Debris-flow and Flooding Hazards Caused by the December 1999 Storm in Coastal Venezuela* (USGS Open-File Report 01-144). US Geological Survey, Reston, VA.

Wolman, M.G. (1954) A method of sampling coarse river-bed material. *EOS Transactions, American Geophysical Union*, **35**(6), December, 951–956.

Zevallos, O., Fernández, M., Plaza, G., and Klinkicht, S. (1993) *Sin Plazo para la Esperanza* (Reporte sobre el desastre de la Josefina). Escuela Politécnica Nacional, Quito.

21

Debris flows caused by Typhoon Herb in Taiwan

Chyan-Deng Jan and Cheng-Lung Chen

21.1 INTRODUCTION

A large number of debris flows occurred in Taiwan after the 1996 Typhoon Herb. Debris flows in Taiwan are mainly triggered by typhoons that bring heavy rains and trigger debris avalanches that transform into debris flows in stream channels. Earthquakes also play a major role in contributing to the accumulation of colluviums on hillslopes and in channels. In addition to typhoons and earthquakes, many other factors related to the land and its inhabitants affect the frequencies and magnitudes of debris flows. To explain how debris flows were triggered by Typhoon Herb and other typhoons, we first address the debris-flow effects of primary factors, such as the geologic and geomorphologic settings, recurrence of typhoons, rainfall conditions, and anthropogenic disturbances.

The severity of debris-flow hazards in Taiwan had not been realized and extensive investigations not undertaken until Typhoon Herb hit Taiwan in 1996 and triggered 52 debris flows. Since 47 debris flows occurred in the watershed of the Chenyoulan stream (i.e., headwaters of the Choshui River) in central Taiwan, we devote a significant part of this chapter to address the causes and effects of debris flows in this watershed. We evaluate the devastation of debris flows caused by Typhoon Herb and its aftermath. We also assess the present status of debris-flow hazards mitigation strategies in Taiwan.

After Typhoon Herb, the combination of a catastrophic earthquake with magnitude 7.3 on the Richter scale (hereafter referred to as the 1999 earthquake) and extremely heavy rains associated with Typhoons Xangsane in 2000 and Typhoons Toraji and Nari in 2001, has caused large and abundant debris flows. In 2001, a total of 192 debris flows occurred in Taiwan. Available resources to mitigate such a large number of debris-flow hazards are limited in Taiwan, and government officials are assessing the adequacy of debris-flow hazards countermeasures presently implemented in Taiwan. Results from debris-flow research should allow for developing cost-effective debris-flow hazards mitigation strategies and measures. This chapter lays the foundation and establishes the viable guidelines for attaining these goals.

21.2 TERRESTRIAL FACTORS TRIGGERING DEBRIS FLOWS IN TAIWAN

21.2.1 Geologic and geomorphologic settings

Taiwan is an island 36,000 km² in size separated from the south-east coast of China by the 175 km wide Taiwan Strait. The island is oval in shape with a length of 394 km and a maximum width of 144 km. Located at the obliquely convergent boundary of the Eurasian Plate and the Philippine Sea Plate, Taiwan was formed by the collision of an island arc with the Asian continental margin. Orogenesis is ongoing, resulting in two-thirds of its area being covered by rugged mountains and hills (Figure 21.1, see color section), steep topography, young (3 million years) and weak geological formations, active earthquakes, and loose soils (Huang, 2002).

The Central Mountain Range stretches along the entire island from north-north-east to south-south-west, forming a watershed divide separating rivers on the eastern and western sides of Taiwan. The Jade Mountain Range lies on the west side with the Jade Mountain peak reaching an elevation of 3,952 m – the highest mountain peak in North-east Asia. Taiwan's average elevation is 765 m and about 31% of the total island area has an elevation exceeding 1,000 m (GIO, 2003). Most mountains are very steep with slope gradients usually exceeding 25° and local relief of 1,000 m.

Significant denudation results from landslides, debris flows, and soil erosion in mountainous terrain. Landslides generate a large amount of loose, weathered materials, which are delivered to creeks, thus forming the dominant sediment supply mechanism for debris-flow-prone channels. Several investigations of landslides in the mountains of the central and eastern parts of Taiwan (Wu et al., 1989; Shieh, 1996) have revealed a spatial density of 0.27 landslides/km² and a landslide area ratio equal to 0.84%.

Landslide causes include weak rock and rapid weathering, while landslide triggers include runoff concentration, undercutting by streams, and road construction (Chang, 1993; Tsai et al., 1996). About 60% of landslides are susceptible to reactivation (Chang and Slaymaker, 2002). The average denudation rate of the Central Mountain Range of Taiwan is about 14 kg/m²/yr, which is equivalent to 5.5 mm/yr if the density of rock is assumed to be about 2.5 g/cm³ (Li, 1976; Dadson et al., 2003). Tectonic activity and associated vigorous denudation supply material to hillslopes and stream channels, thus affecting the frequency and magnitude of debris flows.

21.2.2 Recurrence of typhoons

Typhoons are most frequent from July to October. They usually begin in low latitudes (10°–20°N) in the North Pacific Ocean near the Philippine Islands, move north or north-west, and then pass through the Taiwan Island or turn north-east at high latitudes (20°–30°N) towards Korea or Japan. The western Pacific typhoons have a storm path 80–160 km wide. They travel at about 10–30 km/hr, but the violent winds within a typhoon can cause great destruction, resulting in the loss of life and property due to high winds, suddenly rising waves, and heavy rains. Taiwan has mild weather throughout the year except during a typhoon. Taiwan's subtropical climate is characterized by warm and humid conditions with a mean monthly temperature in the lowlands being about 16°C in winter and between 24°C and 30°C during the rest of a year. The relative humidity averages about 80%.

On average, three typhoons hit Taiwan annually, and people living on the island are used to severe weather accompanying typhoons, especially heavy rains. The amount of rainfall received on Taiwan during the passage of a typhoon is dependent on the speed, size, and intensity of the rain-producing centre of a typhoon and on the topography of the land swept over, thus varying spatially and concentrating on the mountainous interior of Taiwan. Taiwan receives an annual rainfall of 2,500 mm, with maxima exceeding 5,000 mm in some high mountain regions. About 80% of annual rains fall from May to October, especially during typhoons. Rainfall intensity during some destructive typhoons may exceed 100 mm/hr and 1,000 mm/24 hr. The recorded maximum one-hour and 24-hour rainfalls before 1996 were 300 mm and 1,672 mm, respectively. The recorded 24-hour rainfall was broken by the heavy rain accompanying Typhoon Herb and reached 1,749 mm. Rainfall data collected in the Chenyoulan stream watershed during Typhoon Herb will be analysed and elaborated later in this chapter (Section 21.3.1).

21.2.3 Anthropogenic disturbances

Natural conditions to trigger debris flows have been aggravated due to inadequate land-use and engineering practices in mountainous areas. Negligence to enforce restrictions on the use of hillslopes has led to urban sprawl, encroachment onto alluvial fans, and development on steep terraces. Most people in Taiwan live in or near big cities where average population densities exceed $2,000/km^2$. The development and cultivation of hillslopes are often deemed essential to provide food and housing in such densely populated areas. On hillslopes with slopes less than 30°, cultivation is permitted, while mountain forest lands with slopes exceeding 30° are strictly forbidden from cultivation. Of all mountainous lands, 28% have been cultivated. Despite these land-use restrictions, 15% of mountain forest lands have been illegally developed (Lee, 1996; Cheng et al., 2000), leading to a significant increase of the vulnerability of infrastructure and urban development to debris-flow hazards.

21.2.4 Typhoons as agents of debris flows

As a result of heavy rains accompanying typhoons, steep topography, young and weak geological formations, strong earthquakes (25 earthquakes of magnitudes

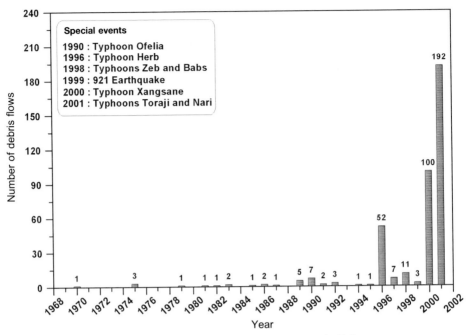

Figure 21.2. Number of debris flows per year in Taiwan.

larger than 6 on the Richter scale in the 20th century), loose soils, and land development in mountainous terrain, many areas in Taiwan are susceptible to landsliding (TGRU, 2001). Among so many causes, typhoons are especially important agents of debris flows in Taiwan. One of the first well documented debris-flow catastrophes occurred on 23 June, 1990 in Tongmen village, located in the west of Hualien City on the east coast of Taiwan, during Typhoon Ofelia, with rainfall intensities exceeding 106 mm/hr (Yu, 1990). This debris flow (with an estimated volume of 56,000 m^3 and an average speed of 8.4 m/s), killed 35 people and destroyed 24 houses (Yu, 1990). Since this event, the number of debris flows in Taiwan has increased, as shown in Figure 21.2. In the following, we detail the impact of Typhoon Herb on the generation of debris flows.

21.3 DEBRIS FLOWS TRIGGERED BY TYPHOON HERB

21.3.1 Rainfall

Typhoon Herb was "born" on 24 July, 1996 at approximately 19.9°N and 148.8°E in the North Pacific Ocean, about 800 km north-east of Guam. It then moved west-north-west at 10–30 km/hr towards Taiwan, and landed on 31 July in Yilan County, north-east Taiwan. From here the typhoon moved across north-western Taiwan with a radius of 350 km, and left on 1 August, 1996 with wind speeds up to 65 m/s. Typhoon Herb was not the strongest typhoon that had stricken Taiwan in the past four decades, considering wind speed and radius of the typhoon. In retrospect,

Typhoon Herb had not been considered a serious threat before striking Taiwan. However, contrary to the prediction, after landing on Taiwan, it caused more damage than any other typhoon because of a record-breaking amount of rain, especially in Taiwan's central region (Chiang, 1996; Yu and Tuan, 1996; F.C. Yu, 1997). Figure 21.3 (see color section) shows the spatial distribution of rainfall in Taiwan for the 96-hour rainfall during Typhoon Herb.

The maximum 10-minute, 1-hour, 12-hour, 24-hour, and 48-hour rainfall near the headwaters of the Chenyoulan stream watershed in Nantou County were 25 mm, 113 mm, 1,158 mm, 1,749 mm, and 1,987 mm, respectively (Chieng, 1998; Lin, 1998). Both the previous 12-hour (862 mm) and 24-hour rainfall (1,672 mm) records were broken. The new 12-hour and 24-hour rainfall records in Taiwan approach the world records (respectively 1,340 mm and 1,870 mm) (TGRU, 2001), as shown in Figure 21.4. The estimated return periods for 6-hour to 24-hour rainfall at Alishan Station well exceeded 200 years (Figure 21.5). In total, 1,987 mm of rain fell at Alishan Station over a period of 43 hours during Typhoon Herb. This total amount equals about 30% of the regional annual rainfall in central Taiwan. An hourly rainfall exceeding 80 mm persisted for 13 hours from 17:00 on 31 July to 06:00 on 1 August 1996 at Alishan Station (Figure 21.6). This record-breaking rainfall and its antecedents are believed to play a major role in triggering debris flows. To analyse the effect of the antecedent rainfall on the debris-flow initiation threshold, the hourly rainfall hydrograph at Alishan Climate Station is plotted in Figure 21.7. This analysis is still underway, but it cannot be elaborated herein until after we find its significant results.

21.3.2 Debris flows generated in Chenyoulan stream watershed

The heavy rains during Typhoon Herb triggered 52 debris flows in Taiwan. Because 90% of them occurred in the Chenyoulan stream watershed (Figure 21.8, see color section), which has an area of 449.5 km^2, our study focuses on this area. Factors controlling debris-flow occurrence include site-specific geomorphic and geologic characteristics and precipitation. Six site factors listed in Table 21.1 include the catchment area upstream of a debris-flow initiation location, ground elevation and gully slope at a debris-flow initiation location, road construction, land ownership, and land use in a debris-flow catchment area, while one precipitation factor listed in Table 21.1 is the total rainfall received by this watershed during Typhoon Herb.

In Table 21.1, Chieng (1998) originally listed only the numbers of debris-flow events for the various ranges (or classes) of the 6 site factors and 1 precipitation factor. An inspection of Table 21.1 reveals that the number of debris-flow events for catchment areas ranging from 50–100 ha, for ground elevations ranging from 500–1,000 m, or for gullies having slopes from 12°–17° is the largest among the potential ranges (or classes) for each of the first 3 site factors mentioned above. Such comparison based solely on the number of debris-flow events is misleading because debris flows triggered in the various ranges (or classes) of a factor did in fact come from different debris-flow initiation areas, which may vary in size. Statistically, the effect of a site factor on debris-flow occurrence cannot be correctly evaluated unless the

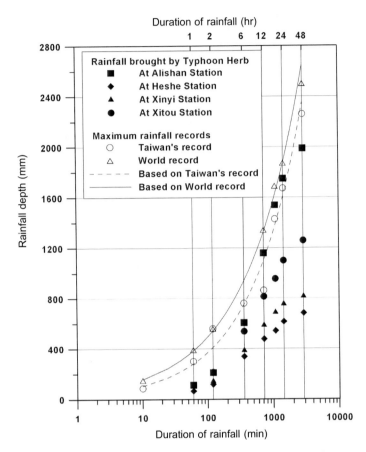

Figure 21.4. Rainfall magnitude–duration graph for Typhoon Herb at 4 rain-gauge stations in the Chenyoulan stream watershed. For comparison, maximum rainfall magnitude–duration relations in Taiwan and the corresponding world records are shown.

numbers of debris flows are normalized by the respective areas in the various ranges (or classes) of the site factor. Unfortunately, we cannot calculate the number of debris flows *per unit area* for all 6 site factors analysed due to the unavailability of data on areas in the potential ranges (or classes) of each factor, as shown in Table 21.1. The 3 site factors lacking such area data are the catchment area upstream of a debris-flow initiation site, roads passing through a debris-flow catchment, and land use in a debris-flow catchment. The remaining 3 site factors for which we can calculate the numbers of debris flows *per unit area* in the potential ranges (or classes) are the ground elevation at debris-flow initiation area, gully slope at debris-flow initiation area, and land classified according to ownership. Analysing a pattern of the numbers of debris flows *per unit area* in the potential ranges (or classes) of these 3 site factors, we can infer a possible trend of debris-flow occurrence in this watershed.

Sec. 21.3] Debris flows triggered by Typhoon Herb 545

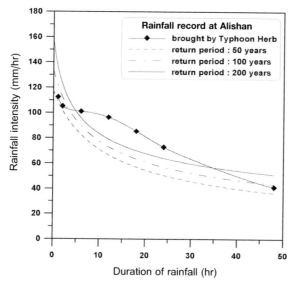

Figure 21.5. Rainfall intensity–duration graph for Typhoon Herb at Alishan Climate Station. For comparison, the corresponding relations for various return periods are also plotted.

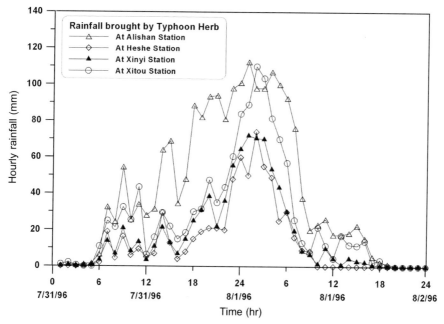

Figure 21.6. Hourly rainfall measured at 4 rain-gauge stations in the Chenyoulan stream watershed during Typhoon Herb. The maximum hourly rainfall at Alishan and Xitou rain-gauge stations exceeded 100 mm.

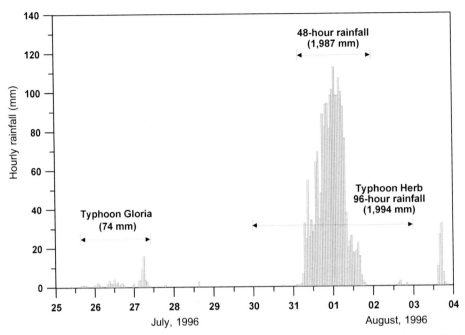

Figure 21.7. Hourly rainfall hydrograph at Alishan Climate Station during Typhoons Gloria and Herb.

As shown in Table 21.1, the number of debris flows *per unit area* for the entire area of the Chenyoulan stream watershed averages $0.105/\text{km}^2$, irrespective of which factor is being investigated. If the number of debris flows *per unit area* in any range (or class) of a site factor is greater than $0.105/\text{km}^2$, it is likely that debris flows occurred above the average in this range (or class) of the site factor. On the contrary, if the number of debris flows *per unit area* is less than $0.105/\text{km}^2$, debris flows occurred below the average. Therefore, the average number of debris flows *per unit area*, $0.105/\text{km}^2$, can be used as a yardstick in measuring the significance of a role that a site or rainfall factor plays in triggering debris flows in this watershed. For example, for gully slopes in the range of 12°–17° is $0.528/\text{km}^2$, which is about 5 times greater than the average number of debris flows *per unit area* ($0.105/\text{km}^2$) in this watershed. This implies the higher likelihood (about 5 times the average), of debris-flow occurrence for gully slopes in the range of 12°–17°. In another example, the number of debris flows *per unit area* for ground elevation between 500 m and 1,000 m is $0.212/\text{km}^2$, which is about 2 times the average $0.105/\text{km}^2$. Although the probability to trigger debris flows for ground elevation in the range of 500–1,000 m is 2 times higher than the watershed average, it is still 2.5 times lower than that for gully slopes in the range of 12°–17°. As for the effect of land ownership on debris-flow occurrence, none of the 4 ownerships generated conspicuously more debris flows than the remaining 3 because the number of debris flows *per unit area* for each ownership did not differ much from the entire watershed average $0.105/\text{km}^2$. Comparing the number of debris flows *per unit area* for experimental forest with

Table 21.1. Site and rainfall factors affecting 47 debris flows, which were caused by Typhoon Herb in the Chenyoulan stream watershed.
After Chieng (1998) but revised based on additional data.

Catchment area (ha) upstream of a debris-flow initiation site	<50	50–100	100–200	200–300	300–400	400–500	>500
Area in specified range (km²)	NA	NA	NA	NA	NA	NA	NA
Number of debris-flow events	8	13	8	6	3	4	5
Number of debris flows/area	NC	NC	NC	NC	NC	NC	NC

Ground elevation (m) at debris-flow initiation area	< 500	500–1,000	1,000–1,500	> 1,500	Total
Area in specified range (km²)	13.1	89.8	111.3	235.3	449.5
Hypsometry (% area in range)	2.9	20.0	24.8	52.3	100
Number of debris-flow events	1	19	16	11	47
Number of debris flows/area	0.076	0.212	0.144	0.047	0.105

Gully slope (degree) at debris-flow initiation area	< 6	6–12	12–17	17–22	22–27	> 27	Total
Area in specified range (km²)	38.6	31.2	34.1	51.4	67.2	227.0	449.5
Number of debris-flow events	0	5	18	15	6	3	47
Number of debris flows/area	0	0.160	0.528	0.292	0.089	0.013	0.105

Roads passing through a debris-flow catchment	NCCI-Highway	Other roads	No road
Area in specified class (km²)	NA	NA	NA
Number of debris-flow events	7	22	18
Number of debris flows/area	NC	NC	NC

Land classified according to ownership	Regular sloped land	Native reserved land	Experimental forest*	National forest	Total
Area in specified class (km²)	17.4	67.1	193.6	171.4	449.5
Number of debris-flow events	2	5	24	16	47
Number of debris flows/area	0.115	0.075	0.124	0.093	0.105

Land use in a debris-flow catchment	Natural forest	Forest having less than 20% of agriculture	Forest having more than 20% of agriculture
Area in specified class (km²)	NA	NA	NA
Number of debris-flow events	33	9	5
Number of debris flows/area	NC	NC	NC

Triggered by debris avalanche?	Yes	No
Number of debris-flow events	44	3

Total rainfall depth (mm)	<600	600–800	800–1,000	1,000–1,200	>1,200	Total
Area in specified range (km²)	0	50.2	305.7	54.4	39.2	449.5
Number of debris-flow events	0	8	28	6	5	47
Number of debris flows/area	IND	0.159	0.092	0.110	0.128	0.105

NA = Data not available. NC = Not yet calculated due to unavailability of data on area in specified range or class. *Owned by National Taiwan University (NTU). IND = Indeterminate.

that for native reserved land, one can only say that the former had the slightly higher probability of debris-flow occurrence than the latter.

In the preceding three examples, we simply analysed the effects of the 3 site factors, one factor at a time, on debris-flow occurrence, using data on the number of debris flows *per unit area*. In practical application, however, the validity of such a simple cause–effect analysis is questionable if one desires to evaluate *simultaneously* the combined effects of two or more factors on debris-flow occurrence. If that is the case, we need a multivariate statistical analysis of all factors involved. The multivariate analysis can pave the way for the development of viable cost-effective debris-flow countermeasures in Taiwan. However, it cannot be undertaken until after data on areas in the specified ranges (or classes) of all factors become available.

Data on rainfall affecting the number of debris flows triggered by Typhoon Herb in the Chenyoulan stream watershed are tabulated using the four bottom rows in Table 21.1. The last row in the table shows a pattern of the numbers of debris flows *per unit area* in the stepwise ranges of the total rainfall received in the watershed. On the one hand, since the numbers of debris flows *per unit area* in all the specified ranges of the total rainfall except for that ranging from 600 mm to 800 mm do not differ much from the watershed average $0.105/\text{km}^2$, suffice it to say that the effect of the rainfall factor on debris-flow occurrence is not very obvious. On the other hand, judging from the number of debris flows *per unit area*, $0.159/\text{km}^2$, in the area where the total rainfall ranged from 600–800 mm, we may infer that this specific rainfall area triggered nearly 1.5 times more debris flows than the average $0.105/\text{km}^2$ in the other areas of the watershed.

Characteristics of nine major debris-flow gullies

Since Typhoon Herb, a large number of analyses of aerial photographs and satellite images as well as field investigations have been conducted in the Chenyoulan stream watershed (Hung, 1996; Jeng and Lin, 1996; Lee et al., 1996; Lin et al., 1996; Tsai et al., 1996; Yu and Tuan, 1996; Chen, 1997; Cheng et al., 1997; Chieng, 1997; Chiu, 1997; Lin, 1997; H.J. Shieh, 1997; F.C. Yu, 1997; L.H. Yu, 1997; Chieng, 1998; Liu et al., 1998; Chen, 1999; Lin et al., 1999; Cheng et al., 2000; Lin et al., 2000; Lin and Jeng, 2000; Chen and Su, 2001). Their studies include the identification of the locations and distributions of debris flows as well as other types of landslides, the assessment of debris-flow hazards, the analysis of debris-flow characteristics, and the correlation between the site factors and the initiation of debris flows and other types of landslides. In Tables 21.2 and 21.3, the characteristics of catchments, gully gradients, rainfall, and alluvial fans are summarized for 9 out of the 47 debris flows caused by Typhoon Herb along the Chenyoulan stream from Nanpingkeng to Shenmu (Figure 21.8, see color section). Note that limits in resources, especially manpower, has so far confined such investigations and analyses to the nine debris-flow gullies only.

Table 21.2 indicates that all the catchment areas of the 9 debris-flow gullies were less than 200 ha, except for the Shenmu debris-flow gully, which exceeds 900 ha. The cultivation rates (i.e., the areas of cultivation over the total area) in

Table 21.2. Characteristics of catchments with 9 major debris flows caused by Typhoon Herb in the Chenyoulan stream watershed. After Yu and Tuan (1996).

Location	Catchment area (ha)	Landslide area (ha)	Cultivation rate (%)	Gully width (m)	Debris-flow gully slope (degree)		
					Initiation	Transportation	Deposition
Nanpingkeng	165.3	7.7	36.5	7	24.0	11.0	—
Junkengkou	86.4	2.1	45.8	5	33.7	16.0	16.0*
Junkengqiao	193.3	5.0	34.5	5	25.2	12.0	9.1
Xinyizhongxin	38.6	—	64.7	4	23.1	18.0	6.0
Shangfengqiu	67.3	—	44.2	5	34.6	16.0	7.0
Fengqiu	186.9	—	37.4	9	27.4	18.0	5.7
Tongfu	83.7	7.0	42.1	6	27.2	17.8	8.5
Longhua	174.6	8.2	40.5	6	28.6	15.0	15.0*
Shenmu	910.1	4.6	0.7	12	29.3	11.2	7.4

*Some debris flows were deposited on the area of high gradient due to the local culvert blockage.

Table 21.3. Characteristics of rainfall and debris flows caused by Typhoon Herb at 9 locations in the Chenyoulan stream watershed.

Location	Total rainfall depth (mm)	Maximum one-hour rainfall (mm)	Debris-flow type	Alluvial fan*				
				Maximum length** (m)	Maximum width** (m)	Area (ha)	Average deposit depth (m)	Volume (10^4 m^3)
Nanpingkeng	810	72	Cobble-gravely	— No alluvial fan —				
Junkengkou	830	73	Bouldery	310	95	0.90	4.0	3.6
Junkengqiao	850	74	Cobble-gravely	228	130	1.45	4.0	5.8
Xinyizhongxin	1100	96	Cobble-gravely	95	60	0.36	3.0	1.1
Shangfengqiu	860	75	Cobble-gravely	76	95	0.52	3.0	1.6
Fengqiu	880	77	Bouldery	400	570	9.10	5.0	45.5
Tongfu	690	75	Muddy	95	230	1.67	3.0	5.0
Longhua	790	78	Muddy	250	190	2.60	4.0	10.4
Shenmu	1300	98	Cobble-gravely	800	90	5.56	4.0	22.2

*Alluvial-fan data were obtained from Yu and Tuan (1996). **Maximum length and width of an alluvial fan were measured respectively along its centerline and perpendicular thereto.

the nine catchments were over 30%, except in the Shenmu catchment, which had only 0.7%. The slopes of debris-flow initiation zones were between 23° and 35°, while the slopes of debris-flow transportation zones were between 11° and 18°. All the debris flows were deposited at areas where the ground slopes were less than 10°, except at Junkengkou and Longhua where debris flows were deposited on the areas with much higher gradients (about 15° ~ 16°) due to local culvert blockage (Table 21.2).

With reference to the heaviness and spatial distribution of the rainstorm brought by Typhoon Herb, rainfall data for the 9 debris-flow locations, as given in Table 21.3, are intended to supplement what has been documented in Table 21.1 and the inset of Figure 21.3. The total 96-hour rainfall depth and the maximum 1-hour rainfall depth that triggered debris flows at the 9 locations during Typhoon Herb were at least 690 mm and 72 mm, respectively (Table 21.3).

If based solely on the sediment-size composition (Lane, 1947), the 9 debris flows triggered in the Chenyoulan stream watershed may be roughly divided into three groups: "bouldery" (rock-rich) debris flows with average particle sizes of 0.256 m or larger, "cobble-gravely" (mixed evenly with rock and matrix) debris flows with average particle sizes between 0.256 m and 2 mm, and "sand-silt-clayey" or "muddy" (matrix-rich) debris flows with average particle sizes of 2 mm or less. According to this classification, debris flows generated at Junkengkou and Fengqiu (Figure 21.9, see color section) were bouldery and, those at Tongfu and Longhua (Figure 21.10, see color section) muddy, while debris flows at each of the remaining 5 locations (Figure 21.11, see color section) were cobble-gravely (Table 21.3). It appears that the geology of colluviums and bedrock in debris-flow initiation areas, along with loose soils accumulated on hillslopes and stream channel side-slopes, interweaves with the rheology (or mobility) of debris flows to control the sediment composition of debris flows, thereby depositing different sizes and shapes of sediment particles along debris-flow gullies and alluvial fans. In fact, these three photos merely reflect the three superficial sediment-size distributions of collapsed colluviums and bedrocks deposited in debris-flow gullies and alluvial fans.

The measured uppermost extents (i.e., maximum lengths and widths) and areas of various alluvial fans formed at the 9 locations are also given in Table 21.3. Debris-flow volume has roughly been estimated by multiplying the measured area of an alluvial fan by the average deposit depth (Table 21.3). It is found that the debris-flow volume is roughly proportional to the catchment area upstream of the apex of an alluvial fan under investigation in 8 of the 9 debris-flow gullies (Figure 21.12). If such a relation between the debris-flow volume and the debris-flow watershed area could be empirically established, it would be useful in the design of a debris basin to control debris flows. Nevertheless, before its application, this relation has yet to be further substantiated using data obtained from sites other than the 9 debris-flow gullies.

In fact, we may draw an inference from the relation of debris-flow volume vs. debris-flow watershed area (Figure 21.12) that the debris-flow volume in each gully did not increase (rather decrease) after the 1999 earthquake, but the total volume of

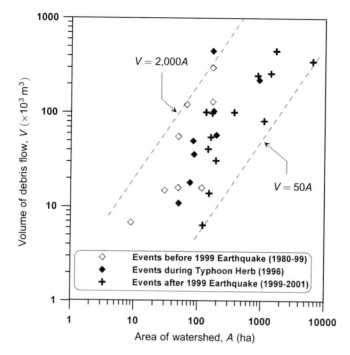

Figure 21.12. Empirical relation between debris-flow volume and debris-flow watershed area. This relation is intended for application in each debris-flow gully, but not the whole Chenyoulan stream watershed.

all debris flows in Taiwan could have increased substantially due to the increased number of newly occurred debris flows (Figure 21.2). This inference remains to be substantiated in the future.

21.3.3 Debris-flow initiation threshold

Hydro-meteorological thresholds of debris-flow initiation can be defined by rainfall intensity (or cumulative rainfall depth) vs. rainfall duration relation (Caine, 1980; Keefer et al., 1987; Wieczorek, 1987). As shown in Figure 21.13, a line (A) relating the rainfall intensity to the rainfall duration on a log–log scale describes an envelope of debris-flow-causing storms before the 1999 earthquake. A definite separation between debris-flow-causing storms and non-debris-flow-causing storms cannot be made based on this relationship. Therefore, we can only claim that the envelope defined in Figure 21.13 is a necessary, but not a sufficient, condition to trigger debris flows in the Chenyoulan stream watershed.

The critical line (A) can be expressed using an empirical rainfall intensity–duration formula, which was originally proposed by Sherman (1931) and later generalized by Chen (1983). A simplified version of the original rainfall

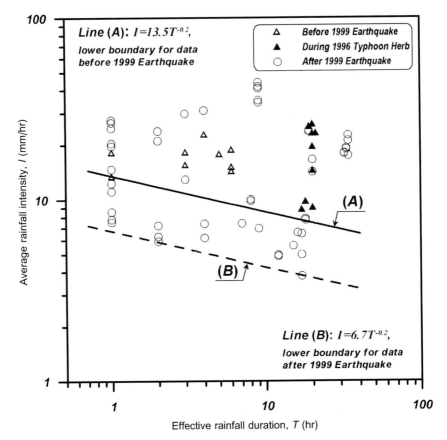

Figure 21.13. Empirical relation between duration and intensity of debris-flow causing rainstorms for the entire Chenyoulan stream watershed, but not for an individual specific debris-flow gully.

intensity–duration formula for debris-flow initiation threshold is derived below using the linear regression analysis:

$$I = 13.5T^{-0.2} \tag{21.1}$$

in which I is the average rainfall intensity in mm/hr and T the rainfall duration in hours.

The 1999 earthquake has significantly lowered the threshold of the rainfall amount required to trigger debris flows. Apparently, the earthquake caused an abundant supply of loose material accumulated on hillslopes and in stream channels, which tended to move more easily than ever before under lighter rains following the earthquake. Therefore, for the same rainfall duration, the lower rainfall intensity to trigger debris flows in the Chenyoulan stream watershed after the 1999 earthquake leads to the formulation of another critical line (B) shown in

Figure 21.13 on a log–log scale. The regression equation of this critical line (B) can be expressed as:

$$I = 6.7T^{-0.2} \tag{21.2}$$

A comparison of (21.1) with (21.2) indicates that for the same rainfall duration, the rainfall intensity needed to trigger debris flows after the 1999 earthquake is only about one-half of that before the earthquake. This critical line (B) reflects the system response of the Chenyoulan stream watershed subject to a combined excitation of the 1999 earthquake and rainstorms accompanying the typhoons. A frequency analysis of available rainfall data has revealed that the lower the rainfall intensity, the higher the frequency of rainfall, as characterized in the rainfall intensity–duration–frequency formulas (Chen, 1983) as well as exemplified in Figure 21.5. Therefore, the decrease in the rainfall intensity to trigger debris flows after the earthquake is equivalent to the increase in the number and hence the frequency of debris-flow occurrence in subsequent years after the 1999 earthquake (Figure 21.2).

Impact of the 1999 earthquake on the hydro-meteorological thresholds of debris-flow initiation was also investigated by Fan et al. (2003), who adopted a similar approach in the derivation of two regression equations for debris-flow initiation threshold, one before the 1999 earthquake and the other after the 1999 earthquake. It should be noted that Fan et al. took into account the effects of some physical parameters in developing their regression equations, plotting the respective threshold lines on a linear scale rather than on a log–log scale. In a sense, their regression equations are *linear* in the parameters, whereas (21.1) and (21.2) are *intrinsically linear* (Draper and Smith, 1981) because both equations expressed in exponential form can be transformed into linear form by taking logarithms of both sides of each equation. To show how the threshold line plotted immediately after the 1999 earthquake has shifted upward in the rainfall intensity–duration graph, Fan et al. elaborated the expression of its shift in terms of the time elapsing after the earthquake. This can be illustrated using Figure 21.13 as follows: the threshold line (B) that was lowered from the threshold line (A) immediately after the 1999 earthquake has gradually moved upward, approaching asymptotically the original threshold line (A) as the time goes by indefinitely, unless it is interrupted by another strong earthquake in the meantime.

21.4 DEVASTATION AND AFTERMATH OF TYPHOON HERB

21.4.1 Casualties and property loss (or damage) caused by debris flows

During Typhoon Herb, 95 people were killed and 463 wounded (Chieng, 1998). Damage caused by Typhoon Herb reached every corner of Taiwan, affecting the national economy, environment, agriculture, and engineering structures (Table 21.4). Property damage included 503 houses completely destroyed, 880 houses severely damaged, 599 ha of crop fields eroded, 1,266 ha of farmlands buried under debris and sediment, and 2,157 ha of farmlands temporarily inundated by seawater in the coastal areas. More than 220,000 metric tons of debris and sediment had to be

Table 21.4. Consequences of Typhoon Herb in Taiwan.
After Chieng (1998).

Category	Sub-category	Casualties, damage, or loss (in 2003 US$)
Human	People	51 killed, 22 missing, and 463 wounded
Housing	Houses	503 completely destroyed and 880 severely damaged
Agriculture	Crop fields	599 ha washed, 1,266 ha buried, and 2,157 ha flooded
	Crops	153,000 ha of crops damaged, equivalent to a loss of US$ 360 million
	Farming	Farming loss of US$ 13 million
	Fishery	6,000 ha of fishing farm flooded, fishery loss of US$ 46 million
	Forest	Forest loss of US$ 19 million
	Structures	US$ 19 million for restoration of damaged soil-conservation-structures, especially in central Taiwan
	Roadways	US$ 54 million for restoration of damaged roadways
Hydraulic facilities	River embankment	160 km of river embankments damaged, especially in Nantou County, central Taiwan
	Coastal embankment	37 km of coastal embankments damaged, and 35,000 ha of area inundated
Transportation	Highway	3,690 sections of highways damaged, especially along NCCI Highway (Highway 21)
	Railway	Numerous sections of railways damaged; 31 cases of landslide-induced damage along Alishan forest railways
	Telecommunication	205,000 telephone lines (2.1% of the national network) severed
Environment	Debris and sediment	220,000 metric tons of debris and sediment deposited
Economy	Total property loss	Over US$ one billion

cleared from buried farmlands, and the total property loss was over US$ one billion (2003). Damage caused by Typhoon Herb was the most severe since Typhoon Ellen on 7 August, 1959. Although Typhoon Ellen did not land on Taiwan, it brought a 1,164-mm three-day rainfall resulting in 1,075 people killed, 295 people injured, 22,426 houses completely destroyed, and 18,002 houses severely damaged (Lin and Jeng, 2000). The total property loss caused by Typhoon Ellen in 1959 was over US$ 100 million (1959) (i.e., about US$ 8 billion (2003)). The casualties as well as property loss and damage caused by Typhoon Ellen were much heavier than those caused by Typhoon Herb because very few structural countermeasures against floods, debris floods, and debris flows existed in Taiwan in 1959. Apparently, such countermeasures, if they existed in 1959, were either ineffective or inferior compared to those implemented in 1996.

During Typhoon Herb, the heaviest casualties and property loss (or damage) were caused by the 47 debris flows in the Chenyoulan stream watershed, Nantou County in central Taiwan. Relatively light casualties and property loss resulted from other 5 debris flows triggered in other counties, one in each of Yunlin, Chiayi, and Pingtung Counties in southern Taiwan and two in Miaoli County in northern Taiwan (Figure 21.3, see color section). To break down the casualties and property loss (or damage), we summarize the consequences of Typhoon Herb in Taiwan (island-wide) in Table 21.4. By the same token, the consequences of Typhoon Herb in 9 exhaustively-investigated debris-flow gullies in the Chenyoulan stream watershed are summarized using Table 21.5 for comparison.

Extreme hydro-meteorological events, such as Typhoon Herb and other heavy rainstorms, and their resultant floods, debris avalanches/flows, landslides, and other types of soil failures are inevitable in Taiwan, but such disastrous events need not have resulted in the huge loss of life and properties if countermeasures had been implemented. It has been found that inadequate engineering construction and land use at least partially contributed to the severe damage (Jeng and Lin, 1996; Hung, 1996; F.C. Yu, 1997; Lin and Jeng, 2000; Chang and Slaymaker, 2002).

21.4.2 Unjustified engineering structures and land use

Typhoon Herb mercilessly exposed inadequacy in the layout, design, and construction of highways, culverts, and bridges. A debris flow at Nanpingkeng destroyed a 4-m-high, 4-m-wide highway drainage culvert, and then discharged into the Chenyoulan stream, while eroding the Nanpingkeng gully bed to 12 m depth. Two highway drainage culverts, which were originally designed to allow floodwater to pass underneath the highway at Junkengkou and Longhua, had an average slope of 16°, but their openings were undersized for delivering debris-flow discharges, thus creating a blockage. Debris flows approaching the blocked culverts were then forced to flow over the highway, destroying and damaging nearby houses (Figure 21.14, see color section). Several bridges were damaged or destroyed due to insufficient conveyance for debris flows (Figure 21.15, see color section) or due to unjustified locations in curved streams or near the confluence of a stream and its tributary.

The New Central Cross-Island (NCCI) Highway was constructed in the 1980s along the Chenyoulan stream, which parallels the Chenyoulan fault of the Central Mountain Range in Taiwan. Hillslopes and tributaries on both sides of the Chenyoulan stream are geologically unstable and prone to landslides, debris avalanches, and debris flows. Prior to the NCCI Highway construction, some geologists had pointed out that hillslopes on the east side of the Chenyoulan stream are more stable than the west side, thus suggesting the east side as being a more appropriate route for the NCCI Highway construction (Wang, 1991; Hung, 1996). However, for unknown reasons, this suggestion was ignored, and the Highway was constructed along the west side. As expected after the completion of the Highway, rainstorms triggered rockfalls, landslides, debris avalanches, and debris flows (Lin et al., 1996; F.C. Yu, 1997; Lin and Jeng, 2000; Chang and Slaymaker, 2002) along the Highway.

Table 21.5. Consequences of Typhoon Herb in the 9 debris-flow gullies in the Chenyoulan stream watershed.
After Lin and Jeng (2000).

Location	Consequences	Remarks
Nanpingkeng	Roadway eroded, check dams damaged, gully eroded, culvert destroyed	The damaged site was located in the debris-flow transportation channel; highway culvert was too small for debris to flow through
Junkengkou	Houses buried, roads buried, 4 people killed	Houses were build too close to a conveying channel of debris flows
Junkengqiao	Houses destroyed, 13 killed, culvert blocked	Culvert was too small for debris to flow through; houses were built too close to debris-flow-prone channels
Xinyizhongxin	Roadways eroded, embankments eroded, bridge piers washed away, bridges destroyed, gully bed eroded/aggraded	Damage was caused by inadequate location and design of Xinyi bridge
Shangfengqiu	Roadways eroded/buried, check dams damaged	The check dam built in 1987 after the 1986 debris flow was partially destroyed by a debris flow due to Typhoon Herb
Fengqiu	Houses buried, 2 killed, farm fields buried	The damaged site was located in the debris-flow deposition section
Tongfu	Houses destroyed, roads culvert blocked	Culvert was too small for debris to flow through
Longhua	Houses buried, school damaged, culvert blocked	Roadway culvert was too small for debris to flow through; damage was attributed to inadequate locations for houses and a school
Shenmu	Roadways damaged, bridge damaged, culvert blocked	Debris flow was retarded and blocked by a bridge built at a junction of streams; damage was attributed to inadequate location of the bridge

In retrospect, if the Highway had been constructed along the east side of the Chenyoulan stream, such damage would likely have been greatly reduced.

Besides the NCCI Highway construction, many roads were built on hillslopes in the 1980s and 1990s to gain access to agricultural lands on terraces, where farmers grew fruit, betel-nut, areca-tea, and vegetables. Such land-use activities have led to more severe damage under heavier rainstorms. A post-typhoon spatial analysis of the sloped area in Chiayi County, central Taiwan by Lin et al. (1999) has revealed that the lower sloped areas seemed to have a higher rate of increase in bared land. They have also found that landscape modification through agriculture significantly lowers slope stability.

21.5 DEBRIS-FLOW MITIGATION STRATEGIES

21.5.1 Vitalization of debris-flow research

The first debris flows documented in Taiwan were the ones triggered by Typhoon Gloria in Taitung County in 1963. At that time, people called it a "sediment disaster" instead of a debris flow. However, many years had elapsed before the public and the government began to realize the economic impact of debris-flow hazards. Systematic studies of debris-flow hazards barely commenced until after the consecutive occurrence of two debris flows in 1985 and 1986 at Fengqiu in the Chenyoulan stream watershed (Figure 21.8, see color section), and the Tongmen debris flow (1990) in Hualien County. Before Typhoon Herb, very few researchers had a complete understanding of the debris-flow process. However, after the Typhoon Herb event, the devastation of debris flows and their hazards were repeatedly displayed on television. Such publicity via the mass media has attracted attention to debris-flow hazards by governmental officials, lawmakers, researchers, and the general public. As a result, the central government in Taiwan, such as the Council of Agriculture (COA) and the National Science Council (NSC), took a leading role in sponsoring debris-flow research and developing a comprehensive program for the mitigation of debris-flow hazards. Because more scientists have been engaged in debris-flow research since Typhoon Herb, the number of debris-flow papers published by Taiwanese researchers has increased considerably, especially in Taiwanese journals and conference proceedings (Figure 21.16).

Being in charge of the central control (e.g., prevention, mitigation, emergency management and evacuation, and rescue operation) of debris-flow hazards in Taiwan, the COA has developed debris-flow hazards mitigation strategies and measures, such as structural and non-structural countermeasures, detailed in the following sections.

21.5.2 Structural debris-flow countermeasures and warning systems

The Soil and Water Conservation Bureau (SWCB) of the COA is responsible for all debris-flow-related engineering projects. The structural countermeasures include the installation of debris barriers, debris breakers, debris basins, deflection berms, slit dams, check dams, and/or Sabo dams in debris-flow gullies and alluvial fans. Photographs of some typical sediment-control structures built after Typhoon Herb are shown in Figures 21.17 and 21.18 (see color section).

In addition to such debris-flow-controlling structures, the countermeasures in this context may embrace the installation of rainfall-based warning/monitoring systems and instrument-operated warning systems. The first rainfall-based warning/monitoring system (hereafter referred to as the monitoring station after installation) was installed in Hualien County, eastern Taiwan in 1992. Six rainfall-based warning/monitoring systems had been installed before the 1996 Typhoon Herb event, and 12 further such monitoring stations were instituted between 1997 and

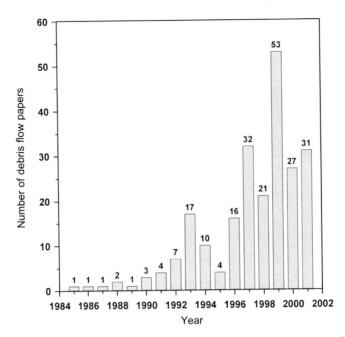

Figure 21.16. Number of debris flow papers published by Taiwanese researchers in Taiwanese journals and conferences proceedings between 1985 and 2001. Not taken into account are those papers published in foreign journals and international conferences proceedings.
Based on data obtained from the website of Science and Technology Information Center, NSC.

1999. Each monitoring station consists of a rain gauge, a geophone, and a video camera. Although the monitoring stations have functioned reasonably well for some subsequent debris-flow events, the overall rainfall-based warning success rate has been only at 30%, thus undermining people's reliance on such stations (C.L. Shieh, 1997; Shieh and Chen, 1999). Debris avalanches and debris flows occurred more frequently after the 1999 earthquake due to heavy rains, especially by Typhoon Toraji in 2001 (Cheng et al., 2003). Therefore, the SWCB installed 10 additional monitoring stations in 2002 and 2003. The new monitoring stations have been equipped with a rain gauge, ultrasonic airborne level-meters, geophones, wire sensors, infrared Charge-Coupled-Device (CCD) cameras, and a satellite communication system to transmit electronically measured data to the Debris Flow Emergency Operation Center of the SWCB. These data can be analysed and the results used in making decisions on how to mitigate debris-flow hazards.

21.5.3 Non-structural debris-flow countermeasures

To establish a database, the COA has commissioned the National Cheng Kung University (NCKU) to investigate the zoning of potential debris-flow hazard areas as well as to identify sites of high debris-flow risk since 1990. The regional and local governments are committed to alert the inhabitants residing in the vicinity of potential debris-flow hazard areas for evacuation when a debris flow occurs. Since 1996, NCKU mapped 485 highly hazardous debris-flow-prone streams based on geomorphologic characteristics and consequences (loss of life, property, and infrastructure). The number of hazardous debris-flow-prone streams has increased to 1,420 following the 1999 earthquake, according to the field investigations conducted by NCKU in 2001 and 2002. The NCKU was further commissioned to implement the zoning and aerial photography of the potential hazard areas along the 1,420 debris-flow-prone streams. In order to enhance public safety and awareness, the COA has installed debris-flow warning signs at the crossing of roads and all the debris-flow-prone streams. A warning sign is posted near a debris-flow-prone stream to alert the public of the danger of debris-flow occurrence (Figure 21.19, see color section).

To avoid culvert and bridge blockage or destruction by debris flows, the design codes of culverts and bridges in mountainous areas have been upgraded. Nowadays, for bridge and culvert design, the design considers floods with a recurrence interval of 50 years rather than the previous standard of 25 years. However, the updated codes have not yet addressed the potential of debris flows and are based on flood flow only, a problem that has also been recognized elsewhere (Jakob and Jordan, 2001). In another example to improve the design codes, a single-span bridge, instead of a multi-span bridge, has been put forth as a standard design for most small highway bridges in mountainous areas. The single-span bridge can avoid the reduction of conveyance in a stream because it does not require a bridge pier in the stream (Lin et al., 2000).

The central government in Taiwan has realized that it is impractical, if not impossible, to completely prevent debris flows. Therefore, the COA has entrusted universities (such as NCKU, National Taiwan University, and National Chung Hsing University) and professional societies (such as the Taiwan Disaster Prevention Society and the Chinese Soil and Water Conservation Society) to teach courses in disaster prevention, thereby increasing public awareness of debris-flow hazards as well as planning evacuation routes and shelters for people residing in high-hazard areas. Under the sponsorship of the COA, 100 mountainous villages have conducted debris-flow disaster prevention and evacuation practices to enhance inhabitants' hazard-mitigation knowledge and emergency-response swiftness as well as their familiarity with evacuation routes and shelters.

21.6 CONCLUSIONS

Catastrophic earthquakes and recurrent debris flows have had significant socio-economic consequences whose impacts on the wealth of Taiwan may eventually exhaust the available resources to cope with the aftermath of such disasters. Although the central government has been making efforts to develop debris-flow countermeasures since Typhoon Herb, many debris-flow mitigation measures and warning systems are still inadequate. For evacuation in the case of emergency, it will be necessary to enhance public awareness of many debris-flow hazards and educate people on how to react to debris-flow hazards. At the same time, more research on debris-flow mechanics, warning systems, and various viable prediction and assessment methods is needed.

Justifiable layout, design, and construction of highways, roads, bridges, and drainage culverts as well as appropriate land use in mountainous areas could offset the poor geologic and geomorphologic settings as well as the extreme hydro-meteorological conditions in Taiwan. For example, to improve culvert and bridge designs, we need to estimate or predict accurately debris-flow/avalanche discharges on hillslopes as well as in gullies and alluvial fans. Unfortunately, the techniques available to estimate or predict such debris-flow/avalanche discharges are still far from adequate. Engineering deficiency can be attributed partly to our insufficient knowledge in coping with debris-flow problems and partly to obsolete design codes used in building and installing sediment-control structures. More fruitful studies on the understanding of debris-flow mechanics, from which we can develop more adequate design codes, are urgently needed in Taiwan.

21.7 ACKNOWLEDGEMENTS

The writers sincerely appreciate the help of J.S. Wang in drawing the figures. Thanks are due to Professor C.C. Wu for providing Figure 21.17, and also to the Disaster Prevention Research Center, NCKU for providing Figure 21.19. Matthias Jakob reviewed earlier drafts of this chapter.

21.8 REFERENCES

Caine, N. (1980) The rainfall intensity-duration control of shallow landslides and debris flows. *Geografiska Annaler*, **62A**(1/2), 23–27.

Chang, J.C. and Slaymaker, O. (2002) Frequency and spatial distribution of landslides in a mountainous drainage basin: Western Foothills, Taiwan. *Catena*, **46**, 285–307.

Chang, S.C. (1993) *Landslides and Their Environmental Impacts in Taiwan* (Multiple Hazards Mitigation Report 81, p. 49), National Science Council, Taipei, Taiwan [in Chinese].

Chen, C-L. (1983) Rainfall intensity-duration-frequency formulas. *Journal of Hydraulic Engineering, ASCE*, **109**(12), 1603–1621.

Chen, H. and Su, D.Y. (2001) Geological factors for hazardous debris flows in Hoser, central Taiwan. *Environmental Geology*, **40**, 1114–1124.

Chen, R.H. (1999) *Integrated Project of Debris-flow Disasters Mitigation* (Report, 24 pp.). National Science Council, Taipei, Taiwan [in Chinese].

Chen, Z.E. (1997) A comparative study of geo-scientific environments about debris flow in Junkeng area, Nantou County. *Proceedings of 1st National Debris Flow Conference, Xitou, Taiwan* (pp. 6–23). National Chung Hsing University, Taichung, Taiwan [in Chinese].

Cheng, J.D., Wu, H.L., and Chen, L.J. (1997) A comprehensive debris flow hazard mitigation program in Taiwan. In: C-L. Chen (ed.), *Proceedings of 1st International Conference on Debris-flow Hazards Mitigation: Mechanics, Prediction, and Assessment, San Francisco, California* (pp. 93–102). American Society of Civil Engineers, New York.

Cheng, J.D., Su, R.R., and Wu, H.L. (2000) Hydrometeorological and site factors contributing to disastrous debris flows in Taiwan. In: G.F. Wieczorek and N.D. Naeser (eds), *Proceedings of 2nd International Conference on Debris-flow Hazards Mitigation: Mechanics, Prediction, and Assessment, Taipei, Taiwan* (pp. 583–592). A.A. Balkema, Rotterdam.

Cheng, J.D., Yeh, J.L., Deng, Y.H., Wu, H.L., and Hsei, C.D. (2003) Landslides and debris flows induced by typhoon Toraji, July 29–30, 2001 in central Taiwan. In: D. Rickenmann and C-L. Chen (eds), *Proceedings of 3rd International Conference on Debris-flow Hazards Mitigation: Mechanics, Prediction, and Assessment, Davos, Switzerland* (pp. 919–929). Millpress, Rotterdam.

Chiang, S.H. (1996) Rainfall associated with Typhoon Herb in central Taiwan. *Proceedings of Typhoon Herb and Engineering Environment, Taipei, Taiwan* (pp. 1–7). National Taiwan University, Taipei, Taiwan [in Chinese].

Chieng, P.W. (editor in chief) (1997) *Review of Hill-slope Hazards Induced by Typhoon Herb* (52 pp.). Soil and Water Conservation Bureau, and the Forestry Bureau of the Department of Agriculture and Forestry of Taiwan Province, Taiwan [in Chinese].

Chieng, P.W. (editor in chief) (1998) *A Record of Sediment Hazards Caused by Typhoon Herb and Their Restorations* (129 pp.). Council of Agriculture, Taipei, Taiwan and the Chinese Water and Soil Conservation Association, Taipei, Taiwan [in Chinese].

Chiu, C.Y. (1997) Investigation of road failure caused by Typhoon Herb. *Proceedings of 1st National Debris Flow Conference, Xitou, Taiwan* (pp. 25–42). National Chung Hsing University, Taipei, Taiwan [in Chinese].

Dadson, S.J., Hovius, N., Chen, H., Dade, W.B., Hsieh, M.L., Sillett, S.D., Hu, J.C., Horng, M.J., Chen, M.C., Stark, C.P., Lague, D., and Lin, J.C. (2003) Links between erosion, runoff variability and seismicity in the Taiwan orogen. *Nature*, **426**, 648–651.

Draper, N.R. and Smith, H. (1981) *Applied Regression Analysis*. John Wiley & Sons, New York.

Fan, J.C., Liu, C.H., Wu, M.F., and Yu, S.K. (2003) Determination of critical rainfall thresholds for debris-flow occurrence in central Taiwan and their revision after the 1999 Chi-Chi great earthquake. In: D. Rickenmann and C-L. Chen (eds), *Proceedings of 3rd International Conference on Debris-flow Hazards Mitigation: Mechanics, Prediction, and Assessment, Davos, Switzerland* (pp. 103–114). Millpress, Rotterdam.

GIO (2003) Geography. *Taiwan Yearbook 2003*. Government Information Office, Republic of China. Available at *http://www.gio.gov.tw/taiwan-website/5-gp/yearbook/chpt01.htm*

Huang, C.Y. (editor in chief) (2002) *Taiwan Tectonics* (210 pp.). Geological Society, Taipei, Taiwan [in Chinese].

Hung, J.J. (1996) Typhoon Herb, the new, central, cross-island highway and slope-land failures in central Taiwan. *Sino-Geotechnics*, **57**, 25–30 [in Chinese].

Jakob, M. and Jordan, P. (2001) Design flood estimates in mountain streams: The need for a geomorphic approach. *Canadian Journal of Civil Engineering*, **28**(3), 425–439.

Jeng, F.S. and Lin, M.L. (1996) Engineering deficiencies exposed by Typhoon Herb. *Sino-Geotechnics*, **57**, 65–74 [in Chinese].

Keefer, D.K., Wilson, R.C., Mark, R.K., Brabb, E.E., Brown, W.M., Ellen, S.D., Harp, E.L., Wieczorek, G.F., Alger, C.S., and Zatkin, R.S. (1987) Real-time landslide warning during heavy rainfall. *Science*, **238**, 921–925.

Lane, E.W. (1947) Report of the subcommittee on sediment terminology. *Transactions of the American Geophysical Union*, **28**(6), 936–938.

Lee, C.Y. (1996) *Watershed Management* (412 pp.). Ruiyu Publications, Pingdong, Taiwan [in Chinese].

Lee, D.H., Tien, K.G., Huang, S.C., and Lin, G.C. (1996) Investigation on slope failure induced by Typhoon Herb at low elevation area at Alishan Highway. *Sino-Geotechnics*, **57**, 81–91 [in Chinese].

Li, Y.H. (1976) Denudation of Taiwan Island since the Pliocene Epoch. *Geology*, **4**, 105–107.

Lin, C.E. (1997) Geological and topographical factors in initiating debris flows: Typhoon Herb. *Proceedings of 1st National Debris Flow Conference, Xitou, Taiwan* (pp. 1–6). National Chung Hsing University, Taichung, Taiwan [in Chinese].

Lin, C.S., Chen, Y.K., and Liou, J.C. (1999) Spatial analysis of bare land increment caused by Typhoon Herb in Chiayi slopeland area. *Journal of Chinese Soil and Water Conservation*, **30**(3), 223–233 [in Chinese].

Lin, M.L., Chuang, M.H., Hung, F.E., Lu, Y.H, Chiang, W.T., Huang, G.T., and Lin, H.H. (2000) *GIS and Characteristics of Debris-flow Rivers in Chenyoulan Watershed* (National Science and Technology Program for Hazards Mitigation 1999 Annual Report, 135 pp.). National Science Council, Taipei, Taiwan [in Chinese].

Lin, M.L. and Jeng, F.S. (2000) Characteristics of hazards induced by extremely heavy rainfall in Central Taiwan: Typhoon Herb. *Engineering Geology*, **58**, 191–207.

Lin, M.L., Jeng, F.S., and Wu, C.C. (1996) Disasters along the new, central, cross-island highway. *Sino-Geotechnics*, **57**, 31–44 [in Chinese].

Lin, S.W. (1998) Report on Typhoon Herb of 1996. *Meteorological Bulletin*, **42**(1), 80–102 [in Chinese].

Liu, K.F., Shieh, C.L., and Ho, C.W. (1998) *Integrated Project of Debris-flow Hazards Mitigation and Its Monitoring* (Report, 24 pp.). National Science Council, Taipei, Taiwan [in Chinese].

Sherman, C.W. (1931) Frequency and intensity of excessive rainfalls at Boston, Massachusetts. *Transactions of the American Society of Civil Engineers, ASCE*, **95**, 951–960.

Shieh, C.L. (1996) *High Potential Creeks of Debris Flow and Landslide in Taiwan* (Report). Council of Agriculture, Taipei, Taiwan [n Chinese].

Shieh, C.L. (1997) Present situation and perspective of debris-flow warning systems in Taiwan. *Proceedings of Hazards Prevention and Mitigation on Hillslopes Development, Tainan, Taiwan* (pp. 27–39). National Science Council Engineering and Technology Promotion Center, Taipei, Taiwan [in Chinese].

Shieh, C.L. and Chen, L.J. (1999) The present and future of debris-flow warning systems. *Proceedings of 2nd National Debris Flow Conference, Hualien, Taiwan* (pp. 308–315). Taiwan Disaster Prevention Society, Tainan, Taiwan [in Chinese].

Shieh, H.J. (1997) Sediment hazards induced by Typhoon Herb and debris-flow control methods. *Proceedings of 1st National Debris Flow Conference, Xitou, Taiwan* (pp. 43–53). National Chung Hsing University, Taichung, Taiwan [in Chinese].

TGRU (2001) *Natural Hazards of Taiwan* (96 pp.). Taiwan Geomorphological Research Unit, Department of Geography, National Taiwan University, Taipei, Taiwan and Council of Agriculture, Taipei, Taiwan [in Chinese].

Tsai, K.J., Sung, Y.M., Wang, H.Y., and Lin, J.B. (1996) GIS/GPS technology applied for investigating disasters caused by Typhoon Herb along the new, central cross-island highway. *Sino-Geotechnics*, **57**, 55–64 [in Chinese].

Wang, S. (1991) *Impact of Road Construction on Hillslopes Stability* (Report). National Science Council, Taipei, Taiwan [in Chinese].

Wang, W.N., Yin, C.Y., Chen, C.C., and Lee, M.C. (2000) Characteristics of slope failures induced by 921-earthquake in central Taiwan. *Proceedings of the Taiwan–Japan Joint Conference on the Investigation and Remediation of the Secondary Disasters after the 921 Earthquake. Taipei, Taiwan* (pp. 79–90). Taiwan Disaster Prevention Society, Tainan, Taiwan [in Chinese].

Wieczorek, G.F. (1987) Effect of rainfall intensity and duration on debris flows in Central Santa Cruz mountains, California. In: J.E. Costa and G.F. Wieczorek (eds), *Debris Flows/Avalanches: Process, Recognition, and Mitigation* (Reviews in Engineering Geology No. 7, pp. 103–114). Geological Society of America, Boulder, CO.

Wu, C.H., Tsai, C.H., and Hu, C.T. (1989) *A Survey Report on Landslide of Slope Land in Taiwan* (139 pp.). Council of Agriculture, Taipei, Taiwan [in Chinese].

Yu, F.C. (editor in chief) (1990) *Investigation of Tongmen Sediment Hazards in Hualien County* (Report, 84 pp.). Council of Agriculture, Taipei, Taiwan [in Chinese].

Yu, F.C. (1997) Case studies of debris flows caused by Typhoon Herb. *Journal of Modern Construction*, **209**, 9–19 [in Chinese].

Yu, F.C. and Tuan, C.H. (editors in chief) (1996) *Preliminary Investigation of Typhoon-Herb-triggered Hazards along the Sides of Chenyoulan Stream in Nantou County* (Report, 88 pp.). Council of Agriculture, Taipei, Taiwan [in Chinese].

Yu, L.H. (1997) Investigation of hazards in Chenyoulan Stream area using GIS and artificial neural network method. MSc thesis, National Taiwan University, Taipei, Taiwan [in Chinese].

22

Jiangjia Ravine debris flows in south-western China

Peng Cui, Xiaoqing Chen, Yuyi Waqng, Kaiheng Hu, and Yong Li

22.1 INTRODUCTION

China is a mountainous country. More than 6.4×10^6 km^2 of the territory consists of complex, tectonically active terrain with a monsoon climate within which debris flows are common and create huge losses every year. Debris-flow activity is concentrated in the west and north-east parts of the mainland, including the provinces of Beijing, Sichuan, Yunnan, Xinjiang, Gansu, Xizang, as well as in Taiwan (Figure 22.1). There are about 50,000 debris-flow sites distributed over 48% of the territory area, threatening or damaging 138 central cities, 1,200 towns and villages, 36 railways, 50,000 km of highways, 1,200 km^2 of farmland, and many mines. River navigation and some large reservoirs, potentially including the Three Gorges Reservoir, are adversely affected by debris-flow sediment. The annual losses from debris flows approach US$ 366 million, and the annual death toll exceeds 100 (Tang et al., 2000). Debris-flow activity also increases soil erosion and ecological degradation, severely restricting the economic development of mountainous areas.

The majority of debris flows in China are triggered by rainfall. The Jiangjia Ravine in Yunnan Province has attracted global attention for the annual occurrence of well-documented episodes of rainfall-triggered debris flows. The site has an unparalleled record of long-term observations and experiments of the Dongchuan Debris Flow Observation and Research Station (DDFORS), a facility of the Institute of Mountain Hazards and Environment, Chinese Academy of Science. Debris flows in Jiangjia Ravine are typical of those in most mountainous areas of China and exhibit a variety of flow types, patterns, and processes under natural conditions. Chinese scientists have studied flow formation, static and dynamic properties, and deposition, and have attempted forecasting, warning, and mitigation of debris flows in Jiangjia Ravine for many years. In this chapter, we summarize some key findings from nearly four decades of data collection and research at the site.

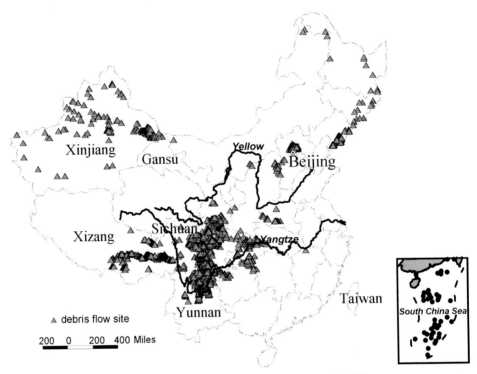

Figure 22.1. Map of debris flows distribution in China.

Jiangjia Ravine watershed with an area of 48.6 km² between N26°13'~N26°17' and E103°06'~E103°13', is located in the northern part of the Yunnan-Guizhou Plateau. The drainage joins the Xiaojiang River 31 km south of the confluence of the Xiaojiang and Jinsha Rivers. The main channel of the Jiangjia Ravine extends from the drainage divide at 3,269 m altitude west to the junction with the Xiaojiang at 1,042 m (Figure 22.2). Slopes steeper than 25° comprise 55% of the basin area with typical local relief of 500 m (Figure 22.3) (Wu et al., 1990). The basin consists of 5 orders of sub-basins (shown in Table 22.1), where the sum of all stream lengths is 124 km, and the stream density is 2.6 km/km². The main channel can be divided into three sections with different morphologic characteristics (Figure 22.4): (1) the erosion section, 10 km in length and 17° of bed slope on average; (2) the debris-flow transport section, 1.3 km in length and 5.1° of bed slope on average, with great cross-sectional variation up to 16 m in depth caused by scouring and silting of debris flows; and (3) the deposition section, 4.2 km in length and 3.7° bed slope on average. Table 22.1 shows the detailed characteristics of the ravine.

The Jiangjia Ravine has developed in the zone of the Xiaojiang fault, of which two branches join in the downstream reaches. The area is characterized by intense tectonism. About 80% of the exposed rocks are highly fractured and mildly metamorphosed. The sandstone and slate can be grouped into two types by color, both of

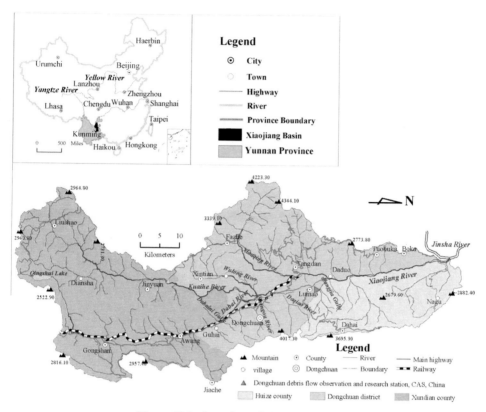

Figure 22.2. Location of Jiangjia Ravine.
From He (2003).

which are weak and easily weathered and broken into fragments. Colluvium and mantlerock are widely distributed on slopes and in channels in some sub-basins. The total volume of accumulated clastic detritus as source material for debris flows is estimated to $1.2 \times 10^9 \, m^3$ (Li and Wu, 1980).

Annual rainfall in the Jiangjia Ravine ranges from 400–1,000 mm and is markedly seasonal. About 85% of the total annual rain occurs between May and October. About 40% rainfall occurs between altitudes of 2,500 m and 3,000 m, but within an area of only 10% of the total watershed. Reflecting both the climate and human activity, the ground cover consists of: forest 4.2%, shrub land 8.4%, grassland 26.9%, and barren terrain 26.3%, while the other 34.2% includes alluvial surfaces and fans, farmland, and other land uses.

These conditions provide an ideal setting for recurrent debris flows. Since 1965, more than 400 episodes of debris-flow activity have occurred, each consisting of many individual surges. As many as 28 episodes have occurred in a single year. Annual sediment yielded in the Jiangjia Ravine is $2.0 \times 10^6 \, m^3$ on average while a maximum of $6.6 \times 10^6 \, m^3$ occurred in 1991.

Figure 22.3. Debris-flow source region of Jiangjia Ravine (Point A is upstream of the Menqian Gully and Point B is the confluence point of the Menqian Gully and the Duozhao Gully, also shown in Figure 22.4. The length from A to B is 2.2 km.)

Photo: Hu, K.H. on 12 July 2001.

Table 22.1. Morphologic characteristics of the Jiangjia Ravine.
From Wu et al. (1990).

Order	Streams			Debris deposited sections		
	%	Length (km)	Slope (°)	%	Length (km)	Slope (°)
I	62.5	77.5	31.8	5.1	1.4	29.7
II	15.8	19.6	23.6	18.9	5.2	12.4
III	11.1	13.8	13.3	17.1	4.7	9.6
IV	4.5	5.6	9.5	36.4	10	7
V	6.1	7.5	4.5	22.5	6.2	3.7
Total	*100*	*124*	–	*100*	*27.5*	–

% = percentage of total watershed area of each sub-watershed order.

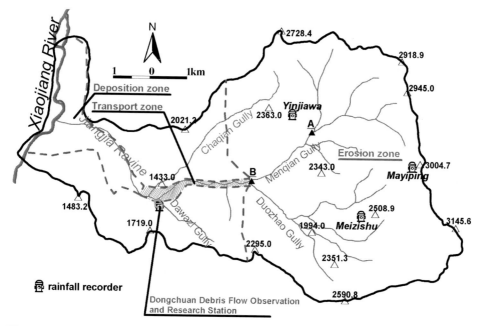

Figure 22.4. Plane view of Jiangjia Ravine. (The number in the figure is an elevation value. Point A and B are the same as the two in Figure 22.3.)

22.2 RELATIONSHIPS BETWEEN RAINFALL AND DEBRIS FLOW

Water is a necessary agent for debris-flow movement through liquefying and mobilizing loose detritus. In the Jiangjia Ravine rainfall is the main source of water while minor sources include snowfall and hail. DDFORS instrumentation in the Jiangjia Ravine includes four rain gauges at different elevations in the watershed. One is located at Mayiping, where debris flows originate; the others are at Meizishu, Yinjiawa, and the station (Figure 22.4). They provide data with which to analyse the relationships between precipitation and debris flow. For example, from 24 July to 8 July 2001, there were only two days with no rainfall but five days with rainfall in excess of 20 mm. The total amount of rainfall over 15 days was 246.7 mm recorded by the gauge at Mayiping (Figure 22.5). Antecedent rainfall saturated loose materials in the source, facilitating the initiation of debris flow. Debris flows occurred in the observations station in the erosion section at about 7:00 on 8 July, after rainfall began at 22:00 on the preceding day and increased abruptly at 6:20 on 8 July. Peak rainfall intensity of 10.8 mm/hr occurred at 8:20, followed by the maximal debris-flow discharge (775.2 m^3/s) at 9:25. Three periods of intense rainfall occurred until 20:20, when the rainfall ended (Figure 22.5). Peak debris-flow activity did not occur until intense rainfall occurred in the middle and downstream reaches (Wei et al., 2002).

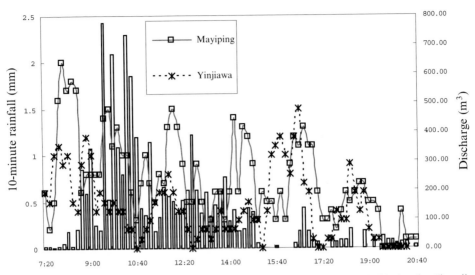

Figure 22.5. 10-min rainfall and debris flow of discharge on 8 July 2001, in the Jiangjia Ravine.

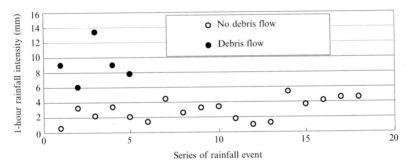

Figure 22.6. 1-hr rainfall intensity and related debris flows.

A single episode of debris flow is usually affected by both antecedent rainfall and concurrent rainfall. The 1-hr rainfall intensity is crucial, as shown in Figure 22.6. However, other rainfall parameters are less positively correlated. For example, the 10-min rainfall intensity appears irrelevant to flow, as shown in Figure 22.7 (Yang, 2002). Therefore, forecasting flows requires more detailed analysis.

Each rainfall parameter appears to play different roles in sub-basins of different orders in the Jiangjia Ravine. For example, according to orthogonal analysis, in first-order tributaries with a slope of 31°, the main component is initiating and antecedent rainfall, accounting for 77.6% of the variation, and the second factor is the rainfall intensity, accounting for only 2.1% of variation. However, for the main stream, antecedent rainfall accounts for 60.2%, and initiating rainfall, 39.8% of variation. Light rain can also trigger debris flows (see rainfall grade in Table 22.2 (according to National Weather Bureau)). Based on the rainfall associated with 26 debris-flow

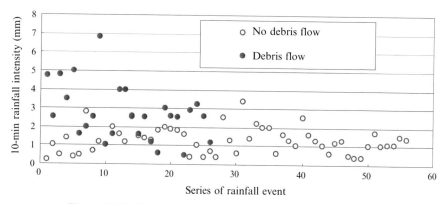

Figure 22.7. 10-min rainfall intensity and related debris flows.

Table 22.2. Rainfall grade.

Type	1-day rainfall (mm)	12-hour rainfall (mm)
Rainstorm	>50	>30
Heavy rain	25–50	15–30
Moderate rain	10–25	5–15
Light rain	<10	<5

episodes, 46.2% were caused by heavy rain, 42.3% by moderate rain, 7.7% by short-duration rain storms, and 3.8% by light rain.

In the Jiangjia Ravine, the rainfall type is also important and can be conceptually analyzed by dividing events into four types shown in Figure 22.8. A-type is concave with its rainfall intensity being smaller at first but increasing later. B-type is convex with its rainfall intensity larger at first but diminishing later. C-type is a reversed S-shape with its rainfall intensity greater first and then decreasing and finally increasing. D-type is S-shaped with its rainfall intensity varying from light to high. Each pattern is associated with a ratio or probability of the occurrence of debris flow, as listed in Table 22.3.

Of course, the minimum daily rainfall necessary to trigger debris-flow activity varies between different basins, shown by minima of 20 mm at Jiangjia Ravine, 50 mm at Heishahe Ravine in Sichuan Province, and 100 mm in Xishan at Beijing (Wu et al., 1993).

22.3 DYNAMICS

22.3.1 Velocity

The velocity field for debris flows is very difficult to measure. Accordingly the velocities provided here are values for the flow fronts which are usually turbulent

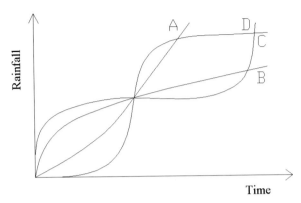

Figure 22.8. Four types of rainfall concerning debris flows in the Jiangjia Ravine.
From Yang (2002).

Table 22.3. Ratio of debris flow events under different rainfall patterns.
From Yang (2002).

Rainfall pattern	A	B	C	D	Other
Event ratio (%)	37	19	22	15	7

(Figure 22.9). In the Jiangjia Ravine, the Reynolds number of debris flows is greater than 105. The velocity at the centre of the front often exceeds 10 m/s. Turbulence varies with velocity as follows: strong turbulence (>8 m/s), turbulence (<8 m/s and >6 m/s), quasi-laminar flow (<6 m/s).

Debris flows in the Jiangjia Ravine range from hyperconcentrated flows with a density below 1.3×10^3 kg/m^3, to debris flows of various viscosities with density between 1.3×10^3 and 2.4×10^3 kg/m^3 and yield stress from 0 to 248 Pa for slurry. Viscous flow moves in the form of multiple surges, commonly followed by continuous hyperconcentrated flow. A surge usually rides on a mud layer, the so-called residual layer left by the preceding surge, which reduces resistance to flow, thus explaining why debris-flow surges move at speeds much higher than water-dominated flows under the same conditions. Compared with other locations in China, the resistance of the debris flow in Jiangjia is lower because the slurry contains a relative fine particle ratio of $d < 0.05$ mm/$d > 0.05$ mm (Rns). Experimentally, the resistance coefficient (i.e., the roughness) n_c is related to the characteristics of a debris-flow slurry by (Wang et al., 2003a):

$$n_c = 0.033 R_{ns}^{-0.51} \exp(0.34 R_{ns}^{0.17}) \ln h \qquad (22.1)$$

where

$$R_{ns} = 4.59 \times 10^{-9} \eta_p^{-2.2} \exp(8.9 \times 10^{-11} \eta_p^{7.99} C_{vf}) \qquad (22.2)$$

where η_p is the viscosity of slurry, C_{vf} is the volume concentration of slurry, and h is

Figure 22.9. Turbulence in a debris-flow front (the front is about 0.5 m high).
Photo: Wei, F.Q. on 19 July 2001.

depth of flow. The velocity V_{cp} can be expressed in the form of the Manning formula that is used widely in hydraulics:

$$V_{cp} = \frac{1}{n_c} h^{2/3} j^{1/2} \qquad (22.3)$$

where h is the flow depth and j the gradient of the channel. Velocity on average is 5–10 m/s, with a recorded maximum of 18.2 m/s on 9 July 1998. The maximum recorded discharge is 2,914 m³/s and five times that recorded for the trunk stream, the Xiaojiang River. The maximum annual sediment yielded from Jiangjia Ravine is 6.59×10^6 m³.

A debris flow occurred on 25 July 1997 in the Jiangjia Ravine. The developing process was observed and the air concentration in the debris flow and its effect on flow resistance were analysed. The results are shown in Table 22.4. The air concentration appears to be closely related to the velocity, as shown in Figure 22.10, an inspection of this figure reveals that the higher the debris-flow velocity (V_{cp}), the higher the air concentration (C_a) in the flow. A relation between C_a and V_{cp} can be approximated by (Jan et al., 2000):

$$C_a = 4.2 \times 10^{-5} V_{cp}^{2.95} \qquad (22.4)$$

Table 22.4. Sediment and entrained air concentrations of debris-flow samples. From Jan, C.D. et al. (2000).

Sample No.	H (m)	V_{cp} (m/s)	C_{sa} (%)	C_s (%)	γ_{ma} (g/cm³)	γ_m (g/cm³)	C_a (%)
1	1.0	10.0	69.2	74.3	2.176	2.263	3.84
2	1.4	8.0	68.4	71.2	2.163	2.211	2.17
3	1.1	6.9	71.0	72.3	2.207	2.229	1.0
4	0.6	5.9	68.9	70.2	2.192	2.192	0.9

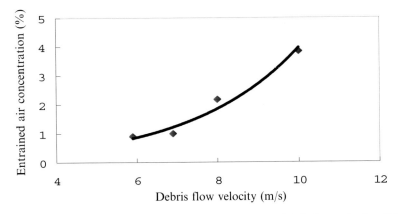

Figure 22.10. Concentration of air entrained in debris flow vs. debris-flow velocity. (Diamond symbol represents data measured from the debris-flow samples in Table 22.4; solid line represents the regression trend of these data points.)

22.3.2 Impact forces

The kinetic energy of debris flows is large because of their high material densities and velocity. When debris flows encounter some obstacles such as a pier and other constructions, the impact force can destroy these structures. In order to measure the impact forces, DDFROS constructed three rectangular, armored concrete pillars in the channel in May 1982. An electric strain sensor was installed inside a steel box in each pillar. After having recorded data from 19 flow episodes, the pillars were destroyed by debris flows in 1983. In July 1985, a new instrument, a piezo-electric crystal sensor, was put in use and measured impact forces for 40 flows. Similar tests were initiated in 2003 (Figure 22.11). The data show that debris flows impact the sensors respectively through individual stones and slurries. The stone impact is intermittent and of short duration, resulting in much greater force than the latter which is continuous and of long duration.

The two kinds of impact often concur and result in a sawtooth-shaped impulsion which is typical for most viscous debris flows. Impact interaction can continuously

Figure 22.11. Armoured concrete pillar for measuring impact force, as used in 2003. (The three steel discs in the photo are pressure sensors with 15 cm diameters.)
Photo: Hu, K.H. on 13 June 2003.

take place within the duration for as long as 500 seconds. Peaks reflect the abrupt impacts of boulders in the surge. With depths of less than 5 m and boulders smaller than 1 m in diameter, the value of dynamic pressure in this pattern is approximately 10^6 Pa. When the flow becomes more uniform, the impact shape becomes rectangular, reflecting relatively steady impact. Values range between 10^5 Pa and 10^6 Pa. The duration of peak impact is very short, usually less than 9 seconds and the maximum dynamic pressure attained is 3.0×10^6 Pa (Wu et al., 1990).

In practice the impact of the slurry is estimated as:

$$P = K\gamma_c V_{cp}^2 \qquad (22.5)$$

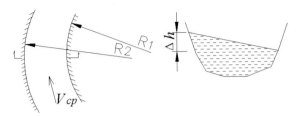

Figure 22.12. Diagram of superelevation in channel bends (see text for explanation).

where P is the dynamic pressure of debris flows, K is a coefficient (from 2.5 to 4.0), γ_c is the unit weight, and V_{cp} is velocity. For large boulders in the flow, impact P_d is:

$$P_d = \gamma_H A V_{cp} C \tag{22.6}$$

where γ_H is specific weight, A is contact area, V_{cp} is velocity, and C is transmission speed of the elastic wave impacting a rock (usually 4,000 m/s).

22.3.3 Superelevation in channel bends

A debris flow has greater inertia than common fluids. In the bends of a channel, the very large kinetic energy allows debris flows to rise to a much greater elevation than water, which often brings about extensive damage. Based on the dynamic equilibrium of a debris flow in a transverse section, the approximate formula for calculating the difference (Δh) between the surface elevations at the concave and convex banks can be written as (Kang, 1990):

$$\Delta h = 2.3 \frac{V_{cp}^2}{g} \log \frac{R_2}{R_1} \tag{22.7}$$

where V_{cp} is the velocity of the debris flow, and R_1 and R_2 are the radii of curvature for convex and concave bank, respectively. Figure 22.12 shows the meaning of each parameter.

Equation (22.7) is agreeable to the field observation. A debris flow occurred on 12 June 1973, and its parameters were determined as: $R_1 = 30\,\text{m}$, $R_2 = 49.8\,\text{m}$, and $C_{cp} = 8.3\,\text{m/s}$. Using (22.7) we obtain $\Delta h = 3.6\,\text{m}$, which is close to the measured value of 4 m.

22.4 STATIC PROPERTIES

The most fundamental static properties of debris flow are density and composition, determined by the sediment concentration in the slurry.

22.4.1 Density and concentration

Debris flows are heterogeneous. The density of the front is greater than that of the other parts of the surge because the largest clasts concentrate within the front. The

Table 22.5. Density of eight samples and their slurries from the debris flows on 8 July, 2001.

No.	Sampling time	Flow pattern	Sample (g/cm^3)	Slurry (g/cm^3) (with grain size <1.2 mm)
01-59	07:38:00	Continuous flow	1.81	1.62
01-60	08:47:20	Body of surge	1.99	1.63
01-61	09:23:41	Tail of surge	2.18	1.71
01-62	10:08:38	Tail of surge	2.22	1.69
01-63	11:38:51	Tail of surge	2.09	1.69
01-64	13:12:40	Tail of surge	1.99	1.66
01-65	14:36:55	Tail of surge	1.94	1.67
01-66	16:33:28	Continuous flow	1.71	1.54

density may change significantly within a given surge and in different surges. In a typical debris flow the first stage is a low-viscous continuous flow, followed by a viscous surge, then by a low-viscous flow, and finally by a debris flood. Different densities of flows are listed in Table 22.5 (sampled on 8 July, 2001). The density of an earlier surge body may be lower than that of a later surge tail such as in sample 01-60. The density of a highly viscous debris flow is remarkably close to the solid density of the bedrock in the Jiangjia Ravine (i.e., 2.65 g/cm^3).

Densities fall in the range between those commonly analysed in studies of sediment concentration and soil mechanics. Based on the analyses of 14 samples it was found that suspended load is related to the hydraulic strength (Wang et al., 2001):

$$C_{vt} = 0.042 f^{0.83} \tag{22.8}$$

where C_{vt} is volume concentration of suspended sediment, and f is hydraulic strength $(\gamma_c V_{cp}^3/(\rho_s - \gamma_f') g h \omega_0)$. The section AB in Figure 22.13 corresponds to hydraulic (dilute) debris flow.

Less viscous debris flows have properties similar to turbulent flow (hence they are called hydraulic debris flows) and subject to the same relation as in (22.8). When the sediment volume concentration is higher than 0.5, plastic features alien to the previous cases will emerge. Like debris flows of high viscosity, the viscoplastic fluid called a gravity flow has a relation as in (22.8) but with a smaller exponent after 19 samples had been analysed (Wang et al., 2001):

$$C_{vt} = 32.94 f^{0.081} \tag{22.9}$$

As shown in Figure 22.13, section BC is due to a gravity (viscous) debris flow. So the exponent in (22.8) is 10 times that in (22.9).

Table 22.6 lists typical bulk concentrations in debris flows measured at different locations at the Jiangjia Ravine.

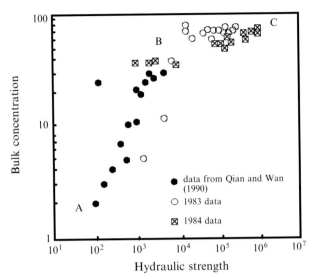

Figure 22.13. Relationship between suspended sediment and hydraulic strength.

22.4.2 Composition

Debris flows of low viscosity are characterized by similar volumes and a narrow range in grain size distribution. For example, dispersion (sorting) of particles (d_{84}/d_{16}) in a dilute debris flow in Jiangjia Ravine is from 0.01 to 0.08, and over 99% of the sediment is sand less than 2 mm in diameter, accounting for 62.5–82% of the total weight. Viscous debris flows have a wide range of grain sizes, with a dispersion of 15–40, some 500–4,000 times that of a flood flow. Grains larger than 2 mm account for 65% of the total solid weight (sand, 18.3%; clay, 18.6%) (Takahashi, 1999).

Cumulative curves of grain-size distribution show different shapes related to concentration of debris flow, reflecting variation in shear stress, cohesive force, and dispersive force of the flow (Figure 22.14). Each curve consists of two segments, representing suspended load and bedload. In the case of a viscous debris flow, this distinction becomes ambiguous, reflecting poor sorting.

As shown in Figure 22.14, eight cumulative curves may be divided into two groups. Five curves (Nos 01-60, 01-61, 01-62, 01-63, and 01-64) are much steeper than the other three in the coarse sizes. The size grading curve (Figure 22.15) has two types: single-peak such as No. 01-63, and double-peak such as No. 01-64. The grain size corresponding to the peak of size grading curve is about 20 mm for Nos 01-61–01-63 samples with greater density. The maximum peak size is 80 mm, and the minimum is 2 mm. These characteristics conform to the evolution of debris flows over time from a low-viscous continuous flow to viscous surges. The weight of grains greater than 2 mm is from 50.5–70% of viscous debris flow, with the remainder considered as fluid load that can suspend particle size up to 0.5 m. The upper limit of grain size of suspended load determines the capacity of transportation.

Table 22.6. Classification for concentration.
From Wang et al. (2001).

Categories	Classification			Density (g/cm³)	Sediment (g/cm³)	Porosity	Moisture (%)	Fluid limit	Flow regime	Flow character
Soil				–	–	0.19–0.44	8–11	Contract limit	Solid	
Debris flow	Viscous DF	High		2.373***	2,180.32	0.19	8.12	Plastic limit	Sub-solid	Non-Newtonian flow
		Middle		2.12	1,778.56	0.34	16.11	Flow limit	Laminar	
		Low		2.04	1,651.5	0.39	19.04			
	Subviscous DF			1.73	1,159.2	0.57	33.09	Slurry limit	Turb–laminar	
	Dilute DF			1.57 (1.24)**	906.75 (380)**	0.66	42.38	Turbid limit	Turbulence	
Hydraulics	Flood			1.17	269.96	0.90	76.09	Turbid		Newtonian flow
	Fluid			1.02	31.76	0.98	96.89	Water		

***Lower limit value; ***Upper limit value; other is mean value.

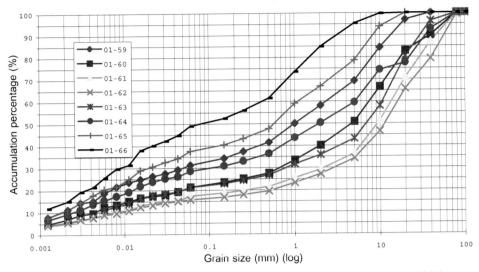

Figure 22.14. Grain-size distribution for eight debris-flow samples (8 July, 2001).

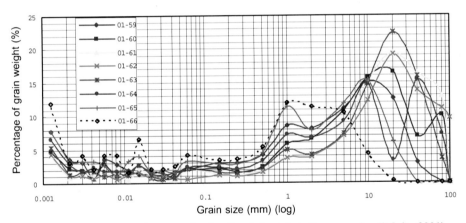

Figure 22.15. Grain weight vs. grain size for eight debris-flow samples (8 July, 2001).

Fine sediment forming the matrix of a debris flow suspends the coarser material. Sediment concentration increases as coarse grains are added and become dispersed in the matrix phase of the flow. When concentration approaches 0.5, coarse sediment can remain in suspension with relative stability. As Takahashi (1999) argued, stone flow occurs on gradient slopes between 14° and 25°. Viscous debris flows sometimes have a density as high as 2.37×10^3 kg/m^3 ($C_{vt} = 0.82$) while coarse grains typically compose 80% of the total weight of the debris flow, and can move on slopes of less than 5.7°. Different fine-grain composition for (22.1) and (22.2) in the matrix of debris flows imposes various influences on motion, flow regime, resistance, and other properties, as shown in Table 22.7.

Table 22.7. Particle size of debris flows and associated slurries, and their rheological parameters and resistance.

No.	Composition (fluid/slurry)				Slurry density (10^3 kg/m^3)	Fine ratio (R_{ns})	τ_{sf} (Pa)	η_p (Pa.s)	Slope	Mean velocity (m/s) (maximum)	Resistance
	>2 mm	2–0.05 mm	0.05–0.005 mm	<0.005 mm							
1	58	24/57	14/35	4/8	1.67	0.76	20.93	0.069	0.06–0.12	5–6	Middle
2	65.2	18.3/52	10.2/29.3	6.3/18.7	1.67	0.92	50.32	0.18	0.05–0.08	8–10 (15)	Low
3	41.1	26.7/45.3	19.0/32.3	13.2/23.4	1.57	0.81	10.91	0.041			
4	53.1	36.8/78.5	6.1/13.0	4.0/8.5	1.66	0.27	7.15	0.09	0.08–0.16	3–5 (6)	High
5	30.5	57.1/82.2	7.2/10.4	5.2/7.1	1.56	0.21	4.2	0.045			

No. 1 = debris flow of HuoShao Gully; No. 2 = viscous debris flow of Jiangjia Ravine; No. 3 = sub-viscous debris flow of Jiangjia Ravine; No. 4 = viscous debris flow of Hunshui Ravine; No. 5 = sub-viscous debris flow of Hunshui Ravine.

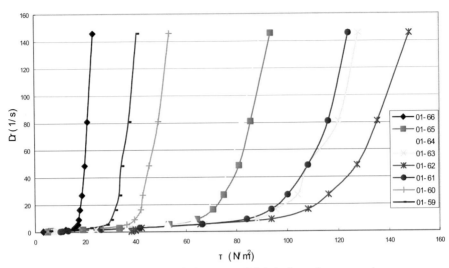

Figure 22.16. Rheological curves of debris-flow slurry samples.

Table 22.8. Yield stress and Bingham number of the slurries taken from the eight samples.

No.	01-59	01-60	01-61	01-62	01-63	01-64	01-65	01-66
Yield stress (N/m^2)	31.77	42.05	101.56	117.52	100.98	71.78	75.02	18.09
Plastic viscosity (N.s/m^2)	0.064	0.081	0.159	0.212	0.194	0.124	0.129	0.038

Debris flows consisting of water and fine-grained soil exhibit partial plasticity and can not be a perfect Bingham fluid. In practice, the Bingham fluid can be considered as a good approximation of sub-viscous and viscous debris flows. Generally, the yield stress of slurry in the Jiangjia Ravine is between 10 and 300 N/m^2 and the plastic viscosity between 0.4 and 15 N.s/m^2. Rheological curves of slurries from the eight samples above measured by a coaxial viscorator are shown in Figure 22.16. The linear least-square-approximation of the standing section of the curve gives the yield stress and plastic viscosity in Table 22.8 (Wei et al., 2002). The immediate conclusion is that yield stress and plastic viscosity increase with the density of slurry ($d_{max} = 1.2$ mm).

22.5 SEDIMENT TRANSPORTATION AND INFLUENCE ON THE MAIN RIVER

22.5.1 Sediment transportation

The annual sediment load caused by debris flows in the Jiangjia Ravine is the sum of sediment load of all the debris-flow events during a year. Debris flows commonly

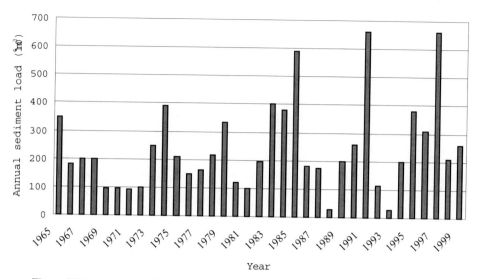

Figure 22.17. Annual sediment load caused by debris flows in the Jiangjia Ravine.
From Cui, P. et al. (1999).

Table 22.9. Sediment transporting capacity of debris flow in the Jiangjia Ravine.
From Cui, P. (1999).

				Maximum discharge and relevant parameters			
Year (Y)	Debris flow event	Flowing duration (hh:mm)	Sediment runoff annually (10^4 m^3/y)	Maximum discharge (m^3/s)	Velocity (m/s)	Sediment content (ton/m^3)	Sediment flux (ton/s)
1994	9	27:45	199.52	2,007.8	11.11	1.988	4,031.27
1995	14	68:15	373.93	1,399.7	13.33	1.847	2,585.25
1996	14	53:00	316.31	2,843.8	12.5	2.008	5,712.34
1997	18	86:21	629.28	1,341.6	12.9	1.926	2,584.59
1998	9	28:41	184.44	2,913.9	16.67	2.008	5,853.15

take place 10–20 times a year in the gully, but occurred 28 times in 1964 alone. The annual sediment loads caused by debris flows in the gully from 1965 to 1999 are shown in Figure 22.17. In the Jiangjia Ravine, 62 debris flows happened from 1994 to 1998 (Table 22.9).

Table 22.10 gives the annual sediment loads of debris flows measured in some of the gullies in China and their erosion module calculated from: $M_C = W_{CS}/A$. Debris flows in the Jiangjia Ravine have a powerful capacity for transporting sediment.

22.5.2 Influence on main river

Debris flows in the Jiangjia Ravine play a vital role in changing the nearby confluence with the Xiaojiang River through the deposition of sediment. Figure 22.18

Table 22.10. Measured annual sediment load and erosion module of debris flows in some gullies.

Gully	Catchment area A (km^2)	Measuring year	Annual sediment load $W_{cs}(10^4\,\text{m}^3)$	Erosion modulus $M_c(10^7\,\text{kg/km}^2\text{a})$	Source of data
Guxiang	26.0	1964–65	461.15	47.00	Wang et al. (1984)
Liuwan	1.97	1963	9.29	12.50	Yang (1984)
Huoshao	2.03	1973	9.66	12.61	Yang (1984)
Niwan	10.3	1965	25.46	6.55	Yang (1984)
Hunshui	4.5	1976–78	63.55	37.43	Zhang and Liu (1989)
Jiangjia	45.1	1965–99	241.58	14.19	Observation
Macao	13.5	1986–88	12.35	2.42	Gao and Qi (1997)
Zhifang	3.73	1959	2.99	2.13	Lanzhou Institute of Glaciology and Cryopedology (1982)

shows one result of the confluence of a debris flow with a main channel through the laboratory simulation that was performed in the Civil Engineering Fluid Mechanics Lab of the Southwest Jiaotong University by Dr. He Yiping from October 2001 to February 2003. The set-up includes one main flume (length, 15 m; width, 0.3 m; height, 0.65 m) and one branch flume (length, 2.5 m; width, 0.1 m; height, 0.4 m). The confluence angle and discharge ratio between the main flume and branch flume are adjustable.

Following the confluence of debris flows and the main channel, the cross section and longitudinal profile of the main river will be changed nearby the confluence spot (Figures 22.19 and 22.20). The upstream profile from confluence has been elevated, and the downstream profile steepened.

On the basis of our observations and experiment, the four modes of confluence between debris flows and the main river are proposed as follows: (a) pounding the opposite bank; (b) partial-complete blockage of main channel; (c) submerging into main channel; and (d) mixing with and submerging into the main channel. Figure 22.21 shows a sketch map of the four modes.

From 4 to 12 July 2001, three episodes of debris flows occurred in the Jiangjia Ravine, which transported large amounts of sediment into the Xiaojiang River (Figure 22.22). The average discharge of the Xiaojiang River in July is between 82 m^3/s and 118 m^3/s, in contrast with a maximum discharge of debris flow on 8 July of 775 m^3/s. The total volume of debris flows was 6.5×10^5 m^3, in addition to sediment discharge of 3.4×10^5 m^3. Figure 22.22 shows the resulting blockage. The mechanics of the confluence of debris flow with main channel is very complex

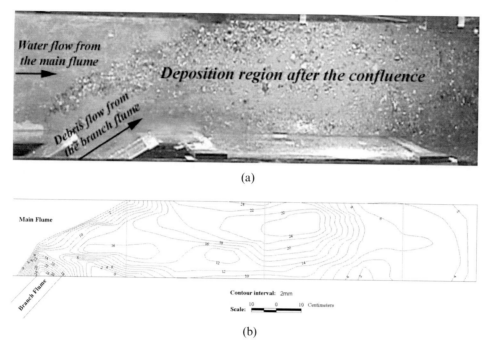

Figure 22.18. Simulation of confluence of branch and main river. (a) Snapshot of deposition after the confluence. (b) Contour map of deposition after the confluence. (The ratio of discharge per unit between the main and branch flumes is 0.354, and the confluence angle is 30°. The value in (b) is deposition thickness.)
From He (2003).

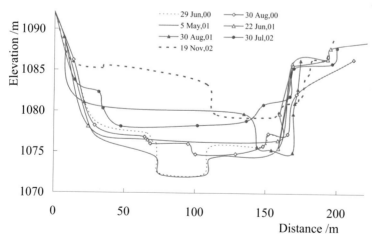

Figure 22.19. Cross section nearby the confluence spot of the Jiangjia Ravine and Xiaojiang River.
From He (2003).

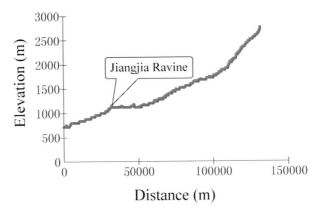

Figure 22.20. The longitudinal profile of the Xiaojiang River.
From He (2003).

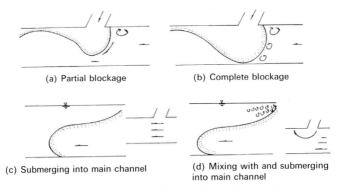

Figure 22.21. Sketch map of the four modes.
From Chen (2000).

because the debris flows are distinguished from the ambient fluid (hyperconcentrated flow) in density, composition, and viscosity. The velocity difference between the two types of fluids is high, so they are mixed, not stratified. At the confluence, we observed that debris flows slid through a short distance toward the opposite bank of the river and then were dumped into the current of the main stream. Shortly after, debris flows rolled up and down in the main channel, and a half-circular wave on the surface of the main flow produced. Upon impacting the opposite bank, the main wave divided into three separate waves. The first reversed its course in an upstream direction and disappeared; the second impacted the bank and flowed downstream; and the third ran directly downstream with a velocity exceeding that of the main flow. Pebbles and boulders were deposited near the point of confluence. The sediments resulted in the water level of the main channel rising to 8.5 m. After the dam was incised by water flow on 11 July, the river stage returned to its normal level.

(a)

(b)

(c)

Figure 22.22. Debris-flow from the Jianjgia Ravine entering the Xiaojiang River. (a) Partial blockage. (b) Complete blockage. (c) Dam break.

Photo: Wei, F.Q. on 8 July 2001.

The height of the dam was 6.5 m and the width was 301 m on that day. The resulting flood inundated fields and destroyed some buildings.

22.6 DEPOSITION

From field observations, debris-flow deposits in China can be described in terms of five classes (Cui, 1996; Scott, 1989):

1. *Normal grading.* Grains are deposited in normal order, with bigger boulders coming to rest before the smaller ones. This occurs in debris flows of lower viscosity.
2. *Disorderly graded bedding.* Grains of various diameters are randomly distributed and poorly sorted, due mainly to high viscosity.
3. *Inverse grading.* Larger grains are deposited first, occurring in the case of sub-viscous flow with a matrix with dispersive force exceeding the viscous force.
4. *Armored deposits.* Coarse sediment remains as finer material in the matrix is eroded.
5. *Basal mud deposits.* A thin layer of mud, consisting of clay and silty sands, is present at the base of the deposit.

The depositional structure of viscous debris flows is dominated by their inner speed and stress distributions. According to observational data from viscous debris flows with hyperconcentration, radical causes of formation for different graded bedding in debris-flow deposits have been analysed, using its rheological properties and the ratio of flow plug. The flow plug ratio (R_H) of the three kinds is equal to the ratio of the mean flow depth (H) to the mean flow plug (H_0), that is (Wang et al., 2003b):

$$R_H = H/H_0 \qquad (22.10)$$

$$H_0 = \tau_{Bf}/(\gamma_c jg) \qquad (22.11)$$

Equation (22.10) determines the speed distribution of every kind of viscous debris flow and total depth condition development by a flow plug of Bingham fluid. That can approach to characteristics of graded bedding texture of every kind for viscous debris flow.

For coarse gravel to deposit disorderly formed hybridize graded bedding structure, because of action of inside resistance by high-viscous medium. It is most graded bedding deposit of viscous or high-viscous debris flow with $R_H = 12$. These deposit structures are observed in many ravines. But when the mean flow plug ratio is high, about 100 in sub-viscous debris flow, there is a lower structure substantial degree (C_{vt}/C_{vm}) for 0.85. So that there are dispersive force and viscous force together in the sub-viscous debris flows, or dispersive force is more than viscous force in sub-viscous debris flow. There is inversely graded bedding structure after debris-flow deposit. There are the structures in most lahar deposits in America, Japan. And similar structures are found in some deposits in China, for example of Dawazhi dully in the branch of Jiangjia Ravine.

When the mean flow plug ratio is about 20–22 in viscous debris flow, they would form the bedding structure of the gravel accumulated at surface, which are the same geomorphologic phenomenon in Major's experiment (Major, 1994). Here that is specially emphasized that the gravel in squirm condition of hyperconcentration viscous flows would tend to upward motion for the effect of Weissenberg, namely viscoplastic principal strain difference, and they are different from rough bedding structure (Wang et al., 2003b).

22.7 HAZARD MITIGATION

Measures for mitigating debris-flow hazards include protective engineering works, forecasting and warning systems allowing for emergency education, and land-use regulation and zoning, as well as evacuating and emergency.

22.7.1 Structured mitigation

Check dams of various types are used in source areas to stabilize slopes, to control channel erosion, and to trap coarse debris. Armored channels are used in downstream areas to convey flows to trunk streams without damaging villages, roads, and fields. Several types of check dams, channels, and structures have been engineered to reduce the damage from debris flows in China. Finally, structures in deposition areas may be required to protect specific facilities. Figure 22.23 shows several main check dams and channels in the Dongchuan area.

A systematic plan for debris-flow prevention has been carried out in the Jiangjia Ravine (Wu et al., 1990). The main engineering works were designed for the upper area – Duozhao gully and lower sections of the ravine (Figure 22.24). In the Duozhao gully, 117 check dams were constructed in the 1980s in order to stabilize slopes and prevent tributaries from incising, and they trap up to $7 \times 10^5 \, \text{m}^3$ of sediment. Most matter came from the Duozhao gully before the 1980s, but after 1985 most of the matter came from the Menqian gully and the mean debris-flow scale decreased.

In the lower reaches of the Jiangjia Ravine, drainage grooves of 1,800 m in length were designed to protect villages and farmland, as well as to control and spread deposition in order to maximize the potential for the subsequent use for rice cultivation. About $0.2 \, \text{km}^2$ in the ravine was planted to restore the vegetation cover.

Debris flows often occur in areas of poor vegetation cover, and restoration of that cover can be effective in reducing debris-flow activities. Measures include forestation, returning cultivated land to forest, banning herding in grasslands, and a combination of these approaches.

Great benefits were obtained from these disaster reducing works. The main benefits include control of debris-flow disasters, recovering ecological environment, reclaiming $7 \times 10^5 \, \text{m}^2$ of farmland, and protecting two bridges and a factory in the upper part of the Xiaojiang river, and so on.

(a)

(b)

Figure 22.23. Two types of drainage groove. (a) Ladder-shaped groove. (b) V-shaped groove.

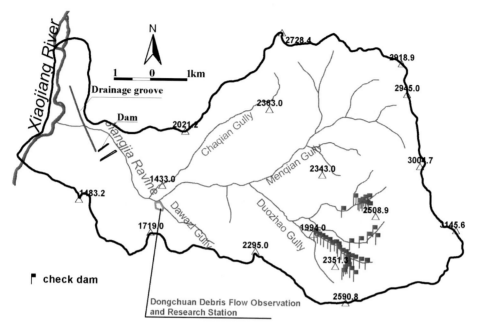

Figure 22.24. Main structure mitigation in the Jiangjia Ravine.

22.7.2 Forecasting and warning system

Debris flow occurs in response to rainfall, so we focus our forecasting on monitoring and forecasting threshold levels of rainfall. Despite the fact that no direct quantitative relation has been found between flow occurrence and precipitation, empirical formulas can estimate critical conditions at which debris flows occur. For instance, one can roughly tell that a debris flow is probable when antecedent and concurrent rainfall intensity reach a certain value. This method is limited by its oversimplification.

On the other hand, the warning system installed in the Jiangjia Ravine functions very well, which includes Acoustic Flow Monitors (AFMs) that function as seismometers; ultrasonic detectors; and infrasound sensors. AFMs sense frequencies ranging from 0.2–1,000 Hz. Signals received and processed by computer can be used to trigger a warning alarm in advance of a debris flow, the delay time being:

$$T = (S_2 - S_1)/V_{cp} \qquad (22.12)$$

where V_{cp} is the mean velocity of the debris flow and $(S_2 - S_1)$ is the distance between the observation sections. Ultrasonic detectors are used because flow depth is typical of the flow magnitude and possible damage, and can give a telemetered alarm 10 km from the source of the debris flow.

In practice, an ultrasonic signal is triggered by a debris flow at its initiation in the source area. The velocity of the ultrasonic wave is about the same as that of audible sound in air (i.e., about 344 m/s at 20°C), much faster than the velocity of a debris flow. Hence, it can warn of the arrival of a debris flow before the flow reaches the lower parts of the gully, if we have a device to receive the wave. The infrasonic observation system for debris flows (Zhang, 2003) has been designed and developed.

22.8 CONCLUSIONS

The Jiangjia Ravine provides most excellent conditions for investigating debris flows. In the last 40 years, scientists from the Institute of Mountain Hazards and Environment, Chinese Academy of Science, have kept observing and monitoring debris-flow occurrences for their formation, motion, deposition, and their action on local landforms and environments. The most conspicuous features of debris flows and their related studies in this very basin are as follows:

1. Debris flows occur at rather high frequencies and have averaged 11 events per year during the last 40 years, allowing long-term and systematic observations to some extent parallel to observations of climate and hydrology, which have been supposed to be interrelated with debris flows.
2. Each debris-flow event in the Jiangjia Ravine consists of tens to hundreds of surges, providing a wide variety of phenomena that are rarely seen in other areas of the world. This calls for systematic research on debris flows.
3. The Jiangjia Ravine has provided an ideal site to explore debris-flow formation and initiation in combination with a background of loose materials and rainfall. This is significant for debris-flow forecasting.
4. Observation data and videos of living debris flows in the Jiangjia Ravine are the most unique resource open to future studies in the field.

22.9 ACKNOWLEDGEMENTS

The authors are grateful to Dr. Scott and Dr. Jakob for their help in reviewing the draft and also to the Dongchuan Debris Flow Observation and Research Station, Chinese Academy of Sciences, for permission to use data and pictures. Dr. James Gardner of the University of Manitoba, Canada, assisted in editing the paper. This research is partly supported by the Chinese Science Foundation for Outstanding Youth (Grant No. 40025103) and National Science Foundation of China (Grant No. 4P831010).

22.10 REFERENCES

Chen, D.M. (2000) Mechanism of confluence between debris flow and the main river. PhD thesis, Chinese Academy of Water Conservancy Science [in Chinese].

Cui, P. (1999) Impact of debris flow on river channel in the upper reaches of the Yangtze River. *International Journal of Sediment Research*, 14(2), 201–203.

Cui, P., Wei, F.Q., and Li, Y. (1999) Sediment transported by debris flow to the lower Jiansha River. *International Journal of Sediment Research*, 14(4), 67–71.

Cui, Z.J. (1996) *Debris-flow Deposition and Environment* (p. 192). Ocean Press, Beijing [in Chinese].

Gao, S.Y. and Qi, L. (1997) Characteristics of debris flow in Macao Gully in Wudu County, Gansu Province. *Mountain Research*, 15(4), 300–304 [(in Chinese].

He, Y.P. (2003) Influence of debris flow on river channel change of mountains. PhD thesis, Chinese Academy of Science [in Chinese].

Jan, C.D., Wang, Y.Y., and Han, W.L. (2000) Resistance reduction of debris flow due to air entrainment. In: G.F. Wieczorek and N.D. Naeser (eds), *Debris-flow Hazards Mitigation: Mechanics, Prediction, and Assessment: Proceedings of the 2nd International Conference on Debris-flow Hazards Mitigation, Taiwan*. A.A. Balkema, Rotterdam.

Kang, Z.C. (1990) *Motion Characteristics of Debris Flow at Jiangjia Gully, Yunnan Province, China*. International Research & Training Center on Erosion & Sedimentation.

Lanzhou Institute of Glaciology & Cryopedology, Traffic Science Institute of Gansu Province (1981) *Debris Flow in Gansu Province* (pp. 83–85, 114–115). Publishing House of People's Transportation, Beijing [in Chinese].

Li, J. and Wu, J.S. (1980) Dynamical data of debris flow in Jiangjiagou Gully (Unpublished) [in Chinese].

Major, J.J. (1994) *Experimental Studies of Deposition at Debris Flow Flume* (USGS Open-file Report 1-28). US Geological Survey, Reston, VA.

Qian, N. and Wan, Z.H. (1990) *Sediment Kinematics*. Science Press, Beijing [in Chinese].

Scott, K.M. (1989) *Origins, Behavior and Sedimentology of Lahars and Lahar Run-out Flows in the Toutle Cowlitz River System, Mt. St. Helens* (USGS Publication PPA19). US Geological Survey, Reston, VA.

Takahashi, T. (1999) Mechanics of viscous debris flow. In: *China–Japan Joint Research on the Mechanism and the Countermeasures for Viscous Debris Flow* (pp. 119–125). Institute of Mountain Hazards & Environment, CAS, and Disaster Prevention Research Institute, Kyoto University.

Tang, B.X., Zhou, B.F., and Wu, J.S. (2000) *Debris Flow in China* (pp. 1–10). Commercial Press, Beijing [in Chinese].

Wang, Y.Y., Jan, C.D., Han, W.L., Hong, Y., and Zhou, R.Y. (2003a) Stress-strain properties of viscous debris flow and determination of velocity parameter. *Chinese Journal of Geological Hazard and Control*, 14(1), 9–13 [in Chinese].

Wang, Y.Y., Jan, C.D., Han, W.L., and Zhou, R.Y. (2003b) A study on the forming mechanism of the bedding structure of gravel accumulated at the surface, and rough bedding structure in deposits of viscous debris flows with hyperconcentration. *Acta Sedimentologica Sinica*, 21(2), 205–210 [in Chinese].

Wang, Y.Y., Jan, C.D., and Yan, B.Y. (2001) *Debris Flow Structure and Rheology*. Hunan Science and Technology Press, Changsha, China [in Chinese].

Wang, W.J., Zhang, S.C., and Wang, J.Y. (1984) *Characteristics of Glacial Debris Flow in Guxiang Gully, Tibet Autonomous Region* (Memories of Lanzhou Institute of Glaciology & Cryopedology, pp. 18–34). Chinese Academy of Science, Beijing, and Science Press, Beijing [in Chinese].

Wei, F.Q., Hu, K.H., Cui, P., Chen, J., and He, Y.P. (2002) Mechanism of blocking of Xiaojiang River by debris flow of Jiangjiagou Gully. *Journal of Water and Soil Conservation*, 16(6), 71–75 [in Chinese].

Wu, J.S., Kang, Z.C., Tian, L.Q., and Zhang, S.C. (1990) *Debris Flow Observation and Research in Jiangjiagou Ravine, Yunnan*. Science Press, Beijing [in Chinese].

Wu, J.S., Tian, L.Q., Kang, Z.C., Zhang, S.C., and Liu, J. (1993) *Debris Flow and Its Comprehensive Control*. Science Press, Beijing [in Chinese].

Yang, K. (2002) Relationship between debris flow and precipitation in Jiangjiagou Gully. MSc thesis, Chinese Academy of Science [in Chinese].

Yang, Z.N. (1984) *Viscous Debris Flow in Wudu Prefecture and Estimation of Their Basic Parameters* (Memories of Lanzhou Institute of Glaciology & Cryopedology, pp. 22–25). Chinese Academy of Science, Beijing, and Science Press, Beijing [in Chinese].

Zhang, S.C. (2003) Detecting infrasound emission of debris flow for warning purposes. In: D. Rickenmann and C-L. Chen (eds), *Debris-flow Hazards Mitigation: Mechanics, Prediction, and Assessment*. Millpress, Rotterdam.

Zhang, X.B. and Liu, J. (1989) *Debris Flow in Dayingjiang Basin, Yunnan Province* (pp. 103–103). Chengdu Cartography Press, Chengdu, China [in Chinese].

Zhang, X.B. and Liu, J. (1989) *Debris Flow in Dayingjiang Basin, Yunnan Provinces* (pp. 103–107). Chengdu Cartography Press, Chengdu, China [in Chinese].

23

Debris flows and debris avalanches in Clayoquot Sound

Terrence P. Rollerson, Thomas H. Millard, and Denis A. Collins

23.1 INTRODUCTION

Clayoquot Sound on the west coast of Vancouver Island is rich in natural resources. The watersheds in Clayoquot Sound are heavily forested with a variety of conifers, and five species of salmon spawn in the streams of the area. Both aboriginal and western cultures harvest the abundance of resources that these streams and watersheds supply. The area supports a major tourist industry as well as commercial fishing and logging operations.

Clayoquot Sound is rugged, with steep mountains rising directly from the ocean, or from a narrow coastal plain that lies in front of the mountains (Figure 23.1). The winter months are wet due to cyclonic storms that originate in the northern Pacific Ocean and track eastward towards the coast of British Columbia (BC). Intense precipitation frequently occurs as the storm fronts rise over the coastal mountains. Debris flows and debris avalanches are common as a result of this intense precipitation.

Until the 1960s, most forest harvesting in Clayoquot Sound occurred at low elevations on the coastal plain and along the valley bottoms. As logging moved onto steeper slopes, landslides became more common; a result of changes in hillslope stability due to both clearcut harvesting and road-building practices that failed to address landslide concerns.

In the 1980s and 1990s, environmentalists began to direct criticism against the forest industry on Vancouver Island and, particularly, in Clayoquot Sound. One of their strongest criticisms was directed towards the effects of forest harvesting on steep slopes that led to debris avalanches and debris flows that delivered large quantities of sediment into fish streams. The visual effect of logged hillslopes scarred with landslides was a potent image (Figure 23.2), and photographs of clearcuts and landslides on west-coast mountainsides were featured in environmental

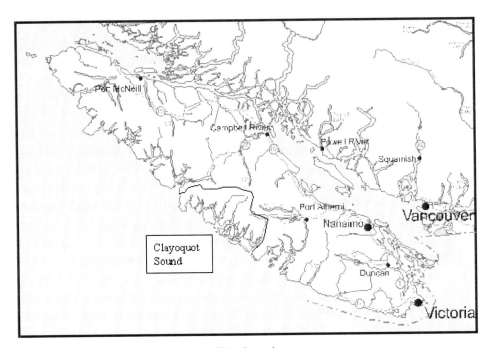

Figure 23.1. Location map.

Figure 23.2. Landslides at Rae Lake, Clayoquot Sound.

publications and in mainstream journals such as the *National Geographic* magazine (September, 1990).

During this time, the BC government and the forest industry began studies to improve their understanding of the effects of logging and road construction on steep slopes. Research was also directed at developing methods that better identified landslide-prone terrain and that minimized landslide hazards and the risks of operating in steep terrain. Much of this research occurred on the west coast of Vancouver Island, with some research centered in Clayoquot Sound. At the same time, extension efforts intensified in the form of workshops, conferences, and handbooks on the management of landslide-prone terrain (e.g., Chatwin et al., 1994).

In 1995 the BC government passed the Forest Practices Code Act of British Columbia (FPC), which contained strict requirements for the assessment and management of landslide-prone terrain. Even greater restrictions were incorporated with the implementation of recommendations from the Scientific Panel for Sustainable Forest Practices in Clayoquot Sound (SPSFP, 1995). Many of the slope stability regulations in the FPC are based on studies that were conducted in Clayoquot Sound, on the west coast of Vancouver Island and elsewhere in coastal BC.

Government and the logging industry recognized the damage caused by logging-related landslides. Shortly before the FPC was introduced, the BC Watershed Restoration Program was implemented to mitigate and minimize further logging-related landslide damage to watersheds throughout BC. An important aspect of this program was the deactivation (or de-building) of old roads that were associated with landslides. In 1996, a particularly long and intense storm in the central and northern portion of Clayoquot Sound as well as widespread areas in Washington, Oregon, and northern California (Robinson et al., 1999) reinforced the need for a watershed restoration program. In the Clayoquot Sound region alone 273 landslides initiated within a four-day period. Of these, two-thirds were related to logging, with landslides from roads accounting for 25% of the total number of landslides.

The west coast of Vancouver Island continues to be difficult terrain in which to conduct forest harvesting. For example, by 1994, more than 80% of the remaining old-growth forest in Clayoquot Sound was located on slopes steeper than 30° (SPSFP, 1995). The industry relies on the advice of terrain specialists to identify the landslide hazards and risks their operations may present. In turn, these experts rely on a locally developed body of research on which to base their assessments.

The remainder of this chapter discusses debris flows and debris avalanches, the most common landslide types in Clayoquot Sound. To a lesser extent, other landslide types are also discussed. Finally, the effects of logging on hillslope stability, and how current forest operations manage landslide risks in Clayoquot Sound are presented.

23.2 LOCATION AND SETTING

Clayoquot Sound is located within the central portion of the Vancouver Island Mountains, a major north-west to south-east-trending physiographic unit that forms the backbone of Vancouver Island and the Estevan Coastal Lowland (Holland, 1964, fig. 1). Elevations range from sea level to 2,200 m. Within the Clayoquot Sound area, the Vancouver Island Mountains can be divided into two sub-units consisting of the North Vancouver Island Ranges and the Vancouver Island Fiordland (Hoadley, 1953; Yorath and Nasmith, 1995).

The North Vancouver Island Ranges are 270 km long and 60 km wide, extending from Quatsino Sound in the north to Barkley Sound in the south. Mountain surfaces were modified by Pleistocene glacial erosion resulting in rounded lower peaks and ridges and steep U-shaped valley cross sections (Howes, 1981). Shallow deposits of colluvium and till mantle many mid to upper valley sides, but bedrock bluffs are common on steeper slopes. In narrow valleys, shallow colluvial and morainal veneers may extend to the valley floor. In wider valleys, lower slopes and valley floors tend to be mantled with thicker deposits of till, and glaciofluvial, fluvial, and debris flow/flood deposits.

The Vancouver Island Fiordland is 260 km long and 60 km wide, extending along the western portion of Vancouver Island from the Brooks Peninsula south to Barkley Sound. It includes both islands and peninsulas bounded by a network of fiords that penetrate inland from the exposed western coast (Figure 23.3). Within this sub-unit, the land rises abruptly from the shoreline to elevations of 900 m beyond which a more gradual slope leads to inland summits. Pleistocene glaciation resulted in rounded summit peaks and ridges. Most of these are densely forested, but isolated areas of alpine tundra do occur. Much like the Vancouver Island Ranges, colluvial materials, bedrock, and thin veneers of till dominate on steep hillsides, ridges, and summits, while deeper till, glaciofluvial, fluvial, and debris flow/flood deposits cover lower hillsides and valley floors.

The Estevan Coastal Lowland is 1–3 km wide extending along most of the west coast of Vancouver Island. Much of the coastal plain is less than 50 m above sea level and is dissected by many inlets and fiords. Where the coastal plain is underlain by soft bedrock or is mantled by Pleistocene sediments, it is smooth; but, where harder rocks are present low hills, hummocks and occasional steep bluffs can be found. In many areas the coastal plain is mantled with glaciomarine or more recent marine sediments; however, till and glaciofluvial deposits are also present. Fluvial and debris flow as well as debris flood fans often cover older glacial sediments along the mountain bases.

23.2.1 Bedrock

The dominant bedrock units in Clayoquot Sound are the Karmutsen Formation basalts and andesites, the Vancouver Island Intrusions (VII) composed of quartz monzonite and granodiorite, and the Coast Plutonic Complex (CPC) diorites and amphibolites. Sicker Group meta-andesites and dacites are occasionally present, as are limestones of the Quatsino Formation (Roddick et al., 1979).

23.2.2 Climate and vegetation

Clayoquot Sound is part of an extensive temperate coastal rainforest that extends from Alaska to northern California. The ecosystems present within Clayoquot

Figure 23.3. Landsat image of Clayoquot Sound study area.
Landsat image data available from U.S. Geological Survey, EROS Data Center, Sioux Falls, SD.

Sound include the Coastal Western Hemlock Zone and the Mountain Hemlock Zone. Western Hemlock (*Tsuga heterophylla*) accompanied by Amabillis Fir (*Abies amabillis*), Western Red Cedar (*Thuja plicata*), and minor amounts of Yellow Cedar (*Chamaecyparis nootkatensis*) dominate the forests in the Western Hemlock Zone. The Mountain Hemlock Zone is dominated by Mountain Hemlock (*Tsuga mertensiana*) and Amabalis fir, and to a lesser extent Yellow Cedar (Green and Klinka, 1994).

In many temperate forests wildfire or windthrow are the major means of natural stand replacement. In parts of Clayoquot Sound and other areas on the west coast of Vancouver Island, however, landslides are one of the major natural processes that initiate the establishment of new forests. For example, in the Clayoquot River valley, landslides initiated stand replacement for 2.4% of the forests in the valley (Pearson, 2000). Red alder (*Alnus rubra*) typically dominates the successional sequence on landslides, advancing slowly upslope from the base of the landslide over periods of up to 15 years (Straker, 1998). Conifer species such as hemlock and red cedar are often early colonizers over the entire surface of many landslides but do not achieve full site occupation until near the end of the initial alder phase (40–60 years). Straker suggests that canopy protection and nutrient additions provided by alder are critical to conifer regeneration success.

Clayoquot Sound like much of the west coast of Vancouver Island is characterized by cool, wet winters and warm, moist summers. Mean annual precipitation increases from about 2,900 mm at sea level on the outer west coast to greater than 4,600 mm inland at sea level at the heads of inlets. Seventy to eighty per cent of the precipitation occurs between October and March (Howes, 1981). Snow is usually restricted to higher elevations and is ephemeral at lower and mid-elevations. The area is subject to occasional rain-on-snow events. The coastline is exposed to Pacific frontal storms of high intensity and long duration, and rainstorms of sufficient magnitude to initiate debris flows usually occur several times a year (Rollerson et al., 1998). Long-term precipitation records for coastal stations show three-day extreme totals with ten-year return periods ranging from 257–330 mm (Howes, 1981). These data, however, reflect low-elevation coastal stations, whereas in the mountains orographic uplift can result in much greater precipitation intensities (Church and Miles, 1987). Marquis (2001) found that a mountain site within Clayoquot Sound located about 2 km inland from the coast experienced individual storm precipitation two to three times as great as a coastal station 15 km away.

Wind data for the west coast of Vancouver Island is scarce, however, lighthouse station records (Cape Scott, Spring Island, and Estevan Point) indicate maximum hourly speeds ranging from 60–120 km. The highest wind velocities usually occur during the winter months (October to March).

23.3 DEBRIS AVALANCHE AND DEBRIS-FLOW INITIATION FACTORS IN CLAYOQUOT SOUND

In January of 1996 an intense, four-day rainstorm resulted in 273 landslides in the Clayoquot Region. An unpublished analysis of the source areas by BC Ministry of Forests staff showed that about 20% of the landslides initiated in unlogged areas, 55% in clearcut areas, and 25% at roads. Many of the landslides associated with roads initiated in large fills located on steep slopes.

An extensive air photo inventory of natural and logging-related landslides carried out in the late 1990s (Jakob, 2000) identified 1,004 landslides in the approximately 2,546 km^2 that comprise Clayoquot Sound. This inventory identified a greater percentage of natural landslides, with 50% initiating in unlogged areas, 29% initiating in clearcut areas, and 20% initiating from roads; 1% were of unknown origin.

Jakob found overall landslide densities[1] in forested areas of 0.22 landslides per square kilometer (ls/km^2) compared with 1.94 ls/km^2 for logged areas, a nine times higher density for logging-related landslides.

The actual ratio between logged and unlogged areas is confounded by a number of factors. Because the inventory used only a single set of air photos it is not certain that the landslides identified in logged and unlogged areas necessarily occurred during the same time period. Natural landslides that occurred in alpine areas and

---
[1] Spatial landslide occurrence or spatial landslide density.

that were obviously older than those in logged areas were excluded from the analysis (the logged areas represented landslides which likely occurred within the 20-year period prior to the inventory). However, some older natural landslides may have been included in the tabulation. Similarly, recent ground-based studies in the Coast Mountains near Vancouver (Brardinoni et al., 2003) and in Oregon (Robinson et al., 1999) found that many smaller landslides in forested areas are missed by air photo interpreters compared to the number of smaller landslides identified on air photos in logged areas. Finally, some landslides in logged areas would likely have occurred even if these areas had not been logged.

Jakob (2000) classified approximately 52% of the 1,004 landslides as debris avalanches, 42% as debris flows, and 6% as rockfalls. Landslides were classified as debris flows if they showed a defined headscarp, channelized flow, a linear, confined track, a well-developed fan, and debris lobes in the runout zone. Debris avalanches by contrast were classified as non-channelized, shallow landslides with width remaining relatively constant or increasing downslope. Our own field observations in the area suggest that a number of the landslides classified as rockfalls by Jacob (2000) were likely complex rockfall–rockslide or rockfall–rock avalanche events.

Some of the debris flows in the Clayoquot Sound area that travel down long, confined channels with gentle gradients become sufficiently water charged to be classified as debris floods (hyperconcentrated flows). These events can be classified on the basis of some deposit sorting and the gentle gradients of their deposits as contrasted to the unsorted and more steeply sloping deposits of debris flows.

The study by Jakob (2000) also found that landslides initiating on open slopes were more frequent than landslides initiating on the sides or headwalls of steep, deeply incised, and confined mountain channels (gullies). About 80% of the landslides inventoried initiated on open slopes, 9% on gully sidewalls, 7% in gully headwalls, and 4% in gully channels.

Jakob (2000) reports that downslope curvature was associated with landslide incidence. In logged terrain 44% of all landslides occurred on concave slopes, 54% on planar slopes, and 3% on convex slopes. In unlogged areas, 70% occurred on concave slopes, 27% on planar slopes, and 1% on convex slopes. Many of the landslides in natural areas occurred on steeper slopes in areas where the forests are not considered merchantable, so the difference in distributions of landslides with slope curvature between logged and unlogged areas may be a function of differences in slope curvature distributions and not the effects of logging.

Jakob (2000) describes an asymmetric landslide distribution by slope aspect. The majority of landslides (56%) had orientations ranging between 90° and 225° or east to south-south-west. The second most common landslide orientations were from 270° through to 45° or west to north-north-east. These orientations correspond well with dominant slope aspects in Clayoquot Sound, and the preponderance of landslides on slopes that face east to south-south-west is thought to be associated with the dominant direction of winter storms that tend to approach from the south-east. This south-east-storm direction is thought to be associated with enhanced precipitation on windward slopes due to local orographic uplift of moist air masses and less precipitation on lee-side slopes. The low landslide rates on north-north-east though east and south-south-west through west-facing slopes likely results from a limited number of slopes with these orientations due to the general alignment of valleys and ridges perpendicular to the main storm tracks. A corresponding trend of decreasing landslide incidence with distance from the coast may be related to higher precipitation rates near the coastal edge of the Vancouver Island Mountains and lower precipitation rates inland.

Identification of landslide-prone terrain for the logging industry requires methods that are capable of assessing extensive areas in relatively short periods of time. Subsurface investigations and slope stability modeling are rarely conducted. A series of regional studies in coastal BC (Rollerson, 1992; Rollerson et al., 1997; Rollerson et al., 1998; Rollerson et al., 2001; Rollerson et al., 2002; Millard et al., 2002) use an effective method of identifying likely post-logging landslide locations. These studies identify terrain attributes that can be used to predict post-logging landslides locations and also determines landslide rates for specific terrain types.

Rollerson et al. (1998) collected data from clearcut areas in Clayoquot Sound and Barkley Sound (immediately to the south of Clayoquot Sound). Each landform sample consists of a field-verified, homogenous terrain polygon. The samples were restricted to areas where 6–15 years had passed since harvesting to ensure that all areas had time for root strength reduction and to experience a number of intense, long duration rainstorms. The data set consists of a total of 1,194 terrain polygons with a total area of 4,812 ha or an average area of 4 ha per polygon. For each sample, terrain attribute data such as slope position, slope gradient, aspect, morphology, lateral curvature, soil drainage, surficial material, bedrock type, and the presence or absence of both natural and post-logging landslides was recorded. Landslides smaller than 0.05 ha were excluded because they could not be consistently identified. The data presented here does not include landslides that initiate from the road prism. Road prism landslides account for about 40% of the logging-related landslides within Clayoquot and Barkley Sounds.

Eighty-five per cent of the 1,194 samples in the data set remained stable after logging. By comparison, 78% of the 760 samples in a data set from the Queen Charlotte Islands (Rollerson, 1992) remained stable for the same 6–15-year period after logging. The overall mean post-harvest landslide density for the study area was 0.06 ls/ha, contrasting with a mean landslide density of 0.17 ls/ha for similar terrain in the Queen Charlotte Islands, and 0.08 ls/ha for a data set representing the entire central and northern west coast of Vancouver Island (Rollerson et al., 1997).

Post-harvest landslide densities showed a statistically significant relationship at the <0.01 significance level for 7 out of the 14 terrain variables analysed (Table 23.1). Another five were significant at the 0.01–0.05 levels. Increases in landslide density tended to correspond to an increase in the percentage of polygons within a group or class that experienced landslides (Table 23.2). The percentage of cases experiencing one or more landslides can be interpreted as an index or ranking of the likelihood of landslide activity following logging on similar terrain in the same climatic region.

Average polygon slope angle categories showed a trend of increasing landslide

Table 23.1. Comparison of terrain features with post-harvesting landslide density.
From Rollerson et al. (1998).

Variable	Significance level	
	Kruskal-Wallis (1)	Chi-square (2)
Slope class	0.000	0.000
Natural landslides	0.000	0.022
Minor natural landslides	0.013	0.003
Landscape position	0.002	0.000
Slope morphology	0.000	0.000
Lateral curvature	0.004	0.025
Soil drainage	0.004	0.006
Slope aspect	0.054	0.055
North vs. south aspect	0.000	0.003
"Combined octants"	0.007	0.008
Elevation	0.021	0.020
Terrain category	0.243	0.231
Bedrock formation	0.504	0.537
Bedrock lithology	0.026	0.051
Bedrock structure	0.049	0.052
Bedrock competence	0.010	0.016

(1) Based on post-logging clearcut landslide density (number/ha).
(2) Based on presence or absence of post-logging clearcut landslides.

density with increasing slope angle up to approximately 46° (Table 23.2). The lower landslide rates on steeper slopes are likely explained by the increasing dominance of bedrock and the limited and discontinuous cover of surficial materials. The highest post-harvest landslide densities were associated with those landforms with slope gradients ranging from 30–40°.

The presence of pre-logging natural landslide scars showed a positive association with post-logging landslide activity. Some of the highest post-harvest landslide densities were associated with these features.

Steep stream escarpments and headwater drainage areas were associated with higher landslide densities, as were areas of highly dissected terrain (frequent gullies) or larger individual gullies. About one-half of the headwater drainage areas were either dissected or were the headward zones of individual, large gullies. About one-third of the areas identified as stream escarpments formed the sidewalls of larger gullies, the remainder of these escarpments formed well-drained, steep, relatively uniform, planar slopes along stream edges. In localized areas, these escarpments were highly dissected and imperfectly drained.

Differences in landslide densities associated with varying surficial materials (terrain categories) were not pronounced; however, cross-tabulations showed some correspondence between different terrain categories and slope class. For example, till, which one would expect to be less stable than angular, bedrock-derived

Table 23.2. Terrain features: clearcut landslide summary statistics.
From Rollerson et al. (1998).

Variable	n	Mean landslide density (ls/ha)	Units with landslides (%)
Slope class (degrees)			
15–19	35	0.00	0
20–25	277	0.02	4
26–30	243	0.05	14
31–35	367	0.10	24
36–40	175	0.07	21
41–46	51	0.07	14
>46	32	0.02	6
Natural landslides			
Absent	1,168	0.06	15
Present	26	0.19	31
Minor natural landslides			
Absent	1,180	0.06	15
Present	14	0.19	43
Landscape position			
Apex	12	0.08	8
Upper slope	397	0.06	15
Mid slope	628	0.06	14
Lower slope	117	0.05	7
Stream escarpment	8	0.26	25
Headwater basin	31	0.20	55
Slope morphology			
Uniform	776	0.06	15
Benchy	32	0.00	3
Dissected	67	0.13	37
Faceted	16	0.01	6
Irregular	221	0.03	7
Single gullies	81	0.18	26
Terrain category			
Morainal (till)	600	0.07	16
Colluvial	75	0.05	17
Glaciofluvial	4	–	–
Marine	3	–	–
Rock	5	–	–
Morainal + colluvial	179	0.06	17
Morainal + glaciofluvial	4	0.03	–
Morainal/rock	202	0.07	15
Colluvial/rock	49	0.03	8
Rock/colluvial (1)	73	0.01	4

Lateral curvature			
Concave	316	0.09	20
Convex	291	0.05	12
Straight	566	0.05	14
Complex	14	0.09	21
Soil drainage			
Rapidly	162	0.03	7
Well	908	0.06	16
Moderately well	119	0.08	15
Imperfectly	2	–	–
Poorly	1	–	–
Slope aspect			
NNE	86	0.03	8
ENE	100	0.04	10
ESE	190	0.07	17
SSE	219	0.08	20
SSW	87	0.09	20
WSW	160	0.08	14
WNW	192	0.05	15
NNW	147	0.04	11
North vs. south aspect			
North	525	0.04	12
South	656	0.08	18
"Combined octants"			
NNW–NNE	233	0.03	10
WNW+ENE	292	0.05	13
WSW+ESE	350	0.07	16
SSW–SSE	306	0.08	20
Elevation (m)			
100	18	0.02	6
101–200	74	0.04	7
201–300	137	0.06	14
301–400	175	0.06	13
401–500	221	0.07	19
501–600	267	0.09	20
601–700	176	0.04	13
700	120	0.03	10
Bedrock formation			
VII/CPC	735	0.07	15
Bonanza	6	–	–
Karmutsen	352	0.05	16
Quatsino	28	0.02	7
Sicker	6	–	–

continued

Table 23.2 (*cont.*)

Variable	n	Mean landslide density (ls/ha)	Units with landslides (%)
Bedrock lithology			
Quartz monzonite	278	0.10	21
Granodiorite	85	0.04	14
Diorite	189	0.03	10
Andesite	61	0.03	12
Basalt	175	0.04	13
Volcanic breccia	21	0.08	13
Gneiss	5	–	–
Greywacke	12	0.04	17
Limestone	29	0.02	10
Bedrock structure			
Massive	154	0.03	10
Fractured	331	0.06	187.7
Sheared	22	0.13	27
Bedded	2	–	–
Bedrock competence			
High	335	0.04	12
Moderate	104	0.08	19
Low	31	0.11	29

(1) Proportion symbols: / = dominant/subdominant; + = either component may be dominant or they may be equivalent.

colluvium, tends to be associated with a gentler range of slope angles than colluvium. Consequently, these two terrain categories exhibit very similar landslide rates. Laterally concave or complex slopes showed higher landslide densities than convex or planar slopes.

Soil drainage showed reasonable correspondence with landslide density. Landslide densities decreased as the soils became drier, indicating an association between soil drainage and slope angle.

When aspect is expressed as exposure to north vs. south, the relationship between landslide density and aspect is significant. Combining aspect octants with similar landslide densities refines the relationship (Tables 23.1 and 23.2). As noted earlier, storms from the south-east or south-west are more common than storms from more northerly aspects. This finding, like the work of Jakob (2000), suggests that local orographic effects may influence landslide rates on south-facing vs. north-facing slopes.

Bedrock lithology, structure, and competence show a statistically significant association with landslide density. However, lithology, structure, and competence could only be documented where bedrock was exposed at the ground surface or in road cuts, typically areas of shallow soil cover.

Decision-tree analysis using CHAID (Magidson, 1993) resulted in the separation of several different classes of terrain, with some classes showing landslides in less than 2% of the samples, while for other classes, more than 50% of the samples experienced post-harvest landslides. The predictor variables identified by the model in order of importance were slope angle, slope morphology, lateral curvature, aspect, and bedrock competence (Figure 23.4).

Debris-flow initiation in gullies (steep, usually deeply incised channels) is an important issue in forestry since gullies are frequently connected to fish habitats. Coastal BC studies (e.g., Van Dine, 1985) often describe mountain channels that experience relatively frequent debris flows. However, many debris flows in logged areas occur in gullies that show limited signs of recent debris-flow activity (Millard et al., 1998). Often these gullies contain trees that are several centuries old, indicating long debris-flow return periods.

Most post-logging debris flows in Clayoquot Sound and other parts of coastal BC occur in gullies when a landslide enters the gully channel. Millard (1999) examined 286 debris avalanches in logged areas in coastal BC of which 75 (or 26%) developed into debris flows. Debris flows initiating by spontaneous entrainment of channel debris were rare, amounting to 2.6% of the total number of debris flows assessed. Similarly, data from the Queen Charlotte Islands indicates that 1–2% of debris flows in gullies may initiate in this manner. Two factors were strongly associated with debris-flow initiation: (1) large debris avalanches were much more likely to initiate debris flows than small debris avalanches when they reached the gully channel, and (2) the planimetric angle at which the debris avalanche entered the channel affected the likelihood of debris-flow initiation. Debris avalanches on gully headwalls showed a very high rate of debris-flow initiation (66%) because the median angle of entry was 0°. In contrast, debris avalanches on gully sidewalls had a median angle of entry of 74°, and only 16% of these transformed into debris flows. Steeper channels were more likely to initiate debris flows, and no debris flows initiated where channels gradients were less than 10°. All these factors are modeled by Bovis and Dagg (1992).

23.4 LANDSLIDE MANAGEMENT IN CLAYOQUOT SOUND

More than 80% of the old-growth forests in Clayoquot Sound are now located on slopes steeper than 30°. Second-growth forests are generally too young to harvest. Logging and road construction in the near future require careful management on those steep slopes still available for development.

23.4.1 Mapping, planning, and avoidance

The BC forest industry commonly uses a two-step process for landslide hazard assessment. Terrain and terrain stability mapping, utilizing a qualitative 5-class hazard ranking system generally at a scale of 1:20,000, identifies land that may experience landslides following forest harvesting or road building. Should a road

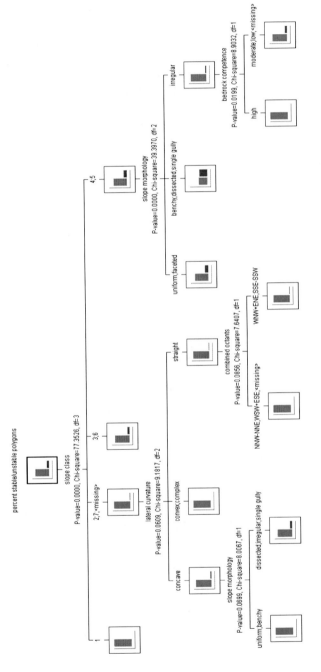

Figure 23.4. Decision-tree diagram for Clayaquot Sound landslides.

or cutblock be proposed in one of the higher hazard areas, a more detailed on-site geologic assessment provides conclusions and recommendations regarding road location and construction techniques, cutblock location and harvesting methods, and the potential effects of these activities on slope stability. Road builders regularly use full-benching and end-hauling or other techniques to minimize landslide potential. Protection forests or reserves and helicopter harvesting are frequently used to decrease landslide probability. Although helicopter logging has been viewed as producing less ground disturbance and therefore fewer landslides, Roberts (2001) shows that helicopter harvesting results in little, if any, reduction in the number of post-harvest landslides. Factors such as hydrologic change or root decay due to forest removal are likely more influential than the extent of ground disturbance. Perhaps the greatest benefit of helicopter logging is the avoidance of roads on steep slopes and the possibility that road related drainage may initiate landslides downslope.

23.4.2 Remote sensing

Landslides and other major sediment sources may now be classified from high-resolution satellite imagery such as the 1-m resolution IKONOS and the 0.61-m resolution QUICKBIRD satellites (Collins et al., 2001; Kliparchuk and Collins, 2003). Types of disturbance, landslide initiation locations, extent of disturbance and runout zones, connectivity to streams and degree of vegetation are identifiable by high-resolution satellite imagery. The high resolution also facilitates estimation of the amount of timber that may be available for salvage from the deposition zones of some landslides (Figure 23.5, see color section).

The following example from Clayoquot Sound shows a landslide associated with road-fill and a rockslide (Figure 23.6, see color section). The area of the landslides can be readily calculated by digitizing the feature boundaries. Near-infrared imagery is useful for differentiating the area occupied by bare landslide material rather than vegetation. The landslide on the left of Figure 23.6 was vegetated with grass seed and alder at one point, but the landslide failed retrogressively and much of the vegetative cover was lost. Isolated areas of vegetation are evident within the landslide track itself. In addition the trace of a deflection berm that was constructed to both control landslide direction and contain landslide material is evident even though it was breached and overwhelmed by subsequent events.

Landslide area can also be determined using the high-resolution satellite imagery. Varying types of landslides (Figure 23.6, see color section) were randomly chosen and digitized to calculate their areas. The resulting areas correspond closely to the areas that have been estimated during site reviews of these landslides and, in fact, notwithstanding requirements for radiometric corrections, may be more accurate because ground surveys of landslides are rarely made using precision equipment.

Image analysis techniques such as Principal Component Analysis, which in fact results in data compression or data merging, may also be used to highlight areas of incipient failure or actively eroding sites which may lead to failure initiation on occasion (Collins et al., 2001).

23.4.3 Road deactivation and landslide rehabilitation

Prior to the FPC, logging roads were largely constructed using cut and fill methods where material excavated from the upslope side of the road was placed on the downslope side to form the road prism and travel surface. Failure of these sidecast fills has resulted in many of the debris flows and debris avalanches seen in the Clayoquot Sound and on the west coast of BC.

Permanent deactivation of older cut and fill roads on moderate to steep terrain has been a major component of the BC Watershed Restoration Program (WRP) that began in the mid-1990s. This work entails professional geoscientists or engineers conducting detailed field assessments prior to developing comprehensive deactivation prescriptions and plans. Hydraulic excavators are used to decompact and cross-ditch old road grades, and to pull back potentially unstable fill slopes thus re-establishing original slope profiles and stream channels. During the 1996 storm event described in Section 23.1, no landslides initiated from roads that had been fully deactivated while approximately 68 landslides started at roads that had not been deactivated or partially deactivated. Many of the roads that were deactivated to the "standard of the day" prior to the WRP had to be reactivated and then properly deactivated to minimize future road related landslides and plantation losses.

Once the roads are deactivated, landslide rehabilitation usually follows. Some of the earliest landslide rehabilitation trials on the BC coast were set up in Clayoquot Sound providing knowledge and skills now in widespread use (Beese et al., 1994). Today, landslide rehabilitation typically involves hydroseeding or dry seeding with grasses and legumes to control surface erosion, followed by alder and/or conifer planting to re-establish forest cover. Occasionally, bioengineering techniques are used to provide better vegetation establishment on very steep or ravelling slopes.

Many of the techniques developed, practiced, and refined by the BC WRP in Clayoquot Sound and elsewhere are provided in a document entitled, *Best Management Practices Handbook: Hillslope Restoration in British Columbia* (Atkins et al., 2001).

23.5 SUMMARY

As with other areas in the Pacific Northwest, many logging-related debris flows and debris avalanches as well as frequent natural landslides exist in the steep coastal forests of Clayoquot Sound. Logging in Clayoquot Sound requires careful assessment and management to avoid landslide damage to its rich resources. In this chapter, case studies and data collected from Clayoquot Sound and adjacent areas were reviewed. Relationships were presented between terrain features and both natural and post-logging landslide occurrences, since identifying terrain susceptible to landslides following logging and road building is critical for ongoing forest management in the Clayoquot Sound area. Techniques now being used to identify landslide hazards, to monitor and minimize landslide occurrence, and to rehabilitate landslide scars were discussed.

23.6 ACKNOWLEDGEMENTS

Over the years a large number of people have worked on and contributed to an understanding of landslides and debris flows in the Clayoquot Sound area, sometimes under difficult and onerous field conditions. We would like to acknowledge Frank Baumann, Gordon Butt, Anthony Collette, Russell English, Pierre Friele, Don MacKinnon, Bruce Maxwell, Matthias Jakob, Denny Maynard, Audrey Pearson, Justin Straker, Gray Switzer, Sid Tsang, and Mel Zwiernk among others.

23.7 REFERENCES

Atkins, R.J., Leslie, M.R., Polster, D.F., Wise, M.P., and Wong, R.H. (2001) *Best Management Practices Handbook: Hillslope Restoration in British Columbia*. Ministry of Forests, Watershed Restoration Program, Resource Tenures and Engineering Branch, Victoria, British Columbia. Available at *http://www.for.gov.bc.ca/RTE/engineering/bmp-handbk-final-mar-2002.pdf*

Brardinoni, F., Slaymaker, O., and Hassan, M.A. (2003) Landslide inventory in a rugged forested watershed: A comparison between air-photo and field survey data. *Geomorphology*, **54**, 179–196.

Beese, W.J., Rollerson, T.P., and Green, R.N. (1994) *Forest Site Rehabilitation for Coastal British Columbia* (Watershed Restoration Technical Circular No. 4). Ministry of Environment, Lands & Parks, and Ministry of Forests, Province of British Columbia, Vancouver.

Bovis, M.J. and Dagg, B. (1992) Debris flow triggering by impulsive loading: Mechanical modeling and case studies. *Canadian Geotechnical Journal*, **29**, 345–352.

Chatwin, S.C., Howes, D.E., Schwab, J.W., and Swanston, D.N. (1994) *A Guide for Management of Landslide-prone Terrain in the Pacific Northwest* (Land Management Handbook Number 18, 2nd edn). Research Branch, Ministry of Forests, Province of British Columbia, Vancouver.

Church, M. and Miles, M.J. (1987) Meteorological antecedents to debris flow in southwestern British Columbia: Some case studies. In: J.E. Costa and G.F. Wieczorck (eds), *Debris Flows/Avalanches: Process, Recognition, and Mitigation* (GSA Reviews in Engineering Geology No. 7, pp. 63–79). Geological Society of America, Boulder, CO.

Collins, D., Kliparchuk, K., Connor, M., and Warttig, W. (2001) *Preliminary Assessment of the Application of IKONOS Satellite Imagery and Its Fusion with RADARSAT-1 Data for Forest Management* (Forest Research Technical Report TR-014). Vancouver Forest Region, British Columbia Forest Service, Nanaimo, British Columbia.

Green, R.N. and Klinka, K. (1994) *A Field Guide for Site Identification and Interpretation for the Vancouver Forest Region* (Land Management Handbook 28). Research Branch, British Columbia Ministry of Forests, Victoria.

Hoadley, J.W. (1953) *Geology and Mineral Deposits of the Zeballos–Nimpkish Area, Vancouver Island, British Columbia* (Memoire 272). Geological Survey of Canada, Vancouver.

Holland, S.S. (1964). *Landforms of British Columbia: A Physiographic Outline* (Bulletin 48). British Columbia Mines and Petroleum Resources, Victoria.

Howes, D.E. (1981) *Terrain Inventory and Geological Hazards: Northern Vancouver Island* (APD Bulletin 5). British Columbia Ministry of Environment, Victoria.

Howes, D.E. and Kenk, E. (1988) *Terrain Classification System for British Columbia* (revised edn). Ministry of Environment, Ministry of Crown Lands, Victoria, British Columbia.

Jakob, M. (2000). The impacts of logging on landslide activity at Clayoquot Sound, British Columbia. *Catena*, **38**, 279–300.

Kliparchuk, K. and Collins, D. (2003) *Using QuickBird Sub-metre Satellite Imagery for Implementation Monitoring and Effectiveness Evaluation in Forestry* (Technical Report TR-026, 24 pp.). Research Section, Vancouver Forest Region, British Columbia Ministry of Forests, Nanaimo.

Magidson, J./SPSS Inc. (1993) *SPSS for Windows CHAID Release 6.0*. SPSS Inc., Chicago.

Marquis, P. (2001) *How Practical Are Precipitation Shutdown Guidelines?* (Streamline Watershed Restoration Technical Bulletin Vol. 5, No. 4. pp. 13–16). British Columbia Ministry of Environment, Vancouver.

Millard, T. (1999) *Debris Flow Initiation in Coastal British Columbia Gullies* (Forest Research Technical Report TR-002). Vancouver Forest Region, British Columbia Forest Service, Nanaimo.

Millard, T., Wise, M.P., Rollerson, T., Chatwin, S.C., and Hogan, D. (1998) Gully system hazards, risks and forestry operations in coastal British Columbia. *8th International IAEG Congress*. A.A. Balkema, Rotterdam.

Millard, T., Rollerson, T.P., and Thomson, B. (2002) *Predicting Post-logging Landslide Rates: Cascade Mountains, Southwestern British Columbia* (Forest Research Technical Report TR-016). Research Section, Vancouver Forest Region, British Columbia Ministry of Forests, Nanaimo.

Pearson, A.F. (2000) Natural disturbance patterns in a coastal temperate rain forest watershed, Clayoquot Sound, British Columbia. Ph.D. thesis, University of Washington.

Roberts, W.B. (2001) A terrain attribute study of helicopter logging related landslides in the southwest Coast Mountains of British Columbia (142 pp.). M.Sc. Thesis, Simon Fraser University, Burnaby, British Columbia.

Robinson, E.G., Mills, K.A., Paul, J., Dent, L., and Skaugset, A. (1999) *Storm Impacts and Landslides of 1996: Final Report* (Forest Practices Monitoring Program, Forest Practices Technical Report No. 4). Oregon Department of Forestry, Salem, OR.

Roddick, J.A., Muller, J.E., and Okulitch, A.V. (1979) Fraser River (Sheet 92). In: R.J.W. Douglas (coordinator), *1:1,000,000 Geological Atlas*. Geological Survey of Canada, Vancouver.

Rollerson, T. (1992) *Relationships between Landscape Attributes and Landslide Densities after Logging: Skidegate Plateau, Queen Charlotte Islands* (Land Management Report No. 76). Ministry of Forests, Victoria, British Columbia.

Rollerson, T., Thomson, B., and Millard, T.H. (1997) Identification of coastal British Columbia terrain susceptible to debris flows. *1st International Symposium on Debris Flows, San Francisco, California, August*. US Geological Survey, Reston, VA, and American Society of Civil Engineers, New York.

Rollerson, T., Thomson, B., and Millard, T. (1998) Post-logging terrain stability in Clayoquot Sound and Barkley Sound. *Proceedings of the 12th Annual Vancouver Geotechnical*

Society Symposium on Site Characterization, Vancouver, British Columbia (13 pp.). Vancouver Geotechnical Society.

Rollerson, T., Millard, T., Jones, C., Trainor, K., and Thomson, B. (2001) *Predicting Post-logging Landslide Activity Using Terrain Attributes: Coast Mountains, British Columbia* (Technical Report 11, 20 pp.). Vancouver Forest Region, Ministry of Forests, Nanaimo, British Columbia.

Rollerson, T., Millard, T., and Thomson, B. (2002) *Using Terrain Attributes to Predict Post-logging Landslide Likelihood on Southwestern Vancouver Island* (Technical Report 11, 15 pp.). Vancouver Forest Region, Ministry of Forests, Nanaimo, British Columbia.

SPSFP (1995) *Sustainable Ecosystem Management in Clayoquot Sound* (Scientific Panel for Sustainable Forest Practices in Clayoquot Sound Report 5). Cortex Consultants, Victoria, British Columbia.

Straker, J. (1998) Soil development and forest productivity on naturally regenerating landslides on Vancouver Island (148 pp.). M.Sc. thesis, University of British Columbia, Vancouver.

VanDine, D.F. (1985) Debris flows and debris torrents in the Southern Canadian Cordillera. *Canadian Geotechnical Journal*, **22**(1), 44–68.

Yorath, C.J. and Nasmith, H.W. (1995) *The Geology of Southern Vancouver Island: A Field Guide*. Orca Books, Victoria, British Columbia.

24

Analysis and management of debris-flow risks at Sörenberg (Switzerland)

Markus N. Zimmermann

24.1 INTRODUCTION

In the Swiss Alps debris flows constitute a well-known threat to many towns and villages, national and international transportation corridors, as well as to infrastructure. Reports about damage caused by debris flows and about how people managed this threat date back several centuries (e.g., Zimmermann et al., 1997a). Surprisingly, in Switzerland systematic debris flows investigations only began following the major flood and debris-flow events in the summer of 1987. Since then numerous field investigations (e.g., Haeberli et al., 1990; Rickenmann and Zimmermann, 1993) as well as laboratory tests were performed (e.g., Rickenmann, 1990a; Tognacca and Bezzola, 1997). In the last few years concern has grown over the potential effects of climate change on natural hazards in general and debris flows in particular (e.g., Zimmermann and Haeberli, 1992). As a result a number of tools were developed to analyse debris-flow processes and to assess the resulting hazards and risks (e.g., Zimmermann et al., 1997b; Rickenmann, 1999). Such information was requested more and more often by the authorities.

The evaluation of the 1987 events provided evidence for a revision of the Swiss legal framework and the approaches to manage natural hazards, on the national and cantonal (state) levels. The new laws, guidelines, and recommendations demand a detailed investigation of the hazards and the establishment of hazard maps before any protection measures are discussed or implemented (*cf.* BWG, 2001). A series of flood and landslide disasters in 1999 provided the opportunity to evaluate the appropriateness of these new approaches. Among the villages and towns affected was Sörenberg, a small tourist resort in the Canton Lucerne (Figure 24.1).

This chapter describes Switzerland's new approach of natural hazard and risk management using the Sörenberg debris flows as an example.[1]

---

[1] Part of this chapter was presented at the First International Conference on Debris-Flow Disaster Mitigation Strategy, Taiwan (Zimmermann, 2002).

Figure 24.1. Tourist resort Sörenberg. The deep-seated landslide complex (approximately 400–500 m wide) is clearly visible. Inset map showing Sörenberg (dot) located at the border range of the Swiss Alps in Central Switzerland.

Photo: M.N. Zimmermann, 29 May, 1999.

24.2 THE SWISS STRATEGY FOR NATURAL DISASTER REDUCTION

24.2.1 Rationale

The 1987 flood and debris-flow disasters in Switzerland caused unparalleled damage to private property and infrastructure (approximately US$ 1 billion) with eight persons losing their lives. These events initiated a paradigm change: for over a century disaster reduction followed the principles of defence against natural hazards and a "maximum safety" approach was followed. Structural measures (check dams, dikes, sediment retention structures, river training works, etc.) were implemented to retain debris, to retard discharge, and to channellize water. In addition, many watersheds were reforested.

Despite these efforts large-scale damage from floods and debris flows could not be prevented as the 1987 events clearly showed. Following these events the Swiss federal authorities developed a new strategy and legal framework for the management of natural hazards (BWG, 2001). The laws regulating disaster reduction (mainly forest and water management) were completely revised. Today, an integrated approach (e.g., Williams et al., 1998) for the management of risks is applied which includes:

1. Identification and assessment of hazards: what type of processes occurs, what is at risk (vulnerability) and what is the probability of occurrence?
2. Definition of protection goals: which elements at risk have to be protected to what extent?
3. Planning of measures to reduce the main risks: what type of measures, structural and non-structural, have to be implemented and what are the priorities?
4. Emergency planning for the residual risks: what organization and what mechanisms are required to cope with disasters?

24.2.2 Hazard assessment

The new Swiss legislation requires the preparation of hazard maps. In the past few years methods were established to identify and assess the various hazardous processes according to a unified system (e.g., Petrascheck und Kienholz, 2003). For each hazard type criteria are defined to classify magnitude and frequency of the processes. Important bases for this step are event inventories, field investigations, and computer modelling.

The hazard maps are produced at a scale of 1:5,000 or 1:10,000 (municipal level). The degree of hazard is determined according to its magnitude (three classes) and its probability of occurrence (four levels: 30-y, 100-y, 300-y, extreme event). Three colors represent the degree of hazard (Figure 24.2, see color section).

The four degrees of hazards and their implications for land-use management are defined as follows:

Red   *High hazard*: heavy damage is expected. People are not safe within buildings. In general, the construction of new buildings is prohibited.

Blue   *Moderate hazard*: damage to buildings or infrastructure may occur. People normally are safe inside buildings. The construction of buildings is permitted under certain conditions only (e.g., flood or debris-flow proofing required).

Yellow   *Low hazard*: damage may occur. No restrictions for private landowners, but restrictions for vital development and infrastructure (e.g., hospitals, schools, telecommunication facilities).

Yellow/white: *Very low probability (>300 years recurrence interval) but high potential magnitude*. This is general information that needs to be taken into account (e.g., for preparedness plans).

24.2.3 Definition of protection goals

During the last decades disaster reduction was seen as either a response activity (involving emergency management) or mitigation activity (implementing protection structures). An integrated approach, however, requires a combination of preventative and response measures, and effective rehabilitation mechanisms. The process is based on well-defined protection goals, which depend on the value of the elements at risk and represent the acceptable level of risk on one side and the lack of safety on the other. The crucial question is: how much safety at what price? It is mainly a political process to define these protection goals. However, it is very difficult for decision makers to set such goals because most people are unfamiliar with probability theory and risk management.

The federal government created a general risk matrix to determine the protection goals for basic land-use classes (Figure 24.3).

24.2.4 Priority for mitigation measures

Federal authorities set clear priorities for mitigation measures: the prime disaster reduction approach includes non-structural measures such as land-use plans based on hazard maps to limit the further increase of damage potential and to reduce the vulnerability of existing structures (mainly through flood, debris flow, or avalanche proofing of buildings). In addition, areas can be delineated where sediment accumulation or flooding is desired to relieve downstream areas.

Structural measures are considered a secondary priority, which deviates from a century-old tradition when torrent control and river training prevailed. In the past, check dams in the source area and sediment retention basins on the fan for debris flows were implemented in numerous places to allow continuing development in the debris-flow hazard zones.

The emergency preparedness as well as the insurance system are considered a third priority. It is accepted that some damage is inevitable. Emergency preparedness must provide the necessary activities to protect people's lives and to limit damage. A number of organizations on a municipal and cantonal level (e.g., fire brigade, ambulance, civil protection) provide disaster relief. Special rescue units of the Swiss armed forces can be called for support during major disasters.

24.3 THE 1999 NATURAL DISASTERS IN SWITZERLAND

24.3.1 Overall conditions

The year 1999 was very unusual with regard to natural disasters. Extreme snowstorms in January and February dropped more than 5 m of snow in many areas of the Alps. More than 1,200 avalanches were triggered causing damage of more than CHF 0.6 billion (US$ 0.45 billion) and 17 fatalities. Prolonged rainfall in May and intense snowmelt triggered a large number of landslides including debris flows (cf. Bollinger et al., 2000). In addition, wide-spread flooding occurred on the rivers and lake shores of the Swiss Plateau. Overall damage reached US$ 0.7 billion, of which the landslides accounted for about 10–15%.

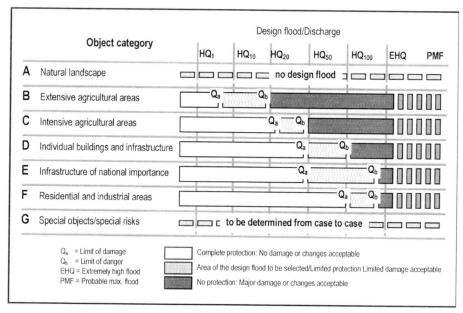

Figure 24.3. Differential safety concept, as developed following the 1987 flood disasters. According to safety objectives a variable design event can be applied for control works. This concept was developed for floods but is being applied for debris flows as well.

24.3.2 The 1999 landslides and debris flows: Causes and effects

At the end of winter 1999 the water equivalent of the snow cover was very high and the months of March, April, and May of 1999 were warmer compared to the average, leading to rapid snowmelt. Large amounts of water saturated the soil. A first series of landslides occurred immediately after the heavy snowfall in February, mainly in the Jura and around the Plateau.

In the eastern Swiss Plateau and the adjacent mountains (1,000 to 2,500 m a.s.l.) 150–200 mm of rain fell from 11–15 May followed by an additional 200 mm between 20–22 May. More than 350 landslides were triggered in central and eastern Switzerland. A clear concentration of events was observed in the area, which received more than 100 mm of rain between 11 and 15 May.

Two main types of landslides were observed during winter and spring 1999 (*cf.* Bollinger et al., 2000):

1. Debris flows and shallow debris avalanches during high-intensity rainfall periods. The transported volumes were of the order of several 1,000 m³.
2. Reactivation of deep-seated landslides, which occurred during snowmelt as well as during the intense rainfall periods. The velocity of the large soil/rock masses accelerated considerably from a few millimetres per month to several centimetres

per week. As a consequence, large-scale secondary collapses transformed into debris flows.

In many cases the landslide areas had previously been delineated, however, the development into large debris flows was not expected at all locations (e.g., at Braunwald, Canton Glarus, cf. Frank and Zimmermann, 2000). The following case study serves as an example to illustrate how one of these major 1999 debris flows developed and how it was managed according to the new Swiss hazard management system.

24.3.3 The Sörenberg case

Sörenberg is a small 40-year-old tourist resort in central Switzerland (cf. Figure 24.1 and Figure 24.4). Before, Sörenberg was a small spa with few buildings situated around the church in a low-hazard zone. Today, about 800 housing units, mainly chalets and tourist apartment buildings are located on old debris-flow deposits. In addition, a number of sports facilities including a public swimming pool were constructed in recent years.

The slopes above the village belong to a large and deep-seated landslide complex. In the 1860s vertical movements of several metres were observed in the area. The unstable flysch series are composed of relatively hard sandstones with interbedded schist layers. The volume of the landslide complex is about 15 million m^3, and has a maximum depth of approximately 100 m. Its surface is completely fractured and, therefore, highly permeable.

On 14 May 1999 the landslide complex showed increased activity. About 250,000 m^3 of rock with a very high content of fine sediments detached from the main mass, transformed into a debris flow and descended some 500 m. The flow stopped 700 m upslope of the first houses of Sörenberg. The fresh deposits were subsequently reactivated as debris flows. A flow of about 30,000 m^3 and many small flows of 500–5,000 m^3 occurred in the days, weeks, and months following 14 May. Initially, the flows resembled very rapid earth flows moving at velocities of several metres per minute. In summer and autumn, during higher rainfall intensities, the debris flows moved at a velocity of 1–2 m/s and followed a small creek into a 150 m wide space between two village districts in an area designated for development.

Surprisingly, no avulsion occurred on the debris-flow fan and the damage in the densely populated village districts was limited to mud accumulations in a few basements and gardens.

24.4 SHORT AND LONG-TERM DEBRIS-FLOW RISK MANAGEMENT

24.4.1 Immediate response

Completely surprised by the first debris flows, the municipal authorities quickly established emergency measures to protect the population, to prepare for further

Figure 24.4. Aerial photograph of the landslide complex (July 1999). The deep-seated landslide (dashed line) has a volume of 15 million m^3. The dotted line marks the extent of the collapse and the subsequent flow (about 250,000 m^3). The black line indicates the flow path of the various debris flows which occurred in 1999.

flows and to manage the emerging crisis. Hours and days after the first debris flows occurred, members of the local emergency services (fire department and civil protection units) were evacuating people living in the vicinity of the creek and implemented minor protection works (sand bags, etc.). In the following weeks more measures were implemented including:

- Establishment of a crisis management committee (one forest engineer employed full-time) to coordinate the immediate response, coordinate between cantonal and municipal authorities and villagers, and coordinate with experts.

- Establishment of evacuation plans (approximately 100 houses). Special consideration had to be given to tourists who were not aware of the crisis.
- Visual monitoring of the debris-flow initiation zone: a number of volunteers were asked to observe the conditions in the headwaters (particularly when heavy rains were forecast). An observation post was installed, equipped with telecommunication equipment and a strong searchlight to illuminate the initiation zone during the night.
- Installation of wire sensors to detect debris flows at two locations.
- Regular geodetic survey of the large moving rock mass (every 4–8 weeks).
- Installation of a continuously recording rain gauge in the headwaters with telemetric data transmission to the crisis centre.
- Implementation of small-scale structural measures such as small dams and channel excavation.

Due to ongoing debris-flow activity from May to October 1999 the above mentioned measures remained in effect throughout the year.

24.4.2 Debris-flow risks assessment

Detailed analysis of the prevailing hazards (i.e., the natural processes) is a first step towards an integrated management of the corresponding risks. The management of the 1999 debris flows, which threatened the village of Sörenberg, included a series of investigations (executed between 1999 and 2002), which are described below:

Process cascade

Mass movements at Sörenberg include creep, falls, slides, and flows that interact and therefore result in composite movements (Figure 24.4). Three main processes can be described:

1. *Deep-seated landslide (rock creep)*. The slow movement of the mass is considered to be the engine for the other processes. Its velocity is approximately 4–10 cm per year (Figure 24.5) and its volume is about 15 million m^3. A shear zone was found through seismic investigations and two boreholes at a depth of 50 m at the edge and about 100 m in the centre of the mass. Inclinometer measurements demonstrate that the movement occurs in this relatively narrow shear zone.
2. *Collapses at the edge of the slowly moving mass*. They constitute the release of tension, which is built up through the main movements. In the past the slides and occasionally the falls had volumes of several 10,000 to about 1,000,000 m^3 (cf. Table 24.1). Collapses can occur on either side of the mass or along its front.
3. *Debris flows*. The collapses directly transform into earth flows (as was the case in 1910) or the freshly deposited material (near its origin) serves as source for subsequent debris-flow activity (as was the case in 1999).

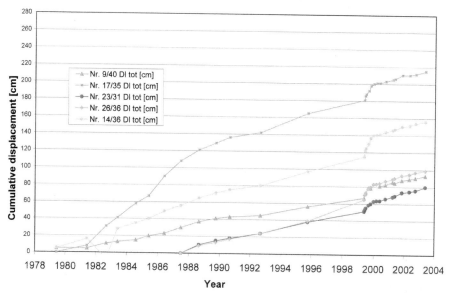

Figure 24.5. Displacement measurement on the Sörenberg landslide complex. Slow movements were recorded for the last 20 years (4–10 cm per year) in a number of locations. The 1999 reactivation caused movements of about 30 cm within 6 months for all measuring points. Today, the velocities are of the same order of magnitude as before. The whole complex is moving at approximately the same rate.

Source: Geotest AG.

Past events

The information about past events (see Table 24.1) is a key element for the assessment of natural hazards. A number of historic documents, such as maps (Figure 24.6), postcards, and a detailed description of the 1910 events by Albert Heim (Heim, 1932), provided an opportunity to reconstruct the sequence of events in the 20th century (Holliger, 2002).

The sequence of events indicates a very active phase at the beginning of the 20th century and a period of inactivity from 1922 to 1999 (if the minor collapse of 1986 is disregarded). The 1910 events have to be regarded as a direct consequence of the large-scale vertical movements of the whole mass at the end of the century. The events of 1910 were about one order of magnitude larger than the 1999 events.

Triggering events

The hydro-meteorological conditions were investigated using the Flühli rain gauge station located about 7 km north of Sörenberg. The Swiss Meteorological Service has operated this station since 1901. The station on the landslide complex was installed in September 1999.

Annual precipitation at Flühli Station is about 1,640 mm (average), with a maximum of 2,212 mm (1999) and a minimum of 1,050 mm (1948). Four of the six

Table 24.1. Past events in the Sörenberg landslide complex.

Year	Primary process	Secondary proces	Volume (very rough estimate)
1860 to 1910	Large vertical movement of the whole complex (>30 m)	No major secondary processes reported	
1902	Soil slips in the western part of the sagging complex	Debris flows, liquid, mobile	Debris flows: 20,000 m^3
1910	Large vertical movement of whole complex (for several days) with 2 major lateral collapses	Individual collapses transformed into major earth flows and slow debris flows; highly viscous, slow movement	Collapses and flows: 3,000,000 m^3
1912	Several lateral collapses	Individual collapses developed into earth flows and debris flows, slow movements	Flows: 250,000 m^3
1922	Several lateral collapses	Individual collapses transformed into earth flows (0.1–1 m/min) and debris flows	Flows: 400,000 m^3
1986	One frontal collapse	Soil movement, falls. Did not reach inhabited areas	Collapse: 25,000 m^3
1999	One lateral collapse	Collapse transformed into an earth flow. Debris flows developed from the mass. First flows slow (0.1–0.5 m/s), later in the year fast movements (1–3 m/s)	Collapse: 250,000 m^3 Debris flows: 50,000 m^3

events occurred in late spring (1999, 1922, 1912, 1910). This suggests that snowmelt and rainfall together are responsible for triggering collapses and subsequent debris flows. Antecedent precipitation (9, 6, and 3 months prior to the event) was analysed and compared to conditions of years without events. In those years a hypothetical occurrence at the end of May was assumed. The results are given in Table 24.2 and Figure 24.7.

The precipitation amounts (snow and rain) recorded 3, 6, and 9 months prior to the respective event are clearly higher than the precipitation of the other years in the 100-year period. Particularly, the ranks for the 6 months period are among the highest 14 years (except for 1986). The daily precipitation on the other hand, does not show extraordinary values except the 120 mm in 1912. This value represents the

Figure 24.6. Topographic map, 1918. The open scars on either side of the mass and the freshly deposited material have been clearly delineated by the cartographer.

highest ever-measured daily rainfall in the past 100 years. However, it is not sure whether this value at Flühli is representative for the Sörenberg area as well.

Simulation of debris flows and scenarios

The preparation of a hazard map, according to the above-mentioned recommendations (Section 24.2.2), requires scenarios based on three probability classes. At Sörenberg, the scenarios were prepared based on the historic context and on field evidence. Volumes were estimated for the collapses as well as for the debris flows

Table 24.2. Total precipitation prior to the 6 events. The rank indicates the position within the 100-year observation period.

Year	Date	9 months prior	Rank	6 months prior	Rank	3 months prior	Rank	Daily rainfall
1999	14 May	1,701 mm	1	1,086 mm	3	586 mm	5	34
1986	9 February	1,277 mm	17	746 mm	40	506 mm	13	1
1922	25 June	1,434 mm	3	1,094 mm	2	599 mm	4	17
1912	13 June	1,310 mm	12	980 mm	6	650 mm	2	120
1910	9 May	1,317 mm	11	919 mm	14	414 mm	37	5
1902	8 August	1,209 mm	25	942 mm	10	545 mm	7	15

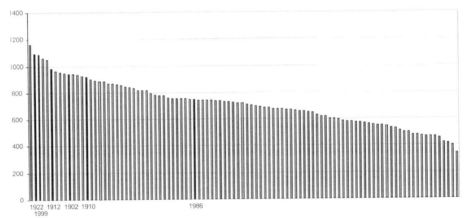

Figure 24.7. Cumulative precipitation amount (mm) for 6 months prior to the event or prior to the end of May. The event-years are indicated with black bars. Out of the 6 events 5 are found among the first 14 years. A threshold for increased movements can be found at about 900 mm/6 months.

Source: Holliger (2002).

arriving in the village. In addition, debris flows were modelled using a digital terrain model, a simple 2-parameter model for the travel distance (Rickenmann, 1990b), and a random walk approach (cf. Gamma 2000) for the analysis of the debris flow on the fan (Figure 24.8, see color section). According to this simulation (and strong evidence in the field) it remains unclear why all the 1999 debris flows followed the small and shallow channel. A diversion at the fan apex could have easily occurred thus affecting inhabited areas. The scenarios for the collapses at the edge of the moving mass and for the subsequent debris flows were established based on the following assumptions:

- Movements of the main rock mass are ongoing. The whole complex is presently moving at the same velocity as before 1999.

- Similar conditions as in 1910, and the years before, are no longer assumed. A fast and large-scale vertical movement seems unlikely to occur.
- Collapses at the edge of the moving mass of various volumes transforms into debris flows. Maximum volumes of about 300,000 m^3 are expected.
- Houses and roads in the village will affect the flows. An accurate forecast of areas affected based on topography alone is not possible.
- Large flows (>100,000 m^3) are relatively slow, small flows can attain velocities of more than 1 m/s.

The scenarios were transferred into so-called intensity maps. For each process (flooding, debris-flow activity, snow avalanches, etc.) the federal recommendations (BWW, BUWAL, RPL, 1997) provide definite criteria to determine the intensity class. The criteria for flooding, sediment accumulation, and debris flows are as follows:

Intensity class	Flooding	Debris flows
High	$h > 2$ m or $h \times v > 2$ m^2/s	$h > 1$ m and $v > 1$ m/s
Moderate	$0.5 > h > 2$ m or $0.5 > h \times v > 2$ m^2/s	$h < 1$ m or $v < 1$ m/s
Low	$h < 0.5$ or $h \times v < 0.5$ m^2/s	No low-intensity class for debris flows

h = flow depth or water depth; v = velocity of the flow.

The hazard map shows the degree of hazards, as outlined in Section 24.2.2, using the magnitude/frequency matrix. At each location the particular intensity/magnitude of the respective probability is determined. For most of the village the degree of hazard is high due to the strong intensity/magnitude of the expected debris flows (Figure 24.9, see color section).

24.4.3 Long-term safety concept for Sörenberg

Protection goals

According to discussions and consultations with state authorities the following protection goals were defined for the residential areas at Sörenberg and are currently being discussed with the local authorities and the population of Sörenberg:

- Casualties are not accepted for any type of event.
- 30-year return period: no damage acceptable.
- 100-year return period: limited damage acceptable.
- 300-year return period: limited protection for assets provided.

Land-use planning

Land-use planning guarantees sustainable development and use of limited space for residential purposes and infrastructure investments. The principal objective is to stop

Table 24.3. Adjustment of the legal framework for future land use in Sörenberg (land-use plan, building code).

Degree of hazard	New buildings	Existing buildings
High (red)	Construction of new buildings prohibited	Normal maintenance of building permitted. Alterations only when number of persons in building is not increased
Moderate (blue)	No new residential areas to be delineated. Existing housing lots: permitted with conditions (local protection of buildings)	Alteration only with increased safety measures (local protection): e.g., elevated entrance, no bedrooms in the basement
Low (yellow)	Local protection recommended; required for essential buildings	No restrictions

the increase of the number and value of elements at risk and to reduce the vulnerability of existing development. This implies adjusting the municipal land-use plan and respective building codes.

A set of regulations in the local land-use plan and building code defines what is permitted in the red, blue, and yellow hazard zones (Table 24.3). The same regulations will be applied to the existing hazard map and an updated map delineating the hazard after the implementation of structural measures. Figure 24.11 (see color section) gives an indication of what this map could look like in the case of all the structural debris-flow mitigation measures being implemented.

In addition to the regulations in the village it is planned to ban cattle grazing in the whole watershed and to increase forest cover.

Monitoring and alert networks

To provide maximum protection for people's lives the existing monitoring and warning network needs to be upgraded and installed permanently. Long-term monitoring should alert the responsible institutions allowing the preparation for short-term activities. The warning devices installed are listed in Table 24.4.

Structural measures

The possibilities for structural measures are limited at Sörenberg. Areas to allow debris-flow deposition are almost completely used for residential purposes (except for one large lot where the 1999 flows accumulated). In addition, the magnitude of possible events is large, even though their probability is low. Therefore, full control of future debris flows by structural measures is considered unlikely.

At present, a system of protection works (Figure 24.10) is planned and discussed that may provide the required safety, together with the non-structural measures. The concept takes into account natural characteristics of debris flows as much as possible. The protection works include the elements in Table 24.5.

Table 24.4. Long-term monitoring and short-term observations.

Activity	Type of measurements/device	Description
Long-term monitoring	Displacement measurements	Regular displacement measurements (every 3–6 months) are performed to obtain information about the velocity of the deep-seated landslide complex
	Rainfall records	Observation of precipitation amounts according to thresholds for 3, 6, and 9-month periods
Observations	Instant rainfall measurements	Intense daily or even hourly rainfall may trigger debris flows. The relevant threshold values are uncertain at the moment
	Wire sensors	Cables across the main debris flow path will alarm the emergency services in case a debris flow is passing
	Visual observation	The authorities constructed a small observation post near the source area where the debris flow starting zone can be observed visually. This post is equipped with a strong light and radio facilities

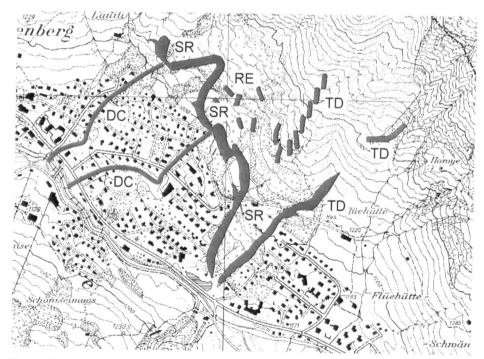

Figure 24.10. Structural measures in the vicinity of Sörenberg Village. TD = training dike; RE = roughness element; SR = sediment retention structure; DC = discharge corridor.

Table 24.5. Structural measures to control debris-flow hazards in the Sörenberg area.

Type of measures	Effect	Reason
Roughness elements (piles of large boulders)	To spread debris flows far upstream of the retention structures and encourage deposition	The retention capacity is limited. In order to increase the natural accumulation capacity of the debris-flow fan a number of roughness elements are constructed in small ravines and channels
Training dikes	To direct the flows into the sediment retention basins	The fan area has a convex surface. Flows may travel in almost all directions
Sediment retention basins	To retain a certain volume immediately upslope of the residential area	To control small events and to provide time for evacuation in case of a major event
Discharge corridor	To direct the overflow of the retention basins through the residential area	The capacity of the retention basins is limited. Flows have to be directed through the residential area. For this purpose a small number of houses have to be relocated

The planned protection works require CHF 14 million (US$ 11 million). The federal and cantonal authorities will contribute 30–40% each, the municipality has to carry about 10–15% and the homeowners another 10%. The final distribution of costs is not yet decided. It will require major efforts to convince the homeowners to contribute to the mitigation effort.

Emergency preparedness

Structural protection works will not guarantee the full safety of the population. Therefore, a preparedness concept was established for Sörenberg. It is divided into an institutional framework, a monitoring concept, and an emergency management plan. Activities were implemented following the 1999 crisis and continuously developed during the past years. Today, the emergency preparedness includes aspects listed in Table 24.6.

24.5 UNRESOLVED ISSUES

Uncertainties. The in-depth investigation of the primary and subsequent landslide processes revealed much new data and information. However, some major uncertainties remain:

- Behaviour of the whole landslide mass: at present it is assumed that rapid vertical movement as it occurred in the end of the 19th century (several tens

of metres) is no longer possible, but the mass will continue to move. Particularly, in the light of predicted climate change the frequency of accelerated movements remains unclear. If major movements occur again resulting in large collapses and earth flows, the relocation of the entire village would have to be considered.
- Lateral collapses: these processes will continue to occur as long as the whole mass is moving. The size can be estimated from what happened in the past. The frequency is dependent on the primary movements.
- Debris flows: depending on soil moisture conditions and sediment composition the flows vary in viscosity. It still remains unclear whether the flows will deposit above the village and avulse or will remain in the small creeks.

Awareness. Risk management aims to prevent fatalities and minimize property damage. In the case of major debris-flow events, lives need to be saved by preparedness measures (warning, alarm, evacuation). The high level of emergency preparedness is already being reduced after only 3 years of inactivity. It will be further reduced after the structural protection measures are in place. However, a certain degree of vigilance must be maintained, but may unfortunately be replaced by complacency over time.

Land use. The enforcement of the land-use regulations in the village depends on the activity of the natural processes. An occasional debris flow or continuous mass

Table 24.6. Emergency preparedness for the debris flow threat at Sörenberg.

Aspect	Type	Description
Organizational structure	Crisis management unit	Maintaining the crisis management team (mayor, one member of the municipal council, members of the emergency services, members of civil protection)
Monitoring and observation		Displacement measurements Precipitation (volume and intensity) Visual observation Wire sensors
Preparedness plan	Alarm	The emergency services are equipped with megaphones in order to warn the population and initiate the evacuation
	Evacuation scheme	A detailed evacuation scheme was established which is now used by the emergency services. Its role is particularly important because of large number of tourists unaware of the situation
	Preparation of resources	In some locations equipment and temporary protection materials (e.g., sand bags, wooden beams) are stored

movements can be used to convince local residents of the need for strict implementation of the rules and regulations. If nothing happens upslope of the village, the pressure from the residents to construct in the red zone will definitely increase and the local authorities may rethink existing regulations.

Public view vs. expert opinion. People's views, at least some of the residents, contrasts the view of the scientists and the cantonal and federal authorities. A number of information campaigns were directed at residents about the hazard and risk analysis and about the planning progress. To relay the message about hazards and risks is particularly difficult if natural processes occur at low frequency and the processes do not behave as expected. So far, the behaviour of the 1999 processes failed to fully convince residents to readily adopting scientifically based recommendations.

24.6 CONCLUSIONS

Debris flows constitute a major risk for the village of Sörenberg. Large debris flows occurred at the beginning of the 20th century, at that time with little downstream consequences. With the economic development of the area (after a relatively long period of geomorphic inactivity) the landslide reactivation and the resulting debris flows transformed a high-hazard process into a high-risk situation. A similar development can be observed in many locations in the Swiss Alps.

The management of these risks follows the new strategy of the federal authorities. Analysis of the hazards revealed that the landslide and debris flows cannot be fully controlled by structural measures. Only a holistic approach can substantially increase the safety of the population with limited funds. In Switzerland such approaches will be increasingly followed, particularly in view of existing uncertainties (e.g., climate change), limited financial resources, and still increasing use of the land in hazardous locations.

The case of Sörenberg shows that complex natural hazards require detailed investigations of the governing processes, of the possible developments, and of the resulting risks for the population and existing assets. The uncertainties of this process have to be communicated to the local authorities and to the population at large. In the future, the residents have to be included in a risk dialogue. This requires an open information policy. This process is very time-consuming and, therefore, more expensive than a pure "top-down" approach. However, for sustainable risk reduction in a particular area such work is urgently required and clearly will likely pay off in the long term.

24.7 ACKNOWLEDGEMENTS

The author is indebted to Gabi Hunziker (Geo7 AG), Markus Liniger (Geotest AG), and Karl Grunder (OekoB AG) for the excellent teamwork and the fruitful discussions during the Sörenberg contract. Local, cantonal, and federal authorities provided valuable support for this work. Ursina Holliger contributed to the project with her master thesis. Without this open collaboration the assessment of risks and the finding of solutions would not have been possible.

24.8 REFERENCES

Bollinger, D., Hegg, C., Keusen, H.R., and Lateltin, O. (2000) Ursachenanalyse der Hanginstabilitäten 1999. *Bull. Angew. Geol.*, **5**(1), 5–38 [in German].

BWG (2001) *Flood Control at Rivers and Streams: Guidelines*. Available at http://www.bwg.admin.ch/service/doku/d/bibvlist.htm Bundesamt für Wasser und Geologie, Biel, Switzerland.

BWW, BUWAL, BRP (1997) *Empfehlungen: Berücksichtigung der Hochwassergefahren bei raumwirksamen Tätigkeiten* [in German and French]. Available at http://www.bwg.admin.ch/themen/natur/d/empfhw.htm

Frank, S. and Zimmermann, M. (2000) Rutschung "Bätschern", Murgänge in der Wüechtenrus (Braunwald und Rüti): Eine Grossrutschung und ihre Folgeprozesse. *Bull. Angew. Geol.*, **1**(5), 131–138 [in German].

Gamma, P. (2000) *df-walk: Ein Murgang-Simulationsprogramm zur Gefahrenzonierung* (Geographica Bernensia, G66). University of Bern, Switzerland [in German].

Haeberli, W., Rickenmann, D., Rösli, U., and Zimmermann, M. (1990) *Investigation of 1987 Debris Flows in the Swiss Alps: General Concept and Geophysical Soundings* (IAHS Publication 194, pp. 303–310). International Association of Hydrological Sciences, Christchurch, New Zealand.

Heim, A. (1932) *Bergsturz und Menschenleben*. Verlag Fretz und Wasmuth, Zürich, Switzerland [in German].

Holliger, U. (2002) Geomorphologische und kulturgeographische Veränderungen im Raume Sörenberg im 20. Jahrhundert. Master thesis, Institute of Geography, University of Bern [in German].

Petrascheck, A. and Kienholz, H. (2003) Hazard assessment and mapping of mountain risks in Switzerland. In: D. Rickenmann and C-L. Chen (eds), *Debris-flow Hazards Mitigation: Mechanics, Prediction, and Assessment* (pp. 25–38). Millpress, Rotterdam.

Rickenmann, D. (1990a) *Bedload Transport Capacity of Slurry Flows at Steep Slopes* (Mitt. VAW ETH, No. 103). ETH, Zurich, Switzerland.

Rickenmann, D. (1990b) *Debris Flows 1987 in Switzerland: Modelling and Sediment Transport* (IAHS Publication No. 194, pp. 371–378). International Association of Hydrological Sciences, Christchurch, New Zealand.

Rickenmann, D. (1999) Empirical relationships for debris flows. *Natural Hazards*, **19**, 47–77.

Rickenmann, D. and Zimmermann, M. (1993) The 1987 debris flows in Switzerland: Documentation and analysis. *Geomorphology*, **8**, 175–189.

Tognacca, C. and Bezzola, G.R. (1997) Debris flow initiation by channel bed failure. In: C-L. Chen (ed.), *Debris-flow Hazards Mitigation: Mechanics, Prediction, and Assessment* (pp. 44–53). American Society of Civil Engineers, New York.

Williams, C.A., Jr, Smith, M.L., and Young, P.C. (1998) *Risk Management and Insurance*. McGraw-Hill, Boston.

Zimmermann, M. (2002) An integrated approach for the management of debris flow risks: A case study from Switzerland. *Proceedings 1st International Conference on Debris Flow Disaster Mitigation Strategy, Taiwan* (pp. 129–150).

Zimmermann, M. and Haeberli, W. (1992) Climatic change and debris flow activity in high-mountain areas: A case study in the Swiss Alps. *Catena*, **22**, 59–72.

Zimmermann, M., Mani, P., and Romang, H. (1997a) Magnitude-frequency aspects of Alpine debris flows. *Eclo. Geol. Helv.*, **90**, 415–420.

Zimmermann, M., Mani, P., Gamma, P., Gsteiger, P., Heiniger, O., and Hunziker, G. (1997b) *Murganggefahr und Klimaänderung: ein GIS-basierter Ansatz* (NFP31 Schlussbericht). Vdf Verlag, Zürich, Switzerland [in German].

25

Engineering for debris flows in New Zealand

Mauri J. McSaveney and Tim R.H. Davies

25.1 INTRODUCTION

The New Zealand debris-flow hazard environment is unusual on several counts. New Zealand's tectonic setting on a convergent plate boundary (Figure 25.1) leads to rapid uplift, and strongly fractured bedrock susceptible to physical weathering. Its mid-latitude position in the oceanic, Southern Hemisphere, athwart the "Roaring Forties" leads to frequent high-intensity rain, and rapid physical weathering. The uplift and weathering lead to steep topography and rapid sediment production. When sufficient sediment is available on steep slopes, heavy rain triggers debris avalanches and debris flows. They occur widely, but low population density and sparse infrastructure in the more susceptible areas allow relatively few opportunities for debris-flow damage. Structural engineering solutions to debris-flow problems are rare; some bridges are built for debris flows to pass beneath, one bridge is intended to withstand debris-flow impact, and there are several diversion walls; one is built to deflect debris flows away from a high-value building; another is intended to stop debris flows from crossing from one catchment to another where there are vulnerable high-value assets downstream. Intentional avoidance is rarely used. Unintentional avoidance is the more usual strategy for debris-flow hazard "mitigation"; the frequent heavy rain leads most mountain torrents to actively discourage close human neighbours and their infrastructure. There is a propensity in New Zealand to neglect debris flows as a hazard, and to be very surprised when they strike. Neither neglect nor surprise is warranted.

Here we describe six debris-flow studies. They mostly are examples from New Zealand's Southern Alps, reflecting a geographical bias in our work, and not in debris-flow distribution. There have been some major debris-flow disasters in North Island, but we have had no specific involvement in their study and can discuss them only briefly. The Walter Peak study (unpublished 2002 report by M.J. McSaveney) treats a common debris-flow problem arising on steep slopes

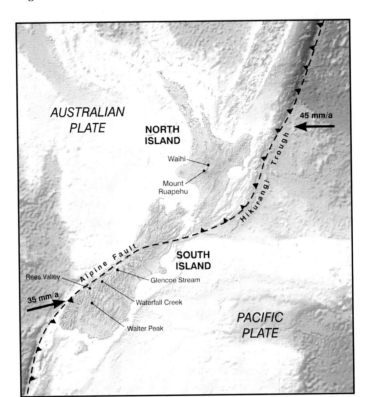

Figure 25.1. Location on the New Zealand continental land mass of the six debris-flow examples discussed in the text. The boundary between the Australian and Pacific tectonic plates through New Zealand is indicated. The oblique convergence between the plates in South Island gives rise to the Southern Alps. With steep slopes, readily eroded rock, and abundant orographic rain, debris flows occur frequently, but they are a generally ignored hazard because they seldom are encountered by people.

underlain by biotite schist in the southern and western Southern Alps. Gully erosion and slumping on the Walter Peak slope in one debris-flow episode raises the issue of conditional probability as debris flows now trigger at a surprisingly low threshold of rainfall intensity that is expected to rise as the scars stabilize. The Rees Valley study (McSaveney and Glassey, 2002) is typical of the New Zealand natural environment, where human involvement was accidental and we provided evidence to a Coroner's inquiry. It illustrates the weather conditions that give rise to debris flows, and the speed with which these can arise and catch people unaware. The Waterfall Creek study is mostly about engineering (McSaveney, 1995), where excavation to speed stream flow through a channel exacerbated an unrecognized debris-flow hazard, "solved" with a "bomb-proof" bridge yet to be tested. The Glencoe Stream study (McSaveney and Davies, 1998; Skermer et al., 2002) presents an unusual case of belated recognition of a debris-flow hazard without on-site historical precedent.

Last, two North Island examples of volcanic debris flows are discussed. The Waihi example (McSaveney and McSaveney, 1998) has killed people in two past events and could do so again. Lahars on Mount Ruapehu are an ever-present danger, causing the loss of 151 lives in 1953, and $NZ 17,000,000 in damage and loss of revenue at an electricity-generating station in 1995 (Taig, 2002). In 1996, the volcano reset itself for a repeat performance of the 1953 lahar, expected in 2005–2006. Preparation for this event brings desires for engineering intervention into conflict with cultural values.

25.2 DEBRIS FLOWS AT WALTER PEAK

The Walter Peak resort property, on the shores of Lake Wakatipu at the foot of Walter Peak (1,800 m), was affected by debris flows (Figure 25.2) during a storm in November 1999. The flows originated on the mid slopes of Walter Peak as many shallow landslides which merged and gathered mass from their channels. One cut a new channel across its fan head when it was diverted by deposition of an early pulse. Damage to buildings was minor, but consequent landscape changes affect

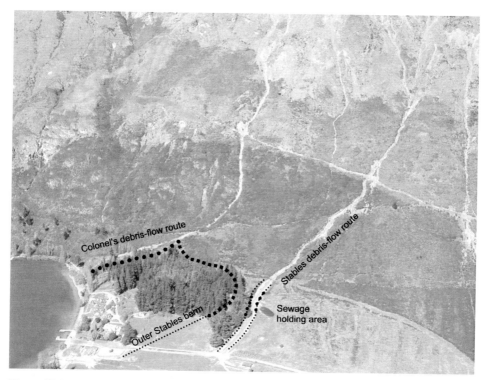

Figure 25.2. Aerial view of the Walter Peak debris flows of November 1999. The Colonel's debris flow on the left passes near to the Colonel's House. The Stables debris flow in the centre passed through the Walter Peak stables in November 1999.

638   Engineering for debris flows in New Zealand   [Ch. 25

the property's safe management. Subsequent mitigation works described below were *ad hoc*, somewhat emergency constructions by an earthmoving contractor in consultation with the resident property manager, without the benefit of professional advice. Unlike professionally designed work described in a later example, they perform as they were intended.

25.2.1 Setting

Walter Peak flanks a broad, formerly glaciated valley. Bedrock is weakly foliated biotite schist. Schistosity dips out of the slope above the lake; a poor alignment for slope stability. Most of the lake-facing slope is a major creeping landslide (Figure 25.3), indicated by "sagging" surface morphology and a dilated rock mass exposed in mid-slope bluffs. This active landslide is ancient, and not in itself considered to be a danger. It has an upper layer of easily eroded, chaotically arranged debris of angular boulders, sand, silt, and clay, a few tens of metres thick, overlying a thicker mass of displaced rock with open joints. Such deep-seated landslides occur widely on schist in South Island, and are source to many debris flows.

Rain and melting snow on the upper Walter Peak slopes feed streams that drain

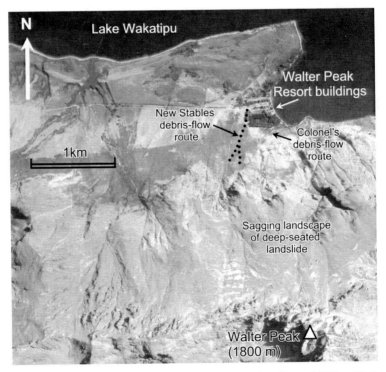

Figure 25.3. Vertical aerial photograph of the Lake Wakatipu face of Walter Peak showing the sagging surface morphology that indicated the presence of a large active landslide which predisposes the Walter Peak slopes to debris flows.

across the "sagging" landscape to gullies cutting through rocky bluffs at mid slope. Flooded mountain torrents tumbling through steep gullies undercut the chaotically arranged debris, and together with high pore-water pressures, initiate shallow landslides into the gullies. There they become debris flows, picking up more material and water along their paths.

The foot slopes are a series of fans consisting of debris from mid slope. The steeper foot slopes are a source of debris-flow material, but only through erosion by flows from above. Flooded streams rework the deposits, carrying sediment across gentler slopes to the lake.

The developed portion of the property is on the lake-shore extremity of a complex of coalesced fans (Figure 25.2). The steeper upper and middle fan portions are deposition zones of flows originating from many sources feeding three major debris-flow channels and one minor one. One channel takes debris flows to the lake edge. This channel near the "Colonel's House" is the "Colonel's debris-flow route", to distinguish it from the "Stables debris-flow route" that passed through (and destroyed) a stable in November 1999. The third major route does not affect the property. There is a minor route between the Colonel's and Stables routes (Figure 25.2).

25.2.2 History

A quick impression of the past behaviour of the fan complex was gathered by walking over it and examining aerial photographs taken in March 1959, and ground-based photographs of the Walter Peak homestead taken in the early 1900s. By 1900, exotic conifers planted *ca.* 1870 on a public reserve were already tall trees. They are on recent debris-flow deposits, with exposed, unweathered boulders. The deposits are from prehistoric debris flows, because there is no debris piled against trees, other than from 1999.

The Colonel's route is where it appears on 1959 aerial photographs. In November 1999, a part of this route temporarily avulsed towards the Colonel's House, but it stopped among trees on a lower fan gradient. Prior to 1999, the Colonel's-route stream had threatened to avulse towards the house. In 1999, some recently completed, small, raised, flood-protection berms helped mitigate damage to buildings. The Stables route was created in the 1999 storm. Before then, for more than a century, the stream took a more westerly route across the fan.

Most of the fans' lower slopes are too gentle for debris to move as a debris flow. There, continued movement of debris in 1999 was by traction bedload movement in floodwater draining passed the debris after primary deposition. This flood flow overturned and partially buried a tractor on the Stables route.

25.2.3 Meteorology and return period of the event

On 15 November, 1999, a stationary front brought heavy rain to southern South Island. Rain continued heavily from north-westly, and later southerly fronts until late on 17 November. Over four days, nearby Queenstown recorded 243.1 mm of rain, with 409 mm at the head of the lake. Lake Wakatipu exceeded its previous highest level (1878 AD) by 10 cm. Rivers in the region had two flood peaks, a day apart, corresponding to major pulses of rain. There was strong convection in the moist air mass, and thunder was widely reported. The Walter Peak debris flows were triggered when a convection cell, with intense rain passed over Walter Peak during the second rain pulse, when the slope was saturated and streams were high. A return period >150 years is likely for this storm, based on Queenstown rainfall and the record lake level. With no historical precedent in more than a century of site occupation, the damage certainly was caused by an exceptional event.

25.2.4 Remedial work undertaken

In cleaning up the property, the 1999 debris-flow deposits were fashioned into *ad hoc* raised berms to divert water and debris away from buildings. A raised berm, varying between 0.5 and 2 m high diverts flows to the lake, away from the Colonel's House. Its effectiveness was tested in September 2002 when a debris flow was contained within the channel (which it filled).

A stream established a new course across the property in 1999 along the Stables debris-flow route. Three raised berms will mitigate future damage (Figure 25.2). A wide, low berm cuts diagonally across the former stable paddock and blocks future floodwater and sediment from encroaching on buildings. Two other raised berms of variable height constrain the perennial stream to pass to the west of the public reserve.

25.2.5 Conditional probability

Rain triggered another debris-flow along the Colonel's debris-flow route around 20 September, 2002. This pulsed flow was smaller than those of 1999, and was barely contained within the channel. One of its pulses reached within metres of the lake shore. The storm was not exceptional; that it triggered a debris flow indicated that the source was still disturbed from 1999, and now is spawning flows at a lower threshold of rainfall intensity than existed until November 1999. The huge amount of thick regolith stored in the Walter Peak slopes assures that debris-flow generation is only "transport limited" (Bovis and Jakob, 1999). Only the passage of time and the growth of plants are available to stabilize the slope and reduce the probability of debris flows in storms at Walter Peak. This dependence of event probability on the history of prior events is called conditional probability. The probability may increase for a time after a major event as in this example, or it may decrease as is likely in the next example where the debris flows are "supply limited".

25.3 THE REES VALLEY DEBRIS FLOWS

On 3 January, 2002, two hikers on a trail had the misfortune to be crossing the channel of a headwater tributary of the Rees River, West Otago (Figures 25.1 and

25.4), when a 2-meter-deep debris flow approached them at high speed. One died; the other narrowly escaped. The event illustrates the speed of debris flows, and how quickly they can arise.

25.3.1 Background

The formerly glaciated, upper Rees Valley is floored by steep fans from tributary basins. The accident occurred on the toe of the largest fan. Large boulders and levees (mostly well vegetated) identify the fans as debris-flow fans (Figures 25.5 and 25.6). On the morning of 3 January, 2002, many debris flows left new levees beside more than half of the 40 or so upper Rees River tributaries. Upslope from the accident site, the mountainside is dissected by narrow gullies. Gravel thinly deposited over alpine tussock beside the stream channel, identified the source of the fatal flow (Figure 25.5). It was a peculiarly elongate basin, about 800 m long and less than 80 m wide, perched on the side of another basin (Figure 25.4).

The local bedrock is well foliated, pelitic, biotite schist. The local schist-derived regolith is of variable thickness (generally less than 5 m) and in many places has been stripped by the recent debris flows to expose bedrock. Deep-seated landslides are not evident in the upper Rees valley, but occur widely down valley.

25.3.2 Meteorology and return period of the event

The warm, moist, unstable, north-west airflow that covered much of New Zealand on 3 January 2002 was caused by a low of < 984 hPa south-west of South Island. A series of cold and occluded fronts were embedded in it. There was light rain in the valley on the morning of 2 January, but the afternoon was sunny. About dusk, rain started again, and lightly continued for most of the night (McSaveney and Glassey, 2002).

In a statement to the New Zealand Police, the survivor recollects "steady" rain through the morning of 3 January. Rainfall intensity increased shortly before the accident, because he noted that streams were up slightly, and discolored, whereas earlier ones crossed had been ankle-deep and clear. He recalled "the lack of heavy rain, more showery easing to drizzle with patches of hail hitting off and on." Although he did not describe thunder in his police statement, he later recollected that it was "rocking and rolling all that morning". Shortly before the accident, he "stopped to admire the awesome show being put on" [by the lightning]. A severe convective storm cell, with localized heavy rain reached the headwaters of Rees River shortly before the witness reached the stream crossing – about when he "stopped to admire the awesome show". The torrential part of this cell however did not cross his path.

Rainfall intensities in the valley were unrecorded. At a site 3.3 km north, 240 mm of rain fell between 8:00 a.m. 3 January and 8:00 a.m. 4 January. This is unexceptional, and likely to occur every 2–3 years on average (Tomlinson, 1980). River flows were not extreme either. McSaveney and Glassey (2002) reason that the duration of heavy rain was short, perhaps only 5–10 minutes. This rain moved in a narrow band

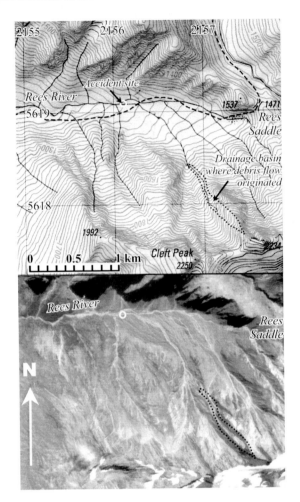

Figure 25.4. Topographic map and vertical aerial photograph of the upper Rees River Valley where a debris-flow fatality occurred in 2002. The dotted line on the map and photograph outlines the drainage basin where the fatal debris flow originated. The grid on the map is a New Zealand map grid at 1-km spacing. The contour interval is 20 m and north is uppermost. The aerial photograph (part of SN3982/18 of 12 March, 1966) is only partly rectified for distortion and topography.

(perhaps only 5 km wide) as a strongly convecting cell was swept along in the westerly air flow. The extent of landscape change in the form of many shallow landslides, debris flows, and massive deposition along channels, is not something very often seen in the upper Rees River Valley. From the sizes of trees and shrubs damaged by the flooding, a storm of this magnitude had not happened there for many decades. If the event was something likely to be seen on average only every 50 to 100 years, the likely 10-minute rainfalls were in the range 30–50 mm (Tomlinson, 1980).

Sec. 25.3]　　　　　　　　　　　　　　　　The Rees Valley debris flows　643

Figure 25.5. The upper fan of the Rees River tributary stream where the debris-flow fatality occurred. The fatal stream crossing was attempted 500 m downstream of this view.

Figure 25.6. Example of the large boulders moved hundreds of metres by the fatal debris flow on the upper fan, Rees River Valley. Here a group of boulders have accumulated as a stable cluster when each became buttressed against the powerful debris-flow current by its surrounding boulders. In the background are other large boulders from a past debris flow, now overgrown by dense sub-alpine plant cover.

25.3.3 The fatal debris flow

The witness travelled alone, but joined with the victim to cross the slightly swollen stream; they linked arms for the crossing. The witness had his parka hood down, and with unobstructed, directional hearing, heard a loud rumbling roar from upstream. He knew instinctively it was not thunder. Without looking to see what might be coming, he scrambled up a 2-m bank. As he climbed, his feet washed from under him. Looking over his shoulder, he saw the victim, parka hood up, first standing facing upstream, then turning to move. He recalls muddy, black water rising up his body as he scrambled up the bank. Looking back on reaching the top, he saw the victim's pack disappear. He never saw him again.

Between the time the two were considering the crossing, and his reaching safety, the water had gone from dirty brown to black indicating an increasing coarse-sediment concentration. The noise was a deafening grinding and banging. Afterwards, he had a "tide mark" of sand on his backpack and clothes. The flow all but ripped off his gaiters and jammed sand and pea-gravel into his boots. It felt like sloppy, wet concrete. Afterwards, the wet debris left on banks was like wet concrete, knee deep, and almost impassable. The matrix of the deposit left a few hundred metres upstream of the accident site contained 63% pebbles, 25% sand, 11% silt, and less than 1% clay (McSaveney and Glassey, 2002).

The coroner found the death to be a rare and unavoidable accident. While more people could be made aware of the possibility of debris flows, and of signs to look for in the landscape and the weather to warn of their likelihood, this has little prospect of saving lives. The event was too rare, too sudden, and too dangerous for knowledge to be particularly useful. The witness survived because of the speed of his instinctive reaction to a loud noise perceived as a danger.

25.4 WATERFALL CREEK DEBRIS FLOWS

The Waterfall Creek episode was more about an engineering misadventure than about debris flows. Waterfall Creek is a minor stream, bridged in a span of 10 m across a deep bedrock channel. Replacing an old, single-lane, wooden bridge with a double-lane, concrete one should have cost ~$NZ 250,000, but has cost >$NZ 1 million to date in a succession of debris-flow mishaps over 20 years. The information comes largely from a review of bridge "maintenance" file records (McSaveney, 1995).

25.4.1 Background

Waterfall Creek is a small perennial mountain torrent draining a 5-km^2 basin cut in weakly foliated biotite schist on the eastern shore of Lake Wanaka (Figure 25.7). It flows steeply for about 3.7 km from an elevation of 1,914 m, through an open upper basin and a lower narrow gorge, to the lake at 279 m. The road crosses the stream some 30 m above the lake, over a plunge pool between two waterfalls (Figure 25.8).

Sec. 25.4] Waterfall Creek debris flows 645

Figure 25.7. Aerial view of the drainage basin of Waterfall Creek. The debris flow of 1983 began mid-basin approximately where two large steep fans reach to the valley bottom from the right of photo. The debris flow of 1994 came from many sites, and most of the fresh landscape damage occurred in the 1994 storm. Note widespread evidence of deep-seated and shallow landsliding in the landscape. The bridge at adjacent Sheepskin Creek (left) was also destroyed in 1994. Lake Wanaka forms the foreground.
Photo: Lloyd Homer.

25.4.2 Bridge #1

Waterfall Creek was bridged with a single-lane structure of wood over steel beams for 50 years without incident. Above the road, the creek passed through two waterfalls to approach the bridge through a narrow slot, passing beneath the bridge through a cross section of $\sim 63\,\mathrm{m}^2$, that easily carried the design flood of

Figure 25.8. Aerial view of the Waterfall Creek bridge site taken in 1995 prior to construction of the "bomb-proof" steel and concrete culvert. Prior to 1982, the waterfall immediately above the bridge was hidden from view in a narrow slot now removed to make room for a wider bridge and to widen the floodway to smooth and speed the passage of floodwater (and debris flows!). The unvegetated area above the road to the left of the bridge is the scar from removal of the Waterfall Creek toppling failure in 1984.

$190 \, m^3/s$. The bridge was always clear of floodwater, but could be reached by waterfall spray. There is no record of debris-flow damage during the life of this bridge.

25.4.3 Bridge #2

Replacement of the 1931 bridge was proposed in 1969 to cope with increasing traffic. Recognizing that a wider bridge might be closer to the falls, designers proposed to widen the channel, and flatten the slope (Figure 25.9) to minimize spray.

Two bridge options were considered using precast, reinforced-concrete beams and decking. The first involved rock excavation to create space to build the bridge in two half widths. With one-half built beside the old bridge, temporary bridging could be avoided. Option two sited the replacement clear of the waterfall, requiring temporary bridging, new part-width approach spans and retaining walls. Reports indicate an overwhelming option-1 bias; a junior team member raised the only

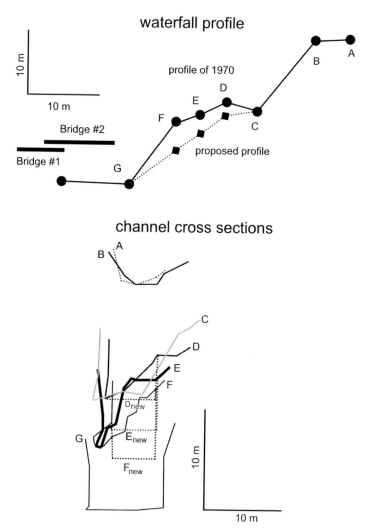

Figure 25.9. Pre-modification survey of the channel of Waterfall Creek, showing the changes as planned in 1970 (and now implemented). The letters on the profile correspond to measured cross sections. Cross sections with subscript "new" represent the equivalent sections after modification.

negative comment: "From an environmental point of view, tampering with the waterfall may not be desirable." Option 2, which did not modify the watercourse, was seen as altogether too complex, and too costly.

Provisional costing was $NZ 267,000 (option 1) vs. $NZ 306,000 (option 2). Option 1 was costed at $NZ9.00/m³ to excavate and dump spoil ($NZ 9.50/m³ for option 2). Option 1's structural costs were quoted at $NZ 430/m² ($NZ 540/m² for option 2). The differences were not explained and were not queried. If costs had been rated equally between the options, the estimates would have been equal, and so savings said to be the deciding factor in selecting options were illusory. Option 2 was built 13 years later for less than either estimate, but the channel geometry was changed by then.

Construction of option 1 started in 1980. Within months, machine operators stopped excavating. The contractor claimed that the rock was not competent to stand at the design batter (4:1 vertical:horizontal, ~76°). The contract supervisor (and designer) noted that "the contractor's heavy handed approach contributed directly to the over-break problem" and that the contractor had "neither the expertise nor resources to complete the works". The designers chose to use their own forces, and hired plant to finish the work, noting "continual minor slipping" after completion. Bridge #2 opened in late 1982.

Bridge #2 was destroyed by a debris flow during the night of 12–13 January, 1983, following rain with a return period of ~3 years. Collapse of about 20,000 m³ of undercut regolith into the stream in the lower part of the upper basin initiated the debris flow. The mode of collapse is unknown. Debris entered the flooded stream and continued as a debris flow, gathering further water and debris. At the crest of the waterfall above the bridge, the channel filled to a height of 6–7 m. A slurry of mostly sandy gravel, with some boulders up to 2 m in diameter poured from the channel to hit the bridge about mid-deck at an angle of 35–40° from the vertical (indicated by remnants of steel linkage bars left projecting from abutments, Figure 25.10). The linkage bars were torn rather than sheared, suggesting that the bridge folded in two lengthwise from the impact. Forces exerted were much greater than the as-built

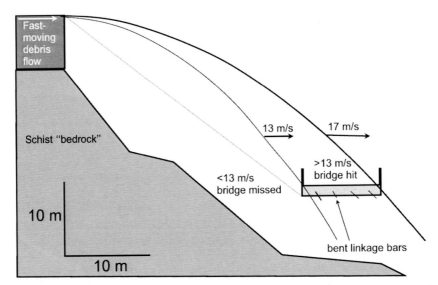

Figure 25.10. Simplified ballistic trajectory of debris flows from the crest of the waterfall to the bridge deck. Scale in metres. Prior to 1982, 50 years of flood flows did not reach the bridge and the channel was too narrow to pass large boulders.

strength. The bridge vanished. It was replaced with a single-lane, temporary, steel Bailey bridge until bridge #3 opened in April 1989.

The bridge reinstatement report states "With the benefit of hindsight, it seems that the single most significant factor that contributed to the loss of the 1982 bridge was shifting it 6 m closer to the waterfall, even though the waterfall channel was modified." With further hindsight and another loss, we now identify the key to be channel changes that launch debris flows to land on the bridge. Excavation also created space to allow half of the bridge to be built between the old bridge and the falls, inhibiting thought of siting the replacement clear of the waterfall.

25.4.4 The Waterfall Creek landslide

In February 1983, cracks in the base of the 76° batter 50 m north of the falls and a headscarp 0.5 m high 100 m above the road indicated an impending landslide of 28,000–40,000 m^3 of fractured rock. The toppling failure was removed in 1985 at a cost of $NZ437,000, reducing the batter to 45° (1:1) (Figure 25.8). Further excavation and rock bolting was required in 1996. Failure was due to the option 1 choice, but the fractured rock indicates that the site is on displaced rock; part of an ancient creeping landslide extending from ridge crest to lake bottom.

While access ramps were available, and plant was on site, the waterfall was further lowered and widened to allow material to pass faster through the waterfall!

25.4.5 Bridge #3

A hydrological report commissioned for the design of bridge #3 noted "considerable evidence for debris flows ... Debris flows are probably a relatively common occurrence and will continue to be so. Bridge design should take both water and sediment into account." These already were obvious to the designers. The report did not suggest how the latter might be done.

Further options were costed and approval given to proceed with the equivalent of option 2 (except for cuts already made), now estimated to cost $NZ225,500. In April 1989, bridge #3 opened to traffic. It was 3.6 m closer to the waterfall than had been the 1931 bridge.

Torrential rain of estimated $\sim$200-year return period in January 1994 caused many debris avalanches and debris flows in the area. Debris flows occurred in all streams bridged by State Highway 6 beside Lake Wanaka. Five bridges, including bridge #3 were buried or destroyed. Their destruction was unobserved.

The 1994 debris flow reached a depth of $\sim$7 m immediately above the roadside waterfall, essentially the same as in 1983, although the top of the waterfall was now steeper and wider. Debris, including boulders more than 2 m in diameter, poured from the top of the waterfall, scouring vegetation from a height of 10 m above one bridge approach.

A vegetation trimline left by the debris flow was superelevated at channel bends. In a section of channel with a slope of about 40°, and curvature of 100-m radius, superelevation of 20° indicates a speed of about 20 m/s. Speed can also be estimated from the trajectory needed for debris to hit the bridge. Neglecting air resistance, a horizontal velocity of <13 m/s is insufficient to hit the deck, and 17 m/s would reach across to the far side (Figure 25.10). An impact speed of ~90–100 km/hr is indicated when the 23 m vertical fall is taken into account. Bridge 3 disappeared into the lake leaving no trace of its mode of failure. It probably folded in two and entered the lake at about 50 m/s (180 km/hr). The creek was temporarily bridged with a single-lane Bailey bridge.

25.4.6 Bridge #4

Waterfall Creek is now bridged with a two-lane, single-span, steel culvert, with concrete replacing conventional earth fill around the steel for added strength. The rationale for the structure is *ad hoc*; it is the strongest structure that could be built in the 10-m gap and still allow for flood water. Will the strongest bridge that could be built in the space be strong enough to survive debris-flow impact? The present design is yet untested.

25.4.7 Was it necessary?

The Waterfall Creek saga is one of an innovative team of enterprising, but inexperienced engineers engrossed in solving day-to-day problems. No one thought to step back to appraise the effects of their work on the goal. The work was carried out to high standard, but some decisions were inappropriate, or at least inopportune. In particular, an adversarial stance between the design team and a contractor clouded judgment and led directly to costly, and ongoing slope instability that was avoidable. A major flaw in design development appears to have been the seemingly trivial and least reviewed decision to alter the waterfall to make it "behave". Had designs been altered to fit the waterfall, at least one of the losses might not have occurred, and some now lost options for mitigation would have been retained. Bridge 4 could have been built without major rock cutting. There was nothing to suggest that cutting might cause a problem, and once implemented, there was no going back.

25.5 GLENCOE STREAM AND AORAKI/MOUNT COOK VILLAGE

The Glencoe stream example from Aoraki/Mount Cook National Park is unusual in New Zealand, because debris-flow mitigation was undertaken without the historical precedent of a destructive debris flow. The hazard was recognized entirely on geomorphic evidence.

Hazards have long been a concern at Aoraki/Mount Cook village. The first hotel was destroyed by a glacier outburst flood in March, 1913. A replacement was built on high ground among the irregular topography of Glencoe fan (for the view and to be clear of flooding from the glacier). The village subsequently developed piecemeal

Figure 25.11. Aerial view of Aoraki/Mount Cook village on Black Birch and Glencoe Fans. Black Birch Fan (BB) forms the foreground, then Glencoe Fan (G) in the center, and the undeveloped Kitchener Fan (K) beyond. Debris flows are only of concern on Glencoe fan. Rockfall of February 1996 produced the dust in the background as it fell from Mount Thompson to Mueller Glacier. Site of the present Hermitage Hotel complex is marked (h). Its former site beside Mueller Glacier is marked (h').
Photo: Lloyd Homer.

about this site (Figure 25.11). Fire destroyed the second hotel in 1957, but it was replaced on the same site. Growth of the village within a National Park puts a premium on development space, and in 1995 an appraisal was sought of the suitability of all land in the village area for "urban" development.

25.5.1 The brief

The brief was to assess potential adverse natural events that might limit sustainability of the village, and asked to:

- identify and quantify natural hazards about the village;
- identify the appropriate standard of protection from natural hazards consistent with relevant New Zealand statutes, principally The Building Act 1991;
- assess the adequacy of existing works and management to meet the standards;
- identify appropriate measures to avoid or mitigate the hazards; and
- identify areas within and adjoining the village which were suitable for development.

The brief was approached by:

- identifying the types and likely extents of hazardous events, based on history and geomorphic evidence;
- identifying and quantifying their possible adverse effects; and
- identifying what could be done to avoid or mitigate these effects.

25.5.2 Historical background

No hazard appraisal of the whole village had hitherto been undertaken. Whitehouse and McSaveney (1990) discuss an earlier appraisal of the part of the village on Black Birch Fan that had been historically active as an alluvial fan, and had been brought into village development through substantial and enduring engineering work.

The village's brief history is too short to sample the range of structurally damaging events that the area can experience. Determination of this range was from evidence in the landscape and deposits underlying it. The principal landforms of the village are fans (Figure 25.11), deposited by side streams on a wide valley floor since glaciers receded more than 8,000 years ago. The larger streams are Black Birch and Glencoe Streams, and Kitchener Creek.

Extreme rainfalls are relatively common at the village. When 491 mm of rain fell in 24 hours in December 1957, the village was severely flooded. The flooding led to minor revetments protecting the hotel built in 1958, but most hazard mitigation went to ensuring adequate water for fire fighting.

Flooding in May, 1978, locally scoured a containment wall beside Black Birch Stream. After repair, a second wall was constructed inside the first. A storm in December 1979 dumped 537 mm of rain in 24 hours on the village. A gully of about $70,000 \text{ m}^3$ eroded 100 m above the fan head to form a large cone in the stream, and forced flood flow against the wall, almost overtopping it. Some $93,000 \text{ m}^3$ of gravel accumulated in Black Birch Stream, which aggraded 7 m at the fan head (Whitehouse, 1982). The containment wall started to scour and fan residents evacuated to safer ground. A minor creek flooded some buildings on Glencoe fan in this event.

On the afternoon and again in the evening of 8 January, 1994, residents left Black Birch fan when a 7,000 m^3 cone of gravel formed at the 1979 site. The stream aggraded 3–4 m at the fan head. In this and a later flood (20–21 January), sections of the containment wall were scoured, but the stream did not break out.

25.5.3 Black Birch fan

Black Birch Fan is a low-gradient alluvial fan. Within the last century, most of its surface has been active. Village growth by the late 1960s needed an upgrade in effluent disposal, and oxidation ponds were built out of sight on Black Birch Fan. This required Black Birch Stream to stay on the southern side of the fan. River-control work gave an illusion of stability, and quickly resulted in village encroachment onto the fan; first, water storage, then storage sheds and a store for hazardous chemicals, then roads and landscaped building sites. Now, most of the village residences are there.

The stream is constrained by massive earthworks to a course it adopted naturally in the 1960s. Where not modified for village uses, the fan has a channelled, boulder-strewn topography. It has a history of fan-head aggradation mentioned above, but no avulsion since 1969.

Gradients on the fan are too low for debris flows to initiate or propagate. Giant boulders above the fan head are believed to be from rockfall and wet-snow avalanching. Snow avalanches historically have reached the fan head.

25.5.4 Glencoe fan

Glencoe Fan is more complex because it partly overwhelms an 8,000-year-old moraine. The village centre is here, along with most capital assets: hotels and motels, park visitor centre, electricity substation, tavern, petrol station, store, and a number of other buildings. Parts of the fan have been altered for roads and building sites.

The fan topography is from episodes of gravel deposition in rare, large events, and long intervals of slower erosion when gravel was redistributed lower on the fan. Included within the Glencoe Fan complex are minor fans deposited by other tributaries: most notable is one small fan which arrived mostly on 26 December, 1957, from a storm-triggered debris avalanche.

The catchment above the Glencoe Fan is small (40 ha), but it is a potential source of much sediment. It contains a crush zone known as the Great Groove Fault, and expanses of weak, shattered rock stand in near-vertical slopes. Rapid runoff, scour, and high pore-water pressures cause crushed rock to fall into the stream. A steep stream gradient, large sediment source areas, and gravel exposed in terraces at the fan head indicate a danger of large debris flows. When the fan head aggrades, the stream can re-route to the western side of the fan, or down the middle; it has done this many times but not since the present hotel complex developed there from 1913.

Severe debris-flow aggradation could happen without warning within minutes, presenting a serious danger to buildings in the vicinity. Evidence on 1960 aerial photographs suggests that significant rapid aggradation occurred in 1957 but it did not re-route the stream. The vegetation stature and lack of surface weathering on boulders on higher terraces at the fan head, indicate that a large debris flow occurred little more than 100 years ago. The annual probability of such a disaster may be in the range 0.5–5%.

Concern for Glencoe Stream was with a large debris flow travelling as a series of pulses, which might build one on top of the other as each slowed to a halt at the fan head. The cumulative effect in a single storm could make safe protection difficult and uncertain. Glencoe Stream thus presents a serious debris-flow hazard, and it was judged that use of some sites on the fan were unsustainable in the long term (100 or so years) without further hazard mitigation (McSaveney and Davies, 1998). It was recognized that political reality precluded abandoning the more hazardous parts of the existing village. An inability to estimate the likely size of the $\sim$500-year return period (10% in 50 years, and 0.2% annual probability) debris flow in Glencoe stream prevented recommendation of hard engineering options. Conceiving of no rationale for compromising current New Zealand building codes, which expect buildings to have a 90% chance of surviving structurally damaging events during their design life (usually taken to be 50 years), McSaveney and Davies (1998) recommended that international consultants with greater expertise in engineering for debris flows be engaged. A large deflection wall is now in place (Skermer et al., 2002). Whether it can mitigate the cumulative effect of a 0.2% annual-probability debris flow remains to be tested.

25.5.5 Kitchener Fan

Kitchener Fan is much steeper, formed by stream aggradation, debris flows, rock falls, and sediment-rich snow avalanches. Boulders to several metres across are carried by snow avalanches, or fall and roll across snow to reach the upper fan. Its extreme hazards are obvious and discourage any development. It threatens the western edge of Glencoe fan, which is liable to impact by debris flows and snow avalanches (including air-blast from the latter). In the 1957 flood, Kitchener Creek trimmed the edge of Glencoe fan. Had a huge wet-snow avalanche in the winter 1995 been a long-runout, dry-snow one with an air blast, a part of the hotel complex could have been damaged. Artificial banks at the fan head currently keep Kitchener Creek on the northern fan edge, but it is an illusory constraint, not expected to persist in the long term. The fan-head banks are easily overwhelmed by wet-snow avalanches.

25.6 SOME NOTABLE NORTH ISLAND DEBRIS-FLOW EVENTS

New Zealand's North Island has a history of notable debris-flow events. One class of events is associated with infrequent tropical cyclones, where the many rainfall-triggered debris avalanches and debris flows add to, but are almost incidental in, the widespread damage from flooding and siltation. Attempted mitigation measures are only through land use (forestry). Another class are associated with the active volcanoes in North Island's central volcanic zone.

25.6.1 Debris flows at Waihi

In 1846, a Māori village was destroyed and more than 65 people killed by a debris flow originating from collapse of a landslide dam near the present settlement of Waihi on the southern shore of Lake Taupo (Figure 25.1). The landslide occurred in hydrothermally weakened rock along a fault scarp. In 1910, a highly mobile landslide from the same site killed another Māori while he was tending crops on the 1846 deposits. Tourist accommodation now is sited on the vulnerable lake shore and concern is mounting about the present and future stability of the steaming fault scarp as hydrothermal alteration continues. Ongoing monitoring of widening cracks for potential warning of collapse is the currently operating hazard mitigation while other measures are debated. Among the uncertainties is whether the next event will be a mobile landslide or a two-stage event with a landslide-dam and subsequent debris flow.

25.6.2 Lahars at Mount Ruapehu

On 24 December, 1953, a lahar (volcanic debris flow) was generated by a partial collapse of the outlet of a crater lake on Mount Ruapehu (Figures 25.1 and 25.12). It scoured piers supporting a railway bridge across Whangaehu River and killed 151 people when a passenger train toppled into the raging lahar. The collapse was the culmination of filling of the lake following an eruption in 1945. Eruptions in 1995 and 1996 have reset the scene for a repeat of the lahar when the present summit crater refills with rain and melted snow to some critical level, expected to occur during 2005, or possibly as late as 2006. Changes around the summit area could now allow a larger lahar than in 1953. The outcome of this predictable future event is to be mitigated mostly by two independent robust warning systems now in place, a radar-based river-level detection system for the rail bridge, and the East Ruapehu Lahar Alarm and Warning System (ERLAWS) in which multiple sets of geophones listen for the rumble of a lahar in the upper valley, giving enough time (~ 90 minutes) to close the road and shut down power-generating turbines in the runout zone (Taig, 2002). Deposition of sediment on the steeper portions of the ring plain surrounding Mount Ruapehu facilitate fast-moving lahars to cross a low-lying drainage divide and enter another drainage system. A "small" lahar when the former crater lake emptied during a 1995 eruption led to greatly accelerated abrasion of turbine blades at the Rangipo hydroelectric power station causing $NZ 7 million damage and $NZ 10 million in lost revenue. It also briefly affected an important trout-fishing river.

Options for mitigating the lahar hazard are limited by several important cultural considerations. The lahars can be viewed as an essential part of the spectrum of natural events within Tongariro National Park and they initiate on a Māori cultural

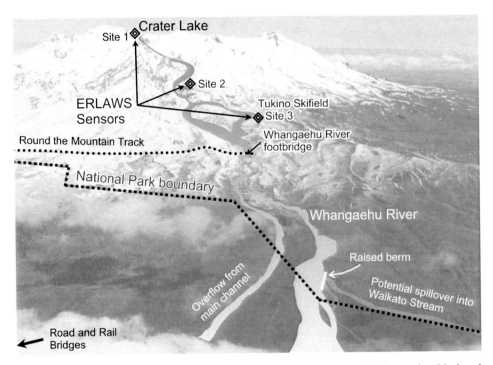

Figure 25.12. View to west of the active volcano, Mount Ruapehu in Tongariro National Park, showing the Whangaehu River Valley lahar route and location of warning sensors (ERLAWS) and raised channel berm to reduce the probability of future lahars entering Waikato Stream and affecting turbines of the Rangipo power station downstream. The location of the present road and rail bridges and site of the 1953 Tangiwai rail disaster are ~15 km off to the right of the photograph foreground.

icon (Mount Ruapehu). They do no damage within the park boundaries. Hence, although it technically may be feasible to eliminate or greatly reduce the anticipated lahar, or to manage its time of occurrence through engineering work at the future lake outlet, such work currently is unacceptable to some major stakeholders. A raised channel berm with a crest 7 m above the present Whangaehu River bed at the park boundary now greatly reduces but does not eliminate the probability of diversion into another catchment (containing the Rangipo and other power stations) in the future. Some people argue that if damage from such natural events can be foreseen and the danger to human life reduced by effective warning, then the events should be left to take their natural course, and affected engineering works (roads, rail, bridges, and power stations) should be designed or managed to cope with them, with appropriate engineering work near the affected site and not at the source. The multifaceted debate has polarized and politicized. It continues with little hint of resolution before the crater lake next overflows.

25.7 CONCLUDING REMARKS

The above examples, though unimpressive by international standards, emphasise some important aspects of debris flows and the hazards they pose.

First, even in a land as sparsely populated as New Zealand, debris flows pose a significant risk to life and property. At Waterfall Creek there was historical debris-flow evidence, but as in many other examples, the evidence was ignored in management decisions. We have found much ignorance of debris-flow hazards among planners and engineers, and this problem is escalating with accelerating development as people increasingly travel, work, play, and live in steeper country. We are aware of sites where dwellings have been permitted on what are plainly debris-flow deposits (some with historical activity, as at Waihi), but the severe hazard to life appears not to be acknowledged. Some people with a will to develop land seem unwilling to accept that some apparently prime sites are just too dangerous. The message is seldom heeded until there is a major disaster. It is a myth that such hazards are only ignored in the underdeveloped world.

Second, we find that the ability to recognize where debris flows have occurred in the past is the prime requirement in debris-flow hazard assessment and mitigation. Ability to analyse or model the events to predict the magnitude of future events comes a long way third, behind inference of the maximum size of past debris flows.

Debris flows are relatively frequent occurrences in steep-land areas of New Zealand, but they mostly occur in areas lacking people and infrastructure. As a consequence, they are generally neglected, and structural solutions to debris-flow problems are relatively rare. In most instances, the engineering solutions have been *ad hoc*, and the few that have been tested generally have been found wanting. Experienced engineering design for debris flows is new to New Zealand, and its products have yet to be tested.

25.8 ACKNOWLEDGEMENTS

The Waterfall Creek study was funded by the New Zealand Foundation for Research, Science and Technology contract CO 5413. The Rees valley study was funded through the GeoNet programme by the New Zealand Earthquake Commission.

25.9 REFERENCES

Bovis, M.J. and Jakob, M. (1999) The role of debris supply to determine debris-flow activity in southwestern B.C. *Earth Surface Processes and Landforms*, **24**, 1039–1054.

McSaveney, M.J. (1995) *Debris Flows and Bridge Losses at Waterfall Creek, SH6 at Lake Wanaka, New Zealand* (Science Report 95/21, 18 pp., 11 figs). Institute of Geological & Nuclear Sciences, Lower Hutt, New Zealand.

McSaveney, M.J. (2002) Assessment of debris-flow mitigation measures at Walter Peak, Queenstown (24 pp.). Unpublished Institute of Geological & Nuclear Sciences Client Report 2002/127 for Locations Management, Queenstown, New Zealand.

McSaveney, M.J. and Davies, T.R. (1998) A hazard assessment for Aoraki/Mount Cook Village. In: D.M. Johnston and P.A. Kingsbury (compilers), *Proceedings of the Natural Hazards Management Workshop, Christchurch 28–29 November* (Information Series 45, pp. 70–75). Institute of Geological & Nuclear Sciences, Lower Hutt, New Zealand.

McSaveney, M.J. and Glassey, P.J. (2002) *The Fatal Cleft Peak Debris Flow of 3 January 2002, Upper Rees Valley, West Otago* (Science Report 2002/03, 28 pp.). Institute of Geological & Nuclear Sciences, Lower Hutt, New Zealand. Available at *www.geonet.org.nz/landslide/report1.pdf*

McSaveney, E.R. and McSaveney, M.J. (1998) Beware of falling rocks: Landslides. In: G. Hicks and H. Campbell (eds), *Awesome Forces: The Natural Hazards that Threaten New Zealand* (pp. 72–97). Te Papa Press, Wellington.

Skermer, N.A., Rawlings, G.E., and Hurley, A.J. (2002) Debris flow defences at Aoraki Mount Cook village, New Zealand. *Quarterly Journal of Engineering Geology and Hydrogeology*, **35**, 19–24.

Taig, T. (2002) *Ruapehu Lahar Residual Risk Assessment* (Report to the Ministry of Civil Defence and Emergency Management, 77 pp.). TTAC Limited. Available at *www.mcdem.govt.nz/memwebsite.nsf/Files/RuapehuReportMainText/$file/RuapehuReport MainText.pdf*

Tomlinson, A.I. (1980) *The Frequency of High-intensity Rainfalls in New Zealand* (Technical publication 19, 36 pp.). National Water & Soil Conservation Organisation, Wellington.

Whitehouse, I.E. (1982) Erosion on Sebastopol, Mt Cook, New Zealand, in the last 85 years. *New Zealand Geographer*, **38**, 77–80.

Whitehouse, I.E. and McSaveney, M.J. (1990) Geomorphic appraisals for development on two steep, active alluvial fans, Mt Cook, New Zealand. In: A. Rachocki and M.A. Church (eds), *Alluvial Fans: A Field Approach* (pp. 369–384). John Wiley & Sons, Chichester, UK.

26

Multifaceted hazard assessment of Cheekye fan, a large debris-flow fan in south-western British Columbia

Pierre A. Friele and John J. Clague

26.1 INTRODUCTION

The Cheekye fan, situated in the southern Coast Mountains 50 km north of Vancouver (Figure 26.1), is one of the largest and most thoroughly studied alluvial fans in British Columbia. This chapter summarizes the substantial body of previous research on the Cheekye fan and illustrates the wide range of approaches that have been applied over the last 50 years to assess debris-flow hazards and risk. Many of the approaches applied to the Cheekye fan are mentioned elsewhere in this book, but their utility in a situation where a fan is under severe development pressure is particularly well illustrated here. No single approach provides answers to all questions that arise in debris-flow hazard assessment, and uncertainties inevitably remain, ultimately necessitating the use of expert judgment. However, combination of several analytical approaches reduces uncertainty and improves confidence in the results.

The Cheekye River basin (Figure 26.2) is located on the west flank of Mount Garibaldi, a Quaternary stratovolcano (Green et al., 1988; Green, 1990). Cool wet winters and warm dry summers characterize the climate of this part of the southern Coast Mountains. The area has been inhabited since the early Holocene by indigenous peoples, whose Salishan name "Chakai", for dirty water, has been applied to Cheekye River. Europeans settled the Squamish River valley in the late 1800s, and today the District Municipality of Squamish extends north from the head of Howe Sound, encompassing the Cheekye fan (Figure 26.2).

The possibility of a disaster on the Cheekye fan is increasing as development spreads across the lower part of the fan. Conditions that predispose this area to a disaster include: the steepness of the Cheekye River basin, with 2,470 m of relief; the presence in the upper part of the basin of the unstable, precipitous, western slope of

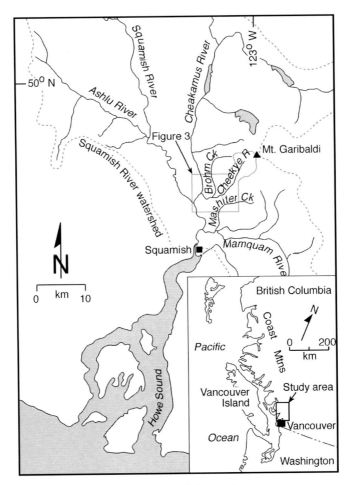

Figure 26.1. Map of the study area.

Mount Garibaldi; and heavy runoff during intense fall rains, rain-on-snow, and spring snowmelt events. This potent mix of landslide-generating phenomena demands a thorough understanding of the hazard before further development of the fan is approved. This compilation, representing a comprehensive analysis of fan geomorphology and debris-flow hazard, provides the scientific basis for risk management.

26.2 SETTING AND EARLY RESEARCH

The first scientific study of the Cheekye fan dates back to the late 1940s and 1950s. William H. Mathews, while a Ph.D. student at the University of California, conducted pioneering work on the volcanic rocks and glaciers in Garibaldi

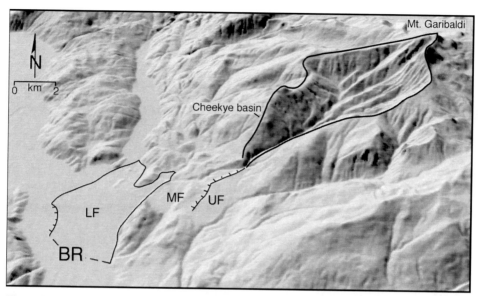

Figure 26.2. Digital elevation model of the Cheekye River basin showing the upper (UF), middle (MF), and lower (LF) Cheekye fans and the community of Brackendale (BR) (view north). Note the well-developed gully system underlying the basin headwall slopes.

Provincial Park and showed that Mount Garibaldi erupted repeatedly during the waning stages of the Fraser Glaciation (Mathews, 1952, 1958). Part of the volcano had formed on and against the late Pleistocene glacier in the adjacent Squamish valley. As the glacier retreated, the west flank of the volcano collapsed to form ice-contact terraces known as the "upper" and "middle" fans (Figure 26.2). The collapsed debris was subsequently incised, reworked and redeposited, with additional material from the Cheekye basin, to form the "lower Cheekye fan", the subject of this study.

The lower Cheekye fan (Figures 26.2 and 26.3) has a radius of about 3.5 km and an average slope of 3° near its apex at 190 m above sea level (a.s.l.), declining to 2° toward the margins. The fan has an area of about 8.3 km². It was formerly much more extensive than today, as its northern and western margins have been eroded by the Cheakamus and Squamish Rivers. Erosion of the toe of the fan indicates that sediment supply has diminished during late Holocene time, and that delivery of sediment to the fan has not kept pace with the transport capacity of the valley-bottom rivers. The head of the fan supports an incised channel, which is 5–20 m deep and extends down the fan to about 120 m a.s.l. There, the channel swings sharply to the north. No abandoned channels are evident on the fan surface.

A debris flow swept down Cheekye River to Squamish River in 1958, alerting the British Columbia Government to the potential landslide hazard in the area (Jones, 1959). In the 1970s, pressure mounted to extend development higher on the fan near the community of Brackendale. The pressure led to a debate as to whether large

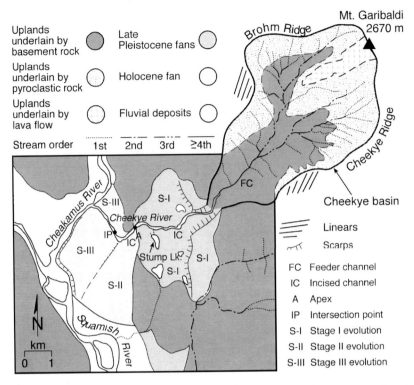

Figure 26.3. Map of the Cheekye basin and fan, showing landforms and other features referred to in the chapter.

debris flows could escape from the existing channel and spread across the fan surface (Crippen Engineering, 1975, 1981).

In 1981, local geologist Frank Baumann discovered fossil wood in an exposure at the eroded margin of the lower Cheekye fan along Squamish River. The wood was collected from a layer 13 m below the fan surface and yielded a radiocarbon age of about 5,900 ^{14}C yr BP (*ca.* 6,800 cal yr BP; Table 26.1), which demonstrated that significant aggradation has occurred on parts of the fan since that time (Eisbacher, 1983). This fact indicated that the fan surface could still be vulnerable to debris flows.

In 1990, excavations at the municipal landfill, in the centre of the fan, exposed numerous logs buried beneath 5 m of diamicton.[1] The diamicton forms a single surface debris flow covering an area of 84 ha (Baumann, 1991). The logs yielded radiocarbon ages between 800 and 1,300 ^{14}C yr BP (800–1,400 cal yr BP; Table 26.1). This discovery confirmed the continuing vulnerability of the fan to debris flows and resulted in a detailed hazard analysis (Thurber Engineering/Golder Associates 1993), the first of its kind on an alluvial fan in British Columbia. All subdivision develop-

---

[1] Diamicton is poorly sorted, massive to weakly stratified sediment ranging in size from clay to gravel.

ment permits were revoked and one large housing project was halted. Since then, no new subdivision approvals have been granted.

More recently, in response to development pressure, the British Columbia Ministry of Water, Lands and Air Protection suggested a mitigation option, consisting of a series of protective dykes on the lower fan. The design debris flow was designated as one with a likelihood of occurring, on average, once in 10,000 years, following a precedent-setting 1973 court ruling at Rubble Creek, 40 km to the north (Berger, 1973). After reviewing the most recent literature and running a 2-D debris-flow-modeling exercise, Kerr Wood Leidal (2003) presented a design for protective works to the Squamish Municipal Council. The economic feasibility of the protective works and the 1-in-10,000-year return design event requirement are currently being evaluated (Squamish Municipal Council Minutes, 2 April, 2003).

The status quo, however, is likely to persist, as the cost of implementation of protective works for the 1-in-10,000-year design event is high. In this type of land-use issue, clear decisions based on sound science and policy are required. The following analysis presents the science; sound policy must come from government.

26.3 BASIN ANALYSIS

The Cheekye River basin (Figure 26.3) has an area of 26 km^2 and an average slope of 25°. The basin headwall exposes pyroclastic flows dipping 10–12° away from Mount Garibaldi's summit and consisting of loose, permeable sediment ranging from silt to blocks many metres in diameter (Mathews, 1958). The headwall, which slopes 43°, truncates the pyroclastic flows. These slopes are actively eroding, deeply gullied (Figure 26.2), and are subject to rockfall and debris slides (Hungr and Rawlings, 1995). Sediment delivery to the head of Cheekye River is very high, and small debris flows are possible whenever rainfall thresholds are exceeded.[2]

A 1-km-long postglacial lava flow clings to the headwall slope directly below Mount Garibaldi (Figure 26.3). The contact between this flow and the underlying pyroclastic deposits has a slope of about 30° into the basin. No evidence has been found for large prehistoric landslides from this area, thus the flow may be strongly welded to the underlying pyroclastic material. If a landslide were to occur, however, it might involve 23–46 × 10^6 m^3 of debris (Hungr and Rawlings, 1995).

Linear cracks occur in pyroclastic rocks on Brohm and Cheekye ridges, adjacent to the Cheekye basin (Figure 26.3). A seismic reflection survey was conducted to investigate the "Cheekye linears" (Thurber Engineering/Golder Associates, 1993). It revealed 100–150 m of pyroclastic breccia overlying a 40-m-thick weathered zone at the surface of the basement rock. The contact between the pyroclastic and basement rocks dips about 15° toward the north and is truncated by a 35–40° scarp slope. The linears are thought to be the surface expression of translational sliding at the base of

---

[2] The Cheekeye River basin in an example of a transport-limited basin (Jakob and Bovis, 1996), where debris-flow initiation is conditional on moisture supply, in contrast to weather-limited basins where initiation depends on debris supply.

Table 26.1. Radiocarbon ages from lower Cheekye fan sediments.

Age (^{14}C yr BP)	Calibrated age (cal yr before AD 2000)[a]	Laboratory number[b]	Dated material	Latitude (°N), longitude (°W)	Comment	Reference
670 ± 50	600–700	GSC-4307	Charcoal	49°47.4', 123°08.3'	Debris-flow deposit along Cheekye River	McNeely and Jorgensen (1992)
810 ± 60[c]	700–800	GSC-6639	Log	49°47.3', 123°08.8'	Sanitary landfill excavation; debris-flow unit, 4–5 m below surface	Clague et al. (2003)
1,080 ± 60[c]	1,000–1,200	GSC-6638	Log	49°47.3', 123°08.8'	Sanitary landfill excavation; debris-flow unit, 4–5 m below surface	Clague et al. (2003)
1,120 ± 70[c]	1,000–1,200	GSC-6636	Log	49°47.3', 123°08.8'	Sanitary landfill excavation; debris-flow unit, 4–5 m below surface	Clague et al. (2003)
1,110 ± 70[d]	1,000–1,300	S	Log	49°47.1', 123°08.3'	Sanitary landfill excavation; debris-flow unit, 4–5 m below surface	Baumann (1991)
1,190 ± 60[d]	1,000–1,300	S	Log	49°47.1', 123°08.3'	Sanitary landfill excavation, debris-flow unit, 4–5 m below surface	Baumann (1991)
1,215 ± 120	1,000–1,400	GX-17892	Charcoal	49°47.0', 123°09.6'	Excavated pit; debris-flow unit, 1 m below surface	Thurber Engineering/Golder Associates (1993)
1,340 ± 65[d]	1,200–1,400	GX-17271	Log	49°47.1', 123°08.3'	Sanitary landfill excavation; debris-flow unit, 4–5 m below surface	Thurber Engineering/Golder Associates (1993)
1,390 ± 65[d]	1,200–1,500	GX-17270	Log	49°47.1', 123°08.5'	Sanitary landfill excavation, debris-flow unit, 4–5 m below surface	Thurber Engineering/Golder Associates (1993)
1,550 ± 80[e]	1,400–1,600	GSC-5100	Charcoal	49°47.8', 123°06.3'	Debris-flow deposit along Cheekye River	McNeely and Atkinson (1996)
1,665 ± 65[e]	1,500–1,800	GX-17891	Charcoal	49°47.4', 123°08.5'	Excavated pit; fluvial gravel, 3 m below surface; maximum age for overlying debris-flow unit	Thurber Engineering/Golder Associates (1993)

Age (^{14}C yr BP)	Calibrated age (cal yr BP)[a]	Lab no.[b]	Material	Location (latitude, longitude)	Comments	Source
2,190 ± 140	2,000–2,400	GSC-5101	Charcoal	49°47.8′, 123°06.3′	Debris-flow deposit along Cheekye River	McNeely and Atkinson (1996)
4,810 ± 80	5,500–5,700	GSC-6293	Sticks	49°47.1′, 123°08.4′	Sanitary landfill excavation; fluvial gravel, 10 m below surface; maximum age for overlying debris-flow unit	Clague et al. (2003)
5,660 ± 175[c]	6,000–6,900	GX-17889	Charcoal	49°47.1′, 123°10.0′	Excavated pit; debris-flow deposit, 8.6 m below surface	Thurber Engineering/Golder Associates (1993)
5,890 ± 100[d]	6,600–6,900	GSC-3256	Charcoal	49°46.2′, 123°10.0′	Bank of Squamish River; silt overlying debris-flow unit that yielded age of 6,595 ± 90 yr BP; maximum age for overlying debris-flow unit	Eisbacher (1983)
5,975 ± 180[f]	6,500–7,300	GX-17893	Charcoal	49°47.0′, 123°09.6′	Excavated pit; weakly stratified sand, 6.2 m below surface; between debris-flow units	Thurber Engineering/Golder Associates (1993)
6,595 ± 90	7,400–7,700	GX-17894	Charcoal	49°46.2′, 123°10.0′	Bank of Squamish River, debris-flow unit, 15 m below surface	Thurber Engineering/Golder Associates (1993)
7,820 ± 95[g]	8,500–9,000	GX-17397	Charcoal	49°46.0′, 123°08.3′	Excavated pit, 0.8 m below surface; minimum age for debris-flow unit on southern sector of fan	Thurber Engineering/Golder Associates (1993)
8,715 ± 100[g]	9,600–10,200	GX-17890	Charcoal	49°46.1′, 123°08.3′	Excavated pit, 4 m below surface; maximum age for debris-flow unit on southern sector of fan	Thurber Engineering/Golder Associates (1993)

[a] Determined from dendrocalibrated data of Stuiver et al. (1998) using the program CALIB 4.2. The range represents the 95% confidence interval ($\pm 2\sigma$) calculated with an error multiplier of 1.0.
[b] GSC, Geological Survey of Canada Radiocarbon Laboratory; GX, Geochron Laboratory; S, Saskatchewan Research Council.
[c] Three ages were obtained from a single log in the "Garbage Dump" debris flow exposed in an excavation at the Squamish sanitary landfill: GSC-6639 on outermost 5 rings, GSC-6638 on rings 181–185, and GSC-6636 on rings 251–256 (innermost five rings).
[d] These four ages were obtained from logs in the "Garbage Dump" debris flow exposed in an excavation at the Squamish sanitary landfill. The analyses were performed by two laboratories. The two Saskatchewan Research Council ages are statistically younger than the two Geochron ages. All four ages are statistically older than GSC-6639, obtained on the outermost rings of a log from the same debris-flow unit.
[e] These two ages are not statistically different and probably date the same debris flow.
[f] These three ages are not statistically different and probably date the same debris flow.
[g] These two ages bracket a debris-flow unit that thus dates to 7,800–8,700 ^{14}C yr BP.

the weathered zone (Hungr and Rawlings, 1995). Trenching of the Cheekye linears indicated that some movement has occurred within the last 3,200 years. Future collapses involving $1-3 \times 10^6 \, m^3$ of debris are possible (Hungr et al., 1984).

The Cheekye basin receives heavy precipitation during Pacific storms. Mean annual precipitation at the valley bottom climate station is about 2,400 mm, with 235 mm as snow, and maximum 24-hr rainfalls ranging from 30–130 mm (Environment Canada, 2001). Fall rains and winter rain-on-snow events are common, and precipitation is enhanced by orographic uplift, so that at high elevation these values may be two to three times higher. A flood with an average recurrence interval of 200 years (the so-called "200-year flood") would be expected to have a peak discharge of $250 \, m^3 \, s^{-1}$ at the apex of the lower fan (Thurber Engineering/Golder Associates, 1993). Maximum snow depth in the basin is of the order of 5 m, and deep deposits of avalanched snow accumulate along the river channel. Runoff during fall rain storms, or later in the season due to rain-on-snow, has the capacity to spontaneously mobilize debris flows up to $1-2 \times 10^6 \, m^3$ in size (Hungr and Rawlings, 1995; Kerr Wood Leidal, 2003).

Cheekye River becomes a fourth-order channel 3 km upstream from the basin mouth, and extends as an incised channel for 2 km to the apex of the lower fan (Figure 26.3). Over this 5-km distance, the channel has an average gradient of 4.7° and a channel flat up to 250 m wide. Debris avalanches blocking the channel may create impoundments that serve as a water source for debris-flow initiation. Impoundment volume (V) can be estimated using the formula:

$$V = h^2 \frac{w}{2s} \qquad (8.1)$$

where h is the height of the dam (in metres), w is the width of the channel (in metres), and s is the slope of the channel (in %) (Jordan, 1987).

Assuming, as in the case of the Cheekye River channel, a 250-m-wide valley flat with vertical sidewalls and a gradient of 4.7°, and assuming 25% water content for debris mobilization, 10, 20, and 30-m-high landslide dams could impound enough water to mobilize 6×10^5, 2.4×10^6, and $5.6 \times 10^6 \, m^3$ debris flows, respectively.

26.4 GEOCHRONOLOGY AND FAN EVOLUTION

Research projects completed since 1993 include a study of the geochronology, sedimentary architecture, and evolution of the Cheekye fan; a ground-penetrating radar (GPR) survey of the fan deposits; and an extension of the hazard analysis of Thurber Engineering/Golder Associates (1993). The pattern and timing of deglaciation of Howe Sound at the end of the Pleistocene have been inferred from studies of ice-contact sediments exposed in the Squamish and Cheakamus valleys upstream and downstream of Cheekye River (Figure 26.4; Friele and Clague, 2002a, b). The highest ice-contact terrace, which dates to about 10,600 ^{14}C yr BP (*ca.* 12,800 cal yr BP), sits at about 520 m a.s.l. at Cheekye River and slopes 4–5° to the south parallel to the reconstructed ice margin. This surface is referred to as the upper Cheekye fan

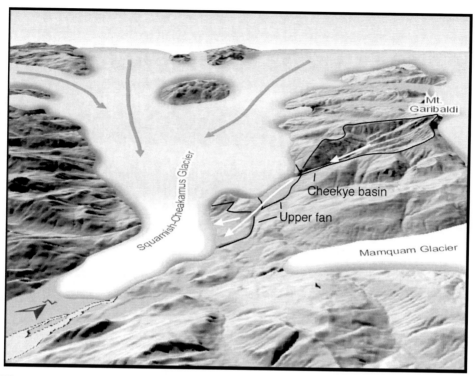

Figure 26.4. Younger Dryas ice margin in Howe Sound about 10,600 ^{14}C yr BP (*ca.* 12,800 cal yr BP). Debris from the Cheekye basin was deflected south along the ice margin at this time (white arrows), forming the upper Cheekye fan (black lines).
From Friele and Clague (2002b).

(Figure 26.3). The middle Cheekye fan, which lies below the upper fan between 200 and 350 m a.s.l., is characterized by kettle lakes, scarps, and raised terrace and fan surfaces. This morphology indicates deposition on top of and against stagnant ice. Radiocarbon ages from basal sediments in Stump Lake (Figure 26.3 and Table 26.2), which is situated on the middle fan, indicate that stagnant ice had melted out, and that Cheekye River had incised to near its present level, by 10,200 ^{14}C yr BP (*ca.* 11,900 cal yr BP) (Friele and Clague, 2002b; Clague et al., 2003).

During the retreat of the Squamish valley glacier, the sea inundated the head of Howe Sound to an elevation of 45–50 m above the present sea level; this marks the level of the highest deltaic surfaces in the area (Friele et al., 1999; Friele and Clague, 2002a). At the time, Howe Sound extended farther up the Squamish River valley (Hickin, 1989), north of the Cheekye River confluence, and the lower fan prograded into the sea. A GPR profile from the central portion of the fan shows steeply dipping reflectors, interpreted to be delta foreset beds (Figure 26.5). The contact between the seaward-dipping foreset beds and the surface-parallel topset beds in this profile drops in a down-fan direction from a maximum elevation of 45–50 m a.s.l. This

Table 26.2. Radiocarbon ages from Stump Lake sediments.

Age (^{14}C yr BP)	Calibraged age (cal yr before AD 2000)[a]	Laboratory number[b]	Dated material	Latitude (N), longitude (W)	Comment	Reference
6,210 ± 60	7,000–7,300	TO-9228	Gyttja	49°46.2', 123°07.2'	Base of gyttja overlying clay layer	Clague et al. (2003)
6,590 ± 130	7,300–7,700	TO-8275	Plant detritus[c]	49°46.2', 123°07.2'	Dated sample brackets tephra	Clague et al. (2003)
7,340 ± 140	7,900–8,500	TO-8276	Conifer needles	49°46.2', 123°07.2'	Top of peaty gyttja	Clague et al. (2003)
9,780 ± 100	10,800–11,600	TO-8723	Twig	49°46.2', 123°07.2'	Layer of plant detritus near bottom of core	Clague et al. (2003)
9,900 ± 120	11,200–12,000	TO-8274	Twig	49°46.2', 123°07.2'	Base of gyttja	Clague et al. (2003)
10,020 ± 80	11,300–12,000	TO-9682	Wood fragment	49°46.2', 123°07.2'	Woody layer near bottom of core	Clague et al. (2003)

[a] Determined from dendrocalibrated data of Stuiver et al. (1998) using the program CALIB 4.2. The range represents the 95% confidence interval (±2σ) calculated with an error multiplier of 1.0.
[b] TO, IsoTrace Laboratory (University of Toronto).
[c] Includes conifer needles.

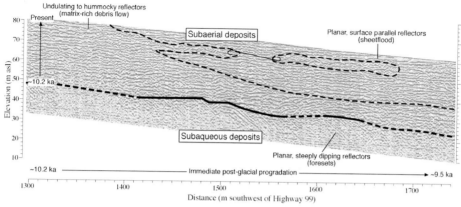

Figure 26.5. Ground-penetrating radar profile trending south-west across the lower Cheekye fan (see Figure 26.6 for location on the fan; after Friele et al., 1999). Three radar facies are identified: a steeply dipping, foreset-bedded deltaic unit; a surface-parallel, planar-bedded fluvial (hyperconcentrated flow, channel) unit; and a surface-parallel, undulating-to-hummocky debris-flow unit.

observation implies that the fan prograded during the period of rapid isostatic uplift immediately following deglaciation, 10,200 ^{14}C yr BP (*ca.* 11,900 cal yr BP) (Figure 26.7(b); Friele et al., 1999). The foreset–topset contact sits −10 m a.s.l. beneath the southern margin of the fan. Because relative sea level was 10 m or more below the present datum 9,000 ^{14}C yr ago (*ca.* 10,200 cal yr BP) (Clague et al., 1982), the southern sector of the fan had reached its modern extent by the early Holocene (Friele et al., 1999). A radiocarbon age from a depth of less than 1 m in a test pit at the southern margin of the fan (Figure 26.6; Thurber Engineering/Golder Associates 1993) indicates that this sector of the fan has been largely inactive since 7,800 ^{14}C yr BP (*ca.* 8,800 cal yr BP; Table 26.1).

Examination of the distribution of radiocarbon ages (Figure 26.6) obtained during excavations on the fan suggests that activity shifted northward after 7,800 ^{14}C yr BP (8,800 cal yr BP) and that an incised channel, now buried, extended in a south-westerly direction from the sharp bend in the modern channel at 120 m a.s.l. In this central fan sector, radiocarbon ages include 6,600 ^{14}C yr BP (*ca.* 7,500 cal yr BP) at 15 m depth, 5,700 ^{14}C yr BP (*ca.* 6,400 cal yr BP) at 8.6 m depth, and 4,800 ^{14}C yr BP (*ca.* 5,600 cal yr BP) at 10 m depth. In contrast, ages from the southern part of the fan are 8,700 ^{14}C yr BP (*ca.* 9,900 cal yr BP) at 4 m depth and 7,800 ^{14}C yr BP (*ca.* 8,800 cal yr BP) at 0.8 m depth.

Development of the paleochannel can be explained by coupling of the fan to the trunk-valley rivers. This is supported by the observation that the paleochannel is oriented perpendicular to the opposite wall of the Squamish valley, and the following reasoning: extension of the fan across the fjord to the opposite valley side would have isolated a lake upfjord of the fan (Hickin, 1989), resulting in drainage of the Squamish and Cheakamus Rivers across the fan toe. This would have marked the

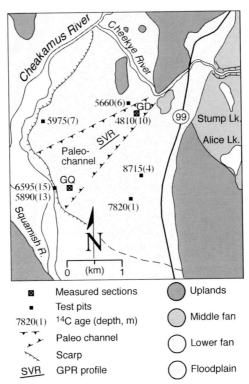

Figure 26.6. Depth and distribution of radiocarbon ages recovered from excavated pits and natural exposures. SVR indicates the location of the GPR profile (Figure 26.5); GD indicates the location of the municipal landfill (see also Figure 26.9(a)); and GQ indicates the location of a gravel quarry (see also Figure 26.9(b)). Highway 99 runs north across the fan. The data suggest that a paleochannel extends south-west across the fan (see text for details).

From Thurber Engineering/Golder Associates (1993) and Clague et al. (2003).

onset of coupling, with the high transport capacity of both rivers exceeding the declining sediment delivery to Cheekye fan, leading to channel entrenchment. Radiocarbon ages suggest coupling occurred between 7,800–6,600 ^{14}C yr BP (8,800–7,500 cal yr BP), and that an entrenched channel extended south-west across the central portion of the lower fan during the period 6,600–4,800 ^{14}C yr BP (7,500–5,600 cal yr BP) (Figure 26.7(b)).

About 1 km below the fan apex is an outcrop of Pleistocene fan sediment with a smooth surface that slopes about 6.5° down the fan (Figures 26.7 and 26.8). This sediment overlies a bedrock hill that extends above the lower fan surface. Modern Cheekye River flows between the Pleistocene sediment outcrop and the bedrock upland to the north, and its channel is incised into bedrock at this site. In the early phase of fan construction, when the southern sector was aggrading, the sediment outcrop probably was connected to the bedrock upland by a sediment fill that extended higher than the paleo-fan surface to the south (Figure 26.7(a)).

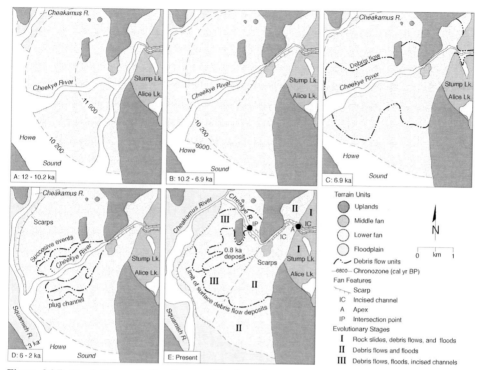

Figure 26.7. Evolution of the lower Cheekye fan. (a) Deltaic progradation, 11,900–10,200 cal yr BP. The active channel of Cheekye River crossed the southern part of the fan during this interval. (b) The Cheekye River channel shifted to the north side of the hummock of Pleistocene debris about 8,800 cal yr BP. (c) The largest Holocene debris flow occurred sometime between 7,500 and 7,100 cal yr BP. (d) Three large debris flows filled the paleo-channel on the central part of the fan between 5,800 and 2,000 cal yr BP. Sometime in the last few thousand years, the active channel shifted to the north, and the northern and western margins of the fan were truncated by the Cheakamus and Squamish Rivers. (e) Surface units and features on the modern fan. The most recent debris flow to spill out of the channel of Cheekye River onto the surface of the fan occurred about 800 years ago.
Clague et al. (2003).

When the southern fan surface attained the level of the fill, the active channel shifted north of the outcrop and was then captured in that position as the active channel incised the fan.

A pit 10 m deep by 40 m long was excavated at the upstream end of the paleo-channel in 1998. It exposed four debris-flow units (GD in Figure 26.6; Figure 26.9(a)). Wood fragments from fluvial sediments at the base of the exposure yielded an age of 4,800 ^{14}C yr BP (*ca.* 5,600 cal yr BP; Table 26.1) (Clague et al., 2003). It thus appears that sometime after 5,600 cal yr BP, the paleochannel became plugged by debris-flow deposits (Figure 26.7(d)) and the active channel shifted north to its present position. The Squamish River delta prograded past the

672 Multifaceted hazard assessment of Cheekye fan [Ch. 26

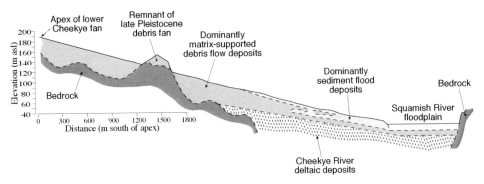

Figure 26.8. Architectural model of the lower Cheekye fan based on a GPR survey and knowledge of the local Quaternary history.

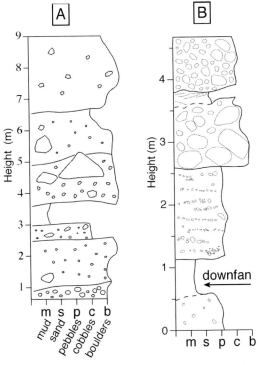

Figure 26.9. Two sections of fan sediment on the lower Cheekye fan (see Figure 26.6 for locations). (a) A sequence of debris-flow units, comprising matrix-supported diamicton with subangular clasts up to boulder size, exposed in an excavation at the municipal landfill. Gravel at the base of the excavation yielded a radiocarbon age of 4,800 ^{14}C yr BP (5,600 cal yr BP). The uppermost debris-flow unit is about 800 years old. (b) Diamicton, stratified hyperconcentrated-flow gravel and sand, and channel boulder lag exposed in a gravel quarry. Vertical casts of standing trees in growth position are preserved in these sediments, indicating that floods enveloped the forest on the fan surface.

mouth of Cheekye River some time between 4,000 and 2,000 cal yr BP (Figure 26.7(d); Hickin, 1989), whereupon the fan-delta became an alluvial fan.

26.5 FAN ARCHITECTURE, SEDIMENTOLOGY, AND HOLOCENE SEDIMENT BUDGET

An architectural model for the lower Cheekye fan (Figure 26.8) was developed from GPR surveys. Traditional methods of stratigraphic investigation allow only shallow exposure: excavation reveals only the uppermost 5–10 m of the fan sediments, and bluffs along Squamish River reveal a maximum height of only 15 m of debris. In contrast, GPR yielded stratigraphic and structural information to depths as great as 50 m (Friele et al., 1999; Ekes and Hickin, 2001; Ekes and Friele, 2003).

Over much of the fan, steeply dipping, linear reflectors are truncated by a sequence of reflectors dipping parallel to the surface (Figure 26.5). The contact between the two radar units separates subaqueous deltaic sediments from subaerial fan sediments. The contact is diachronous and slopes down-fan, indicating that relative sea level fell as the fan grew (Friele et al., 1999).

The subaerial package is divisible into two radar facies: one facies characterized by dominantly undulating to hummocky reflectors, which is found at depth and toward the fan apex; and a second facies with mainly planar reflectors, which is found at the distal margins of the fan. The surface exposures described below suggest that the former facies comprises matrix-rich debris-flow deposits and the latter hyperconcentrated-flow deposits (Figures 26.7 and 26.9).

Excavations in the proximal two-thirds of the fan reveal only diamictic debris-flow deposits (Figures 26.7(e) and 26.9(a)). In contrast, interbedded gravel and sand dominate the distal part of the fan (Figure 26.7(e)). An exposure in the latter materials (Figure 26.9(b)) reveals a basal diamicton overlain by 6 m of planar-bedded gravelly sand. The gravelly sand is capped by an imbricated, subrounded to subangular, cobble-boulder lag, which is conformably overlain by cross-stratified granular sand. The sand, in turn, is overlain, across an erosional contact, by massive, clast-supported, cobble-boulder gravel. With the exception of the basal diamicton, these sediments are interpreted to be hyperconcentrated-flow and channel deposits (Blair and McPherson, 1994).

Ground-penetrating radar surveys and radiocarbon ages allow estimates to be made of changes in sediment delivery to the lower fan during the Holocene (Figure 26.10; Friele et al., 1999). Ninety per cent of the sediment forming the fan was delivered between 10,200 cal yr BP and the onset of coupling 8,800–7,500 cal yr BP, indicating that the Cheekye fan fits the paraglacial model (Church and Ryder, 1972; Church and Slaymaker, 1989). Paraglacial fans are debris-flow-dominated alluvial fans that formed rapidly during and shortly after deglaciation by the reworking of freshly exposed glacial drift.

In an alternate model of fan evolution, Blair and McPherson (1994) describe four stages of development that are defined by a successive decline in surface slope and a shift in the dominant formative processes over time. The precursor stage is

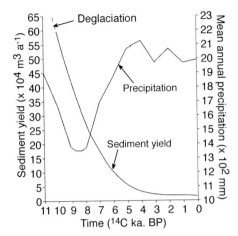

Figure 26.10. Changes in sediment delivery to the lower Cheekye fan (Friele et al., 1999) and precipitation in south-western British Columbia during the Holocene (Mathewes and Heusser, 1981).

marked by rockfalls onto coarse talus. Stage I is characterized by rockfalls, rockslides, debris slides, avalanches, and debris flows. During stage II, the fan surface may be modified by hyperconcentrated flows and then reworked and incised by channel flows. The same processes operate during stage III, but the sediments are finer. The upper Cheekye fan records stage I of the Blair–McPherson model, whereas the lower fan records stages II and III (Figure 26.7(e)). The development of a fourth-order channel in the Cheekye basin also suggests that the fan has reached the later stage, indicating a rapid evolution through 12,000 years of postglacial time.

The decline in sediment supply and changes in climate during the Holocene (Figure 26.10) probably led to shifts in the dominant process affecting the lower Cheekye fan. Blair (1999) describes the contrasting geomorphology of two adjacent fans in Owens valley, California. The source basins share the same climate, but there is a marked difference in their geology. His example illustrates the strong control that geology has on the dominant process affecting alluvial fans. In contrast, the geology of the Cheekye basin has not changed since the fan began to form, whereas sediment supply and climate have varied dramatically (Figure 26.10). During the early Holocene, when climate in British Columbia was warmer and drier than today, abundant available sediment may have favoured debris flows over fluvial activity. Hummocky reflectors noted at the base of the subaerial sequence (Figure 26.5) suggest this is the case. During the middle Holocene, climate cooled and became wetter, while sediment yield decreased. Increased precipitation may have caused larger debris flows, more floods, or both. Confined debris flows in increasingly entrenched channels probably extended the zone of active deposition farther down the fan.

One might suppose that trunk–valley coupling and the substantial decline in sediment supply would have led to more channel entrenchment than is evident.

However, the level of Cheekye River is controlled by bedrock at 200 and 120 m a.s.l., and at its mouth is controlled by Cheakamus River. Further, the level of Squamish River is graded to the sea, which rose during the period 8,000–2,000 ^{14}C yr BP (Clague et al., 1982). Finally, as Squamish River prograde into Howe Sound, its floodplain must aggrade. All these processes prevent significant channel entrenchment.

26.6 DEBRIS-FLOW FREQUENCY AND MAGNITUDE

This section describes three different approaches that have been used to quantify debris-flow hazard on the lower Cheekye fan. The first approach involves stratigraphic analyses of fan sediments and relies largely on excavation of pits. Thurber Engineering/Golder Associates (1993) excavated about 40 pits on the lower fan, ranging from 5–10 m deep. They also reviewed 15 water well logs, and examined and described all significant natural exposures. Using this approach, they produced a stratigraphic model for the surface fan deposits. They calculated that 3.5×10^7 m^3 of sediment have been deposited on the lower fan over about the last 7,000 years. Fifty per cent of this sediment is matrix-supported debris-flow deposits containing 10–30% silt and clay by volume. The only debris-flow deposit they were able to correlate over an extensive area is the surface diamicton at the municipal landfill (the so-called "Garbage Dump" deposit), which was described and mapped by Baumann (1991). Baumann determined that the Garbage Dump debris flow left the channel of Cheekye River at about 120 m a.s.l., just upstream of the landfill, and spread out over 84 ha of the fan. The volume of the deposit has been estimated to be 3–5×10^6 m^3 (Baumann, 1991; Jordan, 1994). Thurber Engineering/Golder Associates (1993) also identified a larger, older debris flow, based on a diamicton unit exposed in a bank near Squamish River and deposits of similar age (5,700–6,600 ^{14}C yr BP, 6,400–7,500 cal yr BP) in two excavations. They termed this deposit the "Squamish River unit".

Thurber Engineering/Golder Associates (1993) produced a probability-of-exceedence plot (Figure 26.11) based on knowledge of the volume of sediment deposited in the last 7,000 years, the proportion of debris-flow material in the near-surface fan, and the estimated magnitudes of the two largest events. They assumed that the largest event, recorded by the Squamish River unit, had a volume of 7×10^6 m^3 (Thurber Engineering/Golder Associates, 1993; Sobkowicz et al., 1995).

The second approach to quantifying debris-flow hazard on the lower Cheekye fan involves analysis of cores recovered from Stump Lake, located just upstream of the fan apex (Clague et al., 2003). Stump Lake is fed by groundwater and has no inlet streams, thus the sediment that normally accumulates in the lake is fine organic matter, or gyttja. However, its outlet sill is less than 10 m above the base of the modern channel, and only 50 m from the lake edge. Thus, the lake could receive sediment from very large debris flows travelling down the Cheekye channel.

Four cores retrieved from the lake provide consistent records (Figure 26.12). Diatom assemblages record an early debris flow that deepened the basin before

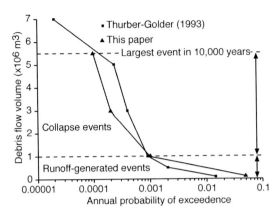

Figure 26.11. Debris-flow probability-of-exceedence plots for the lower Cheekye fan. Such plots indicate the probability that an event of a certain size will be exceeded. The break in slope between small, frequent events and large, infrequent events is interpreted to reflect different initiation mechanisms (see text for details).

10,000 ^{14}C yr BP (*ca.* 11,600 cal yr BP; Table 26.2) and established the sill elevation of the modern lake. A second debris flow entered the lake between 6,500 and 6,200 ^{14}C yr BP (7,500 and 7,100 cal yr BP; Table 26.2). Using two surveyed cross sections extending north from the lake across Cheekye River, Clague et al. (2003) applied a Newtonian flow model (see Chapter 7; Hungr et al., 1984) to estimate velocity and peak discharge of this second debris flow. The analysis yielded a velocity of 15–30 m/s and a peak discharge of 15,000–30,000 m^3/s. By applying an empirical relation between peak discharge (Q) and total debris volume (V) obtained from a global data set of non-granitic debris flows,

$$V = 794.58 Q^{0.849} \tag{8.2}$$

(Mizuyama et al., 1991; Jakob and Bovis, 1996), the average volume of the debris flow was estimated to be 3–5×10^6 m^3 (Clague et al., 2003). Because the feeder channel is 5 km long and has a relatively low gradient (4.7°), it can be argued that the lower bounding velocity and peak discharge are more appropriate (Kerr Wood Leidal, 2003). Using the 95% confidence limit of the relation, a 15,000 m^3/s discharge yields a maximum total debris volume of 5.4×10^6 m^3. The record from Stump Lake thus indicates that the largest event to affect the lower fan during the Holocene was no larger than about 5.5×10^6 m^3. This event probably deposited the Squamish River unit described by Thurber Engineering/Golder Associates (1993).

The third approach uses evidence from damaged living trees adjacent to the modern Cheekye River channel to reconstruct the frequency and magnitude of recent channellized debris flows (Figure 26.13). A sample of 45 tree stem disks and 4-mm-diameter cores, some with several scars of different ages, records nine channelized debris flows over the last 200 years, with return intervals of 10–20 years. Velocity estimates, based on surveyed cross sections, are 7–8 m/s. Estimates of peak discharge range from 800 to 2,800 m^3/s.

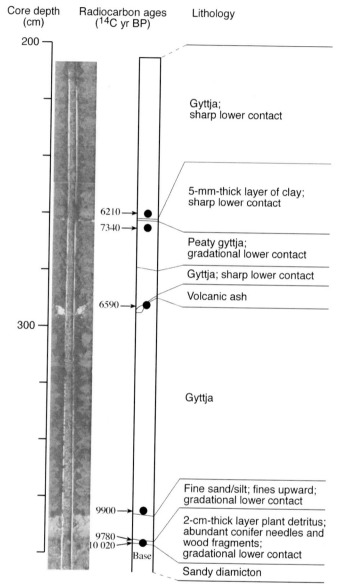

Figure 26.12. Core from Stump Lake showing radiocarbon-dated stratigraphy (Clague et al., 2003). The peaty gyttja and capping clay lamina record a debris flow that occurred 6,500–6,200 ^{14}C yr BP (*ca.* 7,500–7,100 cal yr BP).

A revised exceedence plot is presented in Figure 26.11, based on estimated return intervals for debris flows of different sizes. The Stump Lake event (*ca.* 5.5×10^6 m^3) is assigned a 10,000-year return interval, and channellized flows (*ca.* 50,000 m^3) are given a 20-year return interval. Between the two is the "Garbage Dump" event

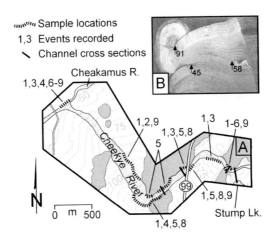

Figure 26.13. Locations of sampled trees scarred by debris flows. Event 1, 1991; 2, 1971–1972; 3, 1962–1963; 4, 1958; 5, 1939–1940; 6, 1920–1921; 7, 1906–1907; 8, 1887–1888; 9, 1878–1879. The example shows a tree scarred by debris flows in 1958 and 1991. The third scar was found on only one sample and is considered erroneous.

documented at the municipal landfill (3×10^6 m^3) with a return interval of about 5,000 years. The three additional debris-flow units at the landfill (Figure 26.8(b)) have thicknesses of 1–2 m, suggesting volumes of 1×10^6 m^3 and recurrence intervals of about 1,000 years.

The revised plot compares favourably with that of Thurber Engineering/Golder Associates (1993). It is non-linear, suggesting that it records two distinct populations. Larger, infrequent events are probably caused by collapse of the steep headwall of the Cheekye basin, whereas the smaller events are triggered by heavy runoff during storms or snowmelt.

26.7 MODELING

Modeling of both rock avalanche and debris-flow travel has been conducted by Thurber Engineering/Golder Associates (1993) and Kerr Wood Leidal (2003). An empirical model relating predicted volume to deposit area (Hungr, 1990) indicates that rock avalanches would not reach the lower fan but rather would stop somewhere on the middle fan (Hungr and Rawlings, 1995).

Debris-flow modeling was carried out by Kerr Wood Leidal (2003) using the commercial software FLO-2D, a 2-D flood routing model. This software is most commonly applied to overbank flooding, but it is also used to analyse unconventional flood problems, such as unconfined flows over complex topography, debris floods, and debris flows.

The model requires a topographic grid, an input hydrograph, and coefficients that govern debris-flow rheology. A 100-m gridded digital terrain model of the fan represented topography. The input debris-flow hydrograph required discharge values and associated sediment concentrations. These values were determined by informed judgment: peak discharges were based on the estimates of Clague et al. (2003), and material properties such as viscosity and effective yield stress of the slurry were estimated by calibrating the model using previously mapped debris-flow deposits.

The model predicts that debris flows with a peak discharge larger than $1,000\,\text{m}^3/\text{s}$ ($\sim 2.75 \times 10^5\,\text{m}^3$) would leave the channel at the fan apex and spread over the fan. Smaller debris flows would remain in the channel, but might travel to the mouth of Cheekye River. A 3×10^6-m^3 debris flow would cover the fan apex, much of the southern sector, and the entire northern sector of the fan (Figure 26.14, see color section). Interestingly, the thick debris-flow deposits at the municipal landfill would deflect the flow to the north and south. Although a 3×10^6-m^3 debris flow does not reach the Brackendale subdivision, a lobe of the 1-in-10,000-year 5.5×10^6-m^3 debris flow does (Kerr Wood Leidal, 2003).

26.8 DISCUSSION

The Cheekye fan provides a good example of the paraglacial concept of Church and Ryder (1972), with 90% of its sediment deposited before about 7,500 years ago. Average annual sediment delivery to the fan is now low; Squamish and Cheakamus Rivers have truncated the margins of the fan, and Cheekye River is entrenched at the fan head. The fan has reached a mature stage in the sense of the evolutionary scheme of Blair and McPherson (1994) – it is entrenched, has a low channel gradient, and depositional processes are now dominated by channellized debris and hyperconcentrated-flow activity. The late evolutionary stage is also indicated by a fourth-order channel network within the Cheekye basin.

Nevertheless, a significant potential exists for a debris flow that would affect the fan surface. Debris flows up to several million cubic metres in volume are possible, with flows exceeding one million cubic metres having an average recurrence interval of about 1,000 years. Although British Columbia lacks landslide hazard legislation, Justice Berger (1973) set a legal precedent by establishing a 10,000-year recurrence interval as the hazard acceptability threshold for rock avalanches and debris flows that might impact a community. Evans (1997) proposed acceptable landslide risk criteria for British Columbia by developing a plot, the so-called F/N plot (Figure 26.15), of landslide frequency vs. number of deaths based on a number of historic fatal landslides in Canada. The plot has a slope of -1, with acceptable risk ranging from 1,000 years for the death of one individual to 1,000,000 years for the death of 1,000 people. A very large debris flow on the Cheekye fan (e.g., $5.5 \times 10^6\,\text{m}^3$) could cause more than 10 fatalities in the community of Brackendale, and in this context Berger's (1973) precedent stands. On the other hand, the British Columbia Ministry of Transport and Highways, which approves subdivision proposals in the province, has established a hazard acceptability level for landslides of 10% in 50 years, or events with approximately a 500-year average return period.

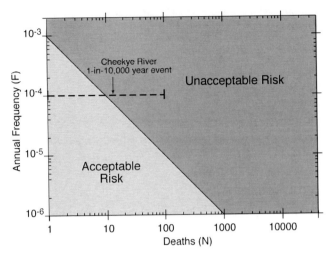

Figure 26.15. F/N plot compiled by Evans (1997). According to Figure 26.11, Cheekye River debris flows with annual return frequencies between 1,000 and 10,000 years would have volumes of 1–$5.5 \times 10^6 \, m^3$. Any of these events could spill onto the fan surface and inundate Highway 99, disrupting a vital transportation corridor. The larger events could impact residential areas, including Brackendale, with the number of deaths of the order of 1–100.

Aside from the lack of clear direction from the province, the probabilistic approach itself presents problems. The long period from which the debris-flow record is derived, and the very nature of the process itself, result in the problem that the phenomenon does not conform to the simple assumptions that underpin the statistical approach, specifically stationarity of environmental controls, events selected from a single population, and independence of events. For example, the rapid growth of the Cheekye fan during the early Holocene was concomitant with a sharp decline in sediment supply and an increase in precipitation. Furthermore, documented debris flows are drawn from at least two populations. Smaller events are produced by high runoff in a transport-limited basin, whereas, larger debris flows are produced by large landslides or landslide dam breaks. Finally, successive events cannot be assumed to be independent, as a large collapse and ensuing debris flow might leave large volumes of debris along the channel that would be available for reworking in subsequent events. Thus, although we have produced an exceedence plot, it serves only to characterize the historical record and has limited predictive value. This poses a problem for jurisdictions like British Columbia that use a hazard acceptability approach.

Debris-flow hazard analysis and mitigation must include judgment, careful and experienced examination of the fan surface (see Kellerhals and Church, 1990), and some combination of geophysical and stratigraphic investigation, radiocarbon and tree-ring dating of event stratigraphy or tree scars, and modern computer modeling. This case study provides one model for debris-flow hazard assessment on alluvial fans in British Columbia. Multifaceted hazard assessments such as the example reported in this chapter can be performed by experienced earth scientists and provide government officials with the knowledge they require to make informed land-use decisions.

26.9 ACKNOWLEDGEMENTS

Our work builds on that of many earth scientists, notably the late Bill Mathews but also Frank Baumann, Csaba Ekes, Oldrich Hungr, Peter Jordan, and Matthias Jakob. Without their contributions, we would not have as clear a picture of the Cheekye fan as we do. Hamish Weatherly of Kerr Wood Leidal kindly provided the modeling results used in Figure 26.14. Our research was supported, in part, by Simon Fraser University research funds and a NSERC (National Sciences and Engineering Research Council, Canada) Research Grant to John Clague.

26.10 REFERENCES

Baumann, F.W. (1991) *Garbage Dump Debris-flow Deposit and Its Relationship to the Geologic History of the Cheekye Fan* (Unpublished report, 25 pp.). Baumann Engineering, Squamish, British Columbia.

Berger, T. (1973) *The Law and Mr. Planner: Reasons for Judgment*. Supreme Court of British Columbia Action No. X4042, August 23, 1973.

Blair, T.C. (1999) Cause of dominance by sheetflood versus debris flow processes on two adjoining alluvial fans, Death Valley, California. *Sedimentology*, **46**, 1015–1028.

Blair, T.C. and McPherson, J.G. (1994) Alluvial fans and their natural distinction from rivers based on morphology, hydraulic processes, sedimentary processes, and facies assemblages. *Journal of Sedimentary Research*, **A64**, 450–489.

Church, M. and Ryder, J.M. (1972) Paraglacial sedimentation: A consideration of fluvial processes conditioned by glaciation. *Geological Society of America Bulletin*, **83**, 3059–3072.

Church, M. and Slaymaker, O. (1989) Disequilibrium of Holocene sediment yield in glaciated British Columbia. *Nature*, **337**, 452–454.

Clague, J.J., Harper, J.R., Hebda, R., and Howes, D.E. (1982) Late Quaternary sea levels and crustal movements, coastal British Columbia. *Canadian Journal of Earth Sciences*, **19**, 597–618.

Clague, J.J., Friele, P.A., and Hutchinson, I. (2003) Chronology and hazards of large debris flows in the Cheekye River basin, British Columbia, Canada. *Environmental and Engineering Geoscience*, **8**, 75–91.

Crippen Engineering (1975) *Geotechnical–hydrogeological Investigation of the Cheekye Fan* (Report). British Columbia Department of Housing, Victoria, British Columbia.

Crippen Engineering (1981) *Cheekye Fan Development Report on Hazard Areas and Protective Works* (Report). Ministry of Environment, Victoria, British Columbia.

Eisbacher, G.H. (1983) Slope stability and mountain torrents, Fraser Lowlands and Southern Coast Mountains, British Columbia. *Joint Annual Meeting, Victoria, British Columbia:*

Field Trip Guidebook (46 pp.). Geological Association of Canada, Mineralogical Association of Canada, Canadian Geophysical Union, Victoria, British Columbia.

Ekes, C. and Friele, P.A. (2003) Sedimentary architecture and post-glacial evolution of Cheekye fan, southwestern British Columbia, Canada. In: C.S. Bristow and H.M. Jol (eds), *Ground Penetrating Radar in Sediments* (Special Publication 211, pp. 87–98). Geological Society of London.

Ekes, C. and Hickin, E.J. (2001) Ground penetrating radar facies of the paraglacial Cheekye fan, southwestern British Columbia, Canada. *Sedimentary Geology*, **143**, 199–217.

Environment Canada (2001) *Canadian Climate Normals 1971–2000* (2001 CDCD West). Meteorological Service of Canada, National Archives and Data Management Branch, Downsview, Ontario. Available at *http://www.climate.weatheroffice.ec.gc.ca*

Evans, S.G. (1997) Fatal landslides and landslide risk in Canada. In: D.M. Cruden and R. Fell (eds), *Landslide Risk Assessment* (pp. 185–196). A.A. Balkema, Rotterdam.

Friele, P.A. and Clague, J.J. (2002a) Readvance of glaciers in the British Columbia Coast Mountains at the end of the last glaciation. *Quaternary International*, **87**, 45–58.

Friele, P.A. and Clague, J.J. (2002b) Younger Dryas readvance in Squamish River valley, southern Coast Mountains, British Columbia. *Quaternary Science Reviews*, **21**, 1925–1933.

Friele, P.A., Ekes, C., and Hickin, E.J. (1999) Evolution of Cheekye fan, Squamish, British Columbia: Holocene sedimentation and implications for hazard assessment. *Canadian Journal of Earth Sciences*, **36**, 2023–2031.

Green, N.L. (1990) Late Cenozoic volcanism in the Mount Garibaldi Lake volcanic fields. Garibaldi Volcanic Belt, southwestern British Columbia. *Geoscience Canada*, **17**, 171–174.

Green, N.L., Armstrong, R.L., Harakal, J.E., Souther, J.G., and Read, P.B. (1988) Eruptive history and K-Ar geochronology of the late Cenozoic Garibaldi Volcanic Belt, southern British Columbia. *Geological Society of America Bulletin*, **100**, 563–579.

Hickin, E.J. (1989) Contemporary Squamish River sediment flux to Howe Sound, British Columbia. *Canadian Journal of Earth Sciences*, **26**, 1953–1963.

Hungr, O. (1990) *Mobility of Rock Avalanches* (Report No. 46, pp. 11–20). National Research Institute for Disaster Prevention, Tsukuba, Japan.

Hungr, O. and Rawlings, G. (1995) Assessment of terrain hazards for planning purposes: Cheekye fan, British Columbia. *Proceedings of 48th Canadian Geotechnical Conference, Vancouver, British Columbia* (pp. 509–517). Bitech Publishers, Vancouver.

Hungr, O., Morgan, G.C., and Kellerhals, R. (1984) Quantitative analysis of debris torrent hazards for design of remedial measures. *Canadian Geotechnical Journal*, **21**, 663–677.

Jakob, M. and Bovis, M.J. (1996) Morphometric and geotechnical controls of debris flow activity, southern Coast Mountains, B.C., Canada. *Zeitschrift für Geomorphologie*, **104**, 13–26.

Jones, W.C. (1959) *Cheekye River Mudfows* (9 pp.). British Columbia Department of Mines, Victoria.

Jordan, P.R. (1987) *Terrain Hazards and River Channel Impacts in the Squamish and Lillooet Watersheds, British Columbia* (Unpublished report). Geological Survey of Canada, Ottawa, Ontario.

Jordan, P.R. (1994) Debris flows in the Southern Coast Mountains, British Columbia: Dynamic behaviour and physical properties (280 pp.). Ph.D. thesis, University of British Columbia, Vancouver.

Kellerhals, R. and Church, M.A. (1990) Hazard management on fans, with examples from British Columbia. In: A.H. Rachocki and M.A. Church (eds), *Alluvial Fans: A Field Approach* (pp. 335–354). John Wiley & Sons, New York.

Kerr Wood Leidal (2003) *Preliminary Design Report for Cheekye Fan Deflection Berms* (Report). District of Squamish, Squamish, British Columbia.

Mathewes, R.W. and Heusser, L.E. (1981) A 12000 year palynological record of temperature and precipitation trends in southwestern British Columbia. *Canadian Journal of Botany*, **59**, 707–710.

Mathews, W.H. (1952) Mount Garibaldi, a supraglacial Pleistocene volcano in southwestern British Columbia. *American Journal of Science*, **250**, 553–565.

Mathews, W.H. (1958) Geology of the Mount Garibaldi map area, southwestern British Columbia, Canada. Part II: Geomorphology and Quaternary volcanic rocks. *Geological Society of America Bulletin*, **69**, 179–198.

McNeely, R. and Atkinson, D.E. (1996) *Geological Survey of Canada Radiocarbon Dates XXXII*. Geological Survey of Canada, Ottawa.

McNeely, R. and Jorgensen, P.K. (1992) *Geological Survey of Canada Radiocarbon Dates XXX* (Paper 90-7, 84 pp.). Geological Survey of Canada, Ottawa.

Mizuyama, T., Kobashi, S., and Ou, G. (1991) Prediction of debris flow peak discharge. *Interpraevent: Internationales Symposium, Bern, Switzerland* (Tagespublikation, Band 4, pp. 99–108). Internationale Forschungsgesellschaft Interpraevent, Klagenfurt, Austria.

Sobkowicz, J., Hungr, O., and Morgan, G. (1995) Probabilistic mapping of a debris flow hazard area: Cheekye fan, British Columbia. *Proceedings of 48th Canadian Geotechnical Conference, Vancouver, British Columbia* (pp. 519–529). Bitech Publishers, Vancouver.

Stuiver, M., Reimer, P.J., Bard, E., Beck, J.W., Burr, G.S., Hughen, K.A., Kromer, B., McCormac, G., van der Plicht, J., and Spurk, M. (1998) INTCAL98 radiocarbon age calibration, 24,000–0 cal BP. *Radiocarbon*, **40**, 1041–1083.

Thurber Engineering/Golder Associates (1993) *The Cheekye River Terrain Hazard and Land-use Study* (Final report). British Columbia Ministry of Environment, Lands and Parks, Burnaby.

27

Debris flows at Mount St. Helens, Washington, USA

Jon J. Major, Thomas C. Pierson, and Kevin M. Scott

27.1 INTRODUCTION

27.1.1 The significance of volcanic debris flows

Volcanic debris flows, also called lahars, pose considerable hazards to populations in many regions around the world, including western North America (see Chapter 10). Expansions of urban, agricultural, and recreational land uses toward many volcanoes, and particularly into adjacent river valleys, have significantly increased the likelihood of disasters from future flows. Arguably, debris flows pose the greatest hazard at many volcanoes. They can flow many tens of kilometres at speeds of tens of km/hr, deposit tens to hundreds of centimetres of coarse sediment on valley floors and floodplains where human settlements are concentrated, destroy or damage all structures along their flow paths, and they can occur with or without an associated eruption. Between 1950 and 2001 volcanic debris flows claimed more than 28,000 lives worldwide (Table 27.1). At some volcanoes, debris flows have choked main channels or blocked tributary channels, impounded lakes, and precipitated secondary floods that endangered people and property when the impoundments breached. In order to develop effective mitigation strategies for protecting the populace at risk from such flows, it is imperative to understand basic physical processes relevant to the generation and behaviour of volcanic debris flows.

The sizes, extents, and hazards of volcanic debris flows can vary widely, especially between glaciated and non-glaciated volcanoes (e.g., Pierson et al., 1990; Carrasco-Núñez et al., 1993; Scott, 1988a; Scott et al., 1995; Newhall and Punongbayan, 1996; Major et al., 2004; Scott et al., in press; see also Chapter 10). Debris flows from glaciated volcanoes, especially those associated with eruptions, generally are larger than flows from non-glaciated volcanoes and non-volcanic debris flows, the physical processes that initiate them are more varied, they can exhibit great complexity downstream, and they commonly flow great distances, as much as many

---

M. Jakob and O. Hungr (eds), *Debris-flow Hazards and Related Phenomena*.
© Praxis. Springer Berlin Heidelberg 2005.

Table 27.1. Fatalities caused by volcanic debris flows between 1950 and 2001.

Volcano	Year	Number of fatalities
New Zealand		
Mount Ruapehu	1953	151
Indonesia		
Agung	1963	200
Kelut	1966–1990	216
Merapi	1969–1986	42
Semeru	1968–1981	>500
Mahawu	1958	1
Awu	1966	~10
Iya	1969	1
Marapi	1979	80
Rinjani	1994	30
Japan		
Sakurajima	1946–1976	12
Mayuyama (Unzen)	1957	13
Usu	1978	3
Philippines		
Binuluan	1952	12
Hibok-Hibok	1954	2
Mayon	1968–1984	>200
Pinatubo	1991–1992	100–200
Parker	1995	~300
USA		
Mount St. Helens	1980	2
Mount Hood	1980	1
Mexico		
El Chichón	1982	1
El Salvador		
San Salvador	1982	300–500
San Vicente	2001	2
Guatemala		
Fuego	1963–1974	10
Santa María (Santiaguito)	1978	1
Nicaragua		
Maderas	1996	6
Casita	1998	2,500
Costa Rica		
Irazú	1963	>30
Chile		
Villarrica	1964–1971	37
Cerro Hudson	1971	3
Colombia		
Nevado del Ruiz	1985	>23,000

Sources: Gallino and Pierson (1985); Major and Newhall (1989); Pierson et al. (1990); Japan Ministry of Construction (1992, 1995); Simkin and Siebert (1994); Smithsonian Institution (1994, 1996); Thouret et al. (1998); Lavigne et al. (2000); Lozano (2001); Major et al. (2004); Scott et al. (in press); R.B. Waitt (U.S. Geological Survey, pers. commun., 2004).

Sec. 27.1]   Introduction 687

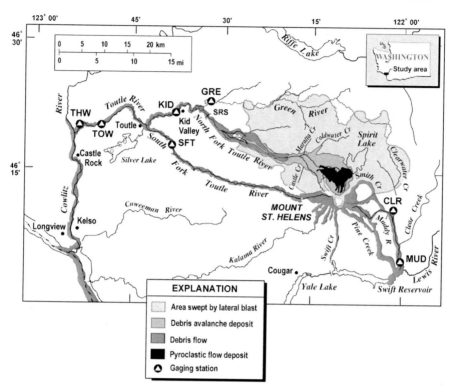

Figure 27.1. Distribution of major volcaniclastic deposits of the cataclysmic 1980 Mount St. Helens eruption and location of gauging stations (e.g., TOW; cf. Figure 27.17). SRS identifies a sediment retention structure. Note the long travel distances of debris flows along major channels draining the volcano.

tens of kilometres. Wide variations in compositions of source sediment, from abundant granular material containing few fines to sediment enriched with hydrothermally altered clays, strongly affect the character of volcanic debris flows. The variations in origins and compositions, distal flow characteristics, and a propensity for long-distance travel make debris flows at snow-clad volcanoes especially hazardous.

Events at Mount St. Helens, Washington (USA) provide signature examples of the variability, complexity, and hazards associated with debris flows from snow-clad volcanoes. The volcano, located in the Cascade Range of south-western Washington State (Figure 27.1), has been exceptionally active, especially within the past 4,500 years (Mullineaux and Crandell, 1981). Numerous modern and ancient debris flows have inundated major tributary and trunk valleys of the Toutle-Cowlitz, Lewis, and Kalama River systems. These three river systems are in turn tributaries to the Columbia River, an economically and ecologically vital commercial and fisheries corridor in the western USA. Debris flows at Mount St. Helens are perhaps the most thoroughly studied anywhere, and several papers have been written about its prehistoric and historical debris flows. In this chapter we synthesize the key findings dispersed throughout that literature. We discuss the variety of debris-flow initiation processes at Mount St. Helens; the characteristics of flows and deposits, and their significance with respect to flow genesis and evolution; the geomorphic, societal, and economic impacts of flows; their frequency of recurrence; and hazard assessments and mitigation measures. Chapter 10 provides a more general discussion of volcanic debris flows.

The stratigraphic record of prehistoric debris flows in the valleys that surround Mount St. Helens, observations and analyses of flowing debris and subsequent deposits associated with the modern eruptions, and analyses of post-1980 debris flows unassociated with eruptions have provided significant insights into volcanic debris flows and their hazards. Debris flows at Mount St. Helens recur frequently, have multiple origins, travel great distances, commonly transform downstream, and leave deposits having characteristics that provide important clues as to their origin and longitudinal development.

27.1.2 Terminology

Rapidly flowing mixtures of water and sediment from volcanoes commonly are called *lahars*. In general, *lahar* refers to an event rather than to a deposit, and that event can include one or more flow phases (Smith and Fritz, 1989; Vallance, 2000; see also Chapter 10). Thus, a single event can encompass a variety of fluid-dynamic characteristics. In the following discussions, we focus on the initiation, behaviour, and consequences of two phases of highly concentrated mixtures of sediment and water. (1) *Debris flow*, defined here as a flowing mixture of approximately equal parts sediment and water in which a broad distribution of grain sizes, commonly including gravel, is vertically well mixed. Debris flow exhibits behaviour that is strongly affected by interactions between the solid and fluid components (see Chapters 6 and 10). (2) *Hyperconcentrated flow*, defined here as a phase of flow that is transitional between debris flow and sediment-laden streamflow in which stresses exerted by the fluid transport the sediment. A hyperconcentrated flow contains volumetrically more water than sediment, but is very sediment-rich compared to normal streamflow (see Chapters 8 and 10). As a result, the coarsest particles settle rapidly and the flowing sediment–water mixture usually contains a narrower distribution of grain sizes than is found in a debris flow; the sediment in many hyperconcentrated flows is predominantly sand.

27.2 ERUPTIONS AND DEBRIS FLOWS AT MOUNT ST. HELENS – AN OVERVIEW

Eruptions and debris flows have been intimately associated with Mount St. Helens throughout its geologic history. The volcano has erupted and discharged debris flows episodically for more than 50,000 years and perhaps as long as 300,000 years (Crandell, 1987; Evarts et al., 2003). The modern phase of activity that began in 1980 and its associated debris flows are merely the latest events in a rich geologic

Table 27.2. Characteristics of deposits from the 18 May, 1980, Mount St. Helens eruption. Lipman and Mulineaux (1981).

Event	Volume of uncompacted deposit (km³)	Area affected (km²)	Deposit thickness (m)
Debris avalanche	2.5	60	10–195
Blast	0.20	600	0.01–1
Debris flows	0.04	50	0.1–3
Pyroclastic flows	0.3	15	0.25–40
Proximal tephra fall	1.1	1,000	>0.05

history. Eruptive styles have ranged from relatively quiescent eruptions that emitted fluid lava flows to violently explosive eruptions that thickly mantled the landscape with volcanic ash (tephra), triggered colossal landslides, and produced pyroclastic flows (dry flows of hot rock and gases) and very large[1] debris flows. Debris-flow and other eruption-related deposits built aprons of clastic sediment around the base of the volcano and partly filled valleys for many tens of kilometres downstream (Mullineaux and Crandell, 1981; Crandell, 1987; Scott, 1988a).

Modern activity at Mount St. Helens highlights the close association between explosive volcanism and debris flows at snow-clad volcanoes. The cataclysmic eruption on 18 May, 1980, consisted of an ensemble of volcanic processes that ravaged several watersheds (Lipman and Mullineaux, 1981; Table 27.2). Within minutes to hours of the onset of the eruption, hundreds of square kilometres of landscape were variously transformed by a voluminous debris avalanche,[2] a directed volcanic blast, debris flows, pyroclastic flows, and extensive tephra fall (Figure 27.1). The eruption began with a colossal failure of the volcano's north flank (Voight, 1981). The resulting debris avalanche deposited 2.5 km³ of poorly sorted rock, soil, ice, and organic debris in the upper North Fork Toutle River valley (Glicken, 1998), buried 60 km² of the valley to a mean depth of 45 m, and disrupted the watershed's drainage pattern (Lehre et al., 1983; Janda et al., 1984). A nearly synchronous directed blast, a type of highly mobile pyroclastic surge, followed the debris avalanche off the volcano, devastated ~600 km² of rugged terrain and blanketed the landscape with up to 1 m of gravel to silt tephra (Hoblitt et al., 1981; Waitt, 1981). Local liquefaction and dewatering of the debris-avalanche deposit spawned the North Fork Toutle River debris flow (Janda et al., 1981; Fairchild, 1987). That debris flow travelled more than 100 km and

---

[1] There is no standard classification system defining relative sizes of debris flows. Chapter 17 proposes one classification scheme. In general, volcanic debris flows are larger than non-volcanic debris flows. However, to avoid superlative appearances for the magnitudes of volcanic debris flows, we use the following arbitrary definitions: small ($<10^5$ m³); moderate (10^5–10^6 m³); large (10^6–10^7 m³); and very large ($>10^7$ m³).
[2] Very large ($>10^7$ m³) slope failures that move rapidly off volcanoes and exhibit some combination of slide and flow-like character commonly are referred to as debris avalanches (e.g., Voight et al., 1981; Siebert, 1984; McGuire, 1996). This usage differs from that in Chapter 2.

deposited tens to hundreds of centimetres of gravelly sand along channels of the North Fork Toutle, Toutle, Cowlitz, and Columbia Rivers (Figure 27.1). On the volcano's western, southern, and eastern flanks, the pyroclastic surge scoured and melted snow and ice, and triggered less voluminous debris flows that travelled up to tens of kilometres and also deposited up to hundreds of centimetres of gravelly sand on valley floors and flood plains (Janda et al., 1981; Fink et al., 1981; Gilkey, 1983; Pierson, 1985; Major and Voight, 1986; Fairchild, 1987; Scott, 1988a; Waitt, 1989).

Episodic eruptions through the mid-1990s (Brantley and Myers, 2000) triggered additional debris flows. Vigorous explosive activity that produced eruption columns and pyroclastic flows occurred in the summer and fall of 1980, but failed to generate debris flows owing to a lack of snowpack. From October 1980 through October 1986, multiple eruptions built a lava dome within the volcano's crater. Minor explosions, collapses of parts of the lava dome, and debris flows accompanied several of these eruptions. Most debris flows occurred when small pyroclastic flows and surges, related to minor collapses of the lava dome, or hot rock hurled against the crater walls by explosions from the dome, rapidly melted snow. The most noteworthy debris flow associated with this phase of activity occurred in March 1982. Between 1989 and 1991 some brief but intense bursts of activity accompanied by small explosions from the lava dome hurled hot rock within the crater, melted snow, and triggered small debris flows. From 1991 to 2000, debris flows at the volcano were triggered solely by heavy rainfalls or glacier outburst floods.

27.3 INITIATION PROCESSES

Debris flows at Mount St. Helens have been triggered by a variety of mechanisms, including transformation from landslides, melting and mixing of snowpack with volcanic debris during eruptions, and by flood surges entraining sediment. Initiation mechanisms have greatly influenced debris-flow volumes, compositions, hydrographs, flow behaviours, and hazards (Table 27.3). The largest and most hazardous flows are related to sediment entrainment by massive flood surges; the smallest and least hazardous are those caused by rainfall- and snowmelt-triggered landslides and by glacier outburst floods.

27.3.1 Transformation from landslides

Like debris flows in other environments (e.g., Iverson et al., 1997), volcanic debris flows can form when landslides transform into rapidly moving flows. Rainfall- and snowmelt-triggered landslides have caused debris flows up to many tens of thousands of cubic meters on the flanks and crater floor of the volcano. On the volcano's east flank, debris flows having discharges up to $50\,m^3/s$ have transformed from rainfall-triggered landslides in saturated debris on the Shoestring Glacier and in the moraine that parallels the outflow channel (Pierson, 1986). In the crater formed by the 1980 eruption, debris flows have started as slumps caused by elevated groundwater pressures augmented by intense rainfalls. For example, during a period of heavy rainfall in September 1997, debris flows having an aggregate volume of about 200,000 m^3 developed from two large failures near the head of Loowit channel on the eastern side of the crater breach; another suite of flows developed from a large circular slump on the western side of the breach (Figure 27.2) (T. Pierson and D. Dzurisin, US Geological Survey, written commun., 2003). Despite their magnitudes, these flows travelled only 2 km or less beyond the mouth of the breach.

In contrast to the relatively small rainfall-triggered events, colossal, eruption-related landslides, known as debris avalanches, have occurred at Mount St. Helens (e.g., Hausback and Swanson, 1990; Glicken, 1998; Figure 27.3), but they have not transformed directly into debris flows. A lack of direct transformation of debris avalanche to debris flow at Mount St. Helens stands in contrast to the behaviour of volcanic debris avalanches elsewhere (e.g., Carrasco-Núñez et al., 1993; Vallance and Scott, 1997; Scott et al., 2001; Chapter 10), perhaps because, owing to the youth of the edifice, the Mount St. Helens debris avalanches contained relatively little hydrothermally altered clay (Glicken, 1998) which is common at many other volcanoes that have shed debris avalanches (e.g., Hyde and Crandell, 1978; Vallance, 1985; Carrasco-Núñez et al., 1993; Vallance and Scott, 1997; Vallance, 1999; Siebert et al., 2004). Instead, debris avalanches at Mount St. Helens have triggered debris flows in at least two other ways: (1) through local liquefaction and dewatering; and (2) by impounding lakes that subsequently breached and formed water floods that entrained sediment and evolved into debris flows (see Section 27.3.3).

A voluminous (10^8 m^3), destructive debris flow that moved through the North Fork Toutle River valley on the afternoon of 18 May, 1980, provides a spectacular example of a flow that developed from liquefaction and dewatering of a debris-avalanche deposit (Janda et al., 1981; Fairchild, 1987). Several thin flows a few metres deep formed when springs developed and discharged, or when water-saturated parts of the avalanche deposit locally slumped and flowed, pooled in closed depressions, and broke out of those depressions as larger, more homogeneous flows (Janda et al., 1981; Fairchild, 1987; Figure 27.4). As these disparate flows merged and entrained sediment, the North Fork Toutle debris flow enlarged. The source of water for this debris flow consisted of groundwater trapped in the avalanche as it swept off the volcano, finely comminuted glacier ice transported in the avalanche and juxtaposed against hot magmatic debris, and locally entrained streamflow.

Downstream movement of the North Fork Toutle debris flow was delayed because of its peculiar origin. Unlike other eruption-related debris flows triggered immediately by snowmelt, the peak of the North Fork Toutle debris flow did not leave its source region on the avalanche deposit until some 5 hours after the eruption began (Janda et al., 1981; Fairchild and Wigmosta, 1983; Fairchild, 1987). The timing of flow initiation matches closely the patterns of seismicity associated with the eruption, and Fairchild (1987) proposed that a succession of harmonic tremors caused by the eruption triggered local liquefaction of the debris-avalanche deposit. From his analyses of melt rates of ice trapped in the deposit and the intensity of the seismicity associated with the tremors, Fairchild (1987) concluded that enough ice

Table 27.3. Characteristics of debris flows and hyperconcentrated flows at Mount St. Helens.

Year	Date	Event	Flow type	Distance from crater (km)	Volume (m³)	Depth (m)	Velocity (m/s)	Peak discharge (m³/s)	Duration (min)	Reference
Snowmelt by hot volcanic debris										
1980	18 May	SFT River	df	4	1.3×10^7	<10	>30	68,000	6.5	3,5,6
				18	1.2×10^7	20–30*	20	30,000	14	
			hcf	45	8×10^6	4	5–10	3,800	72	
				65	8×10^6			2,900	90	
1980	18 May	Muddy R	df	3			40			9
				4		5	~30	~200,000†	<5	
				7		21	28	66,800		
				7.5		5	27	63,900		
				8		2	20	65,000		
				10			23	20,000		
				18			7	7,200		
				28	$1.4 \times 10^{7\text{–}8}$	3–4	3	5,000	120	
1980	18 May	Pine Ck	df	10		15	18	28,600		9
				15		11	21	19,200		
				21		6	12	7,300		
1980	18 May	SW flank	df	5	$1 \times 10^{6**}$	2–3	5	3,000–6,000		7,8
1982	19 March	NFT River	df/hcf	4	$1\text{–}2 \times 10^7$	4–5	15	14,000		4,11,12
				17		3–7	8–10	6,800		
				25		2–3	7	3,700		
				40		3–5	6	960		
				60		2	3–4	450		

Year	Date	Location	Type						Refs
1984	14 May	NFT River	df	8		9	3,400	<10	14
1986	9 May	crater	df	15		5–7	1,000–1,500	<10	14
				3–5	2–4	5–6	870	<2	2

Liquefaction of avalanche deposit

1980	18 May	NFT River	df	15	1.4 × 10⁸	5–8	6–12	7,200	480 (150)‡	3,5,6
				50	1.3 × 10⁸	5–8	6–12	6,600	590	
				70	1.2 × 10⁸	5–8	6–12	6,050	660 (60)	

Lake breakout flood

~2,500 BP		PC1	df	30–50	>10⁹	30	15–20	>200,000		15
				80	>10⁹	~10		50,000		

Rainfall/glacier outburst flood

1980–1983	summer	E flank	df	4	800–4,800	1–3	<1	10–500		1,10
1997	16 Sept	crater	df	2–3	<200,000	7–10	10	4,300		13

* At valley constriction.
† Estimate for broad flow on unchannelled fan that entered multiple tributary valleys.
§ Combined volume of Muddy River and Pine Creek debris flows that entered Swift Reservoir.
** Combined volume of three moderate debris flows.
‡ Values in parentheses represent duration of the discharge peak.

References: (1) Brugman and Post (1981); (2) Cameron and Pringle (1990); (3) Cummans (1981); (4) Dinehart (1999); (5) Fairchild (1987); (6) Fairchild and Wigmosta (1983); (7) Major (1984); (8) Major and Voight (1986); (9) Pierson (1985); (10) Pierson (1986); (11) Pierson (1999); (12) Pierson and Scott (1985); (13) T.C. Pierson and D. Dzurisin (U.S. Geological Survey, written commun., 2003).

Figure 27.2. Oblique aerial view looking south to the crater of Mount St. Helens showing the failure scars of the September 1997 rainfall-triggered debris flows on the crater floor. The head of Loowit channel is in the left-centre of view. Note the rock-avalanche deposit on the north side of the 1-km-diameter lava dome and the glacier growing and enveloping the dome.

Photo: Bergman Photographic Services, 5 October, 2000.

was available and that it melted sufficiently rapidly to saturate part of the avalanche deposit, and that the resulting water-saturated debris could have liquefied under seismic loading. He also concluded that without fine comminution of ice during avalanche transport, the North Fork Toutle debris flow probably would not have formed.

27.3.2 Scour, melting, and mixing of snowpack during eruptions

Debris flows generated by scour and melt of snowpack occur commonly during eruptions of Mount St. Helens, flow away from the volcano immediately, and have short-lived hydrographs. The initial explosions at Mount St. Helens on 18

Figure 27.3. Oblique aerial photograph of the 1980 debris-avalanche deposit. View to the east.
Photo: Austin Post, USGS, 30 June, 1980.

May, 1980, developed into a huge pyroclastic surge that triggered debris flows on all flanks of the volcano, but the exact initiation mechanism is ambiguous. Large debris flows travelled up to tens of kilometers along the South Fork Toutle River, Pine Creek, and Muddy River (Figures 27.1 and 27.5); moderate debris flows swept the volcano's broad southern sector (Figure 27.5).

Eyewitnesses to the 1980 eruption, 8 km west of the volcano, observed the debris flow along the South Fork Toutle River within 10 minutes of the eruption (Scott, 1988a) and reported that its discharge peak lasted but a few minutes. The flow began after the pyroclastic surge scoured and melted snow and ice, and possibly triggered slab snow avalanches (Janda et al., 1981; Fairchild, 1987; Scott, 1988a; Waitt, 1989). Scott (1988a) suggested that the surge may have been partly wet, either from ejected hydrothermal fluid or entrained snow, and that part of it transitioned into a debris flow. Waitt (1989), however, suggested that impacting hot rocks and heat from the

Figure 27.4. Oblique aerial photograph of the debris-avalanche deposit at a constriction in the North Fork Toutle River valley near Maratta Creek. View to the east. Dark-colored areas are deposits of debris flows generated by local liquefaction and dewatering of the avalanche deposit. The debris avalanche is about 1 km wide at constriction.

From Glicken (1996). Photo: Austin Post, USGS, 19 May, 1980.

surge mobilized thousands of small slushflows that eroded and mixed with surge debris and coalesced rapidly into a homogenous debris flow that further scoured channel headwaters. As evidence for this initiation mechanism Waitt (1989) cites: (1) near-source primary deposits of a hot, dry surge overlain sharply by debris-flow deposits, which in turn are overlain by deposits of fine ash that began falling about 20 minutes after the onset of the eruption; and (2) long, streamlined lineations

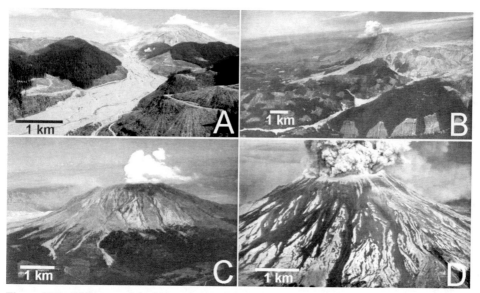

Figure 27.5. Oblique aerial photographs of debris flows triggered by snowmelt on 18 May, 1980. (A) South Fork Toutle River. (B) Muddy River valley. (C) South-west flank. (D) South flank on the morning of 18 May, 1980, showing tracks of small debris flows that descended the broad southern sector of the volcano.

(A) Photo: USGS, summer 1980. (B) Composite image of photographs by Austin Post, USGS, 30 June, 1980. (C) Photo: Austin Post, USGS, 30 June, 1980. (D) Photo: Robert Krimmel, USGS.

along the path of the debris flow, features not observed in areas swept only by the surge. The stratigraphic evidence suggests that leading and trailing edges of the giant pyroclastic surge passed along the headwater channels of the South Fork Toutle valley before arrival of the surge-triggered debris flow and that the discharge peak of the debris flow was a flashy, very short-lived event. Streamlined lineations along headwater paths of the debris flow are indicative of rapid transformation of the hypothesized slushflows to sediment-rich slurries. The great mobility and extraordinarily high velocity of the debris flow in headwater channels (Table 27.3; Fairchild and Wigmosta, 1983; Scott, 1988a; Waitt, 1989) suggest a strong coupling between the debris flow and the surge, and indicate that initiation may also have involved melting and mixing of snow into the basal boundary layer of the surge.

Eruption-triggered debris flows on the volcano's east flank formed in a similar manner (Janda et al., 1981; Pierson, 1985; Brantley and Waitt, 1988; Waitt, 1989). There, the pyroclastic surge triggered several highly energetic debris-flow pulses that swept across fans at the base of the volcano, entered Pine Creek and multiple tributaries of the Muddy River system, and in the Muddy River valley coalesced into a single flow in the trunk channel below its confluence with Smith Creek (Figures 27.1 and 27.5). That debris flow, together with the one in Pine Creek, deposited more than $10^7 \, \text{m}^3$ of sediment into Swift Reservoir (Table 27.3; Janda et al., 1981; Pierson, 1985). Within 5 km of the volcano, discharge peaks of the various pulses of debris flow were very short-lived; as in the South Fork Toutle valley, debris-flow deposits to the east are overlain locally by fine ash that began falling about 20 minutes after the onset of the eruption (Waitt, 1989).

Hours after emplacement of the large debris flows, much smaller debris flows descended the South Fork Toutle and Muddy River valleys (Pierson, 1985; Scott, 1988a; Waitt, 1989). These flows developed when small ashflows and avalanches of hot pumice descended part-way down the volcano's slopes and melted snow and ice high on the volcano. In the Muddy River valley, the later debris flow deposited about 500,000 m^3 of debris into Swift Reservoir (Pierson, 1985).

Small to moderate debris flows triggered by the primary surge formed on the volcano's broad southern sector (Figure 27.5). On the south-west flank of the volcano, three flows having a cumulative volume of about 10^6 m^3 travelled about 5 km from the former summit (Major and Voight, 1986), and comparably sized flows to the south travelled slightly farther along Swift Creek (Figures 27.1 and 27.5).

27.3.3 Sediment entrainment (bulking) and transformation of floods

Floods of water have episodically entrained sediment and transformed into debris flows at Mount St. Helens. This transformation process is known as bulking in the sedimentologic literature, in reference to the associated increase of flow volume that accompanies sediment entrainment (cf. Chapter 7). Transformations of floods to debris flows at Mount St. Helens have been associated with four flood-generating mechanisms: (1) breaching of an eruption-induced meltwater lake; (2) eruption-triggered meltwater floods; (3) breaching of landslide-dammed lakes; and (4) glacier outburst floods.

Breaching of an eruption-induced meltwater lake triggered the largest debris flow at Mount St. Helens between 1980 and 2000. On 19 March, 1982, a minor eruption from the lava dome sprayed hot ballistics that melted and destabilized a meters-thick snowpack on the rear wall of the volcano's crater, triggered snow avalanches, and formed a lake behind the dome when the rate of snowmelt exceeded the rate of meltwater drainage (Waitt et al., 1983). A flood of water and pumice from the lake subsequently discharged simultaneously around both sides of the lava dome, eroded sediment from the crater breach, and emerged as a debris flow that moved across the surface of the debris-avalanche deposit and into the North Fork Toutle River (Figure 27.6; Waitt et al., 1983; Pierson, 1999).

Eruption-triggered meltwater floods spawned most of the post-1980 debris flows that have issued from the crater of Mount St. Helens. Except for the event in 1982, floods developed directly from rapid snowmelt without formation of a lake. The primary processes responsible for destabilizing and melting snowpack include: (1) highly energetic explosive ejection from the dome of hot ballistic fragments that mixed broadly with snowpack and triggered dirty snow avalanches; (2) low-energy explosions that focused hot debris on snow-covered slopes near the vent; and (3) collapses of segments of the lava dome or lobes of newly erupted lava, which triggered hot rock avalanches, pyroclastic flows, or surges that melted snowpack on the crater floor (Waitt and MacLeod, 1987; Mellors et al., 1988; Cameron and

Figure 27.6. Oblique aerial photograph of March 1982 snowmelt-triggered debris flow. View to the south. The flow began as a water flood from the crater entrained sediment along the steep north flank and became a debris flow that moved west (right) down the North Fork Toutle River valley.

Photo: T.J. Casadevall, USGS, 21 March, 1982.

Pringle, 1990; Pierson and Waitt, 1999; Pringle and Cameron, 1999). Examination of several of the post-1980 events within days of occurrence revealed complex interactions of hot volcanic debris with snowpack, which involved combinations of dirty snow avalanches, slushflows, and water floods (e.g., Waitt and MacLeod, 1987; Pierson and Janda, 1994; Pierson and Waitt, 1999). In each case, flows leaving the crater entrained sediment from the north flank of the volcano and emerged onto the valley floor as debris flows. Studies also showed that the initiation and magnitude of snowmelt-generated debris flows were related more to the efficacy of hot-rock interactions with snowpacks than to the relative magnitudes of the eruptions (e.g., Pierson and Waitt, 1999).

The largest debris flows that have been identified at Mount St. Helens occurred in rapid succession in the Toutle River valley about 2,500 radiocarbon years ago, and were associated with successive breaching of one or more landslide dams. A series of four debris flows occurred (named PC1–PC4; Scott, 1988b), the largest (PC1) of which exceeded $10^9 \, m^3$. These debris flows formed as the middle segments of flows that began and ended as flood surges (Scott, 1988b). The largest flow entrained sediment for more than 20 km before it transformed into a debris flow (Scott, 1988b, 1989). A sudden release of water from an ancestral Spirit Lake

(Figure 27.1) is the only possible source of the flood surge that produced the huge debris flow in this series. Deposits of possibly two prehistoric debris avalanches from the north flank of Mount St. Helens have been identified and dated to an eruptive period that occurred between 2,500 and 3,000 radiocarbon years ago (Hausback and Swanson, 1990). These debris avalanches probably blocked the outlet from, and caused enlargement of, Spirit Lake, as did the 1980 debris avalanche. The sequence of prehistoric debris flows that began as large flood waves therefore provides an analogue for what could have happened from breakouts of major lakes formed or enlarged by the 1980 eruption had lake levels not been stabilized by engineering intervention.

Glacier outburst floods triggered small debris flows to $5,000\,m^3$ on the flanks of the volcano during warm weather both before and after the 1980 eruptions (Brugman and Post, 1981; Pierson, 1986). Debris flows triggered by outburst floods occur commonly at Cascade Range volcanoes (e.g., Walder and Driedger, 1995), but they travel relatively short distances and beyond about 10–15 km from a volcano they generally pose little hazard.

27.4 FLOW CHARACTERISTICS

27.4.1 Hydrographs, peak discharges, and flow volumes

Initiation mechanism strongly affects the size and character of, and the degree of hazard associated with, volcanic debris flows. At Mount St. Helens, flow volumes, depths, velocities, runout distances, and hydrographs varied widely as a function of triggering process (Table 27.3). The most hazardous and destructive events are the large, deep ($>2\,m$), fast-moving ($>5\,m/s$) flows that transport great volumes of sediment many tens of kilometers from the volcano (Table 27.3; Figure 27.7). These are the flows that have been triggered by abundant snowmelt by hot volcanic debris, liquefaction of a debris-avalanche deposit, or breakout of a large avalanche-dammed lake. The least hazardous and destructive events are those flows that remain near the volcano. Commonly they are associated with minor snowmelt, intense rainfalls, or glacier outburst floods.

The largest and most hazardous flows at Mount St. Helens were the great prehistoric flows in the Toutle River valley caused by lake breakouts (Scott, 1988b). Deposits of that series of debris flows dominate downstream valley fills in the North Fork Toutle and Toutle River. The largest of those debris flows (PC1) exceeded $1\,km^3$, had a peak depth of about 30 m, and an astounding instantaneous peak discharge of $200,000\text{--}300,000\,m^3/s$ at distances between 30 and 50 km from the volcano (Scott, 1989). At a distance of 80 km the estimated peak discharge was as great as $50,000\,m^3/s$. More than 90% of the flow volume continued on to the Columbia River or has been reworked (Scott, 1988b).

Flows generated by snowmelt during the cataclysmic 1980 eruption can be characterized as intense pulses that had brief but exceptionally great peak discharges near the volcano, but which attenuated rapidly downstream (Table 27.3). About

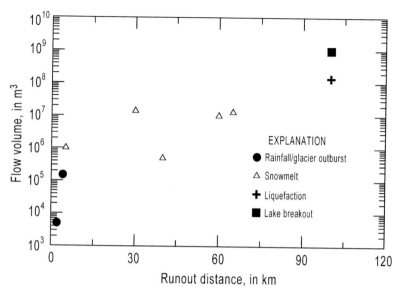

Figure 27.7. Relationships between flow volume and runout distances for volcanic debris flows at Mount St. Helens (see Table 27.3). Runout distances for the lake-breakout and liquefaction generated debris flows are minimum values because at 100 km from the volcano the flows entered the Columbia River.

4 km from the volcano, the South Fork Toutle debris flow had a peak discharge of 68,000 m^3/s and a volume of 13 million m^3, but it lasted only about 6.5 minutes. At the confluence with the North Fork Toutle River (about 50 km from the volcano), the flow had a peak discharge of only 4,000 m^3/s, a volume of 8 million m^3, and it lasted about 1 hour (Figure 27.8; Fairchild and Wigmosta, 1983; Fairchild, 1987). On the east flank, peak discharge across the Pine Creek–Muddy River fan and into tributaries feeding the main channels (Figure 27.1) possibly exceeded 200,000 m^3/s collectively, but decreased exponentially downstream (Pierson, 1985). When the merged Muddy River debris flow reached Swift Reservoir, 30 km from the volcano, it had a peak discharge of less than 5,000 m^3/s and lasted about 2 hours (Pierson, 1985; Table 27.3). The total volume of debris transported through the Pine Creek–Muddy River system exceeded 14 million m^3.

The 1980 North Fork Toutle River debris flow was a much larger, longer event that started later than the other major 1980 debris flows, and it had an unusual hydrograph owing to its peculiar origin (Figure 27.8). The debris flow was highly erosive across the distal avalanche deposit, but eroded little of the North Fork Toutle valley beyond the deposit terminus. Within 4.5 km of source (about 15 km from the volcano), the debris flow had a volume of about 140 million m^3 and lasted nearly 8 hours (Fairchild and Wigmosta, 1983). Its hydrograph rose slowly to peak discharge over more than an hour, remained at peak discharge (7,200 m^3/s) for more than 2 hours, and waned gradually over 4 hours (Fairchild and Wigmosta, 1983). At the mouth of the Toutle River, 70 km from the volcano, flow volume

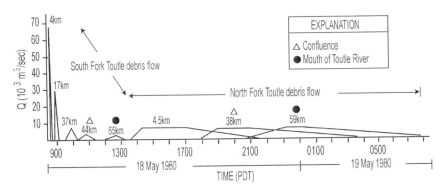

Figure 27.8. Reconstructed hydrographs for the 18 May, 1980, South Fork Toutle and North Fork Toutle debris flows. Distances above each hydrograph indicate distances from source of flow.

From Fairchild (1987), published with permission.

(120 million m^3) and peak discharge (6,050 m^3/s) had decreased by only 15%, but flow duration had increased to 11 hours; peak discharge lasted about 1 hour. The exceptional volume, minimal attenuation, and long duration of both the flow and the near-peak discharge caused extensive damage along the Toutle valley.

Post-1980 debris flows had smaller discharges and attenuated more rapidly than the primary flows triggered by the cataclysmic 1980 eruption (Table 27.3). The largest of the eruption-triggered events between 1980 and 2000, in March 1982, involved a debris flow that transformed downstream into a hyperconcentrated flow. The debris flow had a peak discharge that attenuated rapidly from about 14,000 m^3/s near the crater (Pierson and Scott, 1985; Pierson, 1999) to 3,700 m^3/s 20 km farther downstream. The hyperconcentrated phase near the mouth of the Toutle River had a peak discharge less than 500 m^3/s (Pierson and Scott, 1985; Dinehart, 1999). Most other post-1980 eruption-triggered debris flows were short-duration events that had peak discharges less than 500–1,000 m^3/s at or near the volcano (Table 27.3), and each was assimilated rapidly by streamflow in the North Fork Toutle River (e.g., Waitt and MacLeod, 1987; Cameron and Pringle, 1990; Pierson and Waitt, 1999; Pringle and Cameron, 1999). Debris flows triggered by rainfalls or outburst floods on the flanks of the volcano had volumes less than a few thousand cubic meters within a few kilometers of source and had peak discharges comparable with or smaller than those of the minor snowmelt-triggered flows (e.g., Brugman and Post, 1981; Pierson, 1986). The largest of the rainfall-triggered debris flows, in 1997 along Loowit channel, had an estimated peak discharge of 4,300 m^3/s at the mouth of the crater (T.C. Pierson and D. Dzurisin, US Geological Survey, written commun., 2003).

27.4.2 Sediment entrainment (bulking)

Most debris flows at Mount St. Helens have been formed by meltwater or flood surges that entrained sediment. Depending upon the textural and lithologic characteristics of the entrained sediment, evidence of entrainment may be more or less obvious. Debris flows that form on the slopes of the volcano typically contain a high percentage of angular particles that are composed of volcano-related lithologies. Although these flows may begin as meltwater or flood surges (e.g., 1980 and 1982 debris flows), it is difficult to deduce their origin from textural characteristics of near-source deposits alone, and other diagnostic evidence of origin is required. In contrast, flows that derive most of their sediment from river channels entrain alluvium that typically is rounded and partly composed of lithologies not found on the volcano. The percentages of rounded and exotic particles in deposits at various distances along the flow path can provide information about the degree of sediment entrainment and the likely origin of the flow. Scott (1988a) defined a "bulking factor" as the proportion of demonstrably entrained sediment within a deposit. Using this concept, he showed that about 15–20% of the peak-flow floodplain facies deposits of the 1980 North Fork Toutle and South Fork Toutle debris flows at and beyond the confluence of the forks was entrained alluvium, whereas at least 70% and perhaps more than 90% of the sediment in the gigantic prehistoric lake-breakout debris flow in the Toutle valley was entrained alluvium. On the basis of estimated flow volume and bulking factor, Scott (1988b) concluded that the flood surge responsible for that prehistoric flow eroded about 30 km^2 of the upper North Fork Toutle valley to depths of 10–20 m.

The steeply incised north flank of the volcano has served as a primary sediment source for many of the post-1980 debris flows that have issued from the crater. The March 1982 flow transformed from a water flood to a debris flow over a distance of 3 km as it progressed down the north flank of the volcano (Pierson, 1999). Transformation to debris flow occurred as the flood passed bank-to-bank through a narrow channel bounded by high (>20 m), steep (>45°) banks consisting of non-cohesive sand and gravel, and triggered numerous bank failures. Smaller floods and slushflows likewise entrained sediment from the steep canyons on this flank and transformed into debris flows (Waitt and MacLeod, 1987; Cameron and Pringle, 1990; Pringle and Cameron, 1999). The area has been a persistent source of sediment in the upper North Fork Toutle valley; between 1987 and 1999 it supplied nearly 30% of the sediment eroded from the upper valley (Bradley et al., 2001; US Army Corps of Engineers, 2002).

27.4.3 Distal transitions of flow character

Many debris flows at Mount St. Helens have undergone distal transitions of flow character. The propensity of debris flows to devolve and undergo distal transformations to hyperconcentrated flows and eventually to sediment-laden floods appears to be more common in volcanic than in non-volcanic debris flows. Debris flows that contain less than a few % of clay-sized sediment are more likely to transform distally than are those that contain more clay. At Mount St. Helens, as at other volcanoes in the Cascade Range (e.g., Hyde and Crandell, 1978; Vallance, 1985; Scott et al., 1995; Vallance and Scott, 1997; Vallance, 1999; O'Connor et al., 2001), debris flows that contained more than 3–5% clay (of particles smaller than 32 mm) generally maintained their textural character over many tens of kilometres of travel distance, whereas those that contained less clay-size sediment typically transformed, or began to transform, to hyperconcentrated flow over comparable travel distances (e.g., Scott et al., 1995).

Transformation from debris flow to hyperconcentrated flow, and eventually to sediment-laden flood, involves entrainment of water and loss of sediment mass (e.g., Pierson and Scott, 1985; Chapter 8). When large debris flows at Mount St. Helens moved downstream and diluted, their transport competence decreased and progressively smaller clasts settled out, at first from the head and base of a flow and then progressively upward and into the main body (e.g., Pierson and Scott, 1985; Scott, 1985). Gradually all dense, coarse clasts settled and only pumice gravel, sand, and finer sediment were transported in suspension (e.g., Dinehart, 1999). Eventually, all but those particles within the competence of a sediment-laden flood settled. Flows that contained even a few % clay did not dilute efficiently. In those flows, minimal dilution occurred over travel distances of many tens of kilometres, and both channel and flood-plain deposits at great distances from source are characteristic of debris flows (Hyde, 1975; Scott, 1988a, 1989; Major and Scott, 1988).

Distal transformations of debris flows are well documented for, but are not restricted to, large flows that travelled tens of kilometres along primary channels (Pierson and Scott, 1985; Scott, 1985, 1988a; Major and Scott, 1988; Pringle and Cameron, 1999). Even some small flows that travelled only a few kilometres transformed rapidly (e.g., Cameron and Pringle, 1990). Other small to moderate clay-poor flows that entered channels carrying only ephemeral flow did not transform (e.g., Major and Voight, 1986; Pierson, 1986). Distally transformed flows left characteristic deposits that record the progressive change from debris flow to hyperconcentrated flow (see Section 27.5.3). Recognition of the genesis and significance of those deposits has been used in hazard assessments to extend estimates of the impact and frequency of eruption-related debris flows along primary channels (e.g., Scott, 1989).

27.5 DEPOSIT CHARACTERISTICS

27.5.1 Thicknesses

Debris-flow-deposit thicknesses at Mount St. Helens vary widely, are strongly influenced by flow genesis and topography, and commonly are thin in relation to flow depth. Ancient lake-breakout debris flows produced the thickest and most extensive debris-flow deposits along the Toutle River valley. Those deposits dominate downstream fill in the Toutle valley and form a widespread terrace underlain by as much as 12 m of cobble-rich sediment (Scott, 1988b). Flows formed by snowmelt at the volcano, by liquefaction and dewatering of the 1980 debris-avalanche deposit, or by heavy rainfall or glacier outburst floods produced thinner, less extensive, deposits. Topography, however, exerts a greater control than genesis on debris-flow-deposit thickness. Deposits at Mount St. Helens range in thickness from as little as a few centimetres across unchannelled fans at the base of the volcano (e.g., Pierson, 1985;

Major and Voight, 1986; Waitt, 1989) to as much as several metres on floors, flood plains, and terraces in confined valleys (e.g., Mullineaux and Crandell, 1962; Hyde, 1975; Janda et al., 1981; Fink et al., 1981; Gilkey, 1983; Pierson, 1985; Crandell, 1987; Brantley and Waitt, 1988; Scott, 1988a, b, 1989; Waitt, 1989; Cameron and Pringle, 1990; Pringle and Cameron, 1999). At distances as great as many tens of kilometres from the volcano, debris-flow deposits on flood plains and in terraces are as thick as a few metres. Such great thicknesses at these distances attest to the large size of many debris flows at Mount St. Helens and to the ability of many debris flows to maintain their textural character for great distances. In comparison, the 1980 debris-avalanche deposit in the upper North Fork Toutle River valley ranges from about 10–200 m thick, and has a mean thickness of 45 m over its 20 km length (Glicken, 1998).

Deposit thickness, even at great distances, provides an incomplete picture of hazards associated with a flow, however. In general, deposits of moderate and larger debris flows typically are thin in relation to peak flow depth. Peak flow depths of many of the debris flows triggered by the cataclysmic 1980 eruption were about ten times greater than resulting deposit thicknesses (e.g., Figure 27.9; Janda et al., 1981; Pierson, 1985; Major and Voight, 1986; Scott, 1988a). Comparable relationships hold for ancient debris flows in the Toutle valley (Scott, 1988b)

Figure 27.9. Mudlines on trees along Muddy River valley illustrating the great difference between flow depth and deposit thickness. Note the person for scale in the circle.
Photo: Lyn Topinka, USGS.

and for some of the smaller post-1980 debris flows (e.g., Cameron and Pringle, 1990). Caution should be exercised, however, when inferring relationships between deposit thickness, flow depth, and the nature of the depositional process. The thickness of near-source deposits of small debris flows can overestimate flow magnitude because flows can deposit sediment progressively from head to tail and develop deposits that are thick relative to flow depth, and multiple flows can stack without obvious stratigraphic distinction and construct a deposit that appears to be the product of *en masse* deposition from a single flow (e.g., Major, 1997, 2000). Likewise, large to very large volcanic debris flows can deposit sediment progressively from head to tail during sustained flow and also produce deposits that are thick relative to flow depth (Vallance, 2000).

27.5.2 Megaclasts

Volcanic and non-volcanic debris flows can transport clasts having diameters many hundreds to thousands of times larger than the mean grain diameter of the enclosing sediment (e.g., Johnson, 1970). At Mount St. Helens, dense boulders, breccias, and fragile assemblages of layered deposits up to metres in diameter have been transported as megaclasts within debris flows. The 1980 Muddy River debris flow transported one of the largest known boulders. That boulder, which measured about $10 \times 6 \times 5$ m, was found resting on the 1980 deposit about 5 km below the Clearwater Creek confluence (Figure 27.1; Pierson, 1985). Because the debris-flow depth at that site was only about 4 m, the boulder likely was entrained locally and rolled as bedload.

Metre-scale fragile clasts, such as chunks of debris-flow or debris-avalanche deposits or intact stratigraphic sections, provide important clues regarding the origin and development of some debris flows. For example, Scott (1988b) found deposits of ancient debris flows that contained megaclasts as large as 8 m in diameter composed of intact stratigraphic assemblages and altered dacite breccia (Figure 27.10). The breccia clasts represent blocks of an ancient debris-avalanche deposit, and their abundance strongly suggests the debris-avalanche deposit was perhaps large enough to have dammed the North Fork Toutle River valley. Scott (1988b) used the abundance of those clasts as one piece of evidence to conclude that those ancient debris flows originated as large dam-break floods. Major and Scott (1988) found in the Lewis River valley, 35 km from Mount St. Helens, a debris-flow deposit at least 10 m thick that contains three types of megaclasts as large as 5 m in maximum dimension. One type is composed of intact stratified sand having sedimentary structures such as ripple marks (Figure 27.10(C)); a second type is composed of poorly sorted volcaniclastic breccia (Figure 27.10(D)); and a third type is composed of massive, poorly to moderately sorted sand and granules that probably are eroded masses of hyperconcentrated-flow deposits. Hyde (1975) noted similar megaclasts elsewhere in the valley. These megaclasts are entrained in a deposit that contains abundant rounded gravel. The characteristics of the megaclasts and the abundance of rounded gravel suggest that this deposit resulted from a debris flow having a flood-surge origin much like the largest debris flows in the Toutle

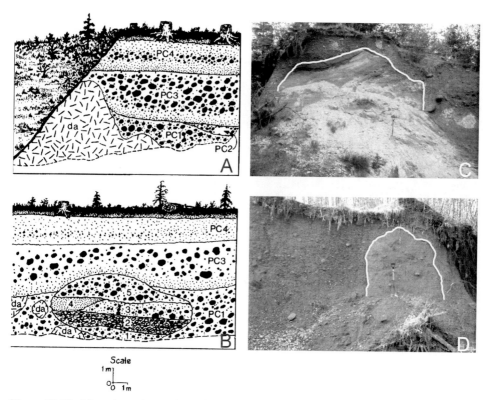

Figure 27.10. Megaclasts in ancient debris-flow deposits. (A) Dacite megaclast, formerly exposed along North Fork Toutle River valley near the SRS, resting on the channel surface and extending above the debris-flow deposit surface. (B) Megaclast composed of the pre-debris flow section of the flood-plain stratigraphy, formerly exposed along the North Fork Toutle River valley near the SRS. The stratigraphic section includes both debris-flow (1,3,5) and alluvial (2,4) deposits (from Scott, 1989). Megaclasts of (C) stratified sandy alluvium and (D) poorly-sorted debris-avalanche deposit in an ancient debris-flow deposit near Yale dam in the Lewis River valley. Clasts are outlined for clarity. Metre-long shovel for scale.
Source: (A) and (B) from Scott (1989); (C) and (D) from Major and Scott (1988).

valley. Evidence of a large volcanic debris avalanche in the Lewis River valley (Newhall, 1982; M.A. Clynne, US Geological Survey, written commun., 2003) and entrainment of debris-avalanche blocks in the debris-flow deposit indicate that the Lewis River at one time breached an avalanche deposit and produced a large flood wave that evolved into a massive debris flow.

27.5.3 Textures

Textures of debris-flow deposits at Mount St. Helens vary widely, but they all share common traits. Deposits of large debris flows commonly can be separated into channel and flood-plain facies; deposits of smaller flows commonly are not subdivided. Channel-facies sediment is deposited by flow within a confined channel; flood-plain facies sediment is deposited by flow that spills over bank. Both facies are composed of deposits that generally are massive (unstratified), poorly sorted, and contain fragmental sediment ranging in size from microns to metres. Particles larger than about 4 mm in the flood-plain facies typically are dispersed and surrounded by distributed smaller grains, but locally they may be in clast-to-clast contact, especially near deposit margins. Channel-facies deposits typically are coarser than flood-plain-facies deposits, are more likely to have a clast-supported framework, and sometimes are mistaken for gravelly flood deposits. Conspicuous irregularly shaped vesicles are sometimes distributed throughout deposits, particularly in the flood-plain facies. Vesicles are interpreted to represent bubbles formed from air entrapped in a nearly water-saturated flow.

Gravel fractions (>2 mm) of both facies of debris-flow deposits exhibit wide variations in size, grading, and degree of dispersal. In some deposits the coarsest gravel is only a few centimetres in diameter; in others it is several tens of centimetres in diameter (Figure 27.11). Gravels may be randomly distributed within deposits, or may exhibit inverse (upward increase in size) or normal (upward decrease in size) grading (Figure 27.11(B,D,E)). In both facies, gravels commonly are inversely graded near the base of a deposit, ungraded in the medial section, and normally graded near the top (e.g., Scott, 1988a). In contrast, particles <2 mm typically are ungraded and uniformly distributed throughout a deposit thickness. In the 1980 debris flows, clasts larger than about 50 cm were more commonly transported as bedload, although some were suspended within the flows. In the large debris flows, decimetre-sized clasts transported as bedload commonly accumulated behind obstructions, on the insides of sharp channel bends (Pierson, 1985), and in medial bars along channel floors (Scott, 1988a; Figure 27.12). In smaller flows coarse suspended clasts accumulated along margins of snouts and in levees (Figure 27.12).

Lithic clasts in the Mount St. Helens debris-flow deposits exhibit variable shapes and compositions indicative of predominant sediment sources. Many deposits contain a high percentage of angular clasts composed of lithologies associated with the volcano (Figure 27.11(A–E)). Those clasts have not been significantly reworked. They represent material entrained predominantly on or near the volcano, or material entrained from older debris-flow deposits along channel corridors. In contrast, some of the debris-flow deposits contain clasts that are predominantly rounded or are broken fragments of rounded clasts, and many of those clasts are composed of exotic lithologies not associated with the volcano (Figure 27.11(F); Major and Scott, 1988; Scott, 1988a, b, 1989). Those clasts represent stream alluvium entrained from channel beds and banks.

Basal contacts and textures of debris-flow deposits at Mount St. Helens show that intensities of basal shear stresses in flows have varied greatly. Surfaces underlying unchannelled fan deposits near the base of the volcano are variably scoured or passively overrun (e.g., Pierson, 1985; Major and Voight, 1986; Waitt, 1989). Surfaces underlying deposits on flood plains and terraces along confined channels exhibit little erosion; 1980 deposits overlie undisturbed forest litter or are in sharp depositional contact with older debris-flow deposits, and ancient deposits locally

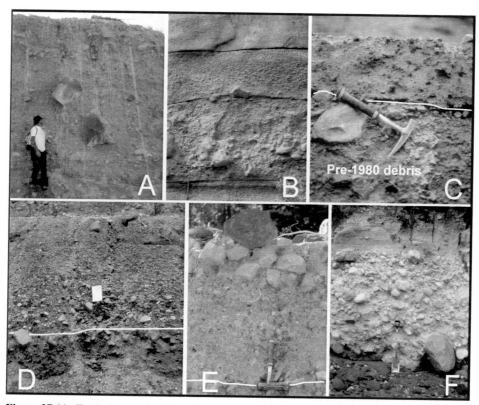

Figure 27.11. Typical textures of debris-flow deposits at Mount St. Helens. (A) Massive, poorly sorted texture of proximal deposit containing angular clasts. (B) Multiple deposits near the confluence of the North and South Fork Toutle Rivers, 50 km from the volcano. Note variations in grain sizes, sorting, and grading of the deposits. Uppermost fine-grained deposit is characteristic of flow transitional from debris flow to hyperconcentrated flow. Axe is about 25 cm long. (C) Massive, poorly sorted 1980 deposit only about 20 cm thick. Mean peak flow depth at this site on the south-west side of the volcano was about 2.5 m. (D) 1980 debris-flow deposit along Pine Creek in which large clasts are normally graded. Notebook for scale. (E) 1980 deposit in which large clasts are inversely graded. (F) Deposit of ancient debris flow in Toutle River valley that contains abundant rounded cobbles. Shovel about 1 m long. Lines in (C), (D), and (E) indicated bases of the 1980 deposits.

overlie delicate, easily erodible tephra (e.g., Fink et al., 1981; Pierson, 1985; Major and Voight, 1986; Scott, 1988a; Major and Scott, 1988). Along channel beds, however, the 1980 debris flows in the Toutle valley locally planed and truncated alluvium (Scott, 1988a).

Compacted basal debris, known as a sole layer, is preserved in some of the largest modern and ancient debris-flow deposits (Scott, 1988a). This layer, apparently unique to large volcanic debris flows, occurs primarily in association with channel rather than flood-plain-facies deposits. Observed sole layers consist of (1)

Figure 27.12. Morphologic features of debris-flow deposits at Mount St. Helens. (A) Channel-margin deposits of small post-1980-eruption debris flows having lobate and digitate morphology. (B) Cobble-boulder levee along a small channel incised into a deposit of 1980 debris flow on the south-west side of the volcano. (C) Accumulation of well-sorted cobbles and boulders, devoid of fine matrix, that was transported near the leading edge of the 1980 debris flow on south-west side of volcano. (D) Medial channel "whaleback" bar deposited by the 1980 South Fork Toutle River debris flow.

Panel (D) from Scott (1988a).

locally foliated, inversely graded sediment that is typically finer grained, texturally more uniform, and strikingly more compacted than the overlying debris (Figure 27.13(A)); or (2) a bed of unusually concentrated, clast-supported, rounded fine pebbles that resemble "ball bearings" (Figure 27.13(B)). The presence of the "ball-bearing" unit on side slopes and in channels and its absence on passively inundated flood plains indicates that it is related to channel-boundary dynamics (Scott, 1988a). The compaction and foliation of the more typical sole layer varies with flow depth and thus with the amount of basal shear exerted by a flow. The "ball-bearing" bed, an extreme product of shear-induced sorting, is found only at the base of the ancient deposit of the fastest, deepest, and coarsest debris flow at Mount St. Helens (Scott, 1988a, b, 1989). These sole layers are useful for distinguishing contacts between debris flows of similar textures and for identifying paleo-inundation limits of prehistoric events.

Deposit textures can change dramatically downvalley if debris flows undergo distal transformation. Over tens of kilometers, flood-plain-facies deposits of many

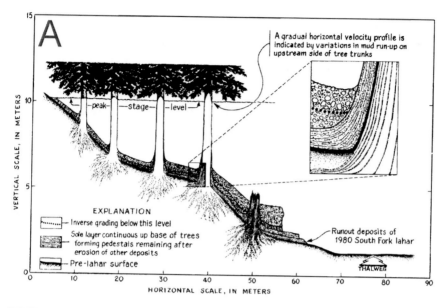

Figure 27.13. Examples of "sole layers" at bases of debris-flow deposits. (A) Compacted sole layer developed at the base of 1980 North Fork Toutle debris flow. (B) "Ball-bearing" unit developed at the base of an ancient debris flow in Toutle River valley.
From Scott (1988a, b).

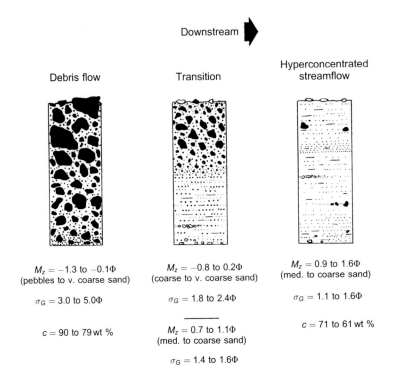

Figure 27.14. Schematic representation of deposit facies of the March 1982 snowmelt-triggered debris flow showing distal transformation from debris flow to hyperconcentrated streamflow. Ranges in mean grain size (M_z), sorting (σ_G), and sediment concentration by weight (C) are indicated. The M_z and σ_G values represent graphic mean and graphic standard deviation (Folk, 1984). Units are given in phi scale.

From Pierson and Scott (1985).

large debris flows at Mount St. Helens are poorly sorted and matrix-supported, and exhibit variable grading of the coarsest clasts. In many deposits, however, the sizes of large clasts decrease longitudinally. Debris flows that mixed significantly with river water dropped their coarsest sediment rapidly and transformed into hyperconcentrated flows and ultimately into sediment-laden river flows (e.g., Pierson and Scott, 1985; Scott, 1988a; Chapter 8). Deposits of distally devolved flows changed longitudinally from massive, poorly sorted gravelly sands that contained dispersed pebbles and cobbles to massive to crudely stratified, moderately sorted sands devoid of dispersed coarse clasts, and ultimately to stratified, well-sorted, sandy alluvium (Figures 27.14 and 27.15).

Grain-size analyses (e.g., Gilkey, 1983; Fairchild, 1985; Pierson, 1985; Major and Voight, 1986; Scott, 1988a) show that most debris flows at Mount St. Helens are very granular and clay-poor, unlike some volcanic debris flows elsewhere in the Cascade Range (e.g., Hyde and Crandell, 1978; Vallance, 1985; Vallance and Scott, 1997; Vallance, 1999). Deposit samples typically can be classified as muddy,

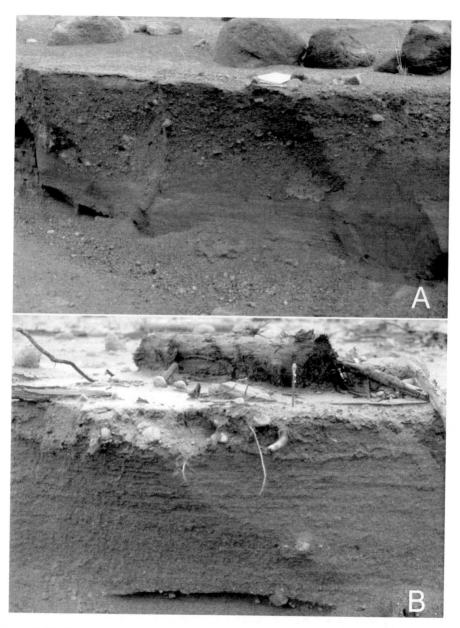

Figure 27.15. Transitional and hyperconcentrated-flow facies of the deposit of the March 1982 debris flow along Toutle River. (A) Transitional deposit showing faintly stratified coarse sand (hyperconcentrated-flow unit) overlain by unstratified gravelly sand (debris-flow unit). The units were deposited by a single flood wave in the process of transforming from debris flow to hyperconcentrated flow. (B) Deposit from fully developed hyperconcentrated flow.

From Pierson and Scott (1985).

sandy gravels or muddy, gravelly sands (Folk, 1984). The poor sorting of debris-flow deposits at Mount St. Helens is reflected in the broad distributions of sample grain sizes (Figure 27.16). More than half of the sediment in many samples is coarser than sand. However, most published grain-size analyses have excluded particles larger than a few centimetres. The coarse ends of the grain-size distributions are thus under-represented. One of the most clay-rich debris flows at Mount St. Helens was the 1980 North Fork Toutle River flow, and only about 2–5 wt% of the sampled deposit finer than 32 mm was composed of clay-sized sediment (Fairchild, 1985; Scott, 1988a).

27.5.4 Relations among deposit textures and initiation and flow processes

Deposit textures, clast shapes and compositions, and the nature of megaclasts provide an array of information regarding debris-flow initiation and transport at Mount St. Helens. The relative percentages of rounded clasts and exotic lithologies have been used to distinguish large debris flows that formed at or near the volcano from those that have formed predominantly as a result of large flood waves that ravaged channels and entrained bed and bank sediment. Debris flows that began at or near the volcano contain abundant angular debris, whereas those caused by large flood surges that moved downstream contain abundant rounded alluvium. Although flows that form at or near a volcano can contain entrained stream cobbles, the percentages of such clasts typically are subordinate. Such criteria for distinguishing initiation mechanisms are somewhat general, however. For example, had the post-1980 Spirit Lake breached and released a large flood surge, that surge would have entrained sediment from the 20-km-long 1980 debris-avalanche deposit – a deposit composed predominantly of angular volcanic debris. Downstream, the deposit from such an event would likely have had a texture distinctly different from that of the ancient lake-breakout debris-flow deposits in the Toutle valley. The abundance of rounded alluvium in the ancient deposits strongly suggests that the ancient debris avalanche(s) that blocked an ancestral Spirit Lake was not as areally extensive as the 1980 debris avalanche (Scott, 1988b).

Clay content has been used to distinguish debris flows that formed at a volcano as a result of large slope failures from more granular flows that likely resulted from eruptions that primarily melted snow and ice. Deposits containing a few to several % clay-sized material (sometimes called "cohesive" debris flows) commonly entrained volcanic debris that had been hydrothermally altered or debris from which the finest particles had not been hydraulically or explosively sorted. Slope failure is the most likely source for such sediment (Scott et al., 1995). In contrast, more granular debris flows containing less than a few % clay-sized material (sometimes called "non-cohesive" flows) are composed of sediment that originally contained few fines or sediment from which the finest particles had been winnowed. Sediment entrainment by meltwater or flood surges, or transformations of relatively small surficial landslides are the most likely causes of these flows. Clay-poor debris flows commonly transform as they move down valley. Although some particularly voluminous "non-cohesive" flows at Mount St. Helens maintained their debris-flow character for many

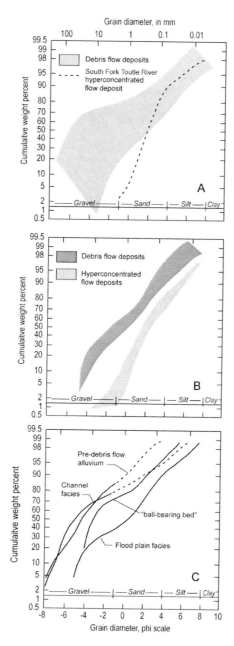

Figure 27.16. Grain-size distribution of debris-flow deposits at Mount St. Helens. (A) Composite distribution envelope of 1980 debris flows along rivers of North Fork Toutle, South Fork Toutle, Toutle, Muddy, Pine Creek, and the south-west flank. (B) Distribution envelopes for 1982 debris flow and hyperconcentrated flow (see Pierson and Scott, 1985). (C) Distribution curves for ancient lake-breakout debris flow in Toutle River valley (see Scott, 1988a).

tens of kilometres, others as large as tens of millions of cubic metres in volume transformed to hyperconcentrated flow over distances of a few tens of kilometres (e.g., 1980 South Fork Toutle flow, Scott (1988a)). Deposits of hyperconcentrated flows have been used to infer the occurrence of clay-poor debris flows upstream, and their relative thicknesses and distances from the volcano have been used to associate the source flows with probable eruptions (e.g., Major and Scott, 1988; Scott, 1988b).

27.6 GEOMORPHIC IMPACTS OF THE 1980 DEBRIS FLOWS

Sediment deposition in the Toutle and Muddy River systems by the 1980 debris flows caused significant channel modification and instability. With the exception of the upper North Fork Toutle valley, gross pre-eruption landforms remained mostly intact. Channels generally were displaced, straightened and smoothed, and changed from gravel-bedded, pool-riffle systems to sand-bedded corridors stripped of riparian vegetation. The debris flows generally filled channels, lowered flow capacity, simplified structure, and reduced roughness. Initial changes to channel cross sections are shown in Janda et al. (1981) and Lombard et al. (1981). Along the lower Cowlitz River, the North Fork Toutle debris flow was mostly confined by levees that lined the channel. The debris flow deposited more than 38 million m^3 of sediment and raised the channel thalweg by about 5 m. This change in bed elevation dramatically reduced channel conveyance capacity. The pre-eruption discharge at flood stage on the Cowlitz River at Castle Rock (Figure 27.1) was about 2,150 m^3/s. After the debris flow, the discharge at flood stage was reduced to about 10% of its pre-eruption magnitude (Lombard et al., 1981). The North Fork Toutle debris flow also deposited about 34 million m^3 of sediment in the Columbia River, and locally raised the channel bed by about 8 m near the confluence with the Cowlitz River.

The style of post-eruption channel adjustments varied with type of disturbance, but generally followed complex cycles of incision, aggradation, and widening. Channel development on the debris-avalanche deposit began during the initial phase of liquefaction and dewatering which triggered fill and spill of small ponds and erosion by the North Fork Toutle debris flow. Channel adjustments subsequently followed a sequence of incision, aggradation, and widening (Meyer and Martinson, 1989). In general, channels on the avalanche deposit incised tens of metres and widened hundreds of metres. Similar, but less dramatic, adjustments occurred in valleys affected by debris flows. Debris-flow-affected channels generally incised up to several metres, widened by tens of metres, and locally aggraded by as much as a couple of metres (Meyer and Janda, 1986; Meyer and Martinson, 1989; Simon, 1999; Hardison, 2000).

Post-eruption erosion and sediment redistribution exacerbated the effects of the 1980 debris flows. Rapid, dramatic, and persistent channel erosion led to extraordinary post-eruption sediment yields. In the Toutle River system, sediment yields were initially as much as several hundred times greater than pre-eruption yields (Figure 27.17), and even after 20 years yields from the upper North Fork Toutle valley remained about 100 times greater than pre-eruption levels (Major et al., 2000).

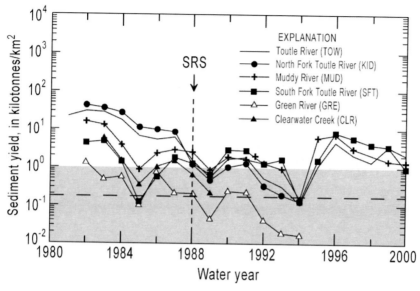

Figure 27.17. Annual suspended-sediment yields at Mount St. Helens from 1980–2000. TOW and KID are gauges that measure transport below the debris-avalanche deposit; MUD and SFT are gauges that measure transport from channels affected by large debris flows; GRE and CLR are gauges that measure transport from basins affected solely by a directed volcanic blast and tephra fall (cf. Figure 27.1). The shaded region depicts the range of, and horizontal dashed line depicts median value of, mean annual suspended-sediment yields from several western Cascade Range rivers. The horizontal dashed line serves as a proxy for pre-eruption sediment yield at Mount St. Helens. The vertical dashed line identifies closure of the SRS (cf. Figures 27.1 and 27.18).

From Major et al. (2000).

Yields from the South Fork Toutle and Muddy River valleys were initially about 20–100 times above pre-eruption levels; they decreased rapidly within five years of the eruption, but even after twenty years remained about ten times greater than pre-eruption levels (Figure 27.17; Major et al., 2000). Such prodigious sediment loads, caused by erosion of freshly deposited sediment as well as of sediment stored in centuries-old terraces, triggered severe channel instability downstream (e.g., Meyer and Janda, 1986) and further reduced channel conveyance capacities.

Post-eruption sediment redistribution occurred largely through fluvial erosion and transport (Major, 2004). The predominant fluvial transport at Mount St. Helens contrasts with predominant post-eruption debris-flow transport at many other volcanoes, especially at those in tropical climates (see references in Major et al., 2000 and Major, 2004). The dearth of post-1980 debris flows at Mount St. Helens is largely a consequence of low to moderate rainfall intensities that characterize the regional climate. However, it also reflects engineering measures undertaken to prevent catastrophic breaching of impounded lakes and the general character of post-1980 eruptions at the volcano, namely relatively quiescent dome building rather than violent explosions. The few explosions that occurred while the volcano was clad in snow triggered the most notable post-1980 debris flows.

Substantial post-eruption channel aggradation is endemic to volcanoes in the Cascade Range. Thick post-eruptive alluvial fills have been found thus far along channels draining Mount St. Helens, Mount Hood, and Mount Rainier (e.g., Crandell, 1987; Major and Scott, 1988; Scott, 1989; Barnhardt, 2003; Rapp et al., 2003; Zehfuss et al., 2003; Pierson, 2004). Persistent geomorphic problems should be anticipated following emplacement of large volcanic debris avalanches and debris flows, and measures designed to mitigate problems related to sediment redistribution need to remain functional for decades (Major, 2003).

27.7 SOCIAL AND ECONOMIC IMPACTS OF THE 1980 DEBRIS FLOWS

The 1980 debris flows destroyed or severely damaged civil works along all of the major river systems that drain the volcano, caused lesser damage along the Cowlitz River, and caused no damage but affected commercial transport along the Columbia River (Schuster, 1983). Fortunately, they caused few fatalities. Of the 57 apparent fatalities caused by the eruption, only two are attributed to debris flow (R.B. Waitt, U.S. Geological Survey, oral commun., 2004). The cataclysmic eruption, including debris flows, caused more than US$1 billion in losses (Foxworthy and Hill, 1982). Along the Toutle valley, the North Fork Toutle and South Fork Toutle debris flows together destroyed or heavily damaged about 200 homes, destroyed or badly damaged 42 bridges, buried about 300 km of highways and roads, partly buried and damaged three large logging camps, and destroyed several privately and publicly owned water-supply and sewage-disposal systems. The Interstate 5 highway and Burlington Northern railway bridges that cross the Toutle River near its confluence with the Cowlitz River, and link Portland and Seattle, sustained only minor damage that was rapidly repaired. To lessen the possibility of subsequent flood damage to these structures, the US Army Corps of Engineers (COE) dredged sediment from the lower 20 km of the Toutle channel through May 1981.

Along the Cowlitz River, the North Fork Toutle debris flow drastically affected operations of the municipal water-supply and sewage-disposal systems of cities and towns located along the flood plain. The debris flow clogged, buried, or removed water intake and treatment structures, pumps, screens, and sediment basins. The flow caused no permanent damage to these municipal systems, but it took weeks before systems and services were fully restored. To restore the original channel and manage future flooding, the COE dredged about 43 million m^3 of sediment from the channel by October 1981 (Figure 27.18; Schuster, 1983).

Sediment deposited by the North Fork Toutle debris flow blocked the navigation channel of the Columbia River and affected regional commerce. Prior to the climactic eruption, about a dozen ocean-going freighters travelled the Columbia River daily to and from Portland, Oregon, and Vancouver, Washington; after the

Figure 27.18. Mitigation of post-1980 sediment transport along the Toutle–Cowlitz River system. (A) Sediment dredging along Cowlitz River. (B) Sediment retention structure constructed on North Fork Toutle River valley.

(A) Photo: Lyn Topinka, USGS. (B) Photo: Bill Johnson, U.S. Army Corps of Engineers.

eruption traffic was halted for about one week until a partial deep-draft channel could be dredged. The full channel was not restored until November 1980. Closure of the shipping channel caused combined daily revenue losses of US$5 million for the ports of Portland and Vancouver (Foxworthy and Hill, 1982). By October 1981, the COE had dredged about 18 million m^3 of eruption-related sediment from the Columbia River. From May 1980 to October 1981, the COE dredged nearly 77 million m^3 of sediment from the Toutle, Cowlitz, and Columbia Rivers (Schuster, 1983).

Along the Pine Creek–Muddy River system, the 1980 debris flows were less devastating owing to their smaller sizes, limited development and the presence of a large reservoir. The flows buried or destroyed 16 bridges and several kilometers of US Forest Service roads. The debris flows passed into Swift Reservoir, but had little impact on the dam or reservoir operation. Prior to the eruption, Pacific Power and Light had lowered the reservoir to accommodate 125 million m^3 of sediment input. Combined, the flows deposited only about 14 million m^3 of sediment, and much of that sediment was deposited at the bottom of the reservoir in a zone of "dead storage". Those flows thus had insignificant impact on the power-production capacity of that facility.

Between 1980 and 1990, the federal government spent more than US$1 billion mitigating problems caused by the 1980 debris flows, the colossal debris avalanche, and post-eruption sediment redistribution. The bulk of the costs entailed channel dredging, design and construction of a large sediment retention structure to trap sediment in the North Fork Toutle valley (Figures 27.1 and 27.18), design and construction of a bedrock tunnel to provide an outlet for Spirit Lake, and temporary pumping of the lake until the outlet tunnel could be completed.

27.8 FREQUENCY AND MAGNITUDE OF LARGE DEBRIS FLOWS

Large debris flows at Mount St. Helens occur frequently. Debris flows large enough to inundate flood plains in the Toutle valley 50 km from the volcano have occurred at least 35 times in roughly 50,000 years (Scott, 1989). At least 26 flows occurred in the past 14,000 years, at least 15 in the past 4,500 years. Unlike floods that generally are considered independent events distributed randomly in time, large debris flows at Mount St. Helens occur during distinct eruptive periods that are separated by dormant intervals lasting a few to many tens of centuries. Furthermore, within each eruptive period debris flows commonly are clustered in time within a few to a few tens of years (Scott, 1989) and variations in triggering mechanisms strongly affect their occurrence. Unlike floods, large volcanic debris flows generally are interdependent, non-random events drawn from substantially variable populations.

Estimating recurrence intervals for large debris flows is conditional upon the state of the volcano. A simplistic frequency analysis based on the sum of flows over a specified time interval provides only a minimum estimate of recurrence. For example, the average recurrence interval of debris flows in the Toutle valley over the past 4,500 years is about 300 years (fifteen events). However, many of these events were clustered, and when the volcano is in a period of eruption, debris flows are more likely to occur. Owing to nonrandom occurrence, a perhaps more useful estimate of the average recurrence interval of large debris flows during periods of eruption can be gleaned by considering only those periods when the volcano was active. Thus, the average recurrence interval of large volcanic debris flows during periods of eruption over the past 4,500 years is 130 years (based on 15 overbank flows during the 1,930 years considered to be within eruptive periods; Scott, 1989).

Estimating time-series distributions of event magnitudes is more difficult because discharges or volumes of ancient flows rarely are well constrained. However, along the Toutle River valley, Scott (1988a, 1989) documented cross-sectional inundation areas for the two largest lake-breakout type debris flows. He also showed that the altitude of the modern channel bed has remained relatively constant over the past 3,000–4,000 years. Therefore, he was able to estimate flow depths and cross-sectional areas for other ancient flows within the past 4,500 years if he assumed that modern relationships between flow depths and deposit thicknesses held for the ancient flows. From inferred paleostages and channel cross sections of several ancient flows he extracted estimated paleodischarges for assumed flow velocities of 5–20 m/s (Scott, 1989).

From his estimates of debris-flow frequency and paleodischarge, Scott (1989) concluded that debris flows sufficient to inundate flood plains throughout the Toutle valley to depths of at least 2 m recur at approximately 100 year intervals when the volcano is active. This depth of inundation corresponds to a debris-flow discharge of at least $10,000 \, m^3/s$ near the confluence of the forks of the Toutle River. The recurrence interval of debris flows of this magnitude is within the typical time frame for long-term land-use and development planning (e.g., multiple generations). Thus large debris flows having discharges of several thousand m^3/s should be anticipated, and perhaps designed for, over the next few centuries given the presently active state of the volcano.

27.9 HAZARD ASSESSMENTS AND MITIGATION RESPONSES

The 1980 eruptions of Mount St. Helens heightened public awareness of explosive volcanism and expanded scientific analyses of eruptive phenomena and assessments of hazards (Tilling, 2000). As a result, volcano-hazard assessments at Mount St. Helens and elsewhere have become more elegant and sophisticated. Prior to the 1980 eruptions, the history and nature of past eruptions of Mount St. Helens were exceptionally well known (Crandell et al., 1975; Mullineaux and Crandell, 1981), but only a generalized hazard-zonation map existed (Miller et al., 1981). That map showed that channels draining the volcano were subject to debris flows, but only in schematic fashion. An updated post-eruption assessment depicted debris-flow hazard zones more meaningfully, but still subjectively, on a topographic base (Wolfe and Pierson, 1995). Newer methods that utilize digital cartography have subsequently been developed for objectively assessing potential areas of debris-flow inundation (Iverson et al., 1998; Chapter 10). In addition to the development of improved hazard-assessment methods, debris-flow detection systems have been developed and refined at Mount St. Helens (e.g., LaHusen, 1996, chap. 12).

Recognition of the origin of the huge ancient debris flow in the Toutle valley as a lake-breakout flood (Scott, 1988b) greatly influenced post-eruption assessments of debris-flow hazards and subsequent mitigation responses to emplacement of the 1980 debris avalanche. With that recognition, the implications for the modern avalanche-dammed Spirit Lake were clear. The potential existed for yet another debris flow transformed from a lake-breakout flood surge, one having a volume several times that estimated for a flood surge alone (*cf.* Dunne and Leopold, 1981; Dunne and Fairchild, 1984). The 20 km of fresh, highly erodible avalanche deposit extending downstream from the blockage increased the likelihood of this outcome. Recognition of this increased hazard stimulated the construction of a pumping operation by the COE in 1982, as well as planning for a long-term solution. The pumping operation, in which lake water was piped across the crest of the debris-avalanche deposit and into the North Fork Toutle River downstream, was designed to draw down Spirit Lake to a level judged at low risk of breakout from even seismically induced failure of the blockage. The long-term solution, completed in 1985, was a 2-km drainage tunnel through bedrock which allowed lake outflow to debouch via a tributary into the river below the main mass of the blockage. Without this engineering intervention, and with continuous inflow but no outflow, the lake level had been projected to reach a level approaching the crest of the blockage by 1984. Even with this intervention, responses to the potential for a catastrophic debris flow included a siren-warning system and an on-site observer in the period before significant drawdown of the lake.

After the 1980 eruptions, the USGS and others assessed the potential for and consequences of failure of natural dams formed by the modern debris-avalanche deposit at the mouths of channels tributary to the North Fork Toutle River. Consequently, stable outlet channels were constructed across some blockages (e.g., Coldwater Lake). For others, the risk of failure declined as sediment gradually filled impounded lakes. At present only Castle Lake at the mouth of Castle creek (Figure 27.1) has a lingering hazard potential. There, the concern involves the stability of the blockage to development of a groundwater table and seismic shock (Meyer et al., 1985; Roeloffs, 1994). The examples of sediment entrainment provided by the series of ancient flows documented the need to incorporate significant entrainment into assessments of potential hazards from breakouts of the tributary-valley lakes. Quantitative estimates of the potential hazards of dam breaches by the USGS, COE, and others (e.g., Dunne and Leopold, 1981; Dunne and Fairchild, 1984) incorporated several-fold increases in volume on the premise that flood surges could readily mobilize sediment from the debris-avalanche deposit and transform into debris flows.

Large-scale sediment entrainment, as first reported at Mount St. Helens, has proven to be a process of general concern at volcanoes elsewhere. After 1980, volume increases caused by entrainment were recognized as an important or even dominant factor in creating large debris flows at other volcanoes. Examples include:

1. A debris flow at Casita volcano, Nicaragua, in 1998. There, a modest landslide triggered by hurricane rainfall devolved into a watery flood and then enlarged into a devastating debris flow (Scott et al., in press). Groundwater released from the failure of fractured bedrock added to surface runoff and yielded a flood surge that swiftly (within 4 km of the base of the volcano) increased in volume several-fold and transformed into a debris flow. The debris flow inundated a kilometre-wide area and killed nearly the entire populations of two towns (2,500 people).
2. A debris flow along Polallie Creek at Mount Hood, Oregon (USA). That flow transformed from an approximately 3,800 m^3 landslide, entrained channel fill, and enlarged and deposited about 76,000 m^3 of sediment at the creek mouth (Gallino and Pierson, 1985).
3. The largest debris flow of the Sherman Crater eruptive period (AD 1843 to present) at Mount Baker, Washington (USA). That flow transformed from a failure of the east side of the crater wall (Scott and Tucker, 2003). Downstream, the debris flow entrained coarse detritus that constitutes as much as 80% of the distal deposit. The total flow volume increased by at least 50% compared to the original failure volume.

27.10 CONCLUSIONS

Analyses of the 1980s eruption-triggered debris flows and studies of ancient eruption-related debris-flow deposits at Mount St. Helens have greatly increased our understanding of debris flows and hazards associated with eruptions of snow-clad volcanoes. Those studies showed that debris flows can form in many ways, even during a single eruption, that the presence or absence of snow plays a key role in generating flows associated directly with eruptions, and that the largest flows can be associated with secondary, post-eruptive processes. Enormous masses of volcanoes can fail as gigantic landslides, called debris avalanches, which can directly or indirectly spawn very large, destructive debris flows. Debris avalanches can transform directly to debris flows, their deposits can locally liquefy and flow, or they can impound lakes that breach and develop secondary floods that entrain sediment, transform into debris flows, and compound downstream hazards. Such a secondary process triggered the largest known debris flow in the history of Mount St. Helens. Many debris flows at Mount St. Helens, especially those associated with eruptions, resulted from meltwater or flood surges that entrained volcanic sediment. Some debris flows began as slushflows or thin sheets of meltwater that rapidly entrained and mixed with volcanic sediment. Others began as pulses of water released suddenly from glaciers, or as surges released from lakes formed by rapid snowmelt in the volcano's crater or impounded by debris-avalanche deposits. Several debris flows, especially those associated with eruptions or lake breakouts, travelled many tens of kilometres. The farthest-travelled flows entered the Columbia River, about 100 km from the volcano.

Small debris flows unassociated with volcanic eruptions have also transformed directly from slope failures and travelled a few to several kilometers. Studies of modern events related to the 1980 and subsequent eruptions documented the significant hazard associated with interactions of hot volcanic debris and snowpack, and they also showed that near-source mass-flow deposits from many events are ephemeral and not easily distinguished after a few years, or even a few months. Therefore, the full extent of mass-flow hazards at snow-clad volcanoes is probably underestimated.

Moderate and larger debris flows at Mount St. Helens typically had short-duration hydrographs that waxed and waned rapidly. Most flows lasted only a few to several minutes at any given location. The 1980 North Fork Toutle debris flow, however, which formed through local liquefaction and dewatering of a debris-avalanche deposit, was a long-duration event that waxed and waned slowly over many hours. Of the modern debris flows, it was the largest and most destructive. Comparisons of the peak stage of that flow with those of several ancient debris flows at comparable distances, however, suggests that past eruption-related debris flows at Mount St. Helens have been substantially larger. When such very large magnitude flows occur again, they will be far more devastating than those of 1980.

Deposit textures at Mount St. Helens revealed that debris flows that contained at least a few % of clay-sized sediment maintained their textural integrity for many tens of kilometres from source whereas "non-cohesive" flows commonly mixed with river flow and diluted, dropped their gravel load, and transformed into sandy flows that progressively devolved into sediment-laden streamflow.

Modern events and ancient deposits at Mount St. Helens showed clearly that large volcanic debris flows are interdependent, non-random events that cluster in time within eruption periods. Therefore, debris-flow hazards are much more acute when a volcano is in a period of eruption than when it is dormant. In the past 4,500 years, a period of apparently extraordinary activity at Mount St. Helens, debris flows recurred with sufficient frequency in the Toutle River system that the hazards posed to distal communities and civil works are within the usual time frame for long-term land-use and development planning. Large debris flows from Mount St. Helens having discharges of several thousands of m^3/s should be anticipated over the next few centuries.

A greater understanding of debris-flow origins, distal transformations, and relationships among deposit textures, flow processes, and initiation mechanisms at Mount St. Helens has improved subsequent assessments of eruptive histories and potential hazards at many volcanoes worldwide. Owing to this improved understanding of volcanic debris flows, studies of volcano hazards in the Cascade Range and elsewhere have evolved from mere academic pursuits to political catalysts that have influenced land-use policy (e.g., Bailey and Woodcock, 2003; Scott, 2004). Technological advancements for mitigating debris-flow hazards at volcanoes have been developed in parallel with scientific understanding of the hazards, and debris-flow detection systems developed and refined at Mount St. Helens have been deployed worldwide.

27.11 ACKNOWLEDGEMENTS

We thank Lee Fairchild, Elizabeth Safran, and Matthias Jakob for perceptive reviews that focused the scope of this chapter and improved its presentation, and Lee Siebert for assistance compiling Table 27.1.

27.12 REFERENCES

Bailey, S. and Woodcock, J. (2003) Mt. Rainiera small event equals a public relations nightmare. *Proceedings of Cities on Volcanoes – 3: Hilo, Hawaii* (p. 8). International Association of Volcanology and Chemistry of the Earth's Interior, Hilo, HI.

Barnhardt, W.A. (2003) Volcanic debris flows and implications for coastal evolution: Subsurface (GPR) imagery of a late Holocene delta at Puget Sound, Washington. *Geological Society of America Abstracts with Programs*, **35**(6), 219.

Bradley, J.B., Grindeland, T.R., and Hadley, H.R. (2001) Sediment supply from Mount St. Helens: 20 years later. *Proceedings of the 7th Federal Interagency Sedimentation Conference* (Vol. 2, pp. X9–X16). U.S. Government Printing Office, Denver, CO.

Brantley, S.R. and Myers, B. (2000) *Mount St. Helens: From the 1980 Eruption to 2000* (USGS Fact Sheet 036-00, 2 pp.). US Geological Survey, Reston, VA.

Brantley, S.R. and Waitt, R.B. (1988) Interrelations among pyroclastic surge, pyroclastic flow, and lahars in Smith Creek valley during first minutes of 18 May 1980 eruption of Mount St. Helens, USA. *Bulletin of Volcanology*, **50**, 304–326.

Brugman, M.M. and Post, A. (1981) *Effects of Volcanism on the Glaciers of Mount St. Helens* (USGS Circular 850-D, 11 pp.). US Geological Survey, Reston, VA.

Cameron, K.A. and Pringle, P.T. (1990) Avalanche-generated debris flow on 9 May 1986, at Mount St. Helens, Washington. *Northwest Science*, **64**, 159–164.

Carrasco-Núñez, G., Vallance, J.W., and Rose, W.I. (1993) A voluminous avalanche-induced lahar from Citlatépetl volcano, Mexico: Implications for hazard assessment. *Journal of Volcanology and Geothermal Research*, **59**, 35–46.

Crandell, D.R. (1987) *Deposits of Pre-1980 Pyroclastic Flows and Lahars from Mount St. Helens Volcano, Washington* (USGS Professional Paper 1444, 94 pp.). US Geological Survey, Reston, VA.

Crandell, D.R., Mullineaux, D.R., and Rubin, M. (1975) Mount St. Helens volcano: recent and future behavior. *Science*, **187**, 438–441.

Cummans, J. (1981) *Mudflows Resulting from the May 18, 1980, Eruption of Mount St. Helens, Washington* (USGS Circular 850-B, 16 pp.). US Geological Survey, Reston, VA.

Dinehart, R.L. (1999) *Sediment Transport in the Hyperconcentrated Phase of the March 19, 1982 Lahar* (USGS Professional Paper 1586, pp. 37–52). US Geological Survey, Reston, VA.

Dunne, T. and Fairchild, L.H. (1984) Estimation of flood sedimentation hazards around Mount St. Helens. *Shin Sabo*, **36**(4), 12–22, and **36**(5), 13–22.

Dunne, T. and Leopold, L.B. (1981) *Flood and Sedimentation Hazards in the Toutle and Cowlitz River System as a Result of the Mount St. Helens Eruption* (Report, 170 pp.). Federal Emergency Management Agency, Region 10, Seattle, WA.

Evarts, R.C., Clynne, M.A., Fleck, R.J., Lanphere, M.A., Calvert, A.T., and Sarna-Wojcicki, A.M. (2003) The antiquity of Mount St. Helens and age of the Hayden Creek Drift. *Geological Society of America Abstracts with Programs*, **35**(6), 80.

Fairchild, L.H. (1985) Lahars at Mount St. Helens (374 pp.). Ph.D. thesis, University of Washington Seattle.

Fairchild, L.H. (1987) The importance of lahar initiation processes. *Geological Society of America Reviews in Engineering Geology*, **7**, 51–62.

Fairchild, L.H. and Wigmosta, M. (1983) Dynamic and volumetric characteristics of the 18 May 1980 lahars on the Toutle River, Washington. *Proceedings of the Symposium on Erosion Control in Volcanic Areas* (Technical Memorandum 1908, pp. 131–153). Public Works Research Institute, Tsukuba, Japan.

Fink, J.H., Malin, M.C., D'Alli, R.E., and Greeley, R. (1981) Rheological properties of mudflows associated with the Spring 1980 eruptions of Mount St. Helens volcano, Washington. *Geophysical Research Letters*, **8**, 43–46.

Folk, R.L. (1984) *Petrology of Sedimentary Rocks* (184 pp.). Hemphill, Austin, TX.

Foxworthy, B.L., and Hill, M. (1982) *Volcanic Eruptions of 1980 at Mount St. Helens: The First 100 days* (USGS Professional Paper 1249, 125 pp.). US Geological Survey, Reston, VA.

Gallino, G.L. and Pierson, T.C. (1985) *Polallie Creek Debris Flow and Subsequent Dam-break Flood of 1980, East Fork Hood River Basin, Oregon* (USGS Water-Supply Paper 2273, 22 pp.). US Geological Survey, Reston, VA.

Gilkey, K.E. (1983) Sedimentology of the North Fork and South Fork Toutle mudflows generated during the 1980 eruption of Mount St. Helens (254 pp.). MSc thesis, University of California Santa Barbara.

Glicken, H. (1996) *Rockslide-Debris Avalanche of May 18, 1980, Mount St. Helens Volcano, Washington* (USGS Open-file Report 96-677, 90 pp.). US Geological Survey, Reston, VA.

Glicken, H. (1998) Rockslide-debris avalanche of May 18, 1980, Mount St. Helens volcano, Washington. *Geological Survey of Japan Bulletin*, **49**, 55–106.

Hardison, J.H. (2000) Post-lahar channel adjustment, Muddy River, Mount St. Helens, Washington (119 pp.). MSc thesis, Colorado State University, Fort Collins.

Hausback, B.P. and Swanson, D.A. (1990) Record of prehistoric debris avalanches on the north flank of Mount St. Helens volcano, Washington. *Geoscience Canada*, **17**, 142–145.

Hoblitt, R.P., Miller, C.D., and Vallance, J.W. (1981) *Origin and Stratigraphy of the Deposit Produced by the May 18 Directed Blast* (USGS Professional Paper 1250, pp. 401–420). US Geological Survey, Reston, VA.

Hyde, J.H. (1975) *Upper Pleistocene Pyroclastic-flow Deposits and Lahars, South of Mount St. Helens Volcano, Washington* (USGS Bulletin 1383-B, 20 pp.). US Geological Survey, Reston, VA.

Hyde, J.H., and Crandell, D.R. (1978) *Postglacial Volcanic Deposits at Mount Baker, Washington and Potential Hazards from Future Eruptions* (USGS Professional Paper 1022-C, 17 pp.). US Geological Survey, Reston, VA.

Iverson, R.M., Reid, M.E., and LaHusen, R.G. (1997) Debris-flow mobilization from landslides. *Annual Reviews of Earth and Planetary Sciences*, **25**, 85–138.

Iverson, R.M., Schilling, S.P., and Vallance, J.W. (1998) Objective delineation of lahar-hazard zones downstream from volcanoes. *Geological Society of America Bulletin*, **110**, 972–984.

Janda, R.J., Scott, K.M., Nolan, K.M., and Martinson, H.A. (1981) *Lahar Movement, Effects, and Deposits* (USGS Professional Paper 1250, pp. 461–478). US Geological Survey, Reston, VA.

Janda, R.J., Meyer, D.F., and Childers, D. (1984) Sedimentation and geomorphic changes during and following the 1980–1983 eruptions of Mount St. Helens, Washington. *Shin Sabo*, **37**(2), 10–21, and **37**(3), 5–19.

Japan Ministry of Construction (1992) *Eruption of Mt. Fugen-dake of Unzen and Volcanic Sabo Projects* (30 pp.). Sediment Control Division, Japan Ministry of Construction, Tokyo.

Japan Ministry of Construction (1995) *Debris Flow at Sakurajima* (Vol. 2, 81 pp.). Ohsumi Works Office, Kyushu Regional Construction Bureau, Japan Ministry of Construction, Public Works Research Institute, Tsukuba.

Johnson, A.M. (1970) *Physical Processes in Geology* (577 pp.). Freeman, Cooper, & Co., San Francisco.

LaHusen, R.G. (1996) *Detecting Debris Flows Using Ground Vibrations* (USGS Fact Sheet 236-96). US Geological Survey, Reston, VA.

Lavigne, F., Thouret, J.C., Voight, B., Suwa, H., and Sumaryono, A. (2000) Lahars at Merapi volcano, Central Java – an overview. *Journal of Volcanology and Geothermal Research*, **100**, 423–456.

Lehre, A.K., Collins, B.D., and Dunne, T. (1983) Post-eruption sediment budget for the North Fork Toutle River drainage, June 1980–June 1981. *Zeitschrift für Geomorphologie, Suppl. Band*, **46**, 143–165.

Lipman, P.W. and Mullineaux, D.R. (eds) (1981) *The 1980 Eruptions of Mount St. Helens, Washington* (USGS Professional Paper 1250, 844 pp.). US Geological Survey, Reston, VA.

Lombard, R.E., Miles, M.B., Nelson, L.M., Kresh, D.L., and Carpenter, P.J. (1981) *The Impact of Mudflows of May 18 on the Lower Toutle and Cowlitz Rivers* (USGS Professional Paper 1250, pp. 693–699). US Geological Survey, Reston, VA.

Lozano, J.R.B. (2001) The tragedy that is Lake Maughan. *Philippine Inquirer*, 23 June.

Major, J.J. (1984) Geologic and rheologic characteristics of the May 18, 1980 southwest flank lahars at Mount St. Helens, Washington. MSc thesis, Pennsylvania State University, State College.

Major, J.J. (1997) Depositional processes in large-scale debris-flow experiments. *Journal of Geology*, **105**, 345–366.

Major, J.J. (2000) Gravity-driven consolidation of granular slurries: Implications for debris-flow deposition and deposit characteristics. *Journal of Sedimentary Research*, **70**, 64–83.

Major, J.J. (2003) Post-eruption hydrology and sediment transport in volcanic river systems. *Water Resources IMPACT*, **5**, 10–15.

Major, J.J. (2004) Post-eruption suspended-sediment transport at Mount St. Helens: Decadal-scale relationships with landscape adjustments and river discharges. *Journal of Geophysical Research*, **109**, F01002, doi:10.1029/2002JF000010.

Major, J.J. and Newhall, C.G. (1989) Snow and ice perturbation during historical volcanic eruptions and the formation of lahars and floods. *Bulletin of Volcanology*, **52**, 1–27.

Major, J.J. and Voight, B. (1986) Sedimentology and clast orientations of the 18 May 1980 southwest flank lahars, Mount St. Helens, Washington. *Journal of Sedimentary Petrology*, **56**, 691–705.

Major, J.J. and Scott, K.M. (1988) *Volcaniclastic Sedimentation in the Lewis River Valley, Mount St. Helens, Washington: Processes, Extent, and Hazards* (USGS Bulletin 1383-D, 38 pp.). US Geological Survey, Reston, VA.

Major, J.J., Pierson, T.C., Dinehart, R.L., and Costa, J.E. (2000) Sediment yield following severe volcanic disturbance – A two decade perspective from Mount St. Helens. *Geology*, **28**, 819–822.

Major, J.J., Schilling, S.P., Pullinger, C.R., and Escobar, C.D. (2004) *Debris-flow Hazards at San Salvador, San Vicente, and San Miguel Volcanoes, El Salvador* (GSA Special Paper 375, pp. 89–108). Geological Society of America, Boulder, CO.

McGuire, W.J. (1996) *Volcano Instability: A Review of Contemporary Themes* (Special Publication 110, pp. 1–23). Geological Society of London.

Mellors, R.A., Waitt, R.B., and Swanson, D.A. (1988) Generation of pyroclastic flows and surges by hot-rock avalanches from the dome of Mount St. Helens volcano, USA. *Bulletin of Volcanology*, **50**, 14–25.

Meyer, D.F. and Janda, R.J. (1986) Sedimentation downstream from the 18 May 1980 North Fork Toutle River debris avalanche deposit, Mount St. Helens, Washington. In: S.A.C. Keller (ed.), *Mount St. Helens: Five Years Later* (pp. 68–86). Eastern Washington University Press, Cheney.

Meyer, D.F. and Martinson, H.A. (1989) Rates and processes of channel development and recovery following the 1980 eruption of Mount St. Helens, Washington. *Hydrological Sciences Journal*, **34**, 115–127.

Meyer, W., Sabol, M.A., Glicken, H.X., and Voight, B. (1985) *The Effects of Ground Water, Slope Stability, and Seismic Hazard on the Stability of the South Fork Castle Creek blockage in the Mount St. Helens Area, Washington* (USGS Professional Paper 1345, 42 pp.). US Geological Survey, Reston, VA.

Miller, C.D., Mullineaux, D.R., and Crandell, D.R. (1981) *Hazards Assessments at Mount St. Helens* (USGS Professional Paper 1250, pp. 789–802). US Geological Survey, Reston, VA.

Mullineaux, D.R. and Crandell, D.R. (1962) Recent lahars from Mount St. Helens, Washington. *Geological Society of America Bulletin*, **73**, 855–870.

Mullineaux, D.R.,. and Crandell, D.R. (1981) *The Eruptive History of Mount St. Helens* (USGS Professional Paper 1250, pp. 3–15). US Geological Survey, Reston, VA.

Newhall, C.G. (1982) A prehistoric debris avalanche from Mount St. Helens. *Eos, Transactions of the American Geophysical Union*, **63**(45), 1141.

Newhall, C.G. and Punongbayan, R.S. (eds) (1996) *Fire and Mud: Eruptions and Lahars of Mount Pinatubo, Philippines* (1126 pp.). University of Washington Press, Seattle.

O'Connor, J.E., Hardison, J.H., and Costa, J.E. (2001) *Debris Flows from Failures of Neoglacial-age Moraine Dams in the Three Sisters and Mount Jefferson Wilderness Areas, Oregon* (USGS Professional Paper 1606, 93 pp.). US Geological Survey, Reston, VA.

Pierson, T.C. (1985) Initiation and flow behavior of the 1980 Pine Creek and Muddy River lahars, Mount St. Helens, Washington. *Geological Society of America Bulletin*, **96**, 1056–1069.

Pierson, T.C. (1986) Flow behavior of channelized debris flows, Mount St. Helens, Washington. In: A.D. Abrahams (ed.), *Hillslope Processes* (pp. 269–296). Allen & Unwin, Boston.

Pierson, T.C. (1999) *Transformation of Water Flood to Debris Flow Following the Eruption-triggered Transient-lake Breakout from the Crater on March 19, 1982* (USGS Professional Paper 1586, pp. 19–36). US Geological Survey, Reston, VA.

Pierson, T.C. (2004) Rapid and persistent post-eruption sedimentation and river-channel aggradation at a temperate Pacific-Rim volcano. *Proceedings of the Geological Society of America Penrose Conference on Neogene-Quaternary Continental Margin Volcanism, January 12–15, Metepec Puebla, Mexico* (p. 61). Universidad Nacional Autónoma de México, Mexico City.

Pierson, T.C. and Janda, R.J. (1994) Volcanic mixed avalanches: A distinct eruption-triggered mass-flow process at snow-clad volcanoes. *Geological Society of America Bulletin*, **106**, 1351–1358.

Pierson, T.C. and Scott, K.M. (1985) Downstream dilution of a lahar-transition from debris flow to hyperconcentrated streamflow. *Water Resources Research*, **21**, 1511–1524.

Pierson, T.C. and Waitt, R.B. (1999) *Dome-collapse Rockslide and Multiple Sediment-water Flows Generated by a Small Explosive Eruption on February 2–3, 1983* (USGS Professional Paper 1586, pp. 53–68). US Geological Survey, Reston, VA.

Pierson, T.C., Janda, R.J., Thouret, J.C., and Borrero, C.A. (1990) Perturbation and melting of snow and ice by the 13 November 1985 eruption of Nevado del Ruiz, Colombia, and consequent mobilization, flow, and deposition of lahars. *Journal of Volcanology and Geothermal Research*, **41**, 17–66.

Pringle, P.T. and Cameron, K.A. (1999) *Eruption-triggered Lahar on May 4, 1984* (USGS Professional Paper 1586, pp. 81–103). US Geological Survey, Reston, VA.

Rapp, B.K., O'Connor, J.E., and Pierson, T.C. (2003) The Sandy River delta, Oregon: A record of late Holocene volcanism at Mt. Hood and sediment into the Columbia River. *Geological Society of America Abstracts with Programs*, **35**(6), 81.

Roeloffs, E.A. (1994) *An Updated Numerical Simulation of the Ground-water Flow System for the Castle Lake Debris Dam, Mount St. Helens, Washington, and Implications for Dam Stability against Heave* (USGS Water-Resources Investigations Report 94-4075, 80 pp.). US Geological Survey, Reston, VA.

Schuster, R.L. (1983) Engineering aspects of the 1980 Mount St. Helens eruptions. *Bulletin of the Association of Engineering Geologists*, **20**, 125–143.

Scott, K.M. (1985) Lahars and flow transformations at Mount St. Helens, Washington, U.S.A. *Proceedings of the International Symposium on Erosion, Debris Flow, and Disaster Prevention, September 3–5, Tsukuba, Japan* (pp. 209–214). The Erosion-Control Engineering Society Japan, Tokyo.

Scott, K.M. (1988a) *Origins, Behavior, and Sedimentology of Lahars and Lahar-runout Flows in the Toutle-Cowlitz River System* (USGS Professional Paper 1447-A, 78 pp.). US Geological Survey, Reston, VA.

Scott, K.M. (1988b) *Origin, Behavior, and Sedimentology of Prehistoric Catastrophic Lahars at Mount St. Helens, Washington* (GSA Special Paper 229, pp. 23–36). Geological Society of America, Boulder, CO.

Scott, K.M. (1989) *Magnitude and Frequency of Lahars and Lahar-runout Flows in the Toutle-Cowlitz River System* (USGS Professional Paper 1447-B, 33 pp.). US Geological Survey, Reston, VA.

Scott, K.M. (2004) Lahars, lidar, and forensic documentation of flow origins, behaviors, and pathways at Mount Rainier, Washington. *Proceedings of the Geological Society of America Penrose Conference on Neogene–Quaternary Continental Margin Volcanism, January 12–15, Metepec Puebla, Mexico* (pp. 111–117). Universidad Nacional Autónoma de México, Mexico City.

Scott, K.M. and Tucker, D.S. (2003) The Sherman Crater eruptive period at Mount Baker, North Cascades – A.D. 1843 to present – Implications for reservoirs at the base of the volcano. *Geological Society of America Abstracts with Programs*, **34**(7), 321.

Scott, K.M., Vallance, J.W., and Pringle, P.T. (1995) *Sedimentology, Behavior, and Hazards of Debris Flows at Mount Rainier, Washington* (USGS Professional Paper 1547, 56 pp.). US Geological Survey, Reston, VA.

Scott, K.M., Macías, J.L., Naranjo, J.A., Rodríguez, S., and McGeehin, J.P. (2001) *Catastrophic Debris Flows Transformed from Landslides in Volcanic Terrains: Mobility, Hazard Assessment, and Mitigation Strategies* (USGS Professional Paper 1630, 59 pp.). US Geological Survey, Reston, VA.

Scott, K.M., Vallance, J.W., Kerle, N., Macías, J.L., Strauch, W., and Devoli, G. (in press) Catastrophic precipitation-triggered lahar at Casita Volcano, Nicaragua: Occurrence, bulking, and transformation. *Earth Surface Processes and Landforms*.

Siebert, L. (1984) Large volcanic debris avalanches: Characteristics of source areas, deposits, and associated eruptions. *Journal of Volcanology and Geothermal Research*, **22**, 163–197.

Siebert, L., Kimberly, P., and Pullinger, C.R. (2004) *The Voluminous Acajutla Debris Avalanche from Santa Ana Volcano, Western El Salvador, and Comparison with Other Central American Edifice-failure Events* (GSA Special Paper 375, pp. 5–23). Geological Society of America, Boulder, CO.

Simkin, T. and Siebert, L. (1994) *Volcanoes of the World* (349 pp.). Geoscience Press, Tucson, AZ.

Simon, A. (1999) *Channel and Drainage-basin Response of the Toutle River System in the Aftermath of the 1980 Eruption of Mount St. Helens, Washington* (USGS Open-file Report 96-633, 130 pp.). US Geological Survey, Reston, VA.

Smith, G.A., and Fritz, W.J. (1989) Volcanic influences on terrestrial sedimentation. *Geology*, **17**, 375–376.

Smithsonian Institution (1994) Rinjani volcano. *Bulletin of the Global Volcanism Network*, **19**(10).

Smithsonian Institution (1996) La Maderas volcano. *Bulletin of the Global Volcanism Network*, **21**(9).

Thouret, J.C., Abdurachman, K.E., Bourdier, J.L., and Bronto, S. (1998) Origin, characteristics, and behaviour of lahars following the 1990 eruption of Kelud volcano, eastern Java (Indonesia). *Bulletin of Volcanology*, **59**, 460–480.

Tilling, R.I. (2000) Mount St. Helens 20 years later: What we've learned. *Geotimes*, **45**, 14–19.

U.S. Army Corps of Engineers (2002) *Mount St. Helens Engineering Reanalysis* (Design Documentation Final Report, 114 pp.). U.S. Army Corps of Engineers, Portland District, Portland, OR.

Vallance, J.W. (1985) Late Pleistocene lahar assemblage near Hood River, OR. *Eos, Transactions of the American Geophysical Union*, **66**(46), 1150.

Vallance, J.W. (1999) *Postglacial Lahars and Potential Hazards in the White Salmon River System on the Southwest Flank of Mount Adams, Washington* (USGS Bulletin 2161, 49 pp.). US Geological Survey, Reston, VA.

Vallance, J.W. (2000) Lahars. In: H. Sigurdsson, B. Houghton, S.R. McNutt, H. Rymer, and J. Stix (eds), *Encyclopedia of Volcanoes* (pp. 601–616). Academic Press, San Diego.

Vallance, J.W. and Scott, K.M. (1997) The Osceola mudflow from Mount Rainier: Sedimentology and hazards implications of a huge clay-rich debris flow. *Geological Society of America Bulletin*, **109**, 143–163.

Voight, B. (1981) *Time Scale for the First Moments of the May 18 Eruption* (USGS Professional Paper 1250, pp. 69–86). US Geological Survey, Reston, VA.

Voight, B., Glicken, H., Janda, R.J., and Douglass, P.M. (1981) *Catastrophic Rockslide Avalanche of May 18* (USGS Professional Paper 1250, pp. 347–377). US Geological Survey, Reston, VA.

Waitt, R.B. (1981) *Devastating Pyroclastic Density Flow and Attendant Air Fall of May 18: Stratigraphy and Sedimentology of Deposits* (USGS Professional Paper 1250, pp. 439–460). US Geological Survey, Reston, VA.

Waitt, R.B. (1989) Swift snowmelt and floods (lahars) caused by great pyroclastic surge at Mount St. Helens volcano, Washington, 18 May 1980. *Bulletin of Volcanology*, **52**, 138–157.

Waitt, R.B. and MacLeod, N.S. (1987) Minor explosive eruptions at Mount St. Helens dramatically interacting with winter snowpack in March–April 1982. *Washington Division of Geology and Earth Resources Bulletin*, **77**, 355–379.

Waitt, R.B., Pierson, T.C., MacLeod, N.S., Janda, R.J., Voight, B., and Holcomb, R.T. (1983) Eruption-triggered avalanche, flood, and lahar at Mount St. Helens: Effects of winter snowpack. *Science*, **221**, 1394–1397.

Walder, J.S. and Driedger, C.L. (1995) Frequent outburst floods from South Tahoma Glacier, Mount Rainier, USA: Relation to debris flows, meteorological origin, and implications for subglacial hydrology. *Journal of Glaciology*, **41**, 1–10.

Wolfe, E.W. and Pierson, T.C. (1995) *Volcanic-hazard Zonation for Mount St. Helens, Washington, 1995* (USGS Open-file Report 95-497, 12 pp.). US Geological Survey, Reston, VA.

Zehfuss, P.H., Atwater, B.F., Vallance, J.W., Brenniman, H., and Brown, T.A. (2003) Holocene lahars and their by-products along the historical path of the White River between Mount Rainier and Seattle. In: T.W. Swanson (ed.), *Western Cordillera and Adjacent Areas* (Field Guide 4, pp. 209–223). Geological Society of America, Boulder, CO.

Index

accelerometers 296
Acoustic Flow Monitor (AFM) 296, 591
active debris-flow mitigation (definition) 446
active mitigation measures
  afforestation 448
  debris-flow breakers 477
  debris-flow bypass 458
  deflection structures 477
  drainage stabilization 452
  flood control reservoirs 458
  integrated watershed management 450
  soil bioengineering 448
  transverse structures 453
  woody debris rake 477
afforestation 448
agroforestry, influence on landslide erosion 399
alluvial fan 413
alpine mudflow 9
alpine permafrost, failure of frozen talus slopes 349
antecedent conditions, different lengths, landslide initiation 334
Antecedent Daily Rainfall Model 337
antecedent hydrologic conditions
  monitoring 292
antecedent rainfall
  earthquakes and volcanic events 334
  regional climate, contribution to debris-flow occurrence and distribution 333
  seasonal, debris flows 332
antecedent soil moisture, time period for build-up 333
Antecedent Soil Water Status model 347
area inundated by a debris flow, correlation to volume 433

bed destabilization, relation to infinite slope stability theory 138
bi-linear fluid model 235
BING 230
Bingham fluid model 235
bog 45
bulking 698, 702

CFX modeling tool 224
channel-facies sediment 708
Cheekye fan 659
  approaches to quantify debris-flow hazard 675
  modeling 678
  proposed mitigation, MWLAP 663
  setting and early research 660
Climate Research Unit (CRU) Global Climate Dataset 343
colluvial fan 413
composite fans 413
contractive behavior 57

cumulative frequency/magnitude (CFM) curves 20

Darcy's Law, pore pressure in soil media 65
debris apron 17
debris avalanche 15, 489
   debris flows, Campania Region (Southern Italy) 489
debris entrainment estimation 425
debris flood 15
   types of floods that qualify as 160
debris flood (definition) 160
debris flow 14
   activity, signs of in transport zone 413
   (DAN) analyses 505
   annual sediment load 582
   anthropogenic disturbance 541
   breakers 477
   bypass 458
   channeling of rainfall, triggering of 511
   channel morphology change, Mount St. Helens 716
   characteristics, Mount St. Helens 700
   China 565
   China, Jiangjia Ravine 566, 592
   clay content as a distinguishing feature 714
   composition 578
   dating
      absolute method 423
      relative method 423
   debris torrent 10
   definition, 688
      channelized v. hillslope 9
      distinction between debris flow and debris avalanche 9
      mudflow 9
      Stiny (1910) 19
   density and concentration 576
   deposition
      deposition area 18
      five classes of 588
   deposit thickness, Mount St. Helens 704
   discharge, correlation 431
   distal transition, Mount St. Helens 703
   early warning systems, important elements 480

earthquake, effect of on rainfall intensity threshold 553
erosion, downstream limits 143
fire-related 350
fire-related susceptibility 371
floods, Latin America 520
and forestry
   importance of gullies 607
   debris avalanches 601
   landslides characterized as debris flows 600
frequency 418
   relation to global climate change 350
hazard
   analysis and mitigation, Cheekye fan 680
   evaluation (Southern Italy) 504
   intensity, velocity of 20
   maps
      definition 436
      using specific return period 437
      Venezuela, 534
   remote sensing 416
history
   China 35
   Japan 34
   Kazakhstan 38
   Spain 39
   Taiwan 36
   United States 37, 38
hyperconcentrated flood 10
influence of forest-harvesting activities 388
influence on main river 583
initiation
   and gullies, important factors 607
   recently burned areas, surface runoff 365
   threshold, rainfall intensity 551
   and transport, Mount St. Helens 714
intensity 418
kinetic energy 574
linkages with landslides in forested terrain 392
magnitude 135, 310, 418, 436
   classification 425
   return period 436
mapping, remote sensing, difficulties in vegetated terrain 279

Index 735

material entrainment
  bed destabilization and erosion 137
  instability of streambanks 138
materials
  definition of "debris" 12
  definition of "earth" 11
  definition of "mud" 11
meltwater lake, eruption induced 698
debris-flow occurrence probability
  supply-limited 418
  supply-unlimited 418
path 16
peak discharge 430
process, post-wildfire effects 364
pyroclastic material, fertility of 511
pyroclastic soils (Southern Italy) 503
relationship with precipitation, China 569
return period 424
risk
  intense urbanization and land-use change 513
  management 343
single-episode
  rainfall and sub-basin order 570
  type of rainfall 570
sources of debris 20
superelevation, channel bends 576
surges 17
  bed stability 141
susceptibility, pyroclastic cover 514
Taiwan 540
Typhoon Herb 542
typhoons as agent of 541
unjustified engineering and land use 555
Venezuela 520
volume 425
  change behavior 310
  relation to watershed characteristics 429
debris-flow danger period (DFDP) 343
Debris Flow Database British Columbia yield rate 150
debris yield rate 17, 310
deflection structures 477
dendrogeomorphology 423
deposition pattern, role of topography 319
dilatant behavior 57
Doppler radar 298

dosing 464
drainage stabilization 452

El Niño-Southern Oscillation (ENSO) 349
eruptions and debris flows, Mount St. Helens 688
Extreme debris-flow danger period (EDFDP) 343

fan deposits, volume estimates 427
Farböschung 221
flood control reservoirs 460
flow anisotrophy 66
flujos de lodos 36
flujos detríticos 36

General Circulation Models (GCMs) 350
GLOF (glacial lake outburst flood) 43
ground vibration sensors 292

halokinesis 209
hazard acceptability level, B.C. Ministry of Transport and Highways 679
hazard acceptability threshold, legal precedent, British Columbia 679
hazard analysis approach, Switzerland 622
hazard intensity 20
hazard assessment
  volcanic debris flows 721
  maps, Switzerland 617
  studies 20
hazard management, integrated approaches, Switzerland 616
hazard mitigation
  debris-flow countermeasures, non–structural 558
  design codes 559
  forecasting, threshold levesl of precipitation 591
  rainfall-based warning/monitoring system 557
  structural countermeasures 557
  structured mitigation 589
hazard planning
  municipal land-use planning, Switzerland 627
  preparedness concept 630
  protection goals 627

hazard potential
  total travel distance 306
  volume of debris flow 306
hazard probability 20
heterogeneous hyperconcentrated flow,
    transition to homogeneous
    hyperconcentrated flow 164
historical method 27
homogeneous hyperconcentrated flow,
    transition from heterogeneous
    hyperconcentrated flow 164
Huaico 36
hyperconcentrated flow
  bed resistance 176
  bedload 172
  bedload transport rates 174
  consideration as a separate process 161
  criteria of a distinct process 191
  defining boundaries 161
  deposition
    by suspension fallout 186
    by traction-carpet accretion 187
  distinction of 159
  distinguishing criteria 162
  effect of bed forms on flow resistance 176
  evolution from debris flows 165
  fundamental characteristic 195
  hazard
    deposition and rapid aggradation 190
    erosivity of 190
    rapid lateral migration of channels 190
  highly concentrated, deposition of 187
  initiation of 164
  sand suspension and settling criteria 164
  semiarid and arid regions 165
  suspended-load transport characteristics 173
  thalwegs 189
  transport of bedload sediment and characteristics of 173
  turbulent 167
  types of suspended load 168
  vertical distribution of suspended sediment 173
  volcanic terrain 178
hyperconcentrated flow (definition) 193, 688

infinite slope stability theory 138
integrated risk management 445
integrated watershed management
    debris-flow hazard mitigation 450

Jokulhlaup 43

lahar 355, 685, 688
  debris-flow phase 248
  definition 248
  edifice collapse 256
  erosion and bulking 257
  genesis 253
  hazard
    appraisal 268
    mapping 269
    mitigation 262
  historical occurrence 40
  hyperconcentrated flow phase 248
  incremental accretion 260
  induced by sudden water release 254
  pyroclastic flow 249
  pyroclastic surge 249
  rheological definition 248
  stream flow phase 248
  transitional flow phase 248
landslide dam 17
landslide incidence
  and forestry
    aspect, relation to 601
    downslope curvature, relation to 601
    slope angle 602
    soil drainage, relation to 606
    steep stream escarpments and headwater drainage areas 603
    surficial materials, relation to 603
  road deactivation 610
landslide management in forestry
  process 607
landslide rehabilitation 610
landslide risk criteria, British Columbia 679
landslides, stress and strength properties of soil
  dilatant soil 62
  formation, speed, and travel distance of 62
landslide-triggered debris flow
  consolidation shear model 82

Index 737

development of 87
effect of rise in ground-water level 101
liquefaction 87
liquefaction initiation under undrained
    conditions 85
ring shear test 93
saturated soil mass, requirement of 101
stress path 83, 84
undrained loading 101
landslide warning system 327
land-use planning, hazard mitigation 480
Leaky Barrel Model 337
LIDAR 282
    advantages 282
llapana 36

main debris-flow danger period (MDFDP)
    343
mass point models, runout distance 311
mazamorra 36
megaclasts 706
moving coil geophones 296
mud flow 14

NIS 228

optical monitoring systems 299

passive mitigation measures 478
perched aquifer, high pore-water pressure,
    trigger debris flows 329
piezometers 293
point source volume estimation 425
pore-water pressure, effect of timber
    harvesting 398
precipitation monitoring 292
prime disaster reduction approach 618

rainfall
    high intensity, triggering of landslides
        326
    intensity-duration, triggering shallow
        landslides 327
    minimum and maximum triggering
        debris flows 327
    pre-storm seasonal, antecedent 332
rainfall threshold
    development of 328

importance of frequency and mean
    annual precipitation (MAP) 327
models 336
prediction and warning 335
real-time remotely sensed data, regional
    slope stability models 347
regional variation 336
triggering of debris flows 335
rapid drawdown 207
recorded events method 27
regional slope stability
    limits of GIS 74
    mapping 74
remote sensing
    analytical techniques
        spectral bands 280
        statistical methods 281
    applications for landslide management in
        forestry 609
    charge-coupled devices (CCDs) 280
    complementary metal oxide
        semiconductors (CMOSs) 280
    LIDAR 282
        advantages of 282
    methods for inventories and
        characterization 284
    multi-temporal aerial photographs 279
    Near Infra-Red imagery 609
    passive microwave sensors 281
    Principal Components Analysis, use in
        forestry 609
    rule for choosing film 277
    satellite imagery for forestry 609
    spaceborne multispectral digital imagers
        280
    spatial resolution 275
    spectral range 276
    synthetic aperture radar (SAR) 283
    terrain mapping 285
remote terminal unit (RTU) 300
rheology (definition), rheological criteria,
    hyperconcentrated flows 162
runout distance
    criteria for deposition 313
    effect of wood debris 313
runout length prediction
    analytical approaches 315
    classification of methods 2
    continuum based simulation model 318

runout length prediction (*cont.*)
  empirical approaches for estimation 313
  lahars 314
  reliability of continuum based models 319
  sediment budget, volume balance 310

Sabo 35
satellite telemetry 302
seepage force, in a porous medium 68
seepage, local, hydraulic heterogeneity and topographic factors 71
SHALSTAB 74
SIMMAP 74
slope stability
  analysis
    infinite slope stability 62
    strength and pullout resistance of roots 64
  effect of forestry roads 400
  benefit of forest vegetation 389
  increase in soil moisture due to logging 397
  partially saturated soils
    estimating factor of safety 73
    shear strength and matric suction 71
SNOTEL aquisition system 344
snowmelt-generated debris flows
  relation to efficacy of hot-rock interactions with snowpacks 699
snowmelt, rain on snow 331
snowpacks, melting of, earthquakes and volcanic events 334
soil bioengineering 448
soil moisture tensiometers 294
soils
  partially saturated
    effective stress 60
    shear strength measurement 61
  water-saturated, effective stress 60
Soil Water Status Model 334, 337
  landslide-triggering threshold 347
  Wellington Region 341
sorting 464
spatial resolution 275
spectral range 276
stream fans 413

stress and strength properties of soil 55
  direct-shear test 55
  triaxial test 55
subaqueous debris flow
  consequences of 237
  difference compared to terrestrial environments 204
  explanatory hypotheses 211
  hazard and risk evaluation 235
  hydroplaning 218
  three main phases 210
  types 204
superelevation equation 430
Supervisory Control and Data Aquisition (SCADA) 300
synthetic aperture radar (SAR) 283

temporary debris deposition
  dosing 464
  sorting 464
terrain mapping 285
Time Domain Reflectometry (TDR) 294
transverse structures 453
travel distance, unobstructed flow 308
TRIGRS 15

ultrasonic range finders 297

Very Small Aperture Terminals (VSATs) 302
VIFFLOW 224, 228
viscosity (definition) 165
volcanic debris flows
  estimating recurrence 720
  glaciated v. non-glaciated 685
  glaciated, snow clad volcanos 687
  scour, melt of snowpack 694
  transformation from landslides 690
wildfire
  effects on debris flows 364
wildfire-related debris flows
  conditions for initiation by discrete failure 370
  conditions for landslide causation 371
  frequency 369
  initiation by discrete failure 369
  process for generating 366

runoff-dominated processes 367
susceptibility, lithology, surface materials 372
wildfire-related debris-flow susceptibility
  function of burn extent and severity 375
  highly correlated factors 378
  threshold conditions 374
woody debris rake 477

yield rate
  determining yield rate, consideration of variance 149
  empirical calculation 149
  estimation of 145
  event return periods 149
yield rate (definition) 142

Colour plates

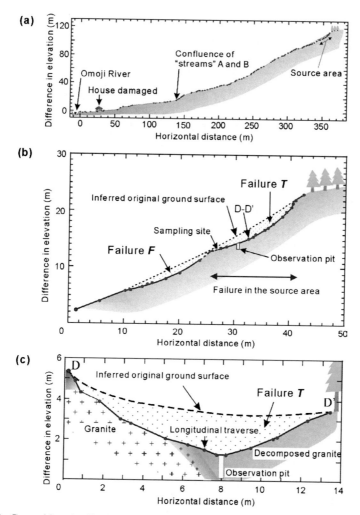

Figure 5.15. Central longitudinal section: (a) from the source area to the Omoji River; and (b) of the source area. (c) Cross section of Failure *T*.

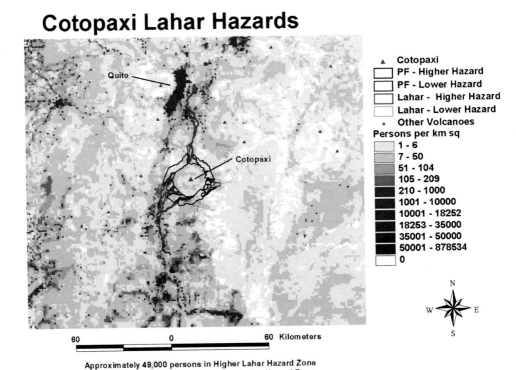

Figure 10.4. Lahar-hazard map of Hall and von Hillebrand (1988) superimposed on Landscan population data courtesy of John Ewert. Landscan data is compiled in 1-km^2 pixels. The combination of population data with lahar-hazard zones follows the volcano population index of Ewert and Harpel (2004) to generate a lahar population index. The generation of Landscan data in 90-m pixels will make the technique more useful with small and moderate-sized lahar-hazard zones.

Figure 10.5. Photograph of Armero after the 13 November, 1985 lahars from Nevado del Ruiz, Columbia.
USGS photograph from Pierson et al. (1990).

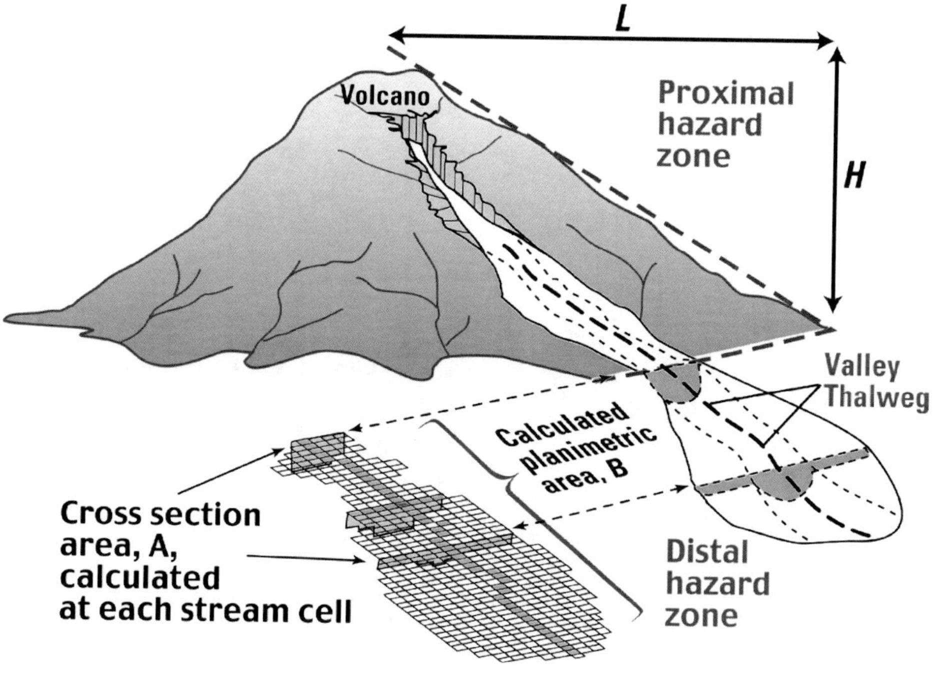

Figure 10.10. Idealized lahar path and geometric relationships between A and B describing the extent of the lahar-inundation hazard zone.
After Iverson et al. (1998).

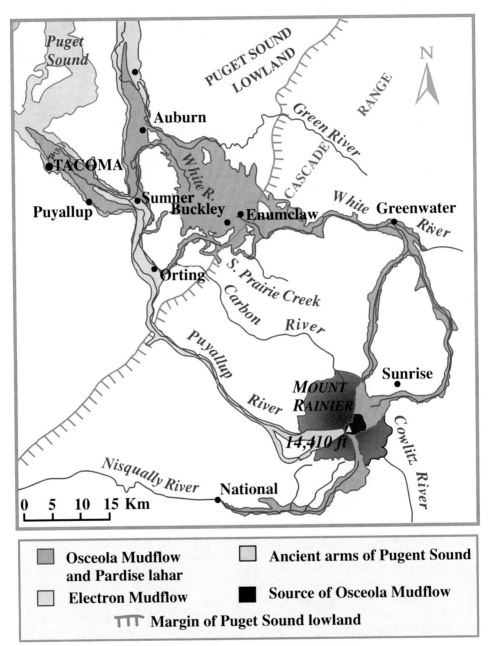

Figure 10.8. Distribution map of the 5,600-year-old, 4-km^3 Osceola Mudflow and coeval Paradise lahar (after Vallance and Scott, 1997) and of the 0.25-km^3, 500-year-old Electron Mudflow (after Crandell, 1971). Edifice collapse of hydrothermally altered weakened rock generated all of these lahars. Eruptions at the summit of Mount Rainier caused the Osceola Mudflow and Paradise lahar, but the cause of the Electron Mudflow may not be related to eruptions.

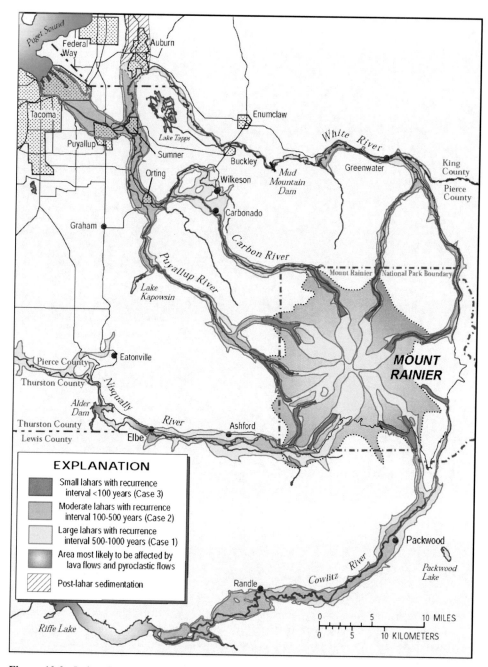

Figure 10.9. Lahar hazard zones for Mount Rainier after Hoblitt et al. (1998). Large lahars (Case 1) will extend farther downstream and will inundate greater areas of the drainage basins than smaller lahars (Case 2). The geologic history of Mount Rainier during the last 10,000 years indicates the approximate frequency with which these lahars occur. The vast majority of these large lahars occur during periods of volcanism. In contrast, the smallest lahars (Case 3) include those that happen nearly every year and are not necessarily genetically related to volcanism.

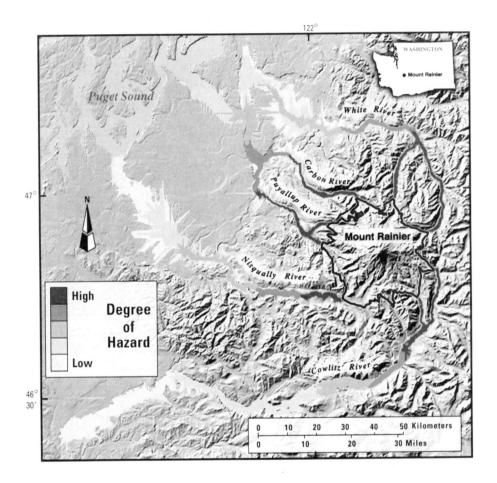

Figure 10.11. Lahar-inundation hazard map constructed by applying the computer model for Mount Rainier and vicinity after Iverson et al. (1998). The dark line encloses a proximal hazard zone, which includes hazards from diverse processes including lahars. Degree of hazard diminishes with simulation volume because large lahars are less likely to recur than small lahars.

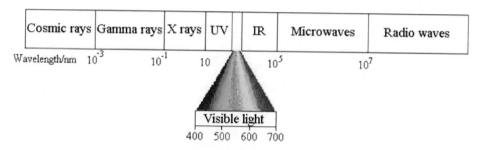

Figure 11.2. Spectral ranges of the electromagnetic spectrum expressed in nanometers.

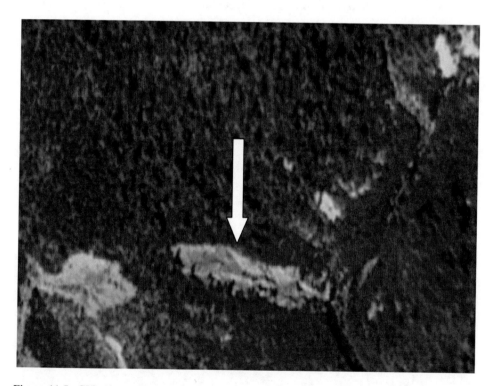

Figure 11.5. CIR photograph showing seepage as dark blue patches in the debris slide scar.

Figure 11.6. Lidar *range image* of a hillslope subject to debris slides and ravelling. The color of a pixel represents the distance from the sensor to that pixel. The colors cycle through the rainbow as pixels become increasingly distant from the sensor.

Figure 11.7. Three-dimensional photograph of the same hillslope shown in Figure 11.6. The shape data came from a lidar sensor and the image data from a color digital camera.

Figure 11.8. Photo showing the use of a ground-based lidar to monitor movement on a debris slide face.
From Paar et al. (2000).

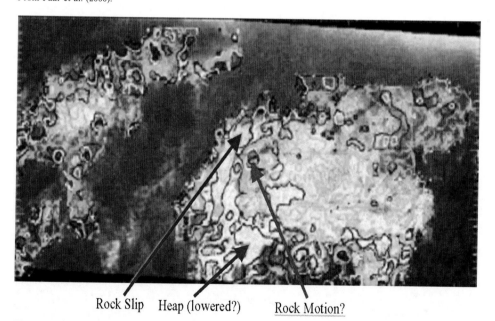

Rock Slip   Heap (lowered?)   Rock Motion?

Figure 11.9. Isoline plot of landslide movements detected by a ground-based lidar system in Austria.
From Paar et al. (2000).

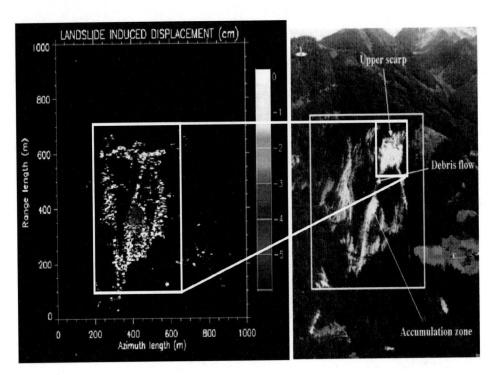

Figure 11.10. Map of the maximum landslide-induced ground deformation provided by a ground-based D-InSAR (*left*). Photo of the debris slide headscarp analysed (*right*).
From Leva et al. (2003).

Figure 14.4. (b) Rainstorm in 2002, Gisborne, East Coast.
Photo: by Michael Crozier.

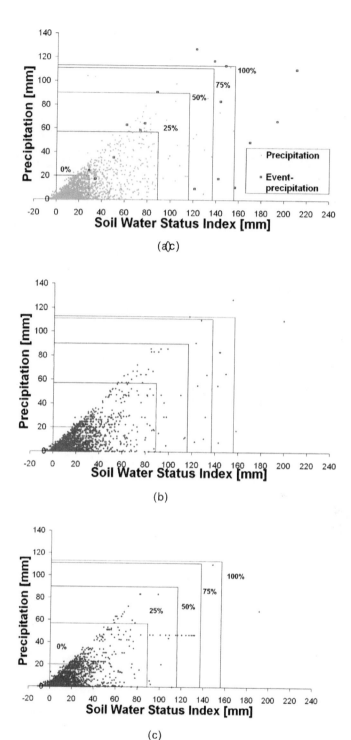

Figure 14.17. The Soil Water Status index calculated for Wellington, New Zealand. Threshold lines refer to calculations from the observation period (Glade, 2000a). (a) Observation period 1950–1979, (b) control run of the downscaled GCM data for the similar period, and (c) climate change scenario based on the GCM run for the period 2070–2099 (refer to Schmidt and Glade (2003) for details). In (a) the event precipitation refers to landslide-triggering rainstorms.

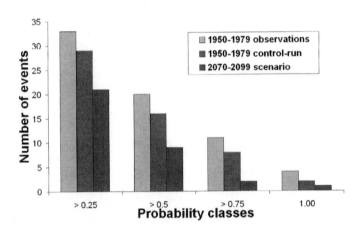

Figure 14.18. Changes of total number of landslide events within each probability class for Wellington, New Zealand. Data refer to Figure 14.17 and give observations, control-run, and scenarios (Schmidt and Glade, 2003). The "number of events" refers to the number of landslide-triggering rainstorm events.

Figure 15.1. (a) Photograph of debris-flow path and (b) deposits generated from a burned basin following the 2002 Coal Seam Fire in Colorado. Pickup truck towing trailer in lower left of (a) for scale.

Photographs by Andrea Holland-Sears, USDA Forest Service.

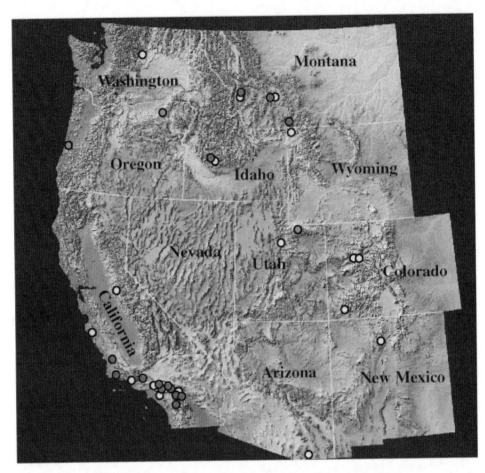

Figure 15.2. Map showing locations of documented fire-related debris-flow events in the western USA. Data for more than one basin exists at most locations. Yellow dots signify locations of basins for which data on either debris-flow volume or peak discharge exist (Gartner et al., 2004); both red and yellow dots show locations of debris-flow producing basins included in Gartner et al. (in press); green dots show additional debris-flow locations reported in the literature.

Figure 15.3. Photograph of debris-flow path generated through the process of progressive bulking of runoff with sediment eroded from hillslopes and channels in the 2002 Coal Seam Fire in Colorado. Levees consisting of poorly sorted, unstratified and matrix-supported material, and mud coatings persisted down the channel from the point indicated by the arrow. Photograph by Andrea Holland-Sears, USDA Forest Service.

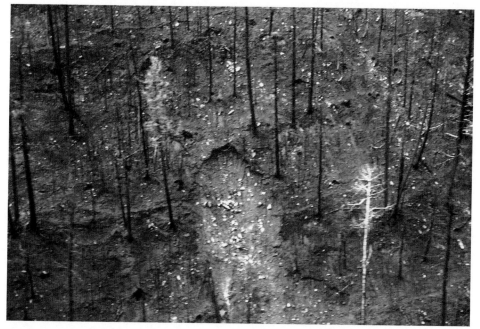

Figure 15.7. Photograph of landslide scar on hillslopes burned by the 2002 Missionary Ridge Fire in Colorado. Material from scar mobilized into a debris flow and travelled only a short distance downslope.

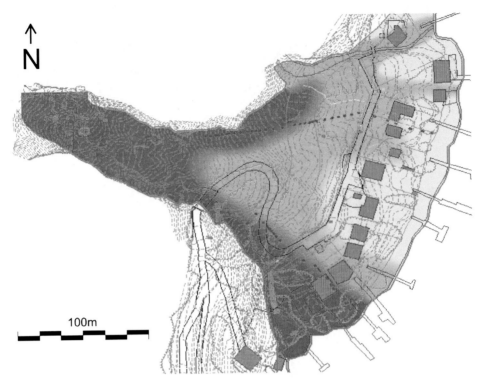

Figure 17.11. Hazard map of Percy Creek at Indian Arm, Vancouver, British Columbia. This map is based on repeated modelling of the 500-year return period debris flow and combined several possible flow paths. High hazard (red) is defined for flow velocities $(v) > 3$ m/s, flow depth $(d) > 3$ m, and maximum boulder size $(D) > 2$ m. Moderate hazard (orange) is defined as $v = 1$–3 m/s, $d = 1$–3 m, and $D = 1$–2 m. Low hazard (yellow) is defined as $v < 1$ m/s, $d < 1$ m, and $D < 1$ m.
KWL Ltd (2003).

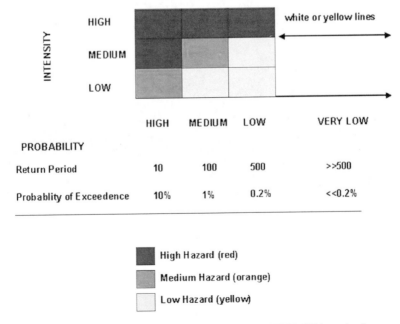

Figure 20.12. Discrete hazard levels (PREVENE 2001 project).

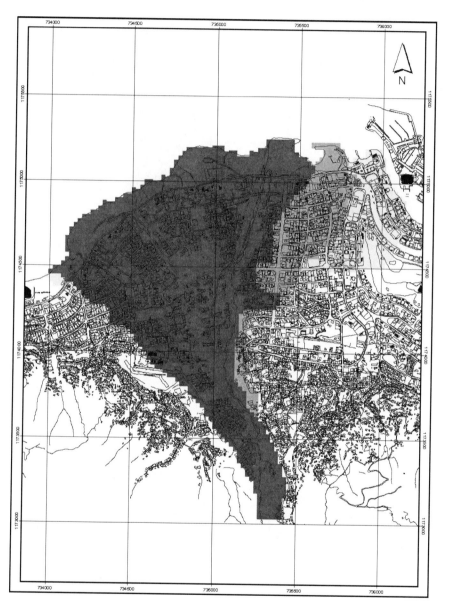

Figure 20.13. Hazard map for San Julian alluvial fan. Squares are 500 × 500 m.

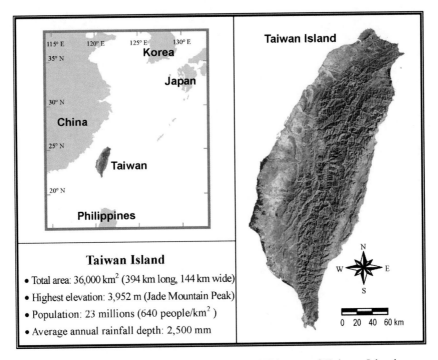

Figure 21.1. Geographical location and SPOT image of Taiwan Island.

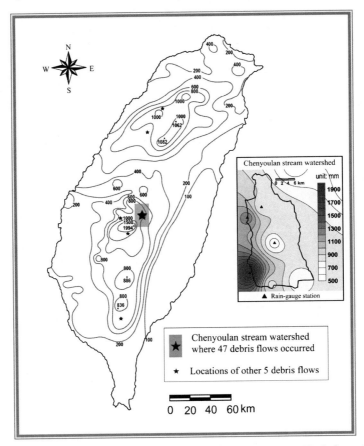

Figure 21.3. Isohyetal map of the 96-hour rainfall during Typhoon Herb from 30 July to 2 August, 1996.

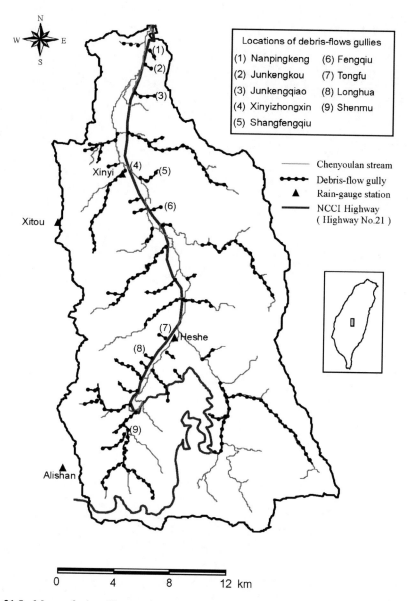

Figure 21.8. Map of the Chenyoulan stream watershed, NCCI Highway, 4 rain-gauge stations, and 47 debris-flow gullies caused by Typhoon Herb.
After Chieng (1998).

Figure 21.9. "Bouldery" 'rock-rich) debris flow at Fengqiu.

Figure 21.10. "Sand-silt-clavey" (matrix-rich) debris flow at Longhua.

Figure 21.11. "Cobble-gravely" (mixed evenly with rock and matrix) debris flow at Shenmu.

Figure 21.14. Cross section of a "cobble-gravely" debris flow deposit at Junkengqiao. The debris flow destroyed one home except for the wall foundation. The thickness of the wall foundation was 0.7 m and the largest boulder in the photograph was 0.4 m. Note the debris avalanche scar in the upper center of the photo, which likely triggered the debris flow.

Figure 21.15. Damaged bridge over the Chushui River at Shenmu. The bridge's clearance was designed for water floods and did not have sufficient conveyance for debris flows. The bridge suffered severe structural damage. The largest boulder transported under the bridge had a diameter of 2.0∼2.5 m.

Figure 21.17. Debris-flow structural countermeasures, such as Sabo dams, debris barriers, and debris basins, at Fengqiu in the Chenyoulan stream watershed. The photo shows a relatively big debris basin having a designed capacity of 80,000 m^3. It was built in 1999, and later divided by a grid-type slit dam (as shown in blue color) in 2000. Photo: Professor C.C. Wu).

Figure 21.18. A slit dam built at the outlet of a debris-flow-prone stream at Jianqing in Hualien County. Note the use of tires on the upstream face of the slit dam to reduce the debris-flow impact. The photo inset shows a series of check dams constructed to stabilize the channel bed and banks along the stream channel upstream of the slit dam.

Figure 21.19. A debris-flow warning sign installed at the crossing of a road and a debris-flow-prone channel.
Photo: Disaster Prevention Research Center, NCKU.

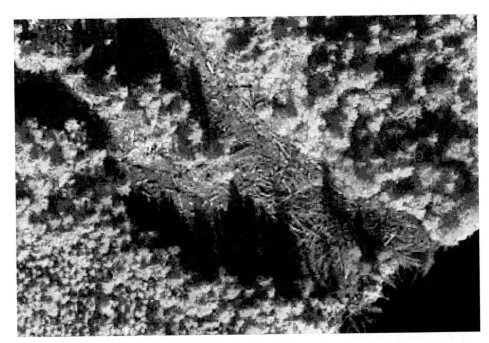

Figure 23.5. Identification of potential timber salvage associated with landslide activity using high-resolution satellite images.
Image copyright of DigitalGlobe.com.

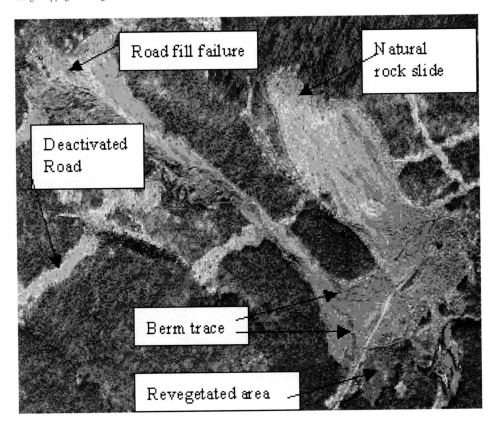

Figure 23.6. Identification of landslide features in Escalante River using near-infrared satellite images.
Image copyright of Spaceimaging.com.

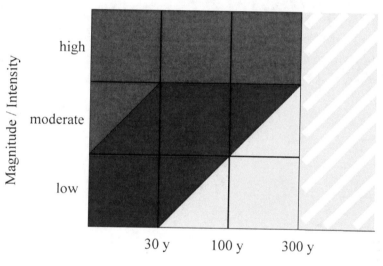

Figure 24.2. Hazard matrix. Specific quantitative criteria (for debris flows: flow depth and velocity) are provided for the magnitude of the processes (*cf.* BWW/BUWAL/BRP, 1997).

Figure 24.8. Simulation of debris flows at Sörenberg using a 2-parameter model for the travel distance and a random-walk approach (using a GIS multiple-flow direction) for the behaviour of the flows on the fan.

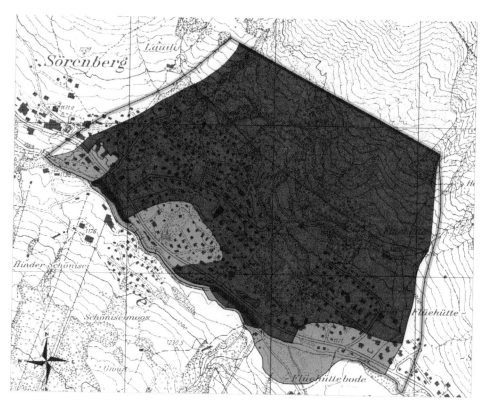

Figure 24.9. Hazard map for Sörenberg. Only the debris flow hazards from the landslide complex are considered (colours according to Figure 24.2).

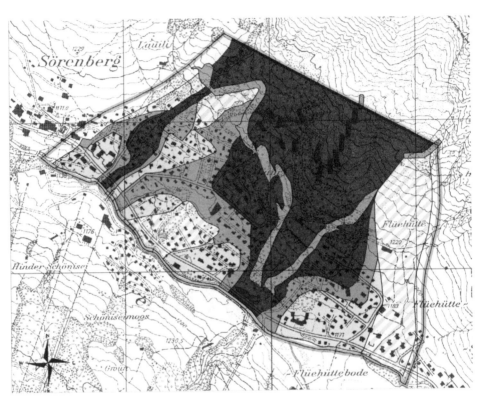

Figure 24.11. Hazard map after the implementation of a set of structural measures as indicated in Figure 24.10 (colours according to Figure 24.2).

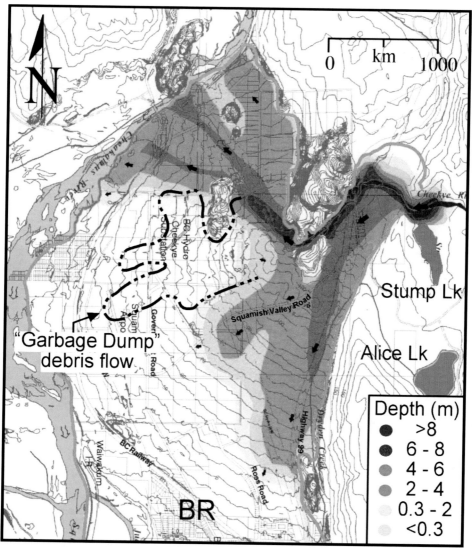

Figure 26.14. Model run for a debris flow with a volume of 3×10^6 m^3 and a peak discharge of 9,600 m^3/s (Kerr Wood Leidal, 2003). According to the exceedence plots in Figure 26.11, this event has a return period of 2,500–5,000 years. Middle and late Holocene debris flows, including the 800-year-old event at the municipal landfill, created a topographic high that splits the modelled event into southern and northern lobes. The thickest materials occur along the modern channel. Note that the debris flow does not reach the residential areas of Brackendale (BR). The contour interval is 5 m.

Printed by Books on Demand, Germany